To my dear friend & colleague, Luis.
Happy reading

Richard

INTERNATIONAL SERIES OF MONOGRAPHS
ON PHYSICS

INTERNATIONAL SERIES OF MONOGRAPHS ON PHYSICS

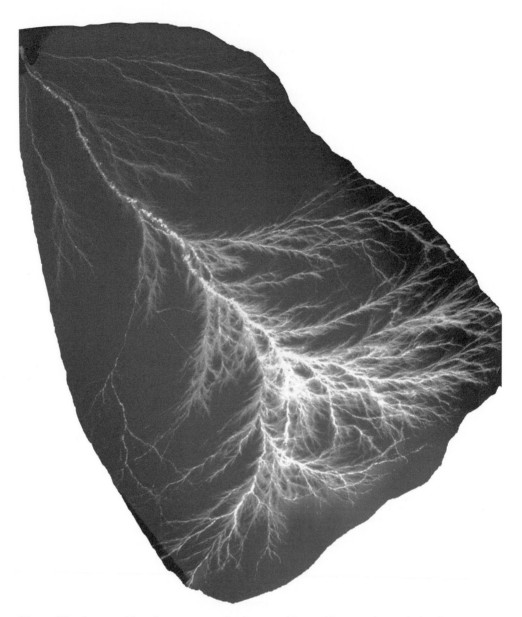

'Beam Tree' created by electromagnetic showers 'frozen' into a piece of plastic.

Calorimetry

Energy Measurement in Particle Physics

Second Edition

Richard Wigmans

Bucy Professor of Physics, Texas Tech University

OXFORD

UNIVERSITY PRESS

OXFORD
UNIVERSITY PRESS

Great Clarendon Street, Oxford, OX2 6DP,
United Kingdom

Oxford University Press is a department of the University of Oxford.
It furthers the University's objective of excellence in research, scholarship,
and education by publishing worldwide. Oxford is a registered trade mark of
Oxford University Press in the UK and in certain other countries

First Edition published in 2000
Second Edition published in 2017

Impression: 1

Published in the United States of America by Oxford University Press
198 Madison Avenue, New York, NY 10016, United States of America

British Library Cataloguing in Publication Data
Data available

Library of Congress Control Number: 2017943897

ISBN 978–0–19–878635–1

DOI 10.1093/oso/9780198786351.001.0001

Printed and bound by
CPI Group (UK) Ltd, Croydon, CR0 4YY

For Nazzi

PREFACE

to the second edition

Calorimeters were originally developed as crude, cheap instruments for some specialized applications in particle physics experiments, such as detection of neutrino interactions. However, in the past 40 years, their role has changed considerably, and calorimetry has become an extremely important experimental technique, especially in experiments that make use of the increasingly powerful accelerators and storage rings that have been built and exploited in this period. Almost every experiment in particle physics today relies heavily on calorimetry. Calorimeters fulfill a number of crucial tasks, ranging from event selection and triggering to precision measurements of the four-vectors of individual particles and jets and of the energy flow in the events (missing energy, *etc.*). This development has benefited in no small part from the improved understanding of the working of these, in many respects somewhat mysterious, instruments.

Much has been learned about calorimetry, thanks to many dedicated R&D projects. This information is contained in a very large number of papers, scattered in the scientific literature. Over the years, many review articles have been published, in which the state-of-the-art is summarized. The ones published in this century include [Ler 00, Fab 03, Wig 08a, Wig 08b, Brau 10, Akc 12a], and a number of book chapters [FF 11, Wig 11]. However, these review papers usually concentrate on a limited number of aspects and were generally not intended as an educational introduction to the topic.

When I first wrote this monograph, I intended it to be not only a reference text, but also a basic introduction for those students (and others) who are confronted for the first time with calorimetric particle detection. Based on the many reactions I received on the first edition, which appeared in 2000, I conclude that it clearly succeeded in that second aspect. The fact that it was rarely available for consulting in the CERN library, which had purchased several copies, also testified to that.

To further strengthen the educational value, I have added in this second edition a new chapter (#9) dedicated to the analysis and interpretation of test beam data, a task that is often assigned to students and postdocs who are new to an experiment. It is a summary of lessons I have learned myself over the years, working with these in many ways highly non-trivial, counter-intuitive detectors. It contains several examples of mistakes that are commonly made in practice, even today. I wish this chapter had existed when I started to learn about these detectors long ago, since it would certainly have prevented several mistakes I have made myself over the years.

Obviously, a lot has happened in the 16 years that have passed since the first edition was published. LEP, the Tevatron and the SLAC colliders have terminated operations, while the LHC experiments, which were in the design stage at that time, have now been running successfully since several years. The Higgs boson, which inspired the design of the ATLAS and CMS calorimeters, as discussed in the last chapter of the first edition, was in fact discovered (2012) and honored with a Nobel prize (2013). Major new experiments prominently featuring calorimeters are nowadays also taking place in Antarctica, in the Mediterranean Sea, on the Argentinian pampa, inside a variety of mountains and deep mines, and in space. Detector R&D is inspired by the prospect of future colliders (ILC, CLIC, FCC,...) that further push the frontiers set by LEP and the LHC. After the compensation issues were settled in the last decade of the previous century, the emphasis in generic R&D has shifted, in view of the perceived needs of future experiments. The completely new chapter #8 describes the recent work in this context.

The basics of calorimetry have not changed and, therefore, the introductory chapters (2–6) are essentially the same as in the first edition, apart from the fact that some new insights have been included, and some more recent experimental data have been chosen to illustrate the issues. Not surprisingly, many of these new data come from the LHC experiments, and particular ATLAS and CMS, in which calorimetry plays such a crucial role. Rereading Chapter 10 of the first edition, I have also been greatly impressed by the enormous developments and successes of the experiments that study natural phenomena (neutrinos, cosmic rays) with calorimetric methods, and I have tried to do justice to this work with a thorough update of the text of that chapter. The remaining chapters (1, 7, 11) have been updated to reflect the developments of the past 16 years as well. I have also used the opportunity to correct a few mistakes that had been pointed out to me over the years, as well as some mistakes I discovered myself in this process. Since this new edition will also be available as an e-book, I have spent a lot of work on improving the figures (which were all bitmapped in the first edition), introducing color where appropriate. I have also discovered that some of the educationally most useful figures can be found in theses by students who obtained advanced degrees working on the various projects discussed in the text. I thank Drs. M.A. Chefdeville [Che 13], Th. Dafni [Daf 07], K. Gümüş [Gum 08], F. Hubaut [Hub 99], Y. Koshio [Kos 98], K. Okumura [Oku 99] and M. Simonyan [Sim 08] for their invaluable contributions, which have allowed me to further improve the educational value of this book.

This book is about calorimetry, not about calorimeters. I have attempted to give a rather complete review of the principles, the possibilities and limitations of calorimetric particle detection. The examples I have chosen to illustrate the important points are primarily derived from projects I am personally familiar with. That does not at all mean that I consider these projects superior or otherwise more worthy of mention than the many other fine calorimeter projects I could have chosen for this purpose. It was just easier for me this way and I apologize in advance to anyone who might feel that his or her work has not been given the credit that it no doubt deserves. I have tried to compensate for this narrow scope in Appendix C, which contains a very brief summary, with references, of all past, present and future calorimeter projects I know of.

Pursuing my educational goals, I have tried to make this book as practical as possible, addressing many issues that those who work with calorimeters encounter in practice. I hope that it will serve as a useful reference guide and manual and that it will find its way into the counting rooms of many experiments. Also, and in particular, I hope that students who are working on their first experiment in particle physics will find it a good introductory source of information.

It would not have been possible for me to write this book without the help of many people. Alessandro Cardini, Ilias Efthymiopoulos, Antonio Ereditato, Don Groom, Ana Henriques, Werner Hofmann, Sungwon Lee, Hans Paar, Giorgio Riccobene, Felix Sefkow, Manoel Seixas, Igor Volobouev and Renyuan Zhu helped me locate material that is included in this new edition and I thank all of them for their contributions. In particular, I thank my long-time collaborator John Hauptman, who proofread this entire book. The credit for a lack of typos and crimes against the English language goes completely to him. I also benefited from the fact that he has written several books himself, which allowed him to make good suggestions on how to deal with controversial issues in the text. Other sections were proofread by Andrea Fontana and Igor Volobouev, and I benefited from their physics insight in several discussions. Sehwook Lee has been the pillar of many RD52 analyses over the years. A good fraction of the figures in Chapter 8 would not exist without him, and he did the work that led to Figures 9.2 and 8.29, which were specially made for this book. Figure 8.3 is the result of simulations that professor Bryan Webber kindly performed with the same purpose. I greatly appreciate these important contributions.

I gratefully acknowledge the hospitality and support of the Department of Physics at the University of Pavia, where I spent nine months on sabbatical leave working on the completion of this project. My host Michele Livan, who has been my friend and collaborator on several calorimeter R&D projects since 30 years and is one of the few people with a deep insight in this topic, helped me by extensive proofreading, and by making valuable suggestions on what to *leave out* of the text of this reference book. In addition, he and his wonderful wife Nadia helped considerably with practical issues such as finding an apartment, and introduced us to many aspects of Italy that few foreigners get to experience.

The frequent interactions with my Italian colleagues/collaborators in Pavia and elsewhere (Pisa, Cagliari, Rome, Cosenza, Como) also led to new insights, which have found their way into this book.

Last but not least, I am indebted to my wonderful wife. Her encouragement and support, as well as her self-sacrifice throughout this project have been absolutely essential for its completion. Her love has been the engine that has kept me going since I first met her 25 years ago. Therefore, this book is dedicated to her.

Ransom Canyon, Texas   R.W.
December 2016

CONTENTS

INTRODUCTION: SEVENTY YEARS OF CALORIMETRY

1.1 Calorimetry in Thermodynamics

The term "calorimetry" finds its origin in thermodynamics. We probably all remember the experiments done in high school or in a freshman physics lab, in which we determined the specific heat of water or other substances, or measured the rate at which the Sun provides energy to the Earth. Calorimeters were the thermally isolated boxes containing the substance of our study. There was always a thermometer sticking out, which provided the experimental information.

Modern, highly sophisticated versions of these instruments are in use in nuclear weapons laboratories, where they are used for the assay of fissionable material. For example, ^{239}Pu produces heat at a rate of 2 milliwatts per gram. Calorimetry can provide an accurate measurement of the amount of plutonium in a sample, in a non-invasive manner.

In nuclear and particle physics, calorimetry refers to the detection of particles, and measurement of their properties, through total absorption in a block of matter, called a *calorimeter*. Calorimeters exist in a wide variety, but they all have the common feature that the measurement process through which the particle properties are determined is *destructive*. Unlike, for example, wire chambers that measure the particle's properties by tracking it in a magnetic field, the particles are no longer available for inspection by other devices once the calorimeter is done with them. The only exception to this rule concerns muons. The fact that these particles may penetrate the substantial amounts of matter represented by a calorimeter is actually an important ingredient for their identification as muons.

In the absorption process, almost all the particle's energy is eventually converted into heat (*calore* in Italian), hence the term calorimetry. However, the units of the energy involved in this process are very different from the thermodynamic ones. The most energetic particles in modern accelerator experiments are measured in units of TeV (1 TeV = 10^{12} eV = 1,000 GeV), whereas 1 calorie is equivalent to about 10^7 TeV! The rise in temperature of the block that absorbs the particle is thus, for all practical purposes, negligible. Therefore, more sophisticated methods are needed to determine the particle properties. These methods, the possibilities they offer and their limitations, are the topic of this book.

1.2 Nuclear Radiation Detectors

Calorimetric particle detection was pioneered in nuclear physics, shortly after World War II, with the advent of scintillation counters. The fact that ionizing particles emitted in nuclear decay may cause fluorescence in certain materials, *e.g.*, ZnS, was already known since the beginning of the twentieth century. However, it was the invention of the photomultiplier tube (PMT) that made it possible to apply this phenomenon for quantitative measurements of particle properties. With the PMT, which converts individual photons into measurable electric signals, it became possible to count the number of scintillation photons created by radioactive decay products, and thus measure the energy deposited in the scintillating material.

The most popular scintillators in those early days included anthracene and (thallium-doped) sodium iodide crystals. Anthracene, an organic compound, is a very bright scintillator with a very short fluorescent decay time, provided the crystal is sufficiently pure. It was mainly used to measure the properties of αs and βs emitted in nuclear decay. Its low density, combined with the fact that it was hard to grow large crystals with the required purity, made it less useful for the detection of γ-rays. High-Z inorganic crystals such as thallium-doped sodium iodide, NaI(Tl), were much better suited for this purpose. A NaI(Tl) crystal with a volume of one cubic inch has already an efficiency of about 50% for detecting typical nuclear γ-rays (with energies around 1 MeV), and it was relatively easy to grow crystals that were sufficiently large to fully contain all the energy released in the absorption of such γ-rays.

NaI(Tl) detectors have played a very important role in nuclear spectroscopy. Until

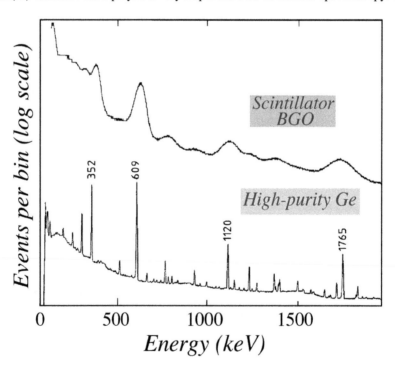

FIG. 1.1. Nuclear γ-ray spectrum of decaying uranium nuclei, measured with a bismuth germaniumoxide scintillation counter (*upper curve*) and with a high-purity germanium crystal (*lower curve*). Courtesy of G. Roubaud, CERN.

the 1960s, they were the instruments of choice in that field. The development of large *semiconductor* crystals (lithium-doped silicon or germanium) ended that role. The reason for that becomes clear from Figure 1.1. This figure shows the γ-ray spectrum of decaying uranium nuclei, measured with a scintillation counter and with a semiconductor crystal. The latter technology offers a spectacularly improved energy resolution. As a result, the measurements with this crystal revealed a wealth of detailed information inaccessible with the scintillator technology.

The history of science is full of quantum leaps that were a direct result of the availability of new, or better, measuring equipment. From the development of the microscope in the seventeenth century and its importance for understanding biological processes, to the Hubble Space Telescope (HST) and the resulting breakthroughs in our knowledge of the Universe, science and technology have always gone hand in hand.

Clearly, the advent of the semiconductor technology mentioned above caused a revolution in experimental nuclear physics, as did the development of calorimeters for experiments in particle physics. And just as in the other examples mentioned in this section, time and again resolution turned out to be *the* crucial feature. The resolution offered by the microscope led to the discovery of bacteria, of the cell structure of living organisms and of viruses. The resolution offered by the HST made it possible, among many other things, to follow the evolution of a supernova explosion that occurred in February 1987, at the incredible distance of more than 10^{18} km. The high resolution offered by the semiconductor nuclear γ-ray counters allowed the unraveling of numerous extremely complicated nuclear level schemes. And in particle physics, in which the smallest details of matter accessible to mankind are being studied, the resolution of the detection instruments has already led to the discovery of intermediate vector bosons, CP violation in the K^0 and B^0 systems, the top quark, and the Higgs boson (Chapter 11). And who knows what other surprises nature holds in store, waiting for discovery until the moment that our supermicroscopes become good enough.

1.3 Calorimetry in Particle Physics

The scientific discipline called nowadays Particle Physics studies the structure of matter and the interactions of its constituents at the innermost level. Over the years, this field has undergone several name changes. In the early days, it was called *Cosmic-Ray Physics*. Later, after the advent of powerful accelerators like the *Bevatron* (Lawrence Berkeley Laboratory, 1954), the discipline became known as *Elementary-Particle Physics* or *High Energy Physics*.

The adjective "elementary" derived its origin from the fact that the newly discovered muons, pions, kaons, Λs, Σs, *etc.* did not fit into the scheme of thinking (the 1950 *Standard Model*) that considered all matter composed of protons, neutrons and electrons (with neutrinos as an accepted, albeit unobserved oddity). Therefore, these new particles had to be different at an elementary level.

However, after the discovery of the quark structure of hadronic matter (1965), one realized that, with the exception of the muon, none of the mentioned new particles was elementary in the sense of *uncompounded*. Also the proton and neutron, until then believed to be elementary building blocks of matter, lost that status. Quarks and leptons became the elementary particles of the new (current) Standard Model. Having learned the lesson that terminology should not be too pretentious, the adjective "elementary" was dropped in describing the discipline.

The experiments in particle physics can be classified as accelerator-based or non-accelerator-based. Most of the accelerators used for these experiments operate at the multi-GeV or even the TeV energy scale, and the term *High Energy Physics* is clearly appropriate to describe the experiments at these machines. However, there are also a number of crucial issues concerning the fundamental structure of matter that are best addressed experimentally at very low energies. Baryon number conservation (proton decay), matter–antimatter symmetry and the neutrino rest mass can be mentioned as examples in this context. *High Energy Physics* is therefore clearly too narrow a description of the field.

It is also not true that the high-energy aspects of particle physics are studied with accelerators and the low-energy aspects in non-accelerator experiments. For example, cosmic-ray phenomena at energies millions of times higher than the current accelerator frontier are studied in a variety of non-accelerator experiments. And experiments at the Anti-proton Decelerator (at CERN) and with ultracold neutron beams at the Institut Laue Langevin (Grenoble, France), Los Alamos National Lab (USA) and the European Spallation Source (under construction in Lund, Sweden) address some of the mentioned fundamental issues at very low energies in an accelerator-based setup.

In summary, experimental particle physics encompasses the study of a wide variety of phenomena, at energy scales that span many orders of magnitude. The experimental approaches chosen to investigate these phenomena are possibly even more diverse. In the following subsections, the (evolution of the) role of calorimeters in some classes of experiments in particle physics is briefly described.

1.3.1 *Shower counters*

At accelerators, the structure of matter is often studied in scattering experiments. In a *fixed-target* geometry, a beam of particles is sent onto a target. The scattering of the beam particles off the target constituents may provide information about the structure of both and about their interaction. In a *colliding-beam* geometry, two beams of particles are brought into collision. This has the advantage that the available energy is used in a more efficient way. In fixed-target collisions, most of the energy carried by the projectiles is used to transfer momentum to the target, while in a colliding-beam setup the center of mass of the collisions may be at rest in the laboratory frame, so that the entire energy carried by the beam particles is available to materialize in any way allowed by the laws of physics.

Until about 1970, almost all experiments in particle physics were of the fixed-target type. In these experiments, one tried to reconstruct the details of the interactions by measuring the four-vectors of all particles produced in these events as best as possible. The charged reaction products were tracked in a magnetic field. From the curvature of the track, the momentum and the charge sign of the particle could be determined. The ionization density (dE/dx) could provide information about the particle's mass.

However, for electrically neutral particles these techniques did not work. These neutral particles include neutrons, photons, K^0s and Λs. For Λs and K^0_Ss, this was not so much of a problem. These particles have a short lifetime and often decayed close to the primary vertex into two charged particles that could be tracked and from which the properties of the parent particle could be derived. Among the remaining neutral particles, photons were by far the most important ones. Most of the photons resulted from π^0 decay ($\pi^0 \rightarrow \gamma\gamma$), and π^0 production occurred in almost every reaction.

In bubble chamber experiments, and in particular the ones using heavy liquids like freon or neon, some fraction of these γs were observed because they converted inside the fiducial volume of the bubble chamber ($\gamma \rightarrow e^+e^-$). The properties of the photons and their π^0 parents could be derived from the momenta of the electrons and positrons in such cases. Usually, the efficiency of π^0 reconstruction (which requires two fully reconstructed γs) was not very high, especially when the bubble chamber was filled with liquid hydrogen, the most widely used liquid (free proton targets!)

In some electronic counter experiments, one tried to get information about π^0 production by installing thin sheets of material with the purpose of converting photons into e^+e^- pairs, whose properties could then be measured in the magnetic field. Also here, the problem was efficiency, combined with precision. In order to convert photons with adequate efficiency, the amount of material needed was such that the electrons and positrons often lost so much energy on their way out that the measurement of their momenta became useless.

Shower counters became a very popular solution to these problems. Inspired by their successes in nuclear γ-ray spectroscopy, and by the relative ease with which large crystals could be grown, NaI(Tl) was the first scintillator to find its way into particle physics experiments. The results were excellent. Photons were measured with 100% efficiency in these devices and thanks to the sub-1% resolutions that were commonly achieved, π^0 reconstruction could be achieved with high efficiency as well, especially at low energies [Chan 78].

These successes started a long tradition of crystal calorimetry for particle physics experiments. One of the disadvantages of NaI(Tl) is its hygroscopicity. Other crystals, *e.g.*, CsI, were found to be less problematic in that respect. New types of crystals were also developed to meet specific needs of experiments, for example in terms of signal speed, radiation hardness, density and/or cost. An overview of inorganic scintillating crystals that have been (or are planned to be) applied in particle physics experiments, including some of their relevant properties, is given in Appendix B (Table B.5).

A totally different type of shower counter that became quite popular in the 1960s was the lead-glass detector. Lead-glass, which exists in a number of varieties, consists of a mixture of SiO_2 and (up to 70%) PbO. This high-density (typically, 4–5 g cm^{-3}), transparent material does not scintillate, but it can be used as a detector for highly relativistic particles through the Čerenkov light that these particles generate in it. The light yield of this type of detector is typically several orders of magnitude smaller than for scintillating crystals. Therefore, the energy resolution is clearly worse. However, because of the instantaneous nature of the Čerenkov mechanism, the signals from these lead-glass counters can be extremely fast, much faster than the signals from scintillating

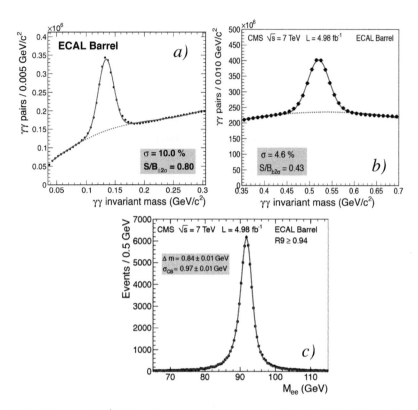

FIG. 1.2. The $\gamma\gamma$ and e^+e^- invariant mass spectra measured by the CMS Collaboration with their PbWO$_4$ crystal calorimeter. The peaks in diagrams a and b correspond to decaying π^0 (mass 135.0 MeV/c^2) and η (547.3 MeV/c^2) mesons, respectively. Diagram c depicts the invariant mass distribution of Z^0 bosons (91.19 GeV/c^2) decaying into electron–positron pairs [CMS 13].

crystals, which are limited by the sometimes very long decay times of the fluorescent processes on which they are based (*e.g.*, 230 ns in sodium iodide). In recent years, also some very fast inorganic scintillators have become available, for example lead tungstate (PbWO$_4$), which is an important detector component in the CMS experiment at CERN's Large Hadron Collider. In Figure 1.2, an example of the performance of crystal calorimeters in particle physics experiments is given. The figure shows the $\gamma\gamma$ and e^+e^- invariant mass spectra of π^0 and η mesons and Z^0 bosons, measured by the CMS Collaboration. Such distributions were used for *in situ* calibration of the detector.

The shower counters mentioned above are all examples of *homogeneous calorimeters*. This means that their entire volume is sensitive to the particles and may contribute to the signals generated by the detector. The functions of absorbing the particles and detecting the signals produced in this process are exercised by the same material, which needs to have a high density in order to perform the first function efficiently.

This feature distinguishes these devices from *sampling calorimeters*, in which the functions of particle absorption and signal generation are exercised by *different* materials, usually called the *passive* and *active medium*, respectively. The passive medium is typically a high-density material, such as iron, copper, lead or uranium. The active medium generates the light or charge that forms the basis for the signals from this calorimeter.

In sampling calorimeters, typically only a small fraction of the energy carried by the entering particles is deposited in the active medium that generates the signals. As a result, the energy resolution of sampling calorimeters is (often considerably) worse than for homogeneous ones, at least for electron and photon detection. However, they are also substantially cheaper. The development of sampling calorimeters was, therefore, primarily inspired by the need for very large detector systems of the types discussed in the following subsections.

1.3.2 *Instrumented targets*

In experiments carried out in the bubble chamber era (1960 – 1975), the bubble chamber served both as the target in which the impinging projectiles interacted, and as the detector with which the properties of the reaction products were determined. With the advent of "electronic" experiments, these two functions were usually separated. The experiment consisted, in these cases, of a separate target plus a variety of detectors that were used

1. to determine if (interesting) interactions were taking place in this target, and
2. to measure the properties of the reaction products in the case of (interesting) events.

However, there were also certain classes of experiments in which the combination of the functions of target and detector was maintained. These experiments were all designed to study very rare phenomena, and required therefore a very large target mass. Examples of such phenomena include neutrino interactions, proton decay and cosmic rays at the very highest energies. Most experiments designed for this purpose consist of a very large, instrumented target.

1.3.2.1 *Neutrino experiments.*

Neutrinos, produced in weak interaction processes such as pion, muon and kaon decay, or in fusion of hydrogen into helium nuclei in the Sun, are extremely elusive particles. For example, the mean free path of a typical solar neutrino amounts to about a lightyear worth of iron. Since the cross section for ν-induced interactions is proportional to the neutrino energy E ($\sigma \approx 10^{-41} E$ cm^2, with E expressed in MeV), the situation is somewhat less extreme for accelerator-produced neutrinos, which have typically energies in the (multi-)GeV range. But even for such neutrinos, the interaction probability in a one-kiloton target/detector is only of the order of 1 in a billion. Intense beams and very massive instrumentation are therefore needed to study the properties of neutrinos and details of their interactions with matter.

Large neutrino experiments in high-energy beams have been operating at CERN and at Fermilab since the early 1970s. The first generation of these experiments, WA1 (CDHS, see Figure 1.3) and WA18 (CHARM) at CERN, and CCFRR at Fermilab, had

FIG. 1.3. The WA1 neutrino detector combination that operated at CERN from 1976 to 1984. Large slabs of absorber material (iron) are interleaved with layers of a plastic scintillator. The rear part of the detector, located at the left-hand side of the picture, is instrumented with wire chambers intended for tracking muons generated in charged current interactions and/or charmed particle production. Photograph courtesy CERN.

an instrumented mass of the order of one kiloton. The absorber material (iron in CDHS and CCFRR, marble or glass in the case of CHARM) was arranged in the form of large slabs perpendicular to the beam. These slabs were interleaved with layers of active material: plastic or liquid scintillators, proportional wire chambers, drift chambers, streamer tubes, spark chambers, *etc*.

A crucial aspect of neutrino interactions is the production of muons. These particles distinguish charged current reactions from neutral current ones. Dimuon events may indicate the production of charmed particles. In order to identify these muons and to measure their momenta, the rear part of the detector (the muon spectrometer) was located inside a magnetic field. A fiducial volume for neutrino interactions was defined in the front part of the detector. Particles produced in interactions in this volume and penetrating into the muon spectrometer were muons by definition and their momenta were measured from their deflection in the magnetic field.

The hadronic part of the interactions was measured calorimetrically, by integrating the signals from the active layers in the fiducial volume. By comparing these signals with those created by hadrons of known energy, which were used to calibrate the detector response, this hadronic energy could be estimated with a reasonable precision, typically of the order of 10%.

These initial experiments contributed greatly to our understanding of nucleon structure and of the weak interaction. They also led to accurate predictions of the masses of the intermediate vector bosons (W and Z), which were discovered shortly afterwards (1982).

Later generations of these experiments focused on more detailed aspects of neutrino physics, for example the purely leptonic scattering processes $\nu_\mu e \to \nu_\mu e$ and $\nu_\mu e \to \nu_e \mu$ (CHARM II). The NOMAD and CHORUS experiments at CERN concentrated primarily on direct detection of neutrinos from the third generation, ν_τ. The instrumented-target setup has been abandoned for this purpose. For example, the CHORUS experiment [Esk 97] used a photographic emulsion target. The rest of the equipment was used to identify candidate ν_τs interacting in this emulsion ("triggering"). After exposing the emulsion to a large number of neutrinos ($\sim 10^{18}$), the stack was developed and scanned for ν_τ interaction vertices in the vicinity of the coordinates indicated by the trigger procedure. The same technique was used in the OPERA experiment, which looked for (and found [Aga 15]) ν_τ interactions in a beam created at CERN and sent over a distance of 730 km through the Earth to the center of the Gran Sasso tunnel in Italy.

Every neutrino experiment ever carried out at an accelerator has been searching for evidence of *neutrino oscillations*, the weak interaction equivalent of the strangeness oscillations observed in the neutral kaon system. If the neutrino rest mass is different from zero, there is no reason why the three neutrino flavors (ν_e, ν_μ, ν_τ, the eigenstates of the weak interaction) should correspond to mass eigenstates. In that case, the wavefunctions of the neutrino flavors would be linear combinations of the wavefunctions of the mass eigenstates. Since each of the latter propagates differently (different masses lead to different velocities, for a given energy), the wavefunction of the neutrino changes as a function of the distance traveled. In this process, the relative contributions of the different mass eigenstates to the neutrino wavefunction are modified, and it is straightforward to show that the neutrino wavefunction at some distance from its source can be written as a linear combination of the different weak interaction eigenstates (see also Section 11.2). When the neutrino is made to interact at a certain distance from where it was produced, there is thus a non-zero probability that the weak interaction selects another eigenstate from the wavefunction and that the original neutrino appears to have changed its flavor.

The probability for such "flavor oscillations" to occur increases (initially proportionally) with the distance traveled. The idea of neutrino oscillations has repeatedly been proposed as a plausible explanation for the "solar neutrino deficit," the lack of observed solar neutrino (ν_e) induced interactions here on Earth. In 1998, the Japanese SuperKamiokande experiment (Figure 1.6) reported an asymmetry in the ν_μ/ν_e ratio between events caused by neutrinos produced in the atmosphere above and below (*i.e.*, across the entire Earth) the detector. This phenomenon is considered the first direct evidence for oscillations and thus implies that neutrinos do have a non-zero rest mass.

More evidence for neutrino oscillations has been provided by several other experiments. The OPERA experiment, mentioned above, used a neutrino beam produced by the CERN accelerator complex. The KAMLAND experiment in Japan [Gan 11] and the Daya Bay experiment in China [An 12] observed the disappearance of $\bar{\nu}_e$s produced by nuclear reactors located at different distances. And the Sudbury facility in Canada

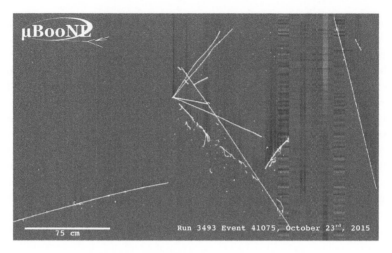

FIG. 1.4. Image of an interaction by a 800 MeV ν_μ produced at Fermilab in the μBooNE liquid argon detector. Image courtesy A. Ereditato.

demonstrated conclusively that the solar neutrino deficit is indeed also due to neutrino oscillations [Ahm 02].

A new generation of "short and long baseline" neutrino experiments is now underway in the United States, where beams of neutrinos produced at Fermilab are or will be sent to large detectors located at different distances, ranging from a few km (μBooNE, at the Fermilab site) to 1,300 km (DUNE, in South Dakota) from the neutrino source. The K2K experiment in Japan is a long baseline experiment in which neutrinos produced at KEK are being sent to SuperKamiokande, 250 km away [Ahn 06]. Interestingly, some of these new experiments use detectors that can be considered genuine "electronic bubble chambers," in the sense that they generate images of the neutrino interactions with bubble chamber quality, but operate completely electronically. They are kiloton liquid argon detectors that work as a Time Projection Chamber (TPC), a technology that was pioneered for the ICARUS experiment [Alm 04] (Gran Sasso laboratory, Italy), and further perfected at the University of Bern, Switzerland [Ere 13]. Figure 1.4 shows an example of a neutrino interaction in the μBooNE detector.

1.3.2.2 *Proton decay/Cosmic-Ray detectors.* The discipline called nowadays experimental particle physics originated from the study of "cosmic rays." Particles such as the muon, the pion and the kaon, which provided the first evidence that the pre-World War II Standard Model (in which protons, neutrons, electrons and neutrinos were considered the elementary building blocks of all matter) was flawed, were all discovered in balloon experiments high up in the stratosphere.

The cosmos is a rich source of information, and cosmic-ray experiments try to exploit this source to the fullest extent possible. In 1987, a Supernova explosion (SN1987a) was observed in the Large Magelhanic Cloud, a galaxy at a distance of about 160,000 lightyears from here. In (core collapse) Supernova explosions, of the order of 10^{57} neutrinos are released when the star undergoes a phase transition in which all available protons in the core of the star are simultaneously transformed into neutrons and the star collapses into an object with the density of nuclear matter. Some of these neutrinos were observed in a number of large underground detectors (installed in the Kamioka mine in Japan, the Morton-Thiokol salt mine in Ohio and under Mount Andyrchi in Russia). Apart from the information that these (25) events provided about the Supernova phenomenon itself, they also were invaluable for particle physics. The neutrinos had traveled for 160,000 years on their way to the detectors. Yet, they all arrived within a time span of about 10 seconds. If neutrinos have a non-zero rest mass, their velocity is smaller than the speed of light. In that case, the time it takes them to travel the distance from their source to the detector is energy dependent. The energies of the detected SN1987a neutrinos varied by an order of magnitude (from 2 to 20 MeV). The Supernova observations put an upper limit of about 6 eV/c^2 on the ν_e rest mass, one of the most restrictive values to date [PDG 14].

Cosmic-Ray experiments usually concentrate on a number of different phenomena. Unexpected surprises of the type just mentioned are of course high on every experiment's wish list, and most experiments are designed to be sensitive to such events. But there are also more common phenomena, with more or less predictable rates, that are of great interest, both from an astrophysics and a particle physics perspective. Among these, we mention

- The production of *atmospheric neutrinos*, mentioned in the previous subsection. Most of these particles are the result of the decay of pions and kaons in the Earth's atmosphere. Most of the muons that reach the Earth's surface and which are the most common manifestation of "cosmic rays," are produced in the same processes.

- The detection of *solar neutrinos*, produced in the nuclear fusion of hydrogen into helium and in some higher-order processes through which the Sun generates energy. These measurements, which have been carried out since the 1960s, provide insight into the processes taking place in the core of the Sun and into details of the weak interaction (in particular the questions of neutrino oscillations and the neutrino rest mass). Detection of solar neutrinos in real time has been reported by the SuperKamiokande and SNO Collaborations.

- The detection of *extremely energetic particles* entering the Earth's atmosphere. The energies of these particles are sometimes in excess of 1 Joule ($6 \cdot 10^{18}$ eV). The mechanisms through which protons and other ions may acquire such exceptionally high energies are still mysterious. The energy spectrum exhibits some interesting, unexplained features, such as the occurrence of "kinks" at energies of $4 \cdot 10^{15}$ eV and $3 \cdot 10^{17}$ eV. Also, the chemical composition of the spectrum changes abruptly at these energies. In the interactions of such particles in the Earth's atmosphere, unusual phenomena may occur, such as the so-called *Centauro events*, in which practically all the available energy is used to produce π^0s. An interesting aspect of these measurements is the question whether the spectrum at the high end will continue to decrease according to a power law, or if there is effectively a maximum energy for the detected cosmic protons. This so-called *GZK cutoff* is the result of photo-pion production off the cosmic microwave background radiation (the omnipresent 2.7 Kelvin photons) and limits the mean free path of protons with energies in excess of $6 \cdot 10^{19}$ eV to a mere 100 Mpc. There seems to be experimental evidence for such a cutoff, although the experiments involved do not agree on the details [Ber 14].

- *Gravitational waves.* Predicted by Einstein in 1916, it has taken exactly 100 years before the first *direct* detection of this phenomenon was reported by the LIGO experiment [Abb 16], a modern (and 360 times larger) version of the original Michelson-Morley interferometer that was used to study the *ether* issue at the end of the 19th century. Indirect evidence for the emission of gravitational waves by a system of two extremely massive objects separated by a short distance and rotating around a common center of mass was obtained by Hulse and Taylor, who observed that the rotation period decreased very slowly, at a rate predicted by Einstein's theory of General Relativity, as a result of the loss of energy through the emission of gravitational waves [Tay 79].

Cosmic-Ray physics has a long tradition and new phenomena are by definition rare. Therefore, a very large instrumented mass is a *conditio sine qua non* for all but the most sophisticated, highly specialized experiments. There are a large number of experiments active, or in various stages of preparation in this very dynamic field. Appendix C contains a list of these experiments, their goals and some of their design characteristics.

Cosmic-Ray physicists were among the pioneers of the use of large sampling calorimeters. Systems consisting of more than 5 tons of lead and iron, interleaved with ionization chambers, were already operating high in the Caucasus mountains in the 1950s [Ale 63]. Nowadays, high mountains on almost all continents host large cosmic-ray detectors. Often, the information from these detectors, which aim to locate the primary vertex and to detect the early part of the cosmic shower as well as possible, is supplemented by data from detectors at much lower altitude. Sometimes, the entire detector is installed at sea level near a major research facility. This offers the possibility to use more sophisticated detection techniques than are usually available in remote locations. An example is the KASCADE-Grande experiment, which operates near Karlsruhe, Germany [KAS 16]. This experiment, shown in Figure 1.5, employs a large calorimeter system based on tetramethylsilane (TMS), a dielectric liquid that requires extraordinary levels of purity and thus needs to be handled in state-of-the-art clean rooms.

Because of the type of phenomena to be studied, many cosmic-ray detectors are located in deep mines, in tunnels under high mountains or deep under water or ice. Some use the extremely low background conditions reigning in these environments for simultaneous studies of another hot topic: proton decay. The fundamental laws of energy and charge conservation do not prevent a proton from decaying into, for example, a positron and a neutral pion, or a μ^+ and a neutral kaon. That these processes do not occur at a rate that has permitted detection so far, is explained by "the law of conservation of baryon number." This law implies that the lightest baryon (the proton) must be stable, just as charge conservation implies that the lightest charged particle (the electron) must be stable. However, conservation of baryon number, which in the framework of the Standard

Model of particle physics is equivalent to the impossibility of a transformation of quarks into leptons or vice versa, is a much more questionable principle than conservation of energy or electric charge. In many theories, this conservation law has to break down at some level, with proton decay as its most dramatic consequence.

If protons decay at all, they are certainly not in a hurry to do so. Current experimental limits have set the partial lifetime for the $p \rightarrow e^+\pi^0$ decay process at more than 10^{32} years, *i.e.*, more than 21 orders of magnitude longer than the estimated age of the Universe!

Particle decay is a statistical process. A lifetime of 10^{32} years implies that in an assembly of 10^{32} protons, on average about 1 decay per year will occur. An example of an assembly containing 10^{32} protons is 300 m^3 of water, the contents of a good-size swimming pool. Therefore, experiments studying proton decay need a very large instrumented volume indeed.

One class of detectors that have greatly contributed to pushing the experimental limits on proton decay to their present values are the *water Čerenkov calorimeters*. An enormous volume of high-purity water is viewed by a large number of photomultipliers that record the passage of relativistic charged particles in this water, through the characteristic blue light that is emitted in this process. When a proton decays into a positron and a π^0, typically five relativistic particles are produced, the positron and the two e^+e^- pairs created in the conversion of the two γs from the π^0 decay. The total energy carried by these particles adds up to the proton rest mass, 938.3 MeV/c^2. Such a decay process should thus give a very characteristic signal, recognizable in this setup.

FIG. 1.5. The KASCADE-Grande cosmic-ray experiment located near Karlsruhe, Germany. The experimental setup consists of a large TMS calorimeter, located in the central building, surrounded by numerous smaller, plastic-scintillator based counters that detect ionizing particles (mainly muons) produced in atmospheric showers. These detectors are housed in the small buildings. Photograph courtesy Kernforschungszentrum Karlsruhe.

FIG. 1.6. Picture taken inside the SuperKamiokande detector, showing part of the 11,200 PMTs that detect the Čerenkov light when the vessel is filled with 55 kilotons of ultrapure water. Note the three technicians standing on the left. Photograph courtesy SuperKamiokande.

Examples of experiments searching for proton decay with such detectors include SuperKamiokande (Japan). Figure 1.6 shows what this detector looks like from the inside. A similar detector, containing 1,000 tons of *heavy water* is operating in a mine near Sudbury (Canada). This detector is primarily intended for studies of solar neutrinos, hence its name: Solar Neutrino Observatory. Examples of experiments that (are planning to) use a *natural environment* to create a gigantic water Čerenkov detector include ANTARES (Mediterranean, near France), NEMO (Mediterranean, near Sicily), NESTOR (Mediterranean, near Greece), and AMANDA / ICECUBE (in the ice under the South Pole).

1.3.3   *4π detectors*

As the energy frontier was pushed higher and higher by new generations of particle accelerators, the shower counters, or *calorimeters* as we prefer to call them nowadays, became gradually the cornerstone of particle physics experiments, and their role in the experiments changed considerably in this process. There were several reasons for this development. First, the precision of the information provided by calorimeters improves with increasing energy, whereas the curvature of charged particles in a given magnetic field, and thus the precision of the momentum measurement, decreases. Calorimetric measurement of the four-vectors of particles other than photons thus became an increasingly attractive alternative. Also, one realized that calorimetric information, in combination with tracking data, allowed the identification of certain particles, in particular electrons. Given the crucial role of electrons in signaling new physics (*e.g.* the discovery of heavy quarks and the intermediate vector bosons), this was a major consideration.

There was another important development that changed the face of experimental particle physics. As the energy increased, and in particular as a result of the transition from fixed-target to colliding-beam geometries, the emphasis of many experiments (especially the ones involving hadron beams) changed from a detailed reconstruction of the four-vectors of *all the particles* produced in the interaction process under study to a measurement of more global event characteristics, usually summarized in the term *energy flow*. Features like missing (transverse) energy and the production of *jets* provided crucial signatures for interesting physics, especially when observed in combination with electron or muon production.

When a beam of particles with mass m and energy E_{beam} strikes a fixed target consisting of particles with mass M, the center-of-mass energy, that is the energy available for the production of new particles in the collision, equals

$$E_{\text{cm}} = \sqrt{m^2 + 2ME_{\text{beam}} + M^2} \qquad (1.1)$$

For high energies, $E_{\text{beam}} \gg m, M$, E_{cm} becomes $\sqrt{2ME_{\text{beam}}}$. The remaining energy, almost all of it for very high values of E_{beam}, is needed to transfer momentum to the reaction products, thus fulfilling the requirements of the law of conservation of momentum.

When two identical particles, each with energy E_{beam} and traveling in opposite directions, hit each other head-on, the center-of-mass frame is the laboratory frame. In that case, the energy available for the production of new particles is equal to the *total energy* involved in the collision, $2E_{\text{beam}}$.

Clearly, the latter method becomes rapidly more favorable for the study of new phenomena at increasing energies. For example, if a 100 GeV proton beam strikes a proton target, only 14 GeV is available in the center of mass, or 14% of the energy available in a head-on collision of two 50 GeV protons. For the 1 TeV case, this fraction drops to 4.5%.

Colliding-beam experiments were pioneered at e^+e^- storage rings, such as ADONE (Frascati, Italy), DORIS (DESY, Hamburg) and SPEAR (SLAC, Stanford) in the 1960s. The first hadron collider was the ISR, which operated at CERN in the 1970s. But the great value of colliders, and the crucial role that calorimeters may play in providing access to new physics, became evident with the operation of the anti-proton–proton collider (Sp$\bar{p}$S) at CERN, in the first half of the 1980s.

In 1982, the UA1 and UA2 (see Figure 1.7) experiments discovered the W intermediate vector boson [Arn 83a, Ban 83], on the basis of the signatures mentioned above: an energetic charged lepton (electron or muon), in combination with missing transverse energy (caused by the neutrino accompanying the charged lepton in the decay $W \to l\nu$). This discovery contributed in no small measure to the change in emphasis mentioned above.

FIG. 1.7. The UA2 calorimeter that operated at CERN's proton–anti-proton collider (Sp$\bar{p}$S) from 1980 to 1986) and played a crucial role in the discovery of the W and Z bosons. Photograph courtesy CERN.

Calorimeters are ideally suited to provide energy flow information. Moreover, they can provide this information, *e.g.* about the total amount of energy detected in an event, about geometric imbalances in this energy (missing transverse energy) and about jet production almost instantaneously, thus offering powerful event selection ("triggering") possibilities.

Since 1980, almost all major new accelerator facilities for experimental particle physics have been colliding-beam machines: the Sp$\bar{p}$S, LEAR, LEP and the LHC at CERN, the Tevatron at Fermilab, PETRA and HERA at DESY, PEP I,II and the SLC at SLAC, CESR at Cornell, TRISTAN and the B-factories at KEK, DAPHNE at Frascati. The experiments at these facilities have put great emphasis on calorimetry, with hermetic coverage usually as one of the most important design criteria.

The *hermeticity* of a calorimeter system indicates the fraction of the 4π solid angle surrounding the interaction vertex that is covered by this system. One hundred percent coverage is not possible in colliding-beam experiments, because of the space requirements of the beam pipe, of signal cables from detectors operating inside the cavity formed by the calorimeter system, *etc.* However, hermeticities in excess of 90% are routinely achieved in modern experiments.

One of the first hermetic calorimeters to operate at a storage ring was the *Crystal*

Ball detector [Ore 82], a 4π detector consisting of NaI(Tl) crystals. This detector was used at the e^+e^- collider SPEAR at SLAC to study radiative transitions within the charmonium family (the $c\bar{c}$ bound states) [Blo 83]. A slice of this detector is schematically shown in Figure 1.8.

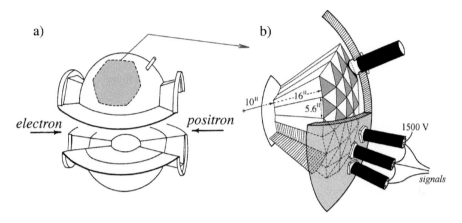

FIG. 1.8. Schematic of the Crystal Ball detector, installed at the interaction point of the electron and positron beams of the SPEAR collider (*a*). The shaded area is blown up and shown in some detail (*b*).

Like in most other experiments at low-energy e^+e^- colliders, this calorimeter focused entirely and exclusively on the detection of electromagnetic showers. Many subsequent experiments at e^+e^- machines have been inspired in their design by the successes of the Crystal Ball. We mention the calorimeters of the CUSB [Bor 80, Mag 81] and CLEO [Kub 92] experiments at CESR, the Crystal Barrel at LEAR [Ake 92], BES at BEPC [Abl 10], BABAR at SLAC [Aub 02] and Belle at KEK [Abe 10].

At the hadron colliders, which operate at much higher energies than the e^+e^- ones just mentioned, the calorimeter designs were inspired by the successes of the UA1 and UA2 experiments, with emphasis on the measurement of energy flow, on detection of jets and on electron identification. A modern version of a calorimeter system at a hadron collider, the ATLAS detector at the LHC, is shown (partially) in Figure 1.11. The electron–proton collider HERA was a unique machine, with special requirements on the calorimetry. For example, some of the crucial physics topics studied at HERA called for the best possible hadronic energy resolution. Results from the calorimeters of ZEUS (one of the experiments at HERA) feature prominently at various places in this book.

1.4   Detection Mechanisms

In this section, the various mechanisms through which calorimeter signals are generated are described. Apart from the cryogenic phenomena discussed in Section 1.4.4, all these mechanisms have been applied in real-life experiments, both in homogeneous and in sampling calorimeters.

1.4.1   *Scintillation*

When charged particles traverse matter, they lose energy through the electromagnetic interaction with the Coulomb fields of the electrons. This energy may be used to ionize the atoms or molecules of which the traversed medium is composed, or to bring these atoms/molecules into an excited state. Scintillation is a phenomenon associated with the latter process.

The excited atomic or molecular states are unstable. Usually, the excited atom or molecule quickly returns to the ground state. In this process, the excitation energy is re-leased in the form of one or more photons. The timescale of this process is determined by the excitation energy, by the number of available return paths, and by the quantum numbers of the states involved (wavefunction overlap). When the energy differences are such that the emitted photons are in the visible domain, this process is called *fluo-rescence* or *scintillation*. Typical timescales range in that case from 10^{-12} to 10^{-6} s, although exceptions in either direction may occur. In general, the timescales get shorter as the molecules get more complex. This can be simply understood from the fact that the density of excited states, and therefore the number of different ways in which an excited molecule can get rid of its excess energy increases sharply with the complexity of the molecule.

Relatively simple scintillating crystals, such as NaI(Tl) and BGO, have decay times of several hundred ns (see Table B.5), orders of magnitude longer than the decay times of complex organic scintillators, such as the plastics anthracene and polystyrene.

Historically, scintillation was the first physics process to be used for the genera-tion of calorimetric signals. And until this day, a large number of calorimeters in a wide variety of particle physics experiments rely upon scintillation light as the prime source of information. Two inventions have played a crucial role in the development of scintillator-based particle detectors in general, and calorimeters in particular:

- *The photomultiplier tube.* Almost 70 years old, the PMT which allows the conver-sion of single photons into electric signals, is still playing a crucial role in many experiments. Although some new devices based on semiconductor applications (such as the Hybrid Photo Detector, the Avalanche Photo Diode and especially the silicon photomultiplier) have more or less successfully addressed some PMT weaknesses, such as the sensitivity to external magnetic fields, the essentially noise-free signal amplification offered by PMTs is still an extremely attractive feature.
- *Wavelength shifters.* These devices absorb the scintillation light and re-emit it at a lower energy (longer wavelength). This development made it possible to ap-ply scintillator-based calorimeters in experiments requiring hermetic coverage, such as the 4π experiments in a colliding-beam setup. The light produced in scin-tillator plates oriented perpendicular to the direction of the incoming particles can be wavelength-shifted and at the same time redirected towards the rear end of the calorimeters, where it can be converted into electric signals. Figure 1.9

shows schematically the readout of scintillator calorimeters with and without wavelength-shifting plates. The price to be paid for these advantages is a loss of light, because of inefficiencies in the process and a longer signal duration, since the wavelength shifters are usually somewhat slower than the scintillators whose light they shift (see Table B.6).

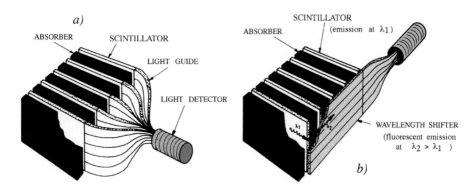

FIG. 1.9. Schematic of the readout systems of scintillator calorimeters without (a) and with (b) wavelength-shifting plates.

Another crucial development, of which the consequences have not been limited to scintillator-based calorimeters alone, concerned the notion that in sampling calorimeters, the active layers do not need to be oriented perpendicular to the direction of the incoming particles (see Section 2.6). Calorimeters may work very well for completely different orientations of the active material, including geometries in which the active layers (*e.g.*, scintillating fibers) run *in the same direction* as the incoming particles. This notion has led to a wide variety of different geometries, including *accordion, tile, lasagna, bayan* and other structures.

The development of plastic optical fibers has also greatly influenced the design of scintillator calorimeters. Scintillating fibers usually consist of a polystyrene core (index of refraction $n = 1.59$), surrounded by one or several layers of cladding with (gradually) lower values of n.

Unlike the optical fibers used for telecommunication purposes, which are designed to *transport light injected along the fiber axis*, the scintillating fibers used in particle physics experiments are both the *source of the light* (generated isotropically) and the medium through which this light is transported to a place where it can be converted into an electric signal. The fraction of the light that is trapped is proportional to the *numerical aperture* $\sqrt{n_{\text{core}}^2 - n_{\text{clad}}^2}$, and most of the light is traveling near the *critical angle*, defined as $\theta_{\text{cr}} = \arcsin\left(n_{\text{clad}}/n_{\text{core}}\right)$.

Apart from these chemically doped optical fibers, undoped plastic fibers are also being applied in particle detectors. For example, clear plastic fibers based on a PMMA core ($n = 1.49$) surrounded by lower-index (fluorinated plastic) cladding material is used to detect Čerenkov light produced in the particle absorption process (see Section

1.4.2).

Optical fibers are being used in many calorimeters, either as the active medium sampling the showers, or as wavelength shifters, converting the scintillation light, *e.g.*, blue light from scintillator plates, to a longer wavelength (*e.g.*, green) and transporting it to light detectors located in a convenient position.

Among the advantages offered by such fibers, we mention

- The perfectly hermetic calorimeter structure that can be achieved,
- The very high sampling frequency (good energy and position resolution!) that can be obtained using fibers as the active medium,
- The high signal speed that can be obtained,
- The arbitrary granularity allowed by the fiber structure,
- The high light yield that can be achieved, and
- The excellent cost/performance ratio

We elaborate on the relevance of these factors, and on the specific advantages of plastic as active material for *hadron* calorimeters in the next chapters.

1.4.2 Čerenkov radiation

When a charged particle travels faster than the speed of light in a certain medium ($v > c/n$, or $\beta = v/c > 1/n$, with n the medium's index of refraction), it loses energy by emitting Čerenkov radiation. This radiation is emitted at a characteristic angle, the Čerenkov angle $\theta_C = \arccos(n\beta)^{-1}$, with the direction of the particle. Therefore, this radiation forms a cone with half-opening angle θ_C. The amount of energy is proportional to $\sin^2 \theta_C$.

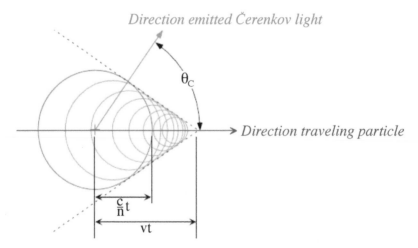

FIG. 1.10. The principle of Čerenkov light emission by a superluminal particle. In a time t, the particle travels a distance vt, while the light it emits travels a distance ct/n. The wavefronts of the light emitted by such a particle form a cone with half-opening angle θ_C.

The spectrum of this Čerenkov radiation exhibits a characteristic $1/\lambda^2$ dependence and, therefore, the visible part of the Čerenkov spectrum is experienced as blue light. This blue light can be abundantly observed in highly radioactive environments, *e.g.*, the moderating liquids in nuclear reactors. It is also a source of light deep in the oceans, where it is created by penetrating cosmic rays.

The emission of Čerenkov light is only a very minor source contributing to the energy loss of the particles. For example, in water, a charged particle with $\beta \simeq 1$ loses about 400 eV/cm in the form of visible Čerenkov photons. That is some four orders of magnitude less than its energy loss through other processes, in particular ionization (2 MeV/cm).

Since the Čerenkov mechanism is sensitive to the *velocity* of particles, it can be used to determine the *mass* of particles of which the *momentum* has been determined by means of deflection in a magnetic field. A variety of devices (threshold Čerenkov counters, differential Čerenkov counters, ring imaging Čerenkov detectors) have been developed to separate electrons, pions, kaons, protons and deuterons from each other, exploiting this effect.

As we will see in later chapters, calorimeters based on the detection of Čerenkov light exhibit some interesting properties, which may be ideal for certain very specific applications, *e.g.*, jet detection very close to the beam pipe in LHC experiments, or dual-readout calorimetry (Section 8.2).

A very important aspect of Čerenkov light is its *instantaneous* character. There are no delaying factors, such as the lifetime of a metastable excited state, which affect the time characteristics of detectors based on scintillation light. Therefore, Čerenkov detectors, including calorimeters, are the instruments of choice for experiments in which ultimate signal speed is required.

1.4.3 *Ionization*

When charged particles traverse matter, they may ionize the atoms of which this matter consists. One or several electrons are released from their Coulomb field in this process, leaving behind an ionized atom. Collection of these liberated electrons is applied as the signal-producing technique in a wide variety of particle detectors. The electrons produced along the trajectory of the ionizing particle may or may not be amplified in this process.

In ionization chambers based on liquid media, no amplification takes place. A potential difference applied over the gap containing the liquid separates the electrons from the ions. The electrons are collected at the anode, the ions at the cathode. In order for this method to work properly, the mean free path of the electrons in the liquid should be long, considerably longer than the size of the gap between the electrodes. Therefore, noble liquids such as argon, krypton and xenon, which have no desire to capture loose electrons wandering around since all the electronic shells of their atoms are filled, are the media of choice in these detectors.

To ensure a sufficiently long mean free drift path for the electrons, very stringent purity standards have to be met in noble-liquid ionization chambers. In particular, contamination by electro-negative elements such as oxygen have to be kept below the 1 ppm (10^{-6}) level. These purity requirements are even more severe in so-called "warm liquids," a class of organic, hydrocarbon-based materials which also exhibit an absence of affinity for electrons. Unlike the noble elements, which require cryogenic operating conditions, these materials are liquid at room temperature. However, they require purity levels in the 1 ppb (10^{-9}) range, which has proved to be a major obstacle for large scale application.

Calorimeters based on noble liquids as active media have been used in particle physics experiments since the 1970s. The technique was pioneered with liquid argon (LAr). Liquid argon is cheap, abundantly available and the required purity levels can be easily achieved and maintained. Among the largest LAr calorimeter systems operating today, we mention the ATLAS experiment at CERN's Large Hadron Collider (Figure 1.11). Previously, the D0 experiment at Fermilab's Tevatron collider and the H1 experiment at the HERA electron–proton collider (DESY, Hamburg) were centered around large LAr sampling calorimeters. A very large *homogeneous* LAr system operating as a Time Projection Chamber was pioneered for ICARUS, in the Gran Sasso laboratory. A 170 kiloton successor (μBooNE) detects neutrino interactions at Fermilab [MB 15] and even much larger devices are being planned for the future.

Other noble liquids, krypton (LKr) and xenon (LXe) are much more expensive and are therefore only used in applications requiring the specific advantages offered by these liquids, such as a higher density, or a higher Z value. For example, the electromagnetic sampling calorimeter of the NA48 experiment at CERN, which studied CP violation in the K^0 system, and its successor NA62, intended for the study of extremely rare kaon

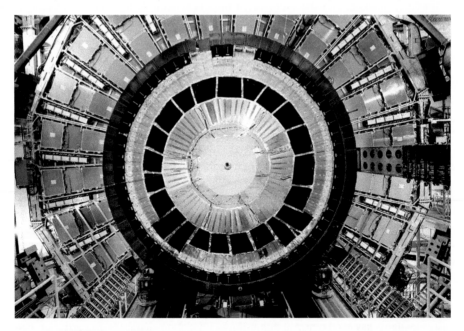

FIG. 1.11. View of one of the endcaps of the ATLAS calorimeter. Photograph courtesy CERN.

decays, is based on LKr as the active medium. This makes it possible to achieve the energy resolution for γ detection required by the sensitivity goals of these experiments [Fan 07].

Another experiment looking for extremely rare decays ($\mu^+ \rightarrow e^+\gamma$) is MEG at PSI (Villigen, Switzerland) [Ada 13a]. Their calorimeter consists of 900 liters of liquid xenon, which was chosen because of its excellent properties for γ detection, a result of the high Z value (54). Apart from the ionization charge produced by charged particles traversing the detector, LXe is also a very bright and fast scintillator, and MEG uses this feature as the detection principle. This property has also been successfully applied before in dedicated nuclear γ detectors, mainly intended for use in high-radiation environments [Gib 84, Apr 87]. The properties of liquid xenon also make it a detector of choice for certain dark matter searches and neutrinoless $\beta\beta$ decay. Elastic scattering of weakly interacting massive particles (WIMPs) gives the heavy Xe nucleus a relatively large recoil energy, which may produce a measurable signal because of the high light yield combined with the sensitivity for ionization charge. And since xenon consists for 8.9% of the isotope ^{136}Xe, which can only decay through the conversion of two neutrons into two protons, LXe can at the same time act as the source *and* the detector for this rare process. Several experiments based on large LXe TPCs are currently operating or being prepared with these goals in mind [Bau 15].

Unlike scintillating crystals, noble liquids are very radiation hard. This property has played a role in the choice of a lead/LAr sampling calorimeter for the electromagnetic shower detection in the ATLAS experiment at Large Hadron Collider.

A totally different class of ionization calorimeters is based on gaseous media. In these devices, which exist in a wide variety, the electrons produced in the ionization process undergo considerable multiplication before being collected at the anode. As they are accelerated in the electric field between the anode and cathode, the electrons may acquire enough energy to ionize other atoms and thus release *secondary* electrons. These may in turn release tertiary electrons, *etc.* The result is an *avalanche* of electrons arriving at the anode and constituting the signal. Since electric fields in the vicinity of a charged object (the anode) strongly depend on the distance r from that charged object, and since the energy acquired by the electrons is proportional to the electric field strength, this multiplication process works best in the immediate vicinity of the anode (small values of r). For this reason, the anode is often made of very thin (30 μm) wires.

Wire chambers may operate in a variety of modes, depending on the type of gas mixture and on the voltage difference between anode and cathode: the proportional mode, the streamer mde, the Geiger mode, *etc.* The time needed for the charge to arrive at the anode may provide information about the spatial coordinates of the particle that caused the signals. This principle is applied in *drift chambers* and in *Time Projection Chambers*. A large number of calorimeter systems rely on wire chambers or tubes of some sort to provide the experimental signals. Especially when very large surface areas have to be covered, this is often the most cost effective solution available.

Thin wires make such detectors very fragile. A short circuit caused by a broken wire may wipe out a large section of the detector system, as has been experienced by several experiments. In the past 20 years, we have witnessed an enormous development of detectors in which the strong electric fields needed for the multiplication of the ionization charge are created in some other way than by means of thin wires. The general term for such devices is "micropattern gas detectors." Thin gaps, tiny holes or some micromesh structure provide the field-shaping geometry, and new developments in photolithography and microelectronics have led to a new class of extraordinary detectors, some good reviews of which can be found in PhD theses [Daf 07, Che 13]. Not surprisingly, these devices have become increasingly popular with designers of tracking systems for new experiments.

Finally, there are also solid state devices that are being used as detectors of charges produced by passing ionizing particles. These are the semiconductors already encountered in Section 1.2. Silicon, germanium and gallium arsenide have all been applied for particle detection since more than 25 years [Bor 89, Ang 90].

The outer-shell atomic levels of these semiconductor crystals exhibit a band structure, consisting of a valence band and a conduction band, separated by a "forbidden" energy gap, in which no energy levels are available. An ionizing particle passing through such a semiconductor excites electrons from the valence band into the conduction band. For every electron that jumps into the conduction band, a *hole* remains in the valence band. This hole is positive relative to the sea of negative electrons in the valence band. Therefore, it acts as a positive charge carrier, and its movement through the semiconductor crystal constitutes an electric current, just as does the movement of the electrons in the conduction band. The electron–hole pairs created by ionizing particles may be collected by means of an electric field.

This technique has several advantages. The energy gap between the valence and conduction bands is very narrow, typically of the order of 1 eV and, therefore, very little energy is required for the production of one electron–hole pair. For example, in silicon every 3.6 eV of deposited radiation energy yields one electron–hole pair. This is typically one order of magnitude less than the energy needed to produce one electron–ion pair in gases and two orders of magnitude less than the energy required for the production of one photoelectron in scintillation counters. Therefore, semiconductor crystals offer the potential of excellent energy resolution in detectors in which fluctuations in the number of primary charge carriers are the limiting factor for resolution. An example of the performance of such detectors is shown in Figure 1.1.

Charge amplification can also be applied in semiconductors. Silicon is an excellent detector of visible light, especially in the longer wavelength region where the quantum efficiency of photocathodes used in PMTs is decreasing. Silicon diodes, which are a common tool in movement detectors, make use of this characteristic. The photoelectrons may also be internally amplified, which is the operating principle of Avalanche Photo Diodes. If the applied voltage is sufficiently large, a detected photon may cause the device to discharge, and thus act like a Geiger counter. Silicon photomultipliers (SiPM) consist of large numbers of extremely small silicon pixels, each of which operates in that way [Dol 06]. Figure 1.12 shows an example of a SiPM that contains 625 pixels on a surface area of only 1 mm^2. As illustrated by the right diagram, these detectors are excellent single-photon counters. The number of photons constituting a given

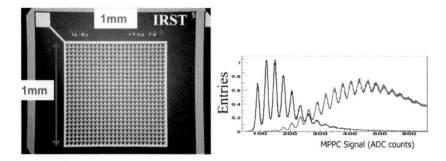

FIG. 1.12. Example of a silicon photomultiplier, and the spectra it produces when exposed to two different sources of visible light.

signal can be precisely determined for up to 20 photons or so. In that sense, as well as in their capability to operate in a magnetic field, they present a clear advantage over PMTs. At the time of this writing, the noise characteristics and the linearity are still considered points of concern. The latter is an issue since a SiPM is essentially a digital detector. When the device from Figure 1.12 is used to detect light pulses from a 1 mm^2 scintillating fiber, it will not produce any signals larger than those from 625 photons, since every pixel will "see" either one or zero photons from that pulse. And when several photons from the light pulse hit the same pixel, that pixel produces the same signal as for one photon. These and other important saturation effects and their consequences are discussed at length in Chapters 8,9.

An additional advantage of semiconductor crystals concerns their response time. Because of their greater density and compact structure, they may be considerably faster than other detectors based on the collection of ionization charge.

1.4.4 Cryogenic phenomena

There is a class of highly specialized detectors that employ calorimetric methods to study a series of very specific phenomena in the boundary area between particle physics and astrophysics: dark matter, solar neutrinos, magnetic monopoles, nuclear double β-decay, *etc*. All these issues require precise measurements of small energy deposits. In order to achieve that goal, the mentioned detectors exploit phenomena that play a role at temperatures close to zero, in the few milli-Kelvin to 1 Kelvin range. These phenomena include:

a) Some elementary excitations require very little energy. For example, Cooper pairs in superconductors have binding energies in the μeV–meV range and may be broken by phonon absorption.

b) The specific heat for dielectric crystals and for superconductors decreases to very small values at these low temperatures.

c) Thermal noise in the detectors and the associated electronics becomes very small.

d) Some materials exhibit specific behavior (*e.g.*, change in magnetization, latent heat release) that may provide detector signals.

Many of the devices that have been proposed in this context are still in the early phases of the R&D process. In many cases, this R&D involves fundamental research in solid-state physics and materials science. However, some devices have reached the stage where practical applications have been successfully demonstrated. Among these, we mention

- *Bolometers*, which are based on principle (*b*). These are calorimeters in the true sense of the word, since the energy deposit of particles (in an insulating crystal at very low temperature) is measured with a resistive thermometer.
- *Superconducting Tunnel Junctions*, in which the quasi-particles and -holes (Cooper pairs) excited by incident radiation tunnel through a thin layer separating two superconducting materials.
- *Superheated Superconducting Granules*, which are based on the fact that certain type I superconductors can exhibit metastable states, in which the material remains superconducting in external magnetic fields exceeding the critical field. These detectors are usually prepared as a colloid of small (diameter 1–100 μm) metallic granules suspended in a dielectric matrix (*e.g.*, paraffin). Heat deposited by an interacting particle may drive one of several granules from the superconducting to the normal state. The resulting change in magnetic flux (disappearance of the Meissner effect) may be recorded by a pickup coil.

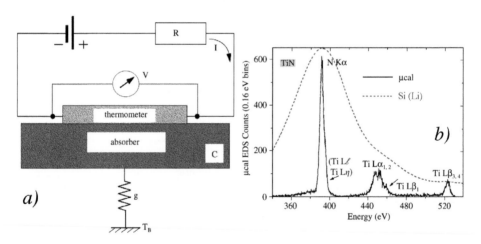

FIG. 1.13. The principle on which a cryogenic calorimeter is based (*a*). Spectrum of X-rays of titanium nitrate, measured with a cryogenic calorimeter, and with a standard Si(Li) semiconductor detector (*b*) [Pre 00].

Figure 1.13a shows the operating principle of a typical cryogenic calorimeter. It consists of an absorber with heat capacity C, a thermometer and a thermal link with a heat conductance g to a reservoir with temperature T_B. The thermometer is typically a thin superconducting strip that operates very close to the transition temperature between the superconducting and normal phases. A small local increase in the temperature may dramatically increase the electric resistance, and thus lower the current flowing through this circuit. The temperature increase needed for this to happen may be caused by phonons created by particles interacting in the absorber, which travel to the surface and break Cooper pairs. The extremely low temperature at which these detectors are operating is needed to reduce the thermal noise, which limits the size of the measurable signals. The use of superconductors as cryogenic particle detectors is motivated by the very small binding energy of the Cooper pairs, ~ 1 meV, compared to 3.6 eV needed to create an electron–hole pair in silicon. Thus, compared to a semiconductor, several orders of magnitude more free charges are produced, which leads to a much higher intrinsic energy resolution (Figure 1.13b).

Because of their sensitivity to very small energy deposits, cryogenic calorimeters are widely used in the search for dark matter, and in particular for low-mass WIMPs. As a matter of fact, the most stringent limits on WIMPs with masses less than 5 GeV/c^2 come from cryogenic experiments such as CDMS, EDELWEISS and CRESST, which all operate bolometric detectors with masses in the 10 kg range at temperatures well below 1 K, in tunnels or deep mines. Plans for upgrades to 100 kg or more exist in all cases. An example of a cryogenic calorimeter looking for neutrinoless $\beta\beta$ decay is CUORE, which is building a detector containing 740 kg worth of TeO$_2$ crystals (*i.e.*, 240 kg of the isotope of interest, ^{130}Te), which will operate at a temperature of 15 mK in the Gran Sasso Laboratory. Superconducting Tunnel Junctions have found useful applications in astronomy, where they are used to detect radiation in the sub-mm wavelength domain.

There is a considerable amount of effort going into the development of these and many related, similarly ingenious devices. However, this highly specialized work falls somewhat outside the scope of this book. The interested reader is referred to reviews of this field that can be found in [Pre 87, Boo 96, Twe 96, Pre 00, Ens 05]. A recent review on dark matter searches is given in [Bau 15]. For all detectors other than calorimeters that are briefly described in this section, [Fab 11] and [Gru 12] are excellent and extensive sources of information.

1.5 Choosing a Calorimeter

In the past decades, detectors measuring the properties of subatomic particles by total absorption (calorimeters) have become crucial components of almost any experiment in high-energy particle physics. Calorimeters exist in a wide variety. The choices of the technology and of the detector parameters for a specific application usually depend on such factors as the physics processes one wants to study, the available budget, the radiation levels and other operating conditions in the environment where the experiment takes place.

It is important to realize that choices made because of one consideration may adversely affect the calorimeter performance in other respects. For example, if one wants the very best energy resolution for electron and photon detection and chooses a calorimeter consisting of scintillating crystals to achieve that goal, then the inevitable consequence of that choice will be very poor resolution for the detection of hadrons, jets and energy flow parameters (missing energy, *etc.*). If that is unacceptable, a reasonable compromise design has to be found, which allows good performance in *all* areas of interest. An example of a project that aims to optimize several aspects of the calorimeter performance simultaneously is discussed in Section 8.2.

The choice of a certain calorimeter system is often based on, or supported by, results from prototype tests in particle beams. The results of such tests have to be interpreted with caution. For example, sometimes the resolution aimed for or claimed on the basis of such test results requires signal integration over a long time, or over a large calorimeter area, while in the practice of the experiment for which the calorimeter is intended, these requirements cannot be fulfilled.

Another reason for caution concerns the fact that in the reality of an experiment often "miscellaneous material" (in the form of support structures, cables, electronics, utilities, *etc.*) is installed in various places in between the interaction vertex and the calorimeter front face. This material affects every particle in a different way, and will in any case cause a degradation in the precision with which the properties of the particles can be determined. One may try to mimick this situation in the test beam setup, but this method clearly has its limits. For example, the effects of material on photons are different from the effects on electrons. Therefore, the results of tests with electron beams cannot be assumed to be also valid for photons, which are not readily available as monoenergetic collimated beams (Section 9.3.4).

But the most important reason for caution in the interpretation of test beam results concerns the fact that the experimental conditions are fundamentally different from those in the experiment for which the calorimeter is intended. A test beam typically consists of particles of a known type and a precisely known energy. This knowledge may be exploited to minimize the width of the total signal distribution for the measured particles, for example by applying weighting schemes for different sections of the calorimeter system, using correlations between signals from different calorimeter sections, *etc.* Also, there is a tendency to eliminate the effects of known sources of fluctuation (*e.g.*, electronic noise, pedestal fluctuations) from the results obtained in test beams.

However, in the reality of a particle physics experiment, only the calorimeter signals are available, without the additional knowledge about the nature and energy of the particle(s) that caused these signals. These signals have to be used to estimate the energy (and the type) of the particle(s). If electronic noise or pedestal fluctuations affect the calorimeter resolution, then this estimate is necessarily and unavoidably less precise than in the absence of such effects. In Chapter 9, these and other aspects of dealing with test beam data are discussed in detail.

In many calorimeters, the total signal depends strongly on the type of particle, *i.e.*, electrons and pions of the same energy cause quite different signals. Therefore, weighting schemes devised for the detection of individual particles of known type and energy basically become useless in situations in which a collection of several particles with unknown composition in terms of particle types and energies enter the calorimeter simultaneously. This is the situation one faces with *jets*. Estimating the calorimeter performance for jet detection, based on single-particle results, is a very non-trivial, and often mishandled issue. We elaborate on this in Chapter 6.

Choosing a calorimeter system involves much more than the choice of the active and

passive materials. Other crucial issues include the segmentation of the system, both laterally and longitudinally, and the readout of the various segments. Also in this respect,
choices made to accommodate one aspect of the calorimeter performance, may jeopardize other aspects. For example, Avalanche Photo Diodes, which are sometimes used
to detect the light from scintillators operating in a strong magnetic field, are extremely
sensitive to charged particles traversing them. They may respond to such particles with
signals that are several orders of magnitude larger than the signal for a scintillation photon. If such devices are used to read out the first section of a longitudinally segmented
calorimeter system, then they may also, by virtue of their location, be traversed by large
numbers of shower particles from hadron-induced showers. As a consequence, the energy of such hadrons could be severely overestimated. An example of this is discussed
in Section 9.1.3.

   In modern particle physics experiments, calorimeters fulfill a number of different
tasks, each with its own specific requirements in terms of detector performance. A
properly designed calorimeter system has all these requirements taken into account,
in accordance with their importance for the experiment as a whole.

2

THE PHYSICS OF SHOWER DEVELOPMENT

When a particle traverses matter, it will generally interact and lose (a fraction of) its energy in doing so. The medium is excited in this process, or heated up, whence the term calorimetry. The interaction processes that play a role depend on the energy and the nature of the particle. They are the result of the electromagnetic (em), the strong and, more rarely, the weak forces reigning between the particle and the medium's constituents.

In this chapter, the various processes by which particles lose their energy when traversing dense matter and by which they eventually get absorbed, are described. We also discuss shower development characteristics, the effects of the electromagnetic and strong interactions, and the consequences of differences between these interactions for the calorimetric energy measurement of electrons and hadrons, respectively. Finally, some Monte Carlo packages that are frequently used to simulate shower development are reviewed.

2.1 Electromagnetic Showers

2.1.1 *Energy loss by charged particles*

The best known energy-loss mechanism contributing to the absorption process is the electromagnetic interaction experienced by charged particles traversing matter. The particles ionize the medium, if their energy is at least sufficient to release the atomic electrons from the Coulomb fields generated by the atomic nuclei. This process also forms the principle on which many particle detectors are based, since the liberated electrons may be collected by means of an electric field and yield an electric signal.

The em interaction may manifest itself, however, in many other ways:

- Charged particles may excite atoms or molecules without ionizing them. The deexcitation from these metastable states may yield (scintillation) light, which is also fruitfully used as a source of calorimeter signals.
- Charged particles traveling faster than the speed of light characteristic for the traversed medium lose energy by emitting Čerenkov light.
- At high energies, energetic knock-on electrons (δ-rays) are produced.
- At high energies, bremsstrahlung is produced.
- At very high energies, the em interaction may induce nuclear reactions.

Already at energies above 100 MeV, and in many materials even at energies considerable lower than that, by far the principal source of energy loss by *electrons and positrons* is *bremsstrahlung*. In their passage through matter, electrons and positrons radiate photons as a result of the Coulomb interaction with the electric fields generated by the atomic nuclei. The energy spectrum of these photons falls off as $1/E$. It extends, in principle, all the way to the energy of the radiating particle, but in general each emitted photon carries only a small fraction of this energy.

In this process, the electron (or positron) itself undergoes a (usually small) change in direction. This is called multiple or Coulomb scattering. This deviation depends on the angle and the energy of the emitted photon, which in turn depend on the strength of the Coulomb field, *i.e.*, on the Z of the absorber material.

These radiative processes, which dominate the absorption of high-energy electrons and positrons, play a role for *any* charged particle traversing matter. However, for heavier charged particles the competition with ionization as the main source of energy loss only starts to play a role at much higher energies. The *critical energy*, ϵ_c, which may be defined as the energy at which the average energy losses from radiation processes equal those from ionization, is higher by a factor $(m/m_e)^2$, where m and m_e are the particle and the electron mass, respectively. The critical energy of the next-lightest charged particle, the muon, is thus about 40,000 times larger than that of the electron.

The energy loss mechanisms for electrons and positrons are governed by the laws of Quantum Electrodynamics (QED) and can be calculated with a high degree of accuracy. The relative importance of ionization and radiation losses at a given energy depends primarily on the electron density of the medium in which the shower develops. This density is roughly proportional to the (average) Z of the medium, since the number of atoms per unit volume is, within a factor of about two, the same for all materials in the solid state.

Results of calculations on the energy loss mechanisms for electrons are shown as a function of energy in Figure 2.1, for three different absorber materials: carbon ($Z = 6$, Figure 2.1d), iron ($Z = 26$, Figure 2.1e) and uranium ($Z = 92$, Figure 2.1f) [Pag 72]. The energy at which the energy losses from ionization equal those from radiation decreases from about 95 MeV for carbon, to 28 MeV for iron, to 9 MeV for uranium.

The Particle Data Group [PDG 14] prefers a slightly different definition of the critical energy (at least for electrons), originally formulated by Rossi [Ros 52]. In this definition, ϵ_c is the energy at which the ionization loss per radiation length (X_0) equals the electron energy[1]:

$$(\Delta E)_{ion} = \left[\frac{dE}{dx}\right]_{ion} X_0 = E \qquad (2.1)$$

This definition would thus be equivalent to the first one if the energy loss to bremsstrahlung were given by

$$\left[\frac{dE}{dx}\right]_{brems} = \frac{E}{X_0} \qquad (2.2)$$

which is true at very high energies, where ionization losses are negligible, but which is only an approximation in the energy regime near ϵ_c. The difference between the two definitions is illustrated in Figure 2.2, for electrons in copper. In this figure, Equation 2.2 is represented by the dashed line. Using this alternative definition, the PDG has fitted

---

[1] See Section 2.1.5.1 for the definition of the radiation length.

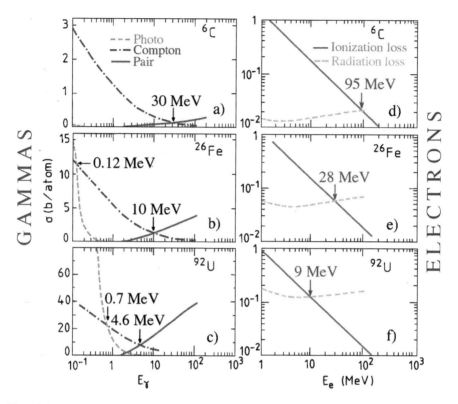

FIG. 2.1. Cross sections for the processes through which the particles composing electromagnetic showers lose their energy, in various absorber materials. To the left are shown the cross sections for pair production, Compton scattering and photoelectric effect in carbon (a), iron (b) and uranium (c). To the right, the fractional energy losses by radiation and ionization are given as a function of the electron energy in carbon (d), iron (e) and uranium (f).

the dE/dx data tabulated by Pages [Pag 72] and recommends the following expressions for the critical energy:

$$\epsilon_c = \frac{610 \text{ MeV}}{Z + 1.24} \tag{2.3}$$

for materials in the solid or liquid phase, and

$$\epsilon_c = \frac{710 \text{ MeV}}{Z + 0.92} \tag{2.4}$$

for gases. These formulae fit the data from the mentioned dE/dx tables to within $\sim 4\%$, with the largest deviations occurring at the highest Z values. For example, for uranium, Equation 2.3 gives $\epsilon_c = 6.54$ MeV, while the data tabulated in [Pag 72] fulfill Rossi's condition at an energy of 6.75 MeV.

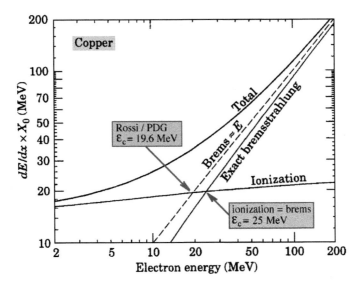

FIG. 2.2. Energy losses through ionization and bremsstrahlung by electrons in copper. The values for the critical energy following from the two definitions discussed in the text are indicated by arrows. From [PDG 14].

The ϵ_c values found in this way are systematically smaller than the ones following from the other definition, where ϵ_c is the energy at which ionization losses equal radiation losses. The differences range from $\sim 15\%$ for carbon to $\sim 35\%$ for uranium (see also Figure 2.2).

In this book, we will use the definition of Rossi and the PDG. The ϵ_c values given in Table B.1 were determined on the basis of the tabulated dE/dx data rather than on Equations 2.3 and 2.4 (which represent approximate fits to these data). They may thus deviate by a few percent from those one would calculate on the basis of these PDG formulae.

2.1.2 Photon interactions

The quantum of the em interaction, the photon (γ) is mainly affected by four different processes: the photoelectric effect, coherent (Rayleigh) scattering, incoherent (Compton) scattering and electron–positron pair production.

2.1.2.1 Photoelectric effect.

At low energies, this is the most likely process to occur. In this process, an atom absorbs the photon and emits an electron. The atom, which is left in an excited state, returns to the ground state by the emission of Auger electrons or X-rays. The photoelectric cross section is extremely dependent on the available number of electrons, and thus on the Z value of the absorber material.

This is illustrated in Figure 2.3, which shows that the cross section scales with Z^n, with the power n between 4 and 5. The photoelectric cross section varies with the photon energy as E^{-3}, so that this process rapidly loses its importance as the energy increases.

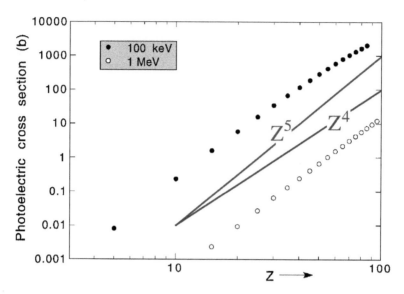

FIG. 2.3. Cross section for the photoelectric effect as a function of the Z value of the absorber. Data for 100 keV and 1 MeV γs.

In uranium, the highest-Z material that can be used for calorimeter construction, the cross section for photoelectric effect is dominating for energies below 700 keV, for iron inelastic scattering already starts to dominate above 100 keV (see Figure 2.7).

2.1.2.2 *Rayleigh scattering.* This (coherent) process is also important at low energies. In this process, the photon is deflected by the atomic electrons. However, the photon does *not* lose energy. Therefore, Rayleigh scattering affects the spatial distribution of the energy deposition, but it does not contribute to the energy deposition process itself.

2.1.2.3 *Compton scattering.* In the Compton process, a photon is scattered by an atomic electron, with a transfer of momentum and energy to the struck electron that is sufficient to put this electron in an unbound state.

Figure 2.4 illustrates this scattering process. Applying the laws of energy and momentum conservation, the relations between the different kinematic variables (energy transfer, scattering angles) can be derived in a straightforward manner. For example, when ζ is defined as the photon energy in units of the electron rest mass ($\zeta = E_\gamma / m_e c^2$), the scattering angles of the electron (ϕ) and the photon (θ) are related as

$$\cot \phi = (1 + \zeta) \, \tan \frac{\theta}{2} \qquad (2.5)$$

In all but the highest-Z absorber materials, Compton scattering is by far the most likely process to occur for γs in the energy range between a few hundred keV and ~ 5 MeV (see Figure 2.7). As we shall see in Section 2.1.4, typically at least half of the total energy is deposited by such γs in the absorption process of multi-GeV electrons, positrons

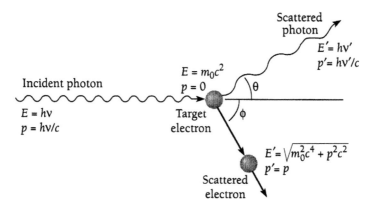

FIG. 2.4. The Compton scattering process.

or photons. Compton scattering is therefore a very important process for understanding the fine details of calorimetry.

The cross section for Compton scattering was one of the first ones to be calculated using Quantum Electrodynamics and is known as the *Klein–Nishina* formula [Bet 59]:

$$\frac{d\sigma}{d\Omega} = \frac{r_e^2}{2} \frac{(1+\cos^2\theta)}{\left[1+\zeta(1-\cos\theta)\right]^2} \left\{ 1 + \frac{\zeta^2(1-\cos\theta)^2}{(1+\cos^2\theta)\left[1+\zeta(1-\cos\theta)\right]} \right\} \quad (2.6)$$

where r_e is the classical electron radius (2.82 fm). In Figure 2.5a, this cross section is shown as a function of the scattering angle θ for photons of 0.1, 1 and 10 MeV. In the limit of zero energy, Equation 2.6 reduces to

$$\frac{d\sigma}{d\Omega} = \frac{r_e^2}{2}(1+\cos^2\theta) \quad (2.7)$$

which is the classical expression for *Thomson* scattering [Jac 74], and would be represented by a parabola in Figure 2.5a. For photons in the energy range in which Compton scattering is the most likely process to occur, this cross section is more or less flat in the backward hemisphere (scattering angles > 90°), and rises to a maximum value for $\theta = 0°$.

The angular distribution of the Compton recoil electrons, shown in Figure 2.5b, exhibits a preference for the direction of the incoming photons ($\cos\phi = 1$), but there is also a substantial isotropic component in the forward hemisphere (the requirements of momentum and energy conservation prevent the electrons from being scattered in the backward hemisphere).

Since the photoelectric effect, in which the photon is absorbed and thus disappears, only plays a role at low energies, many γs in the MeV energy range are absorbed in a *sequence* of Compton scattering processes, in which the photon energy is reduced in a number of steps down to the point where the final absorption in a photoelectric process occurs. In each step, an amount of energy equal to

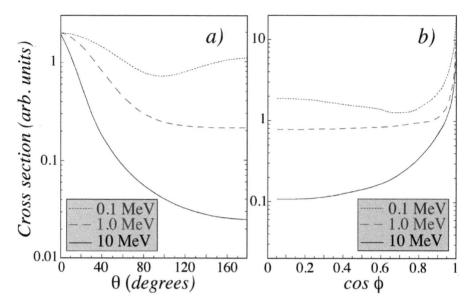

FIG. 2.5. The cross section for Compton scattering as a function of the scattering angle of
the photon (a), and the angular distribution of the Compton recoil electrons (b), for incident
photons of different energies.

$$T = E_\gamma \frac{\zeta(1 - \cos\theta)}{1 + \zeta(1 - \cos\theta)} \tag{2.8}$$

is transferred to the struck electron. In this process, the angular preference still visible for the *first* scattering in this sequence (Figure 2.5b) quickly disappears. Most of the Compton- and photoelectrons produced in this sequential absorption process are isotropically distributed with respect to the direction of the initial γ.

The cross section for Compton scattering is much less dependent on the Z value of the absorber material than the cross section for photoelectric effect. Figure 2.6 shows that the Compton cross section is almost proportional to Z, *i.e.*, proportional to the number of target electrons in the nuclei.

As for the photoelectric effect, the cross section for Compton scattering decreases with increasing photon energy, albeit much less steeply: $\sigma \sim 1/E$. Therefore, above a certain threshold energy, Compton scattering becomes more likely than photoelectric absorption. This threshold ranges from 20 keV for carbon ($Z = 6$) to 700 keV for uranium ($Z = 92$). The values for other elements can be derived from Figure 2.7.

2.1.2.4   *Pair production.*   At energies in excess of twice the electron rest mass, a photon may create, in the field of a charged particle, an electron–positron pair. These particles produce bremsstrahlung radiation as well as ionization along their paths. The electron is eventually absorbed by an ion, while the positron annihilates with an electron. In the latter process, two new photons are produced, each with an energy of 511 keV, the electron rest mass, if the annihilation takes place when the positron has come to rest.

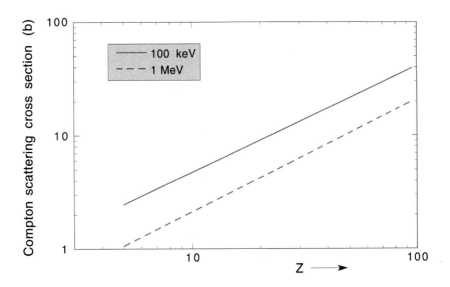

FIG. 2.6. Cross section for Compton scattering as a function of the Z value of the absorber material, for γs of 0.1 and 1.0 MeV.

Typically, more than 99% of the $\gamma \rightarrow e^+e^-$ conversions are caused by *nuclear* electromagnetic fields. For low-Z elements and at high energies, e^+e^- creation in the fields of the atomic electrons also contributes significantly to the total pair production cross section.

The cross section for pair production rises with energy and reaches an asymptotic value at very high energies (> 1 GeV). This cross section is related to the *radiation length* of the absorber material (see Section 2.1.5).

The relative importance of the processes through which photons are absorbed depends strongly on the photon energy and on the electron density ($\sim Z$) of the medium. Figure 2.1 shows the cross sections for these three processes as a function of energy in carbon ($Z = 6$, Figure 2.1a), iron ($Z = 26$, Figure 2.1b) and uranium ($Z = 92$, Figure 2.1c).

Since the cross sections for the photoelectric effect and for Compton scattering decrease with energy, and the cross section for pair production increases, pair production is the most likely process to occur at high energies.

The photoelectric effect dominates at low energies. Compton scattering is the process of choice in some intermediate regime (Figure 2.7). The higher the Z value of the absorber, the more limited the role of Compton scattering in the em absorption process. In uranium, Compton scattering dominates for energies between 0.7 MeV and 4.6 MeV. For iron, this interval is extended to 0.12 MeV on the low-energy side and to 10 MeV on the high-energy side. For low-Z materials such as carbon, pair production only takes over above 30 MeV, while the photoelectric effect is only significant at energies below 50 keV.

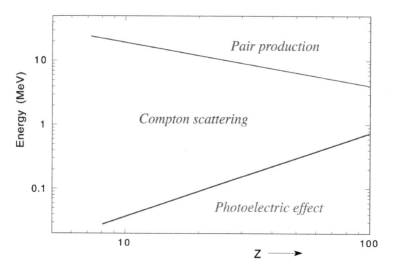

FIG. 2.7. The energy domains in which photoelectric effect, Compton scattering and pair production are the most likely processes to occur, as a function of the Z value of the absorber
material.

Figure 2.1 also shows that the *total cross section* for photon interactions exhibits a
minimum value near the energy where the probabilities for Compton scattering and pair
production are about equal. For high-Z materials, which have the highest total cross
sections and are therefore best suited for shielding against γ-rays, this minimum occurs
for energies around 3 MeV. It may seem quite amazing that it takes less material to
shield effectively against γs of 10–20 MeV than against 3 MeV ones. However, this
peculiarity is by no means unique to the MeV range of the em spectrum. For example,
the Earth's atmosphere is transparent to visible light, but light with a shorter wavelength
(higher energy, *e.g.*, ultraviolet light or X-rays) is effectively absorbed by it.

Photon interactions are fundamentally different from the ones experienced by electrons and positrons. When these charged particles traverse matter, they lose energy in a
continuous stream of events in which atoms or molecules are ionized and photons are ra-
diated away. A multi-GeV electron traversing 1 cm of lead typically radiates thousands
of photons. Some of these photons may have energies in excess of 1 GeV, but the overwhelming majority of these photons are very soft, with energies in the eV–keV–MeV
range.

On the other hand, a multi-GeV photon may penetrate the same thickness of lead
without being affected at all. For such high-energy photon interactions, we may apply
the concept of the *mean free path*, 7.2 mm in the case of lead. Therefore, the probability
that the mentioned photon does interact (*i.e.*, convert into an e^+e^- pair) in 1 cm of lead
is $[1 - \exp(-10/7.2)]$, or about 75%. For electrons, this concept has no meaning. What
we can say is that the original electron has lost, on average, about 83% of its energy after
traversing this material. We will come back to this issue when discussing the concept of

radiation length (Section 2.1.5).

2.1.2.5 *Photonuclear reactions.* At energies in the range 5–20 MeV, a modest role may be played by photonuclear reactions, *e.g.*, γn, γp or photo-induced nuclear fission. The cross sections for such reactions reach a maximum value at the so-called giant dipole resonance, when the photon energy is approximately equal to the *marginal binding energy* of the proton or neutron, *i.e.*, the difference in nuclear binding energy between the target nucleus and the nuclei with one nucleon less. The cross sections for these processes usually do not exceed 1% of the total cross section for the processes mentioned in the previous subsections [Die 88].

2.1.3 *A very simple shower*

Already at very low energies, relatively simple showers may develop. Let us consider, as an example, the γs of a few MeV characteristic for nuclear de-excitation. Such a γ may create an electron–positron pair in the detector. These charged particles lose their kinetic energy through ionization of the medium. When the positron comes to rest, it annihilates with an electron, and two γs of 511 keV are created. These γs may be absorbed in a sequence of Compton scattering processes, ending with photoelectric absorption. The Compton electrons and photoelectrons lose their kinetic energy through ionization. In the sketched process, the total energy of the original γ is absorbed through ionization of the detector medium by a number of charged particles, one positron and several electrons.

The sequence of processes through which a particular γ is absorbed differs from event to event. As an example, we consider γs of 3370 keV, produced in the decay of ^{65}Ga. One possible sequence is shown in Figure 2.8a. Every shower, *i.e.*, every sequence of events such as the one depicted in this figure, is different. In this particular example, the energy of the original γ is deposited in the detector through one positron and six different electrons.

In the γ-ray spectrum measured by this detector, events for which the whole sequence takes place inside the sensitive detector medium yield a signal peak at the energy of the original photon (3369.9 keV, see Figure 2.8b). In small detectors, like the one with which this spectrum was recorded, leakage phenomena may occur. Either one or both 511 keV γs may escape undetected from the detector. This leads to peaks lower by 511 keV (at 2558.9 keV) and by 1022 keV (at 2347.8 keV), respectively. Instead or in addition, one or more of the other γs produced in the absorption sequence may escape, which leads to a continuous background in the recorded spectrum.

In nuclear spectroscopy, one often uses an instrument called an *anti-Compton spectrometer* to improve the signal-to-background ratio of these measurements. In this device, a high-resolution solid-state (Ge) crystal is surrounded by a large scintillating structure (*e.g.*, an assembly of sodium iodide crystals) which detects the escaping γs with high efficiency. By measuring the signals from these two devices in anti-coincidence, only events in which the entire shower is contained in the germanium crystal are recorded. In practice, the Compton background can be reduced by an order of magnitude with such techniques.

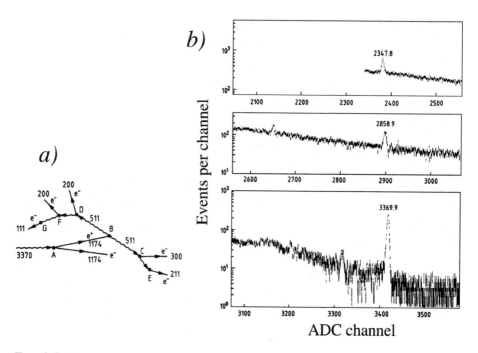

FIG. 2.8. Shower development induced by nuclear γs with an energy of 3370 keV, produced in the decay of ^{65}Ga. In (a), one possible sequence of absorption processes is depicted, with the energies of the positron, the electrons and the photons given in keV. The γ-ray spectrum, measured with a (small) Ge(Li) crystal in which these γs (and others of different energies) interacted is shown in (b). The total-containment peak (3369.9 keV), the single-(2558.9 keV) and double-escape peak (2347.8 keV) and the continuum background reflect the different degrees of absorption that occur in this crystal. See text for details.

2.1.4 *Electromagnetic cascades*

The example in the previous subsection contains many elements that also play a role in the absorption of multi-GeV electrons, positrons and photons: energy deposition by ionization of the absorbing medium (or, in the case of the germanium crystal, creation of electron–hole pairs), conversion of photons into charged particles through pair production, Compton scattering and the photoelectric effect, and the effects of incomplete shower containment (shower leakage). There is one more, crucial, mechanism that determines the absorption characteristics of electromagnetically interacting particles in the GeV domain: *bremsstrahlung*, the radiation of large numbers of photons as a result of the interaction between the high-energy electrons and positrons and the nuclear Coulomb fields.

Bremsstrahlung is by far the principal source of energy loss by high-energy electrons and positrons. As a consequence, high-energy em showers are quite different from the ones discussed in the previous subsection, since an important *multiplication* of shower particles occurs.

A primary, multi-GeV electron may radiate thousands of photons on its way through the detector. The overwhelming majority of these photons are very soft, and are absorbed through Compton scattering and the photoelectric effect. The photons carrying more energy than 5–10 MeV may create e^+e^- pairs. The fast electrons and positrons generated in these processes may in turn lose their energy by radiating more photons, which may create more electron–positron pairs, and so forth. The result is a shower that may consist of thousands of different particles: electrons, positrons and photons.

The shower energy is deposited by the numerous electrons and positrons, which ionize the atoms of the absorber material. Because of the multiplication mechanism described above, the number of electrons and positrons, and thus the amount of energy deposited in an absorber slice of given thickness, initially increases as the shower develops, *i.e.*, with increasing shower depth.

However, as the shower develops, the average energy of the shower particles decreases, and at some point no further multiplication takes place. The depth at which this occurs is called the *shower maximum*. Beyond this depth, the shower photons are, on average, more likely to produce *one* electron in their (Compton or photoelectric) interactions, rather than an electron–positron *pair*. And the electrons and positrons are, again on average, more likely to lose their energy through ionization of the absorber medium, rather than by producing *new* photons (which would in turn convert into *more* electrons) through radiation. Beyond the shower maximum, the number of shower particles, and thus the energy deposited in a detector slice of given thickness, therefore gradually decreases.

All these aspects are illustrated in Figure 2.9, which shows the energy deposit as a function of depth, for 1, 10, 100 and 1,000 GeV electron showers developing in a block of copper. The higher the initial energy of the showering particle, the longer the particle multiplication phase continues. The shower maximum is reached after about 5 cm for 1 GeV electrons. For every order of magnitude in energy, this maximum shifts by ~ 3.5 cm deeper into the detector, to reach a depth of about 16 cm for electron showers of 1 TeV. The amount of copper needed to absorb 99% of the shower energy rises from 23 cm at 1 GeV, via 28 cm at 10 GeV and 33 cm at 100 GeV, to 39 cm at 1 TeV.

The overwhelming majority of the shower particles through which the energy is deposited in the absorber are very soft. This is illustrated in Figure 2.10, which shows the percentage of the energy of 10 GeV electromagnetic showers that is deposited through shower particles with energies below 1 MeV, as a function of the Z of the absorber material. For high-Z materials such as lead and uranium, these soft particles account for about 40% of the energy deposit!

The same figure also shows the percentage of the energy deposited by shower particles faster than 20 MeV. From the relatively small contribution of such fast particles to the absorption process, one might get the impression that pair production is not the most abundant mechanism in em shower development processes, and that most of the energy is deposited through electrons created in Compton scattering and in photoelectric processes. This impression is correct. It turns out that positrons (which are only produced in pair production) are outnumbered by electrons (which are produced in all three photon interaction processes) by a considerable factor. Table 2.1 lists the

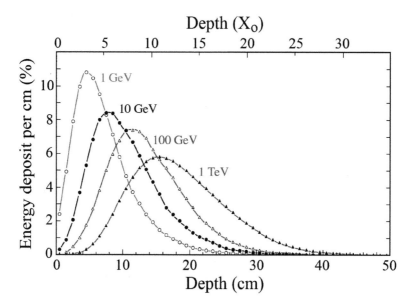

FIG. 2.9. The energy deposit as a function of depth, for 1, 10, 100 and 1,000 GeV electron showers developing in a block of copper. In order to compare the energy deposit profiles, the integrals of these curves have been normalized to the same value. The vertical scale gives the energy deposit per cm of copper, as a percentage of the energy of the showering particle. Results of EGS4 calculations.

Table 2.1 *The numbers of positrons that are generated in em shower development and the fraction of the total energy deposited by these particles. Results of EGS4 simulations.*

Shower energy →	10 GeV		100 GeV	
Absorber ↓	$\#e^+$	E^+/E_{tot}	$\#e^+$	E^+/E_{tot}
Aluminium ($Z = 13$)	191	26%	1750	27%
Iron ($Z = 26$)	285	27%	2920	26%
Tin ($Z = 50$)	427	24%	4330	25%
Lead ($Z = 82$)	554	22%	5730	23%
Uranium ($Z = 92$)	612	23%	5970	23%

numbers of positrons that are generated in the absorption of 10 and 100 GeV electrons in a variety of absorber materials with different Z values, as well as the energy fraction carried by these particles.

The numbers of shower electrons are not listed here, since these numbers depend sensitively on the chosen cutoff values in the simulations. However, they are typically two orders of magnitude larger than the numbers of positrons. The positrons are, on average, considerably more energetic than the electrons and they are thus much less sensitive to these cutoff values.

The number of positrons produced per unit of energy increases by more than a factor

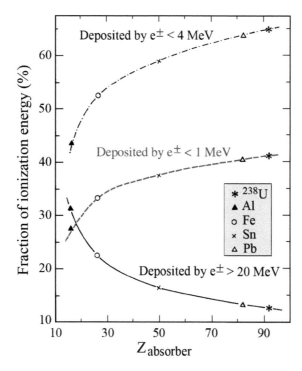

FIG. 2.10. The composition of em showers. Shown are the percentages of the energy of 10 GeV electromagnetic showers deposited through shower particles with energies below 1 MeV (the dashed curve), below 4 MeV (the dash-dotted curved) or above 20 MeV (the solid curve), as a function of the Z of the absorber material. Results of EGS4 simulations.

of three going from aluminium ($\sim 18e^+$/GeV) to uranium ($\sim 60e^+$/GeV). This increase is a consequence of the fact that particle multiplication in showers developing in high-Z absorber materials continues down to much lower energies than in low-Z materials.

Typically, one quarter of the total em shower energy is deposited by positrons, the rest by electrons. This leads to the conclusion that the energy deposited by an average positron ranges from ~ 4 MeV in uranium to ~ 14 MeV in aluminium. This energy difference is commensurate with the difference in the energies at which pair production becomes the dominant interaction process for photons in these materials (see Figure 2.7). Shower electrons, which are predominantly produced by converting bremsstrahlung photons, each deposit typically only a small fraction of 1 MeV.

The spatial profiles of the energy deposited by electrons and positrons are quite different (see Figure 2.11). The contributions from positrons are relatively stronger in areas close to the shower axis (Figure 2.11b) and in the early stages of the shower development, *i.e.*, before the shower maximum is reached (Figure 2.11a). Especially the differences in the lateral characteristics are very substantial: on average, electrons deposit their energy twice as far from the shower axis than do positrons.

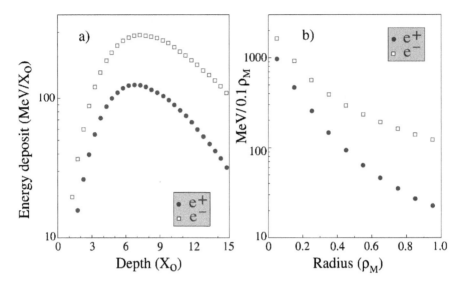

FIG. 2.11. Comparison of the longitudinal (*a*) and lateral (*b*) profiles of the energy deposited
   by electrons and positrons in 10 GeV em showers developing in lead. Note the logarithmic
   vertical scale. Results of EGS4 simulations.

Not surprisingly, the average energy of the shower particles is a function of the depth
inside the detector: the further the shower has developed, the softer the spectrum of its
constituents becomes. This has interesting and important consequences.

For example, as the spectrum of the photons becomes softer, the relative importance
of the three main processes through which photons interact with matter changes. This is
particularly true in high-Z absorber materials, where the photoelectric effect becomes
gradually the dominating process (see Figure 2.1c). As a direct consequence of this, the
relationship between deposited energy and the resulting signal is, in sampling calorimeters, a function of depth. This is discussed in great detail in Section 3.2.

2.1.5 *Scaling variables*

Since the em shower development is primarily determined by the electron density in the
absorber medium, it is to some extent possible, and in any case convenient, to describe
the shower characteristics in a material-independent way. The units that are frequently
used to describe the characteristic shower dimensions are the *radiation length* (X_0) for
the longitudinal development and the *Molière radius* (ρ_M) for the lateral development.

2.1.5.1 *The radiation length.* The radiation length is defined as the distance over
which a high-energy ($\gg 1$ GeV) *electron or positron* loses, on average, 63.2% (*i.e.*,
$1 - e^{-1}$) of its energy to bremsstrahlung. High-energy electrons lose the same *fraction*
of their energy in 18 cm of water ($0.5X_0$) as in 2.8 mm of lead ($0.5X_0$). By expressing
the dimensions of the absorber structure in units of X_0, material-dependent effects are
thus, in first approximation, eliminated.

It can be shown that the asymptotic cross section for *photon interactions* is related to X_0 as

$$\sigma(E \to \infty) = \frac{7}{9} \frac{A}{N_A X_0} \tag{2.9}$$

in which X_0 is expressed in g cm^{-2} and the ratio of Avogadro's number (N_A) and the atomic weight (A) denotes the number of atoms per gram of material. This implies that the mean free path of very-high-energy photons equals $\frac{9}{7}X_0$.

Values of X_0 have been calculated and tabulated by Y.S. Tsai [Tsa 74]:

$$\frac{1}{X_0} = 4\alpha r_e^2 \frac{N_A}{A} \left\{ Z^2 \left[L_{\text{rad}} - f(Z) \right] + Z L'_{\text{rad}} \right\} \tag{2.10}$$

in which α is the fine-structure constant, r_e the electron radius, N_A Avogadro's number, L_{rad} and L'_{rad} parameters with values $\ln(184.15 Z^{-1/3})$ and $\ln(1194 Z^{-2/3})$, respectively (for $Z > 4$), and $f(Z)$ a function that can be parameterized as follows:

$$f(Z) = a^2 \left[(1+a^2)^{-1} + 0.20206 - 0.0369a^2 + 0.0083a^4 - 0.002a^6... \right] \tag{2.11}$$

with $a = \alpha Z$. For approximate calculations, which are accurate to within 3%, the Particle Data Group [PDG 14] recommends the following expression:

$$X_0 = \frac{716.4\ A}{Z\ (Z+1)\ \ln(287/\sqrt{Z})}\ \text{g cm}^{-2} \tag{2.12}$$

Radiation lengths for a number of materials frequently used in calorimeters are tabulated in Table B.1.

The radiation length for a mixture of different materials can be calculated as follows

$$\frac{1}{X_0} = \sum_i V_i/X_i \tag{2.13}$$

in which V_i and X_i are the fraction by volume and the radiation length (expressed in mm) of the ith component of the mixture. Equation 2.13 may, for example, be used to calculate the effective radiation length of a calorimeter consisting of a variety of different materials. Let us, as an example, consider a lead/liquid-argon calorimeter consisting of 5 mm thick lead plates, separated by 3 mm wide LAr-filled gaps. The radiation lengths of lead and LAr are 5.6 mm and 140 mm, respectively, and the fractional volume occupied by these elements is 62.5% for lead and 37.5% for argon. Therefore, we find for the *effective radiation length*: $X_{\text{eff}} = [0.625/5.6 + 0.375/140]^{-1} = 8.75$ mm, only slightly less than the value one would obtain if the argon were replaced by vacuum ($X_{\text{eff}} = 5.6/0.625 = 8.96$ mm).

If the argon were contained in 0.8 mm thick stainless steel containers, separated from the lead plates by 0.2 mm of air, the volume ratio of the different materials in the calorimeter structure would be as follows: lead/argon/iron/air = 5/3/1.6/0.4 = 50%/30%/16%/4%. With the radiation lengths for iron and air being 17.6 mm and 300 m, respectively, the effective radiation length of this structure becomes: $X_{\text{eff}} = [0.5/5.6 + 0.3/140 + 0.16/17.6 + 0.04/30,000]^{-1} = 9.95$ mm.

The radiation length of a *compound* can be calculated in a similar way, using the equation

$$\frac{1}{X_0} = \sum_i m_i / X_i \tag{2.14}$$

in which m_i and X_i are the fraction (by mass) and the radiation length (expressed in g cm^{-2}) of the ith component of the compound. Let us, for example, calculate the radiation length of lead-tungstate crystals (PbWO$_4$). The mass ratio of the elements of which these crystals are composed is as follows: lead/tungsten/oxygen = 207.19/183.85/64.0 = 45.5%/40.4%/14.1%. The radiation lengths of these elements are 6.37, 6.76 and 34.24 g cm^{-2}, respectively. Therefore, we find for the radiation length of lead-tungstate: $X_0 = [0.455/6.37 + 0.404/6.76 + 0.141/34.24]^{-1} = 7.39$ g cm^{-2}.
Since the density of these crystals amounts to 8.30 g cm^{-3}, their radiation length equals 8.9 mm.

2.1.5.2 The Molière radius.

This quantity does not have a physics meaning equal in precision to that of the radiation length. The Molière radius is frequently used to describe the transverse development of em showers in an *approximately* material independent way. It is defined in terms of the radiation length X_0 and the critical energy ϵ_c (Section 2.1.1), as follows:

$$\rho_M = E_s \frac{X_0}{\epsilon_c} \tag{2.15}$$

in which the scale energy E_s, defined as $m_e c^2 \sqrt{4\pi/\alpha}$, equals 21.2 MeV. Typically, ~85–90% of the shower energy is deposited in a cylinder with radius ρ_M around the shower axis (see Figure 2.18 for an example).

Values for ρ_M for a number of materials frequently used in calorimeters are listed in Table B.1. They were calculated with values for the critical energy obtained on the basis of Rossi's definition [Ros 52], using dE/dx values tabulated by Pages and coworkers [Pag 72]. See Section 2.1.1 for details on this point.

The Molière radii for mixtures or compounds of different elements may be calculated in the same way as the radiation length for such mixtures or compounds was obtained, replacing X_i in Equations 2.13 and 2.14 by ρ_i. Let us, for example, calculate the Molière radius of BGO crystals, which have the following chemical composition: Bi$_3$Ge$_4$O$_{12}$. The mass ratio of the elements of which these crystals are composed is as follows: bismuth/germanium/oxygen = (209.0×3)/(72.6×4)/(16.0×12) = 56.5%/26.2%/17.3%. The radiation lengths of these elements are 6.32, 12.25 and 34.24 g cm^{-2}, respectively. For the critical energies, we use Equation 2.3, which gives values of 7.24 MeV for bismuth, 18.4 MeV for germanium and 66 MeV for oxygen. This leads to Molière radii for the different crystal components of 18.5 g cm^{-2} (Bi), 14.4 g cm^{-2} (Ge) and 11.0 g cm^{-2} (O), respectively. When combining all these data, we find for the Molière radius of bismuth-germaniumoxyde:
$\rho_M = [0.565/18.5 + 0.262/14.4 + 0.173/11.0]^{-1} = 15.5$ g cm^{-2}.
Since the density of these crystals amounts to 7.13 g cm^{-3}, their Molière radius thus equals about 22 mm.

The Molière radius is much less Z dependent than the radiation length. This can be seen from Equations 2.3, 2.12 and 2.15. The radiation length scales in first approximation with A/Z^2 (Equation 2.12). If we assume that A is proportional to Z, which is roughly true, the radiation length (expressed in g cm^{-2}) decreases with increasing Z like $1/Z$. The same is approximately true for the critical energy (Equation 2.3). Since the Molière radius is defined as the *ratio* of the radiation length and the critical energy, the Z dependence cancels in first approximation.

This difference in Z dependence may be illustrated by comparing two materials that are frequently used as absorbers in calorimeters, copper ($Z = 29$) and lead ($Z = 82$). The densities are not very different: 8.96 g cm^{-3} for copper *vs.* 11.35 g cm^{-3} for lead. The radiation lengths for these materials reflect the large difference in Z: 14.3 mm for copper *vs.* 5.6 mm for lead, almost a factor of three difference. However, the values for the Molière radii show a completely different pattern: 15.2 mm for copper *vs.* 16.0 mm for lead.

As a consequence, the development of em showers in these two absorber materials has very different characteristics. In the longitudinal direction, it takes about three times as much copper as lead (in cm) to contain these showers. However, laterally, the showers in copper are even *narrower* than those in lead.

2.1.6 *Electromagnetic shower profiles*

Expressed in units of X_0 and ρ_M, the development of electromagnetic showers is, approximately, material independent, since these quantities were defined to eliminate material-dependent effects. Figure 2.12 shows the longitudinal development of 10 GeV electron showers in aluminium, in iron and in lead. The horizontal axis is expressed in units of X_0, *i.e.*, 89 mm for Al, 17.6 mm for Fe and 5.6 mm for Pb. Globally, the profiles look indeed very similar, which means that they roughly scale with the radiation length.

However, there are also some striking differences between the longitudinal profiles for these three absorber materials:

- As Z increases, the shower maximum shifts to greater depth.
- As Z increases, the shower profiles decay more slowly beyond the shower maximum.

As a result of these effects, it takes a larger number of radiation lengths to contain a given em shower in lead than in iron or aluminium. For example, to contain 10 GeV electron showers, on average, at the 99% level, 25 X_0 worth of lead is needed, *vs.* 21 X_0 of iron and 18 X_0 of aluminium.

Figure 2.9 shows that it takes relatively little *extra* material to contain showers of *higher energy* at the same level. This can be understood as follows. A 20 GeV photon travels, on average, $\frac{9}{7}$ radiation lengths before converting into an e^+e^- pair of 10 GeV each. Therefore, it takes only an extra 1.3 X_0 to contain twice as much em shower energy.

This simple example implies a *logarithmic energy dependence* of the longitudinal shower profiles, since a *factor of two* increase in energy increases the material needed to contain the shower by a *constant amount*. In agreement with this notion, parameterizations that have been proposed to describe the (average) longitudinal em shower profile

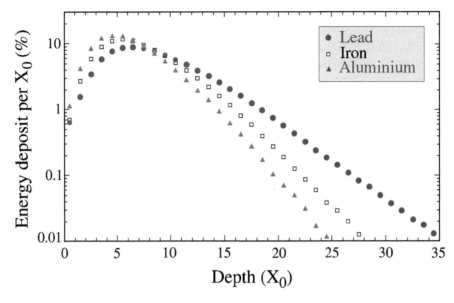

FIG. 2.12. Energy deposit as a function of depth, for 10 GeV electron showers developing in
aluminium, iron and lead, showing approximate scaling of the longitudinal shower profile,
when expressed in units of radiation length, X_0. Results of EGS4 calculations.

[Amal 81, Fab 82] invariably contain a *logarithmic* energy dependence of the depth at
which the shower maximum occurs.

Since there are absorber-specific and, as we will see later, also readout-specific effects that determine details of the longitudinal shower profiles, the practical value of
such parameterizations is limited and, therefore, we do not elaborate on them.

The differences between the longitudinal shower profiles in lead, iron and aluminium (see Figure 2.12) are related to shower characteristics that were pointed out
earlier in Section 2.1.4. Table 2.1 shows that the number of positrons strongly increases
with the Z value of the absorber material. This is because in high-Z materials shower
particle *multiplication*, *i.e.*, the production of e^+e^- pairs by photons and the emission
of bremsstrahlung photons by these charged shower particles, continues down to much
lower energies than in low-Z materials. As the shower develops, the average energy
of the shower particles decreases with each new generation (*i.e.*, each X_0 or $\frac{9}{7}X_0$).
The shower maximum is reached when the average energy of the shower particles
equals the critical energy. The critical energy is only 7 MeV in lead, *vs.* 22 MeV in
iron and 43 MeV in aluminium. At 10 MeV, photons in lead are most likely to convert into e^+e^- pairs, and electrons lose more energy by radiating new photons than
by ionizing the absorber medium. In aluminium, the situation is very different. At 10
MeV, photons interact primarily through Compton scattering, in which process only
one electron is produced, and electrons lose the overwhelming majority of their energy by ionization (see Figure 2.1). Therefore, one may expect the shower maximum,

expressed in units of X_0, to occur at significantly greater depth in lead than in aluminium, in agreement with the results shown in Figure 2.12.

The decay beyond the shower maximum is due to the gradual decrease in the number of shower particles. Photons predominantly produce only one electron in their interactions (Compton scattering, photoelectric effect) and these electrons predominantly range out without generating many new photons. However, in high-Z materials, electrons radiate down to substantially lower energies than in aluminium and iron and, therefore, this decay proceeds more slowly in lead, as seen in Figure 2.12.

The radiation length is defined for very high energies and has no meaning in the MeV range. Details of the shower profiles caused by peculiarities at the MeV level, such as the differences in critical energy, can therefore not be expected to be reproduced correctly in a model that describes shower development in terms of X_0 *alone* (the scaling model).

This is the first example of what will be a recurring theme in this book. Although calorimeters are instruments intended for measuring energies in the GeV and TeV domains, one has to descend to the MeV level (and below) to understand their performance in detail.

The *lateral spread* of em showers is caused by two effects:

1. Electrons and positrons move away from the shower axis because of multiple Coulomb scattering.

2. Photons and electrons produced in more isotropic processes (Compton scattering, photoelectric effect) move away from the shower axis. Also, bremsstrahlung photons emitted by electrons that travel at a considerable angle with respect to the shower axis may contribute to this effect.

The first process dominates in the early stages of the shower development, while the second process is predominant beyond the shower maximum, particularly in high-Z absorber media.

When the shower development in the plane perpendicular to the direction of the incoming particle is discussed, the way data are presented frequently leads to confusion. There are two different ways of presenting such data:

1. The energy density, *i.e.*, the amount of energy per unit volume, is shown as a function of the distance between that unit volume and the shower axis. We will refer to energy distributions and shower profiles of this type as *lateral* energy distributions and *lateral* shower profiles.

2. The energy contained in a radial slice of a certain thickness is shown as a function of the distance between that slice and the shower axis. We will refer to energy distributions and shower profiles of this type as *radial* energy distributions and *radial* shower profiles.

In addition, we will use the term *transverse*, as well as radial or lateral, to describe general, non-quantitative features of shower development in the plane perpendicular to the direction of the incoming particle.

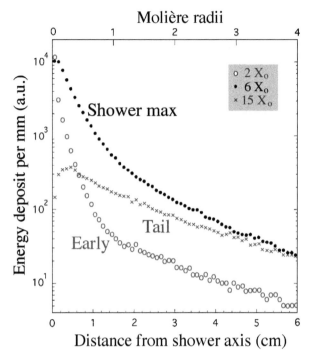

FIG. 2.13. The radial distributions of the energy deposited by 10 GeV electron showers in copper, at various depths. Results of EGS4 calculations.

Of course, distributions/profiles of type #1 are considerably narrower than those of type $2. In practice, both terms are being used for both types of distributions/profiles, and it is sometimes not clear which of these two is shown in presentations. In this book, we use the terminology defined above in a consistent manner.

Both types of distributions may either concern a certain longitudinal slice of the absorbing structure, or the entire absorbing structure. In the latter case, the lateral or radial profiles are said to be *integrated over the full depth*.

Figure 2.13 shows the radial distributions of the energy deposited by 10 GeV electron showers developing in copper, at various depths. The two mentioned components can be clearly distinguished. Both show an exponential behavior (note the logarithmic ordinate in Figure 2.13), with characteristic slopes of ~ 3 mm ($\sim 0.2\rho_M$) and ~ 25 mm ($\sim 1.5\rho_M$), respectively. The radial shower profile shows a pronounced central core (the first component), surrounded by a *halo* (the second component). The central core disappears beyond the shower maximum.

This radial profile, integrated over the total shower depth, is shown in Figure 2.14, together with the equivalent profiles for 10 GeV electron showers developing in lead and aluminium. The distance from the shower axis (plotted horizontally) is expressed in Molière units. Scaling with ρ_M would imply that these three profiles are identical. The figure shows that this is approximately true, but that there are also some clear

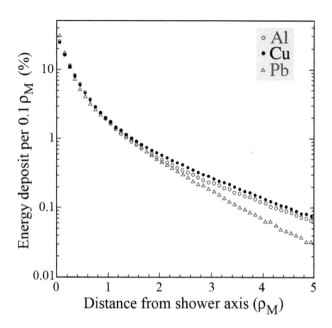

FIG. 2.14. Radial energy deposit profiles for 10 GeV electrons showering in aluminium, iron and lead. Results of EGS4 calculations.

differences between these three profiles. And just as in the case of the *longitudinal* shower development, these differences can be understood from phenomena taking place at the 1 MeV level, for which the Molière radius has no meaning.

The most striking difference between these radial profiles concerns the slope of the halo. This slope is considerably steeper for lead ($\sim 1.1\rho_M$) than for the low-Z materials, aluminium and copper ($\sim 1.6\rho_M$). As we saw above, this halo component of the radial em shower profiles is predominantly due to soft electrons, with energies well below the critical energy, which are produced in Compton scattering and photoelectric absorption. The slope of the halo is thus directly related to the mean free path of the photons causing these processes.

The mean free path of photons in the 1 MeV energy range is typically considerably different from that of GeV-type photons, which convert on average after $\frac{9}{7}X_0$. This mean free path ($\langle l \rangle$, expressed in g cm^{-2}) can be calculated from the total cross section (σ), to which it is related as

$$\sigma(E) = \frac{A}{N_A \langle l \rangle} \tag{2.16}$$

For example, in lead, the asymptotic cross section for photon interactions amounts to 42 barns (Equation 2.9), whereas the cross section drops to 15.6 b for $E_\gamma = 2$ MeV [Sto 70]. Therefore, the mean free path of γs of a few MeV equals $42/15.6 \times 9/7 = 3.5$ X_0. That is why radioactive ^{60}Co sources, which emit γs of about 1.3 MeV, require substantial lead shielding. These γs are among the most penetrating ones available, and therefore the most difficult ones to shield.

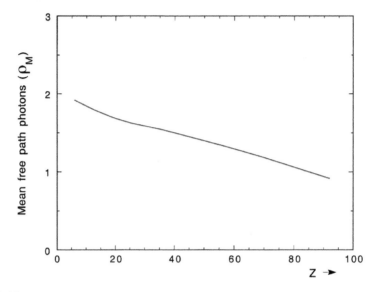

FIG. 2.15. Mean free path, in units of the Molière radius, for photons with energies 1–3 MeV, as a function of the Z value of the absorber material.

We have used Equation 2.16 and the Tables from [Sto 70] to calculate the mean free paths for γs in the energy range from 1–3 MeV for a variety of absorber materials with different Z values and have converted these into units of the Molière radius. The results are given in Figure 2.15. It turns out that this distribution gradually decreases with increasing Z. In the Al–Cu region, the mean free path of these photons amounts to $1.6 - -1.8\rho_M$, while for lead it has dropped to $\sim 1.0\rho_M$. These values are in good agreement with the observed slopes of the halo in Figure 2.14.

Another effect that may contribute to the less steep halo slopes in low-Z materials is that the absorption of low-energy bremsstrahlung photons requires, on average, more steps than in high-Z materials. In Compton scattering, a photon of lower energy is produced and this photon needs to be absorbed in a subsequent process. In the energy range from 0.2–0.5 MeV, Compton scattering is the most likely process in aluminium and iron (see Figure 2.7), while in lead, photons in this energy range are predominantly absorbed in a single-step process, photoelectron production.

All the results shown so far in this subsection were obtained with (EGS4) Monte Carlo simulations. However, available experimental results confirm all the trends mentioned above. Detailed measurements on em shower profiles were carried out by Bathow and coworkers [Bat 70]. They studied the three-dimensional profiles induced by 6 GeV electrons in aluminium, copper and lead. Their experimental data were provided by arrays of tiny (a few mm^3) silver-phosphate dosimeter glasses which were installed at

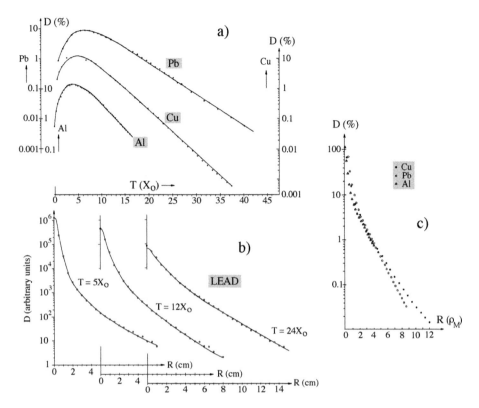

FIG. 2.16. Experimental results on the shower profiles of 6 GeV electrons in aluminium, copper and lead. Shown are the longitudinal profiles in these three materials (a), the lateral profiles in lead, measured at 3 different depths (b), and the energy deposition integrated over depth as a function of the distance to the shower axis, for different absorber materials (c). Data from [Bat 70].

various depths inside the absorber blocks. In these glasses, luminescence centers were formed by ionizing radiation. Afterwards, the accumulated dose could be determined, with a relative precision of about 5%, by exposing the irradiated dosimeters to ultraviolet light.

The absorber blocks with the built-in dosimeters were irradiated with a 6 GeV electron beam provided by the DESY synchrotron, at a rate of $5 \cdot 10^{10}$ particles per second. The dosimeter matrix thus accumulated a three-dimensional image, corresponding to the average shower profile of a very large number of 6 GeV electrons, that was "frozen" into the block and that could be analyzed offline.

Figure 2.16 shows some results of these experiments. In Figure 2.16a, the longitudinal shower profiles in aluminium, copper and lead are given, with the depth expressed in units of X_0. These curves show the same characteristics as the simulated ones in Figure 2.12: approximate scaling with X_0, but as Z increases, the shower maximum shifts to greater depth and the slope beyond the shower maximum becomes less steep.

In Figure 2.16b, the lateral profile in lead is shown, at three different depths. As in the simulated curves (Figure 2.13), there are clearly two components visible, the steepest of which disappears beyond the shower maximum. Figure 2.16c shows the energy deposition, integrated over the full shower depth and plotted as a function of the distance from the shower axis, in units of ρ_M. The latter curves also exhibit the two-component structure, and the long-distance component is steeper for lead than for copper, as in the EGS4 simulations (Figure 2.14).

2.1.7   *Shower containment*

When designing a calorimeter system for a certain experiment, it is important that the detector be sufficiently large to contain the showers of interest at an adequate level. Shower particles escaping from the detector represent a source of fluctuations that may affect the precision of the measurements. In this subsection, we present information on the absorber size needed to contain em showers, *on average*, at a certain level, *e.g.*, 95%. It should be emphasized that this average containment level in itself is no indication of the effects of shower leakage on the energy resolution and other aspects of the calorimeter quality. These effects are determined by *event-to-event fluctuations* about this average. It turns out that 5% longitudinal shower leakage has much larger effects in that sense than 5% transverse shower leakage. In Section 4.5, we elaborate on the relationships between various types of shower leakage and calorimeter quality.

Figure 2.17 shows results for longitudinal shower containment, obtained with EGS4 Monte Carlo simulations. The energy dependence of this containment is shown, as a function of the absorber thickness, in the upper half of this figure, for electrons showering in copper. The absorber thickness needed to contain 95% of the shower energy ranges from $\sim 11X_0$ at 1 GeV to $\sim 22X_0$ at 1 TeV. For 99% containment, the absorber thickness has to be at least $16X_0$ for 1 GeV electrons and $27X_0$ for 1 TeV ones.

The material dependence of the longitudinal shower containment is depicted in Figure 2.17b. Showers initiated by 100 GeV electrons in various absorber materials were simulated with EGS4 for this purpose. This figure shows the same type of deviations from X_0-scaling that became evident earlier (see Figure 2.12). The amount of material needed to contain these showers at the 99% level increases with Z, from $\sim 19X_0$ for aluminium to $\sim 26X_0$ for uranium.

The same figure also shows that γ-induced showers require about one radiation length more material in order to be contained at a certain level than do electron showers of the same energy.

Transverse shower containment results are shown in Figure 2.18, where the energy fraction contained in a cylinder around the shower axis is plotted as a function of the radius of this cylinder. It turns out that the results are almost completely independent of the energy of the incoming particle, and that Z effects are much less pronounced compared to the longitudinal case, in accordance with the radial profiles shown before (Figure 2.14). The figure also indicates that in order to capture 99% instead of 90% of the total energy, the detector mass has to be increased by an order of magnitude. This is of course a consequence of the extensive radial shower tails (Figure 2.14).

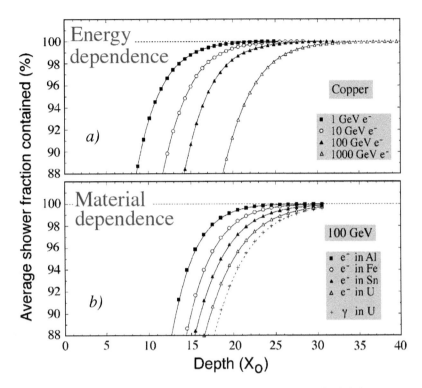

FIG. 2.17. Average energy fraction contained in a block of matter with infinite transverse dimensions, as a function of the thickness of this absorber. Shown are results for showers induced by electrons of various energies in a copper absorber (*a*) and results for 100 GeV electron showers in different absorber materials (*b*). The lower figure also shows the results for 100 GeV γ showers in ^{238}U. Results of EGS4 calculations.

The containment information is in practice very important for the design of calorimeters. The longitudinal containment determines the required depth, while the lateral containment is an important consideration for the chosen cell size (granularity). Typically, one chooses the depth such as to contain, on average, 99% of the shower energy, while the effective radius of the cell size is often chosen to be $1 - 1.5\rho_M$. Figure 2.19 summarizes what these size requirements mean in practice, given the chosen detector material and the energy domain of the experiment.

2.2 Muons Traversing Dense Material

Except at the very lowest energies, the absorption of electrons and photons in matter is a multi-step process (shower development). The phenomenon of particle multiplication in this process leads to the absorption of high-energy particles in relatively small amounts of matter. For example, a small lead brick measuring $5 \times 5 \times 15$ cm^3 (about 4 kg) would absorb more than 90% of the energy carried by the electrons accelerated in the largest synchrotron ever built (LEP at CERN, designed for 100 GeV).

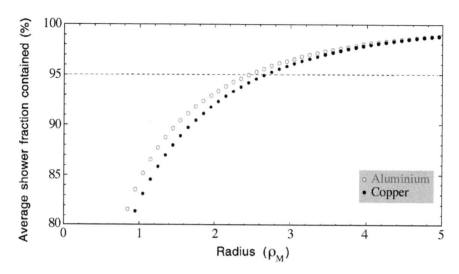

FIG. 2.18. Average energy fraction contained in an infinitely long cylinder of absorber material, as a function of the radius of this cylinder. Results of EGS4 calculations for various absorber materials and different energies.

The other particles subjected to only the em interaction, the muons, behave, at the same energies, in a very different way. Up to very high energies (100 GeV or higher), they lose their energy primarily through ionization and δ-rays. These mechanisms account for energy losses of typically only 1–2 MeV $g^{-1}cm^2$ and, therefore, it takes very substantial amounts of material to absorb high-energy muons.

For this reason, experiments in which cosmic muons constitute a major source of undesirable background have to be located in deep mines or under high mountains, since these muons may sometimes penetrate several kilometers of the Earth's crust. For the same reason, the CERN high-energy neutrino beam that was used for many experiments in the West Area (1963–1998) was equipped with a 300 m long iron shield. The neutrinos were produced from pion and kaon decay ($\pi, K \rightarrow \nu_\mu \mu$) and the muons had to be absorbed in the space between the production target and the neutrino detectors. In iron, muons lose energy at a rate of about 1.1 GeV/m (about a factor of three higher than in the soil of which the CERN site is composed). By virtue of this shield (composed of the Swiss strategic iron reserves), CERN's high-energy neutrino experiments could be installed just inside the perimeter of the Meyrin site.

Higher-order QED processes, such as bremsstrahlung and e^+e^- pair production, do occur in muon absorption. However, compared with electrons, they are suppressed by a scale factor of $(m_\mu/m_e)^2 \approx 40,000$. Therefore, the critical energy at which muons lose, on average, equal amounts of energy through radiation and ionization is at least 200 GeV. Just as for electrons, the contribution of these higher-order QED processes to a muon's energy loss is strongly Z dependent [Loh 85]. For example, the average energy loss of 500 GeV muons in lead increases (with respect to ionization losses) by a factor of 5.8 because of these effects. In iron, this factor amounts to 2.5 and in aluminium 1.8.

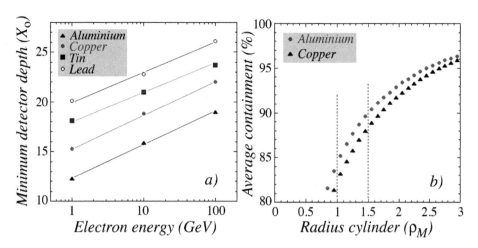

FIG. 2.19. Size requirements for electromagnetic shower containment. The depth of a calorimeter needed to contain electron showers, on average, at the 99% level, as a function of the
electron energy. Results are given for four different absorber media (*a*). Average energy containment of electron-induced showers in a copper and an aluminium based calorimeter, as a
function of the radius of an infinitely deep cylinder around the shower axis (*b*).

At energies below 100 GeV, the average energy loss is primarily determined by ionization, in all absorber materials. The mean energy loss per unit path length, $\langle dE/dx \rangle$,
is given by the well-known Bethe–Bloch formula [Ros 52]:

$$-\langle dE/dx \rangle \; = \; Kz^2 \frac{Z}{A} \frac{1}{\beta^2} \left[\frac{1}{2} \ln \frac{2m_e c^2 \beta^2 \gamma^2 T_{\max}}{I^2} - \beta^2 - \frac{\delta}{2} \right] \qquad (2.17)$$

in which $T_{\max}$ represents the maximum kinetic energy that can be imparted to an electron in a single collision, I is the mean excitation energy of the absorber material, δ a
correction term describing the *density effect*, and the proportionality constant K equals
$4\pi N_A r_e^2 m_e c^2$.

The quantity $\langle dE/dx \rangle$, which is often referred to as the *specific ionization* or the
ionization density, has a characteristic energy dependence, which is governed by the
product of the velocity (β) and the Lorentz factor (γ) of the particles (Figure 2.20).
For relativistic muons, $\langle dE/dx \rangle$ falls rapidly with increasing β, reaches a minimum
value near $\beta = 0.96$, and then exhibits what is called the relativistic rise, to level off
at values of 1–2 MeV g^{-1} cm^2 in most materials. Muons, or other particles with unity
charge such as pions, with an energy corresponding to that at which the $\langle dE/dx \rangle$ curve
reaches its minimum, are called *minimum ionizing particles*, or *mips*,

In relatively thin amounts of material, such as those represented by a typical calorimeter, the total energy loss $\Delta E/\Delta x$ may differ substantially from the value calculated
on the basis of $\langle dE/dx \rangle$. This is because of the relatively small number of collisions
with atomic electrons, and the very large fluctuations in energy transfer that may occur
in such collisions. Therefore, the energy loss distributions measured with (thin) calo-

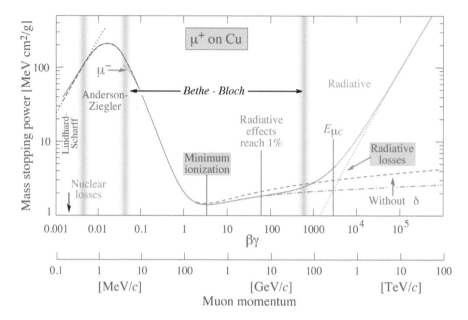

FIG. 2.20. The average energy loss per unit path length $(-\langle dE/dx \rangle)$ for positive muons in copper, given as a function of the product of the Lorentz variables $\beta\gamma$ [PDG 14]. For muon momenta in the range from ~ 5 MeV/c to ~ 50 GeV/c, this energy loss is well described by the Bethe-Bloch formula (Equation 2.17.)

rimeters reach their maximum (*i.e.*, most probable) value in general below the value calculated on the basis of $\langle dE/dx \rangle$ and have a long tail toward large energy losses, the so-called Landau tail [Kop 85]. Only for very substantial amounts of matter, *e.g.*, equivalent to 100 m of water, the energy loss distribution becomes approximately Gaussian.

The various effects discussed above are illustrated in Figure 2.21. This figure shows measured signal distributions from 10, 20, 80 and 225 GeV muon beams traversing the lead-based SPACAL calorimeter [Aco 92c]. The distributions are clearly changing with energy, due to the increasing cross sections of higher-order QED processes, such as hard bremsstrahlung and e^+e^- pair production. Such processes lead to a considerable broadening of the rms width of the energy loss distribution. Also, the average amount of energy lost by the muons in the calorimeter increases with energy. Similar effects were also observed by the RD5 Collaboration [Alb 95].

It should be emphasized that the Bethe-Bloch formula does not only apply to muons, but to the ionization losses of all charged particles. In that sense, it is an extremely important formula for calorimetry in general, since the signals produced by calorimeters are the result of processes in which the atoms of the absorber material are excited. In the next section, we will see examples of the importance of the β^{-2} dependence of the energy loss for non-relativistic nucleons, which dominate the signals from hadron showers.

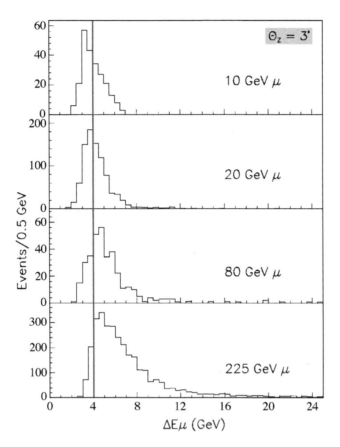

FIG. 2.21. Signal distributions for muons of 10, 20, 80 and 225 GeV traversing the $9.5\lambda_{int}$ deep SPACAL detector at $\theta_z = 3°$. From [Aco 92c].

2.3 Hadronic Showers

So far, we have only considered electromagnetic interactions between shower particles and the (em fields in the) absorbing medium. We now turn our attention to hadronic shower development, in which the strong interactions between the shower particles and the nuclei of the absorbing medium also play an important role.

Because of the nature of the strong interaction, hadronic showers are much more complicated than electromagnetic ones. The variety of processes that may occur, both those at the particle level and those involving the struck nucleus, is much larger.

One immediate consequence of the involvement of the strong interaction, with far-reaching implications for calorimetry, is the "invisible-energy" phenomenon. In em showers, all the energy carried by the incoming electron or photon is eventually used to ionize the absorbing medium, which is something that can be measured. However, as we will see, in hadronic showers a certain fraction of the dissipated energy is *fundamentally undetectable*.

When discussing em showers (Section 2.1), we saw an important difference between the absorption of photons and electrons. Electrons lose their energy in a continuous stream of events, in which atoms of the traversed medium are ionized and bremsstrahlung photons are emitted. On the other hand, photons may penetrate a considerable amount of matter without losing any energy, and then interact in a manner that may change their identity (*i.e.*, the photon may turn into a e^+e^- pair).

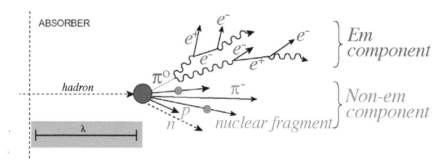

FIG. 2.22. Schematic depiction of a hadron shower. The energy carried by the hadron is typically deposited in the form of an electromagnetic and a non-electromagnetic component. The em component is the result of π^0s and ηs produced in the nuclear reactions. The non-em component consists of charged hadrons, and nuclear fragments. Some fraction of the energy transferred to this component (the "invisible" energy needed to break apart nuclei excited in this process) does *not* contribute to the calorimeter signals.

When a high-energy hadron penetrates a block of matter, some combination of these phenomena may occur (Figure 2.22). When the hadron is charged, it will ionize the atoms of the traversed medium, in a continuous stream of events, in much the same way as a muon of the same energy would do (Section 2.2). However, in general, at some depth, the hadron encounters an atomic nucleus with which it interacts strongly. In this nuclear reaction, the hadron may change its identity dramatically. It may, for example, turn into fifteen new hadrons. Also the struck nucleus changes usually quite a bit in such a reaction. It may, for example, lose ten neutrons and three protons in the process and end up in a highly excited state, from which it decays by emitting several γ-rays.

Neutral hadrons do not ionize the traversed medium. For these particles, nuclear reactions are *the only* option for losing energy. This is in particular true for neutrons, which are abundantly produced in hadronic shower development. As a result, neutrons deposit their kinetic energy in ways very different from those for the charged shower particles, with potentially very important implications for calorimetry.

The particles produced in the first nuclear reaction (mesons, nucleons, γs) may in turn lose their energy by ionizing the medium and/or induce new (nuclear) reactions, thus causing a shower to develop. Conceptually, this shower is very similar to the em ones discussed in Section 2.1. Initially, the number of shower particles increases as a result of multiplication processes, and so does the energy deposited by the shower particles in a slice of given thickness. However, at some depth further multiplication is balanced by the absorption of shower particles. Beyond this shower maximum, the number of shower particles and the energy they deposit in a slice of matter of given thickness gradually decrease.

Despite these similarities, there are also major differences between em and hadronic showers. One of these differences concerns the *scale* of the shower development. Since hadronic shower development is governed by nuclear, instead of electromagnetic, inter- actions, the scales of em and hadronic showers are different to an extent determined by the differences between the cross sections for the em and nuclear reactions.

In this section we have selected aspects of hadronic shower development that have important consequences for calorimetric energy measurements. In the following sub- sections, we first discuss relevant phenomena in the *particle* and *nuclear* sectors of the nuclear interactions. The peculiarities of neutron absorption in matter are the subject of Section 2.3.3, and the characteristics of hadronic shower profiles are discussed in Section 2.3.4.

2.3.1   *The particle sector*

2.3.1.1   *Electromagnetically decaying particles.*   Some of the particles produced in the hadronic cascade, in particular π^0s and ηs, decay through the electromagnetic in- teraction: $\pi^0, \eta \rightarrow \gamma\gamma$. Therefore, hadron showers generally contain a component that propagates electromagnetically. The fraction of the initial hadron energy converted into π^0s and ηs varies strongly from event to event, depending on the detailed processes occurring in the early phase of the shower development, *i.e.*, the phase during which production of these particles is energetically possible.

On average, approximately one-third of the mesons produced in the first interaction are π^0s. In the second generation of nuclear interactions, the remaining hadrons may also produce π^0s if they are sufficiently energetic, and so on. And since the production of π^0s by strongly interacting mesons is an *irreversible* process (a "one-way street" [Gab 94]), the average fraction of the initial hadron energy converted into π^0s gradually increases with that initial energy.

One may argue [Gab 94] that this increase of the average em fraction with energy proceeds according to a power law. This argument goes as follows. If all available shower energy is used to produce mesons, and one-third of the mesons produced in the nuclear reactions are π^0s, then the em content of the shower (f_{em}) amounts, on average, to 0.33 (1/3) after the first interaction, $0.33 + 2/3 \times 1/3 = 0.55$ (5/9) after the second generation of interactions, $0.55 + 4/9 \times 1/3 = 0.70$ (19/27) after the third generation, *etc.* After n generations of reactions, the average fraction of the total shower energy carried by π^0s equals

$$ f_{em} = 1 - \left(1 - \frac{1}{3}\right)^n \qquad (2.18) $$

The non-electromagnetic content of the shower decreases as $(1 - 1/3)^n$. After each col- lision in the hadronic cascade, $(1 - 1/3)$ of the remaining energy is, on average, available for the next generation of collisions. This continues until the point where the available energy drops below the pion production threshold. The number of generations, n, is thus a function of the energy E of the particle that initiated the shower. If we assume that the total number of mesons produced in the shower development is proportional to E, and that the average number of mesons produced per interaction, *i.e.*, the average multiplicity $\langle m \rangle$, is independent of E, then the number of generations (n) increases by one unit every time the energy E increases by a factor $\langle m \rangle$. Under these assumptions, the em fraction of the shower energy changes according to a power law (Equation 2.18) with increasing energy.

In reality, the situation is somewhat more complicated, because of the following factors:

- There are other particles produced apart from charged and neutral pions. Therefore, the factor 1/3 used in the previous discussion, should be considered an upper limit. In reality, this factor, to be called f_{π^0}, will be somewhat lower.
- The average multiplicity $\langle m \rangle$ is not independent of the incident energy E. It is well known that the average particle multiplicity in reactions induced by multi-GeV hadrons increases logarithmically with the energy, which implies that $\langle m \rangle$ slowly increases with energy and that the increase in f_{em} proceeds somewhat less fast with E than indicated by the power law.
- Energy loss by ionization and nuclear excitation of the calorimeter media has been neglected. These losses are strongly media dependent and thus lead to media-dependent em shower fractions.
- Peculiarities, such as the requirement of baryon number conservation, have been neglected. This requirement leads, for example, to smaller em shower fractions in proton-induced showers than in showers initiated by pions of the same energy.

These factors have been studied in detail by Gabriel *et al.* [Gab 94]. The authors give a general expression for the electromagnetic fraction of hadronic showers, as follows

$$f_{em} = 1 - \left(\frac{E}{E_0} \right)^{(k-1)} \tag{2.19}$$

in which E_0 is a scale factor, which corresponds to the average energy needed for the production of one pion, and the exponent $(k - 1)$ is related to the average multiplicity $\langle m \rangle$ and the average fraction of π^0 production in the reactions, f_{π^0}:

$$1 - f_{\pi^0} = \langle m \rangle^{(k-1)} \rightarrow k = 1 + \frac{\ln(1 - f_{\pi^0})}{\ln \langle m \rangle} \tag{2.20}$$

The relationship between the Equations 2.18 and 2.19 can be seen as follows. In Equation 2.18, f_{π^0} is given the value 1/3. In its general form, this equation reads

$$f_{em} = 1 - (1 - f_{\pi^0})^n \tag{2.21}$$

If E_0 corresponds to the average energy needed for the production of one pion, then E/E_0 represents the total number of pions that, *in the absence of an em shower component*, would be produced in the development of a shower initiated by a hadron with energy E. This total number is equal to $\langle m \rangle^n$, so that Equation 2.19 can be written as

$$f_{\text{em}} = 1 - \langle m \rangle^{n(k-1)} \tag{2.22}$$

This equation is equivalent to 2.21, given the relation between the parameters f_{π^0}, $\langle m \rangle$ and k (Equation 2.20).

Inspection of Equations 2.19 and 2.20 shows that the exponent k, which defines the *energy dependence* of the em shower fraction, is determined by two parameters: the average fraction of π^0 production per nuclear interaction, f_{π^0}, and the average multiplicity per nuclear interaction, $\langle m \rangle$. The *overall scale* of the em shower fraction is determined by the number of pions per unit of energy produced in the shower development, *i.e.*, the inverse value of the parameter E_0.

The "slope" parameter k has typically a value around 0.8. This can be easily verified. If we take $f_{\pi^0} = 1/3$ and assume an average multiplicity $\langle m \rangle = 5$, Equation 2.20 gives $k = 0.75$. The value of k rises to 0.82 if f_{π^0} is lowered to 1/4. Changing the average multiplicity $\langle m \rangle$ to 4 or 6 brings the value of k to 0.79 and 0.84, respectively, when f_{π^0} is kept at 1/4.

Gabriel and his coworkers [Gab 94] have made an extensive study of the merits of Equation 2.19, using simulated data on hadronic shower development in a variety of different absorber materials. They arrived at some interesting conclusions:

- In general, the energy dependence of the average em fraction of hadron showers simulated in a given absorber material could be brought in good agreement with Equation 2.19 by choosing appropriate values for the parameters k and E_0.

- The parameter E_0, which describes the overall level of the em shower component was found to be Z dependent in these studies. For example, the average em fraction for showers induced by 100 GeV pions decreased from about 61% for aluminium to about 52% for lead. In order to describe the (simulated) data, the value of E_0 had to be increased by about a factor of two from iron ($E_0 = 0.7$ GeV) to lead ($E_0 = 1.3$ GeV). The values found for k did not significantly depend on Z.

- The em fraction in proton-induced showers is significantly smaller than for pion-induced showers of the same energy. At 100 GeV, the differences were found to be of the order of 15%, and there were no indications for a significant energy dependence of this difference.

It is not easy to verify these predictions experimentally. Yet, there is some supporting experimental evidence. Both the SPACAL [Aco 92b] and the QFCAL [Akc 97] Collaborations have made attempts to measure the electromagnetic components of hadronic showers developing in their detectors.

The SPACAL group measured the em shower fraction by analyzing the lateral profiles of showers induced by pions of different energies (see Figure 2.38). These profiles consisted of a narrow core, surrounded by an exponentially decreasing halo. Assuming that the core represented the em shower component, the em shower fraction was determined by decomposing the measured profiles into the mentioned components.

The results are shown in Figure 2.23. The curves in this figure correspond to calculations based on Equation 2.19, using a value $E_0 = 1.3$ GeV, recommended by Gabriel

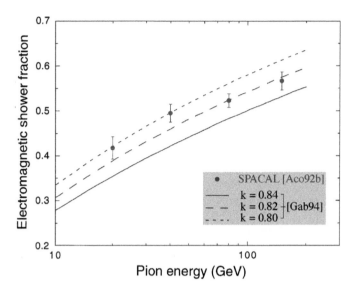

FIG. 2.23. The average em shower fraction in pion-induced showers measured in the SPACAL lead/fiber calorimeter. The curves represent predictions based on Equation 2.19. Experimental data from [Aco 92b].

et al. for lead, the absorber material of the SPACAL detector. The parameter k was varied in these calculations and the three curves represent results for $k = 0.80$, 0.82 and 0.84, respectively. The experimental data favor $k = 0.82$.

The QFCAL group did not measure the values of f_{em} directly. However, because of the extremely *non-compensating* character of their calorimeter, which consisted of quartz fibers embedded in copper [Akc 97], there are rather good indirect ways to estimate these values.

This calorimeter was, for all practical purposes, almost *exclusively* sensitive to the em components of hadron showers [Gan 93]. The signals from the non-electromagnetic component were suppressed by a factor of about seven. Therefore, a measurement of the π/e signal ratio was indicative for the value of f_{em}. In particular, the energy dependence of this signal ratio was found to be very similar to the energy dependence of f_{em}.

The π/e signal ratio measured with the QFCAL calorimeter is shown in Figure 2.24. The dashed line in this figure represents f_{em}, calculated on the basis of Equation 2.19 for copper ($E_0 = 0.7$ GeV [Gab 94]), using the parameter value $k = 0.82$, which gave a reasonable description of the experimental lead data (see Figure 2.23).

The experimental data and this dashed curve show indeed approximately the same energy dependence. The fact that the experimental π/e signal ratios are systematically larger than the f_{em} values is a consequence of the (small) contributions from other particles produced in the shower development to the pion signals. By taking these contributions, which are suppressed by a factor of seven [Gan 95], into account, the π/e signal ratio can be derived from the f_{em} values (see Section 3.5.3 for more details on this point). This leads to the solid curve in Figure 2.24, which is in good agreement with

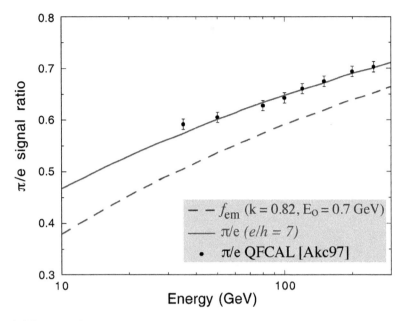

FIG. 2.24. The π/e signal ratio measured with the QFCAL calorimeter. The dashed curve corresponds to the value of the em shower fraction in the pion-induced showers, calculated with Equation 2.19, the solid curve represents the π/e signal ratio derived from this. See text for details. Experimental data from [Akc 97].

the experimental data.

Alternatively, one can thus take the measured π/e signal ratios, correct these for the small contributions from the non-em shower component, and derive a "measured" value of f_{em} on this basis. The result of this exercise is shown in Figure 2.25. The lead data (Figure 2.23) are also included in this figure, as well as the calculated energy dependence of f_{em} (Equation 2.19) based on Gabriel's model, with $k = 0.82$ and $E_0 = 0.7$ GeV and 1.3 GeV for copper and lead, respectively. The experimental results indeed suggest the existence of a systematic difference between the average f_{em} values in copper and lead and most certainly do not contradict the predicted Z dependence of f_{em}.

The experimental evidence in support of the prediction of systematic differences between the average f_{em} values for proton- and pion-induced showers was also provided by the QFCAL group [Akc 98]. Figure 2.26 shows the average calorimeter signal per GeV (the calorimeter *response*), for protons and pions as a function of energy, measured with the QFCAL detector. In the same figure, the response *ratio* is given as a function of energy. The response to protons is systematically smaller than the response to pions of the same energy. The difference ranges from $\sim 13\%$ at 200 GeV to $\sim 8\%$ at 375 GeV. At these high energies, more than 60% of the hadronic energy is already going into the em shower component (Figure 2.25) and the calorimeter response to the remaining (non-em) fraction is strongly suppressed. Therefore, the experimental differences between

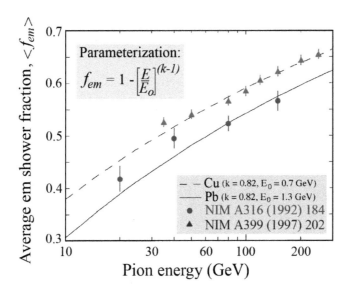

FIG. 2.25. Comparison between the experimental results on the em fraction of pion-induced showers in the (copper-based) QFCAL and (lead-based) SPACAL detectors. Data from [Akc 97] and [Aco 92b].

the pion and proton responses must be due to differences in the average f_{em} values. These results thus confirm the prediction by Gabriel *et al.* [Gab 94] that the em fraction in proton-induced showers is, on average, smaller than in pion-induced ones.

The experimentally observed differences are also *quantitatively* in good agreement with the prediction. According to the authors, f_{em} should be 15% smaller for protons than for pions. When Equation 2.19 is used with the parameter values $E_0 = 0.7$ GeV and $k = 0.82$ (see Figure 2.24), the average f_{em} value for 200 GeV pions in copper is found to be 0.639. If the f_{em} value for 200 GeV protons is 15% smaller, and the non-em components produce signals that are suppressed by a factor of seven, then the π/p signal ratio becomes 0.880, in excellent agreement with the experimentally observed value (0.868±0.020). However, it was not foreseen by Gabriel *et al.* that the proton/pion differences tend to become smaller as the energy further increases.

These phenomena have a simple explanation. To the extent that collisions between high-energy hadrons and atomic nuclei may be described as interactions between constituent quarks, the remaining (spectator) quark(s) from the projectile carry a considerable fraction of the incident energy and emerge as *leading particles* after dressing themselves up to a hadron by picking up an (anti-)quark from the "sea." If the incident particle is a proton, a *baryon* (proton, neutron, Λ, *etc.*) will emerge carrying a large fraction of the initial energy. Baryon number conservation will also apply in subsequent interactions: the leading baryon produces another leading baryon when it collides with a nucleus, and so on.

The requirement of baryon number conservation thus limits the amount of energy available for the production of π^0s, which generate (almost all of) the signal in calo-

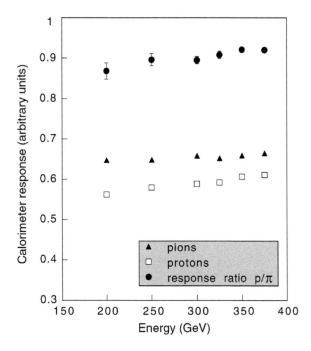

FIG. 2.26. The calorimeter response to protons and π^- mesons, and the ratio of these responses, as a function of energy, measured with the QFCAL detector. Data from [Akc 98].

rimeters of the QFCAL type. In the development of pion-induced showers, there is no such requirement limiting the π^0 production (and thus the signals). As a result, proton-induced showers contain, on average, a smaller em component than pion-induced showers at the same energy.

This explanation is corroborated by several other experimental differences between pion- and proton-induced showers [Akc 98]. For example, the energy resolution was found to be considerably better for protons and the line shape was much more symmetric than for pions (Section 7.5.3). In addition, the leading-particle mechanism leads to significant differences in the shower profiles, which were also experimentally observed.

As the energy increases, the *fragmentation function* which describes the hadronization of the final state particles in nuclear interaction processes changes as well. A frequently used description of this process uses the following parameterization of this function [Hal 84]:

$$D(z) = (\alpha + 1)(1 - z)^\alpha / z \tag{2.23}$$

in which $D(z)$ represents the probability that a hadron produced in this process acquires a fraction z of the available energy. The parameter α slowly rises from values of ~ 4 at LEP energies to ~ 6 at the Tevatron [Gre 90]. This means that the average z value of the final-state hadrons gradually decreases. Therefore, the energy fraction typically carried by the leading particle also gradually decreases as the collision energy increases. This might very well explain why the differences between pion- and proton-induced showers tend to become smaller with increasing energy.

The Z dependence of the em shower fraction, f_{em}, is a consequence of the fact that the secondary hadrons produced in the first nuclear interaction cannot use all the energy they acquire in this process for the production of new hadrons in subsequent reactions. Part of that energy is used to ionize the absorber medium they traverse and another part is transferred to the nuclei with which they interact. Both processes require a larger fraction of the available energy in a high-Z absorber medium than in a low-Z one. As the energy spent per hadron on these processes increases, the *number* of different hadrons produced in the shower development decreases. Therefore, the value of E_0 increases, the effective number of generations (n) decreases and so does the average em shower fraction (Equations 2.18 and 2.19).

2.3.1.2 *Ionization losses by charged hadrons.* The *charged* secondary hadrons produced in the collisions induce a nuclear interaction themselves after having traveled, on average, one *nuclear interaction length* (λ_{int}) in the absorber material[2]. The average energy a minimum ionizing particle loses per unit λ_{int} through ionization in this process is listed for various absorber media in Table 2.2. It amounts to ~ 150 MeV for low-Z

Table 2.2 *The specific ionization energy loss of minimum ionizing particles in various absorber materials, and the average energy lost by minimum ionizing protons over a distance of one nuclear interaction length. Data from [PDG 14].*

Absorber	Z	dE/dx (mip) (MeV g^{-1}cm^2)	λ_{int} (g cm^{-2})	$\Delta E/\lambda_{int}$ (MeV)
Carbon	6	1.742	85.8	149
Aluminum	13	1.615	107.2	173
Iron	26	1.451	132.1	192
Copper	29	1.403	137.3	193
Tin	50	1.263	166.7	211
Tungsten	74	1.145	191.9	220
Lead	82	1.122	199.6	224
Uranium	92	1.081	209.0	226

materials such as carbon, and increases with Z to reach about 225 MeV for high-Z absorber materials, such as lead or uranium. It should be remarked that the values listed in Table 2.2 are for protons. Since pions are smaller, they are less likely to encounter a nucleus. They travel typically a 25% longer distance (and therefore lose 25% more energy) than protons before a nuclear interaction occurs [Kri 99]. Also, the values in Table 2.2 are for *minimum ionizing particles*. In reality, the specific ionization loss of the pions produced in the shower development is typically somewhat larger than the mip value. As a result of these effects, pions produced in hadronic showers developing in uranium

---

[2] See Section 2.3.4.1 for the definition of λ_{int}.

lose, on average, some 300 MeV by ionization, before inducing a nuclear reaction. For iron and aluminium, the losses are 15–25% smaller. These differences constitute a significant contribution to the Z dependence of the number of different hadrons produced in the shower development process and, by consequence, to the Z dependence of the em shower fraction.

In addition to these higher ionization losses, secondary hadrons also transfer a larger fraction of their energy to the *nuclei* with which they interact in high-Z absorber materials than in low-Z ones (see Section 2.3.2 for details on this). This is the other important contribution to the Z-dependent effects discussed here.

2.3.1.3   *Particle multiplicities in hadron showers.*   The description of the particle sector of hadronic showers discussed in the previous subsections also makes it possible to get a feeling for the numbers of hadrons involved in shower development, and for their contribution to the energy absorption process. Let us take, as an example, the case of 100 GeV pions showering in copper. According to Figure 2.25, $\sim 60\%$ of the energy is, on average, deposited by the em shower component. The non-em component takes care of the other 40%. The E_0 value is 0.7 GeV for copper, and thus $40/0.7 \approx 58$ hadrons are involved in that part of the absorption process. Each of these hadrons deposits, on average, ~ 250 MeV in the form of ionization. Ionization by secondary and higher-order hadrons thus accounts for a total energy deposit of about 15 GeV, *i.e.*, 15% of the initial energy and $\sim 35\%$ of the energy carried by the non-em shower component.

The total mass of the hadrons produced in the shower development process corresponds to about 20% of the non-em energy. A small fraction of the charged pions and kaons ($< 1\%$ in dense absorber materials) decay in flight, before undergoing a nuclear interaction. In that case, the (anti-)neutrino and possibly also the muon produced in that process escape the detector. The rest (*i.e.*, almost all) of the mass energy, plus the remaining kinetic energy, are used to excite and dissociate atomic nuclei in the absorber medium. In total, $\sim 65\%$ of the energy comprised in the non-em shower fraction is thus used for that purpose.

The numbers are slightly different when the absorber is lead instead of copper. The showers initiated by 100 GeV pions have, on average, a somewhat smaller em component, carrying 55% instead of 60% of the total energy. For an E_0 value of 1.3 GeV, the total number of hadrons involved in the non-electromagnetic portion of the showers amounts thus, on average, to $45/1.3 \approx 35$. The total ionization loss by these 35 hadrons is about 10.5 GeV, or 23% of the total non-em energy. Since there are fewer hadrons involved, for more available energy, their total mass represents a smaller fraction of the non-em shower energy than in copper, $\sim 10\%$ instead of $\sim 20\%$, The fraction of the total non-em energy used to excite and dissociate atomic nuclei in the absorber medium is relatively similar: 77% for lead *vs.* 65% for copper. However, due to the much smaller number of hadrons, the average energy used for this purpose *per hadron* is considerably larger in lead: 1.0 GeV *vs.* 0.45 GeV for copper.

The effective number of generations of nuclear interactions in hadronic shower development (n, see Equation 2.18) may also be estimated, on the basis of E_0 and $\langle m \rangle$, since $E/E_0 = \langle m \rangle^n$. For 100 GeV pions in copper and an average multiplicity $\langle m \rangle =$

5, one finds $n \sim 3$. That is, charged pions produced by the incoming hadron interact, their interaction products interact too and that is it as far as the production of mesons goes.

Table 2.3 *Characteristics of particle production in pion-induced showers in copper (lead), calculated on the basis of Equation 2.19, with the following parameter choices:* $E_0 = 0.7$ GeV *(copper)*, 1.3 GeV *(lead)*, $k = 0.82$, $f_{\pi^0} = 1/4$.

E_π (GeV)	$\langle f_{em} \rangle$	$\langle \#\pi^\pm, K... \rangle$	$\langle \#\pi^0 \rangle$
10	0.380 (0.307)	9 (5)	3 (2)
20	0.453 (0.389)	16 (9)	5 (3)
30	0.492 (0.432)	22 (13)	7 (4)
50	0.536 (0.482)	33 (20)	11 (7)
80	0.574 (0.524)	49 (29)	16 (10)
100	0.591 (0.542)	58 (35)	19 (12)
150	0.619 (0.575)	82 (49)	27 (16)
200	0.639 (0.596)	103 (62)	34 (21)
300	0.664 (0.624)	144 (87)	48 (29)
400	0.681 (0.643)	182 (110)	61 (37)
500	0.694 (0.657)	219 (132)	73 (44)
700	0.712 (0.678)	288 (173)	96 (58)
1000	0.730 (0.698)	386 (232)	129 (77)

Finally, the number of π^0s produced in hadronic shower development may be estimated on the basis of these considerations, at about one-third to one-half ($[f_{\pi^0}^{-1} - 1]^{-1}$) of the number of hadrons that constitute the non-em shower component. For our "benchmark" showers induced by 100 GeV pions, this gives in copper a total of about 20-30 π^0s, on average. As the shower energy increases, the average number of π^0s increases as well, in a rather complicated manner, governed by Equation 2.19. Choosing values for k and f_{π^0} of 0.82 and 1/4, respectively, we calculated some characteristics of the particle sector in hadronic shower development in copper ($E_0 = 0.7$ GeV) and lead ($E_0 = 1.3$ GeV), for a range of energies of the pions that initiated the showers. These are summarized in Table 2.3. As discussed in Chapter 4, this information is important for understanding the fundamental limitations on the hadronic energy resolution.

2.3.1.4 *Asymptotic consequences.* We finish this section on the particle sector of hadronic shower development by looking at yet another consequence of Equation 2.19, which governs the production of electromagnetically interacting particles in hadronic showers and which is one of the cornerstones for understanding the fundamentals of calorimetry. At present, particles produced in accelerator-based experiments attain energies up to about 8 TeV. However, cosmic-ray events have been reported with energies that are many orders of magnitude higher, up to 10^{20} eV. The equations used in this section for describing hadronic showers in the GeV–TeV energy domain lead to some interesting conclusions when applied to the absorption of such extremely-high-energy hadrons in the Earth's atmosphere (which, as described in more detail in Section

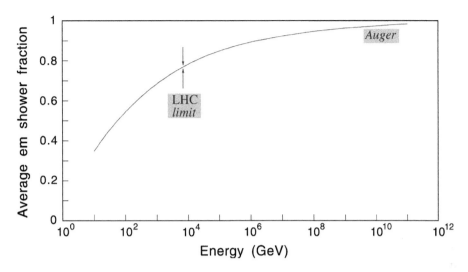

FIG. 2.27. The average fraction of the initial energy carried by the em shower component, as a function of the initial energy, calculated on the basis of Equation 2.19.

10.3, can be used as a calorimeter with some very peculiar properties).

Figure 2.27 shows the average fraction of the energy of the cosmic shower carried by its em component, calculated with Equation 2.19 for the same set of parameter values as above[3]. This fraction continues to rise gradually, and reaches values of 98–99% in the domain of the highest energy cosmic showers on record. It should therefore be no surprise that some of the showers initiated by such extremely high-energetic primaries appear to consist almost entirely of electromagnetic components. Such events have become known as *Centauro* events. At this energy, the number of generations in the shower development (n) reaches values well in excess of 10, and in each generation more energy is siphoned off into the one-way street of π^0 production.

At energies accessible to accelerators, such events, though extremely rare, do also occur. They require the occurrence of processes such as the production of a very-high-z leading π^0 and/or charge exchange reactions (*e.g.*, $\pi^- + p \rightarrow \pi^0 + n$). Such processes form the ultimate limit on the particle identification capabilities of calorimeter systems (see Section 7.6).

2.3.2 *The nuclear sector*

2.3.2.1 *Nuclear spallation reactions.* When an incoming high-energy hadron strikes an atomic nucleus, the most likely process to occur is spallation. Spallation is usually described as a two-stage process: a fast intranuclear cascade, followed by a slower evaporation stage. The incoming hadron makes quasi-free collisions with nucleons inside the struck nucleus. The affected nucleons start traveling themselves through the nucleus and collide with other nucleons. In this way, a cascade of fast nucleons

---

[3] In reality, the em components are probably even larger, because of the smaller E_0 value applicable to air.

develops. At this stage, pions and other unstable hadrons may also be created if the transferred energy is sufficiently high. Some of the particles taking part in this cascade reach the nuclear boundary and escape. Others get caught and distribute their kinetic energy among the remaining nucleons in the nucleus.

The second step of the spallation reaction consists of a de-excitation of the intermediate nucleus. This is achieved by evaporating a certain number of particles, predominantly free nucleons, but sometimes also αs or even heavier nucleon aggregates, until the excitation energy is less than the binding energy of one nucleon. The remaining energy, typically a few MeV, is released in the form of γ-rays. In very heavy nuclei, *e.g.*, uranium, the intermediate nucleus may also fission.

Much experimental information on spallation reactions has been accumulated during the past 70 years. Rudstam [Rud 66] has given an empirical formula, valid within broad limits either of energies (> 50 MeV) or of atomic mass ($A > 20$), which gives a satisfactory description of spallation cross sections. When a particle of energy E hits a target with atomic mass A_T, the relative cross sections σ for the production of spallation products (Z_f, A_f) are given by the relation

$$\sigma(Z_f, A_f) \sim \exp\left[-P(A_T - A_f)\right] \times \exp\left[-R|Z_f - SA_f + TA_f^2|^{3/2}\right] \quad (2.24)$$

in which E is expressed in MeV and the parameters P, R, S and T have the following values:
$P = 20E^{-0.77}$ for $E < 2100$ MeV, $P = 0.056$ for $E > 2100$ MeV,
$R = 11.8A_f^{-0.45}$, $S = 0.486$, $T = 0.00038$.

Figure 2.28 shows the cross sections for nuclides that can be produced from ^{238}U spallation induced by a 2 GeV hadron, computed with this formula. Hundreds of different reactions occur with comparable probability. The largest cross section for an exclusive reaction amounts to only $\sim 2\%$ of the total spallation cross section, and there are about 300 different reactions that contribute more than 0.1% to the total spallation cross section! This example illustrates the enormous diversity of processes that may occur in the nuclear sector of the hadronic interactions.

2.3.2.2 *Nuclear binding energy.*   In these spallation reactions, considerable numbers of nucleons may be released from the nuclei in which they were bound. This was beautifully illustrated by emulsion and bubble-chamber pictures from the early days of high-energy physics, which frequently showed "nuclear stars," nuclei that literally exploded in the collisions.

An example of such a process is depicted in Figure 2.29, which shows (the results of) a collision between a 30 GeV/c proton and an atomic nucleus in a photographic emulsion. About 20 densely ionizing particles, presumably (almost) all protons, are produced in this reaction. The picture shows that these particles are more or less isotropically emitted from the struck nucleus. Several other, much less dense ionization tracks emerge from the collision to the left, *i.e.*, roughly following the direction of the incoming projectile, which enters the picture from the right-hand side. Most likely, these tracks represent pions and fast spallation protons. Not visible in this picture are the neutrons, considerable numbers of which undoubtedly were also released in this interaction.

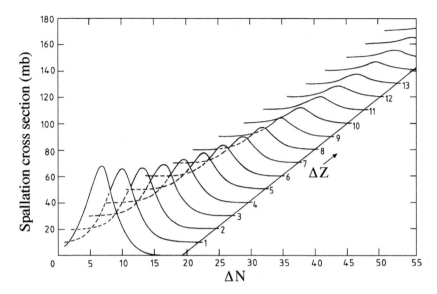

FIG. 2.28. Cross sections for nuclides produced by spallation of ^{238}U, induced by a 2 GeV hadron. The final-state nuclide is defined by the number of protons (ΔZ) and neutrons (ΔN) released from the target nucleus.

The energy needed to release these nucleons, *i.e.*, the nuclear binding energy, is lost for calorimetric purposes, it does not contribute to the calorimeter signal, it is *invisible*. There is a large variety of processes that may occur in hadronic shower development and event-to-event fluctuations in the invisible energy fraction are very large. In one extreme case, an incoming π^+ may strike a neutron and cause the following reaction: $\pi^+ n \rightarrow \pi^0 p$, transferring almost all its kinetic energy to the π^0 (charge exchange reaction). In that case, the invisible energy is almost zero, since the π^0 decays in two γs which decay electromagnetically and the proton loses its (small) energy through ionization of the medium in which the reaction takes place, leaving only the nuclear binding energy of one escaping nucleon unaccounted for. In other extreme cases, invisible energy may consume some 60% of the total available energy. As is shown later on in this subsection, invisible energy accounts, on average, for 30–40% of the non-em shower energy, *i.e.*, energy that is not carried by π^0s or other electromagnetically decaying particles produced in the shower development.

The large event-to-event fluctuations in visible energy have obviously direct consequences for the precision with which hadronic energy can be measured in calorimeters. Because of these fluctuations, which have no equivalent in electromagnetic shower development processes, the energy resolution of hadron calorimeters is usually considerably worse than the em energy resolution. There is, however, one elegant way in which these effects can be limited, by exploiting the correlation that exists between the invisible energy lost when nucleons are released from the nuclei in which they are bound and the kinetic energy carried by these nucleons (see Section 4.8).

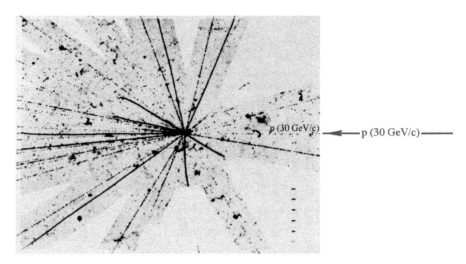

FIG. 2.29. A proton-nucleus interaction in a nuclear emulsion stack. Photograph courtesy CERN.

2.3.2.3 *Spallation nucleons.*   In the spallation reactions with absorber nuclei that take place in the absorption of energetic hadrons, large numbers of nucleons and nucleon aggregates such as α particles are produced. We will now try to get a quantitative feeling of what goes on in this crucial phase of the shower development. In particular, we will try to determine *average values* for the numbers of protons and neutrons produced and for the invisible energy. We will also try to separate the *cascade* nucleons from the *evaporation* ones, on a statistical basis.

Our starting points in this exercise are twofold:

1. Rudstam's equation (2.24) describing the cross sections for the possible nuclear reactions.
2. The conclusion by Gabriel *et al.* [Gab 94] that it takes, on average, 1.3 GeV to produce one pion in the development of high-energy hadron showers in lead.

In Section 2.3.1.1 we saw that the latter conclusion was supported by a number of experimental features of hadronic shower development. We also assume (following the reasoning from Section 2.3.1.2) that such a "typical" pion loses, on average, about 300 MeV by ionizing lead atoms before it interacts with a lead nucleus. The remaining 1.0 GeV, which includes the pion's rest mass, is thus used to excite and dissociate a lead nucleus or, more likely, several lead nuclei.

For simplicity, we assume that the lead absorber consists entirely of $^{208}_{82}\text{Pb}$, its most abundant isotope. The consequences of this simplification are negligible for this analysis. With Equation 2.24, the probability that a particle carrying an energy E converts this nucleus into a final state nucleus (A_f, Z_f) can be computed. In this process, $(208 - A_f)$ nucleons are released, $(82 - Z_f)$ protons and the rest neutrons. This computation can be repeated for all possible reactions that are energetically possible. The relative probabilities of these reactions give the relative probabilities that the corresponding numbers of protons and neutrons are released. The result is a probability distribution of the numbers of protons and neutrons released in this process, which is shown in Figure 2.30a for an incident energy $E = 1,000$ MeV. It should be pointed out that the incident energy E refers to the energy that is transferred to the nucleus. In the case of incident protons or neutrons, E corresponds thus to the *kinetic energy* of the projectiles, while in the case of pions, E represents the *total energy*, *i.e.*, the kinetic energy plus the energy contained in the pion rest mass (139.6 MeV).

At $E = 1,000$ MeV, on average, 2.7 protons and 12.8 neutrons are produced in these spallation reactions. This large discrepancy between the numbers of protons and neutrons released in these reactions is even more striking at lower incident energies. Figure 2.31a shows the average numbers of protons and neutrons as a function of the incident energy E. For energies smaller than 200 MeV, the probability of at least one proton being emitted drops below 50%. However, on average, 7 neutrons still come off at 200 MeV.

This indicates that the protons that are produced in the spallation processes on lead are almost exclusively produced in the fast cascade step. In the evaporation stage of the reaction, almost all emitted nucleons are neutrons. This is not surprising, since the Coulomb barrier for protons in a lead nucleus is ~ 12 MeV. Therefore, in the evaporation stage, where fragments are released with a kinetic energy of typically a few MeV (some fraction of the binding energy per nucleon, which amounts to ~ 7.9 MeV in lead), not many charged particles are expected to emerge from the nucleus.

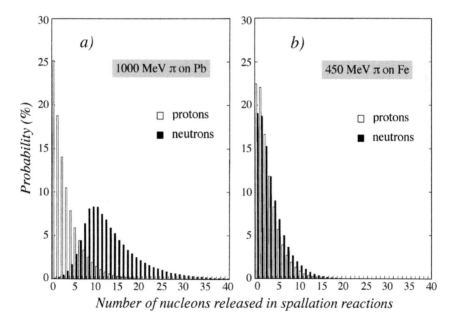

FIG. 2.30. Distribution of the numbers of protons and neutrons produced in spallation reactions induced by 1,000 MeV pions on $^{208}_{82}$Pb (*a*) and by 450 MeV pions on $^{56}_{26}$Fe (*b*).

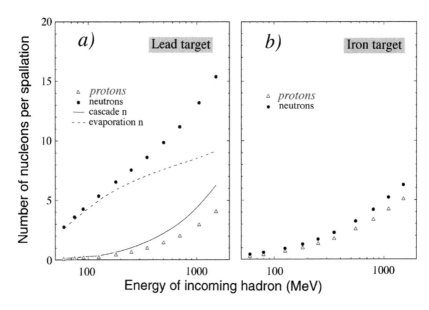

FIG. 2.31. The average numbers of protons and neutrons produced in spallation reactions on $^{208}_{82}$Pb (a) or $^{56}_{26}$Fe (b), as a function of the energy of the incoming hadron. The neutrons are split up in an evaporation and a cascade component.

In the fast cascade step, protons and neutrons are emitted in a ratio that, on average, reflects the numerical presence of these nucleons in the target nuclei. In $^{208}_{82}$Pb, one may thus expect for every cascade proton about 1.5 cascade neutrons (126/82).

These considerations make it possible to split the total nucleon production in the spallation reactions induced by our 1,000 MeV pions ($\sim$ 16 nucleons) into a cascade component and an evaporation component: 9 evaporation neutrons and 7 cascade nucleons (2.8 protons and 4.2 neutrons). The cascade nucleons, in particular the cascade neutrons, are likely to induce themselves new spallation reactions, further increasing the numbers of evaporation neutrons.

The evaporation neutrons are emitted isotropically, but the cascade particles have a dominating momentum component along the direction of the incoming particle. Therefore, the residual target nucleus undergoes a net recoil, in which it acquires a kinetic energy of the order of m/M, where m and M denote the total mass of the cascade nucleons and the residual nucleus, respectively. This recoil energy will, in general, *not* result in a measurable calorimeter signal and therefore has to be considered part of the invisible component of the shower energy.

Let us examine what happens to the energy in the initial "average" spallation reaction. In order to release the 16 nucleons from the lead nucleus, $16 \times 7.9 = 126$ MeV of nuclear binding energy has to be provided. The 9 evaporation neutrons carry a total of $\sim$ 27 MeV of kinetic energy (for $T = 2$ MeV, see Figure 2.32). The remaining $1000 - 153 = 847$ MeV is shared among the target nucleus (30 MeV recoil energy) and 7 cascade nucleons, about 117 MeV each. The range of 117 MeV protons in lead amounts to $\sim$ 2 cm [Jan 82], considerably less than the nuclear interaction length (17 cm). Therefore, these cascade protons are most likely to lose their energy by ionizing lead atoms.

The (on average, 4.2) cascade neutrons, on the other hand, will induce new spallation reactions. According to Figure 2.31a, each of these reactions will be characterized by the production of 5.3 neutrons and 0.25 protons, in total 22.3 neutrons and 1.1 protons. Following the same procedure as before, we can split these up into 21 evaporation neutrons, 1.7 cascade neutrons and 1.1 cascade protons. The energy involved in this stage of the process, $4.2 \times 117 = 490$ MeV is distributed as follows: 185 MeV nuclear binding energy, 62 MeV kinetic energy of the evaporation neutrons and 243 MeV kinetic energy shared by 1.6 cascade neutrons, 1.1 cascade protons and the target nucleus (recoil energy 3 MeV). The cascade nucleons carry thus, on average, a kinetic energy of 89 MeV. As before, the protons deposit this energy in the form of ionization (98 MeV in total), while the 1.6 neutrons cause a third generation of spallation reactions.

Table 2.4 *Destination of the 1.3 GeV total energy carried by an average pion produced in hadronic shower development in lead. Energies are in MeV.*

	Binding energy	Evaporation n (# neutrons)	Cascade n (# neutrons)	Ionization (# cascade p)	Target recoil
Before first reaction				(300) (π_{in})	
First reaction	126	27 (9)	490 (4.2)	328 (2.8)	30
Generation 2	185	62 (21)	142 (1.6)	98 (1.1)	3
Generation 3	59	21 (7)	36 (0.8)	25 (0.5)	1
Generation 4	24	12 (3)			
Total	394	122 (40)		451 (4.4)	34

Following this method, we can continue this process until all energy is deposited in one of four forms: nuclear binding energy, target recoil energy, ionization of lead atoms by cascade protons or kinetic energy of evaporation neutrons. The results are summarized in Table 2.4. The 1,000 MeV used to excite and dissociate atomic nuclei is distributed as follows: ionization by spallation protons 451 MeV, kinetic energy of evaporation neutrons 122 MeV and invisible energy 428 MeV (394 MeV nuclear binding energy + 34 MeV target recoil). On average, 40 evaporation neutrons, 7 cascade neutrons and 4–5 cascade protons are produced in these processes. These numbers refer to a total non-em energy deposit of 1.3 GeV, since the pion that initiated this "nuclear cascade" is assumed to have lost, on its way to the target nucleus, on average 300 MeV by ionizing lead atoms. Since the 1.3 GeV pion is assumed to be the *average* hadron produced in the non-em component of the shower development, this simplistic model allows us to estimate the average composition of this shower component, as well as the way in which the energy contained in this component is deposited in the absorber structure. The results of this estimate are given in Table 2.5.

This table also contains results for hadronic showers developing in iron, obtained with the same model. Starting point for the iron simulations was a 450 MeV pion. According to Gabriel *et al.* [Gab 94], it takes on average 700 MeV to produce one pion in showers developing in iron. These pions are assumed to lose, on average, 250 MeV by ionization before they interact with a nucleus (Section 2.3.1.2), and 450 MeV thus represents the energy typically transferred in these collisions.

The distributions of the numbers of protons and neutrons released from their nuclear environment in collisions between 450 MeV pions and iron nuclei are shown in Figure 2.30b, while the dependence of the *average* numbers of such protons and neutrons on the energy of the incident particle is given in Figure 2.31b. These results are markedly different from the lead ones (Figures 2.30a and 2.31a). Among the most characteristic differences, we mention:

- The strong *asymmetry* between protons and neutrons found in the case of lead is almost absent for reactions with iron nuclei.
- The total *number* of nucleons released in collisions with iron nuclei is considerably smaller than that for collisions with lead nuclei at the same energy.

The proton/neutron asymmetry in lead is a consequence of the Coulomb barrier ($\sim$ 12 MeV), which prevents protons from being emitted by an excited nucleus in the evaporation stage. In iron, this barrier is considerably lower ($\sim$ 5 MeV). Therefore, the probabilities for an excited Fe nucleus to emit a proton or a neutron are not very different from each other.

The differences between the numbers of nucleons released per GeV of deposited energy (40 in lead *vs.* 18 in iron) are due to three effects:

Table 2.5 *Energy deposit and composition of the non-em component of hadronic showers in lead and iron. The listed numbers of particles are per GeV of non-em energy.*

	Lead	Iron
Ionization by pions	23%	35%
Ionization by protons	35%	37%
Total ionization	58%	72%
Nuclear binding energy loss	30%	16%
Target recoil	3%	7%
Total invisible energy	33%	23%
Kinetic energy evaporation neutrons	9%	5%
Number of charged pions	0.77	1.4
Number of protons	3.4	8
Number of cascade neutrons	5.1	5
Number of evaporation neutrons	30.8	5
Total number of neutrons	36	10
Neutrons/protons	10.6/1	1.3/1

1. *The difference in nuclear binding energy.* It takes less energy to release a nucleon from a lead nucleus (7.9 MeV) than from an iron one (8.8 MeV) [Wap 77]. For a given energy, more nucleons are thus released in lead.

2. *The proton/neutron asymmetry.* Protons released from the nuclei typically lose their kinetic energy by ionization, while neutrons interact with other nuclei, and have their entire kinetic energy available for that purpose. In first approximation, kinetic energy carried away by escaping protons is thus lost for the purpose of nuclear excitation. In lead, protons carry $\sim 39\%$ of the energy carried by escaping spallation nucleons (82/208), in iron this fraction amounts to 46% (26/56). Because of the different proton/neutron ratios in the nuclei, a larger fraction of the available energy can thus be used for nuclear excitation in the case of lead absorbers.

3. *Re-interaction within the nucleus.* Spallation protons produced in the initial stage of the nuclear interaction process may transfer some of their kinetic energy to other nucleons they encounter on their way out of the nucleus. These other, accelerated, nucleons may in turn do the same. In each new generation of this *cascade*, some fraction of the available energy is transferred to the nucleus as a whole. In a sense, one may compare this process with the development of a shower in a block of matter, in this case represented by a single atomic nucleus. The larger the nucleus, the smaller will be the energy fraction "leaking out," in the form of spallation nucleons, and the larger will be the fraction of "contained energy," later to be released in the form of evaporation nucleons. And if the latter are almost exclusively neutrons, as in lead, mechanism #2 will lead to a further increase in the number of nucleons released per unit energy.

In first approximation, one would expect the effect of mechanism #3 to be proportional to the path length, *i.e.*, to the nuclear radius, and thus to $\sqrt[3]{A}$. On that basis, the average energy transfer to lead nuclei should thus be 1.6 times larger than for iron nuclei, in collisions with the same projectile. This energy transfer is further increased by mechanism #2, by a factor of 1.2 (46/39), to 1.9. Mechanism #1 translates this difference in energy transfer to the nucleus into a difference in the numbers of released nucleons, a further increase by a factor of 1.1 (8.8/7.9). The difference in the numbers of released nucleons expected on this basis (a factor 2.1 more nucleons released in lead than in iron) is in good agreement with the results listed in Table 2.5 (40/18 $\approx$ 2.2).

This agreement could be further improved by using the *marginal nuclear binding energy* (*i.e.*, the binding energy needed to release a relatively small number of nucleons in the spallation processes) in the calculations, instead of the overall nuclear binding energy per nucleon. For example, to release 10 nucleons from lead, 6.9 MeV per nucleon is needed, whereas we used 7.9 MeV per nucleon in the calculations described above. Values for other elements are listed in Table B.2.

The composition of the non-em shower component and the way in which the energy contained in this component is deposited in the absorber material, as summarized in Table 2.5, were derived on the basis of a few considerations which, in some respects, oversimplify the problem. It is interesting to compare the results with those from elab-

orate, detailed Monte Carlo simulations of hadronic shower development. Such simulations were performed by Gabriel [Gab 85], who studied the absorption of 5 GeV protons in plastic-scintillator calorimeters based on iron or lead as absorber material. Some
relevant results of these simulations are listed in Table 2.6, together with the corresponding numbers (for pure lead and iron) derived on the previous pages.

Table 2.6 *Where does the energy carried by the non-em component of hadronic showers go?*
Handwaving arguments compared with Monte Carlo simulations.

Deposition of non-em energy	Lead Table 2.5	Lead [Gab 85]	Iron Table 2.5	Iron [Gab 85]
Ionization by pions	23%	13.3%	35%	14.4%
Ionization by protons	35%	33.4%	37%	42.5%
Kinetic energy evaporation neutrons	9%	12.2%	5%	7.8%
Excitation γs		2.5%		3.4%
Total invisible energy	33%	38.6%	23%	31.8%

This comparison suggests that our estimates of the energy sharing in the non-em
shower component are accurate at the 10–20% level, with one very notable exception:
the energy deposited through ionization by pions generated in the shower development.
Our assumption was that secondary and higher-order pions would travel, on average,
one nuclear interaction length (λ_π, taken to be $1.25\lambda_{int}$) on their way to the next nuclear interaction. This is clearly an unrealistic oversimplification. The nuclear interaction length is based on the asymptotic total cross section (Equation 2.27) at very high
energies, while the pions produced in the shower development process typically carry
kinetic energies of a few hundred MeV. At those energies, the total cross is affected by
resonance production reactions, such as $\pi^+n \to \pi^0p$, and is considerably larger than in
the multi-GeV regime. Therefore, the mean free path of these shower pions is considerably shorter, and the amount of ionization energy they deposit correspondingly smaller,
than we assumed in our back-of-the-envelope calculation.

The comparison of the results of our calculations and Gabriel's (HETC) model also
confirms some remarkable features, as well as the existence of some significant differences between lead and iron, which will turn out to have important consequences for
calorimetry, for example:

- The relatively minor role of ionization losses by charged pions and other hadronic
  mips in the absorption process, $\sim 20\%$ of the non-em energy.
- The relatively major role of ionization losses by densely ionizing particles, in
  particular protons released from atomic nuclei.
- The large fraction of invisible energy.
- The large numbers of neutrons produced in the shower development, in particular
  in high-Z materials.

Apart from the large difference in the numbers of neutrons (almost a factor of
four, see Table 2.5), there are also some other significant differences between hadron
absorption in lead and iron. In iron, the fraction of the total energy that is deposited by

densely ionizing particles is larger, while the invisible energy and the kinetic energy carried by soft neutrons are smaller than in lead. Because of these differences, the choice of the absorber material for a calorimeter has a number of important consequences, both for the performance characteristics of the detector itself and for the environment in which it has to operate (neutron flux!).

2.3.2.4 *Evaporation neutrons.*   As indicated in Table 2.5, a significant fraction of the hadronic shower energy is carried by large numbers of soft neutrons. The absorption of these neutrons in dense material proceeds very differently from that of the other types of shower particles encountered so far. Electrons, photons, charged mesons and protons are all subject to the electromagnetic interaction. Neutrons depend entirely on the strong (and sometimes the weak) interaction in order to be absorbed in matter. This has very important consequences for calorimetry. These consequences may range from very beneficial to very detrimental [Wig 98] and are extensively discussed in Chapters 3 and 4. In this chapter we concentrate on the neutron spectra and on the absorption mechanisms.

The kinetic energy spectrum of the evaporation neutrons is usually described by a Boltzmann–Maxwell distribution

$$\frac{dN}{dE} = \sqrt{E}\exp(-E/T) \tag{2.25}$$

with a temperature T of about 2 MeV, so that the average kinetic energy of these neutrons amounts to about 3 MeV at production (see Figure 2.32).

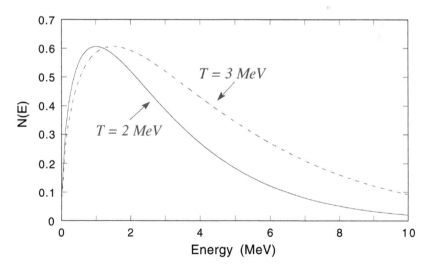

FIG. 2.32. Kinetic energy spectrum of evaporation neutrons, produced according to a Maxwell distribution with a temperature of 2 MeV. For comparison, the spectrum for a temperature of 3 MeV is given as well.

This means that these neutrons have obtained a kinetic energy that is typically of the order of one-third to one-half of the binding energy that confined them to their parent nuclei before the shower development occurred.

The *cascade* neutrons produced in the initial phase of the spallation process are more energetic, but also much less numerous than the evaporation neutrons. Their kinetic energies may extend all the way to the GeV domain. Especially in high-Z materials such as lead, they typically initiate nuclear reactions of the type $^A_Z X \, (n, yn) \, ^{A'}_Z X$, with $A' = A - y + 1$, in which the total number of evaporation neutrons produced in the shower absorption process thus increases by $(y - 1)$. In practice, almost all neutrons that are present in the absorber structure a few nanoseconds after the start of the shower process are thus of the evaporation type.

Experimental measurements have revealed that the numbers of neutrons produced in hadronic shower development are large. Leroy *et al.* [Ler 86] measured the production rates of *thermalized* neutrons in high-energy hadron showers, by analyzing the induced radioactivity resulting from neutron-capture reactions in the absorber material. They found rates of $\sim$ 20 neutrons per GeV of energy in lead and up to 60 neutrons per GeV in ^{238}U, where nuclear fission causes a significant multiplication of the neutron production rates.

Based on knowledge of the spectrum of the neutrons, the average total kinetic energy carried by these neutrons, and the fluctuations in this total kinetic energy may be calculated. Some results of these calculations, for the energy spectrum given in Figure 2.32, are shown in Figure 2.33 .

Because of the Central Limit Theorem and because of the large numbers of neutrons involved, the fluctuations in this total kinetic energy are relatively small. For example, one thousand neutrons, a number typically produced in the absorption of a 50 GeV

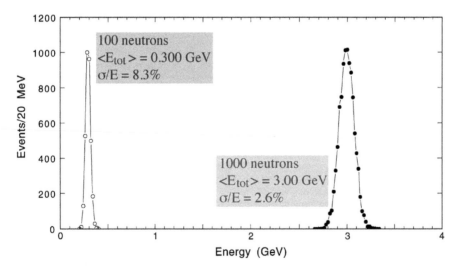

FIG. 2.33. Distribution of the total kinetic energy carried by 100 and 1,000 evaporation neutrons.

hadron in lead, carry a total kinetic energy of 3 GeV ±78 MeV.

Not surprisingly, it turns out that there is a clear correlation between this total kinetic energy carried by the neutrons produced in the shower development and the total amount of *invisible* energy, which is primarily determined by the *number* of target nucleons released in the development of the hadron shower. In Chapters 3 and 4, it is shown how this correlation can be exploited to improve calorimetric performance.

2.3.3   *The interactions of neutrons with matter*

In this subsection we discuss the various mechanisms through which the evaporation neutrons lose their kinetic energy and are eventually absorbed in dense matter.

2.3.3.1   *Elastic neutron scattering.*   At energies between a few eV and approximately 1 MeV, elastic scattering is by far the dominant, if not the only, energy loss mechanism for neutrons. The cross sections for elastic neutron scattering in a number of representative materials are listed in Table B.3, for neutrons in the energy range 1 keV–1 MeV. They are large, usually several barns, which implies mean free paths between collisions of typically a few cm.

The energy fraction f lost by neutrons in collisions with nuclei with atomic number A (approximately proportional to the mass) varies between 0 (for glancing collisions) and $4A/(A+1)^2$, the kinematic limit for central collisions. The average values of f amount thus to 50%, 3.4% and 0.96% for collisions with hydrogen, iron and lead, respectively. It is of course no surprise that, in terms of energy loss, the elastic scattering process is most efficient in hydrogen. A large cross section and a considerable fraction of energy lost in each collision make hydrogen-rich compounds the material of choice for neutron shielding purposes, *e.g.*, in nuclear reactors.

As a result, neutrons in the mentioned energy bracket are sampled very differently from charged particles in calorimeter structures containing hydrogen in the active components. This is further discussed in Chapter 3.

2.3.3.2   *Neutron capture.*   When the neutrons generated in hadronic shower development have lost (almost) all of their kinetic energy in collisions with the target material, one of two things may happen: they decay, or they get captured by an atomic nucleus. Since the timescale for the first process is very long (mean lifetime ~ 15 min) and the cross section for the second process usually large, capture is much more likely to occur. When a neutron is captured by an atomic nucleus, the nuclear binding energy that had to be supplied to the nucleus when the neutron was released (invisible energy) is gained back. The excited "compound" nucleus usually gets rid of this excess energy by emitting γ-rays. In some light nuclei, such as ^{6}Li and ^{10}B, the capture of a neutron may be followed by the emission of an α particle.

The neutron capture process is distinctly different from the processes through which charged particles, such as electrons and protons, get absorbed in the shower development process. After losing their kinetic energy through ionization of the calorimeter materials, these charged particles just become part of the absorbing structure, while the neutrons transform an absorber nucleus into another type of nucleus.

2.3.3.3 *The production of α particles.*   Another process which illustrates that neutrons may be sampled very differently from charged particles takes place at energies
between 3 and 20 MeV. At these energies, neutrons frequently release α particles from
the nuclei with which they interact, for example through (n, α) reactions. This is particularly true for ^{12}C, a key ingredient of organic materials such as plastics or gases used
in wire chambers. Neutrons above 10 MeV may split this nucleus into three α particles,
the reverse process of the one that starts the CNO cycle in aging stars. This process
alone accounts for $\sim 60\%$ of the inelastic n^{12}C cross section. In other light gases such
as oxygen and fluorine, α production is also quite abundant, while in higher-Z materials
like iron or copper, α production takes place in only $\sim 3\%$ of the inelastic reactions (see
Table B.4).

When produced in wire chambers, such αs may give rise to signals that are orders
of magnitude larger than the ones caused by minimum ionizing particles and, therefore,
it is an important process in certain calorimetric applications (see Section 4.8.2).

2.3.3.4 *Inelastic neutron scattering.*   The fact that the energy loss process of neutrons in a given material is extremely dependent on subtleties of that material's nuclear
structure, becomes also clear when we examine the role of inelastic scattering.

In this process, part of the neutron's kinetic energy is used to bring a nucleus in an
excited state. The nucleus releases this excitation energy in the form of one or several
γs, whose (combined) energy corresponds to the energy loss of the neutron. The contribution of this process to the energy loss of the neutrons produced in the calorimeter
depends completely on details of the nuclear level structure.

In some materials, *e.g.*, lead, it becomes insignificant below energies as high as 2.6
MeV (because it takes that much energy to bring the most abundant lead isotope, ^{208}Pb,
from its ground state into the lowest excited state), in other materials it continues to play
a role down to energies well below 1 MeV. For example, the first excited state of the
most abundant isotope of iron, ^{56}Fe, is located 0.85 MeV above the ground state. The
cross section for inelastic scattering processes of the type $(n, n'\gamma)$, in which neutrons in
the energy range of 1–6 MeV lose 0.85 MeV, is more than one barn [Lac 74]. This is one
of the reasons why steel-reinforced concrete is a good shielding material for MeV-type
neutrons.

2.3.4 *Hadronic shower profiles*

2.3.4.1 *The nuclear interaction length.*   Hadronic shower development is based
largely on nuclear interactions, and therefore the shower dimensions are governed by the
nuclear interaction length, λ_{int}. The nuclear interaction length of an absorber medium
is defined as the average distance a high-energy hadron has to travel inside that medium
before a nuclear interaction occurs. The probability that the particle traverses a distance
z in this medium *without* causing a nuclear interaction equals

$$P = \exp\left(-z/\lambda_{\mathrm{int}}\right) \tag{2.26}$$

This definition is thus equivalent to the one for the mean free path of high-energy
photons, which was found to be equal to 9/7 of a radiation length (Equation 2.9). And

just as the mean free path of photons is inversely proportional to the total cross section for photon-induced reactions, λ_{int} is inversely proportional to the total cross section for nuclear interactions:

$$\sigma_{tot} = \frac{A}{N_A \lambda_{int}} \tag{2.27}$$

This cross section is determined by the size of the projectiles and the size of the target nuclei. The cross section of the target nuclei is determined by their radius squared. And since the volume of these nuclei (and thus r^3) scales with the atomic weight A, the cross section scales with $A^{2/3}$. From Equation 2.27, it then follows that λ_{int} scales with $A^{1/3}$, when expressed in units of g cm^{-2} (which eliminates differences in material density).

Table B.1 lists the nuclear interaction lengths for a number of materials used in calorimeters. The smallest values for λ_{int}, around 10 cm, are found for high-density, high-Z materials such as tungsten, gold, platinum and uranium. For frequently used absorber materials such as iron and copper, the interaction length is less than twice as long (a 60–70% increase compared with uranium). This is quite different from the situation encountered earlier for the radiation length, which increases by about a factor of five going from uranium to iron. We will come back to these differences when discussing calorimetric particle identification (Section 7.6).

The nuclear interaction lengths for mixtures of different elements or for a compound can be determined in the same way as discussed for the radiation length and the Molière radius (Section 2.1.5).

The λ_{int} values listed in Table B.1 are for interactions caused by *protons*. The interaction probability, and thus the mean free path of the hadrons, λ_{int}, *also* depend on the size of the hadrons. For example, the total cross section for (fixed-target) pp interactions at 100 GeV amounts to $\sim$ 38 mb, whereas the total cross section for πp interactions at the same energy amounts to $\sim$ 24 mb, about 2/3 of the value for pp reactions [PDG 14]. A similar factor applies to reactions with other targets. Therefore, the interaction lengths for pions are considerably longer than those for protons, which are the ones that are usually listed. This also implies that a calorimeter with a length quoted as $10\lambda_{int}$, represents in fact only some $7\lambda_{int}$ for pions. The so-called *punch-thru* probability, *i.e.*, the probability that a particle traverses this calorimeter without causing a nuclear interaction is thus also very different for protons $(\exp(-10) \approx 5 \cdot 10^{-5})$ and pions $(\exp(-7) \approx 10^{-3})$. Experimental evidence for these proton/pion differences is given in [Aco91b, Akc98, Kri99].

It should also be emphasized that the nuclear interaction length is defined for asymptotic conditions, *i.e.*, for very high energies. In that sense, there is complete similarity with the em scaling variables, the radiation length and the Molière radius. And just as in the case of em showers, essential properties can only be understood from processes that take place at low energy, where these scaling variables lose their validity (*e.g.*, Figure 2.12). For example, the overestimated ionization energy lost by pions produced in the shower development (Table 2.6) was due to the (erroneous) assumption that the mean free path of these soft pions is determined by the nuclear interaction length.

2.3.4.2 *Longitudinal profiles.*   The longitudinal hadronic shower profiles have some
clear similarities with those induced by electrons or photons. The number of shower
particles traversing a thin slice of absorber material, and thus the energy deposited in
this slice, rises initially roughly linearly, reaches a maximum that depends on the particle
energy and on its nature (proton, pion), followed by a decrease that is much less steep
than the initial rise. Figure 2.34 gives a good impression of these characteristics.

The profile depicted in this figure was obtained in a rather unusual way [Ler 86]. A
large stack of 3 mm thick plates of depleted uranium, 250 in total, was exposed to an
intense beam of 300 GeV negative pions. These pions were absorbed in this stack, each
one after developing its own, completely unique shower. A very large number of nuclei
produced in reactions initiated by shower particles were unstable. Most of these nuclei
had half-lives that were much shorter than the total duration of the exposure (one week).
But some fraction was responsible for a significant level of *induced radioactivity* that remained measurable for a long time. The profile of this induced radioactivity represented
the combined effects of nuclear reactions produced in some 100 billion π^- showers of
300 GeV. It is as if the average (three-dimensional) shower profile was "frozen" in the
stack of metal plates with this method.

The horizontal scale in this figure is given in units of λ_{int}. It takes, on average, $8\lambda_{int}$
of uranium, or about 85 cm, to contain these 300 GeV π^- showers at the 95% level.
Containment of 300 GeV electrons at the same level would be achieved with about

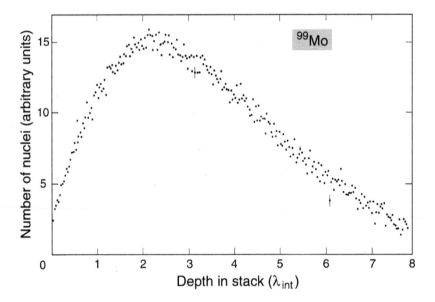

FIG. 2.34. Longitudinal shower profile for 300 GeV π^- interactions in a block of uranium,
measured from the induced radioactivity. The ordinate indicates the number of radioactive
decays of a particular nuclide, ^{99}Mo, produced in the absorption of the high-energy pions.
Data from [Ler 86].

10 cm of uranium. The absorption of hadron showers thus requires considerably more material than the absorption of em showers of the same energy, at least for the type of materials ($Z > 10$) discussed here.

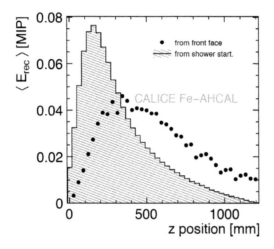

FIG. 2.35. Longitudinal profile of 45 GeV pion showers in iron absorber, relative to the calorimeter front face and relative to the first hard interaction. The integrated profiles have been normalized to unity [Adl 13b].

Longitudinal hadronic shower profiles have also been measured with more traditional calorimeters, in which the energy deposit pattern is recorded event by event. In that case, one may choose to show the profile using either the front face of the calorimeter, or the depth at which the first nuclear interaction takes place, as the origin (Figure 2.35). Examples of both representations can be found in the literature. The second mentioned type of profile can be converted into the first type by convoluting it with an exponential function that describes the longitudinal distribution of the starting point of the shower (Equation 2.26).

Profile measurements extending to a depth of 20 λ_{int} have been carried out with the hadronic section of the ATLAS calorimeter system (TileCal, Figure 2.36a), which consists of an in iron/plastic-scintillator structure [Adr 10]. The results are in good agreement with those obtained 30 years earlier with the CDHS calorimeter (Figure 2.36b). As shown by CDHS (Figure 1.3), the signals measured at very great depths inside the absorber structure are dominated by muons from the decay (in flight) of pions and especially kaons produced in the shower development. The flattening of the profiles beyond 20 λ_{int} reported by ATLAS is consistent with this.

The same ATLAS group has also studied differences in the longitudinal shower profiles between showers induced by protons and pions of the same energy. Figure 2.37 shows that proton showers are shorter than pion ones. This is because the interaction length for pions is 25% longer than for protons [Kri 99], and presumably also for other baryons. That means that the mean free path of a proton is shorter, and that the shower

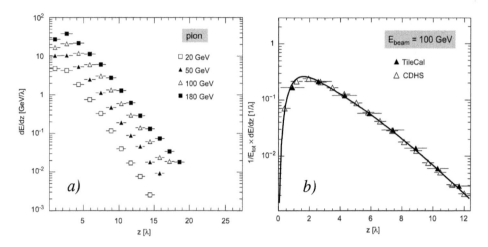

FIG. 2.36. Average longitudinal shower profiles of pions of different energies in the iron/plastic-scintillator calorimeter used in ATLAS (*a*). Comparison of longitudinal shower profiles measured with the ATLAS TileCal and the CDHS calorimeter for 100 GeV pions (*b*). These profiles have been measured with respect to the front face of the calorimeter. Experimental data from [Adr 10] and [Abr 81].

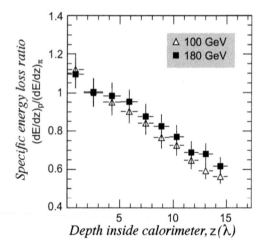

FIG. 2.37. Ratio of the longitudinal shower profiles measured for protons and pions of the same energy in the ATLAS TileCal calorimeter. Experimental data for 100 and 180 GeV from [Adr 10].

thus starts, on average, earlier than for a pion. And since baryon number is conserved, the leading particle in the proton interaction is also a baryon, with a similarly shorter mean free path. As a result, it takes a less thick calorimeter to contain a proton shower at a certain level than a pion shower with the same energy.

2.3.4.3 *Lateral/radial profiles.* Hadron showers do not only start to develop until reaching a much greater depth inside the absorber material, they are also considerably *broader* than electromagnetic showers. The lateral shower profiles exhibit in most materials a narrow core, surrounded by a halo. A representative shower profile, integrated over the full depth of the absorber, is shown in Figure 2.38. This profile was measured with the SPACAL detector [Aco 92b].

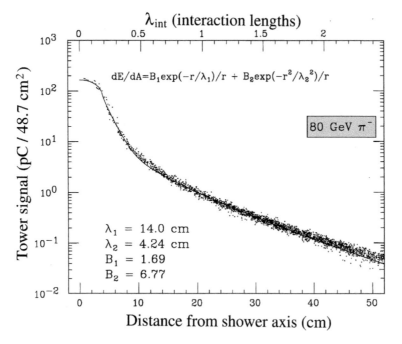

FIG. 2.38. Average lateral profile of the energy deposited by 80 GeV π^- showering in the SPACAL detector. The collected light per unit volume is plotted as a function of the radial distance to the impact point. Data from [Aco 92b].

The narrow core represents the electromagnetic shower component, caused by π^0s produced in the shower development. The halo, which has an exponentially decreasing intensity, is caused by the non-electromagnetic shower component. A detailed comparison of lateral profiles measured with the SPACAL detector showed that the radius of the cylinder around the shower axis needed to contain 80 GeV π^- showers at the 95% level is about 32 cm ($1.5\lambda_{int}$), nine times larger than the 3.5 cm ($1.8\ \rho_M$) radius for containing 80 GeV em showers at the same level [Aco 92b].

Longitudinally, the difference in the amounts of material needed for containing these two types of showers at a certain level is very similar (*i.e.*, a factor of about nine). Measurements showed that the average energy fraction leaking out at the back of the $9.6\lambda_{int}$ deep SPACAL detector amounted to $\sim 0.3\%$, for showers induced by 80 GeV π^- [Aco 91d]. An average containment of 99.7% for 80 GeV electron showers requires

30 X_0 worth of lead (see Section 2.1.7), and $(9.6 \times 170)/(30 \times 5.6) \approx 9$.

Hadron showers are thus larger in *all* spatial dimensions. In the example of SPACAL, it took $9 \times 9 \times 9 \approx 700$ times as much material to contain 80 GeV pion showers at the same level as electron showers of the same energy. This offers multiple possibilities to identify particles on the basis of their shower profile characteristics, both the longitudinal and the lateral ones.

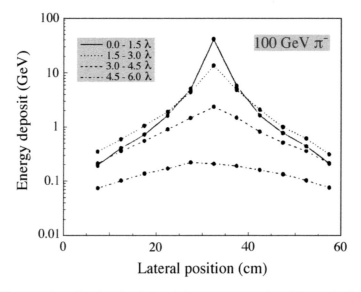

FIG. 2.39. Lateral profiles for pion-induced showers, measured at different depths, with the ZEUS calorimeter. Data from [Bar 90].

Next, we take a look at the *differential* lateral shower profiles, *i.e.*, lateral shower profiles measured at different depths inside the absorber. Not surprisingly, the width of these profiles gradually increases with depth, as shown in Figure 2.39. This figure shows the results of measurements performed by the ZEUS Collaboration on the energy deposit profiles of 100 GeV pions showering in their uranium/plastic-scintillator calorimeter [Bar 90]. The electromagnetic shower core is very prominently present in the initial stages of the shower development, but has completely disappeared beyond a depth of $4.5\lambda_{int}$.

The measurements of shower profiles through the method of induced radioactivity, described above for the longitudinal shower profiles, also revealed interesting features of the lateral profiles.

Figure 2.40 shows the lateral distributions of several radioactive nuclei, measured at a depth of $4\lambda_{int}$ inside the block of uranium. The distributions exhibit considerable differences. The distribution of the ^{239}Np nuclei is much broader than that of ^{99}Mo, which in turn has a larger width than the ^{237}U distribution. These distributions are different because the mechanisms through which these nuclides were produced are different.

Neptunium-239 is the decay product of ^{239}U, which is produced when ^{238}U, the

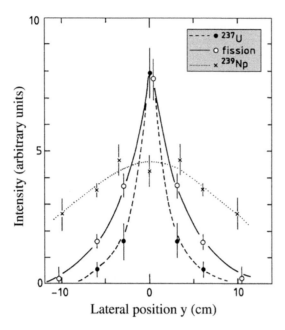

FIG. 2.40. Lateral profiles for 300 GeV π^- interactions in a block of uranium, measured from the induced radioactivity at a depth of $4\lambda_{\text{int}}$ inside the block. The ordinate indicates the decay rate of different radioactive nuclides, produced in nuclear reactions by different types of shower particles. Data from [Ler 86].

absorber material in these experiments, captures a neutron. This is by far most likely to happen for thermal neutrons. In other words, the ^{239}Np distribution is a measure for the spatial density of thermal neutrons at a depth of $4\lambda_{\text{int}}$ inside the block of depleted uranium.

The nuclide ^{99}Mo is a fission product of uranium. The threshold for neutron-induced ^{238}U fission is about 1.5 MeV. Therefore, the distribution of this radioactive nuclide is a measure for the spatial distribution of the non-thermalized, MeV-type neutrons. Obviously, these neutrons have traveled a much smaller distance from their point of origin and are therefore much more concentrated around the shower axis than the thermal neutrons.

Finally, ^{237}U is most likely produced through the reaction ^{238}U (γ, n) ^{237}U. The cross section for this process reaches a maximum value of ~ 0.35 b, at a photon energy of about 11 GeV [Die 88]. Gammas of this energy are abundantly produced in the em showers generated by π^0s. Therefore, one expects to find ^{237}U concentrated in the narrow shower core (Figure 2.38), close to the shower axis.

2.3.4.4 *Fluctuations.* The shower profiles shown so far were all averaged over large numbers of showers. However, it is important to realize that the energy deposit profiles of *individual* hadron showers may deviate substantially from these averages. The electromagnetic core that characterizes the average lateral profile is caused by π^0s produced in the shower development. These π^0s develop em showers, and thus require a much smaller detector volume to deposit their energy than other shower components carrying the same energy. Therefore, the energy *density* is considerably larger in areas near π^0 production, and since most of the π^0s are generated near the shower axis, in the early stages of the shower development, the energy density is, on average, considerably larger in these areas than elsewhere in the absorber.

However, in individual showers, the production of energetic π^0s may occur in completely different regions of the absorbing volume, and this will lead to energy deposit profiles that differ considerably from the average. Figure 2.41 shows a few examples of non-average energy deposit profiles, for pion showers developing in a lead/iron/plastic-scintillator sandwich calorimeter [Gre 94]. The absorber structure consisted of 40 lead plates (3.1 mm thick), followed by 26 iron plates (2.5 cm thick). The energy deposited in this structure was measured in every individual scintillator plate, so that detailed event-by-event information on the longitudinal shower development was obtained.

Figure 2.41a shows a shower with an energy deposit profile that corresponds roughly with the profile averaged over a large number of showers. However, individual beam particles may penetrate deep into the detector before initiating a nuclear reaction (Figure 2.41b,c). Figure 2.41b depicts an event in which a large fraction of the energy was transferred to one or several π^0s in this first nuclear interaction. Also in the event shown in Figure 2.41c, energetic π^0s were produced in the first nuclear interaction. However, in addition, an energetic charged hadron was produced at that point. This particle traveled about one interaction length deeper into the absorber and then transferred almost all its energy to π^0s in a second-generation interaction, leading to a

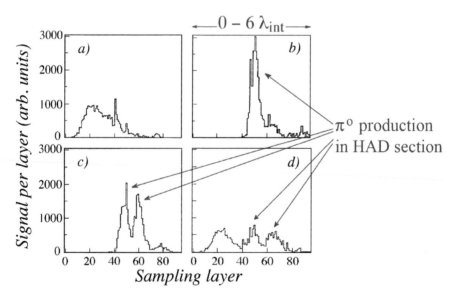

FIG. 2.41. Longitudinal profiles for four different showers induced by 270 GeV pions in a lead/iron/plastic-scintillator calorimeter. Data from [Gre 94].

two-peak structure in the longitudinal energy deposit profile. In Figure 2.41d, even three generations of π^0 production can be distinguished.

Such "stochastic" energy deposit profiles are by no means exceptional in hadronic shower development. They are a consequence of the fact that the π^0s produced in the shower development deposit the energy they carry in a much smaller absorber volume than other shower particles. Therefore, the hadronic energy deposit profiles directly reflect the large event-by-event fluctuations that may occur in both the energy carried by these π^0s and the position in the absorber where they are generated.

2.3.4.5 *Shower profiles in Čerenkov calorimeters.* Only relativistic, charged shower particles contribute to the signals from calorimeters based on Čerenkov light. Their velocity should exceed the Čerenkov threshold: $v > c/n$, where n represents the re-fraction index of the medium in which the particles travel. Media that are frequently used in Čerenkov calorimeters include water ($n = 1.33$), quartz ($n = 1.46$) and various types of lead-glass ($n = 1.5$–1.75). Shower particles that may contribute to the signals include

- Electrons and positrons. For n values of ~ 1.4, these particles emit Čerenkov light when their kinetic energy exceeds 200 keV.
- Charged pions. In typical calorimeter media, these short-lived particles emit Čerenkov light when their total energy is larger than about 190 MeV.
- Protons. These need to carry a kinetic of at least 400 MeV in order to generate Čerenkov light in such calorimeters.

The signals from Čerenkov calorimeters depend completely on the extent to which Čerenkov-capable particles are produced in the shower development. As we have seen before, hadronic shower development involves various processes, each with very differ-ent rates of Čerenkov-capable particle production:

- π^0s produced in hadronic shower development give rise to em showers. Most of the energy of these π^0s is deposited through electrons and positrons above the Čerenkov threshold (kinetic energy larger than 200 keV).
- Of the energy carried by the non-electromagnetic shower component, on average $\sim 20\%$ is deposited by charged pions.
- The rest of the non-electromagnetic energy is deposited by protons and neutrons, almost all of which are non-relativistic, and through release of nuclear binding energy, which leaves no directly measurable signal.

Since electrons and positrons produced in the showers dominate the signals from this calorimeter, hadron showers thus register predominantly through their electromag-netic shower core. This has several important consequences.

One of these consequences concerns the (three-dimensional) hadronic shower pro-files. The instrumented volume needed to contain the Čerenkov-capable shower compo-nent is substantially smaller than that required for full containment of the entire hadron shower. This is true both in depth (longitudinal containment) and in the transverse plane, since the shower tails in all directions are primarily composed of non-relativistic parti-cles (soft nucleons).

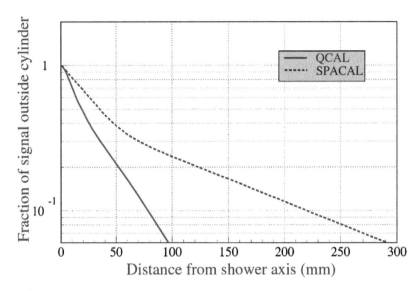

FIG. 2.42. A comparison of the transverse characteristics of 80 GeV π^- showers measured
with a scintillation calorimeter [Aco 92b] and with a Čerenkov calorimeter [Akc 97]. Shown
is the fraction of the signal recorded outside a cylinder with radius R around the shower axis,
as a function of R.

For this reason, hadron showers in Čerenkov calorimeters appear to be considerably
narrower than in other types of calorimeters. This may be an important advantage when
particle densities are very high (for example, in the high-η region of LHC experiments),
since it reduces shower overlap.

This point is illustrated in Figure 2.42, which shows transverse characteristics of
showers, initiated by 80 GeV π^- mesons in the Quartz Fiber Calorimeter (Čerenkov
light, copper absorber) and in the SPACAL calorimeter (scintillating fibers, lead absorber, sensitive to *all* ionizing particles, *cf.* Figure 2.38). The profiles in the Čerenkov
calorimeter are considerably narrower, even though both the Z value and the density of
the absorber in this calorimeter were considerably smaller than in the other one.

This figure illustrates one very important point, which will be emphasized time and
again in this book, namely that calorimeter signals are in general *not proportional* to the
amount of energy deposited in the area from which they are collected. This becomes
extremely clear in the case of the Čerenkov calorimeter. The profile measured with the
Quartz Fiber detector is not at all representative for the energy deposit profile in the
shower development. It just measures the transverse distribution of Čerenkov light produced in this process and thus the transverse distribution of shower particles capable of
emitting such light. Since most of these particles are electrons and positrons generated
in the em shower core, this light is concentrated near the shower axis.

2.3.5   *Shower containment*

When an experiment is designed, one of the most important decisions concerns the total thickness of the calorimeter. Especially in a 4π geometry, this decision has serious consequences for the cost of the experiment. Let us, as an example, consider a spherical calorimeter that surrounds the interaction vertex, starting at a distance of 1 m. Let us also assume that the effective nuclear interaction length of this calorimeter is 20 cm. If we want to make this calorimeter $7\lambda_{int}$ thick, then its total volume amounts to $\frac{4}{3}\pi(2.4^3 - 1)$ = 53.7 m^3. Should we want to add one extra interaction length, this volume would increase by 29% (69.4 m^3), while the surface area of detectors installed outside the calorimeter would increase by 17%. And since the cost of many detectors is more or less proportional to the instrumented mass, a decision to go from $7\lambda_{int}$ to $8\lambda_{int}$ would have major financial implications.

In this subsection, we discuss the absorber thickness needed to contain hadron showers, *on average*, at a certain level. We re-emphasize what was said in Section 2.1.7 for em showers, namely that the effects of shower leakage on the quality of the calorimeter data is determined by *event-to-event fluctuations* about this average, and not by the average shower containment itself. Also in hadron showers, these fluctuations are much larger for longitudinal leakage than for lateral shower leakage, at a given level of shower containment. The reasons for this are discussed in Section 4.5.

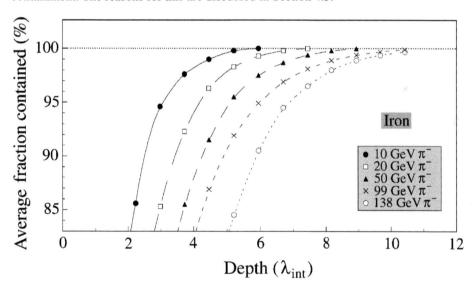

FIG. 2.43. Average energy fraction contained in a block of matter with infinite transverse dimensions, as a function of the thickness of this absorber, expressed in nuclear interaction lengths. Shown are results for showers induced by pions of various energies in iron absorber. Experimental data from [Abr 81].

Because of the practical implications mentioned above, there is plenty of experimental information about hadronic shower containment. Representative results, obtained by WA1 for hadron absorption in iron [Abr 81], are shown in Figure 2.43. The average shower fraction contained in the absorber material is shown as a function of the absorber thickness, for showering pions with energies ranging from 10 GeV to 138 GeV. The absorber thickness needed to contain 95% of the shower energy ranges from $\sim 3\lambda_{int}$ at 10 GeV to more than $6\lambda_{int}$ at 138 GeV. For 99% containment, the absorber thickness has to be at least $5\lambda_{int}$ deep for 10 GeV pions and $\sim 9\lambda_{int}$ for 138 GeV ones.

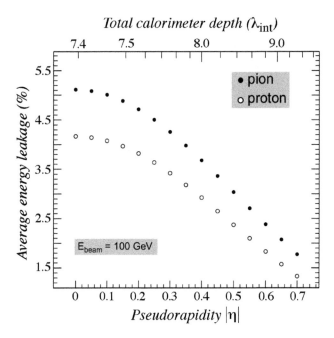

FIG. 2.44. The average shower leakage for 100 GeV pions and protons in the ATLAS calorimeter system, as a function of pseudorapidity. Experimental data from[Adr 10].

The ATLAS Collaboration used the longitudinal profile data shown in Figure 2.36 to determine the shower leakage expected in their calorimeter system. Since the effective calorimeter thickness increases with the pseudorapitidy (η), the leakage decreases with the angle between the particle trajectory and the beam line. Figure 2.44 shows the average energy leakage for 100 GeV pions and protons as a function of η. The corresponding effective calorimeter thickness is plotted on the top axis. The smaller leakage observed for protons is a reflection of the fact that the shower profile for these particles is a bit shorter because of the smaller interaction length, combined with leading particle effects [Adr 10].

Transverse shower containment results are shown in Figure 2.45, where the energy fraction contained in a cylinder around the shower axis is plotted as a function of the

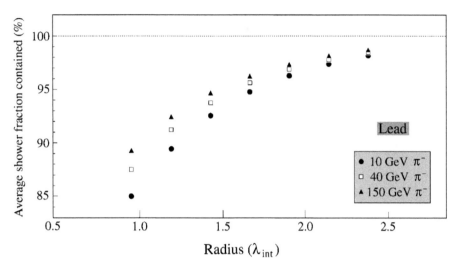

FIG. 2.45. Average energy fraction contained in an infinitely long cylinder of absorber material, as a function of the radius of this cylinder (expressed in nuclear interaction lengths), for pions of different energy showering in lead absorber [Aco 92b].

radius of this cylinder, for pions of several energies showering in lead absorber [Aco 92b]. Unlike for em showers, the results do depend in this case on the energy of the showering particle, in a way that at first sight seems counter-intuitive: the higher the energy of the incoming particle, the narrower the cylinder needed to contain the shower. For example, to contain 10 GeV pions at the 95% level, a cylinder with a radius of $\sim 1.7\lambda_{int}$ is needed, while $1.4\lambda_{int}$ is enough for 150 GeV pions.

This energy dependence is a direct consequence of the energy dependence of $\langle f_{em}\rangle$. The average energy fraction carried by the em shower component increases with energy and since this component is concentrated in a narrow core around the shower axis, the energy fraction contained in a cylinder with a given radius increases with energy as well.

2.4 Properties of the Shower Particles

In the previous sections, we have seen how the absorption process of high-energy particles in a block of matter proceeds, and how the energy carried by the incoming particles is distributed inside the absorber. In this section, we examine in some more detail what happens to the individual particles produced in the shower development.

In electromagnetic showers, for example the ones caused by high-energy electrons or by π^0s generated in hadron-induced showers, the energy is entirely deposited by electrons and positrons. Most of these particles are very soft. This was already illustrated in Figure 2.10, which shows that a considerable fraction, up to 40%, of the energy carried by 10 GeV electrons is deposited by shower particles with energies below 1 MeV, *i.e.*, by particles carrying less than 10^{-4} of the energy of the particle that initiated the shower.

As a consequence, the number of *different* shower particles through which the em
energy is deposited is very large. The overwhelming majority of these particles are *electrons*, they outnumber positrons by a considerable factor (see Table 2.1), thus illustrating
the dominating importance of Compton scattering and the photoelectric effect (in which
electrons, but no positrons are produced) in the energy deposition process.

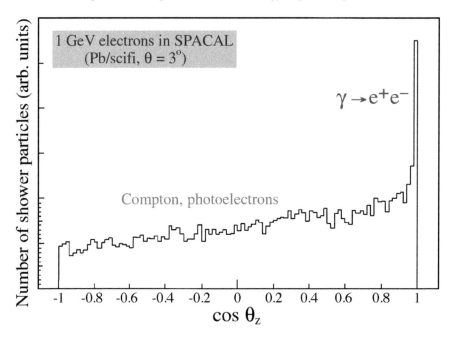

FIG. 2.46. Angular distribution of the shower particles (electrons and positrons) through which
the energy of a 1 GeV electron is absorbed in a lead-based calorimeter. Results of EGS4
Monte Carlo simulations. From [Aco 90].

This also has another important consequence. The shower particles contributing to
the calorimeter signal are to a large extent isotropically distributed, *i.e.*, they have completely "forgotten" the direction of the incoming particle that created the shower. This
is true both for hadronic and electromagnetic showers and is illustrated in Figure 2.46,
which shows the angular distribution of the particles (mostly e^- and a few e^+) through
which the energy of a 1 GeV electron absorbed in a lead-based calorimeter is deposited.
A large fraction of these particles have an angular distribution flat in $\cos \theta$. The electrons are numerously created by Compton scattering and photoelectric processes and
their contribution to the signal of a sampling calorimeter is practically *independent* of
the orientation of the active calorimeter elements (see also Figure 2.5).

The particles through which the energy of em showers is deposited travel, on average, a very short distance in the calorimeter. Figure 2.47 shows their range, as a function
of energy, in a variety of materials. The "benchmark" 1 MeV electrons travel in typical
absorbers that are used in calorimeter construction only a fraction of one millimeter:

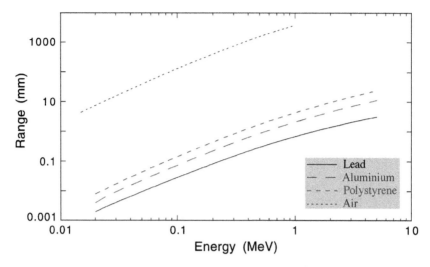

FIG. 2.47. Average range of electrons in various absorber materials, as a function of energy [Pag 72].

0.67 mm in lead, 0.78 mm in iron. Electrons with an energy of 0.1 MeV have a range of only 27 μm in both materials. These distances are much smaller than the distance between active layers typically used in sampling calorimeters. In Chapter 3 we will see that this has important consequences for the performance characteristics of such calorimeters.

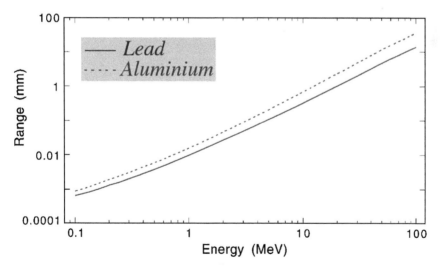

FIG. 2.48. Average range of protons in various absorber materials, as a function of energy [Jan 82].

The distances traveled by the particles through which the *hadronic* energy is deposited are in some cases even smaller than the range of the electrons from em showers. In Section 2.3.2, we saw that a considerable fraction of the non-em shower energy is deposited by protons released from the absorber nuclei. In high-Z absorber materials such as lead, where the Coulomb barrier prevents the evaporation of soft protons, the non-em energy is mainly deposited through spallation protons, but in materials such as iron many evaporation protons may contribute as well.

The spallation protons have typically energies of the order of 100 MeV. Their range in frequently used calorimeter materials such as lead and iron amounts to 1–2 cm (Figure 2.48). On the other hand, 3 MeV evaporation protons travel only 45 μm in lead and 35 μm in iron.

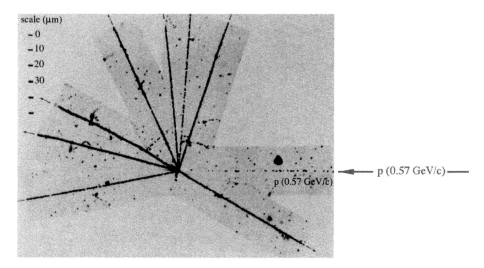

FIG. 2.49. A nuclear interaction induced by a proton with a kinetic energy of 160 MeV in a nuclear emulsion stack. Photograph courtesy CERN.

Figure 2.49 shows a picture of an event in which several soft protons were produced. A proton with a momentum of 570 MeV/c (kinetic energy 160 MeV) has struck a nucleus of a photographic emulsion stack. After developing the emulsion, all the trajectories followed by the charged particles produced in this event appear as dark lines. The larger the specific ionization ($\langle dE/dx \rangle$) of the particles, the higher the density of activated emulsion nuclei and therefore the darker the resulting track has become. The short dense tracks (note the scale of this picture) probably represent evaporation protons released from the emulsion nucleus in this interaction.

The soft evaporation neutrons, which are produced in large numbers in the shower development process, deposit their energy in totally different ways. They do not ionize the atoms of which the absorber material consists, but interact instead with the nuclei. The processes through which they lose their kinetic energy and eventually are captured, are extremely material- and even isotope-dependent. However, in any case, the distances traveled by these neutrons are much longer, by orders of magnitude, than those traveled by their charged counterparts of similar energy. For example, the total cross section for interactions induced by 1 MeV neutrons in lead amounts to $\sim$ 5 b. This corresponds to a mean free path between subsequent interactions of $\sim$ 6 cm, compared with a range of less than 1 mm for electrons and 10 μm for protons of the same energy.

In addition, at these low energies, where elastic scattering is practically the only process through which the neutrons lose kinetic energy, the energy fraction typically lost in one scattering process is small. Therefore, it takes many such scatterings to thermalize the neutrons and make them eligible for nuclear capture (the cross section for capture is roughly inversely proportional to the neutron's velocity). In the capture process, typically 6–10 MeV worth of nuclear γs are emitted (the neutron's nuclear binding energy). As we will see later on (Section 3.2.9), the time involved in this thermalization process is usually too long to be of practical interest for calorimetry.

On the other hand, the *kinetic energy* carried by the evaporation neutrons may be crucially important for calorimetric performance, especially in the case of sampling calorimeters. By choosing a proper combination of active and passive calorimeter materials, a situation may be achieved in which these neutrons are much more efficiently sampled than the charged shower particles. And although the evaporation neutrons carry only a small fraction of the shower energy, typically some 10% of the non-em shower component, this difference in sampling fraction may boost their contribution to the calorimeter *signals* considerably, to the point where the losses due to invisible energy are, on average, compensated for (Section 3.4).

Because of the long mean free path of the evaporation neutrons, they usually dominate the tails of the shower profiles, both laterally and longitudinally. This means that calorimeters for which detecting these neutrons is an essential performance ingredient, *e.g.*, compensating calorimeters, require signal integration over a much larger detector volume (and also a longer time) than calorimeters for which neutron detection is less essential.

2.5   Monte Carlo Simulations

Detailed understanding of the shower development and its dependence on the energy and nature of the showering particles and on the materials in which the processes take place, are crucial in the design of calorimeter systems for particle physics experiments. Analytical parameterizations of the *average* shower behavior are in general not sufficient for this purpose, since the most critical aspects of the system's performance – the precision with which energy and position measurements can be performed – are dictated by event-to-event fluctuations in the absorption processes. For this reason, the simulation of shower development processes by means of Monte Carlo techniques has been developed. In these simulations, which in many cases originated from computer programs developed to study radiation-shielding issues, stochastic models for the elementary electromagnetic and hadronic processes discussed in the previous sections of this chapter are employed to generate individual cascades and to follow their development in considerable detail.

Obviously, the reliability (and thus the usefulness) of these simulations depends on the quality and completeness of the models used to describe the elementary physics processes. In that respect, there is a great difference between the simulations of electromagnetic and hadronic showers.

2.5.1 *Electromagnetic showers*

For electromagnetic showers, a very general and well-documented system of computer codes called EGS4 (Electron Gamma Shower, version 4) [Nel 78], has emerged as the world-wide standard. It is implemented in the GEANT4 framework [Ago 03], in which an arbitrary detector geometry can be specified. Each shower particle is transported, in small steps, through the absorbing media. The interactions undergone by the shower particles in this process are determined on the basis of probability. Random number generators select the various possible processes using tabulated cross sections. The complete history of each generated shower is available for further analysis.

EGS4 has been extensively tested, for a wide variety of applications, and has turned out to be very accurate in its predictions, provided that the shower particles are followed down to a sufficiently low "cutoff" energy. If particles fall below the cutoff energy, they are not followed any further and their remaining energy is deposited at the point where they happen to be in the structure.

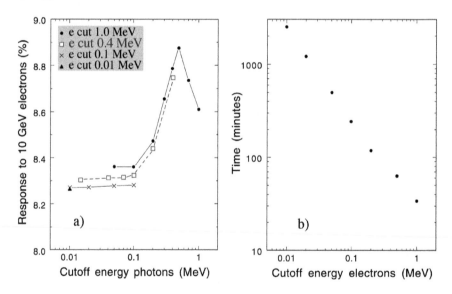

FIG. 2.50. Effects of the cutoff energy in EGS4 shower simulations. Shown are the average signals from 10 GeV electron showers developing in the SPACAL calorimeter, calculated for different values of the energy down to which electrons and photons are tracked in the simulations (*a*), and the computer time needed for the simulation of 1,000 showers, as a function of the cutoff energy (*b*).

A practical limitation of the use of this type of Monte Carlo simulations is the necessary computer time. If one wants to follow the shower development down to a level of 100 keV, each 100 GeV em shower involves tracking of the order of 10^6 shower particles. To go down to 10 keV (which turns out to be necessary for many applications), the number of shower particles, and thus the amount of computer time needed, increases by an order of magnitude. Figure 2.50b shows the computer time needed for the simulation of 1,000 em showers (10 GeV e^-) developing in a lead/plastic-scintillator calorimeter, as a function of the cutoff energy. This time increases from about half an hour for $E_{cut} = 1$ MeV to about 40 hours for $E_{cut} = 10$ keV [4].

Many problems, for example the evaluation of the energy resolution, require simulations of at least 1,000 events to obtain the required precision. And therefore, it is clear that vast amounts of computer time can be easily spent in this way. Obviously, this field has considerably benefited from the tremendous developments in computer hardware over the past decades. Back in the days of the mainframe computers, when the EGS4 code was developed, high-energy calculations were virtually impossible. Nowadays, these simulations can run faster on an iPad, and with better precision, than in the mainframe days.

The effect of the cutoff values on the results is illustrated in Figure 2.50a, which shows the response of the SPACAL calorimeter to 10 GeV electrons, simulated with EGS4 for different choices of the cutoff values for electrons and photons participating in the shower development. This response (which is in this case a measure for the average fraction of the shower energy ending up in the active material of this calorimeter) is plotted as a function of the cutoff energy for photons, for four different values of the cutoff energy for electrons. Several effects are clearly visible.

- For a given choice of the cutoff value for electrons, the response does not change significantly once the cutoff value for photons is chosen smaller than 100 keV, but it does increase quickly for larger photon cutoff values.

- For a given choice of the cutoff value for photons, the response becomes systematically smaller when the cutoff value for electrons is lowered. However, lowering the electron cutoff value below 100 keV does not result in a significant improvement.

Apparently, 100 keV is an important energy for em shower development in lead. Figure 2.51 provides the explanation of these effects.

Figure 2.51a shows that, even at very low energies, shower electrons lose some fraction of their energy through radiation in this detector, but these losses occur almost exclusively in the high-Z absorber material. This radiative energy loss amounts to $\sim 3\%$ at 100 keV and rises quickly to $\sim 17\%$ at 1 MeV. If the electron cutoff value is set at 1 MeV, the entire energy carried by softer electrons is assumed to be deposited at the spot where they were generated, be it in the absorber or in the active detector material (plastic in this case). However, if the electron cutoff value is lowered to 0.1 MeV, electrons in

---

[4] These numbers refer to simulations I did in 1999 for the original version of this book. Surely, the speed has improved considerably since that time. However, this has only affected the absolute value of the numbers on the vertical scale of Figure 2.50b.

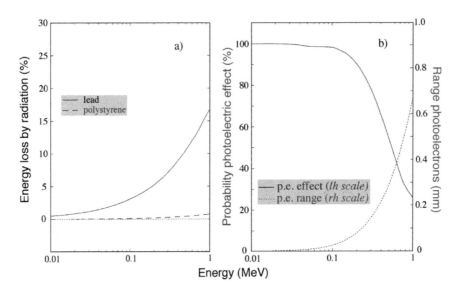

FIG. 2.51. Energy fraction lost to bremsstrahlung by low-energy shower electrons in SPACAL
(a), the probability that low-energy photons interact through photoelectric effect in the absorber material of this detector, and the range of the photoelectrons produced in that process
(b), as a function of energy.

the 0.1–1.0 MeV range are more or less guaranteed to deposit at least some fraction of
their energy, *i.e.*, the fraction lost to bremsstrahlung, in the lead.

Lowering the electron cutoff value thus leads to smaller calorimeter signals. Below
100 keV, this effect disappears, first because the radiation losses become negligibly
small, and second because the range of the shower electrons is now becoming so small
(< 0.2 mm) that they can be contained in the plastic detector elements, so that there
is no longer a guarantee that part of the energy will be deposited in the absorber. This
explains the second of the two mentioned trends that emerged from the Monte Carlo
simulations.

The bremsstrahlung spectrum drops off exponentially to its high-energy limit, *i.e.*,
the energy of the radiating electron. The vast majority of the photons emitted by the
soft electrons mentioned above are thus to be found at energies below 100 keV. These
photons may do one of four things:

1. Photoelectric effect in lead
2. Compton scattering in lead
3. Photoelectric effect in plastic
4. Compton scattering in plastic

Figure 2.51b shows the probability for process #1, photoelectric effect in the absorber material, as well as the range of the photoelectron produced in this process, as
a function of energy. Up to energies of ~ 100 keV, this probability is $> 98\%$, and the
range of the photoelectron is such ($< 27\mu$m), that it is extremely unlikely that it reaches

the active calorimeter material and contributes to the signal. For this reason, one should not expect any significant change in the calorimeter response as a result of lowering the γ cutoff energy below 100 keV.

When the γ cutoff value is increased above 100 keV, the relative probability for photoelectric effect in the absorber material rapidly decreases, and the range of the photoelectron produced in this process increases, so that the likelihood that the shower γs contribute to the calorimeter signals rapidly increases. Raising the γ cutoff value above 100 keV thus results in a larger calorimeter response. This explains the first of the two mentioned trends that emerged from the Monte Carlo simulations.

Absolute response values of calorimeters are hard to determine experimentally. However, experimental features that could be measured for this detector, e.g., energy resolutions for electron detection, were found to be in excellent agreement with the simulations once the low cutoff values were used (100 keV for both electrons and γs).

The lesson learned from this is that the EGS4 predictions are very accurate, provided that the cutoff values are chosen sufficiently low. One can take advantage of this feature to solve very complicated practical problems, such as the intercalibration of the three longitudinal segments of the ATLAS electromagnetic calorimeter [Aha 06], discussed in Chapter 6. But apart from the fact that this quality of the EGS4 simulations allows physicists to optimize the design of a calorimeter system for an experiment in each and every aspect, there is one other major advantage that needs to be mentioned. Since the complete history of every simulated EGS4 shower is available, the simulations may provide important insight into details of the working of the calorimeter in question. Several examples of this insight have been given already in this chapter (e.g., the peculiarities of the longitudinal and lateral shower profiles and their explanation), and many more will follow later on. This is an invaluable benefit of reliable Monte Carlo simulations.

2.5.2 *Hadron showers*

The main reason for the excellent agreement between simulations and reality is the relative simplicity and the detailed fundamental understanding of the physics processes that govern the em shower development. This is very different for hadronic showers and, as a consequence, the Monte Carlo simulations of hadronic shower development are much less accurate than in the electromagnetic case. That does not at all mean that such simulations are not useful, on the contrary. In Section 2.1.3, we saw how Monte Carlo simulations of hadronic shower development have led to important insight in crucial details of the particle sector of this process. Phenomena such as the Z dependence of the average em shower fraction and the difference between proton- and pion-induced showers were predicted and became understood thanks to simulations.

However, the mentioned phenomena are the result of processes that take place in the "easy phase" of the shower development, i.e., the particle production phase. The nuclear sector is much more complicated to describe, because of the enormous variety of processes that may occur. Details of the nuclear level structure of the absorber material may play an important role. For example, a calorimeter based on ^{207}Pb will have very different properties than one using ^{208}Pb as absorber material. The first excited state of the "double-magic" ^{208}Pb nucleus lies 2.6 MeV above the ground state, while for ^{207}Pb the excitation energy is only 0.57 MeV. This has profound consequences for the absorption of evaporation neutrons abundantly produced in hadronic shower development. In ^{208}Pb, the only option for neutrons below 2.6 MeV is elastic scattering, in which process almost no energy is lost. In ^{207}Pb, inelastic scattering $(n, n'\gamma)$ may occur down to ~ 0.6 MeV. As a result, high-energy pions will have a larger fraction of their energy absorbed by a ^{207}Pb block of a given size than by a ^{208}Pb block of the same size. Monte Carlo simulations aiming for an accurate description of the shower development process should include details of this type.

Such details are particularly important for obtaining reliable results on such issues as the hadronic energy resolution and the absolute value of the calorimeter signals. Just as for em showers (see Figure 2.50), these properties are crucially affected by what goes on in the last, very-low-energy stages of the shower development.

The present status of the Monte Carlo packages that are commonly used to simulate hadronic calorimeter performance is still far away from the level of reliability typical for electromagnetic shower detection. Yet, in the 16 years that have past since the first edition of this book was published, hadronic shower simulations have been gradually improving as a result of

1. the availability of more powerful computing resources, and

2. a concerted effort to improve the implementation of the underlying physics processes.

The latter aspect is a direct result of the needs of experiments at current (LHC) and future (ILC) colliders, which also provided many detailed sets of experimental data to guide these efforts.

The commonly used vehicle in this regard is GEANT4 [Ago 03], which is a toolkit for the simulation of the passage of particles through matter. It allows the user to define any detector geometry, choose the details of the physics processes to be used in the simulations and specify the type of requested results. In the latest version of GEANT4 (V10.2 at the time of this writing), there is a wide variety of "physics lists" available for the simulation of hadron showers. The degree of validity, or lack thereof, of these physics lists is continuously tested (and, ideally, improved) by comparing the predictions of the simulations with experimental data sets. This process is called "validation."

The data from ATLAS and CMS beam tests with the combined em and hadronic calorimeters (thick-target) as well as stand-alone hadronic calorimeters have proven to be very important for the validation of the current physics lists. Two different domains (thin- and thick-target) appear critical for further testing of hadronic shower simulations. Measurements of low-energy hadronic interactions in tracker detectors (thin-target) may prove useful to further tune the inelastic hadronic cross sections and final-state models. This region is typically the most difficult to simulate correctly. Hopefully, the collision data from the LHC will shed further light and provide useful input for these efforts.

Elaborate test material for hadronic shower simulations has also been provided by the CALICE collaboration [Sef 15], which has been exploring calorimeters with high granularity, intended for experiments at a future linear collider. Their data provide many details on the sub-structure of developing showers, which may be helpful for further improving the quality of the physics lists. A crucial aspect for the Particle Flow Analysis method (Section 8.3) pursued by the developers of such calorimeters is a correct reproduction of the shower profiles in the simulations, and in particular the profiles in the transverse plane. Figure 2.52 shows a comparison between experimental measurements

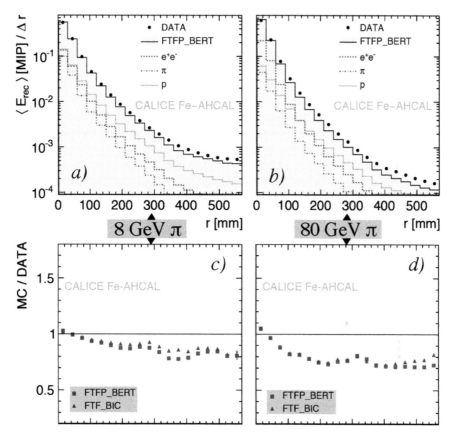

FIG. 2.52. Average radial shower profiles for 8 GeV (*a*) and 80 GeV (*b*) pions measured in the CALICE Fe/scintillator calorimeter. Experimental data are compared to GEANT4 simulations based on the FTFP_BERT physics list. All profiles are normalized to unity. $E_{rec}/\Delta r$ is the average deposited energy in Δr, where r is the radial coordinate. Ratio between the MC values for two different physics lists and the experimental data, for 8 GeV (*c*) and 80 GeV (*d*) pions [Adl 13b].

and Monte Carlo simulations of the radial shower profiles for 8 and 80 GeV pions. The experimental data were obtained with a fine-grained Fe/plastic-scintillator calorimeter, the GEANT4 simulations are based on the FTFP_BERT physics list, which seems to do the best job in this respect [Adl 13b]. The experimental data exhibit a feature we already saw earlier, namely the fact that the shower becomes *narrower* as the energy increases

(Figure 2.45). This feature is (qualitatively) correctly reproduced by the simulations. The overall agreement between the experimental data and the simulated profiles is reasonable, especially close to the shower axis. In the shower halo, differences of $\sim 20\%$ are observed, especially at higher energy (Figures 2.52c,d). Interestingly, the MC results also show the relative contributions of pions, protons and the em shower component to the calorimeter signals. They confirm the dominating role of the em shower component near the shower axis, the increase of the relative contribution of the em component to the signals with energy, and the important feature that the contribution of protons to the hadronic signals is considerably larger than that of pions.

The progress that has been made in the quality of hadronic shower simulations is certainly encouraging, and will hopefully continue. For reasons already mentioned at the beginning of this section, it is important that simulation programs be developed that have the same degree of reliability for hadron showers as EGS4 has for em showers. To achieve this goal, rigorous implementation of all known physics pertaining to shower development in the simulations is a *conditio sine qua non*. The user should *not* be given any options to turn on, switch off or modify (parameters governing) the various physics processes that play a role in hadronic shower development. The only option should concern the cutoff energy.

When testing these programs and comparing the results with experimental data, one should also realize that some types of data are much more sensitive to a correct implementation of the relevant physics than others. For example, a certain simulation program may do a good job in reproducing experimentally observed (average) shower profiles, but at the same time fail miserably in the area of energy resolutions [Fas 93]. The latter are determined by event-to-event fluctuations and thus probe more, and other, aspects of the correct implementation of the physics processes governing the absorption of hadrons in dense matter. For this reason, one should be extremely careful with conclusions about the validity of predictions from such simulation programs: "The appearance of one swallow does not necessarily imply that Summer has arrived."[5]

Very relevant tests of the correctness of hadronic shower simulation programs are provided by those calorimeters in which one particular aspect of the shower development is strongly emphasized. Consider, as an example, Čerenkov calorimeters. The signals from hadron showers in such detectors are, for all practical purposes, completely dominated by the π^0s generated in the absorption process. Therefore, data from such calorimeters provide a very sensitive probe for the correct description of π^0 production, including the event-to-event fluctuations, in hadronic shower development.

Similarly, signals from some lead/plastic-scintillator sampling calorimeters are strongly affected by evaporation neutrons produced in the shower development. Data from such calorimeters are therefore particularly suited for testing that aspect of the simulations.

In this context, it would be very useful to define a series of "benchmark" calorimeter results that can be used to gauge the quality of hadronic shower simulation programs, and to measure the progress achieved in this domain.

---

[5] Dutch proverb.

2.6   Summary of Facts Important for Calorimetry

In this chapter, we have described the processes through which high-energy particles entering dense matter lose their energy and are eventually absorbed. In the next chapters, we will see how signals generated in this absorption process can be used to gain information on the showering particle. The quality of this information and the method(s) by which it can be obtained are in many ways affected by details and peculiarities of this absorption process:

1. The first aspect of shower development with consequences for calorimetry concerns the fact that a large fraction of *the shower particles travel in random directions* with respect to the particle that initiated the shower and whose properties are being measured with the calorimeter (Figure 2.46). This means that the orientation of the active layers of a sampling calorimeter can be chosen as desired, without serious implications for the calorimetric performance of the detector.

   The first generation of sampling calorimeters used in particle physics experiments consisted almost exclusively of instruments of the "sandwich type," *i.e.*, detectors composed of alternating layers of absorber and active material, oriented perpendicular to the direction of the particles to be detected. Although this, from an intuitive point of view, may seem to be the only right choice, the R&D with fiber calorimeters has proven that there is absolutely no need for such a geometry.

   The notion that the active calorimeter layers do not necessarily have to be oriented perpendicular to the direction of the incoming particles, has had a considerable impact on the design of detectors for new experiments. Other orientations may offer considerable advantages in terms of detector hermeticity, readout, granularity, *etc*. Apart from the "spaghetti" type of calorimeters built for a number of experiments, this development is also illustrated by the liquid-argon calorimeters with an "accordion" geometry and the tile/fiber hadron calorimeter of the ATLAS experiment at the LHC [Aad 08].

2. A second aspect of the absorption process with important design implications for calorimeters is the *energy distribution of the particles* through which the energy of the showering particle is deposited. Both in em and non-em showers, very soft particles carry a major fraction of the shower energy: electrons in the sub-MeV range for em showers, spallation protons, nuclear γs and evaporation neutrons in the case of the non-em hadronic shower components. The response to these soft particles and the extent to which they contribute to the signals are, for many calorimeters, major factors that affect the performance.

3. The *differences between the energy deposit mechanisms in electromagnetic and non-electromagnetic showers* have several profound consequences for calorimetry. Some of these consequences represent a serious disadvantage, while others may be exploited with great benefits. The "invisible energy" phenomenon that affects hadronic showers leads, in general, to a hadronic calorimeter response that is smaller than the electromagnetic one and, moreover, depends on the energy. This phenomenon is also responsible for the considerably worse energy resolution for hadronic shower detection that characterizes *all calorimeters*. On the other hand,

the differences in shower profiles may be exploited to identify the particles that caused the signals.

The neutrons produced abundantly in the non-em component of hadron showers may be exploited to obtain ultimate performance, in compensating hadron calorimeters. On the other hand, the same neutrons may seriously jeopardize the usefulness of other types of calorimeters, *e.g.*, through the "Texas tower effect" (Section 4.8.2).

The physics of shower development is complex. As a result, calorimeters, and in particular hadron calorimeters, are non-trivial and often unpredictable devices. On the other hand, this complexity also offers many fascinating options to optimize the detectors, as we will see in the next chapters.

3

THE ENERGY RESPONSE OF CALORIMETERS

After having described in detail the various processes that play a role in the absorption of highly energetic particles in dense material, we now have the tools at hand to exploit this complicated phenomenon to our benefit, in measuring the properties of such particles by means of calorimetric methods.

In this chapter, the emphasis is on the calorimeter *response*. Although this term has an intuitive meaning, and has been used in that sense on several occasions in the previous chapters, it is now time for an explicit and precise definition. In the following, the term **calorimeter response** will be used to indicate

the average calorimeter signal
divided by the energy of the particle that caused it.

Therefore, a calorimeter of which the average signal for the detection of electrons is proportional to the electron energy has an *electromagnetic response that is constant as a function of energy*. Such a calorimeter is also said to be *linear* for em shower detection.

The calorimeter response is thus expressed in units of, for example, the *number of photoelectrons per GeV*, or *picocoulombs per MeV*, or similar, conveniently chosen units, depending on the signal generating mechanism and the data acquisition system.

We will frequently compare the calorimeter responses to different types of particles. Minimum ionizing particles (mips) serve as "benchmark particles" in this respect. If the calorimeter response to particles of type "X" is smaller than the response to mips, *i.e.*, the calorimeter produces, on average, a smaller signal for particles "X" of a given energy than for mips that deposit the same energy, then this will be indicated as follows: the X/mip signal ratio is smaller than 1, or $X/mip < 1$.

In the following, we distinguish between *homogeneous* and *sampling* calorimeters. In a homogeneous calorimeter, the entire detector volume is sensitive to the particles and may contribute to the signals it generates. In a sampling calorimeter, the functions of particle absorption and signal generation are exercised by *different* materials, called the *passive* and *active medium*, respectively. The passive medium is usually a high-density material, such as iron, copper, tungsten, lead or uranium. The active medium generates the light or charge that forms the basis for the signals from such a calorimeter.

The discussion in this chapter is initially limited to calorimeters based on scintillation light or ionization charge as the source of the signals. In the final section (Section 3.5), we discuss the unusual and rather spectacular consequences if Čerenkov light produced in the shower development is used as the source of experimental information.

3.1   Homogeneous Calorimeters

3.1.1   *The response to electrons and photons*

When electrons or photons are absorbed in the calorimeter medium, their entire kinetic energy is used to excite the atoms or molecules of which this medium is composed. In most homogeneous calorimeters, these atoms or molecules emit (part of) this excitation energy in the form of visible light when returning to the ground state and this (scintillation) light forms the basis of the calorimeter signals. In the past 15 years, there has also been a major development of homogeneous liquid-argon calorimeters, which are based on the collection of ionization charge [Alm 04, MB 15]. Multi-kiloton detectors of this type are foreseen for experiments in the context of the Fermilab neutrino program.

Since the entire kinetic energy of the showering electron or photon is used to generate the particles that constitute the calorimeter signals, these calorimeters should be *intrinsically linear* for em shower detection. This is further discussed in Section 3.3.

3.1.2   *The response to muons*

Muons traversing homogeneous calorimeters lose kinetic energy through the same mechanisms as the shower particles produced in em shower development, exciting the atoms and molecules of the detector medium. Usually, the muons easily penetrate these detectors, losing only a small fraction of their energy in the process. For example, mips traversing the BGO crystal calorimeter [Bak 85, Bak 87] that was operated by the L3 experiment at the LEP Collider (CERN) lost on average about 200 MeV (9 MeV cm^{-1}). The specific ionization of muons is slightly higher than for mips, but that affects this number only at the 10% level. Typically, the muons produced in the LEP collisions deposited between 200 and 250 MeV of their energy in the BGO calorimeter.

Because of the similarity between the energy deposit mechanisms, the responses of a homogeneous calorimeter to muons and to em showers are equal. This means that the average signal for a muon that traverses such a calorimeter and loses, for example, 573 MeV in that process is equal to the average signal generated by a 573 MeV electron or photon that is absorbed by shower development in the calorimeter. One may also say

$$e/mip \; = \; 1 \tag{3.1}$$

in which e and *mip* denote the calorimeter responses to em showers and minimum ionizing particles, respectively.

In practice, this means that if a calorimeter of this type is calibrated with em showers, *i.e.*, if the relation between the deposited energy and the resulting calorimeter signal (the "calibration constant") has been established with electrons of known energy, then the signals produced by muons traversing the calorimeter may be converted into the energy lost by these muons in the calorimeter, using the same calibration constant. Although this may seem rather trivial, we will see in the following that this conversion is in general *not* valid for other types of calorimeters, and in particular for sampling calorimeters with high-Z material as absorber medium.

3.1.3   The response to hadrons and jets

The calibration constant derived in the way described above is most certainly not valid for hadrons and jets. Because of the invisible energy phenomenon (Section 2.3.2), only a fraction of the energy carried by these (collections of) particles is used to excite the atoms or molecules of the detector medium. Another fraction is used to dissociate atomic nuclei and does not contribute to the calorimeter signals.

Therefore, if the calibration constant derived from the detection of electrons is applied to the signals generated by pion showers, the energy value comes out too low. In other words, the pion response is smaller than the em one, or

$$\pi/e < 1 \tag{3.2}$$

Pions of a given energy generate signals that are, on average, smaller than the ones generated by electrons of the same energy. And since $e/mip = 1$, one may also say: $\pi/mip < 1$. Pions generate signals that are, on average, smaller than the signals from muons that deposit the same energy.

The response to hadron-induced showers is not only smaller than the electromagnetic one, it is also energy dependent. Homogeneous calorimeters are *intrinsically non-linear* for the detection of hadrons and jets. The reason for this is the energy-dependent em fraction in hadronic showers (see Section 2.3.1). The response to this em component, caused by π^0s produced in the hadronic shower development, equals the response to em showers initiated by high-energy electrons or photons. The response to the non-em shower component is smaller than the em response, because of the invisible energy. Since the average em fraction of hadron-induced showers increases with energy, so does the calorimeter response to such showers. Therefore, *the π/e signal ratio increases with energy*.

In Chapter 2 we saw that the energy in the non-em shower component is partly carried by mesons, spallation protons, evaporation neutrons, recoil target nuclei and nuclear γs and is also partly used to release nuclear binding energy. There is no reason for the distribution of the non-em energy among these various components to be energy dependent. Therefore, the calorimeter response to the non-em component of hadronic showers may be considered constant. We will call this non-em calorimeter response h. Because of invisible energy, h is smaller than the electromagnetic response:

$$e/h > 1 \tag{3.3}$$

A calorimeter for which this relation holds is said to be *non-compensating*. All homogeneous calorimeters are non-compensating. The precise value of e/h indicates the degree of non-compensation. In homogeneous calorimeters, it is only determined by the average fraction of the non-em energy that escapes detection. The e/π signal ratio is *not* a measure for the degree of non-compensation, since part of the pion-induced showers is of an electromagnetic nature. As the energy increases, so does this em fraction. At very high energies, the e/π signal ratio will be close to 1, even in extremely non-compensating calorimeters.

Experimentally, a direct measurement of the e/h ratio requires measuring the em fraction of hadron showers event by event. In Section 8.2.2, we show how this is possible in dual-readout calorimeters (Figure 8.9). Indirectly, the e/h value may be derived from measurements of the e/π signal ratios at a series of energies, preferably spanning as large an energy range as possible. The energy dependence of the average em fraction, $\langle f_{em} \rangle$, needs to be known for this purpose. If e and h denote the calorimeter response to the em and non-em shower fractions, then the response to pions can be written as

$$\pi = \langle f_{em} \rangle \cdot e + (1 - \langle f_{em} \rangle) \cdot h \qquad (3.4)$$

This leads to

$$\pi/e = \langle f_{em} \rangle + \left[1 - \langle f_{em} \rangle\right] \cdot h/e \qquad (3.5)$$

which, when inverted, gives the following relationship between the measured e/π signal ratios and e/h

$$e/\pi = \frac{e/h}{1 - \langle f_{em} \rangle \left[1 - e/h\right]} \qquad (3.6)$$

And since $\langle f_{em} \rangle$ is a function of the energy of the showering hadron, so is the e/π ratio.

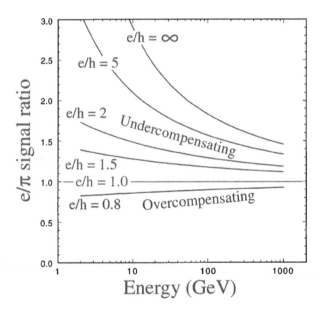

FIG. 3.1. The relation between the calorimeter response ratio to em and non-em energy deposition, e/h, and the measured e/π signal ratios. See text for details.

The relationship between the calorimeter response ratio to the em and non-em shower components (e/h) and the measured e/π signal ratios is graphically depicted in Figure 3.1, for a variety of e/h values ranging from 0.8 to ∞. The energy dependence of the em shower fraction, $\langle f_{em}(E) \rangle$, which one needs to know to derive this relationship, was calculated on the basis of Equation 2.19, with the parameter values $E_0 = 1$ GeV and k

= 0.82 (see Section 2.3.2). This figure illustrates our earlier statement that, in the high-energy limit, the e/π signal ratio tends to approach 1, irrespective of the calorimeter's e/h value.

Let us consider, as an example, what may be expected for $e/h = 2$, a fairly typical value for homogeneous calorimeters. Between 10 GeV and 1 TeV, the energy domain in which almost all calorimeters in accelerator-based experiments are operating, the e/π ratio decreases from 1.49 to 1.17. Since the em response is constant, this implies that the calorimeter response to pions increases by 27% over this energy range. All non-compensating calorimeters, and in particular all homogeneous calorimeters, are thus non-linear for pion detection.

So far, we have only mentioned pions. The situation for jets is very similar. A jet is a collection of particles, resulting from the fragmentation of a quark, a diquark, or a hard gluon produced in the collisions. Especially at high energies, all the particles making up this jet tend to travel in approximately the same direction. Since the four-vector of the jet corresponds to the four-vector of the fragmenting (di)quark, measuring jet properties is in many experiments considered more important than measuring individual hadrons.

For purposes of calorimetry, the absorption of jets proceeds in a way that is very similar to the absorption of individual hadrons. Some fraction of the energy carried by the collection of particles is deposited in the form of em showers, the rest is deposited in non-em form. A (minor) difference is that the em component in hadron-induced showers is the result of π^0s produced inside the calorimeter, while jets usually contain a number of π^0s, or rather the γs from their decay, upon entering the calorimeter. Apart from this "intrinsic" electromagnetic jet component, π^0s produced in the absorption processes of the hadronic jet particles constitute the final em shower component.

The average electromagnetic shower fraction, $\langle f_{em} \rangle$, may be somewhat different for jets than for individual hadrons with the same energy as the jet, due to peculiarities of the fragmentation process. In Section 2.3.1 we saw that $\langle f_{em} \rangle$ differs for showers induced by pions and protons of the same energy, as a consequence of baryon number conservation in the shower development process. Similar effects may also play a role in the fragmentation process. For example, a fragmenting diquark will have a leading baryon, while the leading particle in a quark fragmentation process is most likely a meson. Heavy quarks (c, b) are much less likely to produce leading π^0s than light (u, d) quarks, *etc.*

Because of such details, no precise general statements can be made about differences between the calorimeter response to jets and to individual hadrons. However, the two main effects mentioned above for individual hadrons also apply to jets: the response of homogeneous calorimeters to jets is considerably smaller than to electrons, photons and muons. Moreover, the jet response is energy dependent, which implies significant signal non-linearities.

3.1.4   *Summary of important concepts*

Because of its importance for much of what follows later on, we summarize here the essential concepts introduced in this section:

- The calorimeter *response* is the average signal per unit deposited energy.

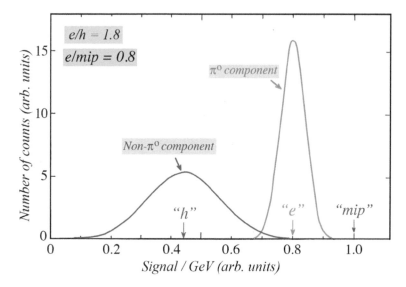

FIG. 3.2. Illustration of the meaning of the e/h and e/mip values of a calorimeter. Shown are distributions of the signal per unit deposited energy for the electromagnetic and non-em components of hadron showers. These distributions are normalized to the response for minimum ionizing particles ("*mip*"). The average values of the em and non-em distributions are the em response ("*e*") and non-em response ("*h*"), respectively.

- A calorimeter is *linear* for the detection of certain particles if its response to these particles is constant, *i.e.*, the average signal is proportional to the particle's energy.

- Every calorimeter is characterized by its e/h value. This is a constant (independent of energy), which relates the response to the em and the non-em components of hadron showers (Figure 3.2). The e/h value can in general *not* be measured directly[6], but can be inferred from the *measured* signal ratios for electrons and pions of the same energy, for a range of energies, using the known energy dependence of the average em shower fraction, $\langle f_{em} \rangle$. Calorimeters with $e/h = 1$ are said to be *compensating*, calorimeters with $e/h > 1$ or $e/h < 1$ are *undercompensating* or *overcompensating*, respectively. The calorimeter from Figure 3.2 is undercompensating, with $e/h = 1.8$. A signal from a pion developing in this calorimeter has a response somewhere in between the "*e*" and "*h*" values, depending on the relative strength of the em component. The larger the $\langle f_{em} \rangle$ value, the closer the response will be to the "*e*" value.

- Every calorimeter is characterized by its e/mip value. This is a constant (independent of energy), which relates the calorimeter responses to electrons and minimum ionizing particles. In homogeneous calorimeters, $e/mip = 1$, in sampling

---

[6]In Section 8.2.2, a method is introduced to measure e/h. However, this method is specific for dual-readout calorimeters (Figure 8.9).

calorimeters it is different from 1. The calorimeter from Figure 3.2 must therefore be a sampling calorimeter, with $e/mip = 0.8$. Just like e/h, this ratio cannot be measured directly, but has to be inferred from e/μ signal ratios measured at different energies.

3.2 Sampling Calorimeters

An important parameter characterizing sampling calorimeters is the *sampling fraction*. We define the sampling fraction of a calorimeter[7] as the energy deposited *by minimum ionizing particles* in the active calorimeter layers, measured relative to the *total energy* deposited by such particles in the calorimeter. Let us consider, as an example, the D0 calorimeter [Abo 89] which operated for more than 20 years at Fermilab's Tevatron collider. The hadronic section of this calorimeter consisted of 3 mm thick depleted uranium plates (^{238}U) interleaved with 5 mm wide gaps filled with liquid argon. Minimum ionizing particles lose, on average, 2.13 MeV cm^{-1} in LAr and 20.5 MeV cm^{-1} in ^{238}U (see Table B.1), or 1.06 MeV and 6.15 MeV in one active and one passive layer, respectively. Therefore, the sampling fraction of this calorimeter amounts to

$$\frac{1.06}{1.06 + 6.15} = 0.147, \text{ or } 14.7\%$$

While this definition of the sampling fraction is extremely simple and straightforward, its connection with experimental data obtained with the calorimeter is not. This is because the response to mips cannot be measured directly. It should be emphasized that a mip is a *hypothetical particle*. As soon as a charged particle with an energy for which dE/dx reaches its minimum value starts traveling through matter, it loses energy and therefore ceases to be a mip.

For all practical purposes, muons are the closest thing nature provides us with in terms of mips. However, even muons with an energy as low as 5 GeV are by no means minimum ionizing particles. Since they are extremely relativistic ($\gamma \sim 50$), the energy loss per unit length is significantly larger than the minimum ionizing value [Loh 85]. The increased specific ionization of muons with energies larger than the minimum ionizing value is due to phenomena such as δ-ray emission (relativistic rise), bremsstrahlung, e^+e^- pair production and, at very high energies, nuclear reactions. The contribution of these effects to the total energy loss is strongly dependent on the muon energy and on (the Z value of) the traversed material.

In practice, the experimental calorimeter response to mips is determined from the signal distributions measured for muons of different energies, by estimating the consequences of the above effects, thereby unfolding the *mip* part of the calorimeter signals [Ake 87, Bern 87, Aco 92c, Aja 97b].

3.2.1 *The response to electrons and photons*

In Section 3.1 we saw that the response of homogeneous calorimeters is the same for all particles that lose their energy exclusively through electromagnetic interactions with the

---

[7]The sampling fraction for (categories of) individual shower particles may be very different from the sampling fraction of the calorimeter as defined here.

absorber material (hence $e/mip = 1$). This is not the case for sampling calorimeters. Experimental data, independently obtained with a large number of different calorimeters, have led to the following conclusion: in sampling calorimeters of which the Z value of the absorber material is larger than the (average) Z value of the active medium, the response to em showers is smaller than the response to minimum ionizing particles ($e/mip < 1$). The larger this difference in Z, the smaller the value of e/mip becomes. We are not dealing with a small effect here. In calorimeters using high-Z absorber materials, such as lead or depleted uranium, e/mip values as low as 0.6 have been measured [And 90, Ake 87, Bern 87, Aco 92c].

The signal from an electron or photon absorbed in a sampling calorimeter is the result of the ionization or excitation of the active layers by all shower electrons and positrons that traverse these layers. Naïvely, one might therefore expect this signal to be equal to the signal from minimum ionizing particles that traverse the detector structure and whose combined energy deposit is equal to the initial energy of the showering electron or photon ($e/mip = 1$).

This expectation is clearly contradicted by experimental results, and this has caused a lot of confusion, at least for what concerns the explanation of the discrepancy. The apparent suppression of the em shower response in sampling calorimeters with high-Z absorber material became historically known under the name *transition effect*. This term was introduced by Pinkau [Pin 65], who argued that the phenomenon should be attributed to effects occurring at the boundary between layers of materials with different Z. He based his argument on the difference in critical energy, ϵ_c, which leads to differences between the shower development characteristics in both materials for the energy region 5–50 MeV. If the shower is considered a collection of particles, each carrying an energy ϵ_c, then the shower tree has more branches in a high-Z (low-ϵ_c) material. Therefore, the *density* of charged particles at a given shower depth is larger in the high-Z absorber material than in the low-Z active layers.

Table 3.1 *Average energy loss of minimum ionizing particles in various absorber materials.*

Absorber	Z	dE/dx (mip) (MeV g^{-1}cm^2)	X_0 (g cm^{-2})	$\Delta E/X_0$ (MeV)
Carbon	6	1.745	42.7	74.5
Aluminium	13	1.615	24.01	38.8
Iron	26	1.451	13.84	20.1
Copper	29	1.403	12.86	18.0
Tin	50	1.264	8.82	11.1
Tungsten	74	1.145	6.76	7.74
Lead	82	1.123	6.37	7.15
Uranium	92	1.082	6.00	6.49

This line of reasoning is perhaps most clearly illustrated in Table 3.1, which lists the average energy lost by minimum ionizing particles traversing one radiation length of material, for different Z values. This energy decreases by more than an order of magnitude going from carbon ($Z = 6$) to uranium ($Z = 92$). On the other hand, the principle of the scaling of em shower profiles with the radiation length (Section 2.1.6) implies that the average energy loss per radiation length *for em showers is material independent*.

Let us imagine a calorimeter structure consisting of $2X_0$ of lead, followed by $2X_0$ of plastic, followed by $2X_0$ of lead. Electromagnetic showers developing in this structure will, in first approximation, deposit equal amounts of energy in these three sections. But according to Table 3.1, minimum ionizing particles deposit about 10 times as much energy in the plastic (carbon) section as in either of the lead sections. Therefore, in this structure the fraction of the em shower energy deposited in the plastic is much smaller than for mips.

Qualitatively, Pinkau's argument is certainly correct. Quantitatively, however, the role played by particles around the critical energy with regards to the e/mip ratio is, in realistic sampling calorimeters, a very minor one. The active layers in realistic sampling calorimeters are not $2X_0$ thick, but rather $\sim 0.01X_0$. When a shower develops across the boundary between layers with different Z values, it only adapts adiabatically to the new situation. Therefore, any significant change in the shower composition as discussed above will only appear after a significant fraction of one radiation length. Since the absorber layers of sampling calorimeters are, measured in units of X_0, typically two orders of magnitude thicker than the active layers, the shower development in these structures is completely governed by the properties of the absorber material and the change in charged-particle density as discussed above is, at maximum, something of the order of a few percent.

In 1985, Flauger [Fla 85] studied the transition effects in some detail, using the EGS4 shower simulation program (see Section 2.5). He subdivided the active and

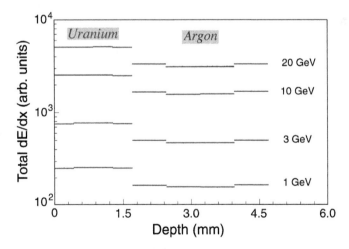

FIG. 3.3. The energy deposition in the different sub-layers of the active and passive sampling layers of a uranium/liquid-argon calorimeter, for showers induced by 1–20 GeV electrons. Results from EGS4 simulations [Fla 85].

passive layers of a fine-sampling calorimeter into sub-layers and found that the energy deposit in the different sub-layers of each sampling layer was practically the same (Figure 3.3). From this, the author concluded that the transition effects occur very rapidly. This indicates that the mechanism suggested by Pinkau does not play a significant role in suppressing the electromagnetic calorimeter response of sampling calorimeters with high-Z absorber material.

The main contribution to this effect is, in fact, a very simple one. It has to do with the way in which low-energy (< 1 MeV) γs interact with matter.

We saw in Chapter 2 that a large fraction of the initial electron energy is used to produce large numbers of low-energy (bremsstrahlung) photons as the shower develops. Therefore, the way in which these photons interact with matter and deposit their energy is absolutely crucial. Analysis of EGS4 data is very revealing in this respect. Figure 2.10 shows that in high-Z materials, such as lead or uranium, $\sim 40\%$ of the shower energy is deposited by electrons softer than 1 MeV. These electrons obtained their kinetic energy from the bremsstrahlung photons. Their range is very short (0.67 mm for 1 MeV in lead, see Figure 2.47), much shorter than the typical thickness of absorber layers used in sampling calorimeters. The sampling of these soft shower electrons is very incomplete, since they travel typically only a small fraction of the distance separating two active layers. Their contribution to the calorimeter signals is thus determined, more than anything else, by the cross sections for the processes through which they are produced: Compton scattering and, especially in high-Z materials, the photoelectric effect (see Figure 2.1). In a calorimeter consisting of high-Z absorber material and low-Z active layers, the Z^5 dependence of the cross section for the latter process (see Figure 2.3) causes the overwhelming majority of the soft photons to interact in the absorber. The photoelectrons produced in this process will only contribute to the calorimeter signals if the conversion takes place very close to the boundary between active and passive media.

Let us consider, as an example, what happens to 511 keV γs in the HELIOS calorimeter [Ake 87], which consisted of 3 mm thick ^{238}U layers, interleaved with 2.5 mm scintillating plastic (polymethylmethacrylate or PMMA). Photons of this energy are abundantly produced in electromagnetic shower development, when positrons annihilate. According to Table 2.1, on average 60 positrons are produced per GeV of energy when em showers develop in uranium. Therefore, 511 keV γs alone account for $\sim 6\%$ of the total shower energy.

The sampling fraction (for mips) of the HELIOS calorimeter was 8.5%. At 511 keV, the total cross section for photon interactions amounts to 76 b in ^{238}U (46 b for photoelectric effect) and 1.7 b for carbon, the main ingredient of plastic [Sto 70]. The ratio of the interaction probabilities of these γs in the active and passive layers is therefore equal to

$$\frac{\sigma_a}{\sigma_p} \cdot \frac{A_p}{A_a} \cdot \frac{\rho_a}{\rho_p} \cdot \frac{d_a}{d_p} = \frac{1.7}{76} \cdot \frac{238}{12} \cdot \frac{1.18}{18.95} \cdot \frac{2.5}{3} = 0.023$$

in which σ, A, ρ and d denote the cross section, the atomic weight, the density and the plate thickness, respectively, and the subscripts a and p refer to the active and passive calorimeter media. Only 2.3% of the 511 keV γs interact thus in the plastic, the rest in the uranium. The range of 511 keV (photo)electrons in ^{238}U is only 0.18 mm. In first approximation, the HELIOS calorimeter's sampling fraction for these γs was thus about a factor of four smaller than the sampling fraction for mips ($\gamma_{511}/mip = 0.27$).

Clearly, the high Z value of the absorber material is the determining factor in this result. Because of this high Z value

1. the cross section for photon interactions in the absorber material is very large, thanks to the contribution of the photoelectric effect, and

2. the range of the electrons produced in the γ interactions is very short, much shorter than the distance between neighboring sampling layers.

In addition, low-energy γs carry a large fraction of the shower energy in high-Z materials, so that the suppression of the calorimeter response to the soft-γ component has a large effect on the overall response to em showers.

Had the 3 mm ^{238}U plates in the HELIOS calorimeter been replaced by iron plates of the same mass, the response to 511 keV γs would have been suppressed by only 17% ($\gamma_{511}/mip = 0.83$), instead of the factor of four found for uranium. Moreover, soft γs carry a considerably smaller fraction of the shower energy in iron absorber than in uranium (see Figure 2.10). For these reasons, the e/mip ratio is expected to be considerably larger in sampling calorimeters with iron absorber, compared with uranium. This is confirmed by experimental data [Bot 81, Dub 86]. For example, in R&D studies for the SLD liquid-argon calorimeter at SLAC, prototypes with ^{238}U absorber were compared with prototypes in which two thirds of the uranium was replaced by iron [Dub 86]. It turned out that the electron/muon signal ratio in these U/Fe modules was larger by about 40%.

The back-of-the-envelope calculation (BOTEC) described above can be refined with Monte Carlo simulations, using the EGS4 package (Section 2.5). Details that were omitted in BOTEC, such as the contributions of electrons produced near the boundary to the calorimeter signals, the effects of electrons produced in the active layers and depositing only part of their energy in those layers, the different energy spectra for electrons from Compton scattering and from the photoelectric effect, the angular distributions of these electrons, *etc*. can be taken into account in a proper manner in such simulations. This was done in [Wig 87], for the HELIOS calorimeter structure. In this study, γs were isotropically emitted from a source homogeneously distributed inside a uranium plate. The energy deposition was followed over 20 X_0 in all directions, using EGS4 cutoff values of 10 keV and 100 keV for photons and electrons, respectively. In this study, it was found that the fraction of the energy deposited in the active calorimeter layers by these 511 keV γs was six to seven times smaller than for mips, rather than the factor of four found in our BOTEC.

The γ/mip ratio was also determined for other photon energies in this study. The results are shown in Figure 3.4, for energies between 100 keV and 1 GeV. At the high-energy end of this range, the γ/mip ratio is, not surprisingly, practically equal to the e/mip ratio, about 0.6 in this calorimeter (experimentally, a value of 0.70 ± 0.05 was reported by HELIOS [Ake 87]). However, below 1 MeV the efficiency for γ detection drops spectacularly, as a result of the onset of the photoelectric effect. At 0.4 MeV, the sampling fraction for γs drops even below 10% of the *mip* value. The dip around

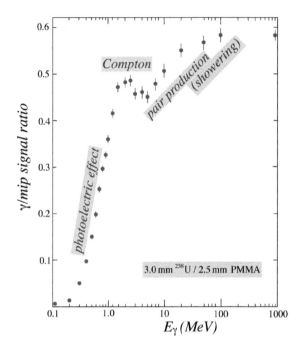

FIG. 3.4. The γ/mip ratio as a function of the γ-ray energy, for a 3 mm U/2.5 mm PMMA
calorimeter. Results from EGS4 Monte Carlo simulations [Wig 87].

5 MeV is the result of the competition between Compton scattering and pair production processes, which have comparable cross sections at that point. As part of the pair
production process, two γs of 511 keV each are produced and, as we saw before, the
contributions of these γs to the signals are strongly suppressed.

These calculations show that the calorimeter response to low-energy photons is
much smaller than the response to mips, in calorimeters consisting of high-Z absorber
and low-Z active material. Because of the dominating role of these low-energy photons
in em shower development in such calorimeters, this effect also leads to e/mip values
that are substantially smaller than 1.0.

The Z values of the active and passive calorimeter materials are the most important
factors determining the e/mip value. This is illustrated in Figure 3.5, which shows
results of EGS4 calculations of the sampling fraction for 10 GeV electron showers in
sampling calorimeters with either plastic scintillator or liquid argon as active material,
as a function of the Z value of the absorber material. The thickness of the absorbers was
chosen to be 1 X_0 in these simulations, the active layers were 2.5 mm thick. The e/mip
ratio, which follows directly from these results, gradually decreases when the Z value
of the absorber increases. The e/mip values are also systematically larger when LAr
($Z = 18$) readout is used instead of plastic scintillator. It is the *difference* in Z values
between active and passive media that determines the e/mip ratio. The simulations
show that this ratio can even be made larger than 1.0 if $Z_{\text{active}} > Z_{\text{passive}}$, *e.g.*, in

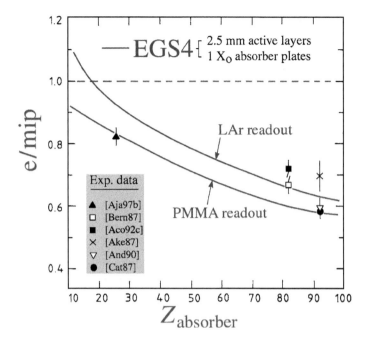

FIG. 3.5. The e/mip ratio for sampling calorimeters as a function of the Z value of the absorber material, for calorimeters with plastic scintillator or liquid argon as active material. Experimental data are compared with results of EGS4 Monte Carlo simulations [Wig 87].

Al/Lar calorimeters.

Experimental results on the value of e/mip are included in this figure. They all concern measurements on calorimeters based on plastic-scintillator readout, and iron [Aja 97b], lead [Bern 87, Aco 92c] or uranium [Ake 87] as absorber material. The results are in reasonable agreement with the simulations.

We conclude that the so-called transition effects, *i.e.*, the suppression of the em shower response, have nothing to do with effects occurring around the critical energy, but are caused by inefficiencies in the sampling of the soft ($E < 1$ MeV) γ component of the em showers. Flauger's observation of the absence of a "transition region" near the boundary between the active and passive calorimeter media [Fla 85] is completely consistent with this conclusion.

Because of the crucial role of the soft γs, it is also clear that reliable predictions of the e/mip value by Monte Carlo simulations such as EGS4 require cutoff values that are low enough to correctly treat the energy deposit by shower particles in the 0.1–1.0 MeV range. Just as for the em energy resolution and the em response (see Figure 2.50), the em sampling fraction (and thus the e/mip ratio) of calorimeters is crucially affected by phenomena that take place at energies corresponding to (less than) 10^{-4} of the energy of the incident particle.

The effects described in this subsection are not only relevant for understanding the electron, photon and muon signals produced by sampling calorimeters. As is shown in Section 3.4.3.1, suppression of the em shower response is an important ingredient to achieve *compensation* in hadron calorimeters. Understanding the details of the mechanism that causes this effect is therefore also crucial in that context.

3.2.2   *Spatial dependence of the electromagnetic response*

In the previous subsection, we saw an example of the invaluable contributions that reliable Monte Carlo simulations can make in the process of understanding the largely non-trivial, difficult to measure, but very important characteristic of sampling calorimeters that is the e/mip ratio. In this subsection we take this process one step further and examine EGS4 predictions that are almost impossible to verify experimentally.

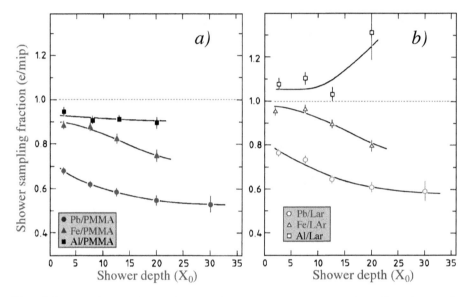

FIG. 3.6. The e/mip ratio as a function of the shower depth, or age, for 1 GeV electrons in various sampling calorimeter configurations. All calorimeters consist of 1 X_0 thick absorber layers, interleaved with 2.5 mm thick PMMA (a) or LAr (b) layers. Results from EGS4 Monte Carlo simulations [Wig 87].

The simulations clearly show that the em sampling fraction, or rather the e/mip ratio, may change as the shower develops. This is illustrated in Figure 3.6, which shows the e/mip ratio as a function of the shower depth, for a variety of sampling calorimeters, based on either PMMA (Figure 3.6a) or LAr (Figure 3.6b) readout. This plot was obtained by comparing the energy deposited in each individual active layer with the average energy deposited in the two absorber layers sandwiching it. The ratio of these two numbers (the "local" shower sampling fraction) is plotted on a scale normalized to the dE/dx ratio for mips traversing an active and a passive layer.

When high-Z absorber material is used, the e/mip ratio clearly decreases as the shower develops. For example, in Pb/LAr e/mip drops from a value of ~ 0.8 in the early part of the shower to ~ 0.6 in the tail. Similar effects are also observed for PMMA readout. For Al/LAr, the opposite effect occurs: the e/mip value rises by about 20% as the shower develops.

These effects can be understood from the changing composition of the showers. In the early phase, most of the energy is deposited by relatively fast shower particles. Pair production is a major source of these particles. As illustrated by the positron production characteristics (Figure 2.11), this process occurs predominantly in the early stages of the em shower development, and near the shower axis.

The soft γs that are absorbed through Compton scattering and through the photo-electric effect, dominate the tails of the showers. Since these γs are responsible for deviations from $e/mip = 1$, their (changing) relative abundance determines the depth dependence of the em sampling fraction.

This phenomenon has important practical consequences when an electromagnetic calorimeter is longitudinally segmented into, for example, sections of 10 X_0 and 20 X_0. Because of the different sampling fractions for the early and late parts of the em showers, the calibration constants (which relate signals to energy deposit) have to be *different* for these two sections, even if the calorimeter structure is exactly the same throughout the detector. This is worked out in detail in Chapter 6.

3.2.3   *Sampling frequency and electromagnetic response*

The soft γs responsible for the suppression of the em calorimeter response in high-Z sampling calorimeters interact overwhelmingly in the absorber layers. The (photo)elec-trons produced in these interactions only contribute to the calorimeter signal if the in-teractions occur in a very thin region, with an effective thickness δ, near the boundary between the active and passive layers.

However, if the sampling *frequency* is increased, *i.e.*, if the absorber layers of the calorimeter are made thinner, then δ represents a larger fraction of the total absorber volume. Therefore, the *fraction* of the shower γs that interact in the δ region increases and so does the sampling fraction for em showers. In the extreme case of a calorimeter consisting of 1 Ångström thick active and passive layers, e/mip will be 1, regardless of the choice of materials.

This effect was also studied with EGS4 simulations [Wig 87]. Results are shown in Figure 3.7, for uranium calorimeters with PMMA or LAr readout. When the thickness of the uranium plates exceeded 5 mm, no significant changes in the e/mip value were observed. Apparently, the contribution of γ interactions in the δ region to the calorimeter signals was completely negligible beyond that point. However, for thinner layers, the e/mip ratio clearly increased by as much as 20% when 0.5 mm thick absorber plates were used.

Experimental evidence for this effect may be derived from the e/mip values pub-lished for two lead/plastic-scintillator calorimeters with very different sampling fre-quencies. The ZEUS Collaboration built and studied a prototype calorimeter consisting of 10 mm thick lead plates, interleaved with 2.5 mm polystyrene-based scintillator. They measured an e/mip value of 0.67 ± 0.03 for this detector [Bern 87]. The SPACAL Col-

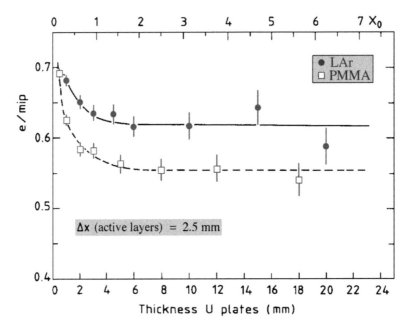

FIG. 3.7. The e/mip ratio as a function of the thickness of the absorber layers, for ura-
nium/PMMA and uranium/LAr calorimeters. The thickness of the active layers is 2.5 mm
in all cases. Results from EGS4 Monte Carlo simulations [Wig 87].

laboration used the same materials, in the same proportion, to construct a fine-sampling
scintillating-fiber calorimeter. The fiber thickness was 1 mm and the fiber-to-fiber spacing 2.22 mm in this device. They measured an e/mip value of 0.72 ± 0.03 [Aco 92c].
The relatively large experimental error bars limit the conclusiveness of these experimental results, but the observed increase in the e/mip ratio with the sampling frequency
is definitely in good agreement with the $\sim 10\%$ effect predicted by EGS4 simulations.

   In the following subsections, we will see why it is so difficult to measure the ratio
of the calorimeter responses to em showers and mips with high precision.

3.2.4   *The response to muons*

Muon detection is an important ingredient of modern experiments in particle physics.
Accurate measurements of the muon momenta require a high-precision magnetic spectrometer. However, such a spectrometer is usually located behind a thick absorber (the
calorimeter) intended, among other reasons, to absorb all other particles and thus to
identify the muons. The muons always lose some fraction of their energy in this absorber. This fraction may fluctuate quite strongly from event to event, especially at high
energies. To identify a charged track as a muon, there has to be consistency between the
measured momentum of the track upstream of the calorimeter and the sum of the energy
loss in the calorimeter and the momentum measured in the downstream muon spectrometer. The strength of this consistency check depends often strongly on the calorimeter

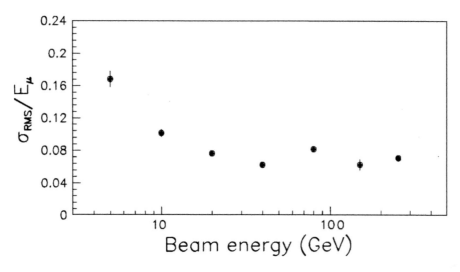

FIG. 3.8. The *rms* spread in the fractional energy loss $\Delta E_\mu / E_\mu$ by muons traversing the SPACAL calorimeter. From [Aco 92c].

measurement.

The necessity for accurate measurements of the energy lost by muons traversing a calorimeter system on their way to the muon spectrometer may be illustrated with data obtained by the SPACAL Collaboration [Aco 92c]. Figure 2.21 shows signal distributions from 10, 20, 80 and 225 GeV muon beams in the SPACAL calorimeter. The distributions are clearly changing with energy, due to the increasing cross sections of higher-order QED processes, such as hard bremsstrahlung and e^+e^- pair production. Both the average amount of energy lost by the muons and the event-to-event fluctuations increase with the muon energy. It is the latter effect, the broadening of the signal distributions, that makes the calorimeter information important for the measurement of the muon properties.

It turned out that the *rms spread* of the fractional energy loss $\Delta E_\mu / E_\mu$ levels off at $\sim 6\%$ in the $9.5\lambda_{\text{int}}$ deep SPACAL calorimeter (Figure 3.8). Therefore, if this calorimeter was just a block of passive absorber, the precision of the muon momentum measurements would be limited by event-to-event fluctuations in the energy loss, rather than by the quality of the downstream muon spectrometer. The fact that the absorber produces signals is thus very important.

Experiments in which accurate muon momentum measurement is an important experimental goal, *e.g.*, the L3 experiment at LEP, may either choose to eliminate events in which muons produce large signals in the calorimeter, or try to derive the energy loss in the calorimeter from these signals. In the latter case, one will need to know the calorimeter response to muons.

The calorimeter response to muons is a complicated issue, since the signals generated by the muons are caused by a variety of different processes, each with its own characteristic response. First, there is the primary ionization, by the muon itself, of the atoms of which the calorimeter structure is composed. There is no reason why the calo-rimeter response to this signal component would be different from the *mip* response.

When a high-energy muon radiates an energetic ($E > 1$ GeV) bremsstrahlung pho-ton, this photon develops an em shower that is sampled in the same way as other em showers in the calorimeter structure. Since e/mip may be significantly different from 1.0, the response to this muon energy loss component may thus be similarly different from the response to the ionization loss component.

When bremsstrahlung photons of lower energies are emitted, the calorimeter re-sponse becomes strongly energy dependent (see Figure 3.4).

The muons also emit energetic knock-on electrons (δ-rays), responsible for the rel-ativistic rise in the dE/dx curves beyond the minimum ionizing energy. For relativistic muons of a given momentum, the ratio of the energy carried by such electrons and the energy deposited as primary ionization is, again, Z dependent: the relativistic rise in-creases with Z. For the soft δ electrons, *i.e.*, those that travel a small fraction of the distance between neighboring sampling layers, this leads to a response that is smaller than the *mip* one, in calorimeters with $Z_{\text{absorber}} > Z_{\text{active}}$. The δ electrons in the 10 MeV range are sampled in a way similar to mips. And the highly energetic ones develop em showers, for which the response is determined by the e/mip ratio.

In summary, this is a very complicated situation indeed. And since it is impossible to disentangle the various contributions to the calorimeter signals event by event, this problem can only be dealt with on a statistical basis. The SPACAL Collaboration did this by comparing the measured muon signal distributions with Monte Carlo ones, and derived the calorimeter response to muons from this comparison.

In Figure 3.9 this response is plotted as a function of the energy of the muons that caused the signals. Not surprisingly, this response depends on the muon energy. For low-energy muons, the signals are dominated by ionization and the response is not very

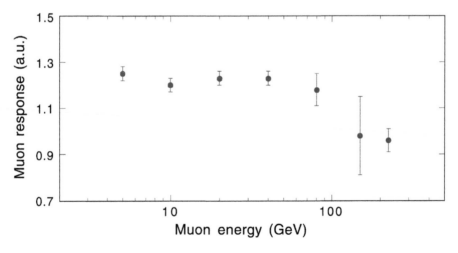

FIG. 3.9. Response of the SPACAL calorimeter to energy deposited by muons traversing it, as a function of the muon energy. See text for details. From [Aco 92c].

different from the response to mips. However, as the muon energy increases, em showers contribute increasingly to the signals. And since $e/mip < 1$ for this calorimeter, the calorimeter response to energy deposited by muons decreases with increasing muon energy by about 30% over the energy range for which these measurements were done (5–225 GeV).

It should be mentioned that the effects discussed here would be smaller than those measured by SPACAL if the calorimeter were made from a lower-Z absorber material than lead. This is because of the Z dependence of the higher-order QED processes that are responsible for the broadening of the energy loss distributions (see Section 2.2), and also because the e/mip ratio is closer to 1, which reduces the impact of these processes on the calorimeter signals.

3.2.5 Experimental determination of e/mip

Because of the complications described in the previous subsection, the experimental determination of a sampling calorimeter's e/mip response ratio is a non-trivial issue. The signal distributions for muons traversing the calorimeter are usually the starting point for this type of analysis. The challenge consists of separating the mip contributions to these signals from the non-mip ones. This is best done at low muon energies, where the non-mip effects are smallest. But it is essential to perform the analysis also at higher energies, to check the internal consistency of the methods that are used: if these methods are correct, they should yield the same e/mip value, regardless of the muon energy.

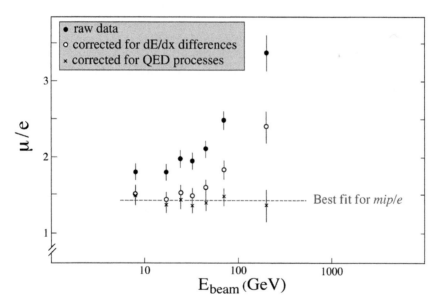

FIG. 3.10. Determination of the e/mip ratio of the HELIOS calorimeter. See text for details. From [Ake 87].

Åkesson and coworkers [Ake 87] determined the e/mip ratio of the HELIOS calorimeter (^{238}U/PMMA) from the muon signal distributions, using analytical correction methods, as illustrated in Figure 3.10. The full circles in this figure represent the average signals measured for muons traversing the calorimeter. The conversion of these signals to energy deposit was done on the basis of the calorimeter calibration with electrons. Based on that conversion, the average energy deposited by the muons in the calorimeter was found to range from about 3 GeV at the lowest energies to about 6 GeV for 150 GeV muons. From the calorimeter dimensions, it was calculated that minimum ionizing particles should deposit 1.802 GeV in this detector. The vertical scale in the figure gives the ratio between the muon energy determined as mentioned and 1.802 GeV, the value expected for mips. Therefore, if muons were mips and the e/mip ratio of the calorimeter was 1, then all measured points should have the value $\mu/e = 1.0$ in this plot.

Next, the authors corrected the measurements for the effects of the relativistic rise in the specific ionization by normalizing the measured energy deposits to the ionization that may be expected from a muon with energy E, instead of the ionization for a mip. That is, the measured energy deposits were not divided by 1.802 GeV, but by a larger, energy-dependent value, derived from Lohmann's Tables [Loh 85]. Since the ionization of the active layers determines the calorimeter signal, the relativistic rise for PMMA was used for this purpose. This procedure led to the open circles in Figure 3.10.

In a third step, corrections for dE/dx losses from higher-order QED processes were applied, using the same Tables. In this case, the calculations were based on the effects of these processes *in the absorber material*, thus assuming that the calorimeter sampled the hard bremsstrahlung photons and the electron–positron pairs in the same way as it sampled em showers. The expected energy deposit was thus increased by a certain fraction, which represented the contribution of bremsstrahlung and pair production to the total energy loss in uranium. The measured muon signals were normalized to the total expected energy deposit. Of course, this correction mainly affected the high-energy points.

The crosses in Figure 3.10 represent the final result. The effects of the relativistic rise and of higher-order QED effects have been eliminated and what remains is the mip part of the muon signals. The mip/e value obtained in this way amounted to ~ 1.5 and was independent of the muon energy, as it should be.

The SPACAL Collaboration evaluated the e/mip value of their calorimeter by comparing the measured signal distributions with Monte Carlo generated ones [Aco 92c]. Some results of that analysis were already shown in the previous subsection. The Monte Carlo program transported the muons through the calorimeter structure, taking into account all the details of the processes mentioned above, and calculated the distribution of the amounts of energy deposited in the scintillating fibers. The e/mip value was a free parameter in these simulations. Its value was found by comparing the measured signal distributions with the Monte Carlo ones. This evaluation of the non-mip part of the calorimeter signals was not only carried out for the *average* amounts of energy lost by the muons, but also for the *most probable* energy losses. The average energy losses are strongly affected by events far out in the tails of the distributions (see Figure 2.21).

By using the most probable values, the sensitivity to systematic effects could be limited.

These analyses also led to internally consistent results, *i.e.*, an e/mip value independent of the muon energy. The e/mip values found using the average and the most probable energy deposits were within 1% the same. The final result of this study, published by the SPACAL Collaboration, was $e/mip = 0.72 \pm 0.03$ [Aco 92c].

Analysis methods similar to those described above were used by Bernardi *et al.* to determine the e/mip ratio of their lead/plastic-scintillator calorimeter [Bern 87], and by Ajaltouni *et al.* for the iron/plastic-scintillator hadron calorimeter for ATLAS [Aja 97b].

The ZEUS Collaboration derived the e/mip value of their uranium/plastic-scintillator calorimeter from the measured calorimeter response to pions and protons with energies near the minimum ionizing value. At these low energies, these particles are more likely to lose their entire kinetic energy by ionization of the absorber medium than through processes involving nuclear interactions. In that respect, they are thus very similar to muons. At a kinetic energy of 0.38 GeV, ZEUS measured the pion response to be $61 \pm 1\%$ of the response to electrons. For protons with kinetic energies of 0.26 GeV and 0.12 GeV, the calorimeter response, normalized to the one for electrons, amounted to 0.60 ± 0.01 and 0.58 ± 0.04, respectively [And 90]. The authors concluded that these numbers were in good agreement with the e/mip value of 0.62 they calculated with EGS4 for their detector. The weighted average of the three mentioned results constitutes the ZEUS point in Figure 3.5, and a systematic error of 0.03 was attributed to that result. In the next subsection, some other interesting aspects of these low-energy measurements are discussed.

3.2.6   *The response to hadrons*

3.2.6.1   *Non-linearity at low energy.*   In Section 3.1.4, the response of homogeneous calorimeters to hadrons was discussed. It was found that the hadronic response of such calorimeters is always smaller than the em response and that, as a result, the hadronic signals from such calorimeters are non-linear: the hadronic response is not constant as a function of energy.

The latter conclusion also applies to sampling calorimeters. It is easy to see why. Consider what happens to charged pions of energy E entering a sampling calorimeter. At low energies, well below 1 GeV, these pions are likely to lose their entire kinetic energy by ionizing or exciting the atoms or molecules in the calorimeter media. They behave very much like mips and therefore, the calorimeter response to such soft pions is not very different from the response to mips.

As the pion energy increases, the likelihood that the particles "range out" rapidly diminishes. Nuclear reactions become an increasingly important aspect of the absorption. As the energy increases, the production of π^0s in these nuclear reactions plays an increasingly important role. And since these π^0s generate em showers, the response of the calorimeter to hadron showers becomes gradually similar to the one for em showers. Figure 3.1 shows that this effect occurs *regardless of the degree of compensation, the e/h value, of the calorimeter*. In other words, at very high energies the e/π signal ratios of any calorimeter approach 1.

In Section 3.2.1, we saw that the e/mip value of practically all sampling calorimeters used in practice is smaller than 1, and that this is particularly true for calorimeters with high-Z absorber material. However, if the response of a calorimeter to hadrons resembles the one for mips at low energy, the one for em showers at high energy, and if $e/mip \neq 1$, then that calorimeter is by definition non-linear for hadrons, *regardless of the degree of compensation.*

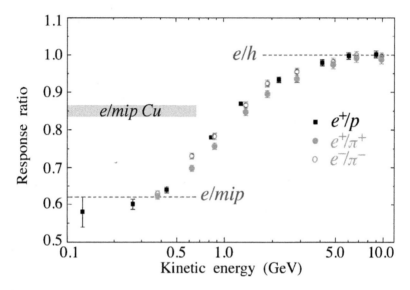

FIG. 3.11. The response of the (compensating) ZEUS calorimeter to low-energy hadrons. Data from [And 90].

This is illustrated in Figure 3.11, which shows the response of the ^{238}U/plastic-scintillator ZEUS calorimeter to low-energy charged hadrons [And 90]. This calorimeter has an e/h value very close to 1.0, but since the e/mip value is about 0.6, the hadronic response increases as the energy decreases below a few GeV, reflecting the increasingly mip-like absorption process.

3.2.6.2   *Non-linearity at high energy.*   The type of hadronic signal non-linearity discussed above is limited to low energies (< 3 GeV), where a significant fraction of the hadrons range out without undergoing a nuclear interaction. At high energies, a completely different mechanism may cause substantial non-linearities.

Figure 3.11 shows that the hadronic response of the ZEUS calorimeter becomes approximately constant for energies above ~ 4 GeV, where all particles develop showers. This is a direct consequence, and one of the most important benefits, of the fact that this calorimeter is compensating ($e/h \approx 1$). Since the responses to the em and non-em shower components are equal, the fact that the (average) energy sharing between these components in hadronic showers changes with the energy is of no consequence for the hadronic calorimeter response.

However, calorimeters for which this condition is not fulfilled, *e.g.*, all homogeneous calorimeters and most sampling calorimeters, are intrinsically non-linear, their hadronic response as a function of energy reflects the changing energy sharing between the em and non-em shower components.

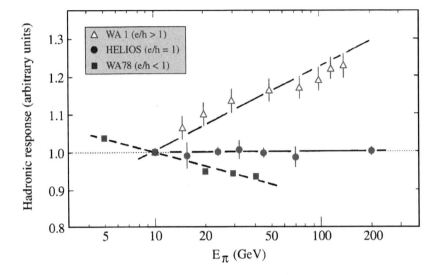

FIG. 3.12. The response to pions as a function of energy for three calorimeters with different e/h values: the WA1 calorimeter ($e/h > 1$, [Abr 81]), the HELIOS calorimeter ($e/h \approx 1$, [Ake 87]) and the WA78 calorimeter ($e/h < 1$, [Dev 86, Cat 87]). All data are normalized to the pion response at 10 GeV.

Figure 3.12 shows the hadronic response of several sampling calorimeters as a function of energy [Wig 88]. The WA1 calorimeter [Abr 81], which was successfully used to study deep inelastic neutrino scattering in the 1970s at CERN, consisted of 5 cm thick iron slabs, interleaved with 2.5 cm thick plastic-scintillator layers, and had an e/h value considerably larger than 1. As a consequence, the response to pions increased by $\sim 20\%$ over the energy range 10–160 GeV, reflecting the increased fraction of the shower energy carried by the em component ($\langle f_{em} \rangle$, see Section 2.3.1).

All homogeneous calorimeters have e/h values larger than 1. Depending on the Z values of the components and on the cross sections for neutron-induced reactions, e/h is somewhere in the range 1.5–2.5. However, in sampling calorimeters, a different and much wider range of e/h values applies.

Most calorimeters used in practice have e/h values larger than 1, *e.g.*, the WA1 calorimeter mentioned above. However, other calorimeters have e/h values very close to 1. They are called compensating calorimeters. An example of such a compensating calorimeter, used by the ZEUS Collaboration for the study of ep scattering at HERA (DESY), was mentioned above. Another representative of this class was exploited by the HELIOS Collaboration for the study of heavy-ion scattering at the CERN Super

Proton Synchrotron. The response of this calorimeter to pions is also shown in Figure 3.12. Because of the compensating nature of this calorimeter, its hadronic response is constant.

There are even examples of calorimeters with e/h values smaller than 1. The WA78 Collaboration [Dev 86, Cat 87] built a uranium/plastic-scintillator calorimeter for their $b\bar{b}$ hadroproduction experiment at CERN's Super Proton Synchrotron. This calorimeter turned out to have an e/h value of about 0.8. As a result, the hadronic response of this calorimeter *decreased* with increasing energy (Figure 3.12).

3.2.6.3 *Non-linearity and e/h.*

The signal non-linearity is thus determined by the e/h value of the calorimeter. In fact, measuring the hadronic signal non-linearity of a calorimeter is one of the methods that can be used to determine its e/h value (see also Figure 3.1). For this purpose, one also needs the energy dependence of the average em shower fraction, $\langle f_{em}(E) \rangle$. Assuming linearity for em shower detection, it follows from Equation 3.5 that the ratio of pion responses at energies E_1 and E_2 is related to the e/h value as

$$\frac{\pi(E_1)}{\pi(E_2)} = \frac{\langle f_{em}(E_1) \rangle + \left[1 - \langle f_{em}(E_1) \rangle\right](e/h)^{-1}}{\langle f_{em}(E_2) \rangle + \left[1 - \langle f_{em}(E_2) \rangle\right](e/h)^{-1}} \tag{3.7}$$

Numerical examples of the application of this relationship are given in Section 7.5.1.

Hadronic signal linearity (at least for $E > 5$ GeV) is only one of many advantages of compensating calorimeters. Other reasons why $e/h = 1$ is a desirable calorimeter property are discussed in Chapter 4. It is, therefore, important to investigate how this desirable property can be obtained.

3.2.6.4 *The non-em calorimeter response.*

Compensation implies that the calorimeter responses to the em and non-em components of hadronic showers are equal. In Section 3.2.1, the em response was discussed. We will now concentrate on the non-em response.

The energy deposition mechanisms that play a role in the absorption of the non-em shower energy are summarized below (see Table 2.5):

- Ionization by charged pions. We will refer to these particles as the *relativistic* shower component, and call the average fraction of the non-em shower energy carried by these particles f_{rel}.
- Ionization by spallation protons. We will refer to these particles as the *non-relativistic* shower component, and call the average fraction of the non-em shower energy carried by these particles f_p.
- Kinetic energy carried by *evaporation neutrons* may be deposited in a variety of ways. The average fraction of the non-em shower energy carried by these neutrons will be called f_n.
- The energy used to release protons and neutrons from calorimeter nuclei, and the kinetic energy carried by recoil nuclei do not contribute to the calorimeter signal. This energy represents the *invisible* fraction f_{inv} of the non-em shower energy.

Using these definitions, the calorimeter response to the non-em shower component, h, can thus be written as follows:

$$h = f_{\text{rel}} \cdot rel + f_p \cdot p + f_n \cdot n + f_{\text{inv}} \cdot inv \qquad (3.8)$$

in which $f_{\text{rel}} + f_p + f_n + f_{\text{inv}} = 1$ and rel, p, n and inv denote the calorimeter responses to the various mentioned components. By normalizing all these responses to the one for mips and by deleting the last term of Equation 3.8 ($inv = 0!$), the e/h value can thus be written as:

$$\frac{e}{h} = \frac{e/mip}{f_{\text{rel}} \cdot rel/mip + f_p \cdot p/mip + f_n \cdot n/mip} \qquad (3.9)$$

Indicative values for f_{rel}, f_p and f_n are given in Table 2.5. Therefore, the e/h value of a given calorimeter can be determined once we know its response to the relativistic (rel/mip) and non-relativistic (p/mip) ionizing shower particles and to the neutrons (n/mip), through which the non-em calorimeter signal is established.

3.2.6.5 *Relativistic charged hadrons.* In Section 2.3.2 we saw that typically about 1 GeV of energy is spent per charged pion produced in the shower development (0.7 GeV in Cu, 1.3 GeV in Pb). Even though they are relativistic, these pions strongly resemble mips in their ionization losses in the calorimeter structure. Therefore, we take $rel/mip = 1$.

3.2.6.6 *Non-relativistic charged hadrons.* A considerable fraction of the energy in the non-em component of hadron showers is deposited through spallation protons (see, for example, Table 2.5). The overwhelming majority of these protons have kinetic energies well below 1 GeV and are thus non-relativistic. Most of them lose their entire kinetic energy through ionization or excitation of the atoms or molecules of which the calorimeter structure consists, without causing new nuclear interactions. In first approximation, one would expect the calorimeter response to such protons to be equal to the response to mips: $p/mip = 1$. However, in practice, the calorimeter response to these protons differs from that to mips, because of (one or several of) the following effects:

- The ratio of the specific ionization in the active and passive calorimeter media, $(dE/dx)_{\text{active}}/(dE/dx)_{\text{absorber}}$, may be different for non-relativistic protons and mips. This leads to different sampling fractions for these particles. The magnitude of this effect depends on the combination of materials and on the proton energy (see Figure 3.13) and may be quite substantial. Especially in lead and uranium calorimeters, the p/mip ratio may considerably *increase* due to this effect.
- Multiple (Coulomb) scattering of protons has the opposite effect. It increases the pathlength of the particles, more so in high-Z materials. Therefore, the fraction of the energy deposited by protons in the (low-Z) active layers *decreases*, and so does the calorimeter response.
- The range of low-energy protons in calorimeter absorber materials is limited. Figure 2.48 shows that the thickness of absorber plates typically used in sampling calorimeters corresponds to a proton range of several tens of MeV. For protons with energies less than ~ 50 MeV, sampling inefficiencies start to play a role. These inefficiencies tend to *reduce* the p/mip response ratio as the energy decreases (Figure 3.14).

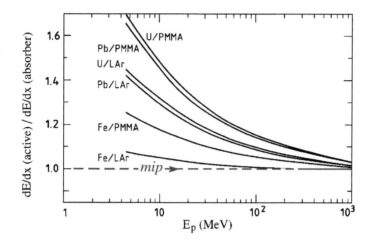

FIG. 3.13. The ratio of energy deposition by non-relativistic protons in the active and passive materials of various calorimeter structures, as a function of the proton's kinetic energy. This ratio is normalized to the one for mips. From [Wig 87].

- This effect may be further amplified by saturation or recombination effects in the active calorimeter medium. Such effects, which occur in scintillators and in liquids operating in the ionization chamber mode (*e.g.*, LAr), become increasingly important as dE/dx increases. They reduce the calorimeter signals and thus the p/mip value.

The combined result of all these effects was studied with a Monte Carlo technique in [Wig 87]. Spallation protons were generated according to a uniform distribution inside an absorber plate of a given sampling calorimeter and transported through the structure, with a step length chosen such that ∼ 0.1 MeV was lost per step. Multiple scattering and dE/dx losses were described on the basis of data tabulated in [Jan 82], and Birks' law [Bir 64] was used to treat saturation or recombination effects (see Section 3.2.10 for details about this point).

Some results of this study are shown in Figure 3.14, for the HELIOS calorimeter: 3 mm ^{238}U plates interleaved with 2.5 mm thick PMMA. The upper curve shows the energy deposited in the active calorimeter layers by the protons. The effect of proton absorption inside the layer where the particle was produced clearly becomes a dominating factor for energies below 50 MeV. The effect of the Z-dependent sampling fraction for densely ionizing particles (Figure 3.13) manifests itself through p/mip values larger than 1 at energies above 50 MeV.

The lower curve shows the amplitude of the calorimeter signals, obtained by correcting the upper curve for saturation effects in the light production by the protons in PMMA. A comparison between the two curves in Figure 3.14 also shows that the saturation effects are very dependent on the proton energy, or rather on the specific ionization dE/dx (see Equation 3.13). At kinetic energies of 10– 20 MeV, the calorimeter signals are reduced by a factor of two due to saturation effects in the plastic scintillator. At 100

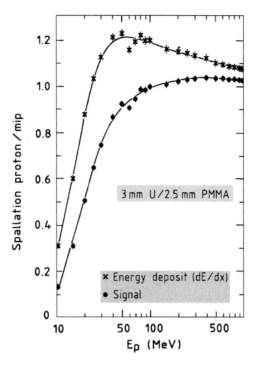

FIG. 3.14. The energy deposit in the active layers (upper curve) and the calorimeter signal
(lower curve) for stopping protons, relative to mips, as a function of the kinetic proton energy,
in a 3 mm ^{238}U/2.5 mm PMMA sampling calorimeter. See text for details. Results from
Monte Carlo simulations [Wig 87].

MeV, the signal reduction due to these effects is $\sim 20\%$ and at 1 GeV it is only a few
percent.

As expected, the p/mip response ratio is strongly energy dependent, and the *spectrum* of the spallation protons generated in the shower development is thus an important
factor. Since many spallation protons have kinetic energies in the 10–100 MeV range,
where saturation plays a major role, the relative contribution of these soft protons to the
energy deposited in the active layers is thus an important factor for the p/mip ratio.
This contribution can be *increased* in two ways:

1. By reducing the thickness of the absorber layers, the softer shower particles are
   sampled more efficiently.

2. By increasing the thickness of the active layers, energy loss in these layers softens
   the spectrum of the spallation protons contributing to the calorimeter signals.

In both cases, the p/mip ratio decreases. This is illustrated in Figure 3.15, which shows
the (simulated) effects of changes in the thickness of the active and passive layers in a
HELIOS-type structure on the p/mip ratio.

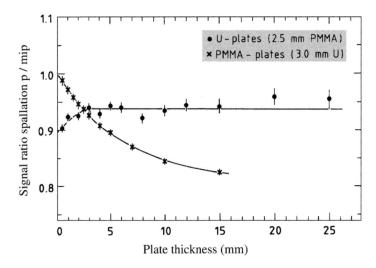

FIG. 3.15. The response ratio p/mip for uranium/PMMA calorimeters as a function of the thickness of the active and passive layers. Results from Monte Carlo simulations [Wig 87].

These simulations were done for protons with an exponentially decreasing energy spectrum, with an average kinetic proton energy of 80 MeV.

In summary, the response of a given sampling calorimeter to the spallation proton component of the shower thus depends on the following factors:

- the spectrum of the spallation protons,
- the Z value of the absorber material,
- the sampling fraction and the sampling frequency of the calorimeter, and
- the saturation properties of the active calorimeter medium.

In [Wig 87], the dependence of the calorimeter response on these factors was studied for a variety of possible sampling calorimeters. For active media that exhibit saturation effects for densely ionizing particles (plastic scintillator, LAr), the p/mip ratio was found to vary between 0.81 (6 mm iron absorber, read out with thick plastic scintillator) and 0.99 (3 mm uranium absorber with thin LAr). If non-saturating active material was used, *e.g.*, silicon, the p/mip ratio could become as high as 1.15 (with uranium absorber).

The dependence of the proton response on certain design parameters of the calorimeter makes it, in principle, possible to affect this response, and thus the e/h ratio, through design choices. For a given absorber medium, the sampling parameters and the saturation properties of the active medium allow for p/mip variations at the level of $\pm 10\%$. However, such changes usually also affect other elements that determine the e/h ratio. This is worked out in detail in Section 3.4.

3.2.6.7 *The response to neutrons.*   According to Table 2.5, 5–10% of the energy in the non-em shower component is deposited by evaporation neutrons. These neutrons are released from the calorimeter nuclei with kinetic energies of typically a few MeV (Figure 2.32). This kinetic energy is deposited through a combination of the following processes:

- *Nuclear reactions* in which the neutron is absorbed by a nucleus, which emits one or several other particles in this process. Examples of such reactions include (n, α), (n, p) and (n, d). Especially in high-Z materials, these reactions cease to play a role below energies of about 5 MeV, mainly because of the Coulomb barrier that prevents low-energy charged particles from escaping the nucleus.

- *Inelastic scattering.* In this process, part of the neutron's kinetic energy is used to bring a nucleus into an excited state. The nucleus releases this excitation energy in the form of one or several γs, whose (combined) energy corresponds to the energy loss of the neutron in this process. The contribution of this process to the energy loss of the neutrons produced in the calorimeter depends completely on details of the nuclear level structure (see Section 2.3.3). In some materials, *e.g.*, lead, it becomes insignificant below energies as high as 2.6 MeV (because it takes that much energy to bring the most abundant lead isotope, ^{208}Pb, from its ground state into the lowest excited state), in other materials (*e.g.*, iron) it continues to play an important role down to energies well below 1 MeV.

- *Elastic scattering.* In this process, the neutrons transfer a fraction of their kinetic energy to a struck nucleus. At energies between 1 keV and 1 MeV, this is by far the most likely process to occur, regardless of the absorber material.

Usually, the thermalized neutrons end their life in the calorimeter through capture by an atomic nucleus. In this process, the marginal nuclear binding energy (*i.e.*, the difference between the binding energies of the nuclei $^A Z$ and $^{A+1} Z$, typically between 6 and 10 MeV) is released, predominantly in the form of γ-rays.

Therefore, the reaction products that may or may not contribute to the calorimeter signals consist of charged nucleons or nucleon aggregates (*e.g.*, αs), nuclear γs and recoil nuclei. For reasons already mentioned above, the first category is in practice insignificant, in terms of its contribution to the energy deposition by (soft) evaporation neutrons. The contribution of γs from inelastic scattering to the energy deposition by the neutrons depends strongly on the choice of calorimeter materials. The calorimeter response to these γs, and also to those abundantly produced in the neutron capture processes, depends strongly on their energy spectrum. All these nuclear γs are sampled in the same way as the γs generated in em shower development, and the γ/mip response ratio discussed before (Section 3.2.1, Figure 3.4) applies in all cases.

In practice, most of the kinetic energy carried by evaporation neutrons is deposited through elastic scattering. The reaction products of elastic scattering are a recoil nucleus and a lower-energy neutron. The efficiency of this process for slowing down the neutrons depends on the mass of the absorber nuclei: the smaller this mass, the larger the average energy transferred in this process. Hydrogen is thus by far the most efficient medium to absorb the kinetic energy carried by these soft neutrons and thus thermalize them.

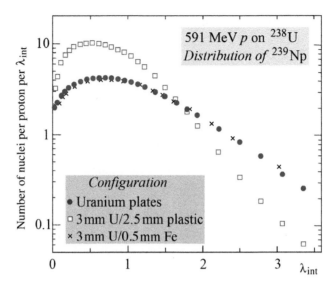

FIG. 3.16. The longitudinal distributions of ^{239}Np nuclei (from neutron capture in ^{238}U) produced by hadron showers in various calorimeter configurations [Ler 86].

The spectacular effects of adding hydrogen to a calorimeter structure become clear
from Figure 3.16. In this figure, the longitudinal distributions of radioactive ^{239}Np nuclei, produced in the absorption of 591 MeV protons in various calorimeter configurations, are shown [Ler 86]. This nuclide is the decay product of ^{239}U, which is produced
by thermal neutron capture in the absorber material, ^{238}U. These distributions are thus
indicative for the longitudinal distributions of thermal neutrons produced in the absorption of 591 MeV protons in these structures.

The data points represented by closed circles and crosses are almost indistinguishable in this figure. These data points describe the neutron distributions in a stack of 3
mm thick ^{238}U plates and a stack in which these plates were interleaved by 0.5 mm
thick iron foils, respectively.

However, the distribution was dramatically affected (note the logarithmic vertical
scale) when these iron foils were replaced by plastic plates with an equivalent thickness, in nuclear interaction lengths. The open squares in Figure 3.16 represent the data
measured for this configuration. Clearly, the neutrons were much more rapidly thermalized in this case and, therefore, must have lost a considerable fraction of their kinetic
energy in elastic collisions with hydrogen nuclei in the plastic.

This conclusion was confirmed by the observation of a significant reduction (by
$\sim 20\%$) in the number of nuclear fissions in this structure, compared with the pure
^{238}U case. Nuclear fission of ^{238}U requires a neutron with a kinetic energy of at least
1.5 MeV, and because of the moderating effect of the plastic, the number of neutrons
capable of inducing fission was smaller in this experiment.

These experimental results show that neutrons lose a disproportionally large fraction of their kinetic energy in collisions with hydrogen when the calorimeter structure

contains this material. With hydrogen as a target, the recoil nucleus is a proton that carries typically half of the kinetic energy of the scattering neutron. If the scattering process occurs in liquid argon, then the recoiling object is an argon nucleus that carries typically only 5% of the kinetic neutron energy. It is not even clear that the argon atom to which this nucleus belongs gets ionized in this process, and in any case the possible contributions of such recoiling argon nuclei to calorimeter signals are completely negligible.

This is very different for the recoil protons produced in hydrogenous calorimeter media. Not only will neutrons which traverse a calorimeter structure that contains hydrogen lose a disproportionally large fraction of their kinetic energy in collisions with this hydrogen, but the recoil protons produced in this process may also substantially contribute to the calorimeter signals. As a result, neutrons in the energy bracket between 1 keV and a few MeV are sampled very differently from charged particles in calorimeter structures that contain hydrogenous active components. This has spectacular calorimetric consequences.

Table 3.2 shows the sampling fractions for neutrons at different energies for three different structures: iron/liquid-hydrogen, iron/liquid-argon and lead/liquid-hydrogen, each with equal volumes of metal and liquid. These sampling fractions were calculated on the basis of the cross sections for elastic scattering (Table B.3) and the assumption (correct if only s-wave scattering plays a role) that the average fraction of the neutron's kinetic energy transferred in elastic collisions with a nucleus containing A nucleons, $\langle f_{el} \rangle$, is given by

$$\langle f_{el} \rangle = \frac{2A}{(A+1)^2} \tag{3.10}$$

From the cross section and the material density, we can calculate the mean free path, $\langle l \rangle$, between collisions. For example, the mean free path of 1 MeV neutrons amounts to 5.5 cm in liquid hydrogen, 24 cm in liquid argon, 4.7 cm in iron and 6.1 cm in lead. In structures consisting of fractions f_i of these elements, the energies deposited in each element are then given by $f_i \langle f_{el}^i \rangle / \langle l_i \rangle$.

Table 3.2 *Sampling fractions for neutrons and mips in various calorimeter structures. See text for details.*

Structure ↓ Particle →	1 keV n	10 keV n	100 keV n	1 MeV n	mip
Fe/LH$_2$ (1/1 vol.)	93.6%	95.9%	95.6%	92.6%	2.4%
Fe/LAr (1/1 vol.)	2.0%	2.8%	11.4%	21.5%	15.7%
Pb/LH$_2$ (1/1 vol.)	99.2%	99.2%	98.8%	98.3%	2.2%

Neutrons at energies between 1 keV and 1 MeV lose almost 100% of their kinetic energy in the hydrogen in the Fe/H$_2$ and Pb/H$_2$ structures, while mips deposit only a few percent of their energy in the hydrogen. Therefore, if hydrogen were the active component of these structures, then neutrons would be much more efficiently sampled than charged particles, the sampling fractions would differ by factors of 40 to 50. In other words, n/mip values would amount to 40–50 in such calorimeter structures. The opposite situation occurs in the liquid-argon case, where mips are sampled more efficiently than soft neutrons.

The n/mip ratio depends sensitively on the fraction of hydrogen contained in the calorimeter structure. This is illustrated in Table 3.3 for the case of lead/hydrogen. Whereas the sampling fraction for mips is almost proportional to the fraction of hydrogen in the structure, the sampling fraction for neutrons changes by less than a factor of three when the fraction of hydrogen is changed by two orders of magnitude. As a result, the n/mip ratio spans a wide domain, from 1.5 in hydrogen-rich structures (99 volume %) to 1630 in hydrogen-poor ones (1 volume %).

Table 3.3 *Sampling fractions for 1 MeV neutrons and mips in Pb structures containing different fractions of liquid hydrogen, assumed to be the active calorimeter medium in these configurations.*

H_2 fraction (vol. %)	1 MeV neutrons	mips	n/mip ratio
1%	36.9%	0.0227%	1630
2%	54.1%	0.0458%	1180
5%	75.3%	0.118%	640
10%	86.6%	0.249%	350
20%	93.5%	0.558%	170
30%	96.1%	0.953%	100
40%	97.5%	1.47%	66
50% (Pb/H_2 = 1/1)	98.3%	2.20%	45
60%	98.9%	3.26%	30
70%	99.3%	4.98%	20
80%	99.6%	8.24%	12
90%	99.8%	16.8%	5.9
95%	99.9%	29.9%	3.3
99%	99.95%	66.6%	1.5

The neutrons are thus much more efficiently sampled than mips in these calorimeter structures, more so if the fraction of hydrogen is reduced. At first sight, it might seem counter-intuitive that it takes a *reduction* in the fraction of neutron-sensitive material to *increase* the relative calorimeter response to these particles (the n/mip ratio). However, this is easily understood from the fact that a change in the hydrogen fraction affects the response to mips much stronger than it affects the response to neutrons. The mips share their energy among the active and passive materials according to the relative abundance of these materials in the structure. If there is a small fraction of active material, only a small fraction of the mip's energy is deposited in this material.

The neutrons, on the other hand, have in practice almost no alternative for depositing their kinetic energy in the active material, even if this active material represents only a small fraction of the total mass. This is because elastic scattering in the lead absorber is an extremely inefficient process for losing energy. On average, the neutrons lose only 0.96% of their kinetic energy in such collisions, *vs.* 50% in hydrogen.

Of course, a reduction of the hydrogen fraction implies that the calorimeter *volume* needed for the absorption of the neutrons increases, since the mean free path between consecutive neutron–proton collisions increases.

The recoil protons produced in elastic neutron scattering off hydrogen nuclei carry typically half the kinetic neutron energy. These recoil protons are very densely ionizing particles. For example, the specific ionization for 1 MeV protons in plastic amounts to 27 keV μm^{-1}, $\sim$ 130 times the minimum ionizing value. For 100 keV protons, the specific ionization increases by another factor of four.

Therefore, if saturation effects play a role in the hydrogen-containing active calorimeter medium, then the n/mip response ratio will be considerably affected. These saturation effects may easily reduce the calorimeter signals from the neutrons, and thus the n/mip response ratio, by a factor of five (see Section 3.2.10) or more. However, in spite of such effects, the n/mip ratios may remain much larger than 1. This means that the neutrons may produce signals that suggest disproportionally large energy deposits in the sampling calorimeter. In Section 3.4 it is shown how this phenomenon can be exploited to achieve compensation.

3.2.7   *Energy dependence of the hadronic response*

In the previous subsection, the calorimeter response to the various shower particles produced in the calorimetric absorption of hadrons was examined. We distinguished three classes of shower particles:

- relativistic hadrons, mainly secondary and higher-order pions
- non-relativistic hadrons, mainly spallation protons
- evaporation neutrons, abundantly produced, with energies of typically a few MeV.

In addition to these particles, produced in the non-electromagnetic component of the shower, there are the electrons and positrons through which the energy of the em shower component is deposited in the calorimeter structure.

The calorimeter response to these different classes of particles determines the e/h ratio (Equation 3.9). Once the e/h ratio is known, the calorimeter response to hadrons of a given type and energy can be inferred from the energy dependence of the average em shower fraction, $\langle f_{em} \rangle$, in the shower development of these hadrons (Equation 3.4). At low energies, an increasing fraction of the hadrons lose their entire energy through the em interaction (ionization or excitation of the atoms or molecules in the calorimeter media) without undergoing nuclear interactions. This leads in general to an increase of the hadronic response (see Figure 3.11).

Figure 3.17 shows the (calculated) effects of the different factors mentioned above on the hadronic response of liquid-argon calorimeters, with iron and lead absorber, respectively. Combining Equations 3.5 and 3.9, the response to pions, relative to the (constant) response to electrons, can be written as

$$\frac{\pi}{e} = \langle f_{em} \rangle + [1 - \langle f_{em} \rangle] \frac{f_{rel} \cdot rel/mip + f_p \cdot p/mip + f_n \cdot n/mip}{e/mip} \qquad (3.11)$$

This response is energy dependent since the average em shower fraction, $\langle f_{em} \rangle$, is energy dependent. In these response calculations, we have used Equation 2.19 to describe

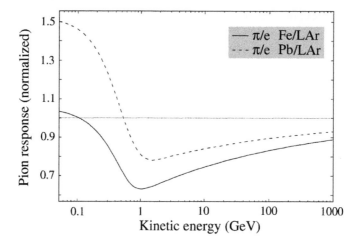

FIG. 3.17. The calculated response of iron/LAr and lead/LAr calorimeters to pions, as a function of the kinetic energy. See text for details.

this energy dependence. The values of the various parameters that were used are listed in Table 3.4.

Equation 3.11 is valid for energies larger than about 3 GeV, when in practice all pions undergo nuclear interactions. At lower energies, the probability that the pions lose their entire kinetic energy by ionization of the calorimeter media rapidly increases. If that happens, the response is likely to be similar to that of a mip. The probability that the particle ranges out, P_{stop}, was taken from data tabulated in [Jan 82], and the pion response at low energies was calculated as

$$\frac{\pi}{e} = P_{\text{stop}}\left(\frac{mip}{e}\right) + \left[1 - P_{\text{stop}}\right]\left(\frac{\pi}{e}\right)_{\text{int}} \tag{3.12}$$

Table 3.4 *Parameter values used to calculate the response of Fe/LAr and Pb/LAr calorimeters to pions of different energies.*

Parameter	Source	Fe/LAr	Pb/LAr
e/mip	EGS4 (Figure 3.5)	0.95	0.65
rel/mip	Section 3.2.6.5	1	1
p/mip	[Wig 87]	0.86	0.94
n/mip	[Wig 87]	0.5	0.5
f_{rel}	Table 2.6	0.15	0.13
f_p	Table 2.6	0.43	0.33
f_n	Table 2.6	0.08	0.12
e/h	Equation 3.9	1.7	1.3
$f_{\text{em}}(E)$	Equation 2.19		
E_0	Figure 2.25	0.7	1.3
k	Figure 2.25	0.82	0.82

where $(\pi/e)_{\text{int}}$ is the response ratio for interacting pions, calculated with Equation 3.11.

The vertical scale in Figure 3.17 is normalized to the calorimeter response for em showers. In both calorimeters, the hadronic response is largest at low energies, where the pions are more likely to range out than to interact with calorimeter nuclei. In the lead calorimeter, the hadronic response becomes even considerably larger than the em one at these low energies, as a result of the suppression of the em response (e/mip is only 0.65 for this detector).

In both calorimeters, the smallest values for the hadronic calorimeter response occur for energies around 1 GeV. A large fraction of the pions develop showers at this energy, and the average em shower fraction $\langle f_{\text{em}} \rangle$ is small, about 10–20%. As the energy increases, so does $\langle f_{\text{em}} \rangle$, and the hadronic response gradually approaches the em response. The Fe/LAr calorimeter is more non-compensating than the Pb/LAr one (e/h = 1.7 vs. 1.3). As a result, its non-linearity for pion showers is clearly worse. Between 5 GeV and 500 GeV, the hadronic response increases by 23% in Fe/LAr, vs. 12% in Pb/LAr.

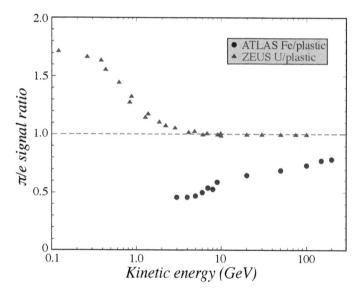

FIG. 3.18. The measured π/e signal ratio as a function of energy over a large energy, for calorimeters based on plastic scintillator as active material and either iron or ^{238}U as absorber. Experimental data from the ATLAS TileCal [Aba 09, Aba 10] and the ZEUS Forward calorimeter [And 90, Beh 90].

All results presented earlier in this subsection were derived on the basis of first principles. Experimental data that may serve as a test for the validity of these results are shown in Figure 3.18. Even though these data were obtained with calorimeters based on plastic scintillator as active material (instead of LAr), and not all data covered the entire energy range of Figure 3.17, the predicted trends are indeed confirmed by these data. Other experimental data in support of the predicted curves are presented in Section 7.5.1 (Figure 7.33).

3.2.8 *Spatial dependence of the hadronic response*

In Section 2.3.4, several experimentally measured hadronic shower profiles are shown, depicting both the longitudinal (Figure 2.34) and the lateral (Figure 2.38) shower development. Since calorimeters are instruments intended to measure energy deposits, it is tempting to interpret these figures as measurements of the (average) *energy deposition profile* in the hadronic shower development. However, this is incorrect.

In the previous subsections, we saw that the hadronic calorimeter signals are caused by a variety of particles generated in the shower development: electrons and positrons in the em shower component; pions, spallation protons and evaporation neutrons in the non-em component. The calorimeter response to these particles varies from species to species, sometimes (*e.g.*, in the case of neutrons) by a considerable factor.

The various types of particles through which the shower energy is deposited in the calorimeter structure also have their own characteristic three-dimensional energy deposition patterns. For example, the electrons and positrons that constitute the em shower component are strongly concentrated near the shower axis, *i.e.*, the path that the particle would have followed in the absence of the absorbing structure. On the other hand, the neutrons behave like a "gas," they propagate more or less isotropically in all directions from the source in which they are produced. They are frequently found at relatively large distances (50 cm or more) from the shower axis.

These phenomena are illustrated in Figures 3.19 and 3.20, which show results from measurements of the spatial distribution of radioactive nuclides produced in the absorption of 591 MeV protons in a stack of 3 mm thick ^{238}U plates, interleaved with 0.5 mm thick iron foils [Ler 86].

In Figure 3.19, longitudinal distributions are given. The crosses show the distribution of ^{48}V, a nucleus that is produced in spallation reactions in the iron foils. This distribution is gradually falling as a function of depth inside this stack. At a depth of $\sim$ 16 cm (about $1.4\lambda_{int}$), the number of ^{48}V nuclei suddenly drops by a factor of seven within 1 cm, to almost zero. This depth corresponds to the range of 591 MeV protons in this ^{238}U/Fe mixture. The Q-values for the reactions in which ^{48}V is produced from ^{56}Fe amount to $\sim$ 80 MeV. Therefore, the spatial distribution of this nuclide in the absorber stack corresponds to the distribution of the particles with energies larger than 80 MeV. These are predominantly the incoming 591 MeV protons, with a few spallation neutrons showing up beyond the range of these protons. Because of the dominance of the primary particles as initiators of the reactions in which ^{48}V is produced, the distribution of this nuclide can be used to derive the interaction length of 591 MeV protons in ^{238}U ($\sim$ 10 cm) and the fraction of these protons that ranged out without causing a nuclear interaction (25%).

Figure 3.19 also shows the distribution of ^{239}Np (the open circles), which has completely different characteristics. This nuclide is produced when thermal neutrons are captured in ^{238}U. The resulting nucleus ^{239}U decays with a half-life of 23.5 minutes to ^{239}Np, which in turn decays to ^{239}Pu ($T_{1/2} = 2.35$ days). This is the main process

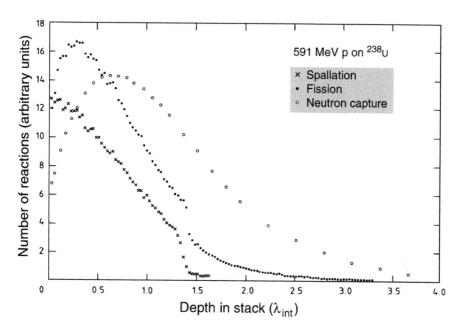

FIG. 3.19. Longitudinal distributions of various radioactive nuclides produced in the absorption of 591 MeV protons in ^{238}U. See text for details. From [Ler 86].

through which plutonium is produced in nuclear reactors. The spatial distribution of ^{239}Np nuclei therefore measures the spatial distribution of thermal neutrons produced in the absorption process of the 591 MeV protons. These neutrons are found in large numbers far beyond the point where these protons range out.

The full dots in Figure 3.19 correspond to ^{103}Ru, a fission product of ^{238}U. Uranium fission may be caused both by the incident protons and by neutrons with energies larger than 1.5 MeV. The distribution of this nuclide shows characteristics of both production mechanisms. The drop in the concentration of this nuclide at a depth of $\sim$ 16 cm was used by the authors to unfold the contributions from protons and fast neutrons to its production. This distribution shows that the neutrons with energies larger than 1.5 MeV have traveled considerably shorter distances in this stack than the thermal ones.

The transverse distributions of these radioactive nuclides make it possible to extend this picture to three dimensions. Figure 3.20 shows radial profiles, measured at a depth of 7 cm. The distribution of ^{239}Np, produced by thermal neutron capture, is clearly much broader than that for nuclear fission products, which can only be produced by neutrons with a kinetic energy of more than 1.5 MeV. This fission distribution is in turn broader than that of ^{48}V, which is almost exclusively produced by the incident protons.

Some shower characteristics revealed by this study also apply to showers initiated by hadrons of much higher energy. Spallation protons of a few hundred MeV are fairly typical shower particles, abundantly produced in the absorption of multi-GeV hadrons (see Section 2.3.2). This study showed, in particular, that the spatial profiles of neu-

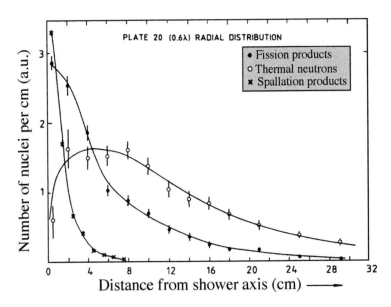

FIG. 3.20. Radial distributions of various radioactive nuclides produced in the absorption of 591 MeV protons in ^{238}U. See text for details. From [Ler 86].

trons and charged shower particles are very different from one another. In addition, the neutron profiles depend strongly on the neutron energy.

In general, neutrons penetrate much deeper into the calorimeter structure, both longitudinally and laterally. Therefore, the calorimeter signals collected from areas far off-axis or at great depth are in practice almost exclusively caused by neutrons. And if the calorimeter response to these neutrons is very different from the response to the particles that were used to set the energy scale (mips or em showers), application of this calibration to the signals from these remote areas leads to a mismeasurement of the energy they represent, sometimes by a considerable factor.

A similar effect may occur very close to the shower axis. Here, the calorimeter signal is primarily due to electrons and positrons produced in the development of the em shower component. To determine the energy represented by the calorimeter signals from this area, calibration constants obtained with em showers thus need to be used. For non-compensating calorimeters, this is in general not done. Calibration constants are rather chosen such as to reproduce, on average, the correct hadronic energy for the entire hadronic shower. As a result of this, the energy contained in the em shower core is mismeasured.

The shower profile characteristics discussed here have one other important calorimetric consequence. In the practice of particle physics experiments, only the signals collected from a well-defined area (usually a cone) around the shower axis are used to determine the energy of the particle or jet that initiated the shower. The relatively wide lateral profiles of the energy deposited by soft shower neutrons, and in particular the energy generated in thermal neutron capture, imply that in practice frequently a considerable fraction of the signals generated by these neutrons is cut.

And if the (thermal) neutron signals survive such geometric cuts, they are likely to be eliminated by constraints on the *duration* of the calorimeter signals. This is discussed in the next subsection.

3.2.9 *Time dependence of the hadronic response*

When discussing the time structure of calorimeter signals, one should distinguish between

1. The time structure of the *shower development process proper.*
   This structure is composed of a collection of δ-functions. Each δ-function represents the passage of an ionizing shower particle through an active calorimeter layer, and is precisely located in time with respect to t_0, *i.e.*, the moment when the incident hadron that initiated the shower entered the calorimeter. The amplitude of each δ-function represents the magnitude of the contribution of that particular shower particle to the total calorimeter signal.
2. *Instrumental effects* that smear this distribution. For example, all scintillators exhibit a characteristic fluorescent decay. The time constant of this process is at least a few ns. This effect gives all δ-functions mentioned above an exponential tail. In calorimeters based on direct collection of ionization charge, the fact that this collection takes time distorts the δ-functions too, in a way that depends on the characteristics of the detector and on the signal handling.

In this subsection the discussion is limited to the first point. Instrumental effects are described in detail in Chapter 5.

Typically, the absorption of hadronic showers takes place in an area with a depth of the order of 1–2 m in calorimeters using dense absorber material. This means that relativistic particles cross the active planes deep inside the calorimeter only a few nanoseconds after the particle that initiated the shower entered the calorimeter. In many cases, the time structure that results from the fact that the shower takes time to develop is narrowed because of the way the calorimeter signals are read out. For example, in scintillation calorimeters, the light is usually collected at the rear end of the detector. This means that light produced late (*i.e.*, deep inside the calorimeter) has to travel a shorter distance to the photomultipliers than light produced in the early stages of the shower development. Remaining time differences between the signals are then only due to the difference between the speed of light in the medium through which the scintillation light is transported ($\approx c/n$) and the speed at which the shower propagates in the calorimeter ($\approx c$).

Tails in the time structure are caused by the non-relativistic shower particles contributing to the calorimeter signals, and in particular the neutrons. In Section 2.3 we saw that the kinetic energy carried by evaporation neutrons may represent some 10% of the energy in the non-electromagnetic shower component. In addition, these neutrons have the potential to release nuclear binding energy when they are captured by calorimeter nuclei. This potential may represent some 30% of the energy in the non-em shower component. However, realizing this potential requires typically a considerable amount of time.

The cross sections for neutron capture are usually only substantial in the thermal regime (kinetic energy in the eV range). The typical timescale for thermalization is of the order of 1 μs, which may be too long for calorimetric particle detection applications. Also, in some materials, *e.g.*, lead, the capture cross section is so small that this process plays no significant role in the calorimeter itself.

The latter statement is definitely not true for calorimeters using ^{238}U as absorber. The cross section for thermal neutron capture in this material is 2.7 b, so that the mean free path of a thermal neutron amounts to only 7.7 cm. Therefore, many of the neutrons produced in hadronic shower development in uranium calorimeters are captured, which leads to the production of ^{239}Pu.

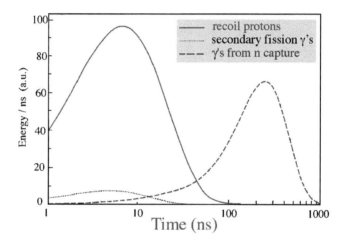

FIG. 3.21. Time structure of various contributions from neutron-induced processes to the hadronic signals of the ZEUS uranium/plastic-scintillator calorimeter [Bru 88].

The timescale of this process is of the order of 1 μs. This is illustrated in Figure 3.21, which shows the time structure of various contributions from neutron-induced processes to the hadronic signals of a uranium/plastic-scintillator calorimeter [Bru 88]. The capture process is considerably delayed with respect to the processes induced by more energetic neutrons, *i.e.*, proton recoil in the plastic scintillator and γs produced in neutron-induced ^{238}U fission. These processes are completed within 100 ns after the start of the shower development (note the logarithmic timescale in Figure 3.21).

The amount of (nuclear binding) energy released in neutron capture may be quite substantial, especially in ^{238}U where, on average, some 60 neutrons are generated for every GeV of hadronic energy deposited in the calorimeter [Ler 86]. This energy is released in the form of γ-rays emitted when the excited compound nucleus (^{239}U*) decays to its ground state.

As a result, the hadronic signals from all uranium calorimeters exhibit significant tails covering a time interval of the order of 1 μs. An example of the consequences of this phenomenon for calorimetry is given in Figure 3.22, taken from the ZEUS experiment that operated from 1992 - 2007 at the electron–proton collider HERA at DESY (Hamburg, Germany). A large ^{238}U/plastic-scintillator calorimeter forms the heart of this experiment [Der 91, Kru 92].

Figure 3.22a shows the ratio of the responses of this calorimeter to 5 GeV/c electrons and pions, as a function of the time over which the calorimeter signals were integrated. When this "gate width" was increased from 70 ns to 600 ns, the e/π signal ratio decreased by about 8%. Since neutron production plays no significant role in the absorption of electron showers, one may conclude from this that the thermal neutrons contributed at least 8% to the signals from 5 GeV/c pions in this calorimeter.

Similar effects were reported by the HELIOS [Ake 85] and WA78 Collaborations [Cat 87], which also operated uranium-based calorimeter systems.

The timescale of the neutron capture process, and the γs that are abundantly produced in this process, is usually too long to be taken advantage of in practice. This is illustrated in Figure 3.22b, which shows another result from the ZEUS experiment [Kru 92]. As the gate width is increased, not only is a larger fraction of the shower signal accumulated, but the "noise" from decaying ^{238}U nuclei is increased as well, in such a way that a net loss in hadronic energy resolution occurs for gate widths beyond ~ 0.2 μs. For electron showers, where all produced scintillation light is collected in a much shorter time (~ 10 ns), the deterioration of the energy resolution becomes already apparent at much shorter gate widths. According to calculations by Brückmann [Bru 88], $\sim 20\%$ of the neutrons produced in ^{238}U are captured within 100 ns, a typical signal integration time for scintillator-based calorimeters.

Another timescale relevant for the neutron contribution to calorimeter signals is the one associated with elastic scattering processes, the dominating (if not the only) mechanism through which neutrons lose their kinetic energy in the keV–MeV range. As pointed out before, this process is particularly interesting for sampling calorimeters with hydrogenous active material. A neutron in this low-energy range loses, on average, 50% of its kinetic energy in a collision with a proton. Therefore, if the time between consecutive collisions is constant, one may expect the contributions from elastically scattering

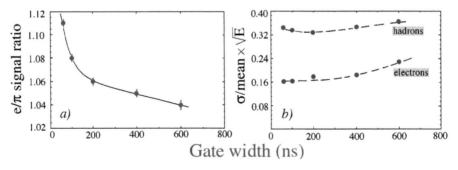

FIG. 3.22. The ratio of the average ZEUS calorimeter signals from 5 GeV/c electrons and pions (a) and the energy resolutions for detecting these particles (b), as a function of the charge integration time [Kru 92].

neutrons to the calorimeter signals to decay exponentially with time. In practice, the velocity of the neutron *decreases* with each collision, but the cross section *increases*. As a result, the mean free path between collisions decreases, but the time between consecutive interactions is, in first approximation, constant.

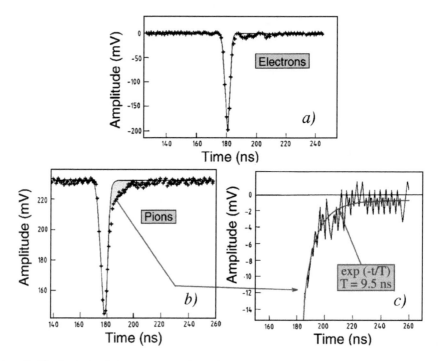

FIG. 3.23. Typical calorimeter signals for 150 GeV electrons (*a*) and pions (*b*) measured with the SPACAL calorimeter. The pion signal exhibits a clear exponential tail with a time constant of ~ 10 ns (*c*). The $t = 0$ point is arbitrary and the bin size is 1 ns. Data from [Aco 91a].

This is illustrated by results obtained by the SPACAL Collaboration [Aco 91a], who constructed and tested an approximately compensating lead/plastic-scintillator calorimeter. This calorimeter consisted of about 20% polystyrene. The cross section for elastic neutron scattering off hydrogen nuclei in this plastic increases from 4.3 b at 1 MeV to 12 b at 100 keV. This translates into a mean free path between elastic collisions with hydrogen nuclei in this structure of ~ 24 cm and ~ 8 cm for 1 MeV and 100 keV neutrons, respectively. The velocities of 1 MeV and 100 keV neutrons amount to $1.4 \cdot 10^9$ cm s^{-1} and $0.44 \cdot 10^9$ cm s^{-1}, so that the *average time* between consecutive elastic neutron–proton collisions is about 17 ns in both cases.

This is in reasonable agreement with the experimental observations by SPACAL. Figure 3.23 shows typical calorimeter signals from 150 GeV electron and pion showers. The latter signal has a clear tail, which is absent for electron showers. After unfolding the (Gaussian) shower shape observed for electron signals, this tail turned out to be well described by an exponentially falling function with a time constant of $\sim$ 10 ns (Figure 3.23c). This tail, whose magnitude (but not the shape) varied strongly from one event to another, is clearly caused by the elastic neutron–proton scattering process discussed here. That the measured time constant is somewhat shorter than the 17 ns estimated above is presumably due to a variety of factors, including energy loss in elastic or inelastic neutron scattering off other nuclei in the calorimeter structure (the remaining 80% of the detector volume consisted of lead, and the plastics contained carbon and oxygen nuclei).

3.2.10 *Material dependence of the hadronic response.*

Many shower particles produced in the absorption of high-energy hadrons are non-relativistic. The nucleons that are released in large numbers in reactions involving the nuclei of which the calorimeter media are composed, have energies ranging from a few MeV (evaporation neutrons) to several hundred MeV (spallation protons).

Non-relativistic charged particles are densely ionizing. The specific ionization, dE/dx, which is typically 1–2 MeV g^{-1} cm^2 for minimum ionizing particles, increases by a factor of about four for 100 MeV protons and by a factor of twenty for 10 MeV protons. The evaporation neutrons may produce charged particles (recoil protons, αs) with even much larger specific ionization values, two to three orders of magnitude larger than for mips.

The contribution of these densely ionizing particles to the calorimeter signals depends crucially on saturation effects that may or may not occur in the active calorimeter medium. Such saturation effects are almost completely absent in gaseous media. A 5 MeV α particle stopping in a gas-filled wire chamber produces a signal that is indeed 1.000 times larger than a mip that loses 5 keV when traversing the same chamber.

The situation is completely different for denser active media, *e.g.*, plastic scintillators or liquid argon. The signals for densely ionizing particles may be considerably reduced due to saturation effects in these materials.

In scintillators, saturation effects are attributed to quenching of the primary excitation by the high density of ionized and excited molecules [Bir 64]. They are usually described by Birks' law:

$$\frac{dL}{dx} = S\frac{dE/dx}{1 + k_B \cdot dE/dx} \tag{3.13}$$

where L is the amount of light produced by a particle of energy E, S a proportionality constant and k_B a material property known as Birks' constant. This constant is typically of the order of 0.01 g cm^{-2} MeV^{-1}, whereas the specific ionization (dE/dx) of a mip is of the order of 1 MeV g^{-1} cm^2. In comparison with mips, the specific light production (photons per unit energy) is thus reduced by a factor of two (eleven) for particles with a specific ionization of 100 (1,000) times the value for minimum ionizing particles. Figure 3.24 shows that this equation gives a reasonable description of experimental data.

Saturation effects of a similar magnitude occur in calorimeters using ionization chamber readout, *e.g.*, liquid argon. In this case, the effects are caused by recombination of electrons and ions along the track of the ionizing particle. This process was first described quantitatively by Onsager [Ons 38]. When an electron originating from

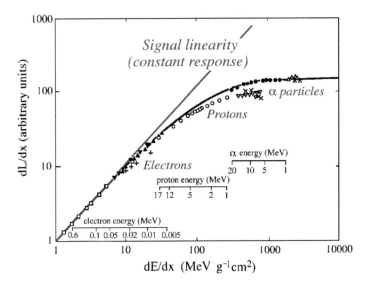

FIG. 3.24. Variation of the specific fluorescence, dL/dx, with the specific ionization loss, dE/dx, in anthracene crystals. The solid curve represents Equation 3.13 with $k_B = 6.6$ mg cm^{-2} MeV^{-1}.

ionization is slowed down to thermal energy and finds itself within a certain distance from its parent ion, recombination of the electron–ion pair is inevitable, according to this theory. If, on the other hand, the thermalization occurs outside that critical sphere, the electron is free from the parent's influence, even in the absence of an external electric field.

An external electric field causes the electron–ion pairs produced in the ionization process to separate. Therefore, saturation decreases when the strength of this field is increased. This is illustrated in Figure 3.25, which shows the saturation characteristics of collected charge vs. electric field for liquid argon and xenon. The experimental points are well described by the curves, based on Onsager's theory, except for very weak electric fields. Typically, the field strength used in LAr calorimeters is of the order of 10 kV cm^{-1}.

It should be emphasized that the data in Figure 3.25 concern *minimum ionizing particles*, conversion electrons of 976 keV produced in the decay of ^{207}Bi. For densely ionizing particles, for example αs produced in neutron-induced reactions, recombination between electrons and ions other than their parent ion also becomes important.

Figure 3.26 shows a comparison between the charge collected from 5.5 MeV αs and mips traversing a liquid-argon gap, as a function of the electric field strength. These data were obtained by the pioneers of liquid-argon calorimetry, Willis and Radeka [Wil 74].

When we compare the open circles and triangles (which were obtained with the same high-purity grade of argon), it becomes clear that the saturation for densely ionizing particles depends even more on the strength of the applied electric field than in the case of mips. The ratio of the electric charge collected for the αs and for mips drops

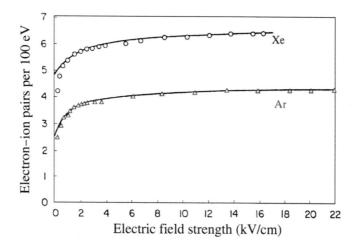

FIG. 3.25. The free-ion yield per 100 eV absorbed energy, as a function of the electric field strength for liquid argon and liquid xenon. The curves are based on Onsager's theory. From [Dok 81].

from about twelve at high field (3 kV mm^{-1}) to five at a field of 0.5 kV mm^{-1}.

But also at high field values, the saturation is substantial. Minimum ionizing particles lose, on average, 0.21 MeV/mm when traversing liquid argon. In the absence of saturation, one would therefore expect to collect 5.5/0.21 = 26 times as much charge from the αs, which deposit their full kinetic energy in 1 mm of argon. However, at high field (3 kV/mm) the charge collected from the αs is only twelve times larger than the charge collected from the βs, which leads to the conclusion that the α signals are suppressed by more than a factor of two, compared with mips. This suppression factor increases to about five when the field strength is decreased to 0.5 kV/mm.

Figure 3.26 also shows another feature of LAr calorimeters. The data points indicated by crosses (αs) and closed circles (mips) were obtained with argon with a higher level of impurities. In particular, oxygen was present at a concentration of $\sim$ 50 ppm. The consequences were dramatic, especially for the α signals, which for low fields even dropped *below* the mip signals. Oxygen is an electronegative substance and electrons strongly tend to attach to oxygen they encounter as they drift towards the anode. The data show that this effect also depends on the electric field strength.

The dependence of the saturation on the strength of the electric field may, to a certain extent, be used to "tune" the response of LAr calorimeters to densely ionizing particles produced in hadronic showers, and thus the e/h value. This is further discussed in Section 3.4.

In scintillation calorimeters, such "tuning" is not available. However, as illustrated by the experimental data shown in this subsection, saturation effects do have large consequences for the contribution of spallation protons and, in particular, shower neutrons to the signals from both types of calorimeters.

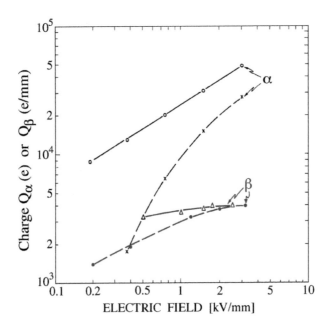

FIG. 3.26. Measured charge as a function of electric field for 5.5 MeV α particles and for
    mips traversing a gap filled with liquid argon of different grades of purity. The open circles
    and triangles were obtained with high-purity (gold-label) LAr, the crosses and closed circles
    were obtained with argon containing a higher concentration ($\sim$ 50 ppm) of oxygen. Data
    from [Wil 74].

3.2.11   *The response to jets*

The statements made in Section 3.1.3 with regard to the detection of jets (in homogeneous calorimeters) also apply to sampling calorimeters. Jets consist of a mixture of γs
(from π^0 decay) and hadrons. The calorimeter signal generated by this mixture is simply the sum of the signals from all particles of which the jet is composed and follows
thus directly from the calorimeter response to these various jet constituents.

   As a result of peculiarities in the fragmentation process, the average electromagnetic
shower fraction, $\langle f_{em} \rangle$, may be somewhat different for jets than for individual hadrons
with the same energy as the jet. Therefore, in non-compensating calorimeters, the jet
response is not necessarily the same as the response to single hadrons. Also, the signal
non-linearity may be somewhat different for single hadrons and jets. However, in compensating calorimeters, where the response to em and non-em energy deposit are the
same, the response to jets is the same as that to electrons and pions of the same energy.

   It should be emphasized that, in the practice of particle physics experiments, jets
are not always clearly identified entities. For example, it is not always clear which of
the many particles observed in complicated interactions do or do not originate from
the fragmenting (di-)quark whose four-vector one wants to measure. Particles not associated with the fragmenting object may happen to travel in approximately the same

direction as this object. And on the other hand, especially for jets produced in collider experiments at large angles with the beam axis (small η values), many soft particles may travel at large angles with respect to the direction of the fragmenting object, because of the transverse momentum (typically 0.3 GeV/c) acquired in the fragmentation process.

These problems are dealt with by means of jet-defining algorithms. In one approach, all energy deposited within a cone in the (η, ϕ) space is attributed to the fragmenting object, *e.g.*, $R = \sqrt{(\Delta\eta)^2 + (\Delta\phi)^2} = 0.3$. In this approach, one aims to choose the size R of this cone such that the energy carried by particles from the "underlying event" that happen to be included in this cone is, on average, offset by jet particles excluded by this algorithm.

3.3 Linearity

A calorimeter is said to be linear if its average signal is proportional to the energy of the particles it detects, *i.e.*, if its response to these particles is constant. Linearity is a very important property for calorimeters, since it is a crucial ingredient for the precision with which the energy of an unknown object that produces signals in the calorimeter can be measured. In this section, we discuss signal linearity, and the causes and implications of non-linearity. Signal non-linearities for em and for hadronic showers are fundamentally quite different. In the previous section, we saw that signal non-linearity for hadron detection is a natural consequence of non-compensation ($e/h \neq 1$). On the other hand, signal non-linearity for em shower detection is usually preventable, and its effects are typically non-recoverable.

3.3.1 *Non-linearity for hadron shower detection*

As we saw in Section 3.1.4, hadronic signal non-linearity is an intrinsic property of all non-compensating calorimeters. It is an inevitable consequence of the energy dependence of the average em fraction of hadron showers. Since $\langle f_{em} \rangle$ increases with the hadron energy, so does the hadronic response of calorimeters with $e/h > 1$ (Equation 3.4). Compensating calorimeters ($e/h = 1$) are not subject to signal non-linearities deriving from the energy dependence of $\langle f_{em} \rangle$. However, they do exhibit a non-linearity at low energies when $e/mip \neq 1$, and an increasing fraction of hadrons range out by ionizing the absorber medium rather than undergoing a nuclear interaction. In the compensating ZEUS calorimeter, the hadronic response gradually increased by more than 50% at energies below 5 GeV as a result of this effect (Figure 3.11).

In longitudinally segmented calorimeters, *i.e.*, calorimeters consisting of different segments in depth, each of which produces a signal, the hadronic signal non-linearity depends sensitively on the intercalibration of these segments. This is discussed in detail in Section 6.2. The non-linearity can also be affected by weighting schemes, in which the signals from different calorimeter segments are multiplied by a weight factor which depends on the event topology, *i.e.*, the measured energy deposit pattern in the calorimeter (Section 6.4).

All the mentioned signal non-linearities for hadron detection are the result of the physics of the shower development process, they do not depend on peculiarities of the calorimeter signals. For that reason, hadronic signal non-linearity does in general not preclude an (on average) correct measurement of the energy of the showering particle on the basis of the observed signals. This is not necessarily true for all the non-linearities that may affect electromagnetic shower detection.

3.3.2   *Non-linearity for electromagnetic shower detection*

For particles that develop em showers, linearity is not only an important property, but also a very fundamental one. Unlike hadrons, these objects transfer their entire energy to the shower particles (electrons and positrons) that constitute the calorimeter signals: a 20 GeV electron generates, on average, twice as many ion/electron pairs or scintillation photons as a 10 GeV electron. Therefore, the calorimeter signal for 20 GeV electrons should be, on average, twice as large as that for 10 GeV electrons (or photons). Intrinsic linearity for em shower detection is thus a very fundamental calorimetric property, not only for homogeneous detectors, but for *all types* of calorimeters. An intrinsically linear calorimeter is in principle very easy to use. Its response serves as the *calibration constant, i.e.*, if the calorimeter is calibrated with 10 GeV electrons, then a particle that produces a signal that is 2.6 times as large as the average signal from 10 GeV electrons should be assigned an energy of 26 GeV. No ifs, ands or buts. It should not depend on the impact point of that particle, or whether it is an electron, positron, γ, π^0, or electromagnetically decaying kaon ($K^0 \rightarrow \pi^0\pi^0 \rightarrow \gamma\gamma\gamma\gamma$) .

However, in practice, deviations from signal linearity may be observed, both in homogeneous and in sampling calorimeters. Signal non-linearity for em showers is either caused by *instrumental effects*, or by *miscalibration*. Depending on the origin of the effect, deviations from signal linearity in electromagnetic calorimeters may either be characterized by a *decreasing* response at increasing energies, or by an *increasing* one. In the next subsections, the different effects that may lead to signal non-linearity are described.

3.3.2.1   *Saturation effects in PMTs.*   In many calorimeters that are based on detecting scintillation light, photomultipliers are used to convert that light into electric signals. These PMTs have a dynamic range over which they are linear, *i.e.*, the integrated charge of the pulses they produce is proportional to the number of scintillation photons. The relationship between that dynamic range and the energy deposited in the calorimeter tower read out by the PMT depends on the high voltage that is used to operate the PMT (See Section 5.4.1). At the high end of the dynamic range, the PMT saturates, which means that adding more light does not lead to more charge. Therefore, the relationship between the energy deposited in the calorimeter and the resulting PMT signal is no longer proportional for signals beyond that maximum. This is illustrated in Figure 3.27, which shows the calorimeter signal as a function of the energy deposited in the calorimeter. Two situations are distinguished, depending on the sharing of the shower energy among different calorimeter cells (towers). If the shower energy is entirely deposited in one calorimeter cell, then the PMT that collects the light from that cell produces a constant signal beyond the maximum light level it can handle as intended, and the result is a constant signal beyond a certain shower energy, 50 GeV in this example (the triangles in Figure 3.27). If the shower energy is shared among several calorimeter cells, only one cell may saturate, typically the one in which the shower axis is located. In that case,

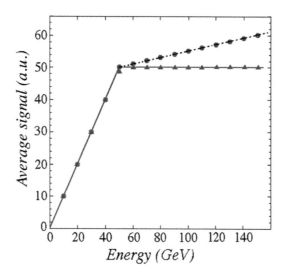

FIG. 3.27. The signal from a scintillation calorimeter in which PMTs are used to convert the light pulses into electric signals, as a function of energy. The shower energy is either entirely deposited in one calorimeter tower (the triangles), or shared between several towers (the circles).

that particular cell produces the same signals for every shower beyond a certain energy, but the other cells will continue to produce signals that are proportional to the fraction of the total energy deposited in them. As a result, the total calorimeter signal, which is the sum of the signals from all participating calorimeter cells, will continue to increase beyond the saturation point of the central cell, albeit more slowly, depending on the fraction of the energy deposited in the saturated cell. This situation is represented by the full circles in Figure 3.27. In Section 5.4.1, ways to deal in practice with the limited dynamic range of PMTs are discussed.

Whereas the effect described above is a straightforward consequence of the high-voltage setting, there may also be more subtle saturation phenomena affecting a PMT based calorimeter. The electric current created in the PMT as a result of the detection of scintillation photons has the effect of lowering the potential differences between the dynodes, in particular in the last amplification stage(s) where this current is largest. As a result, the amplification factor, which is very sensitive to these potential differences, decreases. This effect increases with the current, and thus with the light signal and thus with the shower energy. The result is a response that decreases at increasing energy. This effect is illustrated in Figure 3.28. Figure 3.28a shows a calorimeter response that decreases by about 12% over the energy range from 10 to 150 GeV. This calorimeter thus exhibits a 12% non-linearity for em shower detection in this energy range. Clearly, such effects are extremely undesirable and should be avoided, as much as possible. One way to reduce this effect is to lower the PMT gain. In that way, the current for a given energy deposit is reduced correspondingly. One may also stabilize the voltage

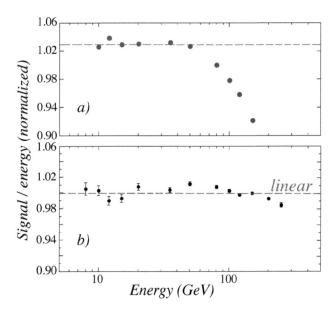

FIG. 3.28. The em calorimeter response as a function of energy, measured with the QFCAL calorimeter, before (*a*) and after (*b*) precautions were taken against PMT saturation effects. Data from [Akc 97].

differences between the PMT dynodes, in particular between the last dynode and the anode, by means of separate power supplies. In that way, the PMT amplification is kept constant, independent of the light level (see also Section 5.4.1). After precautions of this type had been taken, the response of the calorimeter from Figure 3.28 became constant to within $\sim 1\%$ over the energy range 8–250 GeV (Figure 3.28b).

3.3.2.2 *Saturation effects resulting from particle density.* Such effects may play a role in calorimeters in which the ionization charge is collected by gaseous detectors operating in a "digital" mode, such as Geiger counters or streamer chambers. Such detectors are intrinsically non-linear, since each charged particle creates an insensitive region along the struck wire, which prevents other, nearby shower particles from being registered [Engl 83]. And since the density of shower particles increases with energy, the response of such detectors decreases with increasing energy. An example of this effect is shown in Figure 3.29, which contains results obtained in beam tests of a calorimeter read out by means of wire chambers operating in a "saturated avalanche" or "limited streamer" mode. Figure 3.29a shows the average total signal of this calorimeter, as a function of the total deposited electromagnetic energy. The experimental data, located on the dashed line, clearly illustrate the non-linear character of this calorimeter. It should be mentioned that the energy was varied in this experiment by depositing n electrons of 17.5 GeV *simultaneously* in the calorimeter ($n = 1$–10). Therefore, the energy deposit profile was not energy dependent in these measurements. The calorimeter was longitudinally subdivided into five sections. In Figure 3.29b, the average

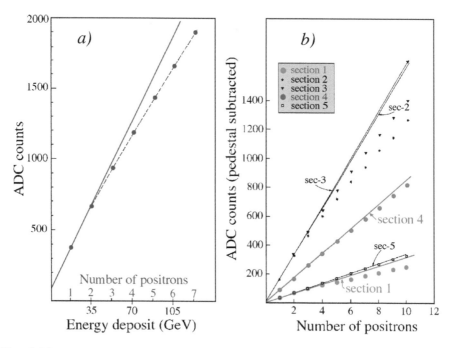

FIG. 3.29. Average em shower signal from a calorimeter read out with gas chambers operating in a "saturated avalanche" mode, as a function of energy [Ata 83].

signals recorded in these sections are shown as a function of the total energy deposited in the calorimeter. This figure illustrates that the observed saturation effect is clearly determined by the *density* of shower particles. The degree of non-linearity is given by the deviation from the straight line. In sections 1 through 5, this non-linearity amounted to 14.5%, 14.8%, 9.3%, 2.4% and 0.5%, respectively, for a total deposited energy of 105 GeV ($n = 6$). Thus, the non-linearity in section 1 was more than a factor of six larger than in section 4, in spite of the fact that the energy deposit in section 1 was less than half of that in section 4. However, the *particle density* in section 1, in the early stage of the shower development, was of course considerably larger than in section 4, located beyond the shower maximum (see also Figure 2.13), and that is what governs this type of non-linearity. A modern example of signal non-linearity caused by a phenomenon of this type is discussed in Section 9.3.3.

Saturation effects resulting from particle density are not limited to calorimeters with gaseous readout. They also play a role in silicon photomultipliers (SiPM), which are devices consisting of a dense array of electrically and optically isolated photodiodes operating in the Geiger mode (pixels). Typically, SiPMs contain 100 to 1,000 of these pixels per mm² (see Section 5.4.1 for more details). If this device is hit by a brief light pulse, *e.g.*, produced by a scintillating crystal, each pixel acts as a digital detector. It responds with a "1" or a "0." Regardless of the number of photons hitting it at the same time, it will produce a signal of "1" photoelectron. The total SiPM signal represents

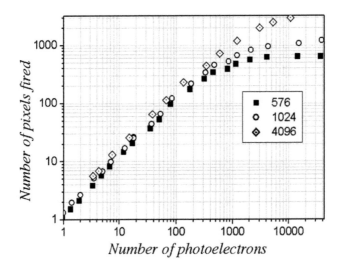

FIG. 3.30. The signal from three different SiPMs, consisting of 576, 1024 and 4096 pixels, as
   a function of the intensity of the light pulse [Ott 06].

the total number of pixels that "fired" (produced a "1") in response to the light pulse.
Depending on the number of photons in the light pulse and the number of pixels in the
SiPM, saturation effects occur, as illustrated in Figure 3.30.

3.3.2.3   *Shower leakage effects.*   As the energy increases, the detector volume needed
to contain the showers increases as well. If the calorimeter signals are collected from
a given detector volume, determined for example by the size of the scintillating crystals, then the fraction of the shower energy contained in that volume decreases with
increasing energy. This results in a response that decreases with increasing energy.

3.3.2.4   *Recombination of ions and electrons into atoms inside the active medium.*   If
this happens, then the ionization charge goes undetected. This process becomes more
likely as the density of ionized atoms increases. This is a well-known phenomenon,
especially in liquid media operating in the ionization-chamber mode. A similar phenomenon plays a role in scintillators, where the primary excitations may be quenched
in an environment with a high density of ionized and excited molecules. These effects
manifest themselves in the form of a reduced response for densely ionizing particles,
such as αs. They may also play a role in the core of em showers. As the shower energy is increased, the density of ionizing particles in the shower core increases as well,
and recombination increasingly reduces the em calorimeter response. These effects are
discussed in more detail in Section 3.2.10.

3.3.2.5   *Instrumental effects that increase the response with energy.*   Whereas all effects discussed so far are resulting in a decrease of the calorimeter response at increasing
energy, there are also effects that may work in the opposite direction and thus cause the
calorimeter response to *increase* with energy. We mention two examples.

1. *Light attenuation.* If the scintillation light produced by the shower particles is strongly attenuated on its way to the light detector (*e.g.*, a PMT), then the energy dependence of shower profiles will cause the fraction of light lost in this process to be energy-dependent as well. If the PMT is mounted at the downstream end of the scintillating crystals, then the light produced in high-energy showers travels, on average, a shorter distance and is thus less attenuated. The result is a response that *increases* with energy. It typically manifests itself as a high-end asymmetric tail in the response function (Figure 3.31).

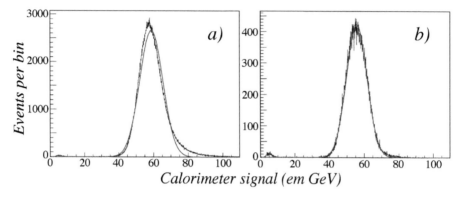

FIG. 3.31. Signal distributions for 60 GeV π^- in a RD52 calorimeter module. Shown are distributions for showers that start deep inside the calorimeter (*a*) and events in which the showers start close to the front face (*b*), where light attenuation effects play much less of a role. Also shown are Gaussian fits to these distributions.

2. A similar effect may occur if the crystals are read out with silicon PIN photodiodes. When these light-detecting devices are traversed by a charged particle (*e.g.*, an escaping shower electron), they produce a signal that may be orders of magnitude larger than the signal generated for a scintillation photon. When the showers are incompletely contained in detectors read out at the downstream end by PIN diodes, the number of escaping shower particles that generate anomalous signals grows disproportionally with the energy. The result is a response that increases with energy. This effect was observed experimentally by the CMS Collaboration in beam tests of $PbWO_4$ crystals. Figure 3.32a shows the signal distribution for 80 GeV electrons measured with 18 cm long crystals equipped with silicon PIN diodes [Pei 96]. The high-energy tail in this distribution was caused by the mentioned effect. This is illustrated by the *absence* of such a tail in the signal distribution shown in Figure 3.32b, where similar crystals were read out with PMTs, which do not exhibit the described phenomena. The small asymmetry that is visible at the *low-energy side* of the distribution in Figure 3.32b is a consequence of small energy losses in upstream material and in shower leakage. At low energies, the high-side tail which is so prominent in Figure 3.32a was practically absent in the photodiode measurements. This tail had the effect of

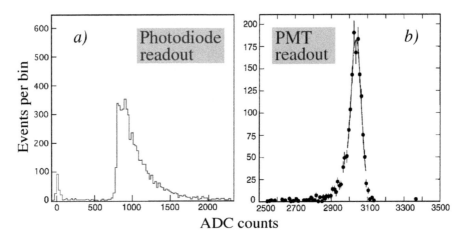

FIG. 3.32. Signal distributions for high-energy electron showers measured with a prototype PbWO$_4$ crystal calorimeter. The calorimeter was read out either with silicon photodiodes (*a*) or with photomultiplier tubes (*b*). Data from [Pei 96].

shifting the mean value of the signal distribution to higher values and thus caused the calorimeter response to increase with energy. The figure illustrates that this effect not only affected the linearity, but it also spoiled the energy resolution of the detector considerably.

In subsequent studies, the CMS Collaboration found that these tails can also be avoided if *Avalanche Photo Diodes* are used to collect and convert the scintillation light produced by high-energy em showers [Ale 97, Auf 98]. However, as discussed in Section 9.1.3, the described phenomena came back to haunt them when detecting hadron showers with these crystals.

3.3.2.6 *Non-linearities resulting from miscalibration.*   An important theme in this book is how to deal with the signals from longitudinally segmented calorimeters. This is not only an issue for hadron showers, which are typically detected in a calorimeter consisting of at least two longitudinal sections (usually called "ECAL" and "HCAL"), but also in em calorimeters that consist of several longitudinal segments. An example is the ATLAS liquid-argon calorimeter, which consists of three longitudinal sections [Aha 06]. Different methods are being used to intercalibrate the signals from these segments. As is shown in Section 6.2, most schemes used for this purpose lead to signal non-linearities. One of the consequences of this is that the reconstructed energy depends on the type of particle that is being detected (Figure 3.33).

3.3.3 *Presenting results from linearity studies*

As illustrated by Figures 3.28 and 3.29, there are two different methods of presenting calorimeter (non-)linearity data. One can either show the *average calorimeter signal* as a function of energy, as in Figure 3.29, or one can show the *calorimeter response* as a function of energy, as in Figure 3.28. Clearly, the latter method is much more sensitive

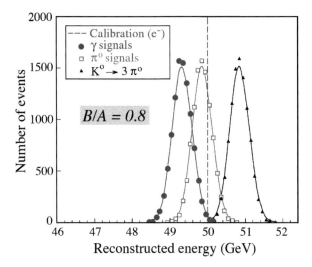

FIG. 3.33. Signal distributions for γs and various hadrons decaying into all-γ final states. All
particles have the same nominal energy and the detector, which consists of two longitudinal
segments and has an intrinsic resolution of 0.5% for em showers of this energy, was calibrated
with electrons such as to optimize the energy resolution for these particles [Wig 02].

to small non-linearities, since it shows the *residuals* of a straight line fit, instead of the
result of the fit itself. Since em calorimeters can be and should be linear at the 1% level,
only results displayed according to the latter method provide in practice meaningful
information.

It is also important to point out that signal linearity implies that the calorimeter
signals are proportional to the deposited energy. In other words, a straight line fit of the
experimental data points relating the measured signal and the deposited energy has to go
through the origin of the plot (zero signal for zero deposited energy). The mere fact that
such data points can be described with a straight-line fit therefore does *not* necessarily
imply that the calorimeter is linear. Figure 3.34 shows an example of a calorimeter
that was mistakenly claimed to be linear because of a situation like this. Even though
the data points were very well described by a straight line, this calorimeter exhibited
actually a 5% non-linearity over the energy range displayed here. More details about
this particular case are given in Section 9.3.5.

3.4   Compensation

The term compensation has already been mentioned several times in previous sections.
We now have all the elements in hand to investigate what is needed for compensation
and how it can be achieved in practice.

3.4.1   *The history of compensation*

The scientific literature is testimony to the fact that the differences between calorimetric
detection of electromagnetic and hadronic showers were initially completely misunder-

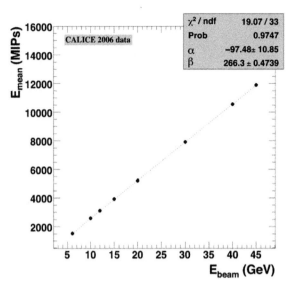

FIG. 3.34. The average signal as a function of energy for positrons detected in a W/Si electromagnetic calorimeter. Data from [Adl 09].

stood. After the first successful use of NaI(Tl) crystals for detecting GeV-type electrons and γs by Hofstadter and his colleagues at SLAC in the late 1960s [Hof 69], it was believed that hadrons could be detected with the same type of precision, provided the detectors were made sufficiently large. To test this idea, the researchers collected as many NaI(Tl) crystals as they could get their hands on and assembled these to a calorimeter with a total mass of about 450 kg (0.12 m^3). They exposed this device to beams of pions of different energies, ranging from 4–16 GeV. Figure 3.35 shows the signal distribution for 8 GeV π^- measured with this detector [Hug 69]. They interpreted the observation that the signals from the pions were, on average, only half as large as those from the electrons as evidence that about half of the energy leaked out of this detector. The fact that Monte Carlo simulations indicated that this leakage was actually much smaller was blamed on a flaw in the simulations. The results obtained at other energies (4, 12 and 16 GeV) were very similar to those shown in the figure: pion signals that were typically only half as large as those of electrons of the same energy, an asymmetric hadronic response function, and a hadronic energy resolution (σ/E) that was approximately *independent* of the energy. Their conclusion from these observations was that the calorimeter was too small to perform well for hadron detection. However, tests of a huge, 60-ton homogeneous calorimeter consisting of liquid scintillator built a few years later at Fermilab revealed that shower leakage was not the dominating problem for the poor hadronic performance. Tests of that detector showed that pion signals were also substantially smaller than electron ones for the same energy, and that the hadronic energy resolution essentially did not improve with increasing energy (Figure 7.11).

We now know that the basic reason for the different hadronic and em calorimeter

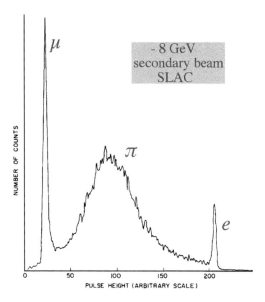

FIG. 3.35. Pulse height distribution for 8 GeV negatively charged particles from a secondary
SLAC beam in a 450 kg array of NaI(Tl) crystals measuring $30 \times 30 \times 120$ cm^3. The momentum resolution of the beam was $\sim 0.8\%$ [Hug 69].

responses lies in the fact that in the absorption of hadronic showers, a significant fraction
of the energy carried by the showering particle is *invisible*, *i.e.*, it does not contribute
to the calorimeter signal. The main source of invisible energy is the energy used to
release nucleons from nuclei, including the nuclear recoil energy. Additional, smaller
contributions come from neutrinos and muons (mainly from π and K decay in flight)
escaping the detector, and from signal saturation for densely ionizing particles.

Non-compensation gives hadron calorimeters a number of undesirable characteristics. In this chapter, we focus on the signal non-linearity caused by it. Degradation of the
energy resolution, discussed in Chapter 4, is another consequence of non-compensation.

The term *compensation* has its origin in the first attempts to circumvent the mentioned problems. By selectively increasing the energy deposited in the calorimeter by
showering hadrons, one would *compensate* for the invisible energy and thus reduce or,
ideally, eliminate the differences between the average calorimeter signals for electrons
and hadrons of the same energy. The mechanism through which this was first believed
to be possible was *nuclear fission* [Fab 77]. By using depleted uranium (^{238}U) absorber
plates, the fission processes induced in the non-em part of the shower development
would contribute extra energy, mainly in the form of nuclear γs and soft (evaporation)
neutrons.

Fabjan and Willis, the proponents of this clever idea, went ahead and successfully
built the first uranium calorimeter [Fab 77]. In this calorimeter, two hundred and fifty
^{238}U plates (with a thickness of 1.7 mm each) were immersed in liquid argon, leaving

2.0 mm wide gaps between the plates. This device was exposed to electrons, pions and protons with energies up to 10 GeV. In order to study the effects of uranium, they also conducted comparative studies in which the uranium plates were replaced by iron ones in the same structure.

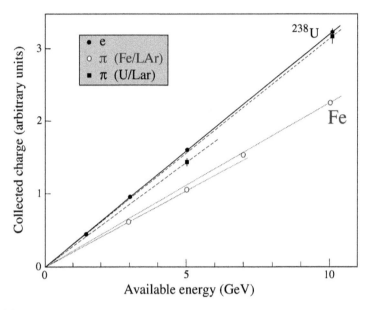

FIG. 3.36. The average charge collected from hadrons and electrons, as a function of energy, in liquid-argon calorimeters with iron or depleted uranium as absorber material. The charge is measured in arbitrary units, which are different for the two calorimeters, but which result in the same response for electrons. Data from [Fab 77].

Some results of these studies are shown in Figure 3.36, where the average collected charge is plotted as a function of the particle energy, for both the Fe/LAr and the ^{238}U/LAr structures. The charge is expressed in arbitrary units, which are different for both calorimeters, but which gave the same results for the electron measurements. The electron measurements thus provided the normalization and made it possible to compare the hadron results for the two different absorber materials. It is clear that the hadronic signals in the uranium structure were much larger than their counterparts in iron. In fact, they were almost as large as the signals for electrons of the same energy, indicating that compensation was almost achieved.

In this type of figure, the calorimeter response as we have defined it is given by the *slope* of the line between the origin and the experimental points. Since the Fe/LAr calorimeter is strongly undercompensating, its hadronic response increases with energy: the slope of the dotted line through the low-energy point is clearly smaller than the slope of the dotted line through the high-energy point. Also quantitatively, this non-linearity is in good agreement with the result derived in Section 3.2.7 (Figure 3.17). The uranium detector also exhibited a small residual non-linearity, as indicated by the dashed lines in Figure 3.36, which suggested that compensation was not completely achieved ($e/h \sim$ 1.1–1.2).

Inspired by this result, the authors designed and constructed a very large uranium-based calorimeter system for the R807 experiment at CERN's Intersecting Storage Rings. This detector used scintillating plastic as active medium. Three mm thick ^{238}U plates were interleaved with 2.5 mm PMMA plastic-scintillator plates in this 4π detector, which comprised some 300 tons of material (mainly uranium). Beam tests of the detector modules, over the energy range 4–200 GeV, revealed that this detector was indeed very linear for pion detection, that its e/h value was very close to 1.0, and that its hadronic energy resolution was considerably better than that achieved with calorimeters based on non-fissionable absorber material [Ake 85, Ake 87].

Following these initial successes, the demand for depleted uranium in the particle physics community soared. At least half a dozen large detector systems based on uranium absorber were planned and/or built, e.g., for the ZEUS and D0 experiments. However, most of these uranium detectors did not meet the excellent performance demonstrated with the device of Fabjan and Willis. The LAr-based uranium calorimeters found the hadronic response to be significantly smaller than the em one. Also, the hadronic energy resolution was considerably worse than expected in these calorimeters [Dub 86].

Another calorimeter, built for the WA78 experiment at CERN, consisted of 1.0 cm thick uranium plates, interleaved with 5 mm thick plastic scintillator [Dev 86, Cat 87]. This device turned out to be strongly *overcompensating*: the signals from pions were *larger* than those from electrons at the same energy. Consistent with this, the hadronic response was found to *decrease* with increasing energy, contrary to the increase found for undercompensating calorimeters (see Figure 3.12).

All these apparently contradictory results were satisfactorily explained in 1987 [Wig 87, Bru 88]. At the same time, it was predicted that compensation could also be achieved with calorimeters using absorber material other than uranium, in particular lead [Wig 87], provided that the right active material was used in the right proportion. Shortly afterwards, this explicit prediction was experimentally tested and found to be correct [Bern 87].

The key to understanding compensation mechanisms lies in the realization that the response of a sampling calorimeter to the various categories of shower particles may be (very) different. The origins of these response differences were extensively discussed in Section 3.2. Compensation is based on the exploitation of these differences, so that equalization of the response to the em and non-em shower components is achieved. There are a number of different ways in which this can be realized. Adding extra (fission) energy to the non-em shower component is neither a necessary, nor a sufficient ingredient.

3.4.2   *The e/h ratio*

In Section 3.1.4, the e/h ratio was introduced as a parameter describing the degree of non-compensation in calorimeters. A calorimeter's e/h value cannot be measured directly. However, it can be derived from experimental measurements of e/π signal ratios, preferably spanning a large energy range. The e/h value is then given by (see Equation 3.5)

$$e/h = \frac{1 - \langle f_{em}(E) \rangle}{\pi/e(E) - \langle f_{em}(E) \rangle} \qquad (3.14)$$

and can be calculated based on the assumed values of the average em energy fraction, $\langle f_{em}(E) \rangle$, in the pion showers. In the following, Equation 2.19 has been used to determine "experimental" e/h values from published e/π data, with parameter values $k = 0.82$ and $E_0 = 0.7$ GeV for iron and copper and 1.3 GeV for lead and uranium. These parameter values were chosen since they give a reasonable description of experimental $\langle f_{em} \rangle$ data (see Figure 2.25).

The calorimeter's e/h value may be defined in terms of the different shower particles that contribute to its em and non-em signals (Equation 3.9):

$$\frac{e}{h} = \frac{e/mip}{f_{rel} \cdot rel/mip + f_p \cdot p/mip + f_n \cdot n/mip}$$

in which f_{rel}, f_p and f_n stand for the average fractions of the energy in the non-em shower component carried by relativistic charged particles, spallation protons and evaporation neutrons, respectively. The calorimeter responses to the various shower particles, normalized to the response to mips, are indicated by e/mip (for electrons and positrons from the em shower component), rel/mip (relativistic charged particles in the non-em component), p/mip (spallation protons) and n/mip (evaporation neutrons).

In a compensating calorimeter, $e/h = 1$. Undercompensating and overcompensating devices are characterized by e/h values larger and smaller than 1, respectively.

In homogeneous calorimeters, the responses to the particles through which the energy from the non-em shower component is deposited are at best equal to the response to the em component. And because of the invisible energy ($f_{rel} + f_p + f_n < 1$), compensation can *never* be achieved in such devices. *All homogeneous calorimeters are undercompensating, with e/h values considerably larger than 1.*

On the other hand, in sampling calorimeters the possibility exists to choose or tune the parameters in Equation 3.9 in such a way that the compensation condition, $e/h = 1$, is achieved. Let us now examine which of the seven parameters can be used for that purpose.

Once the active and passive calorimeter materials have been chosen, the values of f_{rel}, f_p and f_n are fixed. Since the relativistic charged hadrons that traverse the active layers contribute in the same way to the calorimeter signals as do mips, the value of $rel/mip = 1$. Therefore, the only parameters through which the e/h value can be modified are e/mip, p/mip and n/mip.

Because of the invisible energy problem, calorimeters tend to be undercompensating. The e/h value can be brought down either by reducing the em response (e/mip) or by increasing the non-em response (n/mip, p/mip).

3.4.3 *Methods to achieve compensation*

3.4.3.1 *Reducing the em response.* The most effective way to reduce the em response of a sampling calorimeter is to choose high-Z absorber material. In Section 3.2.1, it was shown that the e/mip value of calorimeters using lead or uranium absorber is typically 0.6–0.7. This means that if the calorimeter responded to the particles from the non-em component in the same way as to mips, the suppression of the em response would compensate for 30–40% of invisible energy.

In 1985, Brau and Gabriel [Brau 85] were the first ones to point out that this effect might be an important contributor to the compensation phenomena observed in Willis' uranium calorimeter.

As we saw in Section 3.2.1, the suppression of the em response in sampling calorimeters with high-Z absorber material is caused by the peculiarities in the energy deposition mechanism of the soft-photon component of em showers. Because of the dominating contribution of the photoelectric effect to the cross section, photons below 1 MeV will almost exclusively interact in the passive layers of high-Z sampling calorimeters. The photoelectron created in such a process only contributes to the calorimeter signal if the interaction takes place very close to the boundary layer, so that the electron can escape into the active material.

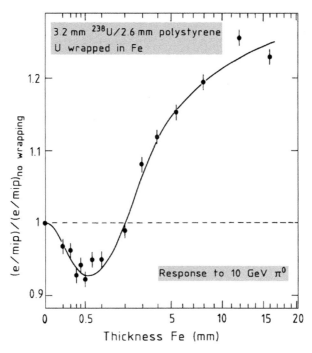

FIG. 3.37. The e/mip response ratio of a uranium/plastic-scintillator calorimeter, in which the 3.2 mm thick uranium plates and the 2.6 mm thick scintillator plates are separated by iron foils of varying thickness. Results of EGS4 Monte Carlo simulations [Wig 88].

One may further enhance the suppression of the em response in high-Z sampling calorimeters by shielding the active layers by thin sheets of passive low-Z material. This is illustrated in Figure 3.37, which shows the e/mip response ratio for a 3 mm ^{238}U/2.5 mm polystyrene structure, with thin sheets of iron inserted between the active and passive layers. The e/mip value is given as a function of the thickness of these foils.

The e/mip ratio reaches a minimum value for iron sheets with a thickness of $\sim$ 500 μm. At that point, the e/mip value is about 8% lower than in the absence of iron foils. Since the iron foils, used in this way, selectively suppress the em calorimeter response, one may also expect a similar reduction in the e/h value.

A thickness of 500 μm of iron corresponds to the range of electrons with an energy of $\sim$ 700 keV. That is approximately the energy at which the cross section for photoelectric effect starts dominating the one for Compton scattering (Figure 2.7). Photoelectrons are thus practically prohibited from contributing to the calorimeter signals by inserting these foils.

At the e/mip minimum, iron represents only a few percent of the total absorber mass. However, when the thickness of the foils is further increased, iron becomes gradually a more sizable component of the calorimeter structure. If iron were the only absorber material, the calorimeter would have an e/mip value around 0.9. Therefore, as the fraction of iron is increased, the e/mip value increases as well.

The results shown in Figure 3.37 concern simulations that were done for the ZEUS experiment. Because of environmental concerns, ZEUS was required to wrap its uranium plates in stainless steel. The simulations show that this requirement had significant effects on the e/h ratio and that the iron thickness was an important design parameter for the performance of the calorimeter.

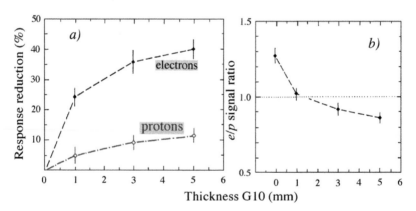

FIG. 3.38. The relative reduction of the response of a uranium/silicon calorimeter that occurs when G10 plates of a certain thickness are inserted between the high-Z absorber material and the active layers (a) and the change in the electron/proton signal ratio (b), as a function of the thickness of the G10 plates. Experimental data from the SICAPO Collaboration [Ang 90].

Experimental evidence for the sensitivity of the e/mip response ratio was reported by the SICAPO Collaboration [Bor 89, Lem 89, Ang 90], which tested generic proto-types for a uranium calorimeter using silicon as active material. The standard support structure for the silicon detectors consisted of two G10 plates, 1.0 and 0.2 mm thick, which sandwiched the active layers. G10 is a type of epoxy that is used as the base ma-terial for printed-circuit boards. The authors studied the signals from this calorimeter not only for this standard configuration, but also for configurations in which *additional* G10 plates were mounted on both sides of the silicon detector planes. These additional plates were 1.0, 3.0 or 5.0 mm thick.

It was found that the compensation characteristics of this instrument clearly de-pended on the thickness of the (low-Z) G10 material, inserted between the high-Z ab-sorber (^{238}U) and the active layers. Figure 3.38a shows the average reduction in the signals from (8–12 GeV) electron and proton showers in this calorimeter that occurred when the mentioned G10 plates were inserted. The signals from electron showers turned out to be much more affected than those from hadron showers [Ang 90]. In additional studies, it was demonstrated that this difference was entirely due to a selective reduction of the em calorimeter response, *i.e.*, the e/mip ratio [Lem 89].

As a result of this effect, the e/h value may be tuned by choosing the thickness of the G10 plates appropriately. This effect is illustrated in Figure 3.38b, which shows the electron/proton signal ratio as a function of the thickness of the G10 layers. How-ever, a word of caution is in place for what concerns the *range* over which e/h can be tuned. Figure 3.38b suggests a large effect, according to the authors the e/p signal ratio changed by $34 \pm 2\%$ when 5 mm G10 were inserted between the uranium and sil-icon layers. However, the detector with which these studies were performed consisted of uranium plates with a thickness of 25 mm (8 X_0). Therefore, the em showers had developed well beyond their maximum in the very first absorber plate, and the silicon layers thus only probed the tail of the em showers. This has two relevant consequences.

1. First, the e/p ratios are *underestimated*. This is easy to see, since in the extreme case of very thick absorber layers, the e/p ratio would become zero.

2. Second, the effects of adding G10 are overestimated. We saw before that these effects are caused by a selective reduction in the sampling fraction of γs with en-ergies below 1 MeV that are absorbed through the photoelectric effect. Such γs are predominantly present in the tails of the em shower distribution, beyond the shower maximum. That is precisely the region that was probed by this calorime-ter. By assuming that this region is representative for the em shower as a whole, the mentioned effects are thus overestimated.

The SICAPO Collaboration used the same silicon structure to collect signals from a fine-sampling electromagnetic calorimeter, with a total depth of $20X_0$ [Lem 89]. The reduction of the e/mip value was in that case determined by comparing the measured signals from em showers with the calculated sampling fraction for mips. It turned out that the e/mip value was reduced by 20% when 5 mm thick G10 plates were added to the structure. This is probably a more realistic figure than the 34% mentioned above.

3.4.3.2 *Boosting the non-em response.* The second, and potentially much more powerful way to affect the e/h ratio of a calorimeter is through its response to neutrons. This method only works if the active material contains hydrogen. In that case, the soft neutrons lose their kinetic energy predominantly through elastic scattering with the hydrogen nuclei, and the recoil protons may contribute directly to the calorimeter signals. Since there is almost no alternative for this very efficient energy loss process, in practice neutrons may end up losing 90% or more of their kinetic energy in this way, whereas the calorimeter's sampling fraction for charged shower particles is typically only a few percent or even (much) less. Even if the signals from these recoil protons are considerably reduced by saturation effects, the net result may very well be a disproportionally large contribution of neutrons to the calorimeter signals.

In Section 3.2.6, some quantitative information was presented with regard to this effect (Table 3.3). It was shown that the n/mip response ratio, and thus the relative contribution of the neutrons to the calorimeter signals, could reach very high values for calorimeters with a very small sampling fraction. As a matter of fact, the n/mip response ratio can be "tuned" to the desired value by choosing the appropriate sampling fraction (for mips), *i.e.*, by choosing the appropriate ratio of active and passive material in the calorimeter.

Although the data from Table 3.3 were obtained for a completely hypothetical active calorimeter material (pure liquid hydrogen), the described mechanism does form the basis on which compensation can be achieved in a wide variety of realistic sampling calorimeter structures. In these instruments, the relative contribution of neutrons to the calorimeter signals can be tuned to the value needed for achieving compensation by choosing the sampling fraction for charged particles appropriately:

The smaller the sampling fraction for charged particles, the larger the relative contribution of neutrons to the calorimeter signals becomes.

The crucial role of hydrogen in the active calorimeter material was experimentally demonstrated in a spectacular way by the L3 Collaboration (LEP, CERN, 1989–2000) [Gal 86]. Figure 3.39 shows the average signal amplitude as a function of energy for electrons and pions showering in their uranium calorimeter, which was read out with proportional wire chambers.

Results are shown for two different gas mixtures that were used in these tests: Ar/CO_2 and isobutane (iC_4H_{10}). The electron signals were barely affected by the gas change, but the pion signals increased by almost a factor of two when the Ar/CO_2 was replaced by the hydrogen-rich isobutane. This gas change altered the calorimeter from undercompensating (e/h may be estimated at ~ 1.3 from these data) to overcompensating ($e/h \sim 0.6$).

Intrigued by these results, the authors tried a whole series of other gas mixtures as well. The results are shown in Figure 3.40, in which the signal ratio between pions and electrons, averaged over energies from 1–5 GeV, is plotted for the different gas mixtures. The horizontal scale in this plot gives the (calculated) average energy deposit in a chamber gap by neutrons. Not surprisingly, this parameter correlates strongly with the relative hydrogen density in the chambers. The π/e signal ratio turned out to be

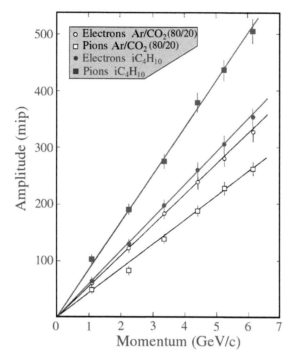

FIG. 3.39. The average signal amplitude for electrons and pions, measured with the ura-
nium/gas calorimeter of the L3 Collaboration, for two different choices of the gas mixture
that was used in the proportional wire chambers [Gal 86].

strongly correlated with this variable and the gas mixture could be chosen such as to
obtain an equal response to pions and electrons, using these data.

From the magnitude of the effects of a change in the hydrogen content on the e/h
value, one may derive that the effective n/mip response ratio in these tests varied from
a value below 1.0 for the hydrogen-free gas to about 10 for isobutane. In a calorimeter
where neutrons carried about 10% of the non-em shower energy, an n/mip value of 10
would change the e/h value by about a factor of three, and the e/π signal ratio at 5 GeV
by a factor of two.

An n/mip response ratio of 10 may seem at first sight very modest compared with
the very large values mentioned before (Table 3.3) for calorimeters with hydrogenous
active material and a small sampling fraction for mips. However, one should realize
that the numbers mentioned in Table 3.3 were calculated under the assumption that the
neutrons are completely absorbed in the calorimeter structure. This was of course not
the case in the detectors L3 tested. For that reason, the term *effective* n/mip response
ratio was used in this context.

As a matter of fact, the total amount of gas in these detectors corresponded to only a
small fraction of 1% of one nuclear interaction length, so that only a very small fraction
of the neutrons interacted at all in the active layers. However, the few neutrons that did

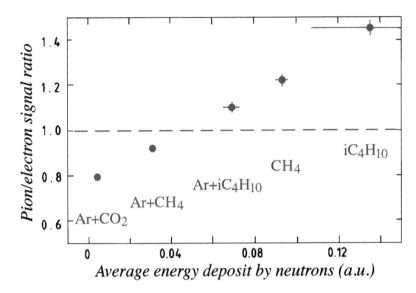

F<small>IG</small>. 3.40. The pion/electron signal ratio, averaged over the energy range 1–5 GeV, measured
for different gas mixtures with the uranium/gas calorimeter of the L3 Collaboration. The horizontal scale gives the (calculated) average energy deposit in a chamber gap by slow neutrons
[Gal 86].

interact deposited in combination ten times more energy in the active calorimeter layers
than the energy that was deposited by a collection of mips carrying the equivalent of the
total energy of all neutrons produced in the shower development combined.

The large contributions of neutrons to the hadronic calorimeter signals of the L3
calorimeter (more than 50% of the signals came from neutrons when isobutane was
used!) became possible thanks to the absence of strong saturation effects in the gases
used for operating the wire chambers. That is very different in dense hydrogenous materials, such as plastic scintillators, where saturation effects may reduce the signals from
typical recoil protons produced in elastic neutron scattering by as much as a factor of
ten (Section 3.2.10). On the other hand, because of the density of the active material,
almost all neutrons generated in the shower development do produce recoil protons in
such calorimeters.

The net result of these effects is shown in Figure 3.41. This figure contains results
of Monte Carlo simulations of the absorption of neutrons in a ^{238}U/PMMA calorimeter
structure [Wig 87]. The neutrons were generated according to a realistic spectrum that
accounted for both the soft evaporation and fission neutrons and the degraded spallation
neutrons. The simulations included all possible reactions the neutrons could initiate,
including uranium fission. Signal quenching because of scintillator saturation was fully
accounted for, using Equation 3.13 with the value $k_B = 9.78$ mg cm^{-2} MeV^{-1} that
was measured for this type of scintillator [Zeu 86].

In Figure 3.41a, the n/mip response ratio is given for ^{238}U/PMMA calorimeters as a
function of the parameter R_d, which measures the ratio of the thicknesses of the passive

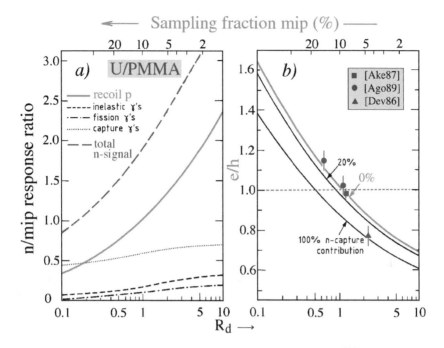

FIG. 3.41. The n/mip response ratio, split up into its components, for ^{238}U/PMMA calorimeters, as a function of R_d, the ratio of the thicknesses of the passive and active calorimeter layers (*a*). The e/h ratio as a function of R_d, assuming that 0%, 20% or 100% of the γs released in thermal neutron capture contribute to the calorimeter signals (*b*). The top axis of both graphs indicates the sampling fraction for mips. From [Wig 88].

and active calorimeter layers. R_d is therefore approximately inversely proportional to the sampling fraction for mips. This sampling fraction is indicated on the top axis of this graph.

The figure distinguishes between the various components contributing to the neutron signals. Not surprisingly, the most dominant contribution comes from recoil protons, and this contribution is indeed strongly dependent on the sampling fraction. If only these recoil protons are taken into account, the n/mip ratio increases from 0.4 for a sampling fraction of 53% ($R_\mathrm{d} = 0.1$) via 1.0 for a sampling fraction of 10% ($R_\mathrm{d} = 1$) to 2.5 for a sampling fraction of 1% ($R_\mathrm{d} = 10$).

Other contributions from the neutrons to the calorimeter signals come from nuclear γs produced in inelastic scattering, in uranium fission or in neutron capture. The latter component represents a considerable amount of energy, 2–3 times the kinetic energy carried by the neutrons. However, this energy is released on a timescale of $\sim 1~\mu s$, too long for practical calorimetric applications (see Section 3.2.9).

In Figure 3.41b, the e/h value of ^{238}U/PMMA calorimeters is shown as a function of the same variable, R_d. Because of the increasing contribution of neutrons to the calorimeter signals, the e/h value decreases with increasing R_d (*i.e.*, e/h decreases with the sampling fraction). The figure shows results for different assumptions concerning the contribution of (γs produced in) neutron capture to the signals: 0%, 20% and 100%. This may give an idea how e/h depends on the charge integration time (the "gate width"). By integrating the calorimeter signals over unrealistically long times and large volumes, the e/h value may be affected by as much as $\sim 15\%$, but the basic trend, e/h decreasing with the sampling fraction, is unaffected by this.

Figure 3.41b also contains some experimental data points, measured for a variety of different ^{238}U/plastic-scintillator calorimeters. Although one should be careful with such a comparison (for example, different types of scintillator with slightly different hydrogen fractions, densities and k_B values were used), these results clearly seem to confirm the predicted e/h dependence on the sampling fraction.

Since the e/h value depends on the sampling fraction, we can choose the sampling fraction such as to achieve compensation: $e/h = 1$. In the case of uranium/plastic-scintillator calorimeters, this condition is fulfilled for calorimeters with a sampling fraction (for mips) of about 10%. That happened to be the choice of Fabjan and Willis when they constructed the first calorimeter of this type. The smaller sampling fraction ($\sim 5\%$) chosen by the WA78 Collaboration for their detector led to a device that was considerably overcompensating [Dev 86, Cat 87].

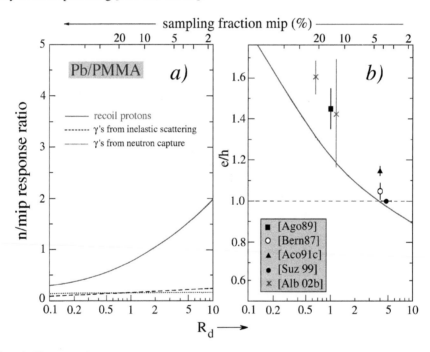

FIG. 3.42. The n/mip response ratio, split up into its components, for Pb/PMMA calorimeters, as a function of R_d, the ratio of the thicknesses of the passive and active calorimeter layers (a). The e/h ratio as a function of R_d (b). The top axis of both graphs indicates the sampling fraction for mips. From [Wig 87].

The calculations of which results are shown in Figure 3.41 may be repeated for any other type of calorimeter configuration. Figure 3.42 shows what happens when the ^{238}U absorber is replaced by lead. The n/mip response ratio looks very insensitive to such a change. The main difference is the absence of γs from neutron-induced fission to the calorimeter signals. But the most important feature, the dependence of n/mip on the sampling fraction for mips (or R_d), appears also to be valid for this configuration.

However, for a given sampling fraction, the e/h of Pb/PMMA calorimeters turns out to be shifted to larger values, compared with those obtained for ^{238}U/PMMA. This is due to two effects (see Equation 3.9):

1. The slightly larger e/mip value, caused by the Z dependence of this parameter.

2. The smaller value of f_n, caused by the absence of neutrons produced in fission processes.

Yet, this does not at all prevent the possibility to achieve compensation. It only means that a Pb/PMMA calorimeter needs to be constructed with a *smaller sampling fraction* than a ^{238}U/PMMA one to achieve this goal. To be precise, the sampling fraction has to be about 3% with lead absorber, *vs.* 10% for uranium.

When this prediction [Wig 87] was made, only lead/scintillator calorimeters with much larger sampling fractions had been built and tested. For example, a 15% device had been built in the framework of the ZEUS prototype program. Its e/h value was measured to be ~ 1.4–1.5 [Ago 89]. Intrigued by the prediction that by doubling the thickness of the absorber plates, combined with halving the thickness of the active plates, a considerable improvement in calorimeter performance could be achieved, ZEUS decided to build another prototype according to these recommended specifications. The test results [Bern 87] confirmed the predictions in great detail, not only for what concerned the compensation (a value $e/h = 1.05 \pm 0.04$ was reported by the authors), but also the hadronic energy resolution was in excellent agreement with the predicted value.

This success, which marked the first time that a detailed prediction (and a very counter-intuitive one, for that matter) with regard to hadronic calorimeter performance turned out to be correct, formed the starting point for SPACAL, a calorimeter R&D project in which compensating lead/plastic-scintillator calorimetry was developed and studied in great detail.

The possibility to achieve compensation in lead/plastic-scintillator structures was reconfirmed by Suzuki *et al.* [Suz 99], who made a systematic study of the performance of this type of calorimeter in which they varied the sampling fraction in small steps. They used pions with energies from 1 to 4 GeV for this purpose.

Figure 3.43 shows some results of their studies. The e/π signal ratio, corrected for the effects of shower leakage, is plotted as a function of the thickness of the lead plates, which was varied in steps of 2 mm from 4 mm to 16 mm. In these tests, scintillator plates with a thickness of 2 mm were used. The authors concluded from these measurements that the compensation condition in their setup was achieved for lead plates with a thickness of 9.1 ± 0.3 mm. That corresponds to $R_d = 4.55 \pm 0.15$ (sampling fraction for mips 3.4%), in excellent agreement with the pioneering results of the ZEUS group. This result is also included in Figure 3.42b.

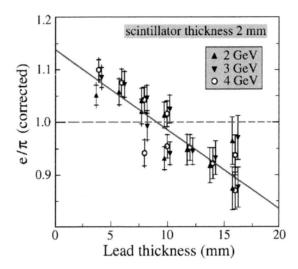

FIG. 3.43. The e/π signal ratio, corrected for the effects of shower leakage, for a lead/polystyrene-scintillator calorimeter with 2 mm thick scintillator plates, as a function of the thickness of the lead plates. The inner (outer) error bars show the combined systematic and statistical uncertainty without (with) the shower leakage corrections. The line in the plot is a result of a linear fit to the experimental data [Suz 99].

Compensation can in principle also be achieved with low-Z absorber materials, such as iron or copper. However, the sampling fraction for which the compensation condition is fulfilled becomes very small in that case, because (*a*) the e/mip value is much larger than for calorimeters with high-Z absorber, and (*b*) neutrons carry a smaller fraction of the total non-em shower energy (see Table 2.5).

Figure 3.44 shows the e/h values for a variety of iron [Hol 78b, Abr 81, Alb 02a, Adr 09] or copper based calorimeters [Ake 85], read out with plastic scintillator, as a function of the ratio of the thicknesses of absorber and active layers ($R_{\rm d}$, bottom axis), or the sampling fraction for mips (top axis). Although compensation was not reached in any of these instruments, the experimental data suggest that e/h will become 1 for a sampling fraction of $\sim 1\%$ ($R_{\rm d} \sim 20$).

The CDHS Collaboration did perform measurements on a detector with these characteristics: 10 (or even 15) cm thick iron plates, interleaved with 5 mm thick plastic scintillator [Abr 81]. For understandable reasons, these measurements did not include electrons, so that no experimental information on the e/π signal ratios was obtained. However, the characteristics of the hadronic energy resolution strongly suggest that the device with the 10 cm thick iron plates had an e/h value close to 1, and that, with 15 cm of iron, the calorimeter became overcompensating. In Chapter 4, we come back to this issue when discussing the relationship between the e/h value and the hadronic energy resolution.

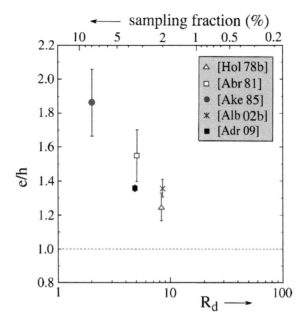

FIG. 3.44. The e/h value for iron/plastic-scintillator calorimeters, as a function of the sampling fraction for mips (top horizontal scale), or the volume ratio of the amounts of passive and active material (bottom horizontal scale).

It is useful to, once again, emphasize the crucial role of saturation of the active calorimeter medium for the compensation issue. Saturation in plastic scintillators reduces the typical neutron signals by a factor of five to ten. If this saturation effect was absent, or much smaller, compensation would be achieved for much larger sampling fractions (smaller R_d values). In gaseous active media, saturation is absent. We saw in Figure 3.39 that this leads indeed to very large neutron signals. However, for reasons to be discussed in Chapter 4, gaseous media are often not the ideal solution for calorimetric needs.

Since 1984, a lot of effort has been spent on the application of so-called "warm liquids" as active calorimeter media. These liquids (for example, tetramethylpentane or TMP) were considered very interesting since they offered the promise of a noncryogenic alternative for liquid argon: a liquid that could be operated in the ionizationchamber mode (collection of ionization charge without internal amplification) at room temperature.

From a compensation perspective, these liquids are also very interesting, not only because they contain a lot of hydrogen, but in particular because they also offer the option of "tunable saturation" through the applied electric field strength [Aub 91, Bac 90], just as liquid-argon calorimeters (Section 3.2.10).

The technical difficulties associated with large-scale applications of these liquids turned out to be extremely challenging and, as far as I know, no experiment in accelerator-based particle physics is using warm-liquid calorimetry. Only the KASCADE-Grande extensive air shower experiment in Karlsruhe uses a large hadron calorimeter ($16\times20\text{m}^2$ surface area, 11.4 λ_{int} deep) as part of the equipment intended for the study of cosmic rays with energies in the PeV domain [KAS 16].

On the other hand, many experiments do use liquid argon as the active calorimeter medium. In these calorimeters, the mechanism of boosting the non-em response discussed in this subsection does not work. Neutrons only contribute to the calorimeter signals through the γs they produce in inelastic scattering. Therefore, the n/mip response ratio is less than 0.5 in these calorimeters. Also, the e/mip values are systematically larger than for calorimeters with plastic-scintillator readout, because of argon's higher Z value (Figure 3.5).

For these reasons, compensation is much harder to achieve in LAr calorimeters, if it can be achieved at all. Typical experimental values for e/h range from $\sim$ 1.8 for Fe/LAr [Fab 77] to $\sim$ 1.4 for Pb/LAr [Dub 86]. In the case of uranium, the e/h value can be reduced by increasing the charge integration time, thus including a larger fraction of the energy released in thermal neutron capture. The D0 Collaboration has reported e/h values ranging from 1.17 for an integration time of 100 ns, to 1.05 for 2 μs [Abo 89].

3.4.3.3 *"Offline compensation."* The need for compensation arises because the calorimeter responses to the em and non-em components of hadron showers are different. The fact that the energy sharing between these components varies from one event to the next and that this sharing, moreover, is energy dependent, causes a variety of undesirable effects, which tend to dominate and deteriorate the performance of non-compensating calorimeters. In this chapter, we have encountered non-linearity of the hadronic calorimeter response as an example of such effects. Several other effects, *e.g.*, effects on the energy resolution and on the line shape of hadron calorimeters, are discussed in the next chapter.

The methods to achieve compensation described so far all aim to equalize the responses to the em and non-em shower components, and thus eliminate the problems at the level of their roots, by making the detector *intrinsically* compensating. There is, however, also an alternative way to eliminate, or at least alleviate, the problems in intrinsically non-compensating instruments.

In this alternative approach, sometimes referred to by the term "offline compensation," one tries to determine the energy sharing between the em and non-em components of hadron showers on an event-by-event basis. If this method is successful, then compensation could, for all practical purposes, effectively be achieved by applying a weight factor e/h to the portion of the signals generated by the non-em shower components.

There are several methods which in principle can be applied for this purpose. We mention two:

1. The *spatial development* of em and non-em showers is usually quite different in a given instrument, especially if high-Z absorber material is used. The spatial development may therefore be used to disentangle the contributions of the two types of components in actual showers. This method was pioneered with some success in the WA1 experiment [Abr 81] and has been used by several other experiments,

in particular H1 [Braun 89] and ATLAS [Aad 08]. The particulars are discussed in Section 4.10.

2. Em showers deposit their energy through electrons and positrons in the absorbing structure. These shower particles are *relativistic* down to energies of a few hundred keV. In the non-em shower component, the energy is predominantly deposited through non-relativistic shower particles, *e.g.*, spallation protons and recoil protons produced in elastic neutron scattering. This property may be used to disentangle the contributions from em and non-em components of actual showers, for example by comparing the Čerenkov and the scintillation light produced in optical calorimeters [Wig 97]. This is the basic principle of dual-readout calorimetry (Section 8.2).

It should be emphasized that it is essential for the success of such methods that the separation of showers in their em and non-em components be addressed at the level of *individual events*. Global weight factors applied to different segments of a calorimeter system have no merit in alleviating the disadvantageous aspects of non-compensating calorimetry. This is discussed in detail in Chapter 6.

3.4.4   *And how about uranium?*

Now that the mechanisms through which compensation is achieved are understood, we briefly come back to the role of uranium. From the discussion in the previous subsection, it has become clear that the use of ^{238}U absorber plates is neither a necessary nor a sufficient condition for achieving $e/h = 1$. However, it may certainly help.

The original argument for using uranium was that the extra energy released in fission processes would compensate for the invisible energy in the non-em shower component. To some extent, this is true. Measurements have shown that in the non-em sector of hadronic shower development in ^{238}U eight to ten fissions take place per GeV deposited energy, depending on the calorimeter configuration [Ler 86]. In calorimeters with plastic-scintillator readout, the number of fissions was found to be $\sim 20\%$ smaller than in LAr calorimeters, because of the moderating effect of the hydrogen on the neutron spectra (a neutron needs a minimum kinetic energy of 1.5 MeV to induce fission in ^{238}U).

In each ^{238}U fission process, a lot of energy is liberated, ~ 200 MeV. However, most of this energy, $\sim 90\%$, goes into recoil of the fission fragments and does not contribute to the measured calorimeter signals, because of the extremely short range (μm) of these fragments. Measurable contributions to the calorimeter signals come from evaporation neutrons produced in the fission process (typically $\sim$three per fission) and from nuclear γs produced in the de-excitation of the nuclear states in which the fission fragments were formed (~ 7.5 MeV per fission) [Bar 69, Gor 80].

The fission neutrons are, on average, considerably softer than the evaporation neutrons from spallation reactions. The Maxwell temperature of the fission neutron spectrum is typically 1.3–1.5 MeV, *vs.* 2–3 MeV for the other neutrons mentioned, which means that their average kinetic energy is about 2 MeV ($\langle E \rangle = 1.5T$).

In total, we may thus estimate the *extra* energy from ^{238}U fission, carried by particles that may contribute to the calorimeter signals, at $10 \times (7.5 + 3 \times 2) = 135$ MeV per GeV of energy in the non-em shower component, or 13.5%. That is equal to about one third of the invisible energy fraction (see Section 2.3.2).

However, more important than the energy carried by these fission products is their impact on the calorimeter signals. In LAr calorimeters, this impact is rather limited. The extra neutrons contribute through γs produced in inelastic scattering. The n/mip response ratio resulting from this process is only ~ 0.2–0.3 in ^{238}U (Figure 3.41). The γ/mip response ratio for nuclear gammas with energies of a few MeV is about 0.5 (Figure 3.4). Therefore, the fission signals increase the non-em calorimeter response by less than 10%, unless a good fraction of the energy released in the capture of the abundantly produced fission neutrons is included in the signals.

The latter can be achieved by increasing the signal integration time. The D0 Collaboration reduced the e/h value of their ^{238}U/LAr calorimeter by 12% in this way, when they increased the signal integration time from 0.1 μs to 2 μs [Abo 89]. However, such an increase may also have adverse effects, $e.g.$, higher noise levels, that outweigh the advantages of a smaller e/h value (see Figure 3.22)

The situation is quite different in calorimeters with hydrogenous active material, since the n/mip signal ratio can be affected through design choices, such as the sampling fraction, in such instruments and, as a result, can be considerably larger than in liquid-argon calorimeters. The neutrons produced in nuclear fission are responsible for a significant increase in the value of the parameter f_n (the kinetic energy carried by neutrons) and therefore the n/mip value required to achieve compensation (Equation 3.9) is considerably smaller than in the absence of these extra neutrons. This is the main reason why the optimal sampling fraction in scintillator calorimeters with ^{238}U absorber is so much larger than for lead absorber.

In conclusion, the extra energy released in fission processes does not represent a fundamental advantage that gives depleted uranium unique properties which distinguish it from other absorber materials. However, this extra energy does make it somewhat easier to achieve or approach the conditions needed for compensation in calorimeters with non-hydrogenous active material. In calorimeters with hydrogenous active material, compensation can be achieved with a larger sampling fraction thanks to this extra energy. This has certain advantages, such as a better energy resolution for em showers. In addition, the natural radioactivity of the uranium absorber material generates a constant signal that can be made to good use for the calibration of the calorimeter, as demonstrated by the ZEUS experiment [Beh 90].

3.4.5 *Compensation and hadronic energy resolution*

In the previous subsections, we have described the various ways in which the e/h signal ratio of a sampling calorimeter can be affected, and be made equal to 1.0, either by means of the design parameters of the calorimeter, or by applying offline criteria to the recorded data in order to effectively achieve the same. The guiding principle behind these efforts is Figure 3.2, which implies that if the response to the em (e) and non-em (h) components of hadron showers are different, event-to-event fluctuations in the energy sharing between these components will introduce a contribution to the hadronic energy resolution that is likely to dominate that resolution, especially at high energies since it does not improve with energy as $E^{-1/2}$. In compensating calorimeters, this effect is eliminated.

However, the underlying phenomenon of non-compensation is the "invisible energy," dominated by nuclear binding energy losses, and fluctuations in invisible energy will continue to affect the hadronic energy resolution in a major way, even in compensating calorimeters. In that context, it is important to realize that different ways of achieving $e/h = 1$ may affect the contributions of fluctuations in invisible energy, and thus the achievable hadronic energy resolution, in different ways. In particular, achieving compensation by properly boosting the response to the shower neutrons is unique in the sense that the kinetic energy carried by neutrons produced in the shower development is strongly correlated to the nuclear binding energy loss, especially in sampling calorimeters with high-Z absorber material (other than ^{238}U). Therefore, in such calorimeters it is actually possible to achieve hadronic energy resolutions that are *better* than the limits set by internal shower fluctuations in "visible energy." This is not possible in calorimeters in which $e/h = 1$ is achieved by other means. This is further discussed in Sections 4.6 and 4.8.

3.5   The Response of Čerenkov Calorimeters

After having discussed the various ways in which the ultimate in calorimeter performance can be achieved (compensation), we finish this chapter with the other extreme of the calorimetric spectrum: the ultimate non-compensating calorimeters, *i.e.*, those based on the detection of Čerenkov light.

Čerenkov light is only emitted by relativistic charged shower particles. A look at Equation 3.9 tells us that of the three categories of particles that constitute the non-em calorimeter signal, two do not meet that criterion: the spallation protons and the neutrons. In first approximation, one would therefore expect

$$e/h = \frac{1}{f_{rel}} \cdot e/mip \tag{3.15}$$

If we take for f_{rel} the value from Table 2.6, and assume that e/mip is the same as for calorimeters based on other detection mechanisms (which is not automatically true), this equation gives e/h values of 5 to 6 when low-Z absorber material is used, and ~ 4 for sampling calorimeters based on high-Z absorber. These are extremely non-compensating devices indeed, with all the disadvantages associated with that property. One thus needs very good reasons to choose this type of detector despite these disadvantages. Such reasons do exist, not least because of the opportunities the detection of Čerenkov light offers in combination with an independent measurement of the energy loss characteristics, dE/dx (see Section 8.2). Considerations of signal speed and radiation hardness provide independent arguments in favor of this detection mechanism as well. Therefore, calorimetry based on the detection of Čerenkov light deserves its own place in this book. Also, because of their extreme properties, Čerenkov calorimeters have provided a lot of insight into fundamental calorimetric issues.

3.5.1 *Electromagnetic showers*

The electromagnetic response of Čerenkov calorimeters has in many ways the same characteristics as the em response of calorimeters based on other detection mechanisms. For example, all Čerenkov calorimeters are extremely linear for electromagnetic shower detection, their em response is constant as a function of energy. This is illustrated by Fig-

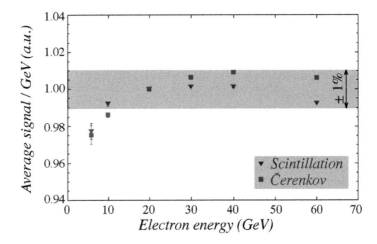

FIG. 3.45. The signal per GeV deposited energy in the scintillation and Čerenkov channels of the RD52 copper/fiber dual-readout calorimeter, as a function of the energy of the electron beam. Data from [Akc 14b].

ure 3.45, which shows the electromagnetic response of the scintillation and Čerenkov signals from the RD52 copper/fiber dual-readout calorimeter. The response is constant within 1%, for both signals, with the exception of the lowest (6 GeV) points, where the effects of upstream material in the beam line caused the beam particles to lose some energy before entering the calorimeter.

Differences with other types of calorimeters are caused by peculiarities of the Čerenkov mechanism, and in particular by the directionality of the emitted Čerenkov light (Section 1.4.2). These differences show up in detectors that are sensitive to the direction in which this light is emitted, *e.g.*, fiber calorimeters. In optical fibers, only light that hits the core/cladding interface at an angle larger than the critical angle is trapped and transported to the light detector. This critical angle is determined by the refractive indices of the core and cladding materials of the optical fibers. For example, the PMMA fibers used in the RD52 calorimeters have indices $n = 1.49$ and 1.40 for the core and cladding, respectively. Therefore, only light hitting the core/cladding boundary at an angle larger than $\arcsin(1.40/1.49) \sim 70°$ is totally reflected. In some detectors that are operating in a high-radiation environment, such as the very forward region of the CMS experiment, high-purity quartz fibers are used as the active medium for calorimeters. This material has an index of refraction of ~ 1.46, and the critical angle is correspondingly larger ($\sim 74°$).

The emission of Čerenkov radiation is a directional process. For particles with $\beta \approx 1$ traveling in quartz, Čerenkov light is emitted in a cone at an angle of about 46° to the direction of the particle. This has led some to believe that this type of calorimeter can only produce meaningful signals if the fibers are oriented at such an angle [Anz 95c]. If that were true, then these detectors would not be useful for collider experiments, where the 4π geometry requires the calorimeter to have a "tower" structure, which can only be achieved with fibers oriented at $\sim 0°$, *i.e.*, pointing to the interaction vertex.

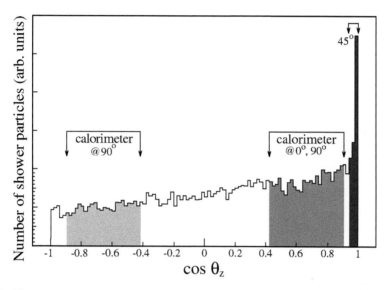

FIG. 3.46. Angular distribution of the shower particles (electrons and positrons) through which the energy of a 1 GeV electron is absorbed in a lead-based calorimeter. Results of EGS4 Monte Carlo simulations. The angular regions contributing to the signals from calorimeters with quartz fibers oriented at 0°, 45° and 90° are shaded and indicated by arrows [Aco 90].

However, as shown in Figure 3.46, the angular distribution of the shower particles through which the energy of em showers is deposited in an absorbing structure contains a sizeable component of more or less isotropically distributed relativistic electrons, *i.e.*, electrons capable of emitting Čerenkov light. These electrons are predominantly produced in Compton scattering. They have "forgotten" the direction of the high-energy particle that initiated the shower and are more or less randomly oriented with respect to that direction. This component leads to a Čerenkov signal in fibers oriented at 0°, or any other angle with respect to the flight path of the showering particles. When the fibers are oriented at the Čerenkov angle ($\theta_C = \arccos(n^{-1}) \approx 46°$) with respect to that flight path, the signals from this type of calorimeter contain, *in addition*, contributions from most of the electrons and positrons produced in the early phase of the shower development, which is dominated by $\gamma \rightarrow e^+e^-$ processes.

Figure 3.47 shows the em response as a function of the angle of incidence of the showering particles (γs in Figure 3.47a, 40 GeV e^- in Figure 3.47b), measured with re-

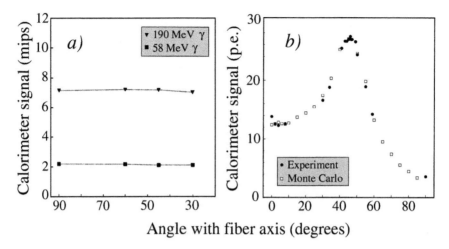

FIG. 3.47. The electromagnetic calorimeter response in fiber calorimeters based on the detection of scintillation light (*a*) or Čerenkov light (*b*), as a function of the angle of incidence of
the particles, measured with respect to the fiber axis. Data from [Ant 95] and [Gan 95].

spect to the fiber axis. Data are given for a scintillating-fiber calorimeter (Figure 3.47a)
[Ant 95] and a quartz-fiber calorimeter (Figure 3.47b) [Gan 95]. Whereas the response
of the scintillating-fiber detector is practically independent of the angle of incidence,
the Čerenkov calorimeter does show an angular dependence. As expected, the highest
response is indeed obtained when the angle of incidence corresponds to the Čerenkov
angle. At that angle, the fibers are sensitive to particles that travel in the same direction
as the incoming particle. If all shower particles that emit Čerenkov light traveled in that
direction, then the response of this calorimeter would be zero except for angles of incidence of $46° \pm \Delta\theta$, where $\Delta\theta$ is determined by the numerical aperture of the fibers and
typically has values in the $10°$–$20°$ range [Akc 03].

However, as mentioned above, a large fraction of the shower particles through which
high-energy electrons and photons deposit their energy in an absorbing structure do not
travel in that direction (see Figure 3.46). The Čerenkov light emitted by such electrons
thus travels at different angles with the fiber axes as well, and the angular dependence
of the calorimeter response provides information about the angular distribution of the
shower particles with energies above the Čerenkov threshold.

For example, the signal measured at $0°$, the orientation needed for calorimeters in
colliding-beam experiments, is caused by shower particles traveling at angles of about
$45°$ with the fiber axes, in the forward direction. The response at this angle is smaller
than the maximum one, by about a factor of two. Even at an angle of $90°$, where the
signal from this calorimeter (partially) depends on particles traveling at an angle of
about $135°$ with the fiber axis, *i.e.*, at $45°$ in the *backward* direction, there is still a
significant response. The different angular regions that contribute to the signals from a
Čerenkov fiber calorimeter, depending on its orientation, are indicated in Figure 3.46.
The experimental response curve (Figure 3.47b) is in excellent agreement with the one

derived from Monte Carlo simulations, which follows from the angular distribution of the shower particles [Gan 95].

A word of caution is necessary concerning results such as those shown in Figure 3.47b. The angular dependence of the em response may in reality be less strong than indicated in this figure. That is because measurements performed at angles larger than 20 degrees typically concern only the first part of the developing shower. It has been shown that the relative contributions of Compton and photoelectrons to the signals are much larger beyond the shower maximum than in the early part [Car 16]. Therefore, the asymmetry between the response at $0°$ and at $90°$ is probably considerably smaller than it appears to be in Figure 3.47b. Nevertheless, the arguments listed above for the angular dependence of the Čerenkov response to em showers are perfectly valid.

3.5.2   *The response to muons*

The directionality of the Čerenkov light turned out to have some very interesting consequences for the detection of muons in Čerenkov fiber calorimeters. As shown in the previous subsection, optical fibers only trap light that is emitted within their numerical aperture. If a relativistic muon travels (approximately) along the axis of an optical fiber, the emitted Čerenkov light it emits does not fulfill that condition. Therefore, the muon does *not* produce a signal, at least not if ionization of the traversed medium was the only energy loss mechanism. However, as the muon energy increases, the contribution of higher-order QED processes becomes gradually a more important component of the energy loss (Section 2.2). Hard bremsstrahlung γs and e^+e^- pairs may develop em showers that do contribute to the signals of the Čerenkov fiber calorimeter. A Čerenkov fiber calorimeter thus only detects the radiative component of the energy loss by muons traversing it in the fiber direction.

This phenomenon was discovered in a spectacular way by the DREAM Collaboration [Akc 04], whose fiber calorimeter detected both the scintillation and the Čerenkov signals from muons traversing it in the direction of the fibers. Figure 3.48 shows the signals. The gradual increase of the response with the muon energy is a result of the increased contribution of radiative energy loss to the signals. The Čerenkov fibers are *only* sensitive to this energy loss component, since the primary Čerenkov radiation emitted by the muons falls outside the numerical aperture of the fibers. The constant (energy-independent) difference between the total signals observed in the scintillating and Čerenkov fibers represents the non-radiative component of the muon's energy loss. Since both types of fibers were calibrated with em showers, their response to the radiative component is equal. This is a unique example of a detector that separates the energy loss by muons into radiative and non-radiative components.

3.5.3   *The hadronic response*

High-energy hadrons deposit their energy through an electromagnetic and a non-electromagnetic shower component. The em component, generated by the π^0s produced in the shower development, consists of large numbers of electrons and positrons and these particles are capable of emitting Čerenkov light down to kinetic energies of a few hundred keV.

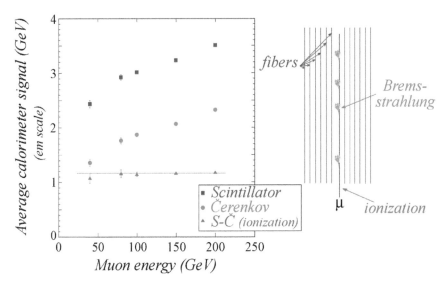

FIG. 3.48. Average values of the scintillator and Čerenkov signals from muons traversing the DREAM calorimeter, as a function of the muon energy. Also shown is the difference between these signals. All values are expressed in units of GeV, as determined by the electron calibration of the calorimeter [Akc 04].

The energy carried by the non-em shower component is predominantly deposited by non-relativistic particles: spallation protons and evaporation neutrons. These shower particles do not leave much of a trace in the form of Čerenkov light, only some γs generated in inelastic neutron scattering might produce relativistic Compton or photoelectrons. In practice, the only other shower particles that contribute to the Čerenkov signals are secondary pions, but these particles carry only a small fraction of the non-em shower energy (see Table 2.6).

Therefore, the Čerenkov signals from hadron showers are typically strongly dominated by the electromagnetic shower core.

3.5.3.1 *Energy dependence.* Since the average fraction of the hadron energy carried by the em shower component, $\langle f_{em} \rangle$, strongly depends on the energy, so does the hadronic response of Čerenkov calorimeters. As a result, Čerenkov calorimeters are the most non-linear detectors of hadronic energy available. This is illustrated in Figure 3.49, which shows the hadronic response of the quartz-fiber calorimeter for the very-forward region of the CMS experiment (LHC, CERN) [Akc 97].

The vertical scale in this figure is normalized to the electromagnetic response of this calorimeter, so that the data points at the same time represent the π/e signal ratio. This ratio rises from 0.592 at 35 GeV to 0.703 at 350 GeV, indicating a 19% signal non-linearity for pions over one order of magnitude in energy.

The figure also shows a curve that represents the prediction of the em shower fraction for pions in copper (the absorber material in this detector), by Gabriel *et al.* [Gab 94].

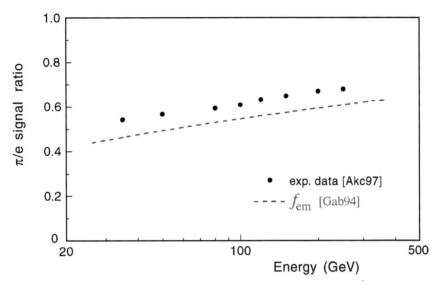

FIG. 3.49. The π/e signal ratio of the Quartz-Fiber calorimeter for the CMS experiment, as a function of energy [Akc 97]. The figure also shows the average em energy fraction predicted for pion showers in calorimeters of this type [Gab 94].

This prediction was derived from Equation 2.19, using the parameter values $E_0 = 0.7$ GeV and $k = 0.84$, suggested by the authors for this case.

The energy dependence of the experimental data follows the predicted one closely. The small, approximately constant difference between the experimental data and the $\langle f_{em} \rangle$ curve, corresponding to $\sim 10\%$ of the em response, is caused by Čerenkov light from the non-em shower component. Since this non-em shower component represents, on average, about half of the total hadron energy, one may conclude from this comparison that the response ratio of the calorimeter to em and non-em energy deposit, e/h, is about 5. This value is much larger than anything we have seen before. Such large e/h values are a typical feature of Čerenkov calorimeters.

3.5.3.2 *Spatial dependence.* In Section 3.2.8 we argued that the shower profile measured by a calorimeter is not identical to the profile of the *energy deposit* in the shower development. This applies *in extremis* to hadronic shower development in Čerenkov calorimeters. Since the hadronic signals in these calorimeters are dominated by electrons and positrons from the em shower core, and since the spallation protons and evaporation neutrons, which populate the tails of the energy deposition profiles, barely contribute to the signals from this type of calorimeter, the measured shower profiles are very different from the energy deposition profiles (see Figure 2.42).

Since the hadron signals from Čerenkov calorimeters are strongly dominated by the em shower component, the measured shower characteristics are dominated by the characteristics of this component. The π^0s produced in hadronic shower development tend to deposit their energy close to the shower axis. Figure 3.50 illustrates this phe-

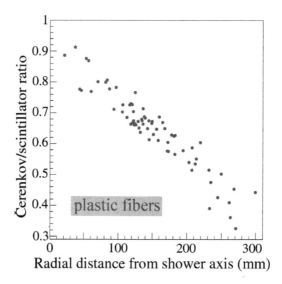

FIG. 3.50. Measurement of the difference between the lateral profile characteristics of 100 GeV π^+ showers in the DREAM dual-readout calorimeter. Each point represents the ratio of the mean values of the Čerenkov and scintillation signal distributions measured in a tower at a radial distance r from the shower axis. The Čerenkov signals were measured with clear plastic fibers [Akc 08a].

nomenon with some results obtained with the DREAM copper/fiber dual-readout calorimeter [Akc 08a]. The ratio of the average Čerenkov and scintillation signals measured with this calorimeter is shown as a function of the distance from the shower axis, for hadronic showers initiated by 100 GeV π^+ in this calorimeter. This ratio decreases by a factor of three over a radial distance of one nuclear interaction length.

Of course, the magnitude of an effect like this also depends on the relative contribution of non-relativistic shower particles to the signals from the scintillation calorimeter. This contribution is significantly larger when lead is used as absorber material instead of copper, because of the much larger neutron production in the absorption of hadrons in lead. Moreover, inelastic neutron scattering is much less likely to occur in lead. The latter feature is a result of the fact that the lowest-lying excited state of ^{208}Pb, the most abundant isotope, is located 2.6 MeV above the ground state. The quantitative differences observed between the results from Figures 2.42 (Pb $vs.$ Cu) and 3.50 (Cu $vs.$ Cu) may be due to such factors.

In hadronic showers, π^0 production is also often limited to the first few generations in the shower development and, therefore, the em shower component rarely extends beyond 7–8 nuclear interaction lengths, even for showers initiated by very-high-energy hadrons. This may translate in differences between the longitudinal profiles of hadron showers measured with Čerenkov light and with dE/dx.

In the longitudinal direction, any shower profile that follows the first nuclear interaction of the showering particle is smeared with an exponential distribution that describes the location of this first interaction. Since the mean free path of a high-energy pion is typically about 1.5 λ_{int}, some 14% (2%) of the pions only start showering after more than 3 (6) nuclear interaction lengths. This effect tends to wash out differences between the details of the shower development proper, such as those discussed here. Therefore, the differences in longitudinal shower profiles between Čerenkov and scintillation calorimeters are less spectacular than the differences between the lateral profiles. Nevertheless, these differences are significant.

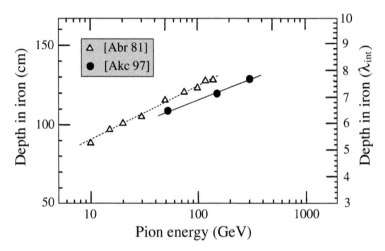

FIG. 3.51. A comparison of the longitudinal shower characteristics in calorimeters based on the detection of scintillation light [Abr 81] and of Čerenkov light [Akc 97]. Shown is the amount of iron needed to limit the average leakage to 5%, as a function of the energy of the showering pion.

Figure 3.51 shows the amount of iron needed to limit longitudinal leakage to 5% of the energy of the showering pion, for calorimeters based on scintillation light and on Čerenkov light, respectively. The experimental data were obtained with the WA1 iron/plastic-scintillator calorimeter [Abr 81] and with the copper/quartz-fiber prototype calorimeter for CMS [Akc 97]. The measurements with the latter instruments were performed by installing increasing amounts of iron upstream of the calorimeter itself and by measuring the effect of this iron on the calorimeter signal distributions in the copper calorimeter. The longitudinal Čerenkov profiles of showers developing *in iron* were thus obtained by using the calorimeter itself as a leakage detector. In the WA1 calorimeter, the differential shower profiles were measured directly.

These experimental data show that hadronic showers in iron were better contained if the calorimeter signals were derived from the Čerenkov light generated by the shower particles than in scintillation-based calorimeters. For 95% containment, the difference is about half an interaction length worth of absorber material.

3.5.3.3 *Time dependence.*   One clear advantage of Čerenkov calorimeters is the extremely short duration of their signals. The emission of Čerenkov light is intrinsically an instantaneous process, and therefore the signals are not affected by the time constants that are typical for scintillators, even for very *fast* ones. Also, tails in the time structure of the signals, which are produced by slow neutrons in other types of calorimeters (see Section 3.2.9), are absent in a detector based on Čerenkov light.

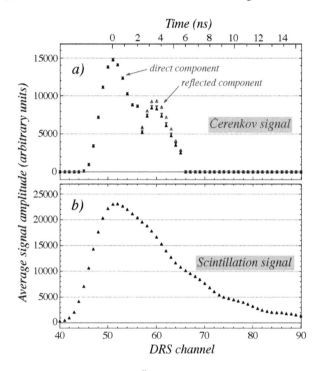

FIG. 3.52. Typical time structure of the Čerenkov (*a*) and scintillation (*b*) signals measured for 80 GeV positrons entering the RD52 calorimeter perpendicular to the fibers. The impact point was located at 30 cm from the mirrored front end of the fibers. The direct and reflected light components of the Čerenkov signals are clearly separated.

As a result of the short duration of the signals, two events that are closely spaced in time can be very well resolved by a calorimeter of this type. This is illustrated in Figure 3.52a, which was obtained with the RD52 dual-readout calorimeter. The figure shows the average time structure of Čerenkov signals produced by 80 GeV positrons entering this detector sideways, *i.e.*, perpendicular to the fiber direction. The Čerenkov fibers were read out by a MCP-PMT on one end, while the other (upstream) end of these fibers was mirrored by means of aluminium sputtering.

The electrons entered the calorimeter at a distance of 30 cm from the mirrored end of the fibers. Thence, they produced two signals, one consisting of light traveling directly to the PMT and the other one consisting of light reflected by the mirrors. These two signals were separated by the time the Čerenkov light needed for the 60 cm roundtrip to the mirrored fiber ends, $\sim$ 3 ns. The figure shown that these two Čerenkov signals were almost completely resolved by the calorimeter.

The RD52 calorimeter also produced scintillation signals for the same events. The scintillating fibers were not mirrored, so that only light traveling directly to the PMT produced a signal in this case. The average scintillation signal from the 80 GeV em showers, shown in Figure 3.52b on the same time scale as the Čerenkov signals, exhibits a characteristic tail, with a decay time constant of about 10 ns.

The time characteristics of the Čerenkov signals are obviously a major advantage in high-luminosity experiments, where they could help to alleviate the pile-up effects that arise when several events are produced during the same bunch crossing.

4

FLUCTUATIONS

In the previous chapter, the *average* signals that calorimeters produce when they absorb high-energy particles were examined in detail. However, in the practice of a particle physics experiment, one will want to use a *given* calorimeter signal to determine the characteristics (*e.g.*, the energy) of the particle that produced it. In order to be able to make a statement about the energy of a detected particle, one needs to know (*a*) the relationship between the measured signals and deposited energy (*i.e.*, the detector calibration), and (*b*) the *energy resolution* of the calorimeter.

The energy resolution determines the *precision* with which the (unknown) energy of a given particle can be measured. It is experimentally determined from the precision with which the energy of particles of *known* energy is reproduced in the calorimetric measurements.

The energy resolution is often considered the most important performance characteristic of a calorimeter. In particle physics experiments, the energy resolution of the calorimeter may be the factor that limits the precision with which the mass of new particles can be determined (*e.g.*, the top quark). It may limit the separation between particles with similar masses (*e.g.*, in the jet–jet decay of the intermediate vector bosons W and Z). And it determines the signal-to-background ratio in event samples collected in almost every experiment.

In this chapter, we discuss the factors that contribute to and limit the energy resolution of calorimeters.

4.1   The Effects of Fluctuations on the Calorimeter Performance

In the shower development that takes place when high-energy particles are absorbed in the block of matter that we call a calorimeter, the energy of the particles is degraded to the level of atomic ionizations or excitations that may be detected. The precision with which the energy of the showering particles can be measured is limited by

1. fluctuations in the processes through which the energy is degraded, and
2. the technique chosen to measure the final products of the cascade processes.

The fluctuations in the shower development process are unavoidable. In electromagnetic showers, they determine the ultimate limit on the achievable energy resolution. However, because of the chosen measurement techniques, the energy resolutions obtained in practice with em calorimeters are usually considerably worse than that.

The situation is quite different for hadron calorimeters. Intrinsic fluctuations in hadronic shower development lead to large event-to-event variations in the fraction of "visible energy," *i.e.*, energy used to ionize or excite the calorimeter's atoms or molecules, and the chosen measurement techniques have often little or no effect on the hadronic energy resolution. However, it turns out that the effect of these intrinsic fluctuations can be (partially) eliminated through a clever design of the readout. In other words, it is possible to obtain hadronic energy resolutions that are *better* than the limits set by the internal shower fluctuations. This can be achieved by exploiting the strong correlation between the invisible energy and the total kinetic energy carried by neutrons produced in the shower development (Section 2.3.2.2), plus the possibility to tune the response of a sampling calorimeter to these neutrons through the sampling fraction (Section 3.4.3.2).

Many, but not all, fluctuations contributing to the energy resolution of calorimeters obey the rules of Poisson statistics. For example, fluctuations in the number of quanta (scintillation or Čerenkov photons, ion–electron or electron–hole pairs, *etc.*) that constitute the calorimeter signals are Poissonian, but shower leakage fluctuations are not.

Let us assume that a particle with energy E creates a signal S that, on average, consists of n signal quanta (*e.g.*, photoelectrons). Event-to-event fluctuations in the signal correspond to Poissonian (or Gaussian, for $n \gtrsim 10$) fluctuations in the number n. The relative width of the signal distribution, $\sigma_S/\langle S \rangle$, *i.e.*, the relative precision of the calorimetric measurement of the energy, σ_E/E, is then equal to $\sqrt{n}/n = 1/\sqrt{n}$.

If the calorimeter is linear, it will produce a signal that consists, on average, of $4n$ quanta when it absorbs a particle with energy $4E$. The event-to-event fluctuations in the detection of such particles correspond to Gaussian fluctuations in the number $4n$. The relative precision of the calorimetric measurement of the energy of these particles amounts to $\sqrt{4n}/4n = 0.5/\sqrt{n}$, *i.e.*, a factor of two ($= \sqrt{4}$) better than for particles with energy E.

For linear calorimeters measuring signal quanta that obey the rules of Poisson statistics, these considerations lead directly to the familiar relationship

$$\sigma_E/E = a/\sqrt{E} \qquad (4.1)$$

We will follow the conventional practice of expressing the energy resolution of the calorimeter, σ_E/E, as a dimensionless number, representing a fraction of the particle energy E, *e.g.*, 3.6%. Unless stated otherwise, we will assume that the particle energy is given in units of GeV. It has become customary to characterize calorimeters, for what concerns the precision with which they can measure the energy of the particles they absorb, in terms of the value of a. In this convention, a thus represents the energy resolution for a 1 GeV energy deposit. In the literature, it is sometimes also expressed as $\sigma/\sqrt{E}$ (*e.g.*, in Figure 4.11), but this variable represents the same dimensionless quantity. Since calorimetry is based on statistical processes (the production of ionization charge, photons, electron–hole pairs, the excitation of Cooper pairs, *etc.*), the relative precision of the energy measurement (σ_E/E) thus *improves* with increasing energy. For example, an em calorimeter for which $a = 10\%$ will measure electrons of 4 GeV with a resolution of 5%, while the resolution will improve to 2% for 25 GeV electrons.

This very attractive feature has greatly contributed to the popularity of calorimeters in particle physics experiments. For other particle detection techniques, the relative precision of the measurements tends to deteriorate with increasing energy. This is most noticeably the case for momentum measurements in a magnetic field, where the size of the spectrometer has to increase proportional to $\sqrt{p}$ to keep the momentum resolution $\Delta p/p$ constant.

However, not all types of fluctuations contribute to the calorimetric energy resolution as $E^{-1/2}$. Some fluctuations are energy independent, *e.g.*, fluctuations resulting from non-uniformities in the calorimeter structure. Other fluctuations may depend on the energy in a different way, *e.g.*, fluctuations resulting from electronic noise (E^{-1}), or from lateral shower leakage ($E^{-1/4}$).

Not all types of fluctuations have a symmetric probability distribution around a mean value. As an example, we mention fluctuations in f_{em}, the em fraction of hadron showers. The probability of finding f_{em} values in excess of $\langle f_{em}\rangle + x$ is larger than the probability of finding values smaller than $\langle f_{em}\rangle - x$, for all values of x (see Section 4.7.1). This phenomenon may lead to an asymmetric *response function*. The response function (sometimes also called the *line shape*) is the calorimeter's signal distribution for particles of a certain type and energy. The mean value of this distribution is the response, and therefore the horizontal scale is expressed in signal units per unit of energy (*e.g.*, pC GeV^{-1}).

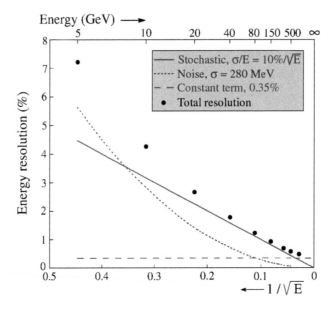

FIG. 4.1. The em energy resolution and the separate contributions to it, for the em barrel calorimeter of the ATLAS experiment, at $\eta = 0.28$. Data from [Gin 95].

In many cases, several sources of fluctuations contribute to the energy resolution of a given calorimeter. For example, Figure 4.1 shows the energy resolution of the em barrel calorimeter of the ATLAS experiment as a function of energy, together with the various contributions to this energy resolution. Since their energy dependence may be different, the relative importance of each of these sources depends on the energy. For example, instrumental effects that cause energy-independent signal fluctuations tend to dominate at high energy, where the effects of Gaussian fluctuations has become very small. On the other hand, electronic noise, an important factor in LAr calorimeters, dominates the energy resolution at low energy, where its E^{-1} dependence overtakes the $E^{-1/2}$ contributions from Gaussian fluctuations.

Most sources of fluctuations that contribute to a calorimeter's energy resolution are mutually uncorrelated. If that is the case, as in Figure 4.1, the uncertainties they cause in the energy of the particles may be added in quadrature. This means that if sources 1, 2 and 3 cause fluctuations with standard deviations σ_1, σ_2 and σ_3 in the measurements of particles with energy E, then the total energy resolution amounts to σ_E/E, with

$$\sigma_E \;=\; \sqrt{(\sigma_1)^2 + (\sigma_2)^2 + (\sigma_3)^2} \;=\; \sigma_1 \oplus \sigma_2 \oplus \sigma_3 \tag{4.2}$$

If the various sources are completely or partially correlated, their effects have to be combined accordingly. Depending on the details, this may result in resolutions that are either better or worse than expressed in Equation 4.2. In this chapter, examples of both situations will be given.

The horizontal scale of Figure 4.1 is one that will be used as the standard throughout this book. The scale is linear in $E^{-1/2}$, with the origin located in the bottom right corner. In that way, the energy increases from left to right, as is common in plots that describe energy dependence, to reach $E = \infty$ for $E^{-1/2} = 0$. Since the horizontal axis is chosen in this way, experimental data that scale with $E^{-1/2}$ (Equation 4.1) will be located on a straight line that extrapolates to the origin (0% resolution at infinite energy). The main reason for choosing this format is that it makes it immediately clear if and what type of deviations from $E^{-1/2}$ scaling are playing a role in the presented data. In general, the energy itself will be plotted on the top axis of plots of this type, which will further help to appreciate their contents.

4.2   Signal Quantum Fluctuations

Fluctuations in the number of detected signal quanta form the ultimate limit for the energy resolution that can be achieved with a given calorimeter. However, in most calorimeters, the resolution is dominated by other factors, discussed in the following sections. In this section, we describe examples of detectors in which fluctuations in the number of detected signal quanta determine the calorimeter resolution.

4.2.1   Semiconductor crystals

The first example concerns the nuclear γ detectors based on semiconductor crystals, such as Ge, Ge(Li) and Si(Li). It takes very little energy to create one electron–hole pair in these crystals, e.g., only 2.9 eV in germanium. The signal generated by a 1 MeV γ fully absorbed in such a crystal therefore consists of some 350,000 electrons. The fluctuations in this number lead to an energy resolution of $1/\sqrt{350,000}$, or 0.17% (at 1 MeV!). In terms of Equation 4.1, this means that $a = 0.005\%$, orders of magnitude smaller than anything we will see for "conventional" calorimeters used in particle physics experiments.

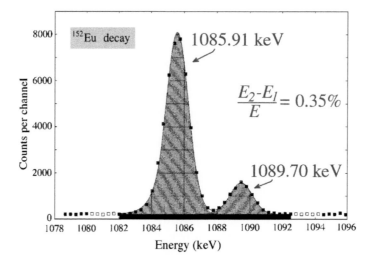

FIG. 4.2. Detection of nuclear γ-rays, from the decay of ^{152}Eu, with a high-purity germanium crystal. The energy resolution of this calorimeter is about 0.1% at 1 MeV. Courtesy of G. Roubaud, CERN.

Owing to correlations in the production of consecutive electron–hole pairs (the so-called Fano factor [Fano 47]), the limit on the energy resolution given by fluctuations in the number of primary processes is even smaller than indicated above. In practice, energy resolutions close to 1.0 keV at 1 MeV are indeed achieved with such detectors (Figure 4.2).

4.2.2 Cryogenic detectors

As a second example, some cryogenic detectors discussed in Section 1.4.4 can be mentioned. *Superconducting Tunnel Junctions* are based on the excitation of Cooper pairs, which then tunnel through a very thin layer separating two superconducting materials. This process takes less than 1 meV per Cooper pair. Therefore, small energy deposits may result in substantial numbers of primary processes and excellent energy resolution. For example, X-rays of 6 keV have been measured with resolutions of about 0.1% in such devices [Pre 87]. This can only be achieved if the number of primary processes is of the order of $\sim 10^6$. In terms of Equation 4.1, this translates into a coefficient a for the stochastic term of the energy resolution of the order of 10^{-6}!

4.2.3 Čerenkov calorimeters

The next example of detectors in which signal quantum fluctuations may play an important role concerns Čerenkov calorimeters. Here, the situation is somewhat less clear-cut than in the previous examples, since in some detectors of this type factors other than Gaussian fluctuations in the number of detected Čerenkov photons may affect the energy resolution.

Detectors of this type include lead-glass em shower counters, water Čerenkov counters (widely used in cosmic-ray experiments and proton-decay studies), as well as a
variety of sampling calorimeters based on quartz as active medium. All these
calorimeters are based on detection of Čerenkov light emitted by relativistic shower
particles with velocities in excess of c/n.

In this discussion, only em showers are being considered. In such showers, the signals are produced by relativistic electrons and positrons generated in the shower development. Typically, the Čerenkov threshold corresponds to a total energy of ~ 0.7 MeV
for these particles (kinetic energy > 0.2 MeV). Figure 2.10 indicates that in practice
most of the em shower energy is deposited by particles capable of emitting Čerenkov
light.

A superluminous particle with charge ze loses energy in the form of Čerenkov photons, per unit track length and in the frequency interval $\nu_1 - \nu_2$, at a rate [Per 00]

$$\frac{dE}{dx} = \frac{z^2}{2}\left(\frac{e^2}{\hbar c}\right)\left(\frac{mc^2}{e^2}\right)\left[\frac{(h\nu_1)^2 - (h\nu_2)^2}{mc^2}\right]\sin^2\theta_C \qquad (4.3)$$

For example, in quartz ($n = 1.46$), this energy loss amounts for particles with $\beta \simeq 1$
to about 500 eV cm^{-1}, for the frequency interval corresponding to visible light ($\lambda = 0.4 - 0.7\mu$m). Such particles thus emit about 200 Čerenkov photons per cm in quartz.
Minimum ionizing particles lose about 4.5 MeV cm^{-1} through ionization in quartz.
If we consider a 1 GeV em shower as a collection of mips, the total track length of
the ionizing shower particles would thus amount to 2.20 m. The total yield of visible
Čerenkov light in a massive quartz detector may thus be estimated as $\epsilon \times 220 \times 200$
photons per GeV. The efficiency ϵ includes effects such as

- the fact that some fraction of the shower energy is deposited by electrons and
  positrons with energies below the Čerenkov threshold ($\beta < 1/n$), and
- the fact that electrons and positrons with $1/n < \beta < 1$ produce less Čerenkov
  light than indicated in Equation 4.3, since the light yield is proportional to
  $\sin^2\theta_C = 1 - (\beta n)^{-2}$.

A reasonable estimate of $\epsilon \approx 0.7$ thus results in about 30,000 visible Čerenkov
photons per GeV of em energy in a detector made of massive quartz. Such detectors
do not exist to my knowledge, but quartz fibers do serve as an active medium in some
sampling calorimeters. We can use the results derived above to estimate the light yield
of such calorimeters. And since this light has to be detected by a device that converts
it into an electric signal, what we are really interested in is the number of photoelectrons generated by em showers in such detectors. This can be estimated by taking the
following factors into account:

- Only some fraction of the shower energy is deposited in the quartz fibers of such
  a sampling calorimeter. In practical devices, this fraction is typically of the order
  of 1%.
- Only some fraction of the Čerenkov light generated in the quartz fibers is trapped
  in these fibers and transported through internal reflection to the light-detecting
  element. If the light was emitted isotropically, then the fraction of trapped light

would only be determined by the indices of refraction of the fiber's core and cladding materials, n_{core} and n_{clad}. For *meridional* light rays, traveling in a plane that includes the fiber axis, this trapping fraction amounts to

$$f_{trap} = \left(\frac{NA}{2n_{core}} \right)^2 \approx \frac{1}{2} \left[1 - \frac{n_{clad}}{n_{core}} \right] \tag{4.4}$$

where the numerical aperture NA is given by $\sqrt{n_{core}^2 - n_{clad}^2}$ [Oko 82, Kir 88, Whi 88]. Typical quartz fibers have a numerical aperture of 0.3–0.4, which means that only of the order of 1% of the Čerenkov light generated in these fibers is trapped and transported to the light detectors.

- The probability that a Čerenkov photon that reaches the light detector is converted into a photoelectron is limited. This quantum efficiency is determined by the wavelength of the photon, and by the properties of the photocathode material and of the window that separates this material from the outside world. In PMTs with bi-alkali photocathodes, the average quantum efficiency is typically of the order of 20–30% [Ham 07].

On this basis, one would thus expect signals consisting of only about 0.5–1 photoelectron per GeV from the type of sampling calorimeters considered here. This is in good agreement with experimental observations. An example of such a detector is the Quartz-Fiber Calorimeter (QFCAL) installed in the very-forward region of the CMS experiment. The light yield of this detector is extremely small, less than 1 photoelectron (p.e.) per GeV. Because of the $1/\lambda^2$ dependence of the Čerenkov light intensity, the precise light yield (and thus the energy resolution) is affected by absorption in the windows of the PMTs that detect the light signals.

This is illustrated in Figure 4.3, which shows results for electron showers measured with two different PMTs, one equipped with a glass window and the other with a quartz window, but otherwise identical. Since the latter window transmitted a larger fraction of the Čerenkov light, the signals were larger and the energy resolution was correspondingly better [Akc 97].

From the measured resolutions, one could actually infer the light yield of this detector. At 1 p.e./GeV, one expects the parameter a in Equation 4.1 to be 1.0 (100%), for 2 p.e./GeV, a becomes $1/\sqrt{2}$ (71%) and for 0.5 p.e./GeV, $a = \sqrt{2}$ (141%). In general, the number of photoelectrons per GeV is given by a^{-2} in this detector. In this way, the authors found a light yield of 0.87 p.e./GeV for the measurements with the PMT equipped with the quartz window and 0.53 p.e./GeV for the glass PMT. These numbers were in excellent agreement with the light yield found in dedicated measurements of this quantity.

One may conclude from these results that contributions to the energy resolution from factors other than fluctuations in the number of photoelectrons are negligible in this detector. The reason for this is the very small light yield, which causes large signal fluctuations, thus dwarfing the effects of other contributions (*e.g.*, sampling fluctuations) to the energy resolution.

The light yield of this calorimeter is very small indeed. At 10 GeV, the measured signals were composed of 5.3 photoelectrons, on average, when read out with a PMT with a glass window. This small number of photoelectrons caused an asymmetric line shape (Figure 4.4a), characteristic for a Poisson distribution P_n for a discrete variable n with a small average value μ:

$$P_n = \frac{\mu^n}{n!}\, e^{-\mu} \tag{4.5}$$

At high energy, the number of photoelectrons was so large that the Poissonian fluctuations led to a symmetric, Gaussian line shape. This is illustrated in Figure 4.4b, for 200 GeV electrons, whose average signal consisted of 106 photoelectrons. The energy resolution σ/E thus corresponded to $1/\sqrt{106}$, or 9.7%, for these high-energy electrons. It improved to $1/\sqrt{174}$ (7.6%) when the calorimeter was read out with a PMT with a quartz window.

Another type of Čerenkov calorimeter used in particle physics experiments is the lead-glass electromagnetic shower counter. This type of detector is homogeneously sensitive and the light yield is thus much larger. Compared with the sampling calorimeters discussed above, light losses resulting from sampling and from the requirement to trap the Čerenkov light within the numerical aperture of optical fibers do not apply. On the other hand, the general level of Čerenkov light production may be somewhat lower than in quartz, because of the larger density of lead-glass.

Let us estimate the size of the signals that may be expected in a typical lead-glass counter, with a density of 5 g cm^{-3} and a refraction index $n = 1.6$. According to

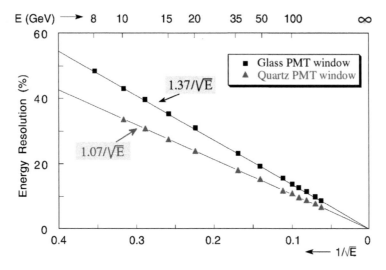

FIG. 4.3. The energy resolution for electron detection with the QFCAL prototype detector, as a function of energy. Results are given for measurements in which photomultiplier tubes with a glass window were used and for measurements in which PMTs of the same type were equipped with a quartz window [Akc 97].

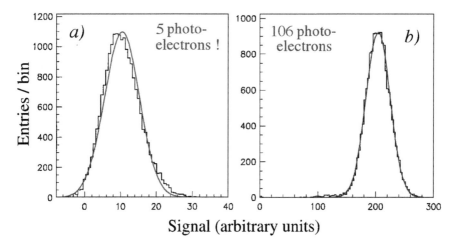

FIG. 4.4. Signal distributions for 10 GeV (*a*) and 200 GeV (*b*) electrons showering in the CMS Quartz-Fiber calorimeter, measured with a PMT with a glass window. The curves represent Gaussian fits to the experimental data [Akc 97].

Equation 4.3, a particle with $\beta \simeq 1$ generates 15% more visible photons per cm track length in this material than in quartz: ≈ 230. The same particle also loses ~ 7 MeV cm^{-1} through ionization. When the same factor is applied to account for non-relativistic inefficiencies as in quartz ($\epsilon = 0.7$), the total light yield may thus be estimated at $230 \times 140 \times 0.7 \approx 23,000$ visible photons per GeV of em energy.

However, in practice the number of Čerenkov photons that reach the photocathode of a PMT mounted behind a lead-glass detector is considerably smaller, because of two effects:

1. Light *absorption* by the detector material.

2. The production angle is such that the light *cannot reach* the photocathode. The critical angle of most types of lead-glass is smaller than 40°, while the Čerenkov angle for particles with $\beta \approx 1$ is typically 50°. In the standard geometry, where the showering particles enter the detector perpendicular to its front/back face, Čerenkov light produced by shower particles traveling along the shower axis thus cannot escape from the detector. Only shower particles traveling at angles $> 10°$ with the shower/detector axis may contribute to the signals in that case.

We have simulated these effects with EGS4, and found that with favorable conditions some 20–25% of the Čerenkov photons produced in the shower development may reach the photocathode. If we assume an effective photocathode efficiency of 20%, as in the example of the sampling calorimeters, this would give signals consisting of $\sim 1,000$ photoelectrons per GeV of deposited energy. Statistical fluctuations in this number would lead to an expected energy resolution (σ/E) of $3.2\%/\sqrt{E}$.

The best lead-glass detector systems have reached energy resolutions of $\sim 5\%/\sqrt{E}$ for electromagnetic showers in the energy range from 1 to 20 GeV [Bro 85, Jef 85].

This might indicate that Gaussian fluctuations in the number of photoelectrons are not the only factor determining the energy resolution of this type of calorimeter, although it is most definitely a very important factor. Additional factors that may play a role include, but are not limited to

- Fluctuations in the energy fraction deposited by shower particles capable of emitting Čerenkov light. On average, some 80% of the energy is deposited by particles with velocities above the Čerenkov threshold, but this fraction fluctuates from one event to the next. It depends, among other things, on the number of positrons produced in the shower development. This is because it takes more shower energy to produce a relativistic positron (through $\gamma \rightarrow e^+e^-$ conversion) than to accelerate an atomic electron to relativistic velocities (Compton scattering).

- Fluctuations in the spectrum of the particles capable of emitting Čerenkov light. Such fluctuations translate into fluctuations in the number of Čerenkov photons because of the strong β- (i.e., θ_C-) dependence of the light yield (Equation 4.3).

- Fluctuations in the spectrum of the Čerenkov photons emitted by the relativistic shower particles. These fluctuations may be important since the quantum efficiency of the light detector is strongly wavelength dependent, with the cutoff occurring in the most densely populated wavelength area.

All these fluctuations obey Poisson statistics and would therefore lead to resolution terms scaling with $E^{-1/2}$, increasing the value of a (Equations 4.1, 4.2). Fluctuations in the absorption of light on its way to the light detectors are unlikely to play a role at this level, since such fluctuations would *not* lead to a $E^{-1/2}$ term in the energy resolution (see Section 4.4.3.4).

However, it is quite possible that in individual detectors of this type losses in light absorption and reflection reduce the signals to levels where photoelectron statistics is, for all practical purposes, the *only* factor determining the energy resolution. It should be pointed out that a resolution of $5\%/\sqrt{E}$, achieved in practice, would correspond to 400 p.e./GeV, if light yield were the only factor determining this resolution. This is only a factor of two to three less than the light yield we calculated in the example above, in which we assumed that the calorimeter's back face was completely covered with photocathode material. In practice, the PMT's are considerably smaller than the crystals. The dominating effect of photocathode coverage on the energy resolution of Čerenkov calorimeters may also be illustrated by the fact that water Čerenkov detectors such as SuperKamiokande, which operates with 40% photocathode coverage, have achieved a light yield of about 7,000 p.e./GeV [Abe 11].

Measurements of the type described in Section 4.2.5 can be used to determine the light yield of individual lead-glass detectors with good precision.

4.2.4 *Scintillation counters*

One type of detector for which the energy resolution is *not* limited by fluctuations in the number of signal quanta, is the scintillation counter. This may be illustrated by the following experimental data. Measurements with NaI(Tl) crystals on 6 keV X-rays have yielded a resolution $\sigma_E/E \approx 15\%$. If we assume that this result is dominated by fluctuations in the number of signal quanta, then this implies that the signals consist on average of ~ 40 photoelectrons. On the basis of this result, one would then for 1 MeV γ-rays expect resolutions of $15\%/(\sqrt{1,000/6}) \approx 1.2\%$. Yet, in reality the best resolutions obtained at this energy are only about 5%. In NaI(Tl) crystals, electromagnetic showers are detected with resolutions of about 1% at 1 GeV, whereas a factor of thirty ($\sqrt{1,000}$) improvement of the 1 MeV result should be expected if photoelectron statistics limited the resolution. Clearly, the energy resolution of this and other scintillation calorimeters is dominated by other factors (see Section 4.4).

4.2.5   *How do we measure signal quantum fluctuations?*

Because of the very small light yield, fluctuations in the number of photoelectrons are almost always the dominating contribution to the em energy resolution of calorimeters that use Čerenkov light as the source of their signals. In Section 8.2.4, we describe how the measurement of the number of Čerenkov photoelectrons in a dual-readout crystal calorimeter is performed in practice.

In general, fluctuations in the number of signal quanta are only one of several factors contributing to the energy resolution. As an example, we describe in this subsection the analysis of the role of photoelectron statistics in the SPAKEBAB calorimeter [Dub 96]. SPAKEBAB was a generic calorimeter R&D project, aimed at the development of a device that would combine the advantages of high resolution and a small e/h value in one instrument. The detector had a sandwich structure. It consisted of a large number of very thin (0.63 or 0.89 mm thick) lead sheets, interleaved with 1 mm thick plastic scintillator plates. The scintillation light was transported to the rear end of the calorimeter by means of a large number of wavelength-shifting fibers, which were spaced by 4 mm and ran through the entire detector structure (Figure 4.5). Because of the very high sampling frequency, this calorimeter had an em energy resolution that was among the very best ever achieved with a sampling calorimeter, $5.7\%/\sqrt{E}$, with a very small deviation from $E^{-1/2}$ scaling (Figure 4.6).

The measurement of the contribution of fluctuations in the number of photoelectrons to the energy resolution of this detector was performed as follows. By installing neutral density filters in front of the PMTs that were used to collect the scintillation light, the number of photoelectrons was reduced by a known factor, f ($f > 1$). This factor could be determined from independent measurements of the properties of the filters or, even better, by simply measuring the ratio of the average calorimeter signals for a given beam of electrons with and without the filters.

By reducing the amount of light, the energy resolution for a given type of showers (*e.g.*, those caused by 10 GeV electrons), deteriorates. By measuring this *change in energy resolution* for showers of different energy and with different filters, the contribution of fluctuations in the number of photoelectrons to the energy resolution could be determined unambiguously.

If x represents the average number of photoelectrons per GeV shower energy (also called the *light yield* of the calorimeter), the average number of photoelectrons from showers initiated by electrons of E GeV equals xE. The standard deviation of fluctuations in this number is $\sqrt{xE}$, and therefore the contribution of photoelectron (p.e.)

statistics to the relative precision of the energy measurement amounts to

$$(\sigma/E)_{\text{p.e.}} = 1/\sqrt{xE} \tag{4.6}$$

The total energy resolution of this calorimeter contained, in addition, contributions from other sources, in particular sampling fluctuations. These fluctuations were also determined by processes governed by Poisson statistics (see Section 4.3), and therefore contributed to the measured energy resolution as (see Equation 4.1)

$$(\sigma/E)_{\text{samp}} = a_{\text{samp}}/\sqrt{E} \tag{4.7}$$

By performing the measurements for low-energy electron showers, the contributions to the energy resolution from other sources, in particular from those contributing in an energy-independent way (*e.g.*, instrumental effects), was made insignificant. In that case, the energy resolution measured *without* filters can be written as

$$(\sigma/E)_{\text{nofilter}} = \sqrt{a_{\text{samp}}^2/E + 1/xE} \tag{4.8}$$

If now the number of photoelectrons is reduced by a factor f, the new energy resolution becomes

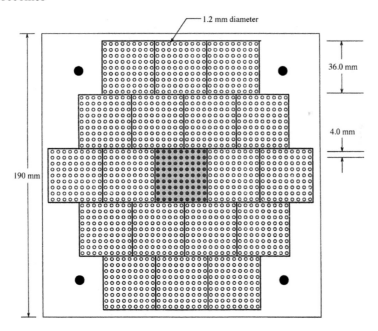

FIG. 4.5. Lateral cross section of the SPAKEBAB calorimeter. The detector consisted of 0.89 mm thick lead plates interleaved with 1 mm thick plastic scintillator plates. Each of the 365 plates had 1539 holes through which ran the WLS fibers that collected and transported the light to the back of the detector. The holes in the corners carried the rods that kept the structure together. A tower structure was achieved by grouping the WLS fibers into 19 cells with 81 fibers each [Dub 96].

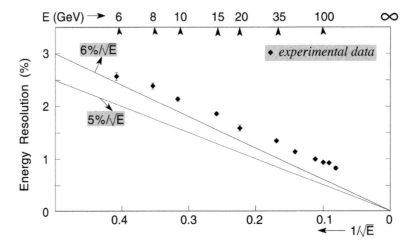

FIG. 4.6. The energy resolution of the SPAKEBAB calorimeter (0.63 mm sampling layers) for electrons as a function of the electron energy. For comparison, the results for $\sigma/E = aE^{-1/2}$, with $a = 5\%$ and 6% (the straight lines) are given as well [Dub 96].

$$(\sigma/E)_{\text{filter}} = \sqrt{a_{\text{samp}}^2/E + f/xE} = \sqrt{(\sigma/E)_{\text{nofilter}}^2 + (f-1)/xE} \qquad (4.9)$$

When f is known, x can thus be derived from the degradation in the energy resolution. This method yields the most accurate results at low energy, where the contributions of non-stochastic processes to the energy resolution are small and the differences between the filtered and unfiltered energy resolutions are large (Equation 4.9).

The measurements were performed with filters that reduced the number of photons by factors (f) of about three and ten, respectively, for electron showers of 6, 8, 10 and 15 GeV. Figure 4.7 shows an example of the results. The measured unfiltered energy resolution at 8 GeV is shown together with curves that show the resolutions that may be *expected* with one and two filters, as a function of x (the number of photoelectrons per GeV). These curves are a graphical representation of Equation 4.9. The experimentally measured energy resolutions with the filters, indicated as data points in the figure, thus correspond to measurements of the value of x, the abscissa in this plot. Other measurements of x, eight in total, were obtained at the other mentioned electron energies.

By combining all results, it was found that this calorimeter produced $1{,}300\pm90$ photoelectrons per GeV deposited energy. Therefore, photoelectron statistics contributed $2.8\%/\sqrt{E}$ to the em energy resolution. Since the total em energy resolution of this detector was $5.7\%/\sqrt{E}$, sampling fluctuations contributed $5.0\%/\sqrt{E}$ ($5.0 \oplus 2.8 = 5.7$).

For sampling calorimeters, this is a high light yield, the result of a very efficient light collection. In many scintillation calorimeters used in particle physics experiments, the light yield is typically one order of magnitude smaller. Therefore, the contribution of photoelectron statistics to the energy resolution of these calorimeters is typically $\sim 10\%/\sqrt{E}$.

In homogeneous, fully sensitive calorimeters based on scintillation light, the light

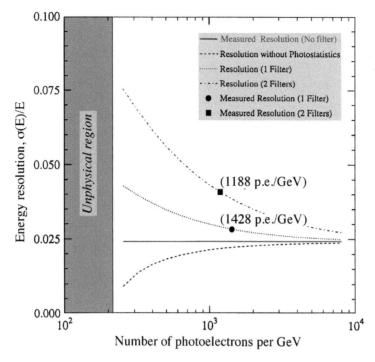

FIG. 4.7. The energy resolution for 8 GeV electrons and the effects of reducing the number of photons by means of filters on this resolution, as a function of the light yield of the SPAKE-BAB calorimeter (in photoelectrons per GeV). See text for details [Dub 96].

yield is much higher than in SPAKEBAB. We already mentioned the 40 photoelectrons that constitute the average signal for a 6 keV X-ray detected in a NaI(Tl) crystal. This translates into 6.7 million photoelectrons for a 1 GeV em shower, and a $0.04\%/\sqrt{E}$ contribution of quantum signal fluctuations to the energy resolution of a NaI(Tl) crystal calorimeter.

The light yield and other relevant properties of a variety of crystals applied in particle physics experiments are listed in Table B.5.

4.3   Sampling Fluctuations

The energy resolution of sampling calorimeters is frequently dominated by the very fact that the shower energy is sampled [Amal 81]. The signal of a sampling calorimeter is the sum of all the signals induced by individual shower particles traversing the active calorimeter layers. Fluctuations in the number of shower particles contributing to the calorimeter signals are thus a component of the energy resolution.

By nature, these sampling fluctuations are governed by the rules of Poisson statistics. Let us suppose, for example, that we have a sampling calorimeter in which a 1 GeV electron shower generates, on average, a signal that is composed of 100 shower particles crossing the active layers. In some of these showers, 90 shower particles contribute to the signals, in others 110. At this energy, sampling fluctuations have a standard deviation of 10% ($1/\sqrt{100}$) and, if no other sources of fluctuations contribute significantly, then the energy resolution of this calorimeter, σ/E, may be expected to be 10% at this energy as well.

If this calorimeter is linear, and for em shower detection this should be the case (Section 3.2.1), then the average signal generated by 10 GeV electrons will be composed of 1,000 shower particles crossing the active layers. The standard deviation of fluctuations in this number, and thus the energy resolution for 10 GeV electrons, corresponds to $1/\sqrt{1,000}$, or 3.2%.

In general, the contribution of sampling fluctuations to the energy resolution of this detector can thus be written as

$$(\sigma/E)_{\mathrm{samp}} = a_{\mathrm{samp}}/\sqrt{E} = 10\%/\sqrt{E} \tag{4.10}$$

with the energy E expressed in GeV. As before, we will express the value of a_{samp} as a dimensionless number (10% in this case) that represents the contribution of sampling fluctuations to the energy resolution at 1 GeV.

Sampling fluctuations thus contribute in the same way to the total energy resolution, σ/E, as signal quantum fluctuations, through a term $a/\sqrt{E}$. In many electromagnetic calorimeters, sampling fluctuations are the main contributing factor to the energy resolution, at least in the energy domain between 10 and 100 GeV, a region of considerable physics interest in modern experiments.

In the following, we examine the factors that determine the value of a_{samp}. Since these factors are different for em and hadronic showers, these two cases are discussed separately.

4.3.1 Electromagnetic showers

The scientific literature contains many papers about the origins of sampling fluctuations in em calorimeters and their dependence on design parameters, such as the choice of passive and active materials, the sampling fraction, and the thickness of the sampling layers.

4.3.1.1 Rossi's Approximation B.
In his classic paper, Amaldi [Amal 81] derives the following, semi-empirical expression for a_{samp} in em sampling calorimeters with dense (non-gaseous) active media:

$$a_{\mathrm{samp}} = 3.2\% \sqrt{\frac{\Delta\epsilon}{F(Z)\langle\cos\theta\rangle}} \tag{4.11}$$

in which $\Delta\epsilon$ (measured in MeV) represents the average energy lost by a mip in one sampling layer, and $F(Z)\langle\cos\theta\rangle$ is a factor that accounts for experimentally observed Z-dependent deviations from a simple $\Delta\epsilon$ dependence.

The basis of this, and other parameterizations proposed to describe a_{samp} in the 1980s (*e.g.*, [Iwa 80, Fab 85]) was the concept of *total track length* developed by Rossi [Ros 52] in his "Approximation B." Rossi considered that an em shower was a collection

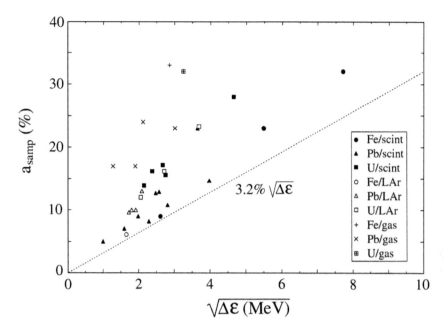

FIG. 4.8. Measured energy resolutions for electrons with sampling calorimeters of different constructions: Fe/scintillator [Sto 78, Hol 78b, Abr 81]; Fe/LAr [Asa 80]; Fe/gas [Lud 81]; Pb/scintillator [Sto 78, Bern 87, Ago 89, Dub 96]; Pb/LAr [Hit 76, Asa 80, Dub 86, Aub 93a]; Pb/gas [And 78, Ata 81, Mue 81, Pri 81]; U/scintillator [Bot 81, Ake 85, Ago 89]; U/LAr [Dub 86, Abo 89]; U/gas [Gal 86]. $\Delta\epsilon$ is the average energy loss in one sampling layer for a mip.

of electrons and positrons each carrying the critical energy ϵ_c. The total track length of all the electrons and positrons through which the shower energy is absorbed in a block of matter of infinite size is, in this approximation, equal to the track length of a minimum ionizing particle carrying the same energy E as the showering electron. The number of shower particles N crossing the active calorimeter layers is then equal to

$$N = \frac{E}{\Delta\epsilon} \tag{4.12}$$

in which $\Delta\epsilon$ stands for the energy lost by a mip in one sampling layer. In this approximation, the energy resolution due to sampling fluctuations is given by the fluctuations in N:

$$(\sigma/E)_{\text{samp}} = \sqrt{\Delta\epsilon} \cdot \frac{1}{\sqrt{E}} \tag{4.13}$$

If E is expressed in GeV and $\Delta\epsilon$ in MeV, then Equations 4.12 and 4.13 become

$$N = \frac{1,000E \text{ (GeV)}}{\Delta\epsilon \text{ (MeV)}} \tag{4.14}$$

and

$$(\sigma/E)_{\text{samp}} = \sqrt{\Delta\epsilon} \cdot \frac{1}{\sqrt{1,000E}} = 3.2\%\sqrt{\Delta\epsilon\ (\text{MeV})} \cdot \frac{1}{\sqrt{E\ (\text{GeV})}} \qquad (4.15)$$

and the contribution of sampling fluctuations to the em energy resolution is thus given by

$$a_{\text{samp}} = 3.2\%\sqrt{\Delta\epsilon\ (\text{MeV})} \qquad (4.16)$$

A representative selection of experimental data from sampling calorimeters consisting of iron, lead or ^{238}U as absorber material, and plastic scintillator, liquid argon or gas as the active medium is compiled in Figure 4.8. Equation 4.16 is represented by the dotted line in this figure. Although some of these experimental energy resolutions may contain contributions from factors other than sampling fluctuations (*e.g.*, photoelectron statistics in scintillation calorimeters), it is clear that the experimental data are not at all well described by these equations.

For most calorimeters, the em energy resolutions are much worse than expected on the basis of Equation 4.15. Only the iron/plastic-scintillator and iron/liquid-argon data, as well as the resolutions of lead calorimeters read out with very thick scintillator plates (> 12 mm [Sto 78]) are reasonably well described by the dotted line. The fluctuations in other calorimeters with high-Z absorber material are clearly larger and the fluctuations in calorimeters with gaseous active material are *much* larger, regardless of the absorber material.

For this reason, Amaldi proposed the mentioned Z-dependent corrections to the simple Rossi-based expression (Equation 4.15). He attributed the observed deviations for gaseous active media to the large event-to-event fluctuations in energy loss by shower particles traversing a gaseous active layer, and extended his expression (Equation 4.11) with what he called *Landau and pathlength fluctuation terms* to account for these.

In Section 2.1.4 we saw that the basic assumption on which this approach is based is not correct. For purposes of sampling calorimetry, electromagnetic showers cannot be considered a collection of electrons and positrons with the critical energy. It was shown that a large fraction of the energy is deposited by shower particles carrying energies that are *much lower* than the critical energy (Figure 2.10).

In Section 3.2.1 it was shown that this fact could explain several unusual features of the calorimeter response, such as the dependence of the e/mip ratio of sampling calorimeters on the Z values of the active and passive calorimeter components (Figure 3.5). The way in which the em shower energy is deposited in the calorimeter structure also has important consequences for the sampling fluctuations.

The concept of relating the sampling fluctuations to $\Delta\epsilon$, the average energy lost by a mip in one sampling layer, is based on the following implicit assumptions:

1. All shower particles contributing to the signals from sampling calorimeters originate from the absorber layers.
2. All shower particles contributing to the signals from sampling calorimeters traverse the active layers completely.

3. All shower particles contributing to the signals from sampling calorimeters traverse the active layers in the same direction, *i.e.*, the direction of the incoming particle.

4. Each shower particle that traverses an active layer loses the same amount of energy in that layer.

5. $e/mip = 1.0$

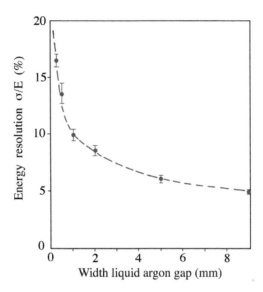

FIG. 4.9. Energy resolution for 1 GeV electrons in Pb/LAr calorimeters, as a function of the width of the liquid-argon gaps. The thickness of the lead plates is 1.9 mm ($\frac{1}{3}X_0$). Results from EGS4 Monte Carlo calculations [Fis 78].

A very clear and early indication that this concept is incorrect for the purpose of describing the contribution of sampling fluctuations to the em resolution of sampling calorimeters follows from Fischer's work, published in 1978 [Fis 78]. Figure 4.9 shows results of his EGS Monte Carlo calculations on the development of 1 GeV electron showers in lead/liquid-argon calorimeters with 1.9 mm thick lead plates. The energy resolution is given as a function of the width of the LAr gaps.

This resolution appears to be very sensitive to changes in the relative amount of argon, it deteriorates by a factor of more than three going from 9 mm to 0.25 mm of argon. At the same time, the value of $\Delta\epsilon$ changes from 4.35 MeV (1.9 mm Pb + 9 mm LAr) to 2.49 MeV (1.9 mm Pb + 0.25 mm LAr), a 43% change *in the opposite direction*.

There is a wealth of experimental data confirming this effect of the thickness of the active layers on the em energy resolution. For example, Stone *et al.* [Sto 78] measured the energy resolution of lead/plastic-scintillator calorimeters, in which they varied the thickness of the scintillator plates in a given absorber structure. For a structure consisting of 4.2 mm thick lead plates, they measured an em resolution of $10.8\%/\sqrt{E}$ when

1.2 cm thick scintillator plates were used as active material. The resolution deteriorated to $12.8\%/\sqrt{E}$ when these scintillator plates were replaced by 0.6 cm thick ones. Obviously, the value of $\Delta\epsilon$ *decreased* as a result of this change.

The resolution due to sampling fluctuations therefore does not at all scale with $\sqrt{\Delta\epsilon}$ when the sampling fraction of a calorimeter with a given absorber structure is changed. It even moves in the opposite direction. It is amazing that after more than 30 years this erroneous concept is still being advocated in review papers [FF 11].

Once again, very soft shower particles determine the calorimeter characteristics, also for what concerns the em energy resolution. Compton and photoelectrons produced in the absorber material may traverse the active layers in any direction, or they may stop in these layers and thus deposit their entire residual energy in the active material. Such electrons may also be *generated* in the active material itself, especially in detectors with a very large sampling fraction. Since these particles typically travel distances that are a small fraction of the thickness of an absorber plate, this thickness determines the *number* of Compton and photoelectrons contributing to the signals. On the other hand, the thickness of the active layers determines the size of the *fluctuations* in the contributions from individual shower particles of this type. Both thicknesses are thus important parameters affecting the sampling fluctuations.

4.3.1.2 *Sampling fraction and sampling frequency.* Because of the reasons mentioned above, a potentially better scaling parameter for a_{samp} than the energy lost by a mip in one sampling layer ($\Delta\epsilon$) is the *sampling fraction, f_{samp}.*

Figure 4.10a shows the em energy resolution as a function of f_{samp} for various representative plastic-scintillator calorimeters, with different absorber materials and plastic scintillator plates with thicknesses varying between 2.5 and 5 mm. The inverse energy resolution, expressed as $1/a_{samp}$, is plotted versus the sampling fraction (for mips), which is shown on a scale that is linear in $\sqrt{f_{samp}}$. The straight line thus corresponds to a situation in which the energy resolution at a given energy is proportional to $\sqrt{1/f_{samp}}$. Therefore, if the sampling fraction is doubled, either by doubling the thickness of the scintillator plates or by reducing the thickness of the absorber plates by a factor of two or in some intermediate way, then the number of shower particles contributing to the calorimeter signals doubles as well, and the resolution resulting from sampling fluctuations is reduced by a factor of $\sqrt{2}$.

This line describes the plotted data reasonably well, but it most certainly does not tell the whole story. Figure 4.10b shows em energy resolutions, plotted in the same way, for calorimeters based on scintillating *fibers*. These fibers are much thinner than the plates, with diameters of typically 0.5–1.0 mm.

For the same sampling fraction, the resolution of these fiber calorimeters is considerably better than that of a calorimeter using scintillator plates. Also, the energy resolution for the detector with the 0.5 mm fibers is better than for a detector with 1 mm fibers and the same sampling fraction, which in turn is better than a detector in which ribbons of 0.9 mm fibers form the active layers. However, in each case where measurements for different sampling fractions were done, the results are well described by a straight line through the origin of the plot, as in Figure 4.10a.

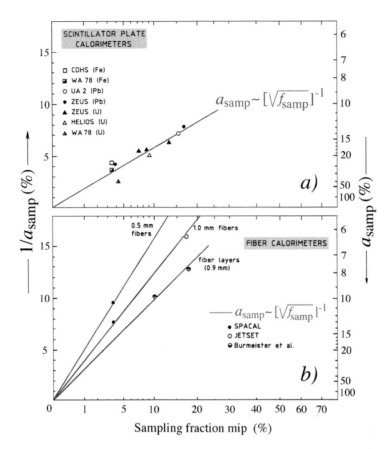

FIG. 4.10. The em energy resolution as a function of the sampling fraction for various representative plastic-scintillator plate calorimeters (*a*) and for scintillating-fiber calorimeters (*b*). From [Liv 95].

In these fiber calorimeters, the number of different active elements in a given calorimeter volume is considerably larger than for the plate calorimeters. More to the point, the total surface of the boundary between the active and the passive calorimeter layers in this volume is much larger.

As we saw before, a large fraction of the shower energy is deposited in the form of electrons that carry less than 1 MeV and thus travel only a small fraction of the distance separating consecutive active layers. By increasing the boundary surface between the active and passive layers, a correspondingly larger fraction of these soft electrons will contribute to the signals, resulting in an improved resolution.

Increasing this total boundary surface can be achieved either by incorporating more active layers of a given type, say with thickness *d*, in the calorimeter volume (increased sampling *fraction*), or by reducing the thickness *d* for a given total amount of active material (increased sampling *frequency*).

In first approximation, the number of *different* soft shower electrons produced in the absorber and contributing to the signals is proportional to the total surface of the boundary between the active and passive components. Therefore, taking only these shower particles into account, the energy resolution from sampling fluctuations may be expected to scale as

$$(\sigma/E)_{samp} = a\sqrt{d/f_{samp}} \cdot \frac{1}{\sqrt{E}} \qquad (4.17)$$

The validity of this relationship may be inferred from Figures 4.10b and 4.11. The fiber data depicted in Figure 4.10b show that for detectors with the same fiber thickness d, the resolution is inversely proportional to the square root of the sampling fraction. For a given sampling fraction, the resolution improves when thinner fibers are used. For example, when lead and plastic are used in the volume ratio 4:1, the resolution is $(12.9 \pm 0.3)\%/\sqrt{E}$ for 1 mm fibers [Aco 91c] and $(9.2 \pm 0.3)\%/\sqrt{E}$ for 0.5 mm fibers [Bad 94a]. The latter result corresponds to an improvement that is about equal to the factor $\sqrt{2}$ expected on the basis of Equation 4.17.

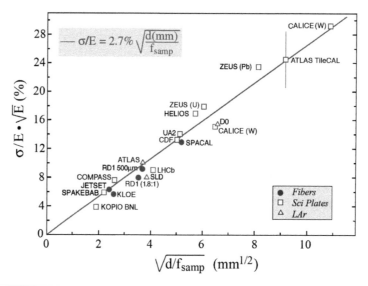

FIG. 4.11. The em energy resolution of sampling calorimeters as a function of the parameter $(d/f_{samp})^{1/2}$, in which d is the thickness of an active sampling layer (*e.g.*, the diameter of a fiber or the thickness of a scintillator plate or a liquid-argon gap), and f_{samp} is the sampling fraction for mips. The error bar for the ATLAS TileCal data point [Kul 06] represents the dependence on the angle of incidence of the electrons.

In Figure 4.11, the em energy resolutions for these and other (*e.g.*, liquid-argon) calorimeters are plotted as a function of the variable $\sqrt{d/f_{samp}}$. The dotted line corresponds to a situation in which the em energy resolution scales with this parameter. We conclude that the experimental data are reasonably well (typically to within $\pm 10\%$) described by

$$a_{\text{samp}} \;=\; 2.7\%\sqrt{d/f_{\text{samp}}} \qquad\qquad (4.18)$$

in which the thickness d of the active layers or fibers is expressed in mm and f_{samp} is the sampling fraction for mips. The agreement could be further improved if one used the sampling fraction for em showers instead of that for mips. However, since the e/mip response ratio also depends on the sampling frequency (see, for example, Figure 3.7), this is not a trivial modification to make.

Equation 4.18 allows a quick and reasonably accurate estimate of the sampling fluctuations in a variety of sampling calorimeters. And since these fluctuations usually dominate the em energy resolution of sampling calorimeters, this resolution may be estimated in the same way. The calorimeters for which Equation 4.18 gives reasonable results include especially those with parameter values likely to be chosen in practical designs, *i.e.*, sampling layers with a thickness in the range 0.1–$1.0X_0$ and sampling fractions in the range of 1–10%.

As an example, we estimate the energy resolution of a lead/plastic-scintillator calorimeter consisting of 5 mm lead plates and 3 mm polystyrene. Using the information from Table B.1, we find for f_{samp} a value $(0.3 \times 2.0)/(0.3 \times 2.0 + 0.5 \times 12.7) = 0.086$. Equation 4.18 gives an estimate for a_{samp} of $2.7\%\sqrt{3/0.086} = 16\%$. Reducing the thickness of the lead plates to 3 mm increases f_{samp} to 0.136 and improves the em energy resolution to about $13\%/\sqrt{E}$.

4.3.1.3 *Correlations.*

So far, we have discussed the effects of two types of shower particles on the sampling fluctuations in em showers:

1. Soft electrons generated in the absorber layers.
2. Shower particles generated in the active calorimeter layers.

There is one more class of shower particles, not mentioned until now: fast e^+e^- pairs created in the conversion of energetic γs. As was pointed out by Willis and Radeka [Wil 74], these particles are always created in pairs and, therefore, the number of statistically independent crossings of active layers is $N/2$ and not N, where N stands for the number of layer crossings from this source. Actually, the number of statistically independent crossings might even be smaller than that. These fast particles may have a range that is larger than the thickness of one sampling layer and in that case they may contribute to the signals in several consecutive active layers.

However, in sampling calorimeters with dense active media, the signal contribution from shower particles traversing more than one sampling layer is usually a negligible effect. As a matter of fact, this was experimentally demonstrated by Willis and Radeka themselves, in their paper describing results of measurements with the first-ever liquid-argon calorimeter [Wil 74]. This calorimeter consisted of a stack of 1.5 mm ($0.1\ X_0$) thick iron plates, separated by 2.0 mm wide liquid-argon gaps. Electronically, this detector was divided into two independent interleaved sections, to be called I and II in the following. In section I, the charge from the odd-numbered LAr gaps was collected, the charge from the even-numbered gaps went to section II. The two sections were operated at the same high voltage, so that in the absence of fluctuations their output signals should be equal.

The authors measured both the sum (I+II) and the difference (I–II) of the signals from these two sections event by event. They found that the two distributions had, within experimental errors, the same relative width (*i.e.*, normalized to I+II): $2.8\pm0.3\%$ (σ_{sum}) vs. $2.8\pm0.1\%$ (σ_{diff}), for 7 GeV electrons. As an aside, we mention that this measured resolution is in reasonable agreement with Equation 4.18 ($f_{\mathrm{samp}} = 0.199 \rightarrow a_{\mathrm{samp}} = 2.7\%\sqrt{(2.0/0.199)}\,(\sqrt{7})^{-1} = 3.2\%$).

We conclude from this result that the two signal distributions from sections I and II were completely uncorrelated. Both sections sampled the same showers, but the shower particles contributing to the signals from section I were not the same as those contributing to the signals from section II. Therefore, the typical shower particle contributing to the em signals from this calorimeter traveled a distance that was short compared with the thickness of one sampling layer (1.5 mm Fe + 2.0 mm LAr) and thus carried an energy of less than ~ 2 MeV, *i.e.*, much less than the critical energy in iron (22 MeV). A similar conclusion can be drawn from the explicit measurements of the sampling fluctuations discussed in Section 4.3.4.

Whereas the contribution of individual shower particles to the signals from more than one sampling layer thus plays no significant role in typical em sampling calorimeters, the sampling fluctuations are strongly affected by such contributions in non-em showers. The consequences of this are further discussed in Section 4.3.2.

4.3.1.4 *Pathlength fluctuations.*

Although Equation 4.17 is useful for a variety of sampling calorimeters, its validity is certainly not universal. In particular, it does not properly describe sampling fluctuations in extreme conditions, *i.e.*, in calorimeters with either a very small or a very large sampling fraction or sampling frequency. This becomes clear when we apply the equation for gas calorimeters. Such calorimeters have extremely small sampling fractions. For example, a calorimeter consisting of 2 mm thick lead plates interleaved with 2 mm wide gaps filled with isobutane has a sampling fraction of $4.7 \cdot 10^{-4}$. Using Equation 4.18, we would estimate $a_{\mathrm{samp}} = 2.7\%\sqrt{(20000/4.7)} \approx 180\%$, which is about a factor of ten larger than the measured value for devices of this type [And 78, Mue 81].

Sampling fluctuations in gas calorimeters are thus not correctly described. We examine the case of gas calorimeters using Figure 4.9 as our starting point. This figure shows the sampling fluctuations in calorimeters consisting of 1.9 mm thick lead plates interleaved with gaps filled with liquid argon. The resolution is given as a function of the width of these gaps. We have repeated and extended the range of these simulations using EGS4. The results are shown in Figure 4.12, for liquid-argon calorimeters with absorber structures of 2 mm or 10 mm thick lead plates. The resolution for 1 GeV electron showers in these calorimeters is plotted as a function of $\sqrt{1/f_{\mathrm{samp}}}$ in this figure, where f_{samp} now stands for the sampling fraction of the showers, and not of a mip.

Equation 4.17 was derived based on the assumption that the sampling fluctuations are determined by the *number* of shower particles that cross the boundaries between the passive and active media in the calorimeter structure and thus contribute to the signals. The fact that the equation describes the experimental data for calorimeters with dense active media as well as it does provides support for this assumption. However, it must

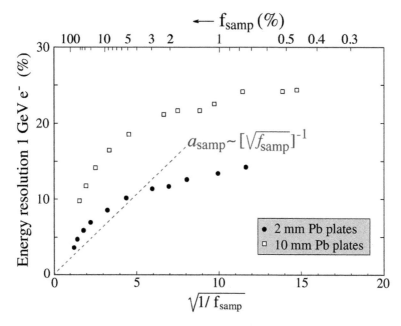

FIG. 4.12. The em energy resolution of Pb/LAr sampling calorimeters with 2 mm thick lead
plates, as a function of the sampling fraction for 1 GeV electrons. This sampling fraction is
varied by means of the thickness of the LAr-filled gaps between the absorber plates. Results
of EGS4 simulations.

be an approximation, since the sampling fluctuations are defined as the fluctuations in
the energy that is sampled by the calorimeter, *i.e.*, *the energy fraction deposited in the*
active material. This approximation breaks down when this energy is either a very large
or a very small fraction of the shower energy. Note that the data points in Figure 4.10
all cluster around sampling fractions of ~ 10%.

Let us first consider the situation for very small sampling fractions, less than a few
%. It is reasonable to assume that in that case the entire signal in this calorimeter is
caused by shower particles produced in the passive (absorber) medium. In first approximation, it should then not make a difference for the sampling fluctuations if the dense
active medium were replaced by a gaseous active medium, since the number of boundary crossings would not be affected by this operation. The same is true for the situation
in Figure 4.12. The number of boundary crossings does not change when the LAr layers are made thinner and thinner, and one would thus not expect to see a change in the
energy resolution if it would be exclusively determined by fluctuations in the number of
shower particles that contribute to the calorimeter signals.

However, if the LAr layers get thinner, or are replaced by gas, the amount of energy
deposited by a single shower particle in such a layer becomes smaller and smaller. For
example, mips traversing a 4 mm LAr gap deposit on average 1 MeV in it. At that
point, the energy deposit by Compton electrons becomes dependent on the *direction* in

which these abundantly produced shower particles cross the boundary. For example, a 2 MeV Compton electron loses about 1 MeV through ionization of the liquid argon if it traverses the gap perpendicularly. If it enters the gap at an angle of 45°, then it loses 1.4 MeV and if the angle is 60° or larger, then the electron will stop in the liquid and deposit its entire 2 MeV energy in it. The contribution of Compton and photoelectrons with energies in excess of 1 MeV to the calorimeter signals thus starts to become dependent on the trajectories these particles follow through the gaps. *Pathlength fluctuations* start to contribute to the sampling fluctuations, and thus to the em energy resolution of the calorimeter. However, a comparison with the $[f_{samp}]^{-1}$ dependence that characterizes the effects of the changing number of participating shower particles (the dashed line in Figure 4.12) shows that the increase in the energy resolution due to these pathlength fluctuations is much slower than that.

When the gap width is further reduced, the pathlength fluctuations become a factor for an increasing fraction of the Compton and photoelectrons that penetrate into the gaps. At a gap width of 2 mm, all shower particles with energies larger than 500 keV are subject to it and at 1 mm, that threshold has dropped to ~ 300 keV. The thresholds for the sampling layers in a typical gas calorimeter are even much lower than that, as can be seen from the curve for air in Figure 2.47. One cm of gas will cause pathlength fluctuations for all electrons with a kinetic energy of more than 30 keV.

However, the sampling fluctuations do not continue to increase indefinitely when the sampling fraction is further reduced. In Section 2.5.1, we saw that the contribution to the signals of lead-based sampling calorimeters is insignificant for shower particles with energies below ~ 100 keV, because the probability that such shower particles escape from the absorber material becomes negligibly small at these energies. For that reason, EGS4 simulations gave stable results once the cutoff values were set at that, or a lower, level. Electrons of 100 keV have a range of 0.15 mm in liquid argon. Therefore, the sampling fluctuations may be expected to reach a stable level when the sampling fraction of our 10 mm Pb/LAr calorimeter drops below $\sim 0.1\%$, *i.e.*, $\sqrt{1/f_{samp}} > 30$. From Figure 4.12, one may expect the sampling fluctuations to stabilize at a level of $\sim 25 - 30\%$ for this calorimeter.[8] For the calorimeter with the 2 mm thick lead plates, the sampling fluctuations are likely to stabilize somewhere in the 15–20% range.

These are thus the resolution levels that could be expected if the liquid argon were replaced by a gaseous active medium. In the literature, measurements with gas calorimeters based on lead absorber structures consisting of 1.27 mm [Mue 81] and 2.8 mm [And 78] thick lead sheets have been described. Both reported an em resolution of $17\%/\sqrt{E}$. Measurements with structures containing thicker absorber plates led to published em resolutions of $23\%/\sqrt{E}$ (7.1 mm Pb, [Pri 81]), $24\%/\sqrt{E}$ (3.5 mm Pb, [Ata 81]) and $32\%/\sqrt{E}$ (4.2 mm U, [Gal 86]). These results are in agreement with the values expected on the basis of the above analysis.

It should be pointed out that, especially in calorimeters with high-Z absorber material, the fraction of the soft shower particles that penetrate the active material may be

---

[8]Extremely thin layers of active material cause some specific technical problems for EGS4. Therefore, we have not explicitly simulated Pb/LAr structures with sampling fractions much smaller than 1%.

strongly affected by the presence of low-Z material between the absorber and the gas (see Section 3.4.3.1). For this reason, the precise energy resolution of such gas calorimeters may be sensitive to details of the design.

Let us now consider the situation for very large sampling fractions ($> 20\%$). At that point, the assumption that only shower particles produced in the absorber layers contribute to the calorimeter signals breaks down. Many shower particles will be created inside the thick liquid-argon layers, and almost all particles generated in the lead and entering the liquid argon deposit their entire kinetic energy in it. As a result, the fluctuations in the energy fraction deposited in the active material rapidly become smaller, and the energy resolution improves faster than proportional to $[f_{samp}]^{-1}$ (Figure 4.12).

The discussion in this section may illustrate that the issue of sampling fluctuations is a complicated one, in which many different factors may play a role. Relatively simple formulae may give a reasonable description for certain classes of detectors, *e.g.*, Equation 4.18 for calorimeters with dense active media. However, such formulae have no universal validity and tend to deviate from reality in extreme conditions, *e.g.*, very large or very small sampling fractions or sampling frequencies.

Fortunately, the EGS4 Monte Carlo package is a very powerful tool for predicting the em performance of specific instruments, provided it is used properly. Del Peso and Ros have made an elaborate study of the validity of EGS4 in simulating the em energy resolution for a large variety of sampling calorimeters [Del 89]. They confirm all the trends mentioned above: the dependence on the thickness of the active and passive layers and on the Z of the absorber material, the importance of pathlength fluctuations for the resolution of gas calorimeters, *etc*.

They emphasize the importance of low cutoff values in the EGS4 simulations. The energy resolution turned out to be very sensitive to the cutoff value for the energy of the shower electrons. For a given choice of this cutoff energy, changes in the cutoff value for the photon energy had little effect on the results. This confirms a trend we mentioned earlier when discussing the simulations of the em response of sampling calorimeters (Section 2.5.1).

4.3.2 *Non-electromagnetic showers*

Hadrons that are absorbed in a calorimeter deposit their energy through an em and a non-em shower component. The processes playing a role in the absorption of the em component are identical to those in the absorption of em showers and, therefore, the sampling fluctuations deriving from this component are identical as well.

The energy contained in the non-em shower component is deposited in the calorimeter structure through ionization by charged hadrons, spallation protons and (especially in calorimeters with hydrogenous active material) particles produced in reactions induced by soft neutrons.

Compared with the sampling fluctuations occurring in em shower development, there are two complicating factors that affect the measurements of the non-em shower component in a sampling calorimeter:

- Individual charged hadrons may traverse a large number of active elements and

thus contribute a correspondingly large number of signal quanta to the total calorimeter signal. This effect might increase the sampling fluctuations considerably.

- Spallation protons and recoil protons from neutron scattering are densely ionizing particles. In spite of the saturation effects that reduce the signals from these
  particles in dense active media such as plastic scintillator or liquid argon (Section
  3.2.10), the signal from such protons may be considerably larger than that of a
  mip crossing a signal plane. As a result, the number of different shower particles
  contributing to a calorimeter signal that corresponds to a given energy deposit is
  smaller than for an em shower of the same energy.

Both factors thus increase the sampling fluctuations for a given energy deposit, compared with em showers. In the following, we examine these effects in a more quantitative
manner.

Sampling calorimeters often use absorber layers with a thickness of the order of 1
X_0. In the case of iron, this corresponds to $\sim 0.1\lambda_{int}$, and for lead to $0.033\lambda_{int}$. Since
pions travel, on average, one interaction length or more before undergoing a nuclear
interaction (Section 2.3.4.1), they may traverse tens of active layers.

If the number of different pions contributing to a calorimeter signal equals N and
each pion traverses, on average, 10 signal planes, then the number of statistically independent crossings amounts to N, while the contribution to the signal is equivalent to
$10N$ particle crossings. Sampling fluctuations are in this case thus increased by a factor
$\sqrt{10}$, compared with em showers in which each shower particle contributes only once.
As indicated above, the precise increase in the sampling fluctuations for this shower
component depends on the thickness of the sampling layers, expressed in units of λ_{int}.

To examine the consequences of the second effect, we consider, as an example, a
100 MeV spallation proton crossing a 3 mm polystyrene-based scintillator plane. The
specific ionization, $\langle dE/dx \rangle$, of this particle in this material amounts to 7.7 MeV cm^{-1},
and the proton thus loses 2.3 MeV in this process. This has to be compared to the specific
ionization and mean energy loss of a mip, 2 MeV cm^{-1} and 0.6 MeV, respectively.
Assuming a k_B value for this type of scintillator of 0.00948 cm MeV^{-1} [Zeu 86], and
applying Equation 3.13 to find the signal reduction due to saturation, we find that the
signal from this spallation proton is typically 3.5 times larger than that of a mip crossing
a signal plane.

A similar calculation yields an increase by a factor of 5.5 (compared with a mip) for
the signal from a 50 MeV proton and by a factor of about ten at 20 MeV.

On the other hand, the signals from recoil protons produced in elastic neutron scattering in the plastic are typically *smaller* than those from mips traversing a scintillator
plane. A 1 MeV recoil proton has a specific ionization of 280 MeV cm^{-1} in polystyrene.
Therefore, its signal is reduced by a factor of about four due to saturation, and is thus
equal to that of a mip depositing 0.25 MeV. Since mips deposit 0.6 MeV in this scintillator plane, the signal from such recoil protons is thus only about 40% of that of a
mip.

The latter effect crucially depends on the dimensions of the active planes. For example, in a scintillator calorimeter with 0.5 mm fibers as active material, the same recoil
protons would produce signals that are *larger* than those of mips (which deposit only

100 keV in that case), by a factor of 2.5.

The purpose of this discussion is to show that the contribution to the energy resolution from sampling fluctuations is larger for hadrons than for electrons, in the same sampling calorimeter. How much larger depends on a variety of details, such as the energy sharing between the em and non-em shower components, the thickness of the sampling layers (in units of λ_{int}), the thickness and the saturation properties of the active calorimeter components, and the relative contribution of recoil protons to the signals from the non-em shower component.

The contribution of sampling fluctuations to the hadronic energy resolution was measured by Fabjan and coworkers for a generic Fe/LAr calorimeter [Fab 77]. The result was quoted as [Fab 85]

$$(\sigma/E)_{\mathrm{samp}} = 0.09\sqrt{\Delta\epsilon(\mathrm{MeV})} \cdot \frac{1}{\sqrt{E(\mathrm{GeV})}} \tag{4.19}$$

in which $\Delta\epsilon$ represents the average energy lost by a mip traversing a sampling layer. The ZEUS group measured the sampling fluctuations for both em and hadronic showers in compensating lead/plastic-scintillator and uranium/plastic-scintillator calorimeters [Dre 90]. They found the hadronic fluctuations to be twice as large as the em ones and quote the results for the hadronic sampling fluctuations in these calorimeters as $a_{\mathrm{samp}} = 0.115\sqrt{\Delta\epsilon(\mathrm{MeV})}$. More details about these measurements are given in Section 4.3.4.

Whereas sampling fluctuations are usually *the* dominant factor determining the energy resolution of em showers in sampling calorimeters, they are often overshadowed by other types of fluctuations in hadronic shower detection. As is shown in Sections 4.6 and 4.7, the hadronic energy resolution of almost all hadron calorimeters is determined by fluctuations in visible energy and (for non-compensating calorimeters) by fluctuations in the em shower content.

There is one notable exception to this rule: gas calorimeters with proportional readout, especially the ones involving hydrogenous gas mixtures. In such calorimeters, densely ionizing particles such as sub-MeV recoil protons (from elastic neutron scattering in the gas) and α particles (from the $(n, 3\alpha)$ reaction on ^{12}C, see Section 2.3.3.3) may produce signals that are two to four orders of magnitude larger than those from mips. This is a result of the absence of saturation effects, which quench the contribution of such processes to the signals of dense active media, such as plastic scintillators. In gas calorimeters, the mentioned processes lead to "hot spots" in the measured energy deposit profile, a phenomenon that has become known as the "Texas tower effect."

In extreme cases, the signal from one readout cell, representing a few cm^3 of the detector volume, may be larger than the combined signals from all other calorimeter cells contributing to the shower. Therefore, such anomalous signals from individual shower particles may lead to very large sampling fluctuations indeed. This effect is discussed in more detail in Section 4.8.2.

4.3.3   *Angular dependence*

In this subsection we address one of the most common misconceptions about calorimetry. In colliding-beam experiments, the central calorimeter system often consists of a
barrel-type structure, with active and passive planes running parallel to the beam line.
In the central rapidity region $(\eta \sim 0)$, the particles produced in the interaction enter the
detector perpendicular to the detector planes, the angle θ between the particles and the
beam line is 90°.

As η increases, the angle θ decreases, and the particles enter the calorimeter in a
skewed direction. Intuitively, one would expect the calorimeter resolution to decrease
as well in this process. Fewer active planes contribute to the signals and therefore the
shower is more crudely sampled, is the reasoning. Since the shower penetration is proportional to $1/\sin\theta$, this is the parameter governing the degradation of the energy resolution. This is the misconception referred to above.

In reality, the energy resolution is barely dependent on the angle θ. This statement is
supported both by experimental data and by Monte Carlo simulations, and can be easily
understood as follows.

Figure 2.46 shows the angular distribution of the shower particles through which the
energy of a 10 GeV electron is deposited in a block of dense matter. A large fraction
of these particles have an angular distribution flat in $\cos\theta$, where θ denotes the angle
with the direction of the particle that initiated the shower. This means that this dominant
shower component is *isotropic*, and thus independent of the direction of the showering
particle.

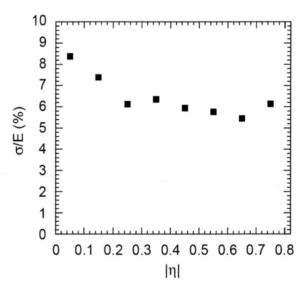

FIG. 4.13. The energy resolution for detection of 180 GeV π^- in the ATLAS TileCal calorimeter, as a function of pseudorapidity [Adr 09].

In a given sampling calorimeter, this dominating shower component, which mainly consists of soft electrons produced in Compton scattering and in photoelectric processes, will produce the same average signal, with the same fluctuations, *regardless of the orientation of the active layers*. In Section 3.5.1 we saw that even in Čerenkov calorimeters, with their rigorous low-energy threshold, this isotropic component represents about half of the em shower signals. In hadron showers, whose signals are even more dominated by isotropically distributed particles (nucleons), this arguments holds *a fortiori*.

Experimental evidence for the independence of the calorimeter resolution on the angle of incidence of the showering particles is presented in Figure 4.13, which shows the energy resolution of the ATLAS hadronic calorimeter (TileCal) for 180 GeV pions, as a function of pseudorapidity η. The η range corresponds to angles of incidence with the calorimeter ranging from $0°$ to $\sim 40°$. There is no indication for a degradation of this resolution when the particles enter the calorimeter non-perpendicularly. If anything, the resolution gets a bit worse for small η values, which may be explained from longitudinal leakage fluctuations, since the calorimeter is least thick in that case.

Because of the limited size of detector prototypes, it is not always possible to perform this type of measurement in an unbiased way. This limitation does not apply to Monte Carlo simulations, where it is of course no problem to define a detector with infinite dimensions.

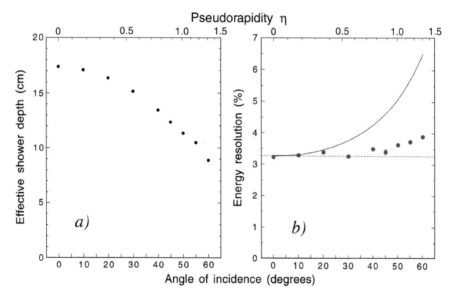

FIG. 4.14. The effective calorimeter depth, *i.e.*, the thickness of the absorber material in which, on average, half of the shower energy is deposited (*a*), and the energy resolution (*b*) for 10 GeV γs in a lead/liquid-argon calorimeter with a sandwich structure (3 mm lead plates, interleaved with 3 mm LAr gaps), as a function of the angle of incidence of the particles ($0°$ is perpendicular to the plates). Results from EGS4 Monte Carlo simulations.

Figure 4.14 shows results of EGS4 simulations, in which 10 GeV γs enter a fine-sampling lead/LAr calorimeter at different angles. For reference purposes, the top scale shows the η value for a colliding-beam experiment with a barrel geometry in which the detector plates run parallel to the beam line. As the angle between the incoming particles and the longitudinal calorimeter axis increases, the effective shower depth decreases, *i.e.*, the shower energy is deposited in a less thick block of absorber material and, therefore, fewer active layers contribute to the signal. This is illustrated by the results in Figure 4.14a. However, the energy resolution is barely affected by this. The increase of the resolution is much smaller than one would expect on the basis of the $\cos \theta$ reduction in the effective number of active layers (indicated by the curve in Figure 4.14b). Between $\eta = 0$ (where the particles enter the calorimeter perpendicular to its front face) and $\eta = 1.5$ (where they enter at an angle of $\sim 65°$ with the longitudinal calorimeter axis), the energy resolution for detecting these γs increases by less than 30%.

The sampling fluctuations in this calorimeter are dominated by fluctuations in the number of particles that cross the boundary between the passive and active layers. The fact that fewer active layers contribute to the signals is largely compensated by the fact that every layer sees, on average, *more* shower particles. As we saw in Section 3.5.1, the response of calorimeters based on ionization or scintillation is independent of the angle of incidence. An increase of the sampling fluctuations could thus only be achieved if fewer shower particles contributed to the signals and that would imply an increase in the average pathlength of the shower particles in the active layers (and thus a larger signal per shower particle). However, such an increase of the average pathlength is barely a factor, because of the largely isotropic character of the shower particles contributing to the calorimeter signals (see Figure 2.46).

This important feature of sampling calorimeters has inspired in recent years a number of calorimeter designs in which the classical "sandwich" geometry of alternating passive and active calorimeters is replaced by a different geometry, optimized for the purposes of the particular application.

Examples can be found in the ATLAS experiment at the LHC. The Pb/LAr em calorimeter has an "accordion" geometry, while the hadronic calorimeter consists of iron and plastic scintillator tiles, which are arranged in such a way that the showering particles enter the calorimeter in a direction *parallel to these planes*.

Another clear example of this trend away from the sandwich structure are fiber calorimeters, in which the particles enter the calorimeter in the "wrong" direction, *i.e.*, along the fibers. This structure allows in principle for a completely hermetic construction, and a very high sampling frequency. Such fiber calorimeters also provide the only exception to the statement made earlier in this subsection that the energy resolution is independent of the angle of incidence of the showering particles. This resolution deteriorates when electrons or γs enter a detector of this type at very small angles with the fiber direction. This phenomenon is a consequence of the fact that in an em shower a large fraction of the energy is deposited in a very narrow cylinder around the shower axis (Figure 2.16). When this narrow cylinder coincides with the position of a fiber, the energy contained

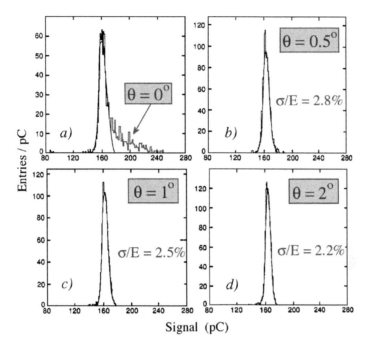

FIG. 4.15. Signal distributions for 40 GeV electron showers measured with the RD1 0.5 mm
fiber calorimeter, for different angles between the particle's direction and the fiber [Bad 94a].

in it is sampled with a very high effciency, resulting in an anomalously large signal.

This is illustrated in Figure 4.15, which shows signal distributions for 40 GeV electron showers, measured with the RD1 fiber calorimeter at different angles between the
particle's direction and the scintillating fibers [Bad 94a]. The anomalous signals mentioned above are clearly visible as a high-energy tail in the signal distribution for $\theta = 0$.
Owing to the small diameter of the fibers in this calorimeter (0.5 mm), this effect is
limited to extremely small angles of incidence. For $\theta = 1°$, the signal distribution is already almost perfectly Gaussian, although the width of the distribution is still somewhat
larger than for larger angles of incidence.

This phenomenon, which is often referred to as *channelling*, was also observed by
the SPACAL (Section 4.4.2) and RD52 (Section 8.2.7) Collaborations. Interestingly, it
is absent for calorimeters in which the fibers only detect the Čerenkov light produced in
the shower development (Figure 8.33). Fortunately, the probability that electrons or γs
enter the calorimeter at the small angles for which this effect plays a role is in practice
very small and can be avoided altogether by the design of the experiment.

4.3.4   *How to measure (effects of) sampling fluctuations?*

If one wants to measure the effects of sampling fluctuations, the recipe must have the
same elements as in case of the signal quantum fluctuations (Section 4.2.5):

- Choose conditions in which sampling fluctuations contribute substantially to the total energy resolution of the calorimeter.
- Change the sampling fluctuations, while keeping everything else the same.
- Derive the contribution of sampling fluctuations from a comparison of the total energy resolution before and after these changes were made.

This procedure was followed by the ZEUS Collaboration, who measured the contribution of sampling fluctuations to the electromagnetic and hadronic energy resolutions of their compensating uranium/plastic-scintillator and lead/plastic-scintillator calorimeter prototypes [Dre 90].

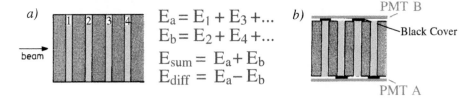

$$E_a = E_1 + E_3 + ...$$
$$E_b = E_2 + E_4 + ...$$
$$E_{sum} = E_a + E_b$$
$$E_{diff} = E_a - E_b$$

FIG. 4.16. Schematic structure of the ZEUS calorimeter (a) and the configuration used to measure the contribution of sampling fluctuations to the energy resolutions for em and hadronic showers (b).

These calorimeters consisted of 2.5 mm thick polystyrene-based scintillator sheets, alternating with either 3 mm thick ^{238}U plates or 10 mm thick lead plates. In order to measure the effects of sampling fluctuations, the calorimeter was split into "two interleaved calorimeters," in the same way as the liquid-argon calorimeter of Willis and Radeka [Wil 74] discussed in Section 4.3.1. This is schematically illustrated in Figure 4.16a. The signals from the odd-numbered scintillator layers were summed up to form "Signal E_A," while "Signal E_B" consisted of the summed signals from the even-numbered layers. In this way, the detector could be considered as two independent sampling calorimeters, A and B, embedded in one and the same instrument.

Technically, this was achieved by covering one side of each scintillator plate with black tape. For the odd-numbered plates, the light from the left-hand side of the scintillators was prevented from reaching the wavelength-shifting plate (and its associated PMT which produced "Signal E_B"). For the even-numbered plates, the light from the right-hand side of the scintillators was prevented from contributing to "Signal E_A." This is graphically illustrated in Figure 4.16b.

For each shower, the signals from A and B were recorded and separate distributions were made of E_A, E_B, and the combinations $E_A + E_B$ (E_{sum}) and $E_A - E_B$ (E_{diff}). It is useful to define the following variables:

$$\sigma_A = \frac{\Delta E_A}{\langle E_A \rangle}, \quad \sigma_B = \frac{\Delta E_B}{\langle E_B \rangle}, \quad \sigma_{sum} = \frac{\Delta E_{sum}}{\langle E_{sum} \rangle}, \quad \sigma_{diff} = \frac{\Delta E_{diff}}{\langle E_{sum} \rangle}$$

which denote the normalized standard deviations of these four distributions.

The resolution of the complete calorimeter is then given by σ_{sum}, while σ_{diff} measures the contribution of sampling fluctuations, σ_{samp}, to this total resolution. This is most easily understood if we consider em showers. In the approximation that only sampling fluctuations contribute to the em energy resolution of sampling calorimeters, one expects $\sigma_{\text{sum}} = \sigma_{\text{samp}}$ and $\sigma_A = \sigma_B = \sigma_{\text{sum}}\sqrt{2} = \sigma_{\text{samp}}\sqrt{2}$, since the sampling fraction in each of the two interleaved calorimeters A,B is a factor of two smaller than that of the total instrument. And if fluctuations in the signals from A and B are completely uncorrelated, $\Delta E_{\text{sum}} = \Delta E_{\text{diff}}$ and, therefore, $\sigma_{\text{sum}} = \sigma_{\text{diff}} = \sigma_{\text{samp}}$.

If other types of fluctuations contribute to the energy resolution, as in the detection of hadron showers, one will find $\sigma_{\text{sum}} > \sigma_{\text{diff}}$ and $\sigma_{A,B} < \sigma_{\text{sum}}\sqrt{2}$. In the extreme case, where the contribution of sampling fluctuations to the total resolution is negligible, σ_A and σ_B would be about equal to σ_{sum}, since sampling fluctuations are the only factor in which these two configurations differ. In general, the contributions of sampling fluctuations can be derived from the measured values of σ_A, σ_{sum} and σ_{diff} by assuming $\sigma_{\text{samp}}(A) = \sigma_{\text{samp}}(A+B)\times\sqrt{2}$.

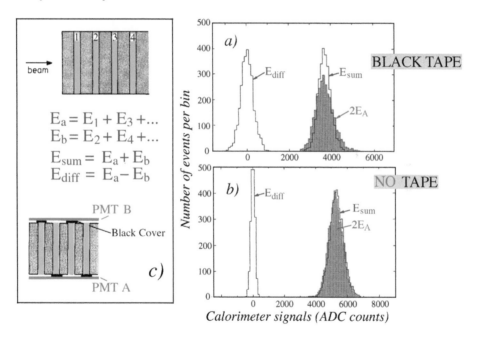

FIG. 4.17. Pulse height distributions for 30 GeV hadrons obtained with the ZEUS lead/plastic-scintillator prototype calorimeter. Diagram a) shows the distributions of E_{sum}, E_{diff} and $2E_A$, measured in the configuration depicted in Figure 4.16b. Diagram b) shows the same distributions measured in the same configuration, but with the black tape removed. See text for details. From [Dre 90].

In practice, the situation was slightly more complicated than described above, because fluctuations in the numbers of photoelectrons and effects of light attenuation in the scintillator plates also contributed measurably to the total energy resolution, especially for em showers. This meant that the difference in em resolution between calorimeter A and the total instrument was not only affected by the change in sampling fluctuations, but also by photoelectron statistics.

This contribution to the resolution was measured by the authors in a way very similar to the one previously described for sampling fluctuations. The only difference was that they removed the black tape (see Figure 4.16b) for this purpose. In this new geometry, the two signals E_A and E_B resulting from a developing shower were thus measured with one and the same sampling device, but each signal contained only (approximately) half of the photoelectrons generated in the event. Therefore, σ_{diff} measured in that case the fluctuations in the number of photoelectrons.

An example of the experimental results obtained in these measurements is shown in Figure 4.17. This figure contains pulse height distributions measured for 30 GeV hadrons with the 10 mm lead/2.5 mm plastic-scintillator prototype calorimeter. Each of the two diagrams shows three distributions: E_{sum}, E_{diff} and $2E_A$. The latter distribution was obtained by multiplying the A signals by a factor of two and thus has the same central value as the distribution of the summed signals of A and B (E_{sum}). Diagram a) shows the results obtained in the configuration depicted in Figure 4.16b, *i.e.*, with black tape, in diagram b) the black tape was removed. This figure exhibits the following features:

- When the tape was removed, σ_{diff} became considerably smaller. This means that fluctuations in the number of photoelectrons were much smaller than the sampling fluctuations.

- When the tape was removed, the distributions of E_{sum} and $2E_A$ were practically identical. Since these two distributions only differ in the average number of photoelectrons constituting the signals, this means that fluctuations in the number of photoelectrons did not contribute significantly to the hadronic energy resolution of this calorimeter.

- When the black tape was in place, the distributions of E_{sum} and $2E_A$ were *not* identical. In this configuration, the sampling fraction was different by a factor of two for these two distributions. Sampling fluctuations were, therefore, a major contribution to the hadronic energy resolution of this calorimeter.

- The distributions of E_{sum} and E_{diff} had practically the same width when the black tape was removed. This means that sampling fluctuations were not only a major contribution to the hadronic energy resolution, but that these fluctuations completely dominated the resolution.

After a careful analysis of all available experimental information, the authors were able to unravel the em and hadronic energy resolutions of these two compensating calorimeters in their different contributing components. The results of this analysis are summarized in Table 4.1.

From these results, the following conclusions can be drawn:

1. The energy resolutions for em showers are strongly dominated by sampling fluctuations in these calorimeters, with a minor contribution coming from photoelec-

Table 4.1 *The contributions of sampling fluctuations and intrinsic fluctuations to the energy resolutions for electrons and pions in compensating uranium/plastic-scintillator and lead/plastic-scintillator calorimeters. Listed are the values of the coefficient* a *(Equation 4.1), expressed in %. Data from [Dre 90].*

Fluctuations	3 mm uranium / 2.5 mm plastic		10 mm lead / 2.5 mm plastic	
(%)	Electrons	Pions	Electrons	Pions
σ_A, σ_B	26.6±1.0	49.5±1.0	36.0±1.0	60.5±1.0
σ_{sum}	18.5±1.0	37.3±1.0	24.5±1.0	43.5±1.0
σ_{diff}	19.2±1.0	32.6±1.0	25.8±1.0	42.3±1.0
σ_{samp}	16.5±0.5	31.1±0.9	23.5±0.5	41.2±0.9
$\sigma_{\text{samp}}/\sqrt{\Delta\epsilon}$	6.1±0.2	11.6±0.4	6.4±0.2	11.3±0.3
σ_{intr}	2.2±4.8	20.4±2.4	0.3±5.1	13.4±4.7

tron statistics (increasing $a_{\text{samp}} = 16.5\%$ to $a_{\text{total}} = 18.5\%$ for the uranium detector).

2. The sampling fluctuations for electrons are in good agreement with the general formula we derived in Section 4.3.1 (Equation 4.18). For the uranium calorimeter, this formula predicted $a_{\text{samp}} = 15.6\%$, while the experimental value was measured to be $16.5 \pm 0.5\%$. For lead, the prediction of 21.9% was also close to the experimental value ($23.5 \pm 0.5\%$). The em sampling fluctuations are about a factor of two larger than expected on the basis of the Rossi's "Approximation B" (Equation 4.16).

3. The energy resolutions for hadrons are also strongly dominated by sampling fluctuations in these calorimeters, especially in the lead one. As a result, the resolutions of calorimeters A and B are, *in all cases*, within experimental errors equal to the resolution of the complete instrument times $\sqrt{2}$.

4. The hadronic sampling fluctuations are about a factor of two larger than the em sampling fluctuations. Expressed in terms of the energy loss by a mip in one sampling layer, $\Delta\epsilon$ (in MeV), a_{samp} amounts to $\sim 11.5\%\sqrt{\Delta\epsilon}$ for hadrons and $\sim 6.2\%\sqrt{\Delta\epsilon}$ for electrons.

5. The contributions of "intrinsic" fluctuations, *i.e.*, the fluctuations that remain after subtracting the contributions from sampling fluctuations, photoelectron statistics and other instrumental effects from the total measured hadronic energy resolution, scale with $E^{-1/2}$. This is typical for compensating calorimeters.

6. The intrinsic fluctuations are smaller in the lead calorimeter than in the uranium one.

The last two points are extensively discussed in Sections 4.6–4.8.

Measurements similar to the ones discussed here were performed by Fabjan and coworkers [Fab 77] on liquid-argon calorimeters (in fact, these measurements served as inspiration for the ZEUS study discussed above). Figure 4.18 shows some results from these measurements, obtained with a calorimeter that consisted of 1.5 mm iron plates separated by 2 mm wide gaps filled with liquid argon. In this figure, the total energy

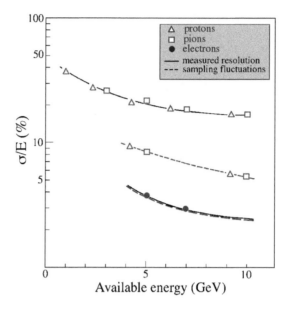

FIG. 4.18. The energy resolution and the contribution from sampling fluctuations to this resolution measured for electrons and hadrons, in a calorimeter consisting of 1.5 mm thick iron
plates separated by 2 mm gaps filled with liquid argon. From [Fab 77].

resolution and the contribution of sampling fluctuations to this resolution are given for
electrons and hadrons, as a function of energy.

There are some similarities between these results and those from Drews *et al.* discussed above, but there are also some striking differences. Both experiments found that
the energy resolution for electrons was completely dominated by sampling fluctuations,
and that a_{samp} was about twice as large for hadron showers as for electron showers.
However, the hadronic sampling fluctuations were only a very minor contribution to the
total energy resolution of the liquid-argon device, while they dominated the hadronic
resolution of the compensating calorimeters. Moreover, as the energy increased, the
relative importance of sampling fluctuations diminished even more in the liquid-argon
calorimeter. At 10 GeV, the hadronic resolution was already more than three times as
large as one would expect if only sampling fluctuations would play a role.

These differences are a direct consequence of the non-compensating nature of the
Fe/LAr calorimeter. They illustrate some of the consequences of non-compensation for
the hadronic energy resolution, which are discussed in much more detail in Sections
4.6–4.8.

4.4   Instrumental Effects

In the previous sections, we have treated calorimeters as ideal instruments, with infinite
dimensions, operating in an environment free of noise and background radiation, with
active elements that respond identically to a given type of energy deposit, no matter

where or in which of the elements it occurs.

In real life, neither the calorimeter construction nor the environment in which it has to operate are ideal, and this has consequences for the calorimeter performance. In this section, we investigate the consequences of some common effects on the energy resolution.

The effects that are discussed have in common that the associated fluctuations do not scale with $E^{-1/2}$. This means that their relative contribution to the total energy resolution is energy dependent. Most effects, namely those that cause energy-independent fluctuations, are dominating the energy resolution at very high energies, where the contributions from the processes governed by Poisson statistics are small. However, the first effect to be discussed is an exception to this rule, since it dominates the resolution at low energies.

4.4.1 Electronic noise

For sampling calorimeters with an active medium based on direct collection of the charge produced in the ionization-chamber (e.g., liquid argon) or proportional mode (wire chambers), the signals typically amount to a few picocoulombs of charge per GeV of shower energy.

The signals produced by these calorimeters correspond to the charge collected during a certain time, which we will call the "gate time" in the following. Since the detector has a certain capacitance, there is inevitably a contribution of electronic noise to the signals. This means that, in the absence of a showering particle, the integrated charge collected during the gate time fluctuates from event to event.

The standard deviation of these fluctuations, σ_{noise}, is measured in units of charge. Since the calorimeter measures the energy of showering particles in the same units, this noise term is equivalent to a certain amount of calorimetric energy, e.g., 100 MeV. The precise value of this energy-equivalent noise contribution (ENC) depends on a variety of factors, such as the gate time, the detector capacitance and the properties of the charge amplification electronics. Willis and Radeka have shown [Wil 74] that this noise term can be minimized with optimal capacitance matching between the detector and the charge amplifier.

Since the standard deviation of the electronic noise fluctuations corresponds to a certain, fixed energy, the contribution of this noise to the energy resolution of calorimetric shower measurements, σ/E, scales like E^{-1}. This is illustrated in Table 4.2 and Figure 4.19, in which the total energy resolution and the contributions to this resolution from noise and from stochastic (e.g., sampling) fluctuations are given for a liquid-argon calorimeter with the following (fictitious, but not unrealistic) properties.

The calorimeter consists of towers, square cylindrical regions in this case. The charge collected from each tower can be integrated separately. The ENC amounts to 200 MeV per tower. The showers deposit typically 90% of their energy in one tower when they enter it in its center. In that case, stochastic fluctuations contribute $15\%/\sqrt{E}$ to the energy resolution. If the signals are integrated over 7 channels, more than 99% of the energy is contained and the stochastic fluctuations are reduced to $10\%/\sqrt{E}$.

The stochastic fluctuations are completely uncorrelated to the noise and the noise in

Table 4.2 *The energy resolution (σ/E, in %) and its contributing terms for a liquid-argon calorimeter of which the properties are described in the text.*

Pion Energy	1 channel			7 channels		
(GeV)	*noise*	*stoch.*	*total*	*noise*	*stoch.*	*total*
1	20.0%	15.0%	25.0%	52.9%	10.0%	53.8%
2	10.0%	10.6%	14.6%	26.5%	7.1%	27.4%
5	4.0%	6.7%	7.8%	10.6%	4.5%	11.5%
10	2.0%	4.7%	5.1%	5.3%	3.2%	6.2%
20	1.0%	3.4%	3.5%	2.7%	2.2%	3.5%
50	0.4%	2.1%	2.1%	1.1%	1.4%	1.8%
100	0.2%	1.5%	1.5%	0.5%	1.0%	1.1%
200	0.1%	1.1%	1.1%	0.3%	0.7%	0.8%

the various electronic channels is completely incoherent (uncorrelated). Therefore the total energy resolution can be written as

$$(\sigma/E)_{\text{tot}} = \frac{0.2}{E} \oplus \frac{15\%}{\sqrt{E}} \qquad (4.20)$$

when only the signals from the central calorimeter tower are used, and as

$$(\sigma/E)_{\text{tot}} = \frac{0.2\sqrt{7}}{E} \oplus \frac{10\%}{\sqrt{E}} \qquad (4.21)$$

when the signals are integrated over 7 towers. As usual, the energy E is expressed in GeV. The results are tabulated in Table 4.2, and graphically shown in Figure 4.19.

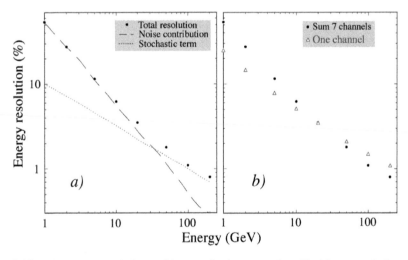

FIG. 4.19. The energy resolution and its contributing terms for a liquid-argon calorimeter of which the properties are described in the text.

This exercise shows several interesting aspects of the noise issue. First, because of its E^{-1} dependence, the noise term dominates the energy resolution at low energy, up to 10 GeV in the case of single-tower readout, up to 20 GeV when the signals are integrated over 7 towers. It also shows that including more shower information does not necessarily lead to a more accurate determination of the shower energy. Adding more towers has the effect that the noise is increased (by a factor $\sqrt{7}$ in this case), which *deteriorates* the energy resolution of this calorimeter for energies up to 20 GeV.

Although these effects mainly play a role in calorimeters based on direct collection of the ionization charge, they may affect other types of calorimeters, in particular scintillator-based ones, as well. In scintillator calorimeters, the PMT signals are digitized and analyzed by means of analog-to-digital converters (ADCs). When no PMT signals are offered, the ADC may still accumulate a certain amount of charge during the gate time, resulting in a "pedestal." To find the calorimeter signals, the pedestal has to be subtracted from the raw signals.

However, the pedestal may exhibit fluctuations, *e.g.*, due to ground loops, improper impedance matching and a variety of other, electronic problems. These pedestal fluctuations play the same role for scintillator calorimeters as electronic noise does for ionization calorimeters. The difference is that pedestal fluctuations can in general be made insignificant, for example by increasing the gain of the PMTs, and thus the size of the signals. In liquid-argon calorimeters, which are based on the collection of *non-amplified* ionization charge, this option does not exist.

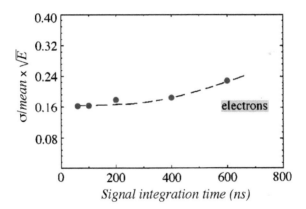

FIG. 4.20. The energy resolutions for 5 GeV/c electrons in the ZEUS calorimeter, as a function of the charge integration time [Kru 92].

A type of electronic noise that could *not* be eliminated by increasing the gain of the PMTs was found by ZEUS, where the radioactive decay of the ^{238}U absorber plates contributed to the signals. As the charge integration time was increased, the energy resolution of detected electron showers deteriorated as a result of this effect (Figure 4.20).

4.4.2 *Variations in sampling fraction*

The second source of instrumental effects that may contribute to the energy resolution may manifest itself in a variety of different ways, depending on the type of calorimeter. The effects discussed in this subsection are in fact so diverse that, at first sight, they may appear unrelated. Yet, they all have the same common origin: event-to-event fluctuations in the sampling fraction (and thus the response) of the calorimeter volume in which the shower develops.

Some of the fluctuations discussed here are *inherent to the structure* of the calorimeters they affect, while others are not. In the absence of a better term, we will call these fluctuations inherent and non-inherent, respectively.

4.4.2.1 *Fiber calorimeters.* The first type of sampling fraction fluctuations that we discuss is inherent to the structure of the calorimeters that it affects. These calorimeters use fibers oriented at $0°$ as active elements. An example of such a device is the Spaghetti Calorimeter (SPACAL), which consists of plastic scintillating fibers (diameter 1.0 mm) embedded in a lead matrix according to a hexagonal pattern, in which each fiber is equidistant (2.22 mm) to its six nearest neighbors (Figure 4.21a).

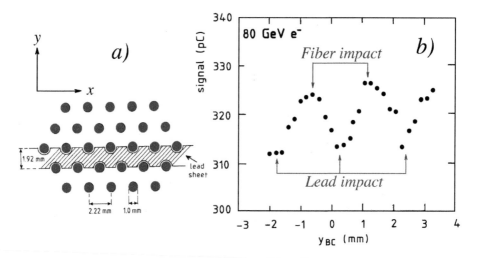

FIG. 4.21. Lateral cross section of SPACAL (*a*). The SPACAL signal as a function of the y-coordinate of the impact point (*b*). Data for 80 GeV electrons. From [Aco 91c].

Figure 4.21b shows the results of a vertical scan performed with 80 GeV electrons across the surface of this calorimeter [Aco 91c]. The electrons entered the detector in the horizontal plane (*i.e.*, the tilt angle $\phi = 0$), at an azimuth angle $\theta = 3°$ between the beam line and the fiber axis. The scan was performed in very small steps, 0.2 mm each. In the figure, the average calorimeter signal is plotted versus the y-position of the impact point. The oscillating pattern has a period of 1.9 mm, which corresponds to the vertical distance between the fiber planes (Figure 4.21a). The amplitude of this oscillation amounts to about 2% of the average signal, and the rms deviation from the mean value is about 1.3%.

This phenomenon is caused by the position dependence of this calorimeter's sampling fraction for narrow showers. Depending on the impact point of the electrons, the part of the shower contained in a narrow cylinder with a diameter of about 1.0 mm around the shower axis is either not sampled at all (for impact points in between two fiber planes) or very efficiently (for impact points in the fiber plane). Since the energy deposit is strongly concentrated around the shower axis in em showers, this difference in sampling fraction for the innermost core causes a non-negligible difference for the sampling fraction of the shower as a whole, resulting in a position-dependent response.

The correctness of this explanation is corroborated by the following experimental observations:

- The effect disappeared when the detector was slightly tilted, *i.e.*, the angle ϕ was made about 3° [Ben 92]. In this way, the possibility that a narrow cylinder around the shower axis did not include any fiber, and that energy contained in such a cylinder thus went unsampled, was effectively eliminated.

- Similar measurements done with a much finer-sampling calorimeter (0.5 mm fibers instead of 1 mm ones, with the same packing fraction) revealed that the effect, although still observable, was considerably smaller, for the same angles of incidence [Bad 94a]. This is understandable, since the volume of the unsampled cylinder around the shower axis was reduced by a factor of four, compared with the SPACAL case.

- The effect was totally absent for hadron detection, at any angle of incidence. This is because the energy fraction contained in the unsampled cylinder around the shower axis is much smaller for hadron showers than for em ones.

The first observation mentioned above also provided an effective recipe for avoiding this phenomenon and its consequences in practice. A calorimeter of this type should be designed in such a way that the particles to be detected cannot enter it at angles θ or ϕ smaller than, say, 2° with the fiber direction. In colliding-beam experiments this is almost automatically taken care of by the magnetic field (for the charged particles) and by the fact that the interactions take place in an extended region of beam overlap and not all at the same point. In addition, the calorimeter may be designed in such a way that the fibers do not point to the beam line, but rather to a cylinder surrounding this beam line.

Figure 4.21b shows that response fluctuations of the type discussed here take place on a scale of a fraction of 1 mm. This is smaller than the position resolution usually achieved for shower detection in calorimeters. In beam tests, the position of the incoming particles can usually be measured with great precision, with the help of dedicated additional detectors (beam telescopes). In that case, these response fluctuations may be accounted, and thus corrected, for event by event, on the basis of the known impact point. This possibility was exploited by SPACAL. Figure 4.22 shows that a considerable improvement was achieved in the em energy resolution when only events were selected in which the beam particles entered the calorimeter in the lead region in between two

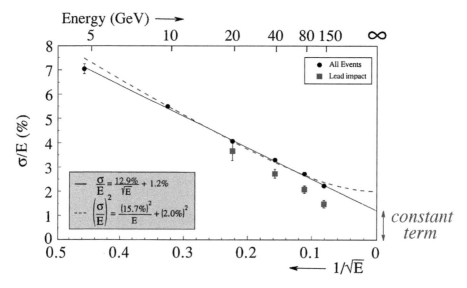

FIG. 4.22. The SPACAL energy resolution for electrons as a function of energy, for $\theta = 3°$ and $\phi = 0°$. The closed circles encompass all events; the open squares are for electrons entering the detector in the lead. From [Aco 91c].

fiber planes.

However, in the practice of particle physics experiments the available position information is usually inadequate for such corrections, and the result is thus an increase in the energy resolution for em shower detection.

Since the lateral em shower profiles are almost independent of energy, the fraction of the shower energy contained in the unsampled cylinder is also energy independent. Therefore, this effect contributes a constant term to the em energy resolution of fiber calorimeters:

$$(\sigma/E)_{\text{tot}} = \frac{a_1}{\sqrt{E}} \oplus a_2 \qquad (4.22)$$

This term a_2 manifests itself mainly at high energy, where the E-dependent terms are small. The straight line through the experimental points (the full circles) in Figure 4.22 does not extrapolate to the bottom right corner of the graph, but to a resolution of 1.2% at infinite energy. This straight line describes the linear sum of the two terms in Equation 4.22.

If the two terms in Equation 4.22 are added in quadrature (which is necessary if both terms are completely uncorrelated, which is debatable in this case), the total energy resolution is described by the dashed line in Figure 4.22.

As was mentioned above, the SPACAL Collaboration used wire chambers installed upstream of their fiber calorimeter to select events in which the electrons entered the calorimeter in a very small area without fibers. The full squares in Figure 4.22 indicate how the energy resolution improved for these event samples. Clearly, the constant term a_2 was greatly reduced in that case.

4.4.2.2 *Accordion calorimeters.*   A second type of calorimeter in which inherent fluctuations in the sampling fraction play a role is the "accordion" Pb/LAr detector developed for the ATLAS experiment at CERN's Large Hadron Collider.

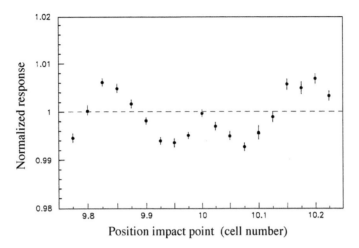

FIG. 4.23. The em response as a function of the impact position of the particles, for 90 GeV electrons in a prototype Pb/LAr accordion calorimeter for the ATLAS experiment [Atl 96a].

In Figure 4.23, the average calorimeter signal from 90 GeV electrons is plotted as a function of the impact point, for a prototype of this device. Also here, a regular pattern is visible, which results from a position-dependent sampling fraction. This pattern closely follows the Monte Carlo prediction of this effect. The constant term in the em energy resolution resulting from this effect amounts to $\sim 0.4\%$ for this calorimeter.

4.4.2.3 *Longitudinal fluctuations.*   In the previous two examples, the fluctuations in the sampling fraction (and thus in the calorimeter response) occurred as the impact point of the particles was moved over the calorimeter surface. We now turn to potentially much more devastating types of fluctuations, those occurring in a calorimeter in which the sampling fraction changes with depth.

One of the most serious mistakes that can be made in the design of a calorimeter is non-uniformity in depth. It is often attractive to make this mistake. We give two examples.

1. One wants to build a hadronic Fe/LAr calorimeter. To save money, the detector is made fine-sampling in the first two interaction lengths, where most of the energy is deposited. As the depth increases, the sampling fraction is gradually decreased. So the detector structure is as follows. Iron plates of 5 mm in the first two interaction lengths, 8 mm in the third, 10 mm in the fourth, 12 mm in the fifth, 15 mm in the sixth and 20 mm in the remaining four interaction lengths. The width of the argon gaps is 5 mm throughout the detector.

2. One wants to build a projective fiber calorimeter, for a 4π experiment. The front

face of the calorimeter is located at 2 m from the interaction point, the rear end at 4 m. To maintain the projectivity as much as possible, all fibers are arranged in projective patterns as well. This means that the fiber-to-fiber spacing gradually increases from 2 mm at the front face to 4 mm in the back.

Both described calorimeters will have a very poor energy resolution as a result of longitudinal fluctuations in the sampling fraction. Let us compare two high-energy pion showers developing in the Fe/LAr calorimeter. These develop completely identically, except that their starting point (*i.e.*, the location of the first nuclear interaction) is different. They deposit 30% of their energy in the first λ_{int} beyond that starting point, 30% in the second λ_{int}, 20% in the third, 10% in the fourth, 5% in the fifth and 5% in the sixth.

The first shower starts right at the front face of the calorimeter, the second pion penetrates $2\lambda_{int}$ before undergoing its first nuclear interaction. The amount of ionization charge created by the first shower ($30/5 + 30/5 + 20/8 + 10/10 + 5/12 + 5/15 = 16.25$) in the described structure is 70% larger than that of the second shower ($30/8 + 30/10 + 20/12 + 10/15 + 5/20 + 5/20 = 9.58$).

One quarter of the pions penetrate at least $2\lambda_{int}$ before undergoing their first nuclear interaction (see Section 2.3.4.1). Therefore, the event-to-event fluctuations in the ionization charge created in this type of detector are very large, at least an order of magnitude larger than the effects of the lateral response fluctuations discussed above.

In calorimeter #2 mentioned above as an example of a flawed design, the sampling fraction varies *gradually* with depth. Assuming that $\lambda_{int} = 20$ cm, the sampling fraction is, at a depth of $1\lambda_{int}$, reduced to 83% of the value at the front face of the calorimeter. At a depth of $2\lambda_{int}$, the sampling fraction is down to 69% of this value, at $3\ \lambda_{int}$ it is 59%, at $4\lambda_{int}$ it is 51%, *etc.* Event-to-event fluctuations in the light yield of identical pion showers, starting at different depths inside this calorimeter, are at the 15% level in this device, independent of the energy.

These effects are not limited to hadron showers. Figure 4.24 shows the results of a Monte Carlo (EGS4) study of the effects of longitudinal detector non-uniformity on the energy resolution for electrons.

In Figure 4.24a, the energy resolution for high-energy γs is plotted as a function of energy in three different Pb/scintillator calorimeters, all equipped with 2.5 mm thick scintillator plates. The first detector contains 2 mm thick lead plates, the plates in the second detector are 5 mm thick. The absorber structure of the third calorimeter consists of fifty 2 mm thick lead plates, followed by 5 mm thick lead plates in the rest of the device.

The two uniform calorimeters behave as expected. Their resolutions scale nicely with $E^{-1/2}$, as indicated by the straight line fits in the figure. The resolution of the calorimeter with the 5 mm plates is larger than that with the 2 mm ones, by a factor of about 1.6 ($\approx \sqrt{5/2}$). Both energy resolutions are as expected on the basis of Equation 4.18.

However, the resolution of the third calorimeter exhibits completely different characteristics. At energies below ~ 5 GeV, the resolution is not very different from that of the calorimeter with the 2 mm plates. This is not surprising, since almost the entire shower energy is deposited in the $18X_0$ deep fine-sampling section at these low ener-

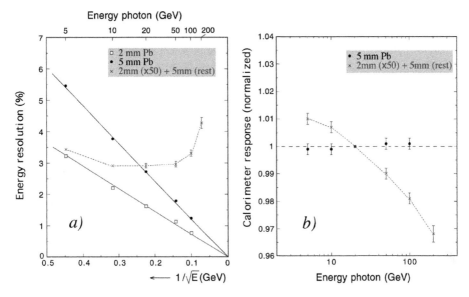

FIG. 4.24. The energy resolution (*a*) and the response (*b*) as a function of energy, for high-energy γs showering in three different Pb/scintillator calorimeters, all equipped with 2.5 mm
thick scintillator plates. Detector 1 uses 2 mm thick lead plates. Detector 2 uses 4 mm thick
lead plates. Detector 3 contains fifty 2 mm thick lead plates, followed by 5 mm thick lead
plates in the rest of the device. Results from EGS4 Monte Carlo simulations.

gies (see also Figure 2.17a). However, as the energy of the γs is increased, the fraction
of the shower energy deposited in the more crudely sampling 5 mm section increases
as well. The sampling fraction in this back section is considerably smaller than in the
front. Therefore, energy deposited in this section of the detector leads to correspondingly smaller signal contributions.

As a result, the response of this calorimeter decreases gradually with increasing
energy and this device is thus intrinsically non-linear (Figure 4.24b). More to the point
for the present discussion is that longitudinal fluctuations in the shower development
increasingly affect the energy sharing between the fine-sampling front section and the
crude-sampling back section. And because of the different sampling fractions for energy
deposited in these two sections, the energy resolution of the calorimeter is affected by
this as well. Figure 4.24a shows that the energy resolution barely improves with energy.
Beyond 100 GeV, where the energy resolution is about the same as at 5 GeV, it rapidly
deteriorates.

Even at these high energies, more than 90% of the shower energy is, on average,
deposited in the first $18X_0$ of the absorbing structure (Figure 2.17a). It is, therefore,
tempting to sample the energy deposited beyond that depth more crudely, since it represents only a small fraction of the total energy. However, this fraction fluctuates strongly
from one shower to the next and that causes the effects described above.

If one wants to subdivide the calorimeter into longitudinal segments with different sampling fractions (and there are often good reasons for doing that), it is thus *absolutely crucial* to record the signals from these segments separately. As illustrated by the above example, failure to do so will lead to a situation where fluctuations in the longitudinal shower development, more than anything else, determine the energy resolution. It should also be pointed out that *any* longitudinal calorimeter segmentation introduces serious, non-trivial problems concerning the intercalibration of the signals from the different segments. These are extensively discussed in Chapter 6.

4.4.2.4 *Cracks.* The effects discussed so far in this subsection are a consequence of the way in which the active and passive detector elements are arranged in the calorimeter. We now turn to the non-inherent fluctuations in the sampling fraction. One type of position dependence, which plays a role in essentially all calorimeters, arises because of engineering considerations. In practice, calorimeters are built according to some modular structure, which has to be supported. The signals have to be transported to the outside world to be measured and analyzed, and utilities (power, cooling, *etc.*) have to be delivered to the detector from the outside world.

As a result, there are usually "cracks" in a calorimeter system, regions that are not (or only partially) sensitive to particles developing showers. An illustrative example of such cracks, and of a solution to deal with them, is provided by ZEUS.

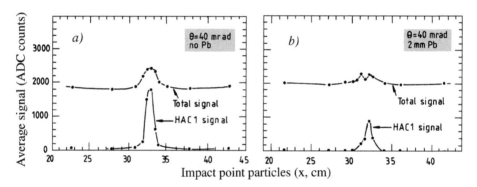

FIG. 4.25. The average signal measured in a position scan with 30 GeV electrons across the surface of the ZEUS calorimeter, before (*a*) and after (*b*) inserting a 2 mm thick lead foil in between neighboring detector modules. The electrons entered the calorimeter at an angle of 40 mrad with its surface in this scan [Beh 90].

The ZEUS forward and central calorimeters consist of alternating uranium and plastic-scintillator plates, oriented (roughly) perpendicular to the direction of the incoming particles. The scintillation light travels sideways and is collected by wavelength-shifting plates that run perpendicular to the sampling layers to the PMTs located at the rear end of the calorimeter. These WLS plates are therefore oriented in approximately the same direction as the one traveled by the incoming particles. The detector modules (20 cm wide) are read out on both sides by three superimposed WLS plates. These plates, together with the module covers, represent $\sim 4\%$ of the total calorimeter surface.

Beam tests in which a 30 GeV electron beam was moved in small steps across the calorimeter surface revealed non-uniformities in the response across the crack (Figure 4.25a). Also the em energy resolution was found to be considerably worse in these boundary regions. The presence of a crack was also revealed by large contributions from the hadronic compartment to the total signals for em showers, a consequence of the absence of dense absorber material upstream.

Interestingly, the signals were *larger* in the crack regions, compared with the rest of the calorimeter. The authors explained this effect as the result of Čerenkov light produced directly inside the WLS plates. This light was converted into wavelength-shifted light with almost 100% efficiency, compared with only a few percent for light produced inside the scintillators.

The authors discovered that the uniformity could be improved by inserting thin metal plates in between the individual calorimeter modules. This is illustrated in Figure 4.25b, which shows the effects of a 2 mm thick lead sheet. The energy resolution also improved as a result of this modification.

4.4.3   *Non-uniformity of active elements*

In the previous subsection, we saw that certain configurations of the active elements of a sampling calorimeter lead to a position-dependent sampling fraction, on a scale that is too small to be resolved on an event-by-event basis. Therefore, this effect causes fluctuations in the calorimeter response which are, in first approximation, energy independent and thus contribute a constant term to the energy resolution.

Similar effects are caused by non-uniformities in the active elements. Unlike the effects discussed before, these non-uniformities are often *avoidable*. They are closely linked to the tolerances maintained in the construction of the detector. Their effects can be minimized by proper quality control.

4.4.3.1   *Liquid-argon calorimeters.*   In liquid-argon calorimeters, non-uniformities may result from variations in the gap width. Such variations may occur over the surface of individual gaps, or from one gap to another. Since liquid-argon calorimeters operate in an ionization-chamber mode, the charge collected from an ionizing particle crossing a gap is directly proportional to the gap width. If a gap with a nominal width of 5 mm varies between 4.5 and 5.5 mm, then the signals from shower particles crossing this gap fluctuate by $\pm 10\%$.

It should be emphasized that shower signals in these calorimeters are usually integrated over many LAr gaps. Therefore, the shower signals are much less sensitive to the mentioned variations than the signals from individual shower particles, provided of course that these variations are random. The worst case of non-randomness occurs when the gaps in the first section of the calorimeter are systematically different (*e.g.*, 4.5 mm) from the gaps deeper inside (*e.g.*, 5.5 mm). In that case, longitudinal fluctuations in the shower's starting point may cause large event-to-event response fluctuations (see Section 4.4.2.3). These considerations illustrate the importance of maintaining tight tolerances in the construction of such calorimeters, and of avoiding systematic effects in the deviations from design specifications.

4.4.3.2 *Scintillating-plate calorimeters.*   Variations in the thickness of the scintillator plates have the same effect for these calorimeters as the gap width in the previous
case. In addition, a number of other effects may play a role. The light produced in the
scintillator plates is somehow guided to the PMTs, where it is converted into an electric signal. This is frequently done by means of wavelength-shifting plates or fibers.
There may be plate-to-plate (or fiber-to-fiber) variations in the efficiency of the process
in which the light is wavelength shifted. The quantum efficiency for the conversion of
light into photoelectrons may also depend on the position where the photons hit the
PMT's photocathode. And finally, there are the effects of light attenuation, which are
discussed separately in Section 4.4.3.4.

Also here, the consequences of these effects may be minimized by paying proper
attention to relevant details in the design and construction of the calorimeter. The ZEUS,
ATLAS and CMS Collaborations, which have built very large calorimeters of this type,
all developed elaborate and sophisticated procedures to deal with these issues.

4.4.3.3 *Fiber calorimeters.*   In scintillating-fiber calorimeters, no separate wavelength
shifters are needed. Therefore, the number of factors that has to be kept under control is
somewhat smaller than for scintillating-plate calorimeters. Fiber-to-fiber thickness fluctuations and position dependence of the quantum efficiency of the PMTs are the major
concerns for these devices.

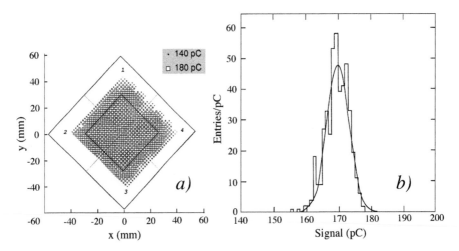

FIG. 4.26. The non-uniformity of 40 GeV electron signals in the RD1 0.5 mm fiber calorimeter.
The size of each square in (*a*) represents the average signal for electrons entering the detector
in that particular area, and serves as one entry in the histogram (*b*) [Bad 94a].

An example of the effect of fluctuations in the fiber thickness is shown in Figure 4.26, which depicts the signal uniformity of the RD1 scintillating-fiber calorimeter
[Bad 94a]. The entire surface of this calorimeter was mapped with a 40 GeV electron
beam. The detector surface was subdivided in some 900 areas with dimensions of 2×2

mm^2. The response for particles entering in each of these areas was measured. For each area, this response is indicated by the size of the square in Figure 4.26a.

This figure shows that the response in the outer regions of the detector was systematically somewhat smaller than average. This is the result of (side) leakage effects. To eliminate these effects, a fiducial area was defined that covered the central region of the calorimeter surface (the inner square in Figure 4.26a). The response of each of the ($\sim$ 500) 2×2 mm^2 areas within this fiducial region was used as an entry for the histogram shown in Figure 4.26b. Analysis of these data showed that fiber-to-fiber response variations contributed an energy independent term of 1.4% to the em energy resolution of this calorimeter.

Since hadronic showers are much broader than em ones, their signals involve many more fibers. Therefore, the hadronic calorimeter signals are much less sensitive to fluctuations in the fiber thickness. In practice, such fluctuations barely affect the hadronic energy resolution of such calorimeters, since the value of the parameter a_1 in Equation 4.22 is large and a_2 is small.

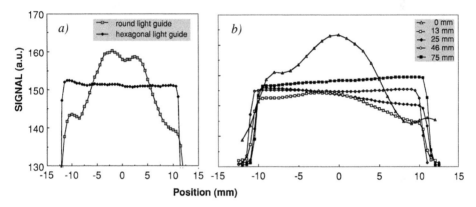

FIG. 4.27. Response to a point-like light source scanning the surface of a light guide/PMT combination. The results concern a circular and a hexagonal cylindrical light guide of the same length, coupled to the same PMT (*a*). Response to a point-like light source scanning the surface of a hexagonal light guide/PMT combination, for various lengths of the light guide (*b*). Note the blown-up vertical scale [Aco 90].

Effects with similar consequences may be caused by non-uniformities in the quantum efficiency of the photocathode, where the light signals are converted into electric ones. This quantum efficiency may vary considerably over the surface of the photocathode, sometimes by factors of two to three. In order to eliminate the effects of this type of non-uniformity on the signals from fiber calorimeters, the light exiting the fibers should, ideally, be uniformly distributed over the entire surface of the photocathode, rather than illuminating a local area of this surface. This can be achieved with a light guide, mounted between the ends of the fibers and the PMT.

It turns out that both the shape and the length of such a light guide are critical parameters for the quality of the light mixing. Figure 4.27a shows the difference in light mixing between two light guides of identical length and composition, but different shape [Aco 90]. A 2 meter long WLS fiber, excited at the far end by means of a UV lamp, served as a point-like light source in these measurements. This source (*i.e.*, the opposite end of the fiber) was moved in small steps across the surface of the light guide, which was optically coupled to a PMT that was equipped with a particularly non-uniform photocathode. Clearly, a much better light mixing was achieved with the hexagonal light guide, compared with the circular one. Dedicated studies by Dukes and coworkers [Duk 90] showed that circular light guides represent the worst possible case, any other shape is better.

Also the length of the light guide is a critical parameter, as illustrated in Figure 4.27b. For very short lengths, the mixing is inadequate, and the non-uniformity profile of the photocathode surface is still visible. Beyond the optimal length, the light guide may act as a lens, producing an inverted, albeit very blurry, image of this non-uniformity profile. The optimal length depends on the numerical aperture and on the transverse dimensions of the light guide. In the SPACAL studies, the optimal length of the hexagonal light guides was found to be $\sim$ 80 mm, corresponding to an aspect ratio (length:apex-to-apex) of 2:1 [Aco 90].

In a dedicated study, Simon and coworkers [Sim 93] showed that a more uniform light mixing could also be achieved by (partially) wrapping imperfect light guides with reflective foil.

4.4.3.4 *Light attenuation.*   A special type of position dependence in the response of scintillation calorimeters is caused by the effects of light attenuation. Light attenuation may be caused by a variety of factors, of which self-absorption and reflection losses are the most common ones. Light attenuation causes the signals to be dependent on the distance the light has traveled between the position where it was generated and the position where it is converted into an electric signal, and thus causes a position dependence in the calorimeter response.

The light attenuation characteristics of all known scintillators change with time, for example as a result of chemical processes induced by the scintillation light itself [Sir 85], but in particular when the scintillator is operated in a radiation field. The mentioned position dependence of the signals is thus, in addition, a function of time.

Light attenuation lengths are typically measured in units of meters. Unlike the position dependence discussed in Section 4.4.2, which takes place on a scale of (a fraction of) 1 mm, the effects from light attenuation can often be corrected for on an event-by-event basis, with the help of information on the impact point of the showering particle. An example of this was provided by the CHORUS Collaboration [Esk 97], who operated a lead/scintillating-fiber calorimeter in their neutrino experiment at CERN.

In this fixed-target experiment, the 3 m long fibers ran perpendicular to the direction of the incoming particles. They were read out on both ends by PMTs. The light attenuation length in the scintillating fibers, λ_{att}, was typically 4.5 m. When the signals from the PMTs on both ends were added, the signal from a given particle traversing the detector in the center was typically $\sim 5\%$ smaller than when the same particle traversed the detector near one of the PMTs. By comparing the arrival times of the light in the two PMTs, the position of the light source could be determined with a precision of a few cm. This made it possible to eliminate the position dependence of the calorimeter response event by event [Buo 94].

Two-sided readout is an efficient way to (a) limit the effects of light attenuation and (b) allow for correction of the remaining effects. As an example, we consider the case of a 1 m long scintillator with $\lambda_{att} = 1$ m. If this scintillator was read out from only one side, and if the signals were normalized to the signal from a particle crossing the scintillator close to the PMT, then the signal from a particle crossing it at the center ($x = 50$ cm from the PMT) would be only $\exp(-x/\lambda_{att}) = 0.6$. If this particle crossed the scintillator at the opposite end ($x = 100$ cm), then the signal would only be 0.37. However, if the same scintillator was read out from both ends, the signal (PMT1 + PMT2) from a particle crossing it at the center would be only 11% smaller than that from a particle crossing near either of the two PMTs ($[2 \times 60]/[100 + 37] = 0.89$).

Many experiments, including ZEUS [And 91], HELIOS [Ake 87] and KLOE [Bab 93] have exploited the advantages of two-sided scintillator readout in their ca-lorimeters and thus strongly mitigated the effects of light attenuation. Especially ZEUS went to great lengths to make the response of their uranium/plastic-scintillator calo-rimeter as uniform as possible. To increase the light yield, and to protect them from

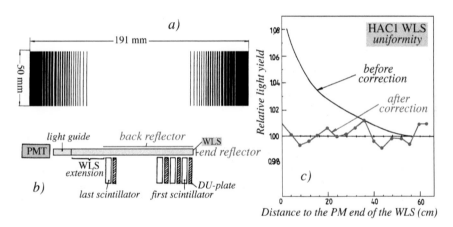

FIG. 4.28. Uniformity correction pattern for scintillating tiles in the em section of the ZEUS Forward Calorimeter (a). Signal uniformity in the ZEUS wavelength-shifting plates as a function of depth (c), before and after applying the correcting reflectors (b). From [And 91].

scratching, every scintillator tile was wrapped in Tyvek paper.[9] On this paper, a black-line uniformity correction pattern was printed (Figure 4.28a). The "bar code" printed on the wrapping paper selectively reduced the response to the average value, i.e., the re-sponse to light produced in the center of the tile. With this technique, the non-uniformity

[9]Tyvek wrapping paper is formed by compressing spun polyethylene. It has the same reflectivity for ultra-violet light as aluminium foil.

of the response was improved from $\pm 7\%$ without the correction pattern to $\pm 2\%$ for the small tiles in the em calorimeter section, and to $\pm 4\%$ in the larger hadronic tiles.

A similar technique was used to eliminate the effects of light attenuation in the wavelength-shifting bars that transported the light to the PMTs. To boost the signals from the WLS region farthest away from the PMT, this end part was equipped with reflectors made of aluminium foil (Figure 4.28b). The effect of this on the uniformity of the response along the wavelength-shifting bar is shown in Figure 4.28c. After applying these reflectors, the WLS response was uniform to within $\pm 2\%$.

By giving great attention to these and other details (see also Figure 4.25), ZEUS managed to achieve very good signal uniformity over the entire calorimeter volume and thus fulfilled a necessary condition for taking advantage of the excellent energy resolution provided by their calorimeter system.

In scintillating-fiber calorimeters, in which the fibers are oriented at $0°$, with the PMTs located at the downstream end of the calorimeter, a similar increase in the effective attenuation length can be achieved by mirroring the upstream fiber ends. For individual scintillating fibers, an increase of the effective attenuation length from 6.8 m to 15.9 m was reported as a result of this procedure [Har 89]. The SPACAL Collaboration combined this mirror technique with the application of filters that selectively absorbed the most attenuated components of the scintillation light, and increased the effective attenuation length of the fibers in their calorimeter to ~ 11 m in this way [Aco 91c].

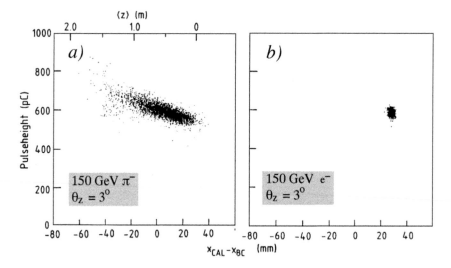

FIG. 4.29. Scatter plot showing the SPACAL signal for 150 GeV π^- (a) and 150 GeV e^- (b) versus the center of gravity of the light production in the showers. The bottom scale shows the lateral displacement of the shower's center of gravity with respect to the particle's impact point. In the top scale, this displacement is converted into the average depth $\langle z \rangle$ of the light production [Aco 91b].

In calorimeters of this type (0° fibers), light attenuation affects the signals as a result of *longitudinal* fluctuations in the shower development, as opposed to the lateral fluctuations that affect the other calorimeters mentioned in this subsection. In general, longitudinal shower fluctuations have much larger consequences for calorimetry than lateral ones. The effects of shower leakage, the topic of Section 4.5, provide a clear example of this. Another example was shown in Section 4.4.2.3, where the effects of a depth-dependent sampling fraction were discussed. Light attenuation in fiber calorimeters such as SPACAL causes effectively a depth-dependent sampling fraction.

In spite of the long attenuation length ($\sim$ 11 m), the effects on the calorimeter signals were still evident. This is illustrated in Figure 4.29, which shows a scatter plot of the signals from 150 GeV pions showering in SPACAL versus the average depth at which the light production took place [Aco 91b]. The latter quantity could be measured thanks to the fact that the detector was slightly rotated in these measurements. The pions entered the detector at an angle $\theta = 3°$ with the fibers. The depth of the light production was determined by comparing the *lateral* position of the shower's center of gravity with the particle's trajectory measured with wire chambers installed upstream.

The figure shows that showers developing deep inside the calorimeter produced, on average, larger signals than showers starting early, commensurate with expectations of the effect of light attenuation in the fibers. The starting point of the hadronic showers fluctuates on a scale of 1 λ_{int} in depth. In first approximation, an energy-independent

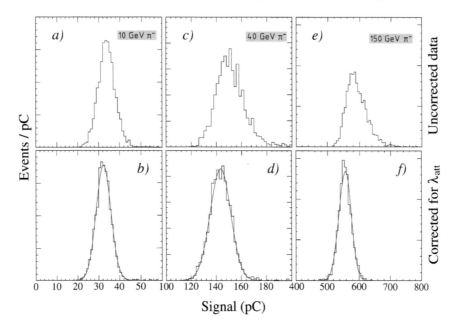

FIG. 4.30. The calorimeter signal distribution for 10 GeV, 40 GeV and 150 GeV π^- before (top row, a, c, e) and after (bottom row, b, d, f) correcting for the effects of light attenuation in the fibers [Aco 91c]. See text for details.

contribution to the hadronic energy resolution of the order of $\lambda_{int}/\lambda_{att}$ may be expected as a result of light attenuation in the fibers, *i.e.*, $\sim 2\%$ in the case of SPACAL. This is in agreement with the experimental observations, as illustrated in Figure 4.30.

When all events shown in the scatter plot of Figure 4.29a were taken together, and the projection was made on the vertical axis, the z-dependence of the fiber response led to an asymmetric calorimeter signal distribution, skewed to the high-energy side. Such asymmetric signal distributions are shown in the top row of Figure 4.30 (diagrams *a, c* and *e*), for the pion signal distributions at 10 GeV, 40 GeV and 150 GeV, respectively.

However, the effect of light attenuation on the calorimeter signals could in this case be eliminated, since the average depth of the light production was measured, event-by-event, and the z-dependence of the fiber response was precisely known. The elimination of the effects of light attenuation was achieved by applying a weighting factor[10] to the measured pulse height. This procedure resulted in the signal distributions shown in the bottom row of Figure 4.30 (diagrams *b, d,* and *f*). At 150 GeV, the (rms) energy resolution improved from 5.5% to 3.5% as a result of this. The correction also made the signal distributions much more symmetric. All distributions in the bottom row of Figure 4.30 were well described by Gaussian functions.

4.4.4 *Other instrumental effects*

There is a variety of other instrumental effects that may affect the energy resolution of calorimeters. In this subsection we briefly discuss two effects that have implications for some practical calorimeters: gain variations and fluctuations in the light collection efficiency.

In calorimeters in which the primary signals are amplified, (*e.g.*, gas calorimeters, or scintillation calorimeters where PMTs or APDs convert the light into electric signals), *gain instabilities* contribute an energy-independent term to the energy resolution. Since the gain is often extremely sensitive to temperature variations, temperature stabilization is an excellent way to minimize these effects. The same holds for the applied electric fields. In PMTs, which are exponential amplification devices, a change of 100 V in the cathode potential causes typically a gain change by a factor of two. Gain stabilization at the 1% level thus requires the high voltage (typically 2 kV) to be stable at the level of 1 V or better. In APDs, the requirements on voltage stability are even more stringent.

In gas calorimeters, the gas gain is usually also very sensitive to the precise composition of the gas mixture, which should thus be maintained as stable as possible.

In Section 4.2.4 we saw that the em energy resolution of scintillating crystals, such as NaI(Tl), CsI and BGO, although very good, is not limited by fluctuations in the number of photoelectrons. Because of their excellent *intrinsic* resolution, these detectors are very susceptible to small effects that might not be noticeable in less precise instruments. As an example, we mention the effects of incomplete shower containment. Leakage effects limit the energy resolution of most crystal calorimeters at high energies to 0.5–1.0% [Bak 85, Kam 94, Auf 96a]. To limit the effects of side leakage at low energies, shower signals usually have to be integrated over a considerable number of

---

[10]The weighting factor was the inverse of the function $I(z)$, which described the z dependence of the fiber signals [Aco 91c].

crystals. This makes the resolution sensitive to crystal-to-crystal variations in the properties of the different calorimeter cells contributing to the shower signals (light yield,
light attenuation, *etc.*), and to the quality of the intercalibration. But even if all crystals
were identical and perfectly intercalibrated, one instrumental effect would still continue
to affect the energy resolution: fluctuations in the light collection efficiency.

The scintillation light produced in such calorimeters is emitted isotropically. It is
collected by a photocathode which is either directly mounted at the back end of the
crystal or connected to it by means of a light guide. The efficiency of this collection
process depends on a number of factors, including:

- the shape of the crystal,
- the reflectivity of the crystal surfaces,
- the light attenuation length (self-absorption), and
- geometric effects, such as the fraction of the crystal surface covered by the photocathode.

In any case, the light collection efficiency depends on the position where the light
is created. The em energy resolution of these calorimeters is thus limited by event-to-
event fluctuations in the light collection efficiency that are the result of fluctuations in
the shower development.

We have studied the fluctuations in electron- and γ-induced showers in CeF_3 crystals with the EGS4 Monte Carlo package [Wig 02]. Some results of this study are shown
in Figure 4.31. In Figure 4.31a, the longitudinal center of gravity of the produced scin-
tillation light ($\langle z \rangle$) is given as a function of energy. As the energy increases, $\langle z \rangle$ moves

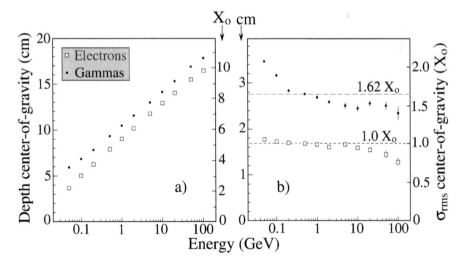

FIG. 4.31. The average longitudinal center of gravity of the light production in em showers
induced by electrons or γs in CeF_3 crystals (*a*) and the standard deviation of the distribution of this center of gravity (*b*), as a function of energy. Results from EGS4 Monte Carlo
calculations [Wig 02].

gradually deeper inside the crystals. This shift in depth is approximately proportional to $\log E$. The light production in γ-induced showers occurs, on average, deeper inside the crystals than for electron-induced showers of the same energy. This is because γs travel, on average, a distance of $\frac{9}{7}X_0$ before they start losing energy, while electrons start losing energy immediately upon their entry into the detector.

This difference between electron- and γ-induced showers also shows up in the event-to-event fluctuations in the location of the center of gravity. The standard deviation of the $\langle z \rangle$ distribution is shown in Figure 4.31b. As expected, this width amounts to $\sim 1X_0$ (1.7 cm) for the electron showers. For the γ-induced showers, this width has to be convoluted with the z-distribution of the shower's starting point. For γs, a standard deviation of $\sqrt{1.0^2 + (9/7)^2} = 1.62X_0$ is thus expected. This is confirmed by the Monte Carlo simulations.

The effects of these fluctuations on the energy resolution depend on details of the calorimeter structure and on the crystals of which the calorimeter is composed. However, it is reasonable to expect an effect of the order of $\sigma_{\langle z \rangle}/\lambda_{\mathrm{att}}$ (see also Section 4.4.3.4).

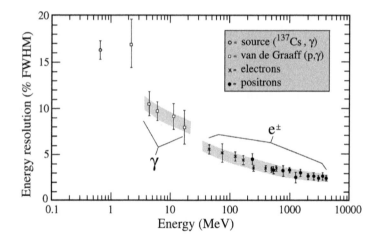

FIG. 4.32. The measured energy resolution of the NaI(Tl) crystals of which the Crystal Ball calorimeter was composed [Chan 78].

These simulations predict different resolutions for γs and electrons of the same energy, for calorimeters in which this effect plays a dominating role. Experimental data, such as the ones measured by the Crystal Ball group [Chan 78] for their NaI(Tl) detector (Figure 4.32) are inconclusive in this respect, since the electron and γ data covered different energy regimes.

The L3 Collaboration, which operated a large BGO calorimeter at LEP, has made an extensive study of various factors that influence the fluctuations in the light collection efficiency (and thus the energy resolution of their instrument). They concluded that a tapered crystal geometry with roughened side surfaces helped in reducing these fluctuations to a level below 1% [Bak 87]. The CMS and ALICE Collaborations operate large electromagnetic calorimeters consisting of $PbWO_4$ crystals. In order to correct for the effects of light attenuation and thus optimize the longitudinal signal uniformity, CMS has deliberately depolished one of the lateral faces of the crystals, which have a truncated pyramidal shape [CMS 08].

4.5 Shower Leakage

Calorimeters are instruments to measure the properties of particles by means of total absorption. However, in practice "total" means 99.9%, or 99%, or even less. When designing a calorimeter system for an experiment, and in particular for a 4π experiment, the choice of the calorimeter depth has important consequences, especially for the cost of the experiment. Therefore, it is important that the decision concerning the degree of shower containment be based on accurate information.

There are two aspects to this problem. First, incomplete shower containment leads to energy-dependent event-by-event fluctuations in the shower leakage, which affects the quality of the calorimetric information. Second, incomplete shower containment means that shower particles escape the calorimeter. These particles may cause signals in other detectors, *e.g.*, the muon system, which may disturb the performance of these detectors.

4.5.1 Effects of leakage on the calorimetric quality

Shower leakage is an energy-dependent effect. The fraction of the energy carried by the showering particle that is *not* deposited in the (fiducial) calorimeter volume depends not only on the particle's energy, but also on the type of particle. In a given calorimeter, electrons of a given energy are better contained than protons of the same energy, which are in turn better contained than pions, on average that is.

Incomplete shower containment has two types of consequences for the calorimetric properties of a given detector. First, the calorimeter response is affected. Since the average undetected shower fraction is energy dependent, the response changes in comparison with the response of a detector that is large enough to make shower leakage insignificant. Although one might intuitively think that the average shower leakage fraction *increases* with energy, this is not automatically true and several calorimeters offer examples of the opposite effect.

Second, there are event-to-event fluctuations in the shower leakage. These fluctuations deteriorate the energy resolution of the calorimeter, in an energy-dependent way.

The calorimetric quality of shower detectors is affected by three different types of shower leakage:

1. *Longitudinal* leakage. When people talk or worry about shower leakage, it is usually this type of leakage. Shower particles escape detection by emerging from the calorimeter's rear end. Considerations about longitudinal shower leakage often drive the design of the calorimeter system, since the depth of the instrumented volume strongly determines its cost. However, it should also be mentioned that some of this type of leakage is in practice unavoidable, since neutrinos as well as muons produced in the decay of pions and kaons will also escape very deep calorimeters.

2. *Lateral* leakage. Although the design of a calorimeter is often driven by considerations about longitudinal shower containment, in practice the effects of lateral
   shower leakage on the calorimeter performance usually dominate. To determine
   the particle's energy, the area surrounding the shower axis over which the calorimeter signals are integrated is typically limited to such an extent that lateral
   losses are easily an order of magnitude larger than the longitudinal ones. From
   this perspective, one might conclude that many calorimeters are probably deeper
   than necessary.

3. *Albedo*, a.k.a. backsplash, *i.e.*, leakage through the front face. Of the three mentioned types of leakage, this is the only one that is fundamentally unavoidable, no
   matter how large the detector is made.

In the following subsections, the effects of these different types of shower leakage
are discussed in some detail.

4.5.2   *Longitudinal vs. lateral leakage*

In Section 4.4.2 we saw that the effects of a depth-dependent sampling fraction on the
energy resolution are much larger than the effects of a sampling fraction that varies
with the impact point of the particles. A similar situation occurs for shower leakage.
At a given (average) leakage level, the effects of shower fluctuations on the energy
resolution are much larger if the leakage occurs longitudinally than when the energy
leaks out sideways.

This was already known a long time ago, as illustrated by Figure 4.33. This figure

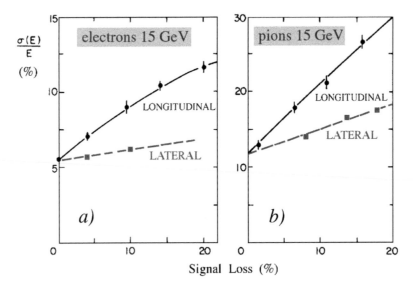

FIG. 4.33. The effects of longitudinal and lateral shower leakage on the energy resolution, as
measured for 15 GeV electrons (*a*) and pions (*b*) by the CHARM Collaboration in a low-Z
calorimeter [Did 80, Amal 81].

shows the effects of longitudinal and lateral shower leakage fluctuations on the energy resolution of 15 GeV electrons (Figure 4.33a) and pions (Figure 4.33b), measured by the CHARM Collaboration for their low-Z (marble absorber, $Z_{eff} \sim 13$) calorimeter [Did 80, Amal 81]. A 10% lateral leakage had a smaller effect on the energy resolution of this detector than a 5% longitudinal leakage, both for em and hadronic showers.

These differences can be qualitatively understood from the very different characteristics of longitudinal and lateral shower fluctuations. Let us first take a look at the 15 GeV pions. On average, these particles penetrated the CHARM calorimeter by some 60 cm before undergoing a nuclear interaction that started the shower. However, for $\sim 15\%$ of the pions, this starting point was located in the first 10 cm of the detector, and for another 15% it was located at a depth of more than 1.2 m. The longitudinal shower fluctuations, and thus the longitudinal leakage fluctuations, are for a very important part driven by these fluctuations in the starting point of the shower, *i.e.*, by the interaction characteristics of *one individual shower particle* (the first one).

As we saw in Section 2.3.4, the lateral tails of hadron showers are primarily caused by neutrons depositing their kinetic energy through nuclear interactions with the calorimeter materials. The number of neutrons that contribute to these processes is large, of the order of 100 for 15 GeV pions in low-Z absorber material (see Section 2.3.2.3). Therefore, event-to-event fluctuations in the lateral shower development, and thus the fluctuations in lateral shower leakage, are determined by the interaction characteristics of *~100 shower particles combined*. As a result, these fluctuations are much smaller than the longitudinal ones.

Similar arguments may be applied to explain the different effects of longitudinal and lateral shower leakage fluctuations on the electromagnetic energy resolution. An incoming high-energy photon converted, on average, after ~ 11 cm ($\frac{9}{7} X_0$) into an e^+e^- pair in the CHARM calorimeter, with 15% converting in the first 2 cm and 15% beyond 22 cm. As in the case of the pions, longitudinal fluctuations in em shower development are driven by this type of fluctuation in the early phase of the shower, caused by a very small number of highly energetic shower particles.

The lateral tails of em showers are almost completely composed of Compton and photoelectrons, produced by numerous soft γ-rays (see Section 2.1.6). Also here, lateral shower fluctuations, and thus fluctuations in lateral shower leakage, are the result of what happens to *a very large number of soft shower particles combined*. As for hadronic showers, lateral leakage has a much smaller effect on the energy resolution than longitudinal leakage.

For a more quantitative assessment of these effects, we should also take into account that there is at least one effect that makes lateral shower fluctuations larger in hadronic showers than in em ones, namely the event-to-event fluctuations in the π^0 content of the shower (the em core). Since this core is usually completely contained in the calorimeter volume over which the hadronic signals are integrated, fluctuations in lateral shower leakage are also determined by event-to-event fluctuations in the em shower content. These fluctuations are large (see Section 4.7). This may well explain why the effects of lateral shower leakage were larger for pions than for electrons in the CHARM detector.

The SPACAL Collaboration has made an extensive study of various aspects of the

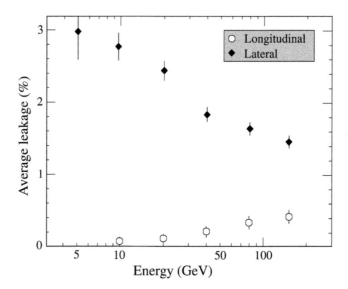

FIG. 4.34. The average lateral and longitudinal leakage out of the SPACAL calorimeter, for
pion showers as a function of beam energy. From [Aco 91d, Aco 92b].

shower leakage issue, for hadron-induced showers. Their lead/scintillating-fiber calorimeter measured $9.5\lambda_{int}$ in depth, with a diameter of $\sim 4.7\lambda_{int}$.

The longitudinal leakage from the back of this calorimeter was on average less than
0.5%, even for the highest-energy pions with which the calorimeter was tested (150
GeV). When the signals from all (155) readout cells were summed together, the lateral
leakage was found to be four times as large as the longitudinal one, at 150 GeV (Figure 4.34). At lower energies, the ratio between lateral and longitudinal shower leakage
rapidly increased, to reach a factor of thirty at 10 GeV [Aco 91d, Aco 92b].

It is interesting to note that the average fraction of the shower energy that leaked out
laterally *increased* at lower energies. Since the longitudinal leakage was small compared
with the lateral one at all energies, this means that high-energy pion showers were better
contained than low-energy ones in this calorimeter (see also Section 2.3.5). This is a
direct consequence of the increase of the (average) energy fraction contained in the
em shower component, $\langle f_{em} \rangle$, with energy. At 10 GeV, this core represents on average
$\sim 30\%$ of the shower energy, rising to $\sim 60\%$ at 150 GeV.

Since lateral leakage only affected the non-em shower component, one may expect
that it was larger at 10 GeV than at 150 GeV, by a factor $(100 - 30)/(100 - 60) = 1.75$.
This expectation is in good agreement with the experimental results (Figure 4.34).

The effects of lateral shower leakage on the hadronic energy resolution were mea-
sured in great detail [Aco 91d]. Figure 4.35 shows the energy resolution for 9.7 and 80
GeV pions as a function of the radius of the area over which the signals were integrated
(Figure 4.35a) and as a function of the average lateral leakage fraction (Figure 4.35b).
In this figure, the resolutions have been normalized to the ones measured for the full
calorimeter.

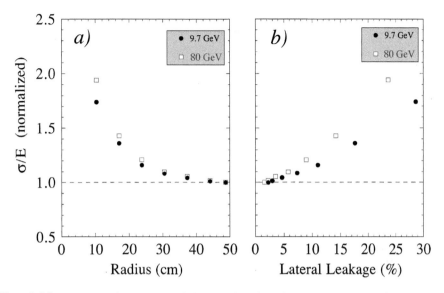

FIG. 4.35. The hadronic energy resolution as a function of the effective radius of the area over which the calorimeter signals were integrated. The energy resolutions have been normalized to the value for the complete SPACAL detector. Results for 9.7 and 80 GeV π^- that entered the calorimeter at an angle of 3° with the fiber axis (*a*). The same data as a function of the lateral shower leakage fraction (*b*). From [Aco 91c].

Shower leakage had a relatively larger impact on the calorimeter resolution as the energy increased. It turned out that the effect of lateral shower leakage on the hadronic energy resolution was reasonably described by a $E^{-1/4}$ term, added in quadrature to the stochastic term ($a_1 E^{-1/2}$) that described the calorimeter resolution in the absence of shower leakage:

$$\sigma/E = \sqrt{\left(\frac{a_1}{\sqrt{E}}\right)^2 + \left(\frac{x}{\sqrt[4]{E}}\right)^2 + \ldots} \qquad (4.23)$$

where x represents the average lateral leakage fraction.

Although the longitudinal shower leakage was *on average* much smaller than the lateral one, the event-to-event fluctuations were much larger. This may be concluded from the effects of longitudinal shower containment cuts on the hadronic energy resolution.

Figure 4.36 shows that at 150 GeV, where the *average* leakage fraction was only 0.4%, event-to-event fluctuations in the leakage had a 10–20% effect on the energy resolution. In order to achieve an effect of the same size for lateral shower leakage, the average leakage fraction should be of the order of 5% (Figure 4.35b). Because of the event-to-event fluctuations, longitudinal and lateral shower leakage thus had a very different effect on the hadronic energy resolution.

Another aspect of longitudinal shower leakage is the distortion of the response function caused by it. Events with significant shower leakage appear typically in the low-energy tail of the signal distribution. By eliminating these events, the signal distribution

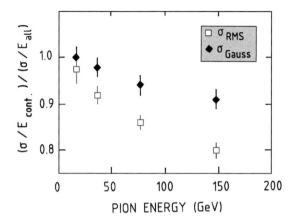

FIG. 4.36. The effects of a containment cut (less than 25 MeV measured in the leakage ca-
     lorimeter) on the energy resolution of the SPACAL calorimeter. Shown are the ratios of the
     energy resolutions for the contained events and for the unbiased event sample, as a function
     of the pion energy. From [Aco 91d].

is thus likely to become more symmetric. This is illustrated by Figure 4.36, which shows
that cutting events in which significant shower leakage occurred had a much larger effect
on the rms standard deviation of the signal distribution than on the width of the Gaussian
fit to this distribution. This means that the cut eliminated predominantly events from the
tail of the distribution.

A systematic study of the effects of longitudinal shower leakage on the hadronic
energy resolution was carried out by ATLAS, with its TileCal calorimeter [Adr 10].
Some results of this study are shown in Figure 4.37, where the degradation of the energy
resolution for pions of different energy is plotted as a function of the calorimeter depth.
This degradation is measured with respect to the resolution that was obtained for a
10.9λ deep calorimeter of this type. Just as for SPACAL, this degradation was found to
be much larger when expressed in terms of σ_{rms} (Figure 4.37a) than when a Gaussian fit
to the signal distribution was considered (Figure 4.37b). This study thus confirmed that
longitudinal shower leakage indeed leads to an asymmetric response function. The study
also found that for a given calorimeter depth, proton showers were better contained than
showers induced by pions (Figure 2.44), and that the effects of a given percentage of
shower leakage on the energy resolution were smaller for protons than for pions. The
shorter interaction length for protons and for the leading baryons subsequently produced
in the shower development, combined with the different characteristics of π^0 production
in the shower development, would indeed lead to effects of this type.

I have used the ATLAS TileCal data [Sim 08] to derive a relationship between the
average shower leakage fraction and the degradation of the hadronic energy resolution.
The contribution of shower leakage, which has to be added in quadrature to the energy
resolution in the absence of leakage, is shown in Figure 4.38 in terms of σ_{rms}, since a
change in the shape of the response function is an integral aspect of this effect. It turns

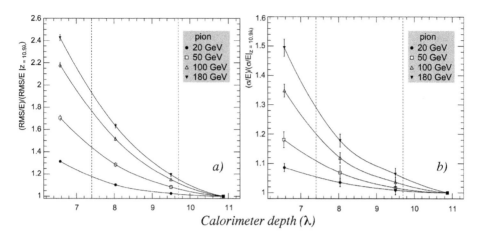

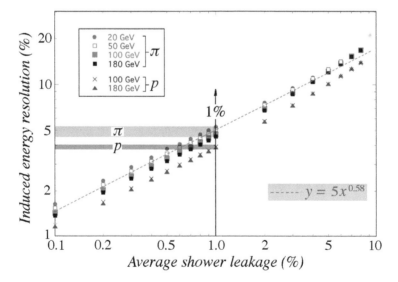

FIG. 4.37. Effects of longitudinal shower leakage on the hadronic energy resolution of the ATLAS TileCal calorimeter. Shown is the degradation of the energy resolution for pions of different energies as a function of the calorimeter depth, measured with respect to the resolution obtained with a 10.9λ deep calorimeter [Adr 10]. This degradation is shown both in terms of $\sigma_{\rm rms}$ (a), and in terms of the σ resulting from a Gaussian fit to the spectrum in the range $(-2\sigma, +2\sigma)$ around the peak (b). The dotted lines indicate the total depth of the ATLAS calorimeter, as well as the depth of the hadronic section, for $\eta = 0$.

FIG. 4.38. The contribution of fluctuations in longitudinal shower leakage to the hadronic energy resolution as a function of the average shower containment, measured with the ATLAS TileCal calorimeter [Sim 08]. Results are shown for pions and protons of different energies. The dashed line through the pion points is drawn to guide the eye.

out that the effect is well described by a power law, and more or less independent of the particle energy. The effects are somewhat larger for pions than for protons.

These results illustrate the importance of making the calorimeter sufficiently deep, especially if one wants to obtain good energy resolution at high energies. This is because

1. The shower penetrates to greater depth as the energy increases, and
2. The effect of an energy-independent term is larger when the resolution in the absence of leakage is smaller.

The figure shows that in order to obtain hadronic energy resolutions at the few % level, longitudinal shower containment at the 99+% level is imperative. At an average leakage level of 1%, event-to-event fluctuations in longitudinal leakage already contribute $\sim 5\%$ to the energy resolution for pions, and $\sim 4\%$ for protons. In the ATLAS measurements from which these results were derived, the TileCal was oriented such that the showers could be followed over a total depth of more than 20λ. It was therefore possible to study the shower containment in great detail, as a function of the depth of the absorber, in the 99+% domain relevant for this purpose. Figure 4.39 shows the average shower

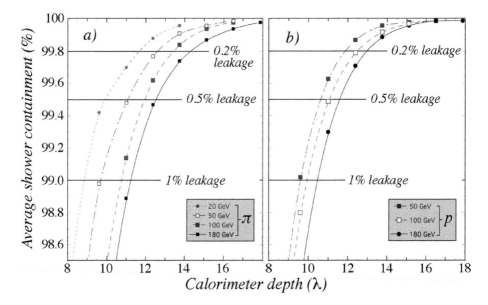

FIG. 4.39. The depth of the ATLAS TileCal calorimeter needed to contain showers initiated by pions (*a*) and protons (*b*) of different energies at a certain level. Experimental data from [Adr 10]. The curves are drawn to guide the eye.

containment for pions (Figure 4.39a) and protons (Figure 4.39b) of different energies, as a function of the depth of the iron/plastic-scintillator calorimeter structure. At the highest energy used in these studies (180 GeV), the calorimeter needs to be about 11λ deep to contain pions, on average, at the 99% level. For protons of this energy, 10λ is sufficient. However, to limit the contribution of fluctuations in longitudinal leakage to the energy resolution to 2%, a calorimeter depth of 14λ would be needed for pions and 12λ for protons. This is more than the typical calorimeter thickness in collider experiments.

However, it is important to consider these results in the proper perspective. At calorimeter depths where the *average* leakage amounts to a fraction of 1% of the deposited energy, a large fraction of the events are actually fully contained, while some other events exhibit sizeable leakage. This results in a large rms value for the event-to-event leakage *fluctuations*, and thus for the induced effect on the energy resolution. Some experiments use a so-called *tail catcher* to deal with the resolution problems caused by the containment requirements. Such a tail catcher, usually a very crude device installed behind the calorimeter system proper, serves primarily to identify and discard events in which particles penetrate deeply inside the calorimeter before undergoing the first nuclear interaction that starts the hadronic shower development. By using the signal from such a device as a veto, either at the data acquisition or the data analysis level, the event samples are limited to those events with showers that are (almost) completely contained in the calorimeter. Figure 4.36 illustrates the improvement of the energy resolution that is achieved by minimal cuts (25 MeV) on the longitudinal leakage energy. If the fraction of events vetoed in this way is small ($\sim 10\%$ or less), then the biases introduced by this veto are minimal, while the benefits in terms of energy resolution may be substantial.

4.5.3 *Albedo*

Shower leakage through the front face of the calorimeter only plays a significant role at very low energies. The shower particles that escape in this way are, by definition, very soft, since they must be produced in scattering processes from which they emerged at large angles. Examples of such processes are Compton scattering and the photoelectric effect in the case of em showers and elastic neutron scattering in hadronic showers. Since the energies of the scattering products are at maximum a few MeV, only low-energy showers are significantly affected.

Experimental evidence for this type of leakage can be derived from Figure 3.19, which shows the longitudinal distribution of a number of different radioactive nuclides, produced in the absorption of protons with a kinetic energy of 591 MeV. The concentrations of the nuclides produced by soft neutrons do not start at zero at the front face of the calorimeter, as one might expect for particles produced in this absorption process, but at a quite elevated level. An extrapolation of the measured curves to the zero-intensity level, upstream of the calorimeter's front face, gives an impression of the fraction of the neutrons making up the albedo in this case.

Figure 2.34 shows a similar distribution for nuclides produced by neutrons generated in the shower development of 300 GeV pions. Also in this case, the concentration is not zero at the front face of the calorimeter, but clearly, the *fraction* of neutrons that escapes through this front face is much smaller than in the previous (591 MeV) case.

The effects of albedo particles may be much more serious than the deterioration of the calorimetric measurements they cause. For example, they may produce signals in upstream detectors that are needed for tracking and particle identification purposes and thus confuse the measurements. Also, they may contribute to radiation damage in such upstream detectors. Albedo neutrons are a serious concern for silicon detectors and silicon-based electronics located inside the calorimeters of 4π experiments at hadron colliders.

4.5.4 *Monte Carlo studies of shower leakage*

The effects of shower leakage on the em energy resolution have been studied at the Monte Carlo level by a number of people. The results of these simulations are typically in good agreement with the experimental data discussed above. For example, a useful parameterization of the effects of longitudinal shower leakage on the em resolution of lead-glass detectors was given by Prokoshkin [Pro 79]:

$$\frac{\sigma}{E} = \left[\frac{\sigma}{E}\right]_{L=\infty} (1 + 4f + 50f^2) \quad (f < 0.1) \tag{4.24}$$

where $L = \infty$ represents the case of an infinitely large calorimeter, and f is the average fraction of longitudinal shower leakage in the actual device. The 40% increase in em energy resolution for $f = 10\%$, predicted by this equation, is also in good agreement with CHARM's experimental result (Figure 4.33). Other parameterizations of this type were compiled by Iwata [Iwa 80] and Amaldi [Amal 81].

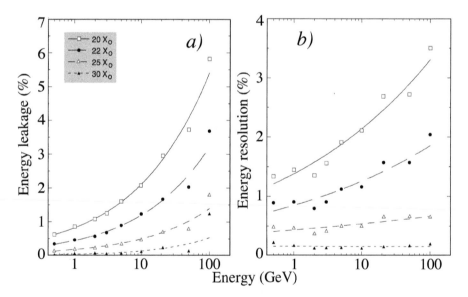

FIG. 4.40. The average fraction of the shower energy carried by particles escaping the calorimeter through the back plane (*a*) and the relative increase in the energy resolution caused by this effect (*b*), as a function of energy, for γ showers developing in blocks of tin with different thicknesses, ranging from $20X_0$ to $30X_0$. The curves are drawn to guide the eye. Results from EGS4 Monte Carlo calculations.

In this subsection, we show some results that emphasize the characteristic features of the effects of shower leakage on the calorimeter performance. These results were obtained with the em shower simulation package EGS4 [Nel 78], and concern em showers developing in a block of tin. This material was chosen because its relevant properties ($Z = 50$, density 7.31 g cm^{-3}) are very similar to those of a variety of em calorimeters used in practice, *e.g.*, crystal calorimeters such as CsI(Tl), NaI(Tl), CeF$_3$, BaF$_2$. The cutoff values used in the simulations were 100 keV for electrons and positrons and 10 keV for photons.

Figure 4.40 shows the effects of longitudinal shower leakage as a function of energy. Calorimeters of four different thicknesses have been simulated, ranging from $20X_0$ to $30X_0$. The average energy fraction that leaks out (Figure 4.40a) and the contribution to the energy resolution caused by this effect (Figure 4.40b) both increase with energy. For calorimeters that are at least $25X_0$ thick, the effects of this are probably acceptable for the energy range considered here. However, for thinner calorimeters, the shower leakage rapidly increases with energy. Around 100 GeV, event-to-event fluctuations in the shower containment correspond already to several percent of the total energy.

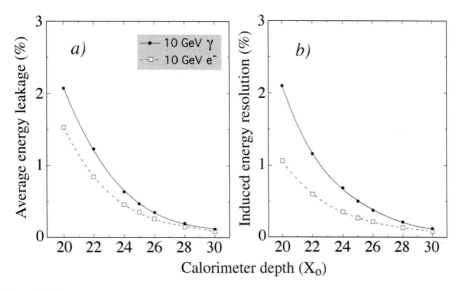

FIG. 4.41. The average fraction of the shower energy carried by particles escaping the calorimeter through the back plane (*a*) and the relative increase in the energy resolution caused by this effect (*b*), for showers induced by 10 GeV electrons and 10 GeV γs developing in blocks of tin with different thicknesses, ranging from $20X_0$ to $30X_0$. Results from EGS4 Monte Carlo calculations [Wig 02].

The results in Figure 4.40 concern showers induced by γs. As illustrated by Figure 4.41, the effects are significantly smaller for showers induced by electrons [Wig 02]. This is caused by two effects:

1. Photon-induced showers may start developing considerably deeper inside the absorber than electron showers. As a result, they are also less contained.

2. Large event-to-event fluctuations in the starting point of photon-induced showers have no equivalent in electron showers. Therefore, the fluctuations in the shower containment are larger for photon-induced showers.

Figure 4.41b shows that these fluctuations are larger by about a factor of two in the simulated case. Laterally, there are no fundamental differences between the development of em showers induced by photons or electrons. Therefore, the effects of lateral shower leakage on the response and the energy resolution are identical for these particles.

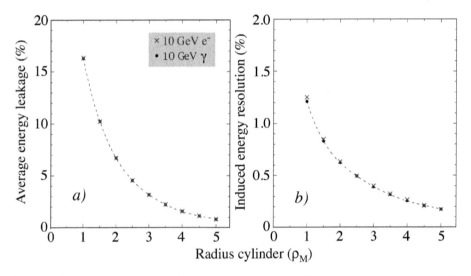

FIG. 4.42. The average fraction of the shower energy carried by particles escaping the calorimeter laterally (a) and the relative increase in the energy resolution caused by this effect (b), as a function of the radius of the calorimeter (an infinitely long block of tin). Results from EGS4 Monte Carlo calculations.

This is demonstrated in Figure 4.42. In these simulations, a cylindrical block of tin with infinite length was used as the absorbing medium. The radius of the cylinder was varied, from $1\rho_M$ to $5\rho_M$. Particles entered this absorber along the longitudinal axis of the cylinder. Figure 4.42a shows the average energy fraction carried by particles escaping sideways. The contribution of event-to-event fluctuations in the shower containment to the energy resolution is given in Figure 4.42b. Both variables are plotted as a function of the radius of the cylindrical tin block. There are no significant differences between electron- and photon-induced showers in this respect.

In Figure 4.43, the energy dependence of the lateral leakage effects is shown. Also in these simulations, the absorber was an infinitely long cylindrical block of tin and the particles entered this block along its longitudinal axis. Results are given for three

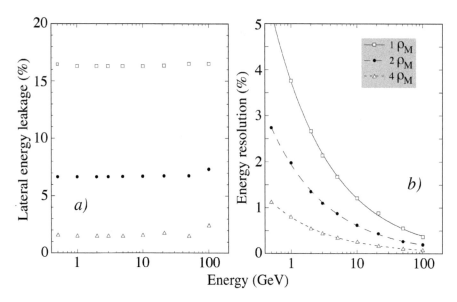

FIG. 4.43. The average fraction of the shower energy carried by particles escaping the calorimeter laterally (*a*) and the relative increase in the energy resolution caused by this effect (*b*),
as a function of energy. The absorber was an infinitely long cylinder of tin, with a radius of
$1\rho_M$, $2\rho_M$ or $4\rho_M$ in these EGS4 Monte Carlo calculations. Electrons entered this absorber
along the longitudinal axis of the cylinder.

different absorber radii: 1, 2 and 4 ρ_M. Comparison of Figures 4.40 and 4.43 reveals
some interesting differences between longitudinal and lateral shower leakage:

- The average lateral leakage fraction is energy independent, while the average
  longitudinal leakage fraction increases with energy. This reflects the fact that the
  lateral em shower profiles are energy independent, while the longitudinal profiles
  do depend on the energy of the showering particles.
- Fluctuations in the energy fraction deposited in a given detector volume *increase*
  with energy when that volume is insufficient to contain the showers in the longitudinal direction (Figure 4.40b), while they *decrease* with increasing energy when
  the detector volume is laterally too small (Figure 4.43b).
- Especially at high energy, the price to pay (in terms of energy resolution) for
  making the detector too shallow is much larger than for making it too narrow. For
  example, at 100 GeV, an average *lateral* energy loss of 16% (incurred when the
  signals are integrated over an area with a radius of $1\rho_M$) translates into an energy
  resolution of only 0.4%. On the other hand, an average *longitudinal* energy loss
  that is four times *smaller* (4%, incurred for a 22 X_0 deep detector) leads to an
  energy resolution that is five times *larger*, i.e., 2% (Figure 4.40). The reasons for
  these differences are discussed in Section 4.5.2.
- The effects of longitudinal leakage are different for electron and γ showers, while
  the lateral leakage effects are the same for both types of particles.

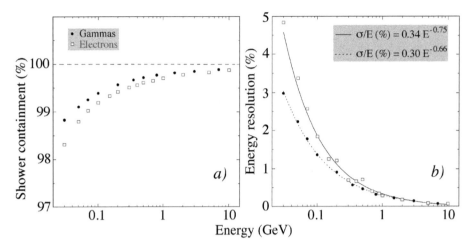

FIG. 4.44. The average fraction of the shower energy carried by particles escaping the calorimeter through its front face (*a*) and the relative increase in the energy resolution caused by this effect (*b*), as a function of energy, for electron and photon showers developing in an infinitely large block of tin. Results from EGS4 Monte Carlo calculations [Wig 02].

The same effects that lead to the different importance of longitudinal shower leakage for electron and photon showers also cause differences for what concerns the consequences of albedo. We studied the production of albedo particles in em showers with the EGS4 Monte Carlo program as well, for electrons and γs entering an infinitely large block of tin. The average fraction of the shower energy carried by the albedo particles as a function of energy is plotted in Figure 4.44a, while the fluctuations in this fraction, and thus the energy resolution induced by this effect, are shown as a function of energy in Figure 4.44b.

The effects of albedo leakage on the calorimeter performance are only significant at low energies. For energies above 1 GeV, the average energy leakage amounts to less than 0.5% and the energy resolution is affected by less than 0.3%. Since showers induced by γs develop, on average, deeper inside the absorber, they are less affected by albedo effects than electron showers of the same energy.

The effects of the different types of shower leakage are compared in Figure 4.45, which shows the energy resolutions resulting from albedo, longitudinal and lateral leakage as a function of the average fraction of the total energy that is carried away by particles escaping from the detector. The longitudinal and lateral leakage data concern 10 GeV γs. The observed difference between the open squares and the closed circles would be even more pronounced at higher energy. For the albedo data, there is no choice of energy: every data point corresponds to photons of a particular (low) energy (see Figure 4.44).

Figure 4.45 shows that a given level of shower non-containment has the least implications for the energy resolution if the leakage takes place sideways. The consequences are worse for longitudinal leakage and worst for albedo. For example, a leakage level

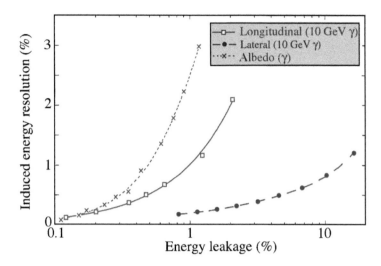

FIG. 4.45. A comparison of the effects caused by different types of shower leakage. Shown
are the induced energy resolutions resulting from albedo, longitudinal and lateral leakage as
a function of the average energy fraction carried by particles escaping from the detector. The
longitudinal and lateral leakage data concern 10 GeV γs, the albedo data are for γ-induced
showers of different energies. Results from EGS4 Monte Carlo calculations [Wig 02].

of 1% translates into an energy resolution of 0.2% when the leakage occurs laterally,
of 1% when the leakage is longitudinal, and of 3% in the case of albedo. However, as
shown above, such albedo levels only play a role for incident particles with energies
below 100 MeV.

4.5.5 *Escaping shower particles*

Even if the calorimeter depth would be adequate to make the effects of longitudinal
shower fluctuations on the energy resolution negligible, the detector might still be too
thin in view of the *rate* at which shower particles escape from the rear and cause interactions in the muon system. In such a situation, installing passive absorber material
between the calorimeter and the muon counters might be an appropriate solution. This
solution was adopted by CMS, who used a substantial amount of polyethylene to absorb
the numerous neutrons that escaped from the calorimeter system, and thus reduce the
high count rates in the muon system, in the high-η regions of the detector system.

The particles escaping a calorimeter after $\sim 10\lambda_{int}$ can be subdivided into three
classes: neutrons, soft charged hadrons and muons. The latter class of particles should
be a source of worry, for the following reason. One of the multiple tasks of calorimeters
in an experiment is to absorb all particles except muons (and neutrinos). Particles that
do penetrate the absorber are identified as muons, based on the assumption that the
calorimeter fulfills this task properly. It is therefore very disturbing if the calorimeter
itself acts as a source of muons. This phenomenon, which is mainly caused by secondary
and higher-order pions and kaons that decay before they (strongly) interact, was clearly

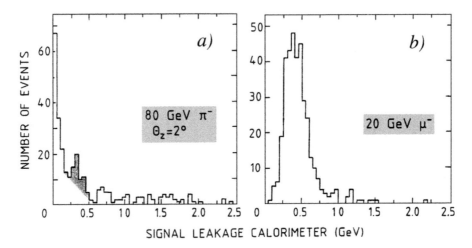

FIG. 4.46. The energy spectrum measured in the leakage calorimeter for 80 GeV π^- entering the SPACAL calorimeter at an angle of $2°$ with the fiber axis (a), compared with the spectrum of the energy deposited by low-energy muons in the leakage calorimeter (b). From [Aco 91d].

identified and quantified by the SPACAL Collaboration [Aco 91d].

The initial evidence gathered for this process from the SPACAL data is given in Figure 4.46a, which shows the energy spectrum measured in a leakage calorimeter backing up the main SPACAL calorimeter. The bump observed at an energy of ~ 0.4 GeV occurred at the same place where muons traversing the leakage calorimeter produced a signal (Figure 4.46b).

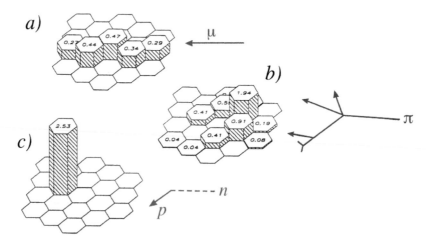

FIG. 4.47. Characteristic energy deposit profiles for muons (a), charged hadrons (b) and neutrons (c) detected in the leakage calorimeter backing up SPACAL. The numbers indicate the pulse heights in the individual calorimeter cells. From [Aco 91d].

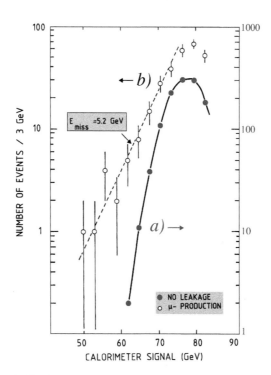

FIG. 4.48. Signal distributions in the SPACAL detector for contained pion showers (a) and for pion showers in which a muon was generated (b). Data taken at 80 GeV. From [Aco 91d].

In this experiment, events in which muons were produced were recognized on an individual basis from the hit patterns produced in the main and leakage calorimeters. Both calorimeters had the same structure, but the leakage calorimeter was oriented with the fibers running perpendicular to the direction of the beam line. In this way, the hit pattern in the leakage calorimeter made it possible to distinguish muons (Figure 4.47a) from charged hadrons (Figure 4.47b) and neutrons (Figure 4.47c).

The probability that an escaping muon is produced in the hadronic shower was measured to be proportional to the shower energy. This probability reached a level of ∼ 2% at 150 GeV. SPACAL also measured the energy distribution of these escaping muons.

Figure 4.48 shows the signal distributions in the main calorimeter for contained pion showers (curve a) and for pion showers in which a muon was generated (curve b). The latter signal distribution contains a low-energy exponential tail, from which the average muon momentum was inferred to lie in the range 2.0–3.7 GeV/c, depending on the relative importance of π and K decay as sources of these muons.

In a follow-up study, the RD1 Collaboration used a dedicated leakage calorimeter, consisting of 10 cm thick iron slabs interleaved with thin-gap wire chambers [Mik 88] to detect escaping shower particles. They confirmed the muon production rates measured by SPACAL and also identified events in which *two muons* were produced in the hadronic showers [Bad 94b]. Such events occurred at a rate of 6% of the single-μ

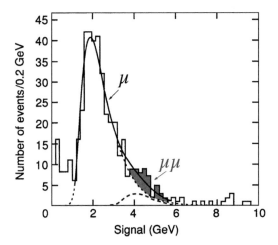

FIG. 4.49. Signal distribution in the leakage calorimeter for events in which μs are produced in
80 GeV pion showers developing in the RD1 fiber calorimeter. The dotted line represents the
Landau response for one μ, the dashed line the response to two muons and the solid line the
response to events in which either one (94% of the time) or two (6%) muons are produced
[Bad 94b].

production rate and were presumed to be due to $\rho^0 \rightarrow \mu^+\mu^-$ decay (see Figure 4.49).

If this is indeed the production mechanism, these muons may be expected to be
considerably more energetic than the ones from π and K decay, where the lifetime
limits the decay to the low-energy members of these species.

SPACAL also measured the rates at which neutrons and soft charged hadrons produced in hadronic showers escaped the calorimeter. These particles outnumbered the
escaping muons by an order of magnitude, over the full energy range at which the
calorimeter was studied.

4.5.6 Leakage signal amplification

In certain types of calorimeters, (longitudinal) shower leakage may affect the signals
in a very unexpected way: instead of being reduced, reflecting the incomplete shower
containment, *the signals are increased*, sometimes by a large factor. Shower leakage is
responsible for a distinct *high-side* tail in the signal distributions measured with these
calorimeters.

An example of this phenomenon was given in Figure 3.32a, where the signal distribution from 80 GeV electrons showering in 18 cm ($20X_0$) deep $PbWO_4$ crystals is
shown [Pei 96]. The crystals were read out with silicon (PIN) photodiodes in the experiment in which these data were collected. The high-side tail in this distribution is the
result of shower leakage.

When a charged particle traverses a silicon PIN diode, it creates electron–hole pairs.
Since mips lose, on average, 3.88 MeV cm^{-1} in silicon, and since it takes 3.6 eV to
create one electron–hole pair, the number of pairs created in a 100 μm thick PIN diode

is typically $\sim 10^4$. The signal from a charged particle is thus four orders of magnitude larger than that produced by a single scintillation photon, which generates only one photoelectron in a PIN diode. The PbWO$_4$ crystals used for the measurements discussed here generated about 2,500 photoelectrons per GeV [Pei 96]. A charged shower particle that escaped from the calorimeter and traversed the PIN diode thus produced a signal equivalent to the signal from all photons generated in the deposit of several GeV of energy inside the calorimeter.

This particular property of PIN diodes made this setup extremely sensitive to shower leakage. It amplified shower leakage dramatically. When ten shower particles escaped from the back of this crystal, this typically meant an energy loss of the order of 20 MeV. However, when these particles traversed the diode, they might easily add the equivalent of 20 GeV to the shower signal.

As with all examples discussed in this subsection, there are of course ways to avoid this problem. One possibility would be to mount two PIN diodes back-to-back and subtract the signal in the downstream diode from that in the upstream one [Kob 87]. Escaping charged shower particles are likely to traverse both diodes in such a geometry, while the scintillation light is only detected by the (upstream) diode which directly faces the light source.

Other possibilities involve the use of Avalanche Photo Diodes, in which the signals from photoelectrons are internally amplified. The silicon conversion layer in these devices is much thinner than in silicon PIN diodes and therefore the amplification factor for charged shower particles is correspondingly smaller. The CMS Collaboration has built an APD-based readout system for their em crystal calorimeter [Ale 97, Auf 98], which is discussed in somewhat more detail in Section 5.2.1.

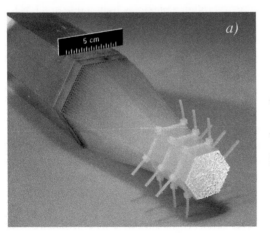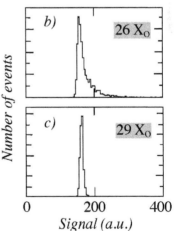

FIG. 4.50. Fibers exiting the rear end of a prototype of the SPACAL calorimeter (a). Signal distributions for 40 GeV electrons showering in detectors of this type, with lengths of $26X_0$ (b) or $29X_0$ (c). From [Aco 92c].

Another example of leakage signal amplification comes from scintillating-fiber calorimeters. Figure 4.50 shows signal distributions from 40 GeV electrons showering in
a lead/scintillating-fiber calorimeter [Aco 92c]. The fibers in this calorimeter were oriented along the direction of the incoming particles. The fibers emerging from the back
of the calorimeter were bunched together and coupled via a light guide to a PMT (Figure
4.50a).

The sampling fraction of this calorimeter for em showers was about 2%. Shower
particles produced inside the calorimeter volume therefore deposited 2% of their kinetic
energy in the fibers. However, shower particles escaping from the back of this calorimeter traversed a 15 cm long region occupied by large numbers of fibers and nothing
else. In this region, the sampling fraction for these leakage particles was thus, in first
approximation, 100% instead of 2%. Further amplification of this leakage signal was
caused by the fact that the fraction of scintillation light trapped in the fibers increased
strongly close to the PMT, due to the participation of the cladding in the light trapping
process [Har 89].

Remarkably, the high-side tail in the signal distribution (Figure 4.50b) disappeared
almost completely when the calorimeter length was extended by an extra $3X_0$, to $\sim$
$29X_0$ (Figure 4.50c). This illustrates the need to contain showers at the level of 99.9% in
calorimeters in which leakage signal amplification occurs, while 99% might be adequate
for calorimeters where this phenomenon is not an issue.

A third calorimeter in which signal leakage amplification was observed contained
clear plastic fibers as active elements [Win 99]. Čerenkov light produced in these fibers
formed the basis of the shower signals. This (7 λ_{int} deep hadronic) calorimeter was read
out in the same way as the scintillating-fiber one mentioned above. The fibers exiting at
the rear end of the calorimeter were coupled through a plexiglas light guide to a PMT.

In this case, the production of Čerenkov light in this light guide itself, by escaping
relativistic shower particles, was the most likely mechanism responsible for the high-
side tail that characterized the hadronic signal distributions from this calorimeter. This
may be concluded from the observation that such tails were notably absent in calorimeters of the same type and length, in which the fibers were coupled to the PMTs by
means of air light guides.

Obviously, readout techniques prone to leakage signal amplification should be
avoided in the first (electromagnetic) section of a longitudinally segmented calorimeter
system. When hadrons start to develop showers somewhere in this first section, it is
very likely that large numbers of shower particles traverse the very region where this
amplification process takes place. This may result in substantial random disturbance of
the hadronic calorimeter signals. Interestingly, this prediction from the first edition of
this book (1999) was confirmed in a spectacular way in the CMS calorimeter system
(2012). The reader is referred to Section 9.1.3 for details on this.

4.6   Fluctuations in "Visible Energy"

In any calorimeter the energy resolution for hadrons is worse than for electrons of the
same energy. This means that event-to-event fluctuations in the signals from hadron

showers are larger than in the signals from em showers. One important reason is that, in addition to the fluctuations that determine the em energy resolution, other types of fluctuations contribute to the hadronic energy resolution. Prominent among those hadron-specific effects are the fluctuations in "visible energy."

In Section 2.3.2, the origins of the invisible-energy phenomenon were discussed in detail. A sizeable fraction of the shower energy is used to release nucleons from nuclei in the hadronic interactions. This nuclear binding energy represents typically 30–40% of the energy carried by the non-em shower component (see Table 2.6), with considerable fluctuations about this average.

The energy carried by the em shower component is overwhelmingly deposited through processes in which nuclear binding energy loss plays no role at all. Therefore, the fluctuations in visible energy are strongly correlated with fluctuations in the em content (f_{em}) of the hadron showers. The latter fluctuations are discussed in the next section. In this section, we concentrate on the consequences of fluctuations in the nuclear binding energy losses for the visible energy fraction of the non-em shower component alone.

In Figure 4.51a, the distribution of the nuclear binding energy loss that may occur in spallation reactions induced by protons with a kinetic energy of 1 GeV on ^{238}U nuclei is shown. This distribution was obtained from a Monte Carlo simulation involving the cross sections for all possible reactions that may occur (Equation 2.24). More than 300 different reactions contributed each more than 0.1% to the total spallation cross section, and the largest contribution of any individual reaction amounted to only 2% of this total cross section [Wig 87].

On average, 90 MeV was lost, but the fluctuations about this average were huge, with a σ_{rms} of $\sim 70\%$ of the mean value. Under the assumption that these fluctuations were governed by Poisson statistics, they could be described as

$$\frac{\sigma}{\Delta E_B} = \frac{21\%}{\sqrt{\Delta E_B}} \qquad (4.25)$$

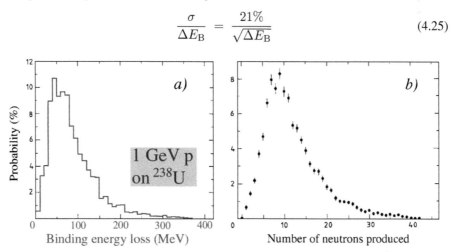

FIG. 4.51. The nuclear binding energy lost in spallation reactions induced by 1 GeV protons on ^{238}U nuclei (a), and the number of neutrons produced in such reactions (b). From [Wig 87].

where ΔE_B represents the average binding energy loss, in GeV.

However, it is important to realize that most of the nuclear binding energy losses occur in reactions induced by particles that are considerably less energetic than 1 GeV. This may be illustrated by the example worked out in Section 2.3.2.2. In this example, which concerns shower development in lead, about 70% of the binding energy is lost in reactions induced by protons and neutrons, which themselves are produced in spallation reactions.

We repeated these Monte Carlo simulations for reactions induced by protons with kinetic energies of 200 MeV, 100 MeV and 50 MeV on a lead target. In each case, we convoluted a large number of reactions, so that the combined binding energy losses amounted to several GeV, and a Gaussian distribution around the mean value was obtained. It turned out that the fluctuations in nuclear binding energy loss became relatively smaller as the energy of the incoming particles decreased. We found that $\sigma/\Delta E_B$ decreased from $17\%/\sqrt{\Delta E_B}$ for 200 MeV protons via $14\%/\sqrt{\Delta E_B}$ at 100 MeV to $12\%/\sqrt{\Delta E_B}$ for reactions induced by 50 MeV protons. Based on the relative contributions of particles of different energies to the total nuclear binding energy losses (see Table 2.4), we estimate that in high-Z materials such as lead or uranium, the fluctuations in nuclear binding energy losses are described by

$$\frac{\sigma}{\Delta E_B} \approx \frac{15\%}{\sqrt{\Delta E_B}} \tag{4.26}$$

In these high-Z absorber materials, most of the nucleons released in nuclear reactions are neutrons (see Section 2.3.2). The distribution of the number of neutrons released in the spallation reactions mentioned above is shown in Figure 4.51b. On average, twelve neutrons are released when a 1 GeV proton interacts with a ^{238}U nucleus (not counting the neutrons released in subsequent reactions, e.g., nuclear fission, induced by the spallation products). The fluctuations about this average number of neutrons are similar to those in the case of nuclear binding energy losses.

Not surprisingly, there may be a strong correlation between the kinetic energy carried by these neutrons and the nuclear binding energy loss. This has important consequences for the energy resolution achievable with compensating calorimeters. This is further discussed in Section 4.8.

4.7   Fluctuations in the Electromagnetic Shower Content

The most important and the most consequential of all fluctuations that play a role in hadronic shower development are the event-to-event fluctuations in the em shower content, f_{em}. These fluctuations are large and they are *not* governed by Poisson statistics.

4.7.1   *Pion showers*

When a high-energy pion interacts with an atomic nucleus, a certain number of secondary hadrons are produced. Approximately one third of these secondaries are π^0s, the rest are other hadrons, predominantly charged pions. These secondary hadrons produce tertiary ones in their interactions with calorimeter nuclei. Also here, approximately one third of the reaction products are π^0s. The production of π^0s, which constitute the em component of the hadron shower, is irreversible: charged pions may produce π^0s, but the reverse process does not occur. This feature leads to an asymmetry in the distribution of the energy fraction carried by the em shower core, in the following way.

At high energy, the particles produced in the nuclear reactions may be considered the products of a fragmenting quark excited in the interaction process. The particle containing this quark, *i.e.*, the leading particle, carries a large fraction of the total energy. This highly energetic particle may either be a π^0 or some other hadron. If it is a π^0, then its energy goes into the em shower component. If it is another particle, then this particle re-interacts, and the leading particle produced in that process may in turn either be a π^0 or some other hadron.

Because of the irreversibility of the production of π^0s and because of the leading-particle phenomenon, there is an asymmetry between the probability that an anomalously large fraction of the energy is contained in the em shower component and the probability that a similarly large fraction of the energy is contained in the non-em shower component. When the leading particle in the first reaction is a π^0, the em shower component is guaranteed to be large. When the leading particle is not a π^0, the em shower component is not guaranteed to be small, since the leading particle produced in the second generation (or even later) may be a π^0. Overall, there is thus a larger probability for the em shower fraction to be anomalously large than anomalously small.

As a result of this asymmetry, the distribution of the em shower fraction, f_{em}, tends to be skewed to the high side. This is confirmed by experimental data, as illustrated by Figure 4.52. Figure 4.52a shows results of a SPACAL experiment in which an attempt was made to measure the f_{em} distribution, exploiting the characteristics of the lateral development of pion showers in this lead/scintillating-fiber calorimeter [Aco 92b].

The lateral energy deposit profile exhibits two distinct components (see Figure 2.38).

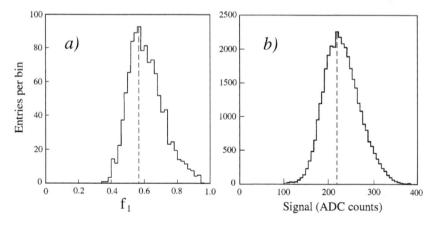

FIG. 4.52. The distribution of the fraction of the energy of 150 GeV π^- showers contained in the em shower core, as measured with the SPACAL detector (*a*) [Aco 92b] and the signal distribution for 300 GeV π^- showers in the CMS Quartz-Fiber calorimeter (*b*) [Akc 98].

The non-em shower component appears as an exponentially decreasing halo that surrounds a narrow, Gaussian em shower core. The em component is thus strongly concentrated around the shower axis.

This SPACAL measurement was possible thanks to the fine-grained tower structure of this calorimeter (155 hexagonal cells with a cross section of ~ 49 cm^2 each). The measurement was performed with 150 GeV pions, which were sent into the center of the central calorimeter cell. Since almost all em energy was deposited in this central cell and since most of the non-em energy was deposited in the other 154 calorimeter cells, the fraction of the total calorimeter signal recorded in the central cell ($E_1 / \sum_{i=1}^{155} E_i = f_1$) may be considered a reasonable approximation of f_{em}. The f_1 distribution, shown in Figure 4.52a indeed exhibits the expected asymmetric shape.

Another possibility to study f_{em} and its fluctuations is offered by Čerenkov calorimeters. As we saw in Section 3.4, the hadronic signals from such calorimeters are completely dominated by the contributions of the em shower components ($e/h \sim 5$). Therefore, the response function of these instruments may be expected to reflect event-to-event fluctuations in f_{em}. Figure 4.52b shows the signal distribution for 300 GeV π^- showers in a prototype module of the Quartz-Fiber Forward Calorimeter of the CMS experiment [Akc 98]. This signal distribution exhibits the same asymmetric shape as the f_1 distribution measured with SPACAL.

A comparison of the two distributions in Figure 4.52 also reveals the limitations of the approximation $f_1 \approx f_{em}$. Since the central SPACAL cell also recorded a significant fraction of the energy contained in the non-em shower component, f_1 tended to be systematically somewhat larger than f_{em}, especially in events where f_{em} was small. This effect led effectively to a low-side cutoff in the distribution, since values $f_1 < 0.2$ were essentially impossible, even for $f_{em} = 0$. Such a cutoff is clearly absent in the signal distribution from the Čerenkov calorimeter (Figure 4.52b).

This effect should be taken into account when interpreting f_1 data, since

- $\langle f_1 \rangle$ is systematically larger than $\langle f_{em} \rangle$, more so at low energy. This was experimentally confirmed by the SPACAL Collaboration itself, who measured $\langle f_{em} \rangle$ independently, by unfolding the average lateral shower profiles [Aco 92b]. The differences between $\langle f_{em} \rangle$ and $\langle f_1 \rangle$ ranged from $\sim 10\%$ at 150 GeV to $\sim 20\%$ at 20 GeV.

- The event-to-event fluctuations in f_1 are systematically smaller than those in f_{em}, more so at low energy. Evidence for this effect may be derived from a comparison with event-to-event fluctuations in the Čerenkov calorimeter data (Figure 4.53)

The rms width of the f_{em} distribution for 150 GeV pions (Figure 4.52a) was found to be $\sim 18\%$. The event-to-event fluctuations in the fraction of the hadronic energy carried by the em shower component were thus substantial. At lower energies, the width of the f_1 distribution increased further, in a way that seemed to scale with the logarithm of the energy, rather than the familiar $1/\sqrt{E}$ dependence (Figure 4.53a).

The energy resolution for pions detected with the quartz-fiber calorimeter is shown as a function of energy in Figure 4.53b. The measured energy resolutions are indicated by the closed circles. The triangles show the contribution of photoelectron statis-

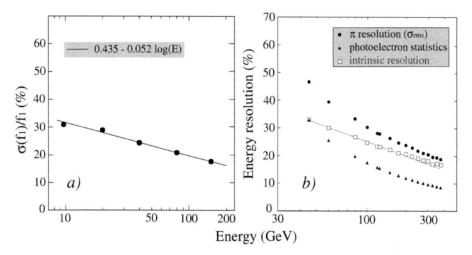

FIG. 4.53. Fluctuations in the em fraction of pion showers. Diagram (a) shows the rms width of the f_1 distribution in SPACAL, as a function of energy [Aco 92b]. Diagram (b) shows the hadronic energy resolution, σ_{rms}/E, of a quartz-fiber calorimeter (the full circles). Also shown in this diagram are the contribution to this resolution from photoelectron statistics (the triangles) and the resolution obtained after subtracting this p.e. contribution in quadrature from the measured values (the squares) [Akc 97]. The lines are drawn to guide the eye.

tics to this resolution and the squares represent the *intrinsic* resolutions, obtained after subtracting the contribution of photoelectron statistics in quadrature from the measured resolutions [Akc 97]. This intrinsic resolution exhibits approximately the same type of energy dependence as the width of SPACAL's f_1 distribution.

The fact that the width of the f_{em} distribution does not scale with $E^{-1/2}$ is a consequence of the non-Gaussian nature of the fluctuations in f_{em}, just like the asymmetric shape of the f_{em} distribution. The width does become smaller as the energy increases, but not as fast as one would expect if only fluctuations in the number of different π^0s contributing to the signals would play a role. The energy sharing between those different π^0s varies wildly, in a way that is governed by the peculiarities of the fragmentation processes and other details of the nuclear reactions. Increasing the shower energy does not make much difference in that respect and, therefore, the improvement in the width of the f_{em} distribution proceeds slower than one would expect on the basis of $E^{-1/2}$ scaling.

However, a logarithmic energy dependence cannot be the correct description of this improvement, simply because it would lead to an unphysical situation (a negative resolution) at very high energies. In addition, close inspection of Figure 4.53b reveals that the squares, representing the intrinsic resolution, do not really follow a straight line in the logarithmic plot. Just as the energy dependence of the average value of f_{em}, the energy dependence of the event-to-event fluctuations in this variable is more correctly described with a power-law function.

In Section 2.3.1 we saw that the average value for $f_{\rm em}$ is related to the average number of pions produced in the shower development as (Equation 2.19)

$$\langle f_{\rm em} \rangle = 1 - \left[\frac{E}{E_0} \right]^{(k-1)}$$

in which k is a parameter related to the average particle multiplicity in the nuclear interactions and E_0 represents the average energy needed for the production of one pion. When we tried to describe the energy dependence of the width of the $f_{\rm em}$ distribution in a similar way

$$\frac{\sigma(f_{\rm em})}{\langle f_{\rm em} \rangle} = \left[\frac{E}{E_0} \right]^{(l-1)} \tag{4.27}$$

we found that the best results were obtained for an E_0 value that was very close to the one that produced the best results for the energy dependence of $\langle f_{\rm em} \rangle$ in copper (the QFCAL absorber material): 0.7 GeV. As illustrated in Figure 4.54, Equation 4.27 with the parameter values $E_0 = 0.7$ GeV and $l = 0.72$ resulted in a good description of the experimental QFCAL resolution data.

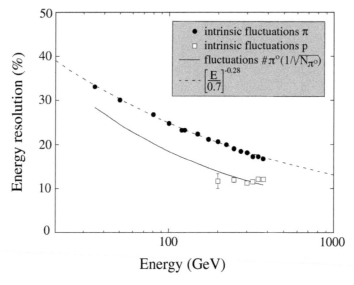

FIG. 4.54. The intrinsic energy resolution for pions (full circles) and protons (open squares) of the Quartz-Fiber Calorimeter prototype for the high-η region in the CMS experiment. The dashed line represents a fit to the pion data (Equation 4.27). The solid line represents the fluctuations in the number of different π^0s contributing to the signals from this calorimeter. Experimental data from [Akc 97] and [Akc 98].

If it takes, on average, 0.7 GeV to produce one pion in the non-em component of pion-induced showers in copper, then we can also estimate the number of π^0s that contributed to the signals. At 35 GeV, the lowest experimental point in Figure 4.54, the average value of the em shower fraction amounted to 0.51 (Equation 2.19). This means that the non-em shower component comprised some 25 charged pions ($0.49 \times 35/0.7$). If the probability for π^0 production is equal to that for π^+ or π^-, the 35 GeV showers contained thus, on average, contributions from 12–13 π^0s. Gaussian fluctuations in this *number* of π^0s would lead to a standard deviation equal to $\sim 28\%$ of the mean value. This may be compared with the intrinsic resolution of the QFCAL detector, which was measured to be $33.1 \pm 0.7\%$ at this energy [Akc 97]. We repeated this calculation for the other energies at which the intrinsic resolution was measured. The results are indicated by the solid curve in Figure 4.54.

This exercise serves to provide a rough idea about the magnitude of the fluctuations in the em content of hadron showers. However, there are large differences in the energy carried by the individual π^0s contributing to the calorimeter signals. And since f_{em} represents the energy carried by all π^0s combined, one may expect the event-to-event fluctuations in this variable to be larger than those derived on the basis of the *number* of π^0s alone. In other words, the solid curve in Figure 4.54 represents a lower limit to the fluctuations in f_{em}, and thus to the resolution of a Čerenkov calorimeter.

Calorimeters based on Čerenkov light are extreme examples of devices in which the hadronic energy resolution is dominated by fluctuations in the em shower content, because there is such a large difference between the calorimeter responses to em and non-em energy deposit ($e/h \sim 5$). However, also in other non-compensating calorimeters, the contribution of these fluctuations may very well be the single most important component of the hadronic energy resolution. This is especially true at high energies, where the contributions of stochastic fluctuations (the $E^{-1/2}$ terms) vanish.

It is a well-known fact that the hadronic energy resolution of non-compensating calorimeters does not scale with $E^{-1/2}$. The hadronic energy resolution of a non-compensating calorimeter is often described as

$$\frac{\sigma}{E} = \frac{a_1}{\sqrt{E}} \oplus a_2 \qquad (4.28)$$

where the value of the constant term, added in quadrature to the stochastic term, is completely determined by the degree of non-compensation, $|e/h - 1|$. However, from the previous discussion it follows that it is more accurate to describe the contribution of fluctuations in f_{em} to the hadronic energy resolution of non-compensating calorimeters with a term that depends on the energy, rather than with a constant term. Such a term should be added in quadrature to a stochastic, $E^{-1/2}$ term which includes the effects that affect the resolution at low energy (sampling fluctuations, photoelectron statistics, etc.):

$$\frac{\sigma}{E} = \frac{a_1}{\sqrt{E}} \oplus a_2 \left[\left(\frac{E}{E_0} \right)^{l-1} \right] \qquad (4.29)$$

As was shown above, the QFCAL data (copper) are best described with the parameter values $E_0 = 0.7$ GeV and $l = 0.72$. The value of a_2 depends on the e/h ratio. It varies between 0 for compensating calorimeters and 1 for extremely non-compensating ones ($e/h = \infty$).

At high energies, where the effects governed by Poisson statistics become vanishingly small, the fluctuations in em shower content start dominating the resolution and the energy dependence characteristic for this resolution component becomes the dominant feature.

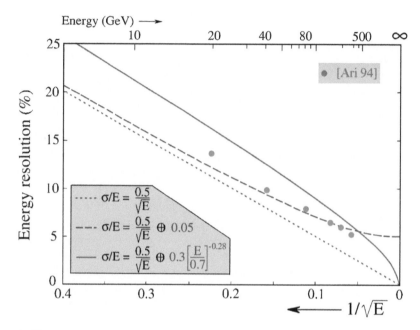

FIG. 4.55. Comparison between Equations 4.28 (the dashed curve) and 4.29 (the solid curve). The dotted line represents the stochastic term in these equations. The experimental data were taken from [Ari 94]. See text for details.

The difference between Equations 4.28 and 4.29 is illustrated in Figure 4.55. The dashed and solid curves describe the energy resolutions calculated with Equations 4.28 and 4.29, respectively. The stochastic term has been chosen the same in both cases: $a_1 = 0.5$. For the constant term in Equation 4.28, we have chosen $a_2 = 0.05$, while the non-compensation parameter in Equation 4.29 has been given the value $a_2 = 0.3$. The parameters E_0 and l were given the QFCAL values mentioned above. Since the horizontal scale is proportional to $E^{-1/2}$, the stochastic term is represented by a straight (dotted) line through the bottom right corner in Figure 4.55.

The difference between the two expressions becomes mainly apparent at energies in excess of 100 GeV, where the second term starts to dominate the resolution. However, there are also differences at low energies. Since a constant term contributes very little to the total resolution at low energies, experimental low-energy data should scale approximately with $E^{-1/2}$ if Equation 4.28 described reality, even in non-compensating calorimeters. On the other hand, in Equation 4.29 the second term also contributes significantly to the energy resolution at low energies. Therefore, a deviation from $E^{-1/2}$

scaling at low energies constitutes important experimental information.

We have included in Figure 4.55 some representative experimental data that were reported for a prototype of the iron/plastic-scintillator hadron calorimeter of ATLAS [Ari 94]. The experimental points are located on a line that is approximately parallel to the dotted line. This suggests that the resolution ought to be described by a *linear sum* of a stochastic term and a constant term:

$$\frac{\sigma}{E} = \frac{a_1}{\sqrt{E}} + a_2 \tag{4.30}$$

This feature is quite common for non-compensating calorimeters, and several authors have fitted their experimental data to an expression of this type. However, this expression suggests that there is complete correlation between the fluctuations that contribute to the two different terms, which is nonsense.

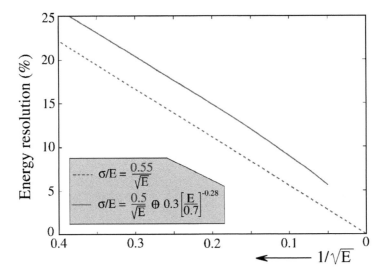

FIG. 4.56. The energy resolution calculated with Equation 4.29 for energies up to 400 GeV (the solid line), and calculated with a sole stochastic term with a slightly larger a_1 value (the dotted line). See text for details.

Figure 4.56 provides the solution of this apparent mystery. The solid line in this figure is exactly the same as in Figure 4.55, except that it stops at $E = 400$ GeV, which is the highest pion energy for which experimental data have been reported. This curve runs almost parallel to the dotted line, which represents a stochastic term with a coefficient $a_1 = 0.55$ (in the solid curve, $a_1 = 0.50$). This means that it is practically impossible to distinguish between fits such as

$$\frac{\sigma}{E} = \frac{50\%}{\sqrt{E}} \oplus 30\% \left[\left(\frac{E}{0.7} \right)^{-0.28} \right]$$

and

$$\frac{\sigma}{E} = \frac{55\%}{\sqrt{E}} + 3.5\%$$

in the energy range for which experimental data are available. Experimental data at very high energies would be needed to make that distinction. The observation that experimental resolution data tend to be better described by a linear sum of a stochastic term and a constant term rather than by a quadratic sum of such terms may thus be interpreted as support for Equation 4.29.

In summary, it looks as if experimental data do indeed favor Equation 4.29 over Equation 4.28 and that fluctuations in f_{em} thus contribute an energy-dependent term to the energy resolution, rather than a constant term.

4.7.2   Proton showers

In Section 2.3.1 we saw that the average value of f_{em}, and thus the response of a non-compensating calorimeter, is not the same for all types of hadrons. In particular, $\langle f_{em} \rangle$ is smaller for proton-induced showers than for pion-induced ones, by $\sim 15\%$ in copper. This is a consequence of the requirement of baryon number conservation in hadronic shower development. This requirement prohibits the production of leading π^0s in proton-induced showers and thus reduces the energy fraction contained in the em shower component, in comparison with pion-induced showers where no such limitations exist.

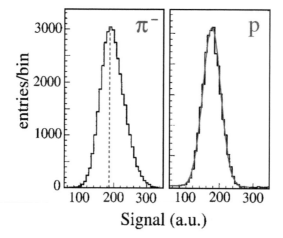

FIG. 4.57. Signal distributions for 300 GeV pions (a) and protons (b) detected with a quartz–fiber calorimeter. The curve represents the result of a Gaussian fit to the proton distribution [Akc 98].

This effect also has consequences for the signal distributions. These consequences are twofold:

1. The energy resolution for protons is better than for pions of the same energy. This is illustrated in Figure 4.57, which shows signal distributions for 300 GeV pions

and protons, recorded with the CMS quartz-fiber calorimeter. The rms width of the proton distribution was measured to be $\sim 20\%$ smaller than that of the pion distribution. When the contributions from fluctuations in the (small) number of photoelectrons to the resolutions were subtracted, the difference became even larger, $\sim 30\%$. The intrinsic resolutions for protons are included in Figure 4.54, where they are represented by the open squares. Interestingly, the proton results are in good agreement with the naïve expectation, symbolized by the solid curve in Figure 4.54, that the fluctuations in f_{em} are determined by fluctuations in the *number* of different π^0s contributing to the shower signals.

2. The response function for protons does not exhibit the asymmetric shape charac- teristic for pion-induced showers. It is much more symmetric and well described by a Gaussian fit (Figure 4.57b). As we saw before, the high-side tail in the pion signal distributions is populated by events in which a leading π^0 is pro- duced in the early stages of the shower development. Such events do not occur in proton-induced showers, where the leading particles tend to be baryons. The em component of proton-induced showers is typically populated by π^0s that share the energy contained in this component more evenly.

Since π^0s are very unlikely to be leading particles in proton-induced showers, the large differences between the energies of individual π^0s that were caused by leading- particle effects and that led to a substantial broadening of the signal distributions for pions, are absent in this case. As a result, the fluctuations in f_{em} are smaller for protons, the energy resolution is better, and the response function is more symmetric than for pions. This conclusion applies to *all* non-compensating calorimeters, but the extent of the proton–pion differences depends of course on the degree of non-compensation.

The origin of the observed differences between proton and pion showers strongly suggests that the mentioned effects are not limited to these particles. In particular, one may expect to see significant differences between kaon and pion showers as well. Just as the baryon number is conserved in proton showers, the strangeness quantum number is conserved in the strong interactions that take place in kaon-induced showers. The strange (anti-)quark contained in the incident particle is likely to be transferred to a highly energetic particle in each generation of the shower development. The production of π^0s in kaon showers is therefore limited by a mechanism very similar to that in proton showers, and the results may be expected to be similar as well: a smaller response, a better energy resolution and a more symmetric response function than for pion-induced showers. Apart from the CMS measurements of the kaon response (Figure 7.38), I am not aware of any experimental data that may serve to test this prediction.

We started this chapter with the statement that the energy resolution determines the precision with which the energy of a showering particle can be measured. However, the issues discussed in this subsection illustrate that there are many unsuspected aspects to that problem. In non-compensating calorimeters, the precision with which the energy of a showering hadron can be determined is not only affected by fluctuations in the em shower content, but also by uncertainty about the *type* of particle that caused the shower. The latter effect leads to systematic mismeasurements. For example, in the case of the quartz-fiber calorimeter discussed above, the energy of protons would be systematically underestimated if this device was calibrated with pions.

4.7.3   Jets

Showers induced by jets are similar to those induced by hadrons in the sense that the showers consist of a mixture of em and non-em components and that the energy carried by each of these components may fluctuate wildly from one event to the next. As for single hadrons, these fluctuations become more and more important for the jet energy resolution of non-compensating calorimeters as the energy increases and the contributions of effects governed by Poisson statistics to the resolution become small.

The em component of a shower initiated by a jet contains contributions from π^0s, or rather γs, produced in the fragmentation process that generated the jet, plus contributions from the em cores of individual showers initiated by the hadronic jet fragments.

The difference between jets and single hadrons is most apparent in the early stages of the shower development. When a 100 GeV pion undergoes its first nuclear reaction in the calorimeter, typically fewer than 10 secondary particles are produced, a combination of π^0s, charged pions, kaons and other hadrons. The energies of these particles are affected by the requirement of momentum conservation in the collision.

A 100 GeV jet is the result of the fragmentation of a 100 GeV (di-)quark or gluon. The number of particles generated in this process is, in general, larger than the number of secondaries produced in the fixed-target collisions of single hadrons at the same energy. For example, the average charged-particle multiplicity in e^+e^- collisions at LEP ($\sqrt{s} = 90$ GeV) was measured to be ~ 21 [Bet 92]. Including π^0s, the average multiplicity was thus more than 30 and these particles belonged usually to one of two 45 GeV jets ($e^+e^- \rightarrow q\bar{q}$). Therefore, in the fragmentation of a 45 GeV (anti-)quark, typically some 15 particles were produced. The fastest of these particles (the leading particle, which contained the fragmenting quark) carried, on average, $\sim 30\%$ of the initial energy. For comparison, the average particle multiplicity in 45 GeV proton–proton collisions in a fixed-target geometry is only ~ 5 [Per 00].

Because of this difference in the multiplicity of the initial state, the event-to-event f_{em} fluctuations in showers initiated by jets may be somewhat smaller than in those initiated by single hadrons. On the other hand, there are some specific additional sources of fluctuation which affect the jet resolution and which thus tend to make the jet resolution larger than that for single hadrons of the same energy. We mention two:

1. Fluctuations resulting from the hadronic particle composition of the jet. In the previous subsection, we saw that there are systematic differences between the $\langle f_{em} \rangle$ values for showers induced by pions and protons of the same energy, and that it is likely that similar differences exist between pions and kaons. Therefore, the response of a non-compensating calorimeter to these particles is not the same and the jet response depends thus on the composition of the non-em jet component.

2. Fluctuations resulting from the jet-defining algorithm. In practical experiments at colliding-beam machines, a jet is defined as a collection of particles contained in a well-defined region of the η–ϕ space, usually a cone with radius $R = \sqrt{(\Delta\eta)^2 + (\Delta\phi)^2}$ around the jet axis. This jet axis connects the interac-

tion vertex with the center of gravity of the jet's energy deposit in the calorimeter. Typical values for R are in the 0.3–0.4 domain. As a result of this algorithm, some particles that belong to the jet may be excluded when the jet's energy is determined. On the other hand, some other particles, unrelated to the jet but traveling by accident within the acceptance of the cone, may be included. The effects of these fluctuations on the energy resolution are thus similar to those induced by a combination of lateral shower leakage (Section 4.5.2) and noise (Section 4.4.1). The "noise" (or "underlying-event") part becomes rapidly more important as the collision energy and/or the luminosity increase.

In most experiments, the jet energy resolution has a greater practical importance than the single-hadron resolution. However, this jet resolution is in practice not as easy to measure, since test beams of jets with precisely known energy and composition are not available at accelerators.

One way of dealing with this problem is to use Monte Carlo simulations to establish the calorimeter properties for jet detection. However, because of the unreliability of Monte Carlo simulations of hadronic shower development, this is not a very meaningful and satisfactory solution.

An alternative method was developed by CMS [Gum 08]. They used experimental signal distributions for em and hadronic showers as the basis for estimating the properties of their calorimeter for jet detection. They generated jets, of the type they wanted to study, according to one or several of the well-established fragmentation models (*e.g.*, the Lund model, the basis for several popular event generators [Ben 86]). Every jet consisted of a series of physical particles, with precisely known four-vectors. The calorimeter signal from each of these jet particles was derived from the experimental signal distributions [Abd 09], as follows.

If the jet contained, for example, a 13.3 GeV π^0, its signal was drawn at random from the experimental signal distribution for 12 GeV electrons, and the result was multiplied by 13.3/12. The signal from a 73.2 GeV π^- was taken from the experimental 80 GeV pion distribution, and the randomly drawn signal from this distribution was multiplied by 73.2/80, *etc.*

In this way, the total signal from a 1 TeV jet could be constructed from the signals caused by each and every particle constituting the jet. By repeating this procedure, a signal distribution for 1 TeV "tagging jets," indicative for the production of Higgs bosons and an important benchmark for this particular calorimeter, could be derived. More details about this procedure are given in Section 9.3.4.

A similar method was used earlier to predict the jet energy resolution of the CDF Plug Upgrade calorimeter [Alb 02b] and of quartz-fiber calorimeters [Gan 93, Gan 95]. In these studies, jets were generated according to the following fragmentation function

$$D(z) = (\alpha + 1)(1 - z)^\alpha / z \tag{4.31}$$

which gives a reasonable description of high-energy Tevatron jets when the parameter $\alpha = 6$ [Gre 90]. Every jet particle drawn from this z-distribution was given a 1/3 possibility of being a π^0 and a 2/3 probability of being a charged pion. The contributions of the individual particles constituting the jet were drawn from the envisaged signal

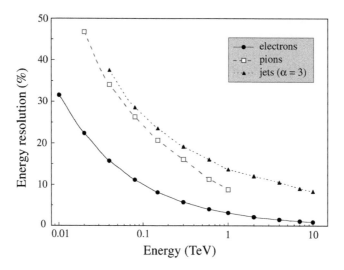

FIG. 4.58. Predicted energy resolution for single electrons, single pions and jets in a quartz–fiber calorimeter [Gan 95].

distributions for electrons and pions. The electron distributions were assumed to be Gaussian and exclusively determined by photoelectron statistics (assuming a light yield of 1 p.e./GeV), while the pion signal distributions were the f_1 distributions measured by SPACAL (see Section 4.7.1).

Figure 4.58 shows the resolution for these jets in the quartz-fiber calorimeters as a function of energy. For comparison, the resolutions for single electrons and pions are shown in the same figure. The jet resolution turned out to be systematically somewhat worse than the resolution of the single particles, even without the fluctuations resulting from the jet algorithm and the contributions of protons and kaons, which were not included in these simulations.

As in the case of single hadrons, the *average* em shower fraction in jets depends on the *type* of fragmenting object. For example, the leading particle of a fragmenting diquark is a baryon. Therefore, the average energy carried by π^0s in such jets is smaller than for jets resulting from the fragmentation of u or d quarks, where the leading particle may very well be a π^0. Also jets from heavy quark (s, c, b) and gluon fragmentation contain, on average, less em energy.

In non-compensating calorimeters, the energies of jets from fragmenting heavy quarks, diquarks and gluons thus tend to be systematically smaller than the energies from u or d quark jets, for the same reason as the energy of protons is systematically underestimated in a calorimeter calibrated with pions.

4.8   Fluctuations in a Compensating Calorimeter

4.8.1   *The intrinsic limit to the hadronic energy resolution*

In Section 3.3 we saw that only compensating calorimeters are linear for hadrons and jets. In addition, the dependence of the calorimeter response on the type of particle or jet that initiated the shower disappears in compensating calorimeters. Compensation also has important consequences for the energy resolution of hadrons and jets:

1. The energy resolution may be considerably better than for non-compensating calorimeters, especially at high energies, since a major source of fluctuations is eliminated: fluctuations in the energy sharing between the em and non-em shower components.

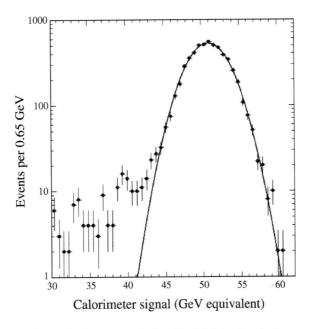

FIG. 4.59. Signal distribution for showers induced by 50 GeV pions in the compensating ZEUS calorimeter, together with the result of a Gaussian fit to the experimental data. To limit the effects of shower leakage, only events for which less than 10% of the total signal was recorded beyond a depth of 4.1 λ_{int} were retained for this plot. Note the logarithmic vertical scale [Beh 90].

2. The eliminated fluctuations are non-Gaussian. Therefore, the asymmetric line shape that characterizes the hadronic signal distributions of non-compensating calorimeters (Figure 4.52) may be eliminated as well. This is illustrated in Figure 4.59, which shows that the right-hand side of the signal distribution for 50 GeV pions in the ZEUS calorimeter is very well described by a Gaussian shape, over three decades. At the left-hand side, deviations from this shape are observed at the 1% level. These deviations are the result of shower leakage.

With this major source of fluctuations eliminated, one may wonder which remaining factors determine and limit the hadronic energy resolution of compensating calorimeters. These factors include:

- Sampling fluctuations
- Signal quantum fluctuations
- Fluctuations in the visible energy

The first two types of fluctuations also determine and limit the *electromagnetic* energy resolution of sampling calorimeters. These fluctuations can in principle be reduced by increasing the sampling fraction and the sampling frequency of the calorimeter.

Fluctuations in the visible energy are unique to hadronic showers. The elimination of fluctuations in f_{em} does of course not mean that fluctuations in the visible energy are eliminated. As a matter of fact, fluctuations in the visible energy constitute the ultimate limit to the energy resolution of compensating calorimeters.

The relative importance of such fluctuations depends on the way in which compensation is achieved. If compensation is realized through proper amplification of the signals from neutrons released in the shower development process, then the effect of fluctuations in nuclear binding energy losses on the hadronic energy resolution may be reduced considerably. By definition, nuclear binding energy loss coincides with the release of nucleons from the calorimeter nuclei. Especially in high-Z materials, the overwhelming majority of these nucleons are (evaporation) neutrons (see Table 2.5). Therefore, one may expect the kinetic energy carried by these neutrons to be correlated to the nuclear binding energy loss.

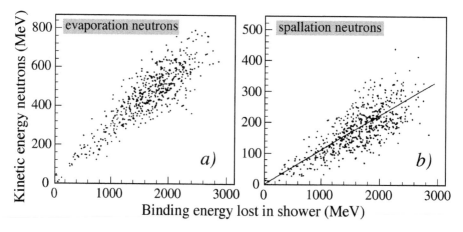

FIG. 4.60. Scatter plots showing the correlation between the kinetic energy carried by soft neutrons ($E < 20$ MeV) and the nuclear binding energy lost when 5 GeV π^- mesons are absorbed in depleted uranium (^{238}U). The distributions are shown separately for neutrons that originated from evaporation processes (*a*) and from nuclear spallation (*b*). Results from Monte Carlo simulations with the HETC/MORSE package [Brau 89].

This was confirmed by dedicated Monte Carlo calculations performed by Brau and Gabriel [Brau 89]. Figure 4.60 shows some results of these calculations, for 5 GeV π^- showering in depleted uranium. The scatter plots show the kinetic neutron energy versus the nuclear binding energy lost in the shower development. In the simulations, a distinction was made between neutrons that originated from evaporation (Figure 4.60a) and from spallation (Figure 4.60b). There is a clear correlation between nuclear binding energy losses and kinetic neutron energy in both cases. However, the fluctuations are somewhat larger for the spallation neutrons, since *protons* may also carry a considerable (and varying) fraction of the available energy in spallation processes (see Section 2.3.2.3). The figure also shows that most ($\sim$ 70%) of the kinetic neutron energy is carried by neutrons originating from evaporation processes, where there is almost no competition from protons.

However, when assessing the effects of fluctuations in visible energy on the hadronic energy resolution of a compensating calorimeter, the crucial factor is not so much the correlation between the nuclear binding energy loss and the kinetic neutron energy, but rather the correlation with the *signals* produced by the latter in the active calorimeter material.

Brau and Gabriel studied this correlation for two different calorimeter configurations. In the first configuration, 3 mm ^{238}U plates were interleaved with 3 mm plastic-scintillator plates. In the second one, 0.4 mm thick silicon layers, sandwiched between two layers of G10, served as active material in a sampling calorimeter consisting of 5 mm ^{238}U absorber layers. The correlation between the nuclear binding energy lost in the shower development and the total signal generated by neutrons in the active material of these two calorimeters is shown in Figures 4.61a and 4.61b, respectively. The poor sampling of the neutron component by the silicon leads to significantly larger fluctuations, and thus to a worse energy resolution, even though both configurations might be perfectly compensating.

In compensating calorimeters, the fluctuations in visible energy, and thus the limits on the achievable hadronic energy resolution, are thus determined by

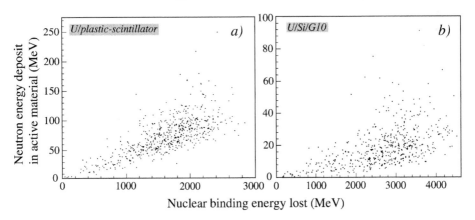

FIG. 4.61. Scatter plots showing the correlation between nuclear binding energy losses and the signals generated by soft neutrons ($E < 20$ MeV) in uranium calorimeters based on plastic-scintillator (*a*) or silicon (*b*) readout. Results from Monte Carlo simulations of pion-induced showers with the HETC/MORSE package [Brau 89].

1. the correlation between nuclear binding energy losses and the total kinetic energy carried by neutrons released in the shower development process, and

2. the correlation between this kinetic energy and the signals generated by these neutrons in the active calorimeter material.

The second factor causes the difference between the hadronic resolutions achievable with calorimeters where compensation is achieved by suppressing the em response (*e.g.*, U/Si/G10) and those where neutron amplification in hydrogenous active material is the mechanism through which the em and non-em calorimeter responses are equalized. Neutrons deposit a major fraction of their kinetic energy in the active layers of the latter (see Table 3.3) and variations in this sampling fraction are inconsequential for the hadronic energy resolution.

In these hydrogenous compensating calorimeters, the ultimate limits on the achievable hadronic energy resolution are thus determined by factor #1. This factor has two components:

1*a*. The correlation between nuclear binding energy losses and the *number* of neutrons produced in the shower development.

1*b*. The correlation between the number of neutrons and the *total kinetic energy* carried by these particles.

Especially in high-Z materials such as lead, almost the entire nuclear binding energy loss is caused by the release of neutrons in nuclear reactions. Table 2.5 shows that for every proton 10 neutrons are released when hadron showers develop in lead. Therefore, there is a very strong correlation between the nuclear binding energy loss and the number of neutrons in this case, and the contribution of fluctuations in the neutron/proton production ratio to the hadronic energy resolution is small. According to Table 2.5, on average four protons are released per GeV hadronic energy. A fluctuation by one standard deviation (two protons) would change the proton/neutron ratio dramatically, but it would change the nuclear binding energy loss per released neutron only by 5%. This effect is small compared with that caused by fluctuations in the total kinetic neutron energy (factor 1*b*).

The *irreducible* event-to-event fluctuations in the signals from compensating lead-scintillator calorimeters are thus expected to be dominated by fluctuations in the *total kinetic energy* carried by a given number of neutrons. In the following, we estimate the contributions of such irreducible fluctuations to the energy resolution for hadrons with energy E, from first principles.

When high-energy hadrons are absorbed in a calorimeter, the non-em shower energy, E_h, can be subdivided into three components:

- Energy deposited in the absorbing structure by charged shower particles, E_c.
- Energy deposited in the absorbing structure by neutrons, E_n.
- Invisible energy, mainly caused by nuclear binding energy losses, ΔE_B.

In Section 2.3.2.3 we saw that ΔE_B represents, on average, $\sim 40\%$ of E_h, in lead-based calorimeters. The kinetic energy carried by neutrons accounts, on average, for $\sim 10\%$ of E_h, and the charged particles take care of the remaining 50% (Table 2.6).

In Section 2.3.2.4, the kinetic energy spectrum of the evaporation neutrons was discussed. This kinetic energy was assumed to be distributed according to a Maxwell spectrum with a temperature of 2 MeV (Equation 2.25, Figure 2.32). Based on these assumptions, we calculated the total kinetic energy carried, on average, by a large number of such neutrons. Results of these calculations, for $N = 100$ and $N = 1,000$, are shown in Figure 2.33. The fluctuations in the total kinetic energy (E_n) carried by N neutrons were found to scale as

$$\frac{\sigma}{E_n} = \frac{83\%}{\sqrt{N}} \tag{4.32}$$

If we assume that the neutrons produced when hadron showers develop in lead are distributed according to this Maxwell spectrum (and thus have an average kinetic energy of 3 MeV), one should thus expect, on average, 33 neutrons per GeV of non-em energy ($E_h = 1$ GeV), and 80 neutrons per GeV nuclear binding energy loss ($\Delta E_B = 1$ GeV). Perfect correlation between nuclear binding energy losses and the number of neutrons implies that the latter rate is fixed, it does not fluctuate about an average:

$$\frac{N}{\Delta E_B (\text{GeV})} = 80$$

For a given value of E_h, downward fluctuations in ΔE_B imply a correspondingly reduced number of neutrons, while the number of neutrons increases in the case of upward fluctuations in ΔE_B. However, the kinetic energy carried by these N neutrons can of course fluctuate, in ways governed by Equation 4.32. The kinetic energy distribution of the neutrons has a relative width

$$\frac{\sigma}{E_n} = \frac{83\%}{\sqrt{80 \Delta E_B}} = \frac{9.3\%}{\sqrt{\Delta E_B}}$$

In Section 4.6 it was shown that the fluctuations in the nuclear binding energy losses in lead have a standard deviation of about $15\%/\sqrt{\Delta E_B}$ (Equation 4.26).

Based on these assumptions, we wrote a BOTEC Monte Carlo program to calculate the ultimate limits on the hadronic energy resolution of compensating lead/plastic-scintillator calorimeters. First, the fluctuations in the calorimeter signals for a fixed non-em energy deposit (10 GeV) were determined. The value of ΔE_B was chosen from a Gaussian distribution centered around 4 GeV, with a fractional width $\sigma/\Delta E_B = 15\%/\sqrt{4} = 7.5\%$. This ΔE_B value also determined the number of neutrons, $N = 80\Delta E_B$. The neutron energy E_n was chosen from a Gaussian distribution centered around $3N$ MeV, with a fractional width given by Equation 4.32. The values of E_n and ΔE_B automatically determined the value of E_c.

To determine the calorimeter signal, the two components contributing to it (E_c and E_n), were given different weights. The total (non-em) signal is thus given by

$$S_h = E_c + w_n E_n \tag{4.33}$$

The fractional width σ/S_h of the total signal from 10 GeV non-em energy deposit in a compensating lead/plastic-scintillator calorimeter, based on the assumptions described

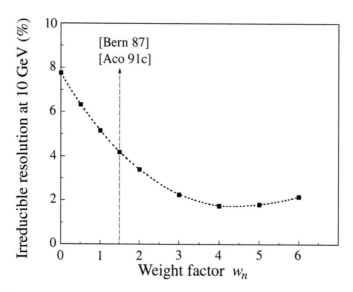

FIG. 4.62. The (calculated) contribution of fluctuations in the total kinetic neutron energy to the hadronic energy resolution at 10 GeV of compensating Pb/plastic-scintillator calorimeters, as a function of the weighting factor for the neutron signals that has to be applied to achieve compensation in such calorimeters. In practice, n/mip values of ~ 1.5 are needed for this purpose, as indicated by the dashed line. See text for details.

above, is given as a function of the relative weight w_n of the neutron signals in Figure 4.62.

As expected, this width gradually decreased when the weight given to the neutron signals was increased. The best results were obtained for $w_n \approx 4$, at which value the energy equivalent of the neutron signals equalled the nuclear binding energy loss (for $\Delta E_B = 1$ GeV, there were 80 neutrons with a total kinetic energy of 240 MeV). This is where the assumption of perfect correlation between ΔE_B and E_n had the largest impact on the total signal distribution.

From Equation 3.9 and Table 3.4, we see that the weighting factor w_n is approximately equal to the n/mip ratio. In practice, the choice of w_n is thus not arbitrary, but it is dictated by the compensation requirement (Equation 3.9). From Figure 3.42, we see that compensation in lead/plastic-scintillator calorimeters is achieved when $n/mip \sim 1.5$. At that point, the irreducible energy resolution is about 4% for a 10 GeV non-em energy deposit.

This figure clearly illustrates the benefits of exploiting the correlation between the neutron signals and the nuclear binding energy losses. If this correlation were absent, or if the neutrons would not contribute to the calorimeter signals, then the (ultimate limit on the) resolution would be worse by almost a factor of two, compared with the compensating lead/plastic-scintillator case.

The fluctuations discussed above only affect the non-em component of hadronic showers. To determine the effect on the *total* calorimeter signals, we included the energy sharing between the em and non-em shower component in our BOTEC Monte Carlo program. We used the experimental data on f_{em} and fluctuations therein, measured for pion-induced showers in lead by the SPACAL Collaboration [Aco 92b], for that purpose. For each event, first the energy sharing between em energy (E_{em}) and non-em energy (E_h) was chosen. This was done by selecting a value for the em shower fraction from a distribution around the mean value for the energy of the showering hadrons. For example, for 10 GeV pions, the mean value for f_{em} was 0.393. When we limited ourselves to compensating calorimeters, the precise choice of the distribution around that mean value had no significant consequences. We chose a Gaussian distribution with a fractional width of 20%, but when this width was changed to 10%, the results were not significantly different.

The total energy E of the showering particles was split into em and non-em fractions: $E = f_{em}E_{em} + (1 - f_{em})E_h$. The non-em part E_h was treated as described above and the total calorimeter signal was determined as

$$S_{cal} = w_{em}E_{em} + E_c + w_n E_n \qquad (4.34)$$

Since we were only interested in the ultimate limits of the hadronic energy resolution, only compensating calorimeters were considered. For these, $w_n = 1.5$, while $w_{em} = 0.65$ (see Table 3.4 for lead). The fractional widths of the resulting S_{cal} distributions, σ_{intr}/E, are listed in Table 4.3 for pions with energies ranging from 5 to 150 GeV, together with the average value of f_{em}, the average energy of the non-em shower component, the average number of neutrons produced in the showers, and the energy resolution for (hypothetical) pions that do not produce π^0s in their shower development, $(\sigma/E)_h$.

Table 4.3 *The effect of fluctuations in the kinetic energy carried by neutrons generated in hadronic shower development on the energy resolution for pions in lead-based calorimeters. See text for details.*

Pion energy (GeV)	f_{em}	$\langle E_h \rangle$ (GeV)	# neutrons	$(\sigma/E)_h$	σ_{intr}/E	$\sigma_{intr}/\sqrt{E}$
5	0.327	3.37	108	5.91%	4.85%	10.9%
10	0.393	6.07	194	4.18%	3.26%	10.3%
20	0.418	11.64	372	2.96%	2.26%	10.1%
40	0.495	20.20	646	2.09%	1.49%	9.4%
80	0.523	38.16	1221	1.48%	1.02%	9.1%
150	0.566	65.10	2083	1.08%	0.71%	8.7%

It is interesting to note that the fractional width of the total signal distribution, σ_{intr}/E does *not* scale with $E^{-1/2}$. Over the energy range from 5 to 150 GeV, σ_{intr}/E gradually *improves* from $10.9\%E^{-1/2}$ to $8.7\%E^{-1/2}$. This reflects the gradually increasing fraction of the energy carried by the em shower component.

The ZEUS Collaboration has measured the contributions of irreducible fluctuations to the hadronic energy resolution of compensating lead/plastic-scintillator and uranium/plastic-scintillator calorimeters (see Section 4.3.4). For lead, they obtained the following result (Table 4.1):

$$\frac{\sigma_{\text{intr}}}{E} = \frac{(13.4 \pm 4.7)\%}{\sqrt{E}} \tag{4.35}$$

This result is graphically represented by the shaded area in Figure 4.63. The (calculated) values of σ_{intr}/E listed in Table 4.3 are indicated by the dots in Figure 4.63.

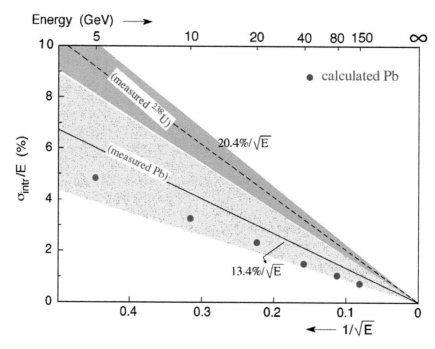

FIG. 4.63. The (calculated) contribution of fluctuations in the total kinetic energy carried by neutrons to the hadronic energy resolution of a compensating Pb/plastic-scintillator calorimeter, as a function of energy (the closed circles). The measured value of the irreducible fluctuations in such a calorimeter is indicated by the shaded area. The dashed line represents the measured value of the irreducible fluctuations in a compensating [238]U/plastic-scintillator calorimeter.

It should be emphasized that the calculations described above were by no means intended to predict or reproduce these ZEUS measurements. They were an attempt to estimate the effects of one particular type of fluctuations, which are irreducible and which we believe to constitute the ultimate limit in the energy resolution that can be achieved with any type of hadron calorimeter. The agreement between the results of the calculations and the experimental ZEUS data (within experimental errors) indicates that these fluctuations are indeed the dominating ingredient for this limit. In our calculations, a number of simplifying assumptions were made. These assumptions had in common that they *underestimate* the fluctuations and thus the energy resolution. We mention the most significant ones:

- All neutrons were assumed to result from evaporation. In practice, this is not true. Spallation neutrons slowed down after having initiated nuclear reactions of the type (n, xn) have an energy distribution that is quite different from the Maxwell spectrum used in our calculations. Therefore, Equation 4.32 underestimates the fluctuations in the kinetic energy carried by a given number of neutrons.
- The effects of saturation in the scintillator signals were neglected in our calculations. This has the same consequences as the previous issue, since it affects the effective energy spectrum of the neutrons. In polystyrene, which has a k_B value of ~ 0.1 g cm^{-2} MeV^{-1}, the signals from 1 MeV recoil protons are suppressed by a factor 3.5, compared with mips. The suppression is only a factor 1.8 for the signals from 5 MeV recoil protons (Equation 3.13).
- We assumed perfect correlation between the nuclear binding energy losses and the *number* of shower neutrons produced in the event.

The last assumption is definitely very crucial. As argued above, this assumption is quite reasonable for the case of showers developing in lead, where fluctuations in the number of protons were estimated to affect the resolution by only $5\%/\sqrt{E_h}$, compared with $13\%/\sqrt{E_h}$ for the fluctuations in kinetic energy for a given number of neutrons ($83\%/\sqrt{40} = 13\%$, Equation 4.32). However, the situation is completely different for showers developing in uranium.

In uranium, a large fraction of the neutrons are created in nuclear fission processes. The fission probability and the number of neutrons produced in the fission process are largely independent of details of the preceding spallation process and, therefore, the number of neutrons produced in showers in uranium is much more weakly correlated to the invisible energy than for showers developing in lead. As a consequence, the hadronic energy resolution achievable with lead calorimeters should be considerably better than for calorimeters based on fissionable absorber material.

This was confirmed by the measurements of the ZEUS team mentioned above. Using the same procedure as for their compensating lead/plastic-scintillator calorimeter, they also measured the contribution of irreducible fluctuations to the hadronic energy of a compensating ^{238}U/plastic-scintillator calorimeter (Table 4.1). The result was significantly larger than for lead:

$$\frac{\sigma_{\text{intr}}}{E} = \frac{(20.4 \pm 2.4)\%}{\sqrt{E}} \qquad (4.36)$$

This resolution, represented by the dashed line in Figure 4.63, is about twice as large as the value we calculated for a compensating lead/plastic-scintillator calorimeter. Dedicated calculations showed that such a degradation could be explained if we assumed that, on average, half of the neutrons (*i.e.*, the fission neutrons) were completely uncorrelated to the nuclear binding energy losses, while the other half were perfectly correlated, as in lead. Another factor that clearly contributed to the degradation of the energy resolution turned out to be the considerably smaller value of the weighting factor for the neutron signals (w_n, see Figure 4.62). In compensating uranium/plastic-scintillator calorimeters, $n/mip \sim 1.0$, vs. 1.5 for lead/plastic-scintillator, see also Figure 3.41).

4.8.2   *The Texas tower effect*

Not all compensating calorimeters offer superior hadronic energy resolution. Calorimeters using gaseous active material are a notable exception.

The measurements on ^{238}U/gas calorimeters described in Section 3.4.3.2 not only revealed the "tunability" of the pion response (and thus of the e/h value) through the composition of the gas mixture, but also demonstrated another important aspect of the contributions of shower neutrons to hadronic calorimeter signals. Detailed analysis of the energy deposit profiles showed the occurrence of "hot spots." Individual calorimeter cells would sometimes produce signals that were considerably larger (sometimes by orders of magnitude) than those in all neighboring cells combined. These hot spots seemed to occur at more or less random positions, sometimes very far away from the shower axis, sometimes very close. In the latter case, their anomalous character was usually less striking, since it is fairly typical that large numbers of different shower particles contribute to the signals from cells close to the shower axis, whereas cells far away from that axis are often empty.

4.8.2.1   *Origin of the effect.*   These hot spots are very likely caused by neutrons interacting in the wire-chamber gas. To understand this phenomenon, one should realize that a sampling calorimeter of the type discussed here has a very small sampling fraction for charged particles, typically of the order of 10^{-5}. This means that when a 100 GeV hadron is absorbed in this structure, only of the order of 1 MeV of energy is deposited in the form of gas ionization by *all the charged shower particles combined*. This sets the energy scale of the calorimeter.

A particle for which the charged shower components generate a total ionization of 0.5 MeV in such a calorimeter will thus be attributed an energy of 50 GeV. If now a MeV-type (evaporation) neutron produced in the shower development underwent elastic scattering off a hydrogen nucleus in the gas and transferred 0.2 MeV of kinetic energy to that proton, then the ionization of the gas caused by this recoil proton would generate a signal equivalent to that of a 20 GeV showering hadron, assuming that the recoil proton stopped in the gas.

This is a reasonable assumption since the range of a 0.2 MeV proton in the gas mixtures used in wire chambers is typically 2.0–2.5 mm, less than the typical gap width of such chambers. Since elastic neutron–proton scattering is a local event, the result is a large signal in only one calorimeter cell. And since shower neutrons usually undergo many elastic scattering processes before being captured, the scattering process taking place in the gas may occur anywhere in the calorimeter. In other words, this type of event has precisely the characteristics of the "hot spots" described above.

Let us take a more quantitative look at this example. Suppose that the calorimeter contains equal volumes of ^{238}U and hydrogen gas at atmospheric pressure, each about 40 cm in depth. Measurements have shown that on average 60 neutrons are produced per GeV of non-em energy when hadronic showers develop in depleted uranium [Ler 86].

A 100 GeV shower thus generates typically $\sim 3,000$ neutrons in this device.

The mean free path of a 1 MeV neutron in hydrogen gas is 43 m, for a 0.1 MeV neutron it is about 15 m. Neglecting energy losses in the absorber and assuming that all neutrons travel in a straight line, this means that the n–p interaction probability for neutrons in this energy range varies from about 1–3%. In other words, of the 3,000 neutrons, 30 to 90 may be expected to create "hot spots," or "spikes," as the authors of [Gal 86] called them. Their experimental observation of 0.6 spikes/GeV is in good agreement with this estimate.

The mentioned hot spots were also observed by the CDF Collaboration in its Forward Calorimeter system, which originally was also based on wire-chamber readout [Bran 88, Cih 88, Fuk 88]. CDF named it the *Texas tower effect*, because of the similarity with oil rigs sticking out in the Texas desert and because of the home institute of the investigators who discovered it. Since neutron production in iron (the absorber material used in the CDF detector) is about a factor of ten smaller than in ^{238}U, the occurrence of the effect was also an order of magnitude less frequent than in the uranium detector discussed above. Although the effect could be recognized and dealt with to some extent in offline event analysis, it disturbed the search for events with anomalous missing transverse energy at the trigger level to such an extent that it was decided to replace the gas calorimetry by a system based on plastic-scintillator readout [Apo 98a, Apo 98b].

Before making this decision, an extensive investigation was undertaken to see if the described effects could be eliminated through modifications of the gas calorimeters and in particular, inspired by the results of the L3 group [Gal 86], if a different choice of gas could provide an adequate solution.

Figure 4.64 shows some quantitative results from this study [Cih 89]. In Figure 4.64a, energy spectra of Texas towers, measured with the CDF forward calorimeter, are given for two different gas mixtures. These mixtures (argon/ethane 50/50 and argon/CO_2/methane 90/7.5/2.5) differed by a factor of twenty in their relative content of hydrogenous gas. The vertical scale is normalized for the same integrated number of hadrons. For energies above ~ 10 GeV, the spectra were very similar, but the frequency of Texas towers was almost an order of magnitude larger for the hydrogen-rich gas mixture. In Figure 4.64b, the average frequency of Texas towers is shown as a function of the energy of the showering hadron, again for both gas mixtures. This figure reconfirms the dependence of the Texas tower production on the hydrogen content of the gas and shows that this production was approximately proportional to the shower energy, and thus to the neutron yield. These data also illustrate that Texas tower production was by no means an exceptional phenomenon. At 250 GeV, the effect was observed in more than half of the events.

Figure 4.64c shows the radial distribution of the Texas towers. Just as in L3, many of the isolated hot spots occurred at a large distance from the shower axis. Distances in excess of 1 m were by no means exceptional.

The Texas tower effect was also observed when the calorimeter was exposed to neutrons from a Pu-Be source. These MeV-type neutrons indeed generated calorimeter signals that were interpreted as GeV-type energy deposits. Again, the frequency at which this effect was observed turned out to be strongly dependent on the hydrogen content

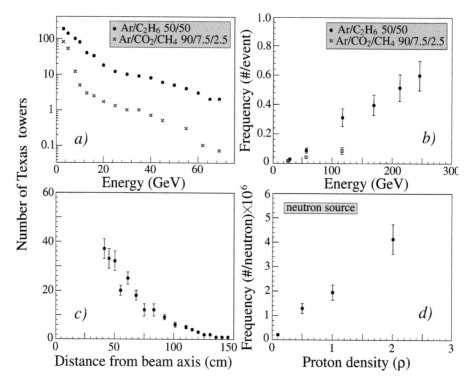

FIG. 4.64. The energy spectra of Texas towers in the CDF forward calorimeter systems, measured for the same number of showering hadrons with two gas mixtures that differed considerably in hydrogen content (*a*). The frequency of Texas towers observed for these two gas
mixtures as a function of the energy of the incident hadrons (*b*). The distribution of Texas
towers as a function of the radial distance to the shower axis of the incident hadron (*c*). The
frequency of Texas towers observed in an exposure of this calorimeter to a neutron source as
a function of the hydrogen density of the gas mixture (*d*) [Cih 89].

of the gas (Figure 4.64d). All these characteristics identified shower neutrons beyond a
shadow of a doubt as the source of the observed effects.

The CDF group found that the effects were strongly reduced, but did not completely
disappear, when hydrogen-free gas mixtures, *e.g.*, ArCO$_2$, were used. The remaining
component is, at least partly, due to α production as discussed in Section 2.3.3.3.

Using the cross section data from Table B.4, we calculated that the probability for a
neutron in the energy bracket 3–20 MeV to produce one or more αs in the active part
of a 10 λ_{int} deep calorimeter consisting of iron and ArCO$_2$ in a volume ratio of 10:1 is
a few times 10^{-4}. Given the production rates of such neutrons in high-energy hadronic
shower development [Ler 86, Gab 85], the probability that one or more αs are produced
in neutron-induced reactions with nuclei of the wire-chamber gas molecules is thus of
the order of a few percent, for a 100 GeV pion shower. Therefore, Texas towers caused
by α particles produced by neutron interactions in wire-chamber gas, or in material

located very close to this gas, do not constitute a very exceptional phenomenon.

Of course, the production of protons, α particles or heavier nucleon aggregates in spallation reactions that take place in the wire-chamber gas (*e.g.*, Figure 2.29) also contributes to the Texas tower effect. In first approximation, the rate of such reactions is governed by the sampling fraction of the calorimeter. For example, if the development of a certain hadronic shower in a calorimeter with a sampling fraction of 10^{-5} involves 100 spallation reactions of the type in which the mentioned particles are produced, then the probability that at least one of these reactions occurs in the wire-chamber gas is of the order of 0.1%.

In summary, the Texas tower effect is the result of a combination of three factors:

1. The very small sampling fraction for charged particles.

2. The very large specific ionization of recoil protons and α particles produced by MeV-type neutrons.

3. The absence of saturation effects that could limit the signals from such particles when they are produced in wire chambers.

Such saturation effects, combined with the much larger sampling fraction, and the correspondingly smaller energy equivalent of one single recoil proton, prevent the occurrence of Texas towers in calorimeters based on plastic-scintillator or liquid-argon readout. However, silicon based calorimeters are certainly not immune to this phenomenon, as illustrated in Section 4.9.

4.8.2.2 *Experimental consequences.* The Texas tower effect has a profound impact on the response function of all calorimeters with gaseous active media, and in particular also on the response function of compensating calorimeters of this type. In compensating calorimeters, signals from neutrons generated in the shower development usually contribute substantially to the hadronic response. It is not unusual that shower neutrons contribute 30–40% of the non-em response of such detectors.

It is important to realize that compensation is always achieved *on average*. The *average* calorimeter signal per GeV deposited energy is the same for the em and non-em shower components. Fluctuations about this average determine the hadronic response function. In gaseous calorimeters, (sampling) fluctuations in the signals from the neutrons tend to be the dominating factor in that respect. The fact that only a small fraction of the neutrons contribute to the response and that the signals from individual neutrons may vary by a large factor has important consequences for the hadronic response function of compensating calorimeters.

Figure 4.65 shows signal distributions recorded with the L3 hadron calorimeter (^{238}U/gas), for 6 GeV electrons and pions [Gal 86]. The wire chambers were filled with two different gas mixtures in these measurements. The solid histograms were obtained with an Ar/CO$_2$ mixture, while the dashed histograms show the signal distributions that were obtained when the wire chambers were operated with pure isobutane. For the electron signals, the gas choice made little difference (Figure 4.65a), but for the hadron signals the gas change had a tremendous effect (Figure 4.65b). When Ar/CO$_2$ was replaced by isobutane, the hadronic signals approximately doubled. This was the result

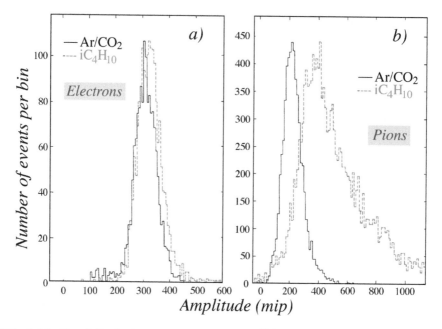

FIG. 4.65. Signal distributions recorded with the L3 ^{238}U/gas calorimeter, for 6 GeV electrons (a) and 6 GeV pions (b), with two different gas choices. The solid histograms were obtained with a mixture of argon (80%) and CO_2 (20%), the dashed histograms with isobutane [Gal 86].

of the contribution of recoil protons produced by elastic neutron scattering in the wire-chamber gas (see also Figure 3.39).

Figure 4.65b shows that the hadronic signal distribution also became more asymmetric as a result of this change. The relative rms width (σ_{rms}/mean) increased from 0.32 to 0.48. The authors found that when a "compensating" gas mixture was chosen instead of (the undercompensating) Ar/CO_2 or (the overcompensating) isobutane, the response function exhibited a substantial high-side tail as well. The relative rms width amounted to 0.40 for 6 GeV pions in that case.

The peculiarities of the signals caused by a small number of shower neutrons thus dominated the performance of this (compensating) calorimeter. There is a striking similarity with the effects of fluctuations in f_{em}, which dominate the resolution for pions in non-compensating calorimeters and which are responsible for the asymmetric pion response function in such detectors (Section 4.7.1). The reasons for these effects are essentially the same, both in the non-compensating calorimeters and in the gas-based compensating ones. In both cases, the calorimeter signals are determined by a small number of shower particles, π^0s in one case, recoil protons from elastic neutron scattering in the other. The wide range of possible energies for π^0s generated in pion-induced shower development leads to a high-side tail in the response function of non-compensating calorimeters. In gas calorimeters, the wide range of possible signals from individual neutrons causes the same effect. The asymmetry arises from the fact that one individual neutron may cause an anomalously large calorimeter signal, but not an anomalously small one.

We performed a dedicated Monte Carlo study to quantify the implications of the Texas tower effect [Wig 98]. In this study we generated signals caused by soft neutrons in an iron/isobutane calorimeter with a volume ratio iron/gas of 10/1 (sampling fraction for mips $\approx 2 \cdot 10^{-5}$). The gas gap was chosen to be 5 mm wide. Neutrons from a Maxwell spectrum (Equation 2.25) with a temperature of 2 MeV (see Figure 2.32) were made to interact at random positions inside this gas gap. The recoil protons produced in these processes were assumed to travel at a random angle with respect to the normal to the gas gap, which is reasonable since the neutrons that initiated the reactions were generated isotropically. The energy fraction transferred to the recoil protons was chosen as a random number between 0 and 1 (see Section 2.3).

Most of the recoil protons ($\sim 60\%$) deposited their entire energy inside the gap in which they were produced, the remaining 40% reached the wall of the wire chamber and thus deposited only part of their energy in the gas.

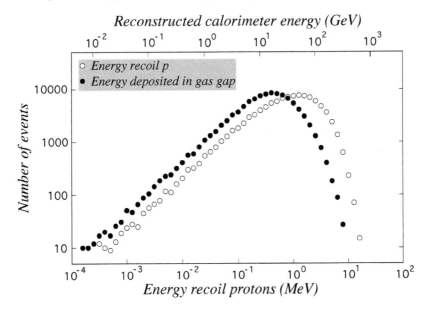

FIG. 4.66. Distribution of the energy carried by recoil protons produced by neutron scattering, and of the energy deposited by these protons in a 5 mm wide gap filled with isobutane. See text for details.

This is illustrated in Figure 4.66, which shows the distributions of the energy carried by the recoil protons and of the energy deposited in the gas by these protons. On average, the kinetic energy of the recoil protons amounted to 1.5 MeV. Protons with this energy travel typically ~ 20 mm in isobutane and, therefore, many of them reached the walls and deposited only part of their energy in the gas.

The average energy deposited by a recoil proton in the isobutane-filled gap amounted to ~ 0.6 MeV. However, since the energy scale of the simulated calorimeter was set by the response to mips, the proton signals were, in the absence of saturation effects, interpreted as energy deposits that were $(2 \cdot 10^{-5})^{-1}$ higher than in reality. For example, the average energy deposit of 0.6 MeV was interpreted as a calorimetric energy deposit of 30 GeV. The energy scale for these "hot spots" in this calorimeter is given on the top axis of Figure 4.66.

This figure also illustrates another aspect of the Texas tower effect that was mentioned above, namely the large range of energies covered by this phenomenon. Note that the horizontal scale is logarithmic and that the displayed recoil proton spectrum covers some five orders of magnitude in energy, for a probability range of three orders of magnitude. A given recoil proton produced by elastic scattering of a neutron in the wire-chamber gas may thus have yielded a "spike" that was reconstructed by the calorimeter as an energy deposit somewhere between 10 MeV and 1 TeV. Whereas about half of the protons produced calorimeter signals that corresponded to less than 15 GeV, there was also a small fraction that produced much larger signals. For example, one out of 700 neutron-induced recoil protons (0.15%) mimicked an energy deposit of 300 GeV or more.

This may cause severe problems in practical experiments, in particular if collisions involving hadrons are studied. In modern hadron colliders, the interaction rates are very high, for example of the order of 1 GHz at the LHC. This is necessary because the cross sections for the interesting processes (*e.g.*, deep inelastic quark–quark scattering) are very small and events in which such processes occur thus represent only a tiny fraction of all interactions. Experiments intended to extract interesting new physics from these interactions thus have to be extremely selective. Among the various event selection criteria, large amounts of energy flowing into the plane perpendicular to the colliding beams, or a large directional imbalance in the energy flowing into this plane (so-called *missing transverse energy*), are considered prime indicators of interesting physics.

The effects discussed above may easily mimick such signatures, at rates that are orders of magnitude above levels that are acceptable in view of the required event selectivity. A conservative estimate puts the production of evaporation neutrons in an iron-based calorimeter, by particles emerging from the collisions at angles between 45° and 90° with the beam line of the LHC, at 1 GHz. If each and every one of these neutrons had a 0.15% probability of mimicking an energy deposit in excess of 300 GeV, then events with a missing transverse energy of this magnitude would occur at a rate of ~ 1 MHz, five orders of magnitude beyond the rate at which events can be retained for offline analysis in these experiments.

For these reasons, gas-based calorimeters should thus not be used in such experiments, at least not as detectors intended for measuring the energy flow in the collisions, and thus playing a prime role in the event selection.

4.9   Catastrophic Effects Caused by a Single Shower Particle

The "Texas tower effect" discussed in the previous section illustrates how one individual shower particle, in this case a neutron, can seriously deteriorate the performance of a calorimeter, and make it in particular unsuitable for triggering purposes. The phenomenon was caused by

1. The very small sampling fraction of the calorimeters in which it occurred,

2. The possibility that a recoil proton from elastic neutron scattering in an active (gas) layer could be stopped in that layer, and

3. The absence of signal saturation for such a recoil proton.

Because of this, such a scattering process by an MeV-type neutron could mimic a multi-GeV energy deposit. The impact of this effect on the triggering capability caused the CDF Collaboration to replace the forward calorimeter of their experiment, which was initially using wire chamber readout, by a plastic-scintillator based one [Alb 02a].

There are several other situations in which one individual shower particle could wreck similar havoc. One example, discussed elsewhere in this book (Section 9.1.3), occurs in the CMS calorimeters, where one nuclear interaction close to the APD that converts the scintillation light from a $PbWO_4$ crystal into an electric pulse may cause a signal that is equivalent to hundreds of GeVs when the densely ionizing reaction products of this event traverse the silicon layer of the APD (Figure 9.7). These "spike" events may be recognized and eliminated offline on the basis of their topological features (Figure 9.16) and/or their time structure (Figure 9.17). However, as in the case of CDF, they have an adverse effect on the triggering capability of the experiment.

The third example is hypothetical, since it is not (yet) supported by experimental evidence. However, based on our understanding of the two cases described above, anonymously large signals may also be expected when hadronic showers develop in a silicon based calorimeter system such as the one adopted by the CMS Collaboration for replacing the crystal-based endcap in the High-Luminosity LHC era [CMS 15b]. The sampling fraction of this detector is very small, $e.g.$, $6 \cdot 10^{-4}$ in the central region of the 3.5λ deep FH section, which consists of twelve 5 cm thick brass absorber plates interleaved with 100 μm silicon layers. In high-energy hadron showers, nuclear reactions such as the ones depicted in Figures 2.29 and 2.49 are commonplace. Note that the event in Figure 2.49 was caused by a proton with a kinetic energy of only 160 MeV, and that the range of the reaction products is less than 100 μm. Therefore, if such a reaction took place in a silicon layer, an estimated 50 MeV could be deposited in that layer (see Figure 2.48). Because of the absence of saturation effects, such an event would be interpreted as an energy deposit of $0.05/(6 \cdot 10^{-4}) = 83$ GeV in the calorimeter. Shower particles such as the one that caused this particular event occur abundantly in hadronic shower development, they are as a matter of fact the dominating component of the non-em signals in hadron showers (Table 2.6). One should therefore not be surprised if this CMS upgrade would lead to yet another "spike" problem, and it would be prudent to keep this in mind when testing prototype modules.

These examples might create the impression that it is always a bad idea to install silicon based detectors inside a calorimeter. This is incorrect, since silicon photomultipliers are immune to the effects described above. The authors of [Ber 16] even performed dedicated tests that demonstrated the absence of such effects. The reason is of course that a SiPM is a pixelated device operating in the Geiger mode. Therefore, a passing ionizing particle creates the same signal as one scintillating photon in these detectors.

4.10   Offline Compensation (Myths)

So far, we have encountered the following beneficial properties of (non-gaseous) compensating calorimeters:

- The hadronic response is constant, *i.e.*, the calorimeter produces signals that are, on average, proportional to the hadron or jet energy.
- The hadronic response is independent of the type of particle or jet.
- The signal distributions for hadrons and jets are Gaussian, since the non-Gaussian fluctuations in the em shower content, which lead to asymmetric response functions, are eliminated.
- The elimination of the fluctuations in the em shower content, which usually dominate the hadronic resolution of non-compensating calorimeters, makes it possible to achieve superior hadronic energy resolution. In particular, the deviation from $E^{-1/2}$ scaling that is observed at high energy in all non-compensating calorimeters is avoided. Therefore, the energy resolution is especially superior in the high-energy domain.

However, most calorimeters used in particle physics experiments are non-compensating, and e/h values in excess of 1.5 are by no means uncommon. This limits the achievable hadronic energy resolution in practice to values of 5–10%, at best.

In addition, systematic differences in the average em shower fraction make the response (and thus the mean value of the signal distribution) of non-compensating calorimeters dependent on the *type* of particle or jet that caused the signal (Section 4.7). This leads to systematic mismeasurement of the energy of certain types of particles or jets.

The asymmetric pionic response function, characteristic for non-compensating calorimeters, may have serious experimental consequences, such as trigger biases. For example, if one uses the calorimeter signals to select events with a minimum (missing) transverse energy from a steeply falling distribution, then the event sample is likely to be strongly dominated by events in which the actual value of this energy was smaller than the trigger level, but in which upward fluctuations pushed it beyond that level. An asymmetric response function makes it extremely difficult to deal with this problem in a correct way.

Among the strategies used for dealing with these problems in practice, we mention:

1. The problems are denied and/or ignored
2. One applies "offline corrections"
3. One applies online weighting schemes of one type or another

These strategies are often based on a very common myth about compensation, which can be phrased as follows: "*Compensation is achieved when the average calorimeter signals for electrons and pions of the same energy are equal,"* or $e/\pi = 1$ in shorthand. In a longitudinally segmented calorimeter system, this condition is then achieved by giving appropriate weighting factors to the signals from the different segments, exploiting the differences in the longitudinal shower profiles of electrons and pions.

In Chapter 6, which deals with calibration issues, the merits of such schemes are discussed in detail. However, at this point we want to emphasize that such schemes do **not** lead to the beneficial effects of compensation listed above, because of the following reasons:

i) The energy resolution and the response function are determined by *fluctuations*, not by *mean values*. A non-compensating calorimeter responds differently to em energy deposits (*e.g.*, the absorption of high-energy photons) and non-em energy deposits (ionization by spallation protons, charged pions, recoil protons from neutron scattering, *etc.*). Electromagnetic and non-electromagnetic energy deposits take place throughout the calorimeter, in all segments. The application of weighting factors has no effect on the event-by-event fluctuations in the em shower content, which determine the hadronic energy resolution. Nor do these factors change the asymmetric line shape, characteristic for non-compensating calorimeters.

ii) The application of weighting factors does not change anything regarding the dependence of the average energy fraction carried by the em shower fraction on the type of particle or jet. Weighted or unweighted, the differences in calorimeter response remain.

iii) And finally, the "compensation" achieved in this way is only valid for one particular energy, when one set of weighting constants is used. Since the longitudinal shower profiles are energy dependent, the condition $e/\pi = 1$ requires different weighting constants for each energy.

The validity of these arguments may be illustrated by a rather trivial, but straightforward, example taken from practice [Fer 97]. The QFCAL group performed measurements of the performance of a strongly non-compensating quartz-fiber calorimeter, in which this detector was preceded by various amounts of "dead" material (iron). This iron had a larger absorbing effect on electron showers than on hadronic ones. As a result, the e/π signal ratio measured with the calorimeter installed behind the absorber gradually decreased as the amount of absorber was increased.

For an absorber thickness of about 10 X_0 (17 cm iron), the average signals from 80 GeV electrons and pions were about equal: $e/\pi = 1$, the compensation condition was achieved (Figure 4.67a). Yet, the hadronic energy resolution was significantly worse than without the "dummy" iron section (Figure 4.67b), both for single pions and for multi-particle "jets" produced by high-energy pions that interacted in a thin target placed upstream of the absorbing structure.

This is of course no surprise, since the signals were collected from only part of the block of matter in which the shower develops. Fluctuations in the fraction of the energy deposited in the part from which the signals were collected added to the ones that determined the resolution in the absence of the dummy section and thus deteriorated the resolution.

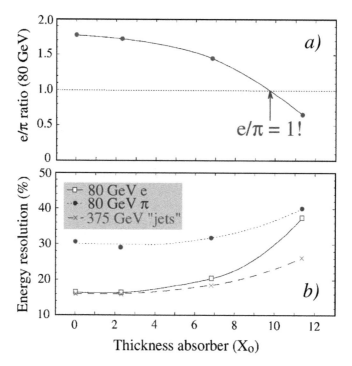

FIG. 4.67. The e/π signal ratio at 80 GeV (*a*) and the energy resolution (*b*) of a quartz-fiber
    calorimeter preceded by dead material (iron), as a function of the thickness of this material
    [Fer 97]. The energy resolution is given for 80 GeV electrons and pions, and for multi-particle
    "jets" generated by 375 GeV pions in an upstream target.

Although this is maybe a somewhat extreme example, it does illustrate that there is
no magic in the e/π signal ratio. The resolution of a non-compensating calorimeter is
determined by the event-to-event fluctuations in the em shower content and as long as
nothing is done to reduce (the effect of) these fluctuations *event by event*, no improvement in the hadronic energy resolution may be expected.

The key for possible success of such weighting methods lies thus in the *event-by-
event* aspect. There are some examples in the literature of more or less successful attempts to improve the hadronic energy resolution through a determination of the em
shower content event by event. These methods were pioneered by the WA1 Collaboration [Abr 81], following a suggestion by Dishaw *et al.* [Dis 79].

Their calorimeter consisted of 2.5 cm thick slabs of iron, interleaved by 5 mm thick
plastic scintillator plates. Each scintillator plate was read out individually on both sides.
Figure 4.68a is a scatter plot of the total signal measured for 140 GeV pion showers
(vertical), *vs.* the largest signal observed in any single scintillator plane (horizontal).
This plot shows a clear correlation between these two quantities.

This calorimeter had an e/h value of about 1.6. Therefore, events with a large value

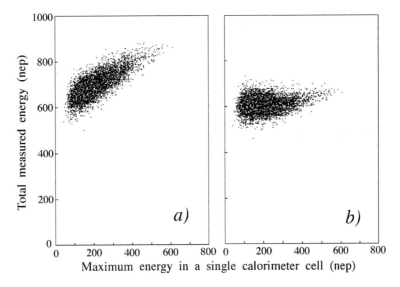

FIG. 4.68. WA1 results on offline compensation, showing the correlation between the total measured signal and the maximum signal observed in one individual calorimeter segment. Results are given for 140 GeV pions before (*a*) and after (*b*) applying a weighting factor, based on the signals observed in the individual calorimeter segments [Abr 81].

of f_{em} produced large signals. However, since the em shower core was highly localized, such events were also characterized by large energy deposits in individual scintillator planes. This was the basis of the observed correlation.

The authors then used the observed correlation to reduce the effects of the fluctuations in the em shower fraction on the energy resolution. They assumed that the value plotted on the horizontal axis was a measure of the em shower fraction and reduced the signal observed in each individual scintillator plane by a factor determined by the signal value in that plane:

$$S_i' = S_i \left[1 - \frac{C}{\sqrt{E}} S_i \right] \qquad (4.37)$$

The optimum value for the constant C was determined empirically from data at a wide range of energies. The result of this procedure is shown for the 140 GeV pions in Figure 4.68b. In Figure 4.69a, the projections of the scatter plots from Figure 4.68 on the vertical axis are shown. The described event-by-event corrections clearly made this signal distribution considerably narrower. Also, the signal distribution became much more symmetric as a result of this procedure.

The hadronic energy resolution is shown in Figure 4.69b, before and after the described corrections were applied. The resolution improved considerably, especially at high energies. Before the corrections, the measured energy resolution leveled off at about 7%. After the corrections, the energy resolution was observed to scale as

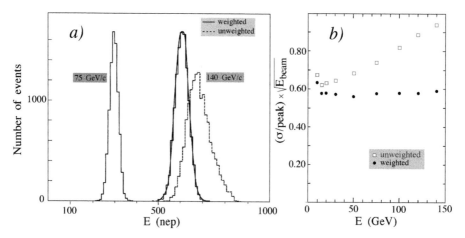

FIG. 4.69. WA1 results on offline compensation. The signal distributions for 140 GeV pions (*a*) and the hadronic energy resolution as a function of energy (*b*), before and after the weighting procedure described in the text was applied to the experimental data [Abr 81].

$$\frac{\sigma}{E} = \frac{0.58}{\sqrt{E}}$$

Also the H1 [Braun 89] and ATLAS [Akh 00] experiments have developed schemes in which the energy deposit pattern of individual events is used to recognize em shower components and thus make corrections for the non-compensating nature of their calorimeter systems. Whereas such schemes do seem to improve the calorimeter performance for beam tests, in which the properties and the trajectories of the detected particles are precisely known, the benefits in a real-life collider experiment are less evident, especially for what concerns the detection of jets (*i.e.*, collections of particles showering simultaneously). It does not help that some of the correction factors employed in these schemes depend on the energy of the object to be detected.

With procedures of this type, aimed at determining the em shower content *event by event*, some (but not all) of the benefits of an intrinsically compensating calorimeter may be restored in an intrinsically non-compensating calorimeter. To be successful in this respect, the calorimeter needs to be sufficiently fine-grained, as in the WA1 case, where each active plane was read out separately.

The experimental WA1 data showed that procedures of this type work especially well at high energies, where π^0s and non-em forms of energy deposit led to distinctly different signals in individual active elements of their detector. At low energies, where these differences were much less spectacular, the improvements in energy resolution were found to be marginal, at best. When given the choice, an intrinsically compensating calorimeter is thus definitely much more attractive.

5

INSTRUMENTAL ASPECTS

Equipped with the theoretical background needed for understanding the principles on which calorimetric detectors are based and the factors that determine their performance characteristics, we now turn to the practical aspects of building and operating these instruments.

Of course, each calorimeter system is unique. It carries the signature of the people who designed and built it, guided by the environment in which it had to operate and by the tasks it had to fulfill.

In this chapter, we describe *general principles* that apply to all calorimeters or to certain classes of calorimeters (*e.g.*, those operating with liquid argon as active material). If and when appropriate, examples taken from practice are used for illustrative purposes.

This chapter covers a broad spectrum of issues, ranging from the mechanical construction of the detectors to details of the data acquisition systems. The final section deals with radiation damage, a topic that, in itself, could easily fill a book the size of this one and that is increasingly important for the design and operation of calorimeters.

5.1 Construction Principles

5.1.1 *The tower structure*

A calorimeter is a block of matter in which energetic particles are absorbed. In order to be able to determine relevant particle properties, the calorimeter is segmented. This can be done in a variety of different ways. For example, the detector could be subdivided into small, identical cubes. When a particle is absorbed in the calorimeter, its properties can then be reconstructed from the signals produced by all the cubes involved in the shower development process.

In this example, each cube acts as an independent detector. The energy deposited in this cube is measured individually, through its own readout electronics. If light is the source of information in this calorimeter, the readout chain may consist, for example, of a photomultiplier tube, an amplifier and an analog-to-digital converter (ADC). For the data acquisition system, this cube is an individual readout channel.

The number of readout channels is limited by the budget of the experiment. Therefore, it is important to choose the segmentation in the most economic way commensurate with the physics goals of the experiments. The cube scheme mentioned above is unlikely to meet that criterion. Many cubes, especially those located deep inside the absorber, may never produce a significant signal. Others may be too large to measure the position of the detected particles with the required precision. Also, it may not be easy to find out which cubes actually contributed to the signal generated by a given showering particle.

To address the latter issue, it is important to choose the segmentation in such a way that the volume in which the showers develop corresponds to *a small number of cells*, and that *most of the energy is deposited in one individual cell*. At the same time, the cell size should be chosen sufficiently small, so that two particles separated by a distance corresponding to their typical lateral shower dimensions can still be recognized as a pair. And finally, the total number of readout channels should be kept at a minimum. When all these criteria are met simultaneously, the calorimeter is segmented in an efficient way.

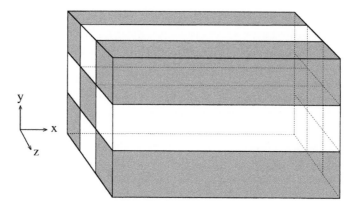

FIG. 5.1. Example of a tower structure.

When the particles to be measured with the calorimeter enter the detector in a non-random direction, as is the case in accelerator-based experiments, this type of efficient segmentation can be realized with a so-called *tower structure*. An example of a tower structure is shown in Figure 5.1.

The detector volume is subdivided into long rectangular parallelepipeds, whose axes are oriented in the direction of the incoming particles, in this case a broad beam traveling parallel to the x-axis. Each parallelogram forms a readout cell, *i.e.*, all the energy deposited in this cell is integrated in the signal that it produces. The lateral cross section of the cells, which is square in Figure 5.1, may also be hexagonal or triangular. The size is chosen such that a particle entering a cell in its center deposits not more than about 90% of its energy in this one cell. In that case, the (y, z) coordinates of the impact point can be determined with adequate accuracy from the energy sharing between different cells.

It should be emphasized that the calorimeter from Figure 5.1 only has a tower structure when it is oriented as shown with respect to the incoming particles. If this detector were rotated over 90°, so that the tower axes became oriented perpendicular to this particle beam, then the advantageous properties mentioned above would disappear: the shower energy would be shared by a large number of readout cells, and it would be impossible to determine the coordinates of the impact point without additional, external information.

In return, very detailed information could be obtained about the longitudinal development of the individual showers. However, the physics value of such information is usually very limited.

5.1.2 Projective vs. non-projective tower structures

The basic information that one wants to extract from the calorimeter data concerns the momentum vector, the energy and the nature of the particles that caused the signals. The calorimeter structure should be chosen so that this information can be obtained in a natural way, using simple and straightforward procedures.

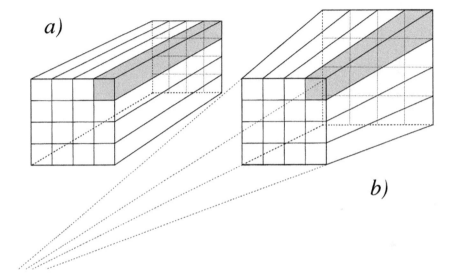

FIG. 5.2. Schematic structure of calorimeters consisting of non-projective (a) or projective (b) towers.

This is best achieved in calorimeters with a *projective tower structure* (Figure 5.2b). In such a calorimeter, the cells in which the detector is subdivided have the shape of truncated pyramids. The interaction vertex, in which the particles to be measured are produced, is the common summit of all these pyramidal calorimeter cells. In other words, all boundary lines between the calorimeter cells point to the interaction vertex.

The advantage of such a structure is that the calorimeter signals immediately reveal the production angle of the detected particles. Each cell represents a certain angular region, in three-dimensional space. In collider experiments, it is customary to express the production angle of the particles in terms of the parameters η (the *pseudorapidity*, a measure[11] for the polar angle θ in the plane formed by the measured particle and

---

[11] The pseudorapity η is defined as $-\ln[\tan(\theta/2)]$.

the beam line) and the azimuthal angle ϕ (measured, for example, with respect to the horizontal plane). These angles are indicated in Figure 5.3, which shows cross sections of the calorimeter in a colliding-beam experiment, along the beam line (Figure 5.3a) and perpendicular to the beam line (Figure 5.3b).

A given cell may cover, for example, the area $[0.7 < \eta < 0.8,\ 30° < \phi < 40°]$. The nearest neighbors of this cell cover the areas $[0.6 < \eta < 0.7,\ 30° < \phi < 40°]$, $[0.8 < \eta < 0.9,\ 30° < \phi < 40°]$, $[0.7 < \eta < 0.8,\ 20° < \phi < 30°]$ and $[0.7 < \eta < 0.8,\ 40° < \phi < 50°]$.

In this example, the *granularity* of the calorimeter, defined in terms of the angular dimensions of the cells, amounts to: $\Delta\eta = 0.10$, $\Delta\phi = 0.17$ (radians) = $10°$.

From the energy sharing between these different cells, the production angle of a certain particle can then be determined, for example as $[\eta, \phi] = [0.76\pm0.02, 37°\pm2°]$. The smaller the granularity, the smaller the experimental uncertainty in these numbers becomes.

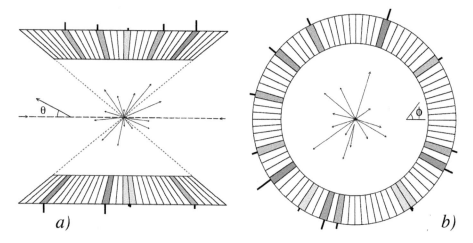

FIG. 5.3. Example of an event taking place in the interaction vertex of a colliding-beam experiment, and the reaction products detected in a calorimeter with a projective tower structure. The beam line is indicated by the dashed line, the particles produced in this interaction are indicated by arrows, whose lengths indicate the particle energies. The particle energies are also indicated by the lengths of the stubs in the calorimeter cells in which they shower. The event is shown in two projections, one in a plane along the beam line (a) and the other in a plane perpendicular to the beam line (b).

In this type of calorimeter structure, every readout cell thus has a precise angular location with respect to the interaction vertex and the direction of the beam particles. This has important advantages for the determination of kinematic variables, such as the transverse energy, $E_\perp$. When a certain amount of energy E_i is deposited in cell i, located at an angle θ_i with respect to the interaction vertex and the direction of the beam particles (Figure 5.3a), the transverse energy measured in this cell, $E_\perp^i$, is defined as

$$E^i_\perp = E_i \sin\theta_i \tag{5.1}$$

The total transverse energy in an event of the type shown in Figure 5.3 can then be determined as

$$E_\perp = \sum_{i=1}^{N} w_i \cdot E_i \tag{5.2}$$

i.e., a weighted sum of the actual energy deposits in each of the N calorimeter cells. The weighting factors w_i (equal to $\sin\theta_i$ in this case) are uniquely determined by the cell positions. Because of the simplicity of this algorithm, it is possible to obtain the values of $E_\perp$ (and other energy flow parameters) almost instantaneously at the trigger level, thus facilitating the event selection process. This is further discussed in Section 5.5.

A major advantage of the projective tower structure is that the described measurements are *independent* of the longitudinal energy deposition characteristics of the shower development. Whether the particle is an electron that starts to shower immediately and deposits all its energy in the first 20 cm of the detector, or whether it is a pion that penetrates the structure half a meter before undergoing its first nuclear interaction, the $[\eta, \phi]$ values can be directly derived from the energy sharing between the towers contributing to the signals.

This is not the case for other, non-projective, tower structures. For example, in the structure shown in Figure 5.2a, the energy sharing between the various calorimeter towers clearly depends not only on the angle of the entering particle with the calorimeter surface, but also on the longitudinal energy deposit profile of the shower that develops inside the calorimeter.

This is illustrated in Figure 5.4. Showers a and b are both initiated by 100 GeV pions entering the detector at the same angle θ. The center of gravity of shower a, in which most of the energy is deposited in the first few interaction lengths, is located somewhere near the boundary between towers 5 and 6. On the other hand, the center of gravity of shower b, in which the particle penetrated several λ_{int} before undergoing a nuclear interaction, is shifted to a position somewhere in tower 3.

This example shows that the towers in a non-projective geometry do not cover a precise angular region in the same unambiguous manner as in a projective geometry. If one assigned $[\eta, \phi]$ values to the towers in Figure 5.4, in the same way as described for the projective geometry above, then the reconstructed values found for showers a and b would be different (and both wrong!), even though the particles entered the detector at the same angle.

These problems become of course more and more serious as the angle of incidence with respect to the tower axis increases. In a projective geometry, this angle is by definition zero. In a geometry of the type shown in Figure 5.2a, the angle depends on the position of the tower. If the mentioned angle is only a few degrees, then the measurement errors are in general at an acceptable level, especially for the detection of jets, where the longitudinal event-to-event fluctuations are much smaller than for individual hadrons.

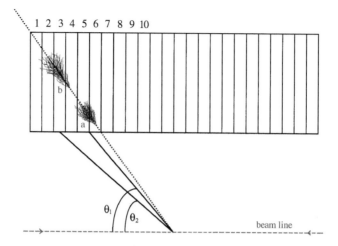

FIG. 5.4. Mismeasurement of the particle direction (θ) in a calorimeter with a non-projective
tower structure. The dashed line represents the beam direction, the dotted line the direction
traveled by the particle. Depending on the shower characteristics, this direction will be measured as θ_1 (shower a, with its center of gravity in tower 5/6) or θ_2 (shower b, center of gravity
in tower 3).

On the other hand, it should be pointed out that for single hadrons, angles as small
as a few degrees may give rise to clearly measurable effects. In Section 4.4.3.4, measurements are described in which the lateral displacement of the shower's center of
gravity made it possible to determine the average depth at which the shower energy
was deposited (Figure 4.29a), and thus correct for the effects of light attenuation on the
calorimeter signals (Figure 4.30). These measurements were performed with hadrons
entering the detector at an angle of $3°$ with the tower axis.

5.1.3 *Longitudinal segmentation*

The tower structure described above constitutes the *lateral* segmentation of the calorimeter. For readout purposes, these towers may or may not be *longitudinally* subdivided
into several segments. In most calorimeters, the towers consist of an em and an hadronic
section. Sometimes, these sections are further subdivided, as for example in ATLAS,
where both the em and hadronic calorimeter section consist of three longitudinal segments each.

The em section, typically 20–30 radiation lengths deep, completely contains the
showers initiated by photons and electrons. Hadrons typically deposit about half of their
energy in this section, with very large fluctuations about this average. For example, if
the depth of the em section corresponds to 1 λ_{int}, 36% $(= e^{-1})$ of the protons entering
this detector traverse this section without inducing a nuclear interaction.

Hadron showers are usually considerably broader than electromagnetic ones. In a
calorimeter with a granularity designed to match the lateral dimensions of the showers
developing in it, the hadronic section is thus more crudely segmented than the em sec-

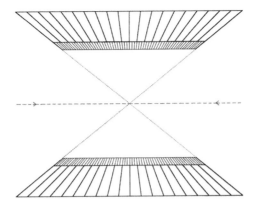

FIG. 5.5. Schematic structure of a calorimeter with projective towers and a different granularity for the em and hadronic sections.

tion. An example of such a structure is shown in Figure 5.5. In this example, nine em towers correspond to one hadronic one. If the granularity of the em calorimeter section is $[\Delta\eta, \Delta\phi] = [0.05, 0.05]$, the granularity of the hadronic section is $[0.15, 0.15]$.

Many calorimeter systems are equipped with a so-called *tail catcher*, a calorimeter section installed downstream of the rest of the calorimeter structure [Kud 90, CMS 08]. The main purpose of this tail catcher is the recognition of showers generated by particles that penetrated the calorimeter very deeply before undergoing their first nuclear interaction. For example, the fraction of protons that start showering beyond a depth of $5\lambda_{int}$ in a given calorimeter amounts to 0.7% ($= e^{-5}$). Because of the larger value of the interaction length, this fraction may be about 2% for pions.

These late starting showers are most likely only partially contained in the calorimeter proper, and may thus be expected to produce significant signals in the tail catcher. The purpose of the tail catcher is to recognize these events, so that they can be discarded from the event samples. In CMS, where the barrel hadron calorimeter is relatively short, since it had to fit inside the magnet coil, the tail catcher (installed outside the magnet coil) also serves to measure the hadronic shower energy that leaks out [CMS 08].

5.1.4   Hermeticity

Calorimeters are intended to measure the properties of particles they absorb. One instrumental aspect that may systematically disturb these measurements is the *non-hermeticity* of the calorimeter system, *i.e.*, the degree to which its spatial coverage is non-hermetic.

5.1.4.1   *The effects of non-hermeticity.*   To appreciate the effects of non-hermeticity on the measurements, let us imagine a calorimeter that consists of two modules, which are separated by a few centimeters (Figure 5.6).

If a particle enters one of these modules and develops a shower in this system, then some of the shower particles may "fall through the crack," *i.e.*, escape from the absorber system without depositing the energy they carry (shower particle #1 in Figure 5.6). They may also re-enter the calorimeter at a point much deeper inside the structure. If this

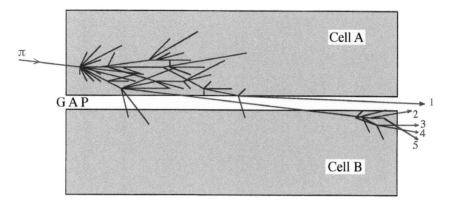

FIG. 5.6. Effects of non-hermeticity. See text for details.

happens to a highly energetic shower particle, then the result may be increased shower leakage (particles #2,3,4,5), since the effective depth of the calorimeter is reduced.

In both cases, the shower energy deposited in the calorimeter is reduced and therefore the showering particle's energy is, on average, underestimated.

The extent of these effects depends on the following factors:

- The impact point of the particle. The closer the impact point is located to the crack, the larger the fraction of the shower energy that is, on average, lost in the crack. In an extreme case, the incoming particle may traverse the crack and not produce a calorimeter signal at all.

- The type of shower that develops. Hadron showers in which a large fraction of the energy is carried by the em shower component are considerably narrower than showers with a small value of f_{em}. If two identical hadrons enter the calorimeter at the same impact point, then the energy loss depends on the number and the energy of the shower particles traversing the gap, and these depend on the type of shower that develops and, more precisely, on the value of f_{em}.

- The width of the "crack." Obviously, the larger the crack, the larger its effects.

- The trajectory of the incident particle. When the particles enter the calorimeter system at a non-zero angle with the direction of the crack, the impact point dependence of the energy losses caused by the crack is smeared out. The incoming particles can no longer "channel" through such *non-projective* cracks. On the other hand, even particles with an impact point located as much as 1 m away from the position of this crack, may experience its effects in this case. The net result is thus a less position-dependent calorimeter response, more so as the angle of incidence of the particles with the direction of the crack is increased.

- The "content" of the crack.

The latter point is of great practical importance. In the practice of designing and constructing a calorimeter system, cracks in the otherwise hermetic calorimeter structure usually have a function. They are there, for example, to contain the cables needed for transporting detector signals to the outside world (*i.e.*, the counting room). Or, they are needed to contain structural elements: carbon-fiber slots to hold the crystals of a lead-glass or scintillating-crystal calorimeter, a steel frame needed to hold the modules of a lead/plastic-scintillator sampling calorimeter together, the cryostat containing the cryogenic liquid of an ionization-chamber calorimeter, *etc.*

Air-filled cracks are, therefore, an exception rather than the rule. In practice, the cracks are thus at least partially filled with materials of the mentioned types. From a calorimetric perspective, these materials are passive.

In fully sensitive calorimeters (*e.g.*, those consisting of scintillating crystals or lead-glass blocks), shower energy deposited in these passive materials has the same effect as shower leakage: it is lost for detection and thus does not contribute to the calorimeter signals. In sampling calorimeters, these materials represent a *sampling anomaly*. The effects are similar to those that affect the performance of fully sensitive devices. However, the same amount of passive material usually has a smaller effect in sampling calorimeters than in fully sensitive ones.

5.1.4.2 *Types of non-hermeticity.* We define non-hermeticity as the condition in which material that is not part of the calorimeter structure proper is present in areas where particles to be detected by the calorimeter deposit (part of) their energy.

When defined in this way, various types of non-hermeticity that may affect the calorimetric measurements can be distinguished:

- *Lateral* non-hermeticity. The calorimeter structure contains "cracks" such as the ones encountered above (Figure 5.6).
- *Longitudinal* non-hermeticity. In this case, dead material is installed at some depth inside the calorimeter structure. A special case of longitudinal non-hermeticity occurs when this material is placed *upstream* of the calorimeter.
- *A combination of lateral and longitudinal non-hermeticity.* An example of this situation occurs in cryogenic detectors (*e.g.*, LAr), in which the calorimeter or segments of it are contained in a cryostat. The cryostat wall upstream of the calorimeter represents a longitudinal non-hermeticity, while the walls separating different sections of the calorimeter system, *e.g.*, the barrel and endcap sections, represent a lateral non-hermeticity.

In the following subsection we examine the relative importance of the effects caused by these different types of structural imperfection.

5.1.4.3 *Minimizing non-hermeticity and its effects.* Calorimeters operate in a non-ideal world. They are subject to gravity and have to be held in position somehow. The light or charge they generate in response to the absorption of energetic particles has to be collected and the signals have to be transported to a place where they can be processed.

Therefore, a certain degree of non-hermeticity is unavoidable. However, by properly designing the calorimeter system, the effects of this non-hermeticity can be minimized. For example, an iron I-beam needed to support the calorimeter structure is much more detrimental to the calorimeter performance when it is installed near the shower maximum of typical particles the calorimeter should detect than when it is located somewhere in the tails of such showers. Also, the needed mechanical strength of the support element may be achieved with less mass when a different material is chosen. The Z-value of the dead material may also make a big difference for the magnitude of the effects.

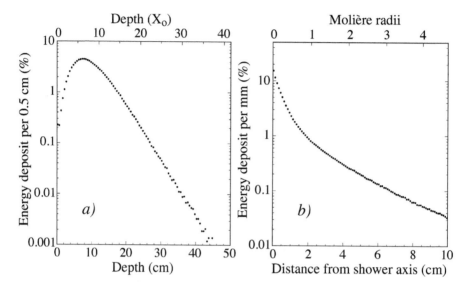

FIG. 5.7. Typical longitudinal (a) and lateral (b) shower profiles for 100 GeV electrons absorbed in tin ($Z = 50$). Results from EGS4 simulations.

The following general rules may be helpful in minimizing the effects of non-hermeticity on the calorimeter performance:

1. Structural elements should be distributed as much as possible. Ten rods of 1 cm thickness each are better than one rod of 10 cm thickness.

2. The effects caused by structural elements placed perpendicular to the direction of the incoming particles are much smaller than those caused by structural elements oriented along the direction of the incoming particles.

3. Structural elements should be avoided as much as possible in areas where a large fraction of the shower energy is deposited, *i.e.*, in the first 10 X_0.

4. Cracks should be non-projective. This point has already been discussed before.

5. Low-Z dead material is less harmful than high-Z material, for the same mass.

6. Dead material upstream of the calorimeter becomes more harmful as the distance to the calorimeter's front face increases.

7. The installation of some active material sandwiching the dead material may limit the detrimental effects.

To be more quantitative on these points, we performed a series of Monte Carlo simulations for em showers, using the EGS4 package (Section 2.5.1, [Nel 78]). Some of the results are given below. These results are given in terms of the *signal loss. i.e.*, the fraction of the shower energy that goes undetected as a result of the non-hermeticity, and in terms of the *induced energy resolution*. The induced energy resolution is a measure for the deterioration of the energy resolution as a direct result of the non-hermeticity. In order to find the actual energy resolution for detecting electrons and γs in the calorimeter, the induced energy resolution has to be added in quadrature to the resolution for these particles in the absence of the non-hermeticity.

First, we simulated the development of 100 GeV electron showers in a material with $Z = 50$, representative for a variety of crystal calorimeters (*e.g.*, CsI, BaF$_2$, NaI(Tl), CeF$_3$). Figure 5.7 shows the average energy deposition profiles, both in a longitudinal and in a lateral projection, when these particles are absorbed in this material. In these simulations, the absorber was considered to extend infinitely in the lateral and longitudinal directions.

The energy resolution, only determined by fluctuations in shower leakage (albedo) for this geometry, amounted to only 0.02%. Next, we investigated the effects of introducing "dead" material into this structure.

5.1.4.4   *Dead material against the calorimeter's front face.*   First, we studied the effects of placing an iron plate with a variable thickness in front of the calorimeter. Figure 5.8 shows the effect of such a plate on the signals (Figure 5.8a) and the energy resolution (Figure 5.8b) for 10 GeV electrons and photons showering in this device.

It turned out that one may pile quite some material in front of the detector before these effects become disturbingly large. For example, 3 cm of iron absorb, on average, only $\sim 2\%$ of the shower energy and affect the energy resolution only by $\sim 1\%$. This reflects a characteristic feature of the shower profiles, *i.e.*, the fact that only a very small

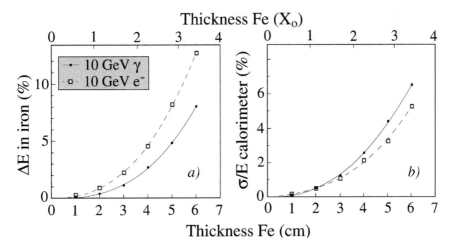

FIG. 5.8. The effects on the average calorimeter signal (*a*) and the energy resolution (*b*) of the $Z = 50$ calorimeter for 10 GeV electrons and γs, as a function of the thickness of an iron plate placed directly upstream of the calorimeter. Results from EGS4 simulations.

fraction of the shower energy is deposited in the first few radiation lengths (Figure 5.7).

Figure 5.8 also shows significant differences between showers induced by electrons and photons of the same energy. As we have seen before (Figure 4.31a), photon showers deposit their energy, on average, deeper inside the calorimeter than electron showers, because their starting point (*i.e.*, the point where the first conversion $\gamma \rightarrow e^+e^-$ takes place) may be located at a depth of several radiation lengths. Also, fluctuations in that starting point, which have no equivalent for electron showers, lead to larger longitudinal shower fluctuations. This causes, for example, larger effects of longitudinal shower leakage on the energy resolution for photons (see Figure 4.41).

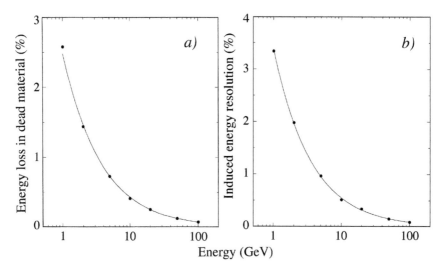

FIG. 5.9. The effect of a 2 cm thick iron plate, placed directly upstream of a $Z = 50$ calorimeter, on the average calorimeter signal (*a*) and the energy resolution (*b*) for γ-induced showers, as a function of energy. Results from EGS4 simulations.

The same phenomena are responsible for the differences between electron and photon showers observed in Figure 5.8. Electrons entering a 2 cm ($1.14X_0$) thick iron plate start radiating immediately and lose, on average, 68% of their energy when traversing the plate. However, when a γ enters the same plate, it may or may not interact (convert into an e^+e^- pair) in it. In the given example, the probability that the γ traverses the plate unnoticed is $\sim 41\%$ ($\exp[-1.14 \times 7/9]$, see Section 2.1.5). On the other hand, the γ may also convert in the very beginning of the plate (the probability for that to happen in the first 2 mm is $\sim 9\%$). In that case, a 5 GeV electron and a 5 GeV positron traverse the rest of the plate, and the γ thus loses, on average, almost twice as much energy and produces twice as many shower particles as a single 10 GeV electron. Therefore, the *fluctuations* in the number of secondary shower particles and in the energy deposited by these particles in the dead material are larger for the photon-induced showers.

The energy dependence of the effects of a 2 cm thick iron plate immediately preceding the calorimeter are shown in Figure 5.9, for γ-induced showers. As the energy increases, both the average energy fraction deposited in this plate and the fluctuations in this fraction decrease.

5.1.4.5   *Dead material at a depth z inside the calorimeter.*   Whereas the effects of this 2 cm iron plate on the calorimeter performance are remarkably small when the plate is installed directly upstream of the calorimeter, its impact changes dramatically when it is placed in an area where typically a substantial fraction of the shower energy is deposited.

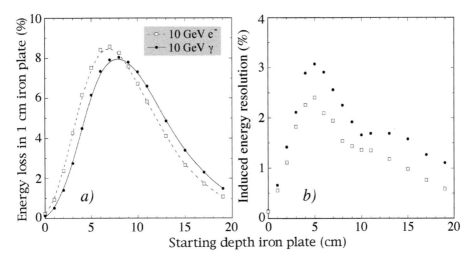

FIG. 5.10. The average calorimeter signal (*a*) and the energy resolution (*b*) of the $Z = 50$ calorimeter for 10 GeV electrons and γs, as a function of the depth at which a given amount of dead material is installed inside the calorimeter. The dead material is represented by a 1 cm thick iron plate, placed perpendicular to the direction of the incident particle. Results from EGS4 simulations.

This is illustrated in Figure 5.10, where the effects of a plate half as thick (1 cm iron, representing 7.85 g cm^{-2} or $0.57X_0$) are displayed as a function of its depth (z) inside the calorimeter. The plate was oriented perpendicular to the direction of the incoming particles in these studies. Figure 5.10a shows the effects on the average calorimeter signal and Figure 5.10b the effects on the energy resolution, as a function of z. The largest signal losses occur, obviously, near the shower maximum, around $z \sim 8$ cm. However, the largest effects on the energy resolution occur when the plate is located around $z \sim 5$ cm, where the second derivative of the shower profile is zero. Figure 5.10b exhibits another local maximum in the energy resolution near the other point where the second derivative is zero, on the trailing edge of the longitudinal shower profile.

The differences between showers induced by electrons and photons discussed above also affect the results shown in this figure. Showers induced by photons develop, on average, deeper inside the detector (Figure 5.10a) and exhibit larger longitudinal fluctuations (Figure 5.10b).

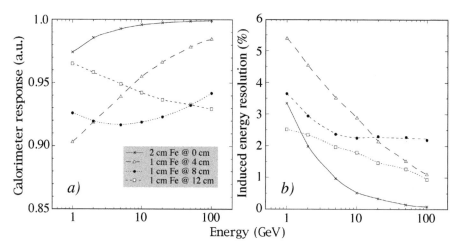

FIG. 5.11. Effects of iron plates installed at different depths inside the $Z = 50$ calorimeter on the average calorimeter signal (a) and on the energy resolution (b), as a function of the energy of the incoming particles (γs). The plates are oriented perpendicular to the direction of the incident particle. Results from EGS4 simulations.

The energy dependence of the effects of an iron plate inserted somewhere inside the absorber structure on the calorimeter performance is shown in Figure 5.11. The effects of a 1 cm iron plate installed at depths of 4 cm, 8 cm and 12 cm on the calorimeter response and on its energy resolution are given as a function of the energy of incoming γs. For comparison, the effects of an iron plate twice as thick, installed upstream of the calorimeter, are shown in the same figure. The results reflect the energy dependence of the longitudinal shower profiles. For example, the largest signal losses occur in the plate at $z = 4$ cm for 1 GeV γ showers, in the plate at $z = 8$ cm for 10 GeV and in the plate at $z = 12$ cm for 100 GeV. This corresponds to depths of 3.3, 6.6 and $9.9X_0$, respectively, the approximate depths of the shower maxima at these energies.

To study the material dependence, these simulations were repeated with plates made of lead or aluminium, with the same mass as the 1 cm iron plate mentioned above. The lead plate thus had a thickness of 0.69 cm (corresponding to $1.24X_0$), while the aluminium plate measured 2.91 cm ($0.33X_0$). The plates were inserted at a depth $z = 8$ cm, perpendicular to the direction of the incoming γs. The results, shown in Figure 5.12, illustrate point 5 from our list: for a given mass, the effects caused by high-Z dead material are much worse than those caused by low-Z material.

5.1.4.6   *Dead material upstream of the calorimeter.*   Structural materials should thus, preferably, have low Z values and be installed in areas of relatively small energy deposit. The area directly upstream of the calorimeter is such an area. Only a very small fraction of the shower energy is deposited in the first radiation length, typically only $\sim 0.3\%$ in the case of 100 GeV electrons in material with $Z = 50$ (Figure 5.7).

However, a radiation length worth of material may have very severe consequences

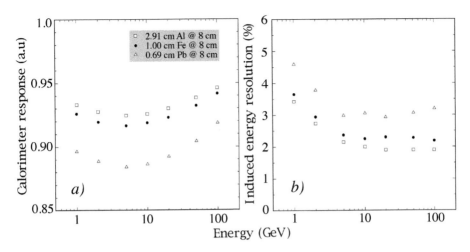

FIG. 5.12. Effects of plates of different materials installed at 8 cm depth inside the $Z = 50$ calorimeter on the average calorimeter signal (*a*) and on the energy resolution (*b*), as a function of the energy of the incoming particles (γs). The plates all had the same mass (7.85 g cm^{-2}) and were oriented perpendicular to the direction of the incident particles in these EGS4 simulations.

for the calorimeter performance when it is installed at a certain distance upstream. The reason for this is as follows.

An energetic electron traversing this material loses typically more than 60% of its energy in this process by radiating a large number of photons. Many of these photons convert in the absorber, producing electrons and positrons. Therefore, a large number of particles (electrons, positrons and photons) emerge from this $1X_0$ thick absorber. When this absorber is placed against the front face of the calorimeter, all these particles seamlessly enter into the calorimeter. However, when there is a certain distance between the absorber and the calorimeter, many of these shower particles may be lost.

First, many shower particles are traveling at relatively large angles with the direction of the incoming electron, in particular the Compton- and photoelectrons. As the distance between absorber and calorimeter is increased, these particles are increasingly likely to miss the area over which the calorimeter signals are integrated to determine the energy of the showering particle.

Second, many shower particles may not reach the calorimeter surface at all when a magnetic field is applied in the region upstream of the calorimeter.

5.1.4.7 *Magnetic field effects.* We studied these effects for our generic $Z = 50$ calorimeter. Dead material, also with $Z = 50$, was placed at a distance d in front of the calorimeter. The shower detected by the calorimeter was defined as the collection of particles that deposited energy within a cylinder with a radius of $3\rho_M$ around the direction of the incoming particle. The entire setup was assumed to be placed in a magnetic field oriented perpendicular to the trajectory of the incoming particles.

Figure 5.13 shows the effects of particle losses as described above on the energy

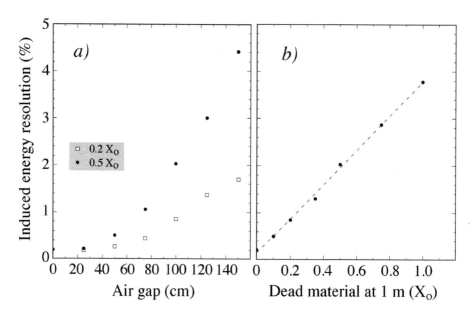

FIG. 5.13. Effects of dead material placed upstream of a $Z = 50$ calorimeter, operating in a transverse magnetic field, on the energy resolution for 100 GeV electrons. Diagram a shows the effects of $0.2X_0$ and $0.5X_0$ worth of material as a function of the distance between this material and the calorimeter's front face. In diagram b, the induced energy resolution caused by dead material placed 1 m from the front face is shown as a function of the thickness of this material. The magnetic field strength was 2 T and the calorimeter signals were integrated over a cylinder with a radius of $3\rho_M$ around the shower axis in these EGS4 simulations.

resolution for 100 GeV electrons. In Figure 5.13a, the effect of the distance d is shown, for fixed amounts of dead material. The energy resolution was found to increase approximately quadratically with d. Next, the distance d was fixed and the amount of dead material was varied. The results (Figure 5.13b) indicate that there is an approximately linear relationship, represented by the dashed line, between the amount of material and the energy resolution it induces.

The influence of the magnetic field is shown in Figure 5.14. Increasing the magnetic field in a given geometry means that charged shower particles generated in the dead material upstream of the calorimeter are increasingly likely to miss the fiducial volume around the shower axis that is used to define the showers. The magnetic field strength defines the cutoff energy for such shower particles: the stronger the field, the higher this cutoff. Figure 5.14 shows an approximately linear relationship between the magnetic field strength and the induced energy resolution.

The energy dependence of the effects caused by dead material placed upstream of a calorimeter operating in a magnetic field is shown in Figure 5.15. The simulated conditions were similar to the ones used before. The calorimeter and the dead material were both made of $Z = 50$ material. The dead material was $0.2X_0$ thick and was placed 1 m

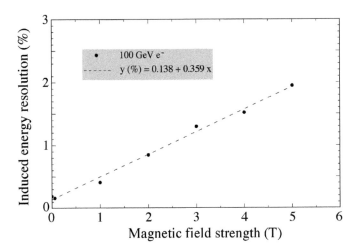

FIG. 5.14. Effects of dead material placed upstream of a $Z = 50$ calorimeter, operating in a transverse magnetic field, on the energy resolution for 100 GeV electrons. The effect on the energy resolution is shown as a function of the strength of the magnetic field. The $0.2X_0$ thick layer of dead material was placed 1 m from the calorimeter's front face and the calorimeter signals were integrated over a cylinder with a radius of $3\rho_M$ around the shower axis in these EGS4 simulations.

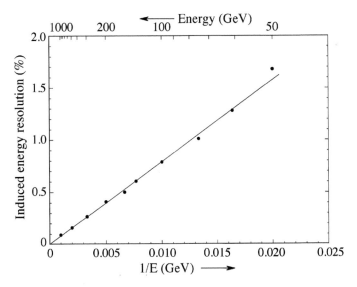

FIG. 5.15. Effects of dead material placed upstream of the $Z = 50$ calorimeter, operating in a 2 T transverse magnetic field, on the energy resolution for electrons, as a function of energy. The calorimeter signals were integrated over a cylinder with a radius of $3\rho_M$ around the shower axis in these EGS4 simulations.

in front of the calorimeter's front face. The magnetic field had a strength of 2 T and was oriented perpendicular to the direction of the incoming electrons. It turned out that the energy resolution induced by shower particles deflected outside the fiducial volume that was used to define showers in the calorimeter was approximately inversely proportional to the energy of the incoming electrons: $\sigma/E \approx 0.8/E$ (GeV).

All results shown in Figures 5.13, 5.14 and 5.15 concern electron showers. It should be emphasized again that the problems caused by dead material placed at a certain distance upstream of the calorimeter are fundamentally different for photons. Photons may or may not convert in this material. If they do not, it is as if the dead material were not there, no energy is lost and no shower particles are deflected by the magnetic field. However, if the photons do convert, they continue as an electron–positron pair. Depending on the track density and the quality of the experiment's tracking system, this phenomenon may or may not be recognized[12]. Especially in experiments at high-luminosity hadron colliders, the overwhelming majority of such conversions go unnoticed. This translates into a *detection inefficiency* for γ-rays produced in the interaction vertex. For electrons, on the other hand, upstream dead material leads to a systematic mismeasurement of the energy and to a degradation of the energy resolution.

5.1.4.8   *Lateral non-hermeticity.*   The next set of simulations concerned the effects of *lateral* non-hermeticity. The iron plate was now installed longitudinally in the calorimeter structure, *i.e.*, perpendicular to the calorimeter's front face. Figure 5.16 shows

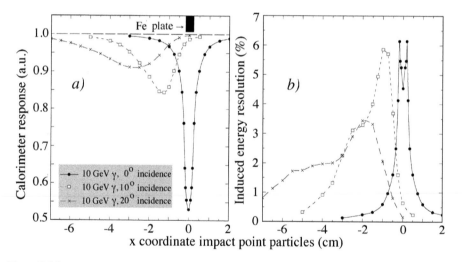

FIG. 5.16. The average calorimeter signal (*a*) and the induced energy resolution (*b*) of the $Z = 50$ calorimeter for 10 GeV photons as a function of the impact point of the particles. The calorimeter contained a 4 mm thick iron plate oriented perpendicular to its front face. The particles entered the calorimeter at angles of 0°, 10° or 20°. Results from EGS4 simulations.

---

[12] An impressive example of recognizing γ conversions was given by the CDF Collaboration [Abe 95], who managed to make a detailed map of the material installed upstream of their calorimeter with this information.

the average calorimeter signal and the induced energy resolution for 10 GeV photons as a function of the impact point of the particles. The figure also shows the importance of the direction of the incoming particles. The photons entered the calorimeter at three different angles: 0° (*i.e.*, perpendicular to the front face), 10° and 20°. The thickness of the iron plate was 4 mm, considerably thinner than the plates that were used to study the effects of longitudinal non-hermeticity. The first observation is that the effects of this plate on the calorimeter's performance were much larger than in the situation where dead material was installed perpendicular to the direction of the incoming particles, even in the worst possible position, *i.e.*, around the shower maximum. For example, 10 GeV photons that entered the 4 mm plate in its center lost as much as half of their energy in it, while the same particles lost, at maximum, only 3% of their energy if the same plate were oriented perpendicular to the direction of the incoming particles.

The explanation of this phenomenon is simple. The lateral shower dimensions are much smaller than the longitudinal ones. Therefore, the energy fraction that may be contained in a longitudinally oriented plate of given dimensions is much larger than when the same plate is oriented perpendicular to the direction of the incoming particles. Also the fluctuations in this fraction, which determine the energy resolution, are much larger.

It is thus no surprise that the effects of the lateral "crack" decreased rapidly when the particles entered the calorimeter at a non-zero angle, *i.e.*, when the crack became gradually less projective. Figure 5.16 shows the results for incident angles of 0° (projective), 10° and 20°. In a sense, the plate installed at a constant depth (*i.e.*, at a 90° angle with the incoming particles) may be considered the ultimate non-projective crack.

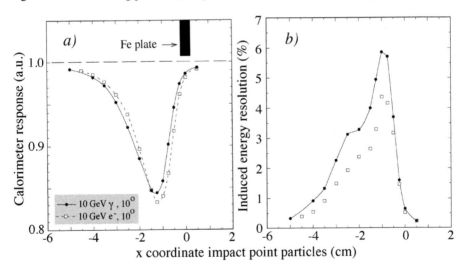

FIG. 5.17. The calorimeter response (*a*) and the induced energy resolution (*b*) of the $Z = 50$ calorimeter for 10 GeV photons and electrons as a function of the impact point of the particles. The calorimeter contained a 4 mm thick iron plate oriented perpendicular to its front face. The particles entered the calorimeter at an angle of 10°. Results from EGS4 simulations.

This example also confirms point #2 from our list of general rules for minimizing the effects of non-hermeticity on the calorimeter performance (Section 5.1.4.3): structural elements placed perpendicular to the direction of the incoming particles are much less disturbing than those placed parallel to that direction. Or, in other words, lateral non-hermeticity is worse than longitudinal non-hermeticity.

The differences between the effects of lateral non-hermeticity on the calorimeter performance for electron and photons are shown in Figure 5.17. Not surprisingly, these differences are very similar to those observed for the longitudinal non-hermeticity (Figure 5.10). When the particles entered the calorimeter at a non-zero angle (θ), the 4 mm plate acted as a (thicker) region of dead material covering a certain depth region. This region was determined by the impact point of the particles (x, see Figure 5.17) and θ, as $x \sin^{-1} \theta$. Therefore, the maximum in the energy resolution observed near $x = -1$ cm in Figure 5.17b corresponds to the maximum near $z = 5$ cm in Figure 5.10b ($\sin^{-1} 10° \approx 5.8$). Similarly, the secondary maxima near $x = -2.5$ cm (Figure 5.17b) and $z = 13$ cm (Figure 5.10b) correspond to each other.

As a final result of these simulations, the lateral non-hermeticity effects are shown as a function of the thickness of the iron plate for photons of three different energies. These photons entered the detector at an angle of 10°, at a position 1 cm away from the center plane of the dead material. These results show that plates as thin as 1 mm may already affect the energy resolution by more than 1% and cause signal uniformities of several percent. This figure also shows that the effects caused by a given plate of dead material tend to become smaller at higher energies.

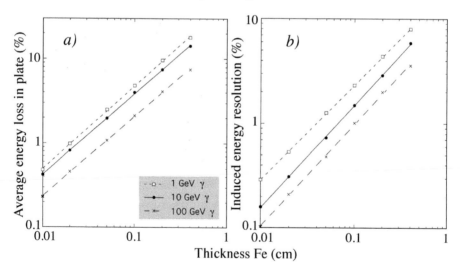

FIG. 5.18. The average energy loss (a) and the induced energy resolution (b) of the $Z = 50$ calorimeter resulting from an iron plate installed inside the calorimeter, perpendicular to its front face. Results are given as a function of the thickness of the plate, for γs of three different energies: 1, 10 and 100 GeV. These particles entered the calorimeter at an angle of 10° at a position 1 cm away from the plate's center plane. Results from EGS4 simulations.

Figure 5.18a shows that the energy loss in the dead material is approximately proportional to the thickness of this material. If, for example, one plate with a thickness of 1
mm were replaced by two plates of 0.5 mm, spaced by about 5 cm, then the signal non-
uniformity resulting from energy losses in these plates could thus be cut in half. This
illustrates point #1 from our list in Section 5.1.4.3. Similar benefits, derived from dis-
tributing the structural elements needed for the mechanical strength of the calorimeter,
also apply to material oriented at any other angle, and in particular also at 90°.

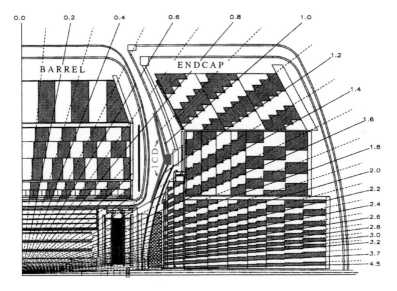

FIG. 5.19. Schematic cross section of one quadrant of the D0 calorimeter system, which was
contained in one barrel and two endcap cryostats. The Intercryostat Detector (ICD), sampled
energy deposited in the 2.5 cm thick stainless steel cryostat walls. The straight lines indicate the value of the pseudorapidity for particles generated in the interaction vertex. From
[Aba 94].

In many calorimeter systems, a certain degree of lateral non-hermeticity is unavoidable. In those situations, the installation of dedicated active material in the vicinity of
these cracks may be very helpful. In Section 4.4.2.4 it is shown how the ZEUS Collaboration effectively minimized the effects of lateral non-hermeticity in this way.

Another successful example of an experiment in which the effects of calorimeter
non-hermeticity were considerably reduced through this technique is D0, which operated at Fermilab's Tevatron collider [Aba 94]. The D0 calorimeter system comprised
several cryostats, which housed the barrel and endcap sections. The cryostat walls were
typically 2.5 cm thick, and the different cryostats also needed to stay clear from one
another for various operational reasons.

This resulted in a considerable non-hermeticity, both laterally and longitudinally.
Figure 5.19 shows a schematic longitudinal cross section of one quadrant of the D0

calorimeter system. It can be seen that particles traveling in the pseudorapidity interval from 0.8 to 1.4 encountered substantial amounts of dead material located at critical depths inside the absorber structure, *i.e.*, upstream of the sensitive volume and/or near the shower maxima.

To minimize the problems resulting from this, the D0 Collaboration decided to install a *crack detector* in the problematic boundary region between the barrel and endcap cryostats ($0.8 < |\eta| < 1.4$). This *Intercryostat Detector* (ICD), indicated in Figure 5.19, consisted of a scintillating-tile structure that sampled the energy deposited in the stainless steel cryostat walls. It clearly alleviated the mentioned problems.

5.2 Readout of Calorimeters Based on Light Detection

In the previous section the segmentation of calorimeter structures was discussed, and the arguments for choosing certain segmentation schemes (*e.g.*, a projective tower structure) were reviewed. However, we have not yet addressed the question of *how to realize* this segmentation in practice. It is the task of the calorimeter's *readout system* to make sure that all the energy deposited in a particular area of the calorimeter that is defined as a readout cell (*e.g.*, a projective tower with $\eta = 0.45 \pm 0.05$ and $\phi = 1.25 \pm 0.05$) gets collected and translated into an electronic signal that can be handled by the experiment's trigger logic and data acquisition system.

In this and the following subsection we describe some general features of calorimeter readout systems. A good readout system meets the following general requirements:

- Its contribution to the non-hermeticity of the calorimeter system is negligible.
- Its effects on the precision of the calorimetric energy, position and time measurements are negligible.
- It does not increase the system's vulnerability to radiation.

In this section the readout systems of calorimeters based on light detection are discussed.

5.2.1 *Homogeneous detectors*

All calorimeter systems of this type operating at particle accelerators are optimized for the detection of *electromagnetic* showers. As was discussed in Section 3.1, there is no particular advantage in using these expensive detectors for the detection of hadrons or jets, on the contrary.

With few exceptions, these calorimeters consist of individual crystals, in which the particles to be detected generate scintillation or Čerenkov light. Typically, these crystals have a length of 20–30 X_0 and a lateral cross section of 0.5–1.0 $(\rho_M)^2$. In recent years, liquid xenon has appeared as an alternative, for example in the MEG experiment, which searches for evidence of the extremely rare $\mu^+ \rightarrow e^+\gamma$ decay process and uses a detector based on an 800 liter LXe calorimeter for that purpose [Saw07].

In 4π geometries at collider experiments, the crystals are tapered. The front face is smaller than the back face, by a factor determined by the crystal length and by the distance to the interaction point, so as to maximize the hermeticity of the structure.

Ideally, non-hermeticities are only caused by the material needed to isolate the individual crystals optically from their neighbors and to make them light-tight. In addition, some structural elements, usually made of strong, low-Z materials such as carbon fiber, hold individual crystals or groups of crystals together.

Each crystal is read out from the back, by a PMT or some other element that converts light into electric pulses, such as a silicon photodiode [Bar 99], a vacuum phototriode [Akr 90], a Hybrid Photon Detector (HPD) [Anz 95a] or an APD [Lor 94]. Sometimes, several readout elements are connected to each crystal, *e.g.*, in the CMS experiment, where each crystal is read out by two APDs.

CMS faced a particular problem in designing their em calorimeter (ECAL), which had to operate in a very strong (4 T) magnetic field. This precluded the use of pho- tomultipliers for reading out the PbWO$_4$ crystals, which had been selected because of their presumed radiation hardness. Besides vacuum triodes, which are used to read the signals from the endcap crystals, where the magnetic field is less of a problem since it is oriented approximately along the crystal axes, solid state based light detectors were the only practical option. They offer a good quantum efficiency for the conversion of photons into photoelectrons in the relevant wavelength range (400–800 nm), and are insensitive to magnetic fields. However, standard silicon PIN photodiodes were also ruled out because of their extreme sensitivity to charged shower particle leakage (see Figure 3.32). These considerations led to the choice of avalanche photo diodes as light detectors for the barrel ECAL (Figure 5.20). Unlike PIN photodiodes, APDs have an

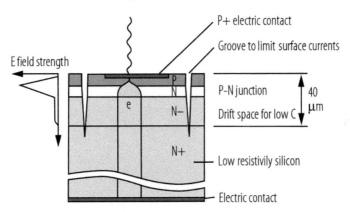

FIG. 5.20. Schematic diagram of an avalanche photo diode.

internal gain, of the order of 10 to 100, but this charge amplification comes at the cost of additional electronic noise, an important difference with the noise-free charge am- plification in PMTs. The detection of scintillation light takes place in a very thin (few μm) active layer, and the amplification of the charge created in this process takes place over a distance of $\sim 40\mu$m. This reduces the sensitivity to charged particles, compared to PIN photodiodes, by a factor of 20 to 200 [Lor 94]. Yet, it turned out that the "nu- clear counter effect," as the detection of charged particles by the light sensors is usually called, was by no means negligible. The small light yield of the PbWO$_4$ crystals, less than 1% of that in NaI(Tl), was of course an important factor in this.

Figure 5.21 clearly illustrates the effects caused by charged shower particles leaking out of the PbWO$_4$ crystals. A broad beam of 225 GeV muons was sent through a PbWO$_4$ crystal equipped with an APD, parallel to the crystal axis. The trajectory of each muon was precisely measured with additional tracking detectors. For each muon, it was thus known whether or not it traversed the APD. If it did not, then the APD only recorded the scintillation light produced in the process. If it did, then the electron–hole pairs produced in the silicon diode by the muon itself gave an additional contribution to the signals.

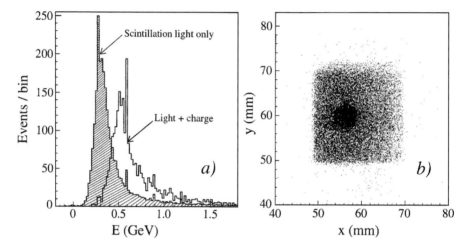

FIG. 5.21. Muon signals in the PbWO$_4$ crystal detector of CMS. Diagram (a) shows the signal distributions for muons that did and did not traverse the surface of the APD that recorded the signals. Diagram (b) shows the impact points of the particles that produced a signal larger than 0.4 GeV [Ale 97].

These two distributions are displayed in Figure 5.21a. The muons lost, most likely, 285 MeV on their way through the crystal. This value corresponds to the peak of the dashed ("scintillation light only") distribution. By comparing the two signal distributions, it was found that the additional contribution from the charge generated by the muon itself in the silicon entrance window of the APD corresponded to $\sim$ 320 MeV ($\sim$ 100 MeV μm^{-1}). Figure 5.21b shows the distribution of the impact points of those muons that generated signals with an energy equivalent larger than 400 MeV. The position, the size and the shape of the APD are very clearly visible in this plot.

When an APD of the type used by CMS is traversed by an ionizing particle, it generates a signal that is typically two orders of magnitude larger than the signal from a scintillation photon. CMS realized that the effects of this could be mitigated by covering a larger fraction of the exit surface of their crystals with APDs. Since APDs with a sufficiently large surface area did not yet exist at the time these decisions were made, each crystal was equipped with two APDs. In doing so, the number of detected photons per GeV was increased and the energy equivalent of a charged particle leaking out of the crystals and traversing an APD was thus reduced [Ale 97, Auf 98], by a factor of two, to ~ 160 MeV for a mip.

In order to save money, the two APDs connected to one crystal were electronically treated as one (larger) sensor. This turned out to be an unfortunate mistake. In Section 2.3.2, we saw that hadron showers are characterized by large numbers of nuclear reactions, such as the ones shown in Figures 2.29 an 2.49. The nuclear fragments, mainly protons, visible in these reactions have typical ionization densities of 100 to 1,000 times that of a mip. If such a reaction took place sufficiently close to an APD, these particles could thus create signals equivalent to hundreds of GeVs. This is precisely what CMS observed once they started to operate their calorimeter system in the LHC environment, and the ECAL was exposed to countless high-energy hadrons which developed showers in it. The anomalous events, in which a single crystal produced a signal that was an order of magnitude (or more) larger than the signals from neighboring crystals, became known as "spike events." An example, as well as the way in which CMS deals with this problem, is given in Section 9.1.3 (Figure 9.7).

If CMS had decided to read the signals from both APDs connected to a single crystal separately, these spikes could have been easily recognized, and their effects avoided. Because of the very short range of the nuclear fragments, they would be detected only in one of the two APDs reading out each crystal, and the signal from the other APD could in that case still be used for determining the energy equivalent of the scintillation light. Based on the explanation of this phenomenon, one should expect the frequency of these unusable "spike" events to increase proportionally with the lunimosity, as well as the collision energy at the LHC.

Experiments that have operated a scintillating-crystal calorimeter at a collider experiment include CLEO (Cornell, [Beb 88]), BaBar (SLAC, [Bar 99, Sta 98]) and BELLE (KEK Japan, [Ohs 96, Ahn 98]), which all used CsI(Tl) crystals, and L3 (LEP, CERN), which was based on BGO crystals [Bak 87]. An example of a fixed-target experiment that used a calorimeter consisting of scintillating crystals (CsI) is KTeV (Fermilab, [Kes 96]). Current and future experiments with crystal-based em calorimeters include BELLE-II, which will soon (2017) start taking data at the upgraded SuperKEKB asymmetric $e^- e^+$ B-factory, BESIII at the Beijing τ-charm factory and PANDA, an experiment planned for the Facility for Anti-proton and Ion Research (FAIR), under construction in Darmstadt (Germany). The calorimeter for BELLE-II consists of CsI(Tl) crystals, readout by PIN photodiodes in the barrel region, and of faster ($\tau \sim 30$ ns, vs. $1\mu s$ for the thallium doped version) pure CsI crystals, readout by vacuum photopentodes in the endcaps [Abe 10]. The BESIII calorimeter also consists of CsI(Tl) crystals, read out by photodiodes [Don 08], while the PANDA calorimeter will be based on $PbWO_4$ crystals, read out by APDs and vacuum phototriodes/tetrodes [Kav 11]. The heavy-ion collision experiment ALICE at the LHC also has a calorimeter consisting of $PbWO_4$ crystals read out by APDs [Aam 08].

Crystals based on the production of Čerenkov light (lead-glass) were used in the OPAL experiment (LEP, CERN, [Akr 90]). This material has also been very popular in a variety of fixed-target experiments carried out since the 1960s at all major accelerator centers. Examples of experiments using this type of calorimeter include NOMAD at CERN [Aut 96], HERMES at DESY [Ava 96], E852 at Brookhaven [Cri 97] and E760 at Fermilab [Bart 90].

A unique experiment based on a homogeneous electromagnetic calorimeter takes place at PSI near Villigen, Switzerland, where the MEG experiment looks for evidence of the lepton-flavor violating process $\mu^+ \rightarrow e^+\gamma$. The telltale signature of this process is the production of a back-to-back positron and γ ray by a stopping muon, each with an energy of 52.8 MeV. In order to reduce background, precise measurements of energy, emission angle and time are crucially important. An 800 liter LXe calorimeter is used to measure γ-rays with the required precision. Liquid xenon is a very bright and fast scintillator (Table B.7), but the emission spectrum of the scintillation light peaks at 174 nm, *i.e.*, in the VUV region. For this reason, the 846 light sensors are immersed in the liquid, so that the photons can be detected with a minimum of transmission losses. Special PMTs equipped with quartz windows and designed to operate at cryogenic temperatures were developed for this purpose. Beam tests showed that 50 MeV γs could be detected with an energy resolution of slightly more than 1% in this calorimeter [Saw07].

Finally, the NOVA experiment, designed for the study of neutrino oscillations in in Fermilab's NuMI beam, operates two huge detectors based on liquid scintillator [Pat 13]. The 300-ton Near Detector is installed at the Fermilab site, 500 m from the neutrino production target, the 22 kiloton Far Detector is located 810 km away, in Minnesota.

5.2.1.1 *Non-accelerator experiments.*

In these experiments, the calorimeter also serves as the target in which the particles to be detected are produced. The tasks of the calorimeter system therefore include in that case determining the position of the vertex, something that is usually not needed in accelerator-based experiments.

The light-based calorimeter systems operating at non-accelerator experiments are of an entirely different type than those at accelerator-based experiments. They consist, for example, of a large volume of (liquid or frozen) water, typically thousands of cubic meters. The Čerenkov light produced in this water by relativistic charged particles created in interactions induced by cosmic rays or by some other process (*e.g.*, neutrinos produced at one or several nuclear reactors, radioactive decay) is detected by large PMTs.

The light produced by individual particles is emitted at an angle $\theta = \arccos\left[(n\beta)^{-1}\right]$ with the direction in which the particle travels (the Čerenkov cone). From the signals detected in the various PMTs, the trajectories of the particles emitting this light can be reconstructed. Experiments of this type include SuperKamiokande (Japan), AMANDA (South Pole) and ANTARES (Mediterranean Sea, near Toulon, France). Several of these experiments are featured in Chapter 10.

5.2.2 *Sampling calorimeters*

The first generation of calorimetric detectors used in particle physics experiments consisted almost entirely of detectors of this type: metal plates interleaved with layers of scintillating plastic. However, in the past three decades, these calorimeters have undergone a complete metamorphosis, which was mainly inspired by the need for hermetic devices.

Sampling calorimeters based on scintillating plastics are attractive because of their relative simplicity and low cost. However, it is non-trivial to extract the light from the area in which it was generated and convert it into electric signals without affecting the hermeticity of the detector.

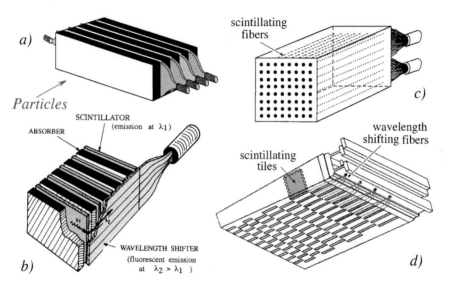

FIG. 5.22. Schematic representation of frequently used readout techniques for scintillation calorimeters: scintillator plates individually coupled to PMTs (a). Scintillator plates read out by
WLS plates (b). Scintillating fibers coupled via a light guide to a PMT (c). Scintillating tiles
oriented in the direction of the incoming particles and read out by WLS fibers (d).

The first generation of sampling scintillation calorimeters [Eng 69] operated in fixed-target experiments, for example at the CERN SPS or the first proton accelerator at Fermilab. In these experiments (*e.g.*, the neutrino experiments WA1 [Abr 81] and WA18 [Did 80] at CERN), each scintillator plate was individually coupled to a PMT. In many cases, each plate was read out by two PMTs, one at each end of the plate (Figure 5.22a).

The development of wavelength-shifting (WLS) optical elements made it possible to construct reasonably hermetic calorimeter systems for collider experiments, which require a 4π detector geometry. The technique of coupling individual scintillator slabs through a WLS plate to a photon detector located behind the calorimeter (Figure 5.22b) has been applied in a variety of experiments at, for example, the CERN Intersecting Storage Rings (R807), the *ep* collider HERA at DESY (ZEUS), and the *pp̄* colliders at CERN (UA2) and Fermilab (CDF).

This technique allows the creation of a tower structure in a natural way, since all active elements that are part of a certain tower are read out by the same WLS plates. Therefore, all the light produced in this tower is detected by the same PMT(s).

Disadvantages of this technique include:

1. The reduction of the signal speed as a result of the wavelength-shifting process. The time constant of this process is usually considerably longer than that of the scintillation process, typically at least 10 ns. One of the main attractive features of scintillation-based calorimeters is affected in this way.

2. The possibilities to create a highly granular structure are limited. The WLS plates represent a non-hermeticity of the worst possible kind, since they are oriented (approximately) perpendicular to the direction of the incoming particles (Section 5.1.4.3). Typically, they occupy a space of the order of 1 cm between the individual calorimeter cells. The cell sizes in the mentioned calorimeter systems employing this technique are typically 20×20 cm^2, so that non-hermeticity introduced by the WLS plates is a relatively minor problem. However, if the calorimeter cells measured 10×10 cm^2, the non-hermetic region would already account for 10% of the total calorimeter surface. This fraction rapidly increases further for even smaller cell sizes.

These problems were solved with the introduction of scintillating and wavelength-shifting optical fibers, and with the notion that the active material in sampling calorimeters does not necessarily have to be oriented perpendicular to the direction of the incoming, showering particles.

Scintillating fibers used as active material oriented in the same direction as the incoming particles eliminate the need for separate wavelength shifters. In addition, there are no significant restrictions on the lateral granularity of such a structure: if desired, every fiber could be read out individually.

On the other hand, longitudinal segmentation of such a calorimeter structure is not trivial. Two possible schemes have been proposed and tried out for this purpose (see Figure 5.23). In the first scheme, the fibers of the first calorimeter segment (the em sec-

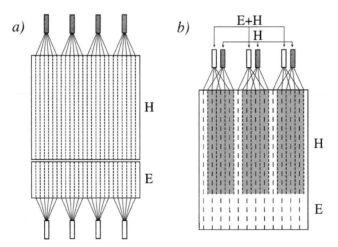

FIG. 5.23. Longitudinal segmentation in a scintillating-fiber calorimeter. See text for details.

tion, E) are read out from the *upstream end* (Figure 5.23a). At first sight, this may seem a very unfavorable arrangement. However, one should keep in mind that scintillation light is emitted isotropically and, therefore, it does not matter from which side the fibers are read out. If desired, the downstream end of the fibers may be aluminized to increase the light yield. However, in that case it is very important that the reflection coefficients of the mirrors on the different fibers are approximately equal, because inequalities may cause signal non-uniformities and degradation of the resolution.

In this scheme, the fibers of the hadronic calorimeter section (H) are read out from the downstream end, as in unsegmented fiber calorimeters. The two segments can be mounted with adjacent surfaces very close to each other.

In the second scheme, a longitudinal segmentation is effectively achieved by having some fraction of the fibers start at a certain depth (*e.g.*, 1 λ_{int}) inside the calorimeter. These fibers are read out separately from those that run all the way from the front face to the rear end of the calorimeter (Figure 5.23b).

In this scheme, the signals from these two sets of fibers do not represent the signals from the em and hadronic calorimeter sections, as in conventionally segmented devices. The "short" fibers represent the hadronic section, while the "long" fibers give the total signal (E+H). Information about the energy deposited in the em section thus can be obtained by subtracting the signal from the short fibers from that in the long fibers. It has been demonstrated that the results obtained with this method, for example in terms of electron identification capability, are not much worse than those obtainable with a "classically" segmented device [Bad 94b].

This second method lends itself in a very natural way to achieving longitudinal segmentation in a fiber calorimeter consisting of tapered modules, which are typically needed in a 4π geometry. If the fibers are running parallel to each other, which is essen-

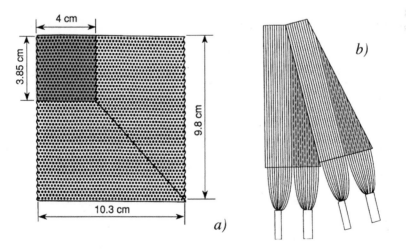

FIG. 5.24. Longitudinal segmentation in a projective fiber calorimeter. Shown is a detail of the front face of a tower (*a*) and a schematic impression (not to scale) of the longitudinal cross section of two neighboring modules (*b*). From [Bad 94b].

tial for achieving a constant sampling fraction (see Section 4.4.2.3), then some fraction of the fibers necessarily have to start at a certain depth inside the calorimeter. These fibers can be read out separately, to form the hadronic section of such a device (Figure 5.24).

Optical fibers may also be used as wavelength shifters, reading out the light from calorimeter structures consisting of absorber and scintillator plates or tiles. This has several advantages compared with the use of WLS plates for this purpose:

- The fibers are much smaller than the plates. Typically, WLS fibers used for this purpose have a diameter of 1 mm, while plates are at least three times as thick. This means that the non-hermetic region between neighboring calorimeter modules can be considerably reduced.

- Unlike the WLS plates, the fibers are flexible. As a result, fibers allow the light to be collected and transported to the photon detectors in ways that are impossible with plates. As an example, we mention the so-called σ-tiles used in the forward calorimeter of the CDF detector [Aot 95, Apo 98a] and in the hadronic calorimeter of CMS [CMS 97b]. The latter calorimeter consists of copper absorber plates interleaved with plastic-scintillator tiles, oriented perpendicular to the direction of the incoming particles. Each tile contains a WLS fiber, embedded in a keyhole-shaped groove (see Figure 5.25a,b). This fiber collects the scintillation light produced in the tile and re-emits it at a longer wavelength. The WLS fiber is optically coupled to a clear, undoped optical fiber, which bends over 90° and transports the light to the rear end of the calorimeter, where the light detectors (HPDs) are located. This scheme allows a great flexibility in readout options, since the signals from each individual tile are available for readout. Signals from different tiles can be combined into the same HPD as desired.

- When combined with very thin scintillator plates, WLS fibers allow arbitrarily fine lateral granularity. This option was explored in the SPAKEBAB project [Dub 96], where a very high sampling frequency was achieved by means of very thin (1 mm) scintillator plates that were interleaved with even thinner (0.89 mm) lead absorber plates. Scintillation light is very strongly attenuated in such thin scintillation plates. To avoid response non-uniformities and maximize the light yield, a large number of WLS fibers, spaced by only 4 mm and distributed uniformly over the entire detector volume, were used to transport the light to the rear end of the calorimeter. By combining these fibers into bunches that were connected to light sensors, a tower structure was realized, as can be seen in Figure 4.5. The light attenuation in the thin scintillation plates was such that there was almost no crosstalk between these towers, even though they were not separated by physical boundaries. The near absence of non-hermeticity achieved in this way eliminates one of the main disadvantages of the WLS plate technique, while the advantage of easy and straightforward longitudinal segmentation is maintained.

Almost every sampling calorimeter system built in the past 25 years that is based on scintillation light uses optical fibers, either as active material in which the light is generated (e.g., CHORUS [Esk 97], KLOE [Bab 93]) and/or as the medium through

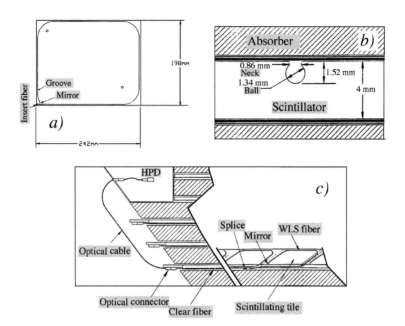

FIG. 5.25. Schematic representation of the light collection in the hadronic calorimeter of CMS. Scintillation light is produced in tiles which are read out by means of WLS fibers (*a*) embedded in a keyhole-shaped groove (*b*). The wavelength-shifted light is transported out of these tiles through these fibers, which are then spliced and connected to clear (undoped) fibers which transport the light signals to the light detectors, HPDs in this case (*c*). From [CMS 97b].

which the light constituting the signal is transported to the detectors (*e.g.*, the ATLAS [Atl 96c] and CMS [CMS 97b] hadron calorimeters, and the Near and Far detectors of Fermilab's MINOS neutrino oscillation experiment [Mic 08]).

5.2.3 *Operational aspects*

Every light-based analog particle detector suffers from two instrumental inconveniences:

1. The detector is inherently non-uniform.

2. The detector is inherently unstable.

The first aspect is a result of the inevitable losses that occur in the transportation of the light between the position where it is produced and the location where it is converted into an electric signal. These attenuation losses are either due to self-absorption or to reflection losses at the interface between media with different indices of refraction.

The non-uniformities of some practical calorimeters in which light is the origin of the signals, as well as the techniques that are being employed to minimize these effects, are discussed in some detail in Sections 4.4.3.4. Three methods were mentioned in this context:

1. *Reduction*: Reading and summing the signals from both ends of the optical elements substantially reduces the effects of light attenuation. Making one end reflective has the same benefit.

2. *Correction*: Determining the position where the light is produced makes it possible to correct for the effects event by event.

3. *Elimination*: By selectively limiting the generation of light in a position dependent way, the response can be made independent of the position where the light was created, as in the elegant "bar code" method used by ZEUS (Figure 4.28).

However, one should realize that the light attenuation characteristics of optical elements such as scintillating plates or fibers also tend to change over time, thus limiting the practical value of the last method. The enormous development in fast electronics is offering new possibilities concerning method #2. It turns out that precision measurements of the time structure of showers make it possible to determine the depth at which light is produced in a fiber calorimeter. This is because the light in the fibers travels at a speed c/n, whereas the particles that produce that light typically travel at a speed c. Examples of application of this feature are given in Section 8.2.7.2.

The second mentioned inconvenience (detector instability) has several components:

- Scintillation is usually a very complex process, sensitive to a variety of factors that may change with time, *e.g.*, temperature, environment, the presence of a magnetic field. As these factors change, so does the amount of light produced per unit of energy deposited in the scintillating material.

- The transparency of most materials that are used to transport the light from the production point to the photon detector changes with time as a result of chemical processes. A well-known example is the phenomenon that transparent plastics used outdoors invariably turn yellow after some time as a result of a photochemical process initiated by the ultraviolet component of sunlight.

- A very specific type of instability is caused by ionizing radiation. The effects of such radiation on scintillating materials and on materials used to transport scintillation or Čerenkov light are discussed in detail in Section 5.10.

- All detectors that are used to convert light produced in the calorimeter into electric signals are intrinsically unstable. The gain of photomultiplier tubes is sensitive to temperature variations, especially if the number of stages is large. In a 12-stage tube, a gain change by a factor of two corresponds typically to a change in applied high voltage of only 50–100 V. This means that a change of 1 V (on 2 kV applied voltage) causes a gain change of 1–2%. Avalanche Photo Diodes are even more sensitive than PMTs to changes in temperature and voltage [Kob 95].

These features imply the necessity for elaborate monitoring systems. All calorimeters based on light detection are equipped with such, sometimes very sophisticated systems. The general features of such systems are discussed in Section 6.6.1.

5.3   Readout of Calorimeters Based on Charge Collection

Charge-collection readout is the second widely used method. Ionization charge liberated in the active medium forms the basis of the signals. The active medium may be gaseous, liquid or solid. In some very sophisticated applications, the calorimetric equivalent of a Time Projection Chamber (TPC) is achieved.

5.3.1   Gaseous active media

Charge collection in gases, usually followed by some degree of internal amplification, is used as the signal-generating mechanism in many diverse calorimeter systems.

When a charged particle traverses a gas-filled volume, a discrete number of ionizing collisions take place. In this process ion–electron pairs are created. The electron ejected in this process may have enough energy to further ionize the medium and thus produce secondary ion–electron pairs. The total number of electron–ion pairs produced per unit of length by a mip is listed in Table 5.1 for a number of gases used in wire chambers. A typical value is 100 pairs per centimeter, or one ionization every $100 \ \mu m$.

Table 5.1 *Properties of gases used in wire chambers. Listed are for each gas the density, the specific energy loss by ionizing particles, the specific ionization yield and (for some gases) the mobility of the positive charge carriers. Data are given for atmospheric pressure. The $\langle dE/dx \rangle$ and n_T data concern minimum ionizing particles.*

Gas	Density $(mg \ cm^{-3})$	$\langle dE/dx \rangle$ $(keV \ cm^{-1})$	n_T $(ions \ cm^{-1})$	μ^+ $(cm^2 V^{-1} s^{-1})$
H_2	0.08	0.34	9.2	13.0
He	0.17	0.32	7.8	10.2
N_2	1.17	1.96	56	
O_2	1.33	2.26	73	2.2
Ne	0.84	1.41	39	
Ar	1.66	2.44	94	1.7
Kr	3.49	4.60	192	
Xe	5.49	6.76	307	
CO_2	1.86	3.01	91	
CH_4	0.67	1.48	53	
C_4H_{10}	2.42	4.50	195	

Table 5.1 also lists the specific energy loss, $\langle dE/dx \rangle$, and from this one can derive that the effective average energy needed to produce one electron–ion pair is about 30 eV.

When an electric field is applied across the gas volume, the electrons and ions start drifting to the anode and cathode, respectively. The drift velocity, $\vec{v}_d$, is proportional to the strength of the electric field, $\vec{E}$:

$$\vec{v}_d = \mu \vec{E} \tag{5.3}$$

The proportionality parameter μ is called the *mobility*. The value of the mobility is different for the electrons and the ions. Some typical values for the ion mobility in different gases are listed in Table 5.1. A value of 1 (cm/s)/(V/cm) (= 1 cm^2 V^{-1} s^{-1}) means that an applied field with a strength of 10 kV cm^{-1} gives the ions a velocity of 10,000 cm s^{-1}, so that it takes them 50 μs to cross a gap with a width of 5 mm.

The ion mobility is determined by collisions with the gas molecules. Therefore, it depends on the sizes of the ions and the molecules, and of course also on the gas pressure. The lower the pressure, the higher the ion mobility.

Unlike the ions, the mobility for electrons is not constant. Therefore, Equation 5.3 is an oversimplification for the case of electrons, except for very low fields. Due to their small mass, electrons can substantially increase their kinetic energy between collisions with the gas molecules under the influence of the electric field. If the field strength reaches values above a few kilovolts per cm, then the electrons may receive enough kinetic energy between two collisions to cause excitation and even ionization of the molecules with which they collide.

In a wire chamber, where the electric field strength is inversely proportional to the distance r from the center of the anode wire, this phenomenon may lead to avalanche formation at very small values of r. Details of the processes occurring in the region close to the anode wire depend on the applied voltage. This is illustrated in Figure 5.26.

At zero voltage, the electron–ion pairs recombine as a result of their electrostatic attraction, and no charge is collected at the electrodes. As the voltage is raised, these electrostatic forces are overcome and more and more electrons and ions reach the anode and cathode, respectively. At some voltage, all the electron–ion pairs are collected and a further increase of the voltage does not result in an increased current (region II of Figure 5.26). A detector operating in this regime is called an *ionization chamber*.

If the voltage is increased beyond a certain value (250 V in the example of Figure 5.26), the electrons produced in the primary ionization processes may acquire enough energy to be able to ionize molecules themselves. The electrons liberated in these secondary ionizations are also accelerated and produce in turn more ionizations: an ionization *avalanche* develops. This charge multiplication process takes place almost entirely at a very small distance from the anode wire, the only place where the electric field reaches values that make it possible.

The increased number of electron–ion pairs gives rise to an increased current. This increase is proportional to the number of primary ionization processes, with the proportionality constant depending on the applied voltage. The voltage region in which this proportionality condition is fulfilled extends to about 600 V in the example given in Figure 5.26. This example also shows that the charge multiplication factor of the *proportional chambers* operating in this regime can reach substantial values (10^4–10^6, depending on the type of particle that caused the primary ionization).

If the voltage is further increased, the proportionality condition begins to be lost. The space charge created in the avalanche process distorts the electric field. Photons emitted in the decay of excited molecules may also contribute to the ionization of the gas. Eventually, the counter operates in a discharge, or *Geiger* mode, in which the output current varies little with the applied voltage.

It is important to realize that the signals from wire chambers operating in the proportional mode are primarily generated by *induction* due to the moving charges rather than

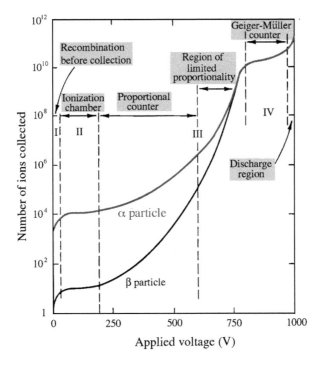

FIG. 5.26. The number of ions collected as a function of the applied voltage for α and β particles traversing a single-wire gas chamber.

by the *collection* of these charges at the electrodes [Sau 77]. Also, since the electrons typically travel distances that correspond only to a few anode wire radii ($< 100 \ \mu$m), their contribution to the signals is much smaller than that of the ions, which traverse the entire chamber gap. For all practical purposes, the signals from these wire chambers are thus primarily caused by ions that drift from the anode to the cathode and induce a voltage drop on the anode wire in this process.

Most gaseous detectors used as active elements in calorimeters are Multi-Wire Proportional Chambers (MWPCs), invented by Charpak [Char 68]. In devices of this type, many closely spaced wires act as independent proportional gas counters. By giving each anode wire its own electronic readout, and comparing the signals induced by the same ionizing particle on different wires, MWPCs can be fruitfully used as position sensors, in addition to their role in calorimetric energy measurements.

5.3.1.1 *Operational aspects.* The internal amplification of the charge produced by ionizing particles in the gas facilitates the signal processing, since it converts the relatively small number of electrons produced by the ionizing particle traversing the gas into an electric pulse with a usable amplitude, *i.e.*, in the μV–mV range. However, at the same time this internal amplification is the source of important systematic uncertainties, which affect the precision of the experimental data collected with proportional wire chambers as active calorimeter elements. The value of the proportionality constant, which determines the amplitude of the signals generated by these chambers, and thus the reconstructed energy values of the showering particles, is very sensitive to a number of parameters, for example:

- The diameter of the anode wire. Since most of the charge multiplication takes place in the immediate vicinity of this wire, where the electric field (inversely proportional to the distance from the wire axis) reaches very high values, small variations in the wire dimensions may have a large impact on the size of the signals.

- The composition of the gas mixture. Since the signals are dominated by the induction caused by drifting ions, variations in the drift velocity, as brought about by changes in the composition of the gas mixture, would have a direct effect on the signals. Also, small changes in the effective ionization energy caused by changes in the gas mixture could have large effects on the multiplication factor, and thus on the signal size.

- The pressure and the temperature of the gas. These quantities determine the mean free path between collisions. The multiplication factor is very sensitive to changes in these parameters

In practice, it is very difficult to keep the multiplication factor of the wire chambers, and thus the energy scale of the calorimeter in which they operate, constant to better than $\sim 10\%$. In experiments requiring high-precision calorimetry, with energy resolutions at the few-percent level, other readout techniques should therefore be preferred.

5.3.2   Liquid active media

Calorimeters using liquified noble gases such as argon, krypton or xenon are also based on direct collection of the ionization charge produced by charged shower particles. Since these liquids are relatively dense, there is no need for charge amplification in an avalanche process, as in the gas-based detectors discussed above.

These liquid-based calorimeters all operate in the ionization-chamber mode, in which the electrons produced in the ionization processes are collected and form the signals [13]. This offers major advantages compared with the gas-based detectors, since the gain stability problems that make the latter hard to operate are avoided in this way.

The basic active element of a noble-liquid calorimeter consists of two parallel metal plates, between which a potential difference ΔV is applied. The gap between these plates is filled with the liquid sampling medium (Figure 5.27). When a charged shower particle traverses this gap, it ionizes the atoms in the liquid. Because of the presence of an electric field, the electrons and ions created in this process drift to their respective electrodes. The charge collected on these electrodes forms the calorimeter signal, in contrast to the gaseous detectors described above, where the signals correspond to the current *induced* on the electrodes by the drifting charges (ions).

---

[13] In LAr TPCs, additional wire planes also provide extra information based on induction.

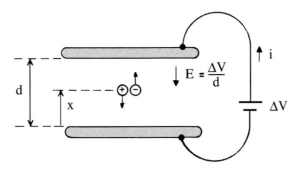

FIG. 5.27. Schematic drawing of the basic element of a noble-liquid calorimeter. Two parallel metal plates, between which a voltage (ΔV) is applied, are immersed in a liquified noble gas. From [Wil 74].

As in the case of the gaseous detectors, the drift velocity of the electrons in noble liquids is determined by the electric field strength and by the (medium-dependent) mobility. However, because of the much higher density of the liquid, the mean free path between collisions is orders of magnitude smaller than in gases. Therefore, acceleration of the electrons between collisions to the point where they acquire enough energy to cause ionizations themselves and thus initiate avalanche development, does not play a role here.

In the most widely used medium in calorimeters of this type, liquid argon, the electron mobility amounts to $475 \text{ cm}^2 \text{ V}^{-1} \text{ s}^{-1}$. For a gap with a width d of 3 mm, and an applied voltage of 300 V (*i.e.*, the field strength equals 1 kV cm^{-1}), the drift velocity of the electrons in LAr thus amounts to $4.75 \cdot 10^3 \text{ m s}^{-1}$. At this speed, it takes the electrons 0.63 μs to cross this gap.

However, in practice, more time than that is required for this process. Even though intra-collision electron acceleration is negligible, Equation 5.3 still represents an over-simplified description of the relationship between the strength of applied electric field and the resulting drift velocity for electrons in noble liquids. It turns out that the electron mobility (μ_e) decreases as the field strength is increased, which leads to a saturation of the drift velocity.

This is illustrated in Figure 5.28, which shows the drift velocity for electrons in liquid argon as a function of the electric field strength. The dotted line corresponds to the electron mobility mentioned above. Only for very small field values does the slope of this line correspond to the one of the dashed line, which represents the case of pure liquid argon. At higher field values, the slope of the dashed line (*i.e.*, the "effective" electron mobility) decreases considerably and to obtain the mentioned drift velocity of 4.75 km s^{-1} a field strength in excess of 10 kV cm^{-1} is required. The electron drift velocity eventually reaches a maximum value of $\sim 8 \text{ km s}^{-1}$ for extremely strong electric fields.

Figure 5.28 also shows that the electron drift velocity can be considerably increased by adding a small percentage of hydrocarbons to the liquid. For example, 0.2% of ethylene approximately doubles the electron drift velocity for fields in the 4–10 kV cm^{-1}

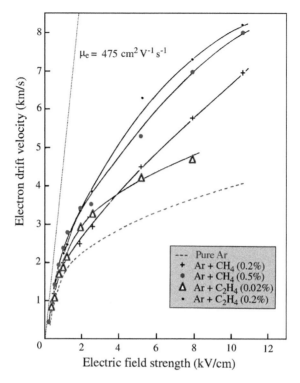

FIG. 5.28. The electron drift velocity as a function of the electric field strength in pure liquid argon (the dashed line) and the effects of small admixtures of methane or ethane on the saturation characteristics. The dotted line represents the unsaturated v_d–E curve. From [Shi 75].

range. This seems to be a consequence of the "cooling" of the liberated electrons in collisions with the hydrocarbon molecules, which can be excited into a variety of vibrational states. In this way, the diffusion coefficient of the electrons is reduced and their mobility is increased [Shi 75]. However, there is a price to be paid for this increased electron drift velocity, in the form of smaller signals and increased radiation sensitivity.

The drift velocity of the ions is three to five orders of magnitude smaller than the drift velocity of the electrons. Therefore, the ions do not significantly contribute to the signals, for typical charge collection times of 1 μs or less.

The number of ionizations per unit length is of course considerably larger than in gases. In first approximation, the energy needed per ionization process is the same in the liquid and the gaseous phases, so that the number of ionizations per unit length is larger by a factor equal to the density ratio.

In the case of argon, the density of the liquid is about 800 times larger than the density of the gas at room temperature and atmospheric pressure. On that basis, one may expect the number of electron–ion pairs produced by a mip traversing a LAr gap to be $\sim$ 70,000 per cm. Therefore, if all the charge produced in these processes is collected, signals of the order of 10 fC may be expected. In practice, the signals also depend on the strength of the electric field that is applied across the gap, which affects the recombination rate (see, for example, Figure 3.25).

Just like plastic scintillators, liquid dielectrics exhibit saturation phenomena. In Section 3.2.10, it was shown that the collected charge depends on the specific ionization, $\langle dE/dx \rangle$, or more to the point, on the density of ion–electron pairs. These saturation effects depend on the strength of the applied electric field: the stronger the field, the smaller the recombination probability (and thus the saturation effects). This was illustrated in Figure 3.26, which showed that the ratio of the signals from 5.5 MeV αs and mips in liquid argon increases considerably with the strength of the electric field.

It turns out that the addition of small amounts of hydrocarbons, intended to increase the electron drift velocity, considerably increases the probability for recombination and thus amplifies the saturation effects. This results in smaller signals, especially for densely ionizing particles, such as αs or recoil protons produced in neutron scattering. These effects were studied in detail by Anderson and Lamb [Ande 88], who

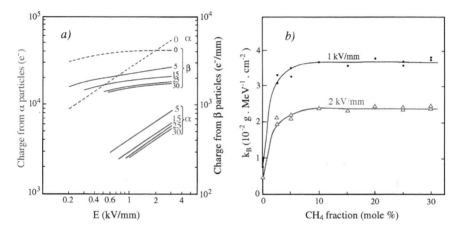

FIG. 5.29. Charge liberated by 5.5 MeV α particles and the charge induced per mm by β particles in liquid argon doped with methane, as a function of the electric field strength, for different molar fractions (5%, 15%, 25% and 30%) of methane. The dashed curves correspond to pure (undoped) LAr (a). The value of Birks' constant as a function of the methane concentration, for two different values of the electric field strength (b). From [Ande 88].

measured the signals from 5.5 MeV α particles and energetic nuclear βs (mips) in a LAr cell doped with methane. Figure 5.29a shows the collected charge as a function of the strength of the electric field applied over this cell. Results are given for pure argon and for three different methane concentrations.

The results for pure argon are in good agreement with those of Willis and Radeka (Figure 3.26). At a field strength of 3 kV mm^{-1}, adding 5% of methane to the argon reduced the mip signals by $\sim$ 35%, but the α signals shrunk by more than a factor of six. In other words, the α/mip signal ratio was reduced by about a factor of four. Further increasing the methane concentration did not change this result much.

Anderson and Lamb assumed that the saturation effects in LAr could be described by the same formula that was proposed by Birks to describe saturation effects in scintillators (Equation 3.13). When they interpreted the observed effects in terms of Birks' constant (k_B), they found that adding a small fraction of methane raised k_B by a factor of four to five, not far from its asymptotic value (Figure 5.29b). The authors concluded that liquid argon with a small fraction of methane mixed into it apparently behaves very similar to pure methane, where the signals from αs and other densely ionizing particles are mostly due to the (lightly ionizing) δ-rays that are produced in the stopping process. The heavily ionizing signal component, due to direct ionization of the liquid, almost completely disappears as a result of recombination.

5.3.2.1 *Operational aspects.*   Because of the absence of internal amplification, which causes major stability problems for gases, liquid ionization detectors are in principle very stable. This stability may be jeopardized by the presence of electronegative impurities, predominantly oxygen, which may attach drifting electrons. To avoid this problem, purification of the liquid to extremely high levels is needed. For example, in liquid argon, the contamination of oxygen has to be kept at the level of 1 ppm or lower.

Electronegative impurities reduce the mean distance over which the electrons can drift in a given electric field before they are absorbed, or in other words the average *electron lifetime*. As we saw above, the typical drift time of electrons in LAr-based sampling calorimeters amounts to less than 1 μs (for a gap width of a few mm and a field strength of 1 kV cm^{-1}). The purity level needed to make electron absorption in such LAr systems insignificant can nowadays be routinely obtained with relatively simple purification systems.

More challenging are fully sensitive liquid-argon detectors, such as the ones used in the Fermilab neutrino experiments μBooNE (Figure 1.4) and DUNE, in which the electrons are required to drift over distances of several meters. However, even these problems are not insurmountable. The results of small scale prototype tests for the ICARUS detector [Cen 94, Alm 04], in which this technique was pioneered, indicated that an electron attenuation length of $\sim$ 7 m, at an electric field strength of 1 kV cm^{-1}, might be achievable [Apr 85]. Ereditato and co-workers demonstrated experimentally that electrons can drift over a distance of up to 5 m, at an electric field strength of 250 V cm^{-1}, provided that the oxygen content of the argon is kept sufficiently low, 0.15 ppb ($1.5 \cdot 10^{-10}$) in their tests. The lifetime of the electrons produced in the ionization processes was measured to be longer than 2 ms [Ere 13]. Figure 5.30 shows some examples of cosmic ray events recorded in the 5 m long tube filled with liquid argon that was used for these studies. Figure 5.30a depicts an interaction by a charged particle in which three charged particles are produced, plus possibly some neutral objects. Several δ-rays are visible along the track of the charged particle that reinteracts after about 2 m. Figure 5.30b shows a textbook example of an em shower.

The high stability that can be achieved in LAr systems is very attractive for calorimeter applications, since it greatly simplifies the calibration efforts. However, to take full advantage of this feature, it is important that very tight tolerances be maintained in the construction of the system. Since no internal charge amplification takes place, the

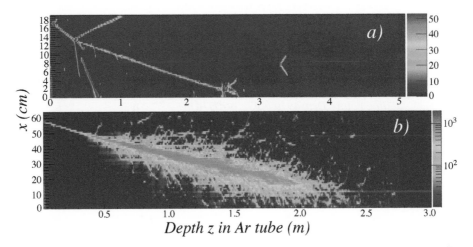

FIG. 5.30. Examples of cosmic ray events in the 5 m long liquid argon tube installed at the University of Bern. See text for details. From [Ere 13].

signals are directly proportional to the width of the liquid-argon filled gap. For example, if this gap is nominally 2.0 mm wide, the signals from particles crossing a gap with a width of 1.9 mm are 5% smaller than the average value. If one aims for high-precision calorimetry, variations in the gap width should thus be rigorously minimized.

Calorimeters based on liquified noble gases need to be operated at cryogenic temperatures ($87°$K for LAr, $120°$K for LKr, $165°$K for LXe). Therefore, such calorimeters are necessarily enclosed in cryostats. That includes the readout electronics, which makes a huge difference for the signal/noise ratio (Section 5.4.2). The cryogenic system required to operate such calorimeters is the source of many constraints imposed on other detectors in the experimental setup, especially in the 4π geometry required in a colliding-beam environment. Also, this cryogenic system usually implies major compromises with regard to the hermeticity of the calorimetric coverage in such experiments (see, for example, Figure 5.19).

In the 1980s, the desire to circumvent these problems triggered several R&D projects in which the feasibility to replace LAr by *warm liquids* was investigated. These liquids, *e.g.*, tetramethylsilane (TMS), tetramethylpentane (TMP) or tetramethylgermane (TMGe), would not need to operate at cryogenic temperatures.

5.3.2.2 *Warm liquids.* The possibility that ions may drift over long distances in ultrapure hydrocarbon liquids has been known since a long time. The first indication for this may be derived from J.J. Thomson's observation that the electric conductivity of vaseline oil increased under the action of X-rays [Tho 97]. In 1969, Schmidt and Allen studied the liquid tetramethylsilane (TMS, $C_4H_{12}Si$). They found a high free-ion yield and a high mobility of charge carriers [Schm 69]. The observation that the high mobility concerned not only the ions, but also the electrons, opened the possibility of operating ionization chambers at room temperature [Schm 77].

Table 5.2 *Properties of liquid dielectrics used as active material in sampling calorimeters. Listed are for each liquid the boiling point at atmospheric pressure, the density of the liquid, the specific energy loss by ionizing particles, the specific ionization yield (in the absence of an electric field) and the electron mobility. Data are given for the boiling point at atmospheric pressure. The $\langle dE/dx \rangle$ and n_T data concern minimum ionizing particles.*

Liquid	Boiling point (°C)	Density at b.p. (g cm^{-3})	$\langle dE/dx \rangle$ (MeV cm^{-1})	n_T (ions cm^{-1})	μ^{-} (cm^{2}V^{-1}s^{-1})
Argon	-186	1.40	2.13	$9.1 \cdot 10^4$	475
Krypton	-153	2.41	3.23	$16 \cdot 10^4$	
Xenon	-108	2.95	3.79	$24 \cdot 10^4$	2200
TMS	27	0.645	1.36	$0.86 \cdot 10^4$	100
TMP	123	0.72	1.58	$1.2 \cdot 10^4$	30
TMGe	43	1.006	1.78	$1.1 \cdot 10^4$	90
TMSn	78	1.314	2.09	$1.3 \cdot 10^4$	70

A number of different candidate liquids have been studied for this purpose [Hol 85]. Some of their relevant properties are listed in Table 5.2, together with those of the cryogenic liquified noble gases discussed before. The table shows that both the free-ion yield and the electron mobility of the warm liquids are smaller than for the noble liquids. On the other hand, a reduction of the "effective mobility" with increasing field strength, characteristic for the noble liquids (Figure 5.28), does *not* occur in the warm liquids. In other words, the drift velocity is much better described by Equation 5.3 for these liquid hydrocarbons than for their noble counterparts.

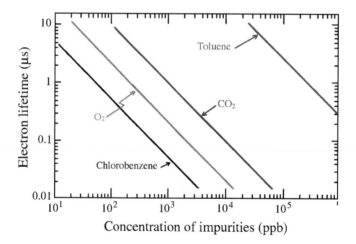

FIG. 5.31. Dependence of the electron lifetime in TMS on the concentration of various electronegative impurities [Rad 92].

As a result, above a certain field strength, the electron drift velocity in the hydrocarbons becomes even larger than in noble liquids. For example, TMS is faster than LAr for field strengths larger than 4 kV cm^{-1} [Rad 92]. The increase of the drift velocities that was observed when methane was added to liquid argon (Figure 5.28) is consistent with this tendency.

The main problem with the warm liquids turned out to be a technological one. The sensitivity of the electron lifetime in TMS to small concentrations of electronegative impurities is illustrated in Figure 5.31. In order to obtain the level of stability that makes liquid-argon calorimeters so attractive, this lifetime needs to be of the order of 100 μs. The figure shows that this requires the impurity levels to be controlled at the (sub)-ppb level, several orders of magnitude smaller than in LAr.

In small-scale test setups, this could eventually be achieved [Eng 86, Alb 88]. However, for some large-scale applications, such as the upgrade of the UA1 calorimeter system [Aps 91], the problems associated with this requirement were prohibitive. In addition, it turned out that the liquids considered here are quite susceptible to radiation damage [Hol 90]. As a consequence, the interest in these liquids has gradually diminished. The only experiment that successfully operates a large calorimeter system of this type is the KASCADE-Grande cosmic-ray experiment [KAS 16].

The central detector of this experiment, whose primary goal is to measure the chemical composition of the high-energy ($\sim$ 1 PeV) cosmic rays entering the Earth's atmosphere, is a 4,000 ton Fe/TMS calorimeter, with a surface area of 20×16 m^2 and a depth of about 10 λ_{int}. The active medium, TMS, is contained in 10,000 ionization chambers, each of which measures $50 \times 50 \times 1$ cm^3 and contains four electrodes of 25×25 cm^2 [Mie 95, Ple 06]. This fine segmentation makes it possible to recognize shower particles separated by as little as half a meter as separate entities. This detector has been in operation for about twenty years and has performed in general very satisfactorily.

For an extensive review of warm liquids and their application as the active media in sampling calorimeters, the reader is referred to [Rad 92].

5.3.3 Solid active media

Many of the problems associated with liquid or gaseous media sampling the ionization charge produced by shower particles may be avoided when a semiconductor such as high-resistivity silicon is used as active calorimeter material. The basic charge carriers in such a device are electron–hole pairs created along the path of the ionizing particle. This process requires only 3.6 eV per pair, which is about an order of magnitude less than the creation of an ion–electron pair in liquid or gaseous media. This property gives silicon sampling calorimeters obvious advantages over other unity-gain devices such as LAr calorimeters:

- For the same sampling fraction, the signals are an order of magnitude larger. The signal/noise ratio is correspondingly larger as well.

- Signals and signal/noise ratios comparable to the ones in LAr calorimeters may be achieved with much smaller sampling fractions. Silicon based sampling calorimeters can thus be considerably more compact.

- In such compact calorimeters, the charge collection also takes place in a much shorter time than in LAr ones, since the electrons have to drift only over very short distances to reach the electrodes. Typical depletion layers in Si junctions are 0.2 mm thick, compared with 2 mm gap widths in LAr calorimeters. In such thin layers, all the charge may be collected in 5–10 ns.

Other advantages of silicon calorimetry include:

- The drift velocity can be considerably higher than in LAr. The drift velocities of both the electrons and the holes increase with the applied electric field strength, to saturate at a value of ~ 100 km/s, which is an order of magnitude higher than in LAr.

- Saturation effects are virtually absent. The k_B value is, for all practical calorimetric purposes, zero in silicon.

- No cryogenic operation is required and no elaborate purification systems are needed.

However, these advantages come with a price tag, in the most literal sense of the word. Although the cost of high-resistivity silicon diodes has come down considerably in the past decades, it is still extremely expensive to equip large multi-ton calorimeter systems with this type of readout. Yet, the CALICE Collaboration has built a proto-type tungsten/silicon calorimeter with a large-scale application in mind [Rep 08]. More details about this project are given in Section 8.3.3.

Much of the credit for developing silicon sampling calorimetry goes to the SICAPO Collaboration, which built and tested several generic prototypes [Bor 89, Ang 90]. Until now, practical applications of Si-based calorimetry have been limited to relatively small devices. Several such detectors have served as luminosity monitors in the experiments in which they were used [Alm 91, Bed 95]. In one case, a detector of this type has served as payload in a balloon-borne cosmic-ray experiment [Boc 96].

Apart from the cost issue, the radiation sensitivity of silicon [Fre 96] is also con-sidered a disadvantage, especially for applications as the active medium inside a calo-rimeter structure where the fluences of neutrons (to which Si is particularly vulnerable) reach their maximum levels. This is discussed in more detail in Section 5.10.4.2. For a more detailed specific review of silicon calorimetry, the reader is referred to [Ler 11].

5.3.4   *Imaging calorimeters*

Measurements based on direct collection of the ionization charge created by charged particles have a long history in experimental physics. Many tracking systems in modern particle physics experiments are based on this technique. Especially in 4π detectors, with their inherent limitations and hermeticity requirements, the tracking of charged particles produced in the collisions has been developed into an art and many ingenious detectors have been developed for this purpose. Prominent among these is the Time Projection Chamber (TPC), invented by David Nygren [Nyg 81].

In a TPC, the ionization charge drifts over a long distance (order 1 m) in a very homogeneous electric field to a detector (usually a MWPC) that converts the charge in electric signals. This is illustrated in Figure 5.32, which shows a schematic of a TPC

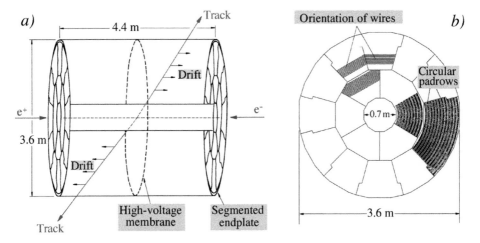

FIG. 5.32. Schematic of the TPC of the ALEPH experiment at LEP (*a*), and a detailed view of one of the end planes where the ionization charge is converted into electric signals (*b*) [Ame 86].

used in a colliding-beam experiment. The drift field is oriented along the beam line. The charge drifts towards two detector planes oriented perpendicular to the beam line.

With such a detector, it possible to reconstruct the origin of the detected charge in space. The sense wires and cathode strips of the detector plane (see Figure 5.32b) provide two coordinates, while the drift time provides the third one. Several experiments (*e.g.*, ALEPH [Ame 86], DELPHI [DEL 91], CDF [Sni 88] and ALICE [Aam 08]) have demonstrated in practice that a TPC may provide a highly detailed, three-dimensional image of the particles produced in interactions in a colliding-beam experiment.

One could use the same principle in calorimeters, and thus create an *imaging calorimeter*, which provides very detailed information about the production pattern of the ionization charge in the shower development process. Either gaseous or liquid active media may be used for this purpose. An example of a gas-based detector of this type was the High-density Projection Chamber (HPC), which served as the em calorimeter in the DELPHI experiment at LEP [Cat 85, Fis 88]. The electric field cage of this detector was formed from close-packed lead "wires," which also served as the calorimeter's absorber material. The ionization charge produced in the shower development drifted along the channels formed by the lead wire arrays and was collected in a MWPC at the end.

This configuration provided detailed information on the shower development. Along the drift direction, the position resolution was 2.4 mm. The resolution in the two perpendicular coordinates was determined by the size of the cathode pads, which varied from 2 to 6 cm in this instrument.

The TPC technique was also applied in ICARUS, a 600-ton homogeneous liquid-argon detector, which studied interactions induced by a beam of neutrinos produced at CERN and sent through the Earth to the detector, located in the Gran Sasso Laboratory in Italy, 730 km from the source of the particles [Alm 04]. In an instrument of this type, the detected ionization charge may be produced at any given point in the detector volume. Therefore, this device is a true "electronic bubble chamber" that produces three-dimensional images of shower development or other types of events that take place in it. Figure 5.33 shows an example of a neutrino interaction in the T600 ICARUS detector.

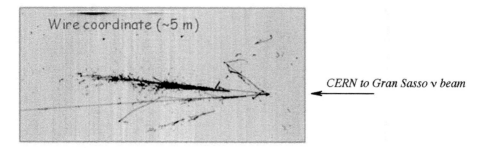

FIG. 5.33. Example of a neutrino interaction in the ICARUS T600 detector [Ant 14].

An important difference with the HPC described above is that no internal amplification takes place in the detection of the ionization charge. This has an advantage and a disadvantage. The disadvantage is that the signals are typically very small and require very sensitive, low-noise preamplifiers (e.g., J-FETs, see Section 5.4.2) to be detected. The advantage is that the drifting electrons induce signals on a number of different sense wires, and thus provide more information for the three-dimensional reconstruction of the events.

The ICARUS T600 detector operated from 2010 - 2014 in the Gran Sasso Lab. During the periods when there was no neutrino beam available, the detector was also used to study important physical parameters (e.g., free electron yield, recombination probability, drift velocity, electron lifetime) and their dependence on the strength of the electric field and the liquid purity in great detail. It achieved impressive performance. For example, the argon purification system managed to reduce the electronegative impurity level to 20 ppt ($2 \cdot 10^{-11}$), which led to a free electron lifetime of 15 ms, at the standard electric field strength of 500 V cm^{-1} [Ant 14]. This result means that the ionization charge can indeed drift over long distances in this medium, and that it is therefore realistic to envisage very large detectors of this type, with a mass in the multi-kiloton range.

Such multi-kiloton detectors are an important component of the new, upgraded neutrino program at Fermilab. One of these detectors will be ICARUS, which is refurbished at CERN before being shipped to the USA. Another detector of this type is μBooNE, which is already taking data in the Fermilab neutrino beam (Figure 1.4).

The ICARUS detectors also served to test a variety of solutions to the many technical problems that have to be solved. The realization of a TPC structure in a cryogenic environment is very challenging indeed. For example, thousands of signal cables have to be brought from inside the dewar to the outside world. Other feedthroughs are needed to supply the high voltage (150 kV in the case of ICARUS) needed for the drift field to electrodes inside the liquid. All these feedthroughs have to be completely reliable, none of them should leak, the materials used should not contaminate the liquid and, last but not least, the chosen solution has to be cheap. The latter requirement may eventually limit the size of the ultimate instruments of this type.

These liquid-argon TPCs are arguably among the most sophisticated and powerful detectors operating in a non-accelerator environment.

5.4 Front-End Signal Electronics

With increasing demands on the performance of calorimeters, the associated electronics has developed into the most advanced and complex analog signal processing circuitry in instrumentation for particle physics. This electronics has to process the signal charge delivered by a photomultiplier tube, a semiconductor based light detector, a wire chamber or an ionization chamber. Typically, it has to meet specifications of the following type:

- A *very large dynamic range*. For example, in the experiments at the Large Hadron Collider, this dynamic range should span the signals originating from minimum ionizing particles (muons, which typically loose 0.1–1 GeV in the individual calorimeter segments) to multi-TeV energy deposits by hadrons or relativistic ions. The required dynamic range is therefore typically 4–5 orders of magnitude, which translates into 15–16 bits of effective ADC resolution ($2^{16} = 64$K).

- It must be possible to perform *precision measurements of low-level calibration signals*, e.g., those generated by radioactive sources or cosmic muons.

- The signal electronics has to make it possible to extract adequate *timing information* on the events. This means that the electronics has to keep up with the rate at which events occur (*e.g.*, 40 MHz at the LHC). Also, the time available for providing information needed for first-level trigger decisions (see Section 5.5) is limited.

- The signal electronics has to be organized in a system of a large number (10^4–10^5) of independent channels.

In addition, other, detector-specific requirements may apply. In the following, some examples are given of systems that have been developed to cope with these challenges.

5.4.1 *Light-based calorimeters*

5.4.1.1 *Photomultiplier tubes.* The oldest and still frequently used light detector in light-based calorimeters is the PMT. The sensitivity to single photons and the essentially noise-free amplification offered by these devices are very attractive features. A PMT consists of a photocathode, in which the photons convert into electrons through the photoelectric effect, followed by an electron collection system, an electron multiplier section and an anode from which the final signal can be taken. For an excellent introduction into the operation of PMTs and their use in connection with scintillation counters, the reader is referred to [Leo 87].

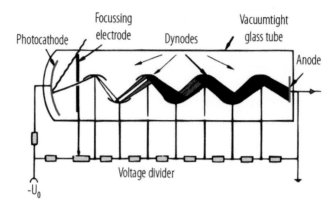

FIG. 5.34. Schematic diagram of a photomultiplier tube.

In conventional PMTs, the electron amplification takes place in a string of dynodes. In recent years, alternative devices in which an array of very narrow capillaries, a so-called "micro-channel plate" (MCP), is used for this purpose have become available. in Figure 5.34, a schematic diagram of a dynode based PMT is shown . All elements are located in an evacuated glass envelope. At the photocathode an electron is extracted through the photoelectric effect. A voltage difference accelerates this electron towards the first dynode, out of which several electrons are extracted by secondary emission. This process is repeated several (typically ten) times, until the electrons reach the anode. The voltage difference between the anode and cathode is typically 1 - 2 kV. With a large enough gain in the first acceleration step, the fluctuations in the number of electrons in the final charge pulse are dominated by Poisson fluctuations in the number of photoelectrons. With bi-alkali photocathodes (Cs-K), the efficiency of the photoelectric process can reach 25%, for photons in the 450 nm wavelength region. For short wavelengths, this *quantum efficiency* is determined by the transparency of the entrance window. Quartz, CaF_2 or LiF windows are necessary for detecting near-UV light.

PMTs have exponential gain characteristics. At each dynode of the multiplier section the number of electrons is multiplied by a factor δ, so that the total amplification of the initial photoelectron reaches a value δ^n after n amplification stages. The secondary emission factor (δ) depends on the potential difference between the dynodes, so that the PMT gain depends on this potential difference to the nth power. Therefore, the gain is extremely sensitive to small variations in the applied voltage. For example, if one wants to maintain the gain of a 10-stage tube constant to within 1%, then the voltage supply must be stable to within 0.1%.

The useful dynamic range of PMTs (*i.e.*, the energy range over which the PMT is sufficiently linear) is limited on the low-energy side by the requirement that a signal should consist of at least one photoelectron. On the high-energy side, it is limited by space charge saturation. The emitting electrodes are surrounded by a cloud of electrons, which reduces the effective electric field in the region and thus prevents the acceleration of emitted electrons towards the next dynode. When the potential differences between

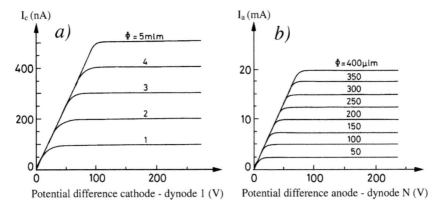

FIG. 5.35. Current–voltage characteristics of a PMT cathode (*a*) and anode (*b*), for different intensities of the incident light [Scho 70].

the electrodes are increased, this space charge is swept away and all of the emitted current may be collected. This is crucial, since PMT linearity requires that the current at each amplification stage be completely collected by the next receiving electrode.

Figure 5.35 shows the typical dependence of the cathode (a) and anode (b) currents on the voltage difference with the neighboring dynode, for various light intensity levels at the photocathode. Below a certain potential difference, space charge limits these currents, but if the potential difference is increased all the emitted current is collected. To ensure linearity, the cathode, dynode and anode currents should thus all be in the flat portion of these characteristic $\Delta V - I$ curves. Figure 5.35 shows that this requires larger and larger voltage differences as the light intensity (*i.e.*, the energy of the showering particle) increases.

Maintaining the potential differences between the various electrodes at levels required for linearity also becomes more of a problem as the energy increases, since the current in the tube increases as well. Especially in the last amplification stages, this effect tends to lower the potential differences between the electrodes. And even though clever tricks may be applied to extend the high-energy reach of the PMT as much as possible, eventually the linearity breaks down. In practice, the dynamic range of PMTs typically comprises four orders of magnitude.

The most common way to deliver well regulated voltages to the electrodes of a PMT is to use a stabilized high-voltage supply in conjunction with a voltage divider. An example of such a system, for a 12-stage tube, is shown in Figure 5.36. It consists of a chain of resistors which are chosen to provide the desired potential values to the various electrodes. Since the last few amplification stages are the most critical ones for the linearity, the potential differences in these stages are usually made larger than those in the earlier stages. In this way, the operating point on the $\Delta V-I$ curve (Figure 5.35b) is shifted to the right, thus increasing the safety margin.

In order to limit variations in the dynode potentials as a result of changing currents in the tube, it is important that the current in the resistor chain (the "bleeder" current) be

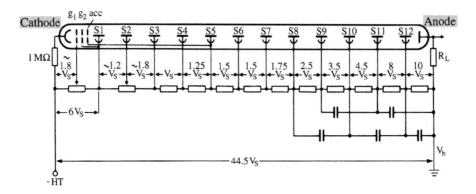

FIG. 5.36. Example of a voltage divider network for a 12-stage PMT. This network is optimized for signal linearity [PDH 90].

large compared with the tube current. This is achieved by properly choosing the values of the load resistor (R_L) and the other resistors in the voltage divider chain.

To avoid sudden potential drops during peak currents, the last stages in this scheme are maintained at a fixed potential by means of decoupling capacitors, which can provide the necessary charge during a short time. The same effect could be achieved by replacing the resistors with Zener diodes in these last stages. In very-high-current situations, it may be necessary to maintain the voltage differences between the electrodes in the last amplification stage(s) with separate power supplies. This technique was used to restore linearity in the example shown in Figure 3.28.

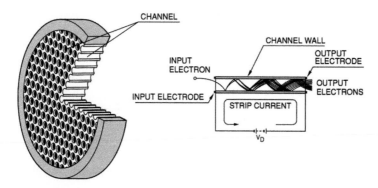

FIG. 5.37. Schematic structure of an MCP-PMT and its principle of electron multiplication.

Figure 5.37 shows the schematic structure of an MCP-PMT, and its operating principle. The multi-channel plate consists of a two-dimensional array of a great number of glass capillaries (channels) bundled in parallel and formed into the shape of a thin disk. Each channel has an internal diameter ranging from 6 to 20 μm, with the inner wall processed to have the proper electrical resistance and secondary emissive properties. Each channel thus acts as an independent electron multiplier. The photoelectrons are produced in the same way as in a conventional PMT, *i.e.*, by means of photoelectric effect in a photocathode. When such a primary photoelectron impinges on the inner wall of a channel, secondary electrons are emitted. Accelerated by the electric field created by the voltage V_D applied across both ends of the MCP, these secondary electrons bombard the channel wall again to produce additional secondary electrons. This process is repeated many times along the channel and as a result, a large number of electrons are released from the output end.

The distance between the photocathode and the MCP is much shorter than the distance to the first dynode in a conventional PMT. Moreover, the voltage difference is larger and, therefore, MCP-PMTs are much less sensitive to magnetic fields, a notorious weak point of conventional PMTs. Also, the signals are considerably faster because of the much shorter transit time, and variations therein, as the charge travels from the photocathode to the anode. Full pulse widths of less than 1 ns and rise times of less than 0.5 ns are common for commercially available devices. Unfortunately, these excellent and desirable characteristics come (literally) at a high price, typically at least five times that of a conventional PMT with the same photocathode surface. However, if the calorimeter signals are intrinsically fast enough, as in the example shown in Figure 7.28, the fast response time offered by MCP-PMTs may well be worth the cost.

An interesting alternative to provide high voltage to PMTs is to equip them with a *Cockroft–Walton base*, in which the required voltage is generated by induction. This idea was first proposed by Lu, Mo and Nunamaker [Lub 92] and has been applied in the ZEUS calorimeter. A 60 kHz oscillator is powered by a 24 V dc power supply. The pulses are stepped up to ~ 100 V by a ferrite transformer. This rf power drives a Cockroft–Walton accelerator, which consists of a chain of rectifiers and capacitors and produces the desired high voltage locally on each individual PMT base. This scheme has the following advantages:

- There is no need for large numbers of high-voltage cables running through the detector from the power supplies to the individual PMTs. Such cables form a safety hazard and occupy premium space.

- The heat production is considerably lower than in conventional, resistive bases. This saves power and alleviates cooling problems.

- Gain stability is much easier to maintain under high counting rate conditions.

On the other hand, detectors using such PMT bases are more sensitive to noise pickup. They are also less flexible for fine-tuning the delivered voltages.

5.4.1.2 *Hybrid Photon Detectors.*

A very large dynamic range is one of the attractive features of Hybrid Photon Detectors (also called Hybrid Photomultiplier Tubes), which have been proposed as a feasible alternative to PMTs in high-energy particle physics experiments [Arn 94]. The HPD tube is conceptually a very simple device. It consists of a planar silicon diode that faces a photocathode. The photoelectrons are accelerated by a potential difference that is typically of the order of 10 kV. They generate electron–hole pairs in the bulk of the reversely biased diode. Since it takes only 3.6 eV to create one electron–hole pair in silicon, gain factors of the order of several thousand can be achieved in this way.

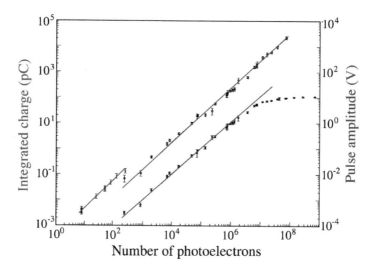

FIG. 5.38. Linearity measurement of the integrated charge (left scale, the two leftmost curves) and the signal amplitude (right scale, right curve) for a Hybrid Photon Detector. The two charge distributions do not line up because of differences in the measurement setup and in the calibration for these two sets of measurements [Arn 94].

Figure 5.38 shows the results of linearity measurements with such a device. The figure displays both the integrated charge and the amplitude of the signals as a function of the intensity of the light pulses sent onto the photocathode surface, measured for an accelerating voltage of 7 kV and 95 V diode bias. The pulse amplitude, represented by the right-hand scale and the rightmost set of data points, saturated at values of 5–10 V. However, the integrated charge turned out to be proportional to the light intensity over the full seven orders of magnitude for which the measurements were performed. The measurements of very small light pulses were carried out with the help of a charge amplifier and were normalized differently, so that the two (leftmost) distributions representing the results of the charge measurement do not exactly line up in Figure 5.38.

The excellent linearity and the large dynamic range of HPDs are a consequence of (a) the single-step amplification process and (b) the absence of saturation in silicon. Other advantages of the HPD include the gain stability (because of the linear voltage–gain relationship), the possibility to introduce position sensitivity by pixelizing the anode, the small power consumption (no need for a bleeder current here) and the possibility of operating it in a magnetic field, especially if this is oriented axially ($\vec{B} \parallel \vec{E}$, see Section 5.8). HPDs also distinguish themselves from conventional PMTs by their excellent photon counting characteristics. Figure 5.39 shows that events producing a different number of photoelectrons are easily resolved. The resolution of a conventional PMT is much worse due to low statistics in the electron multiplication at the first dynode and

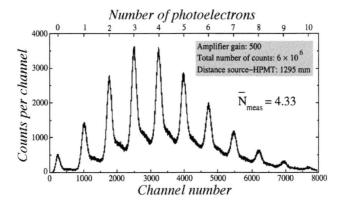

FIG. 5.39. Photoelectron distributions measured with a Hybrid Photon Detector [Ambr 03].

gain fluctuations in the dynode chain [Ambr 03].

The only major experiment that has chosen HPDs as light detectors for a calorimeter is CMS, which uses them in their hadronic calorimeter. The capability to operate these devices in a magnetic field was an important consideration for this choice. They are also used in some other detectors, such as the Ring Imaging Čerenkov detector of the LHCb experiment at the LHC [Eis 14], and in X-ray cameras.

One important drawback of this novel technology is the so-called Ion Feedback problem (Figure 5.40), which is a result of a gradual degradation of the vacuum inside the tube. This vacuum has to be much better than in a conventional PMT, because of the risk that gas trapped inside is ionized by the very energetic photoelectrons. The ions will then be accelerated back to the photocathode, where they may release many additional electrons. The result is a nasty big cluster on the silicon diode. This problem is the main

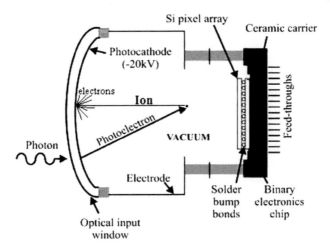

FIG. 5.40. The Ion Feedback mechanism in a Hybrid Photon Detector [Eis 14].

reason why CMS has decided to change the readout of its hadronic calorimeter, and switch to silicon photomultipliers during the next major shutdown of the LHC.

5.4.1.3   *Semiconductor based light detectors.*   Starting in the 1980s, semiconductor based light detectors have become a very popular alternative to the photocathode based devices discussed above. Almost all of these use silicon as the active material, although devices based on other materials (Ge, InGaAs, HgCdTe) are available as well. However, silicon has the largest sensitivity to photons in the visible domain, and is also the cheapest, hence its appeal for applications in calorimetry. These applications became possible thanks to several technological breakthroughs, in particular the use of high ohmic silicon and ion implantation, which allowed the fabrication of very shallow p^+ layers with low series resistance and low leakage currents. Apart from their high quantum efficiency in the relevant wavelength domain, the main advantages of silicon (PIN) photodiodes over PMTs are their compactness, complete insensitivity to magnetic fields, high stability due to unity gain, and mass production potential. In addition, high voltage supplies are not needed to operate them either.

A disadvantage of the absence of internal charge amplification is that the signals need to be amplified, which is exacerbated by the fact that the sensors themselves are also noisy to begin with. Therefore, even though PIN photodiodes are inherently very fast, one needs special filter amplifiers with rather long time constants to reduce the noise to acceptable levels. Yet, for calorimeters based on crystals such as NaI(Tl), CsI(Tl), BaF$_2$ and BGO, which scintillate with time constants of the order of $1\mu s$, this is not a show stopper, and many experiments have used and are still using PIN photodiodes to read out their crystal calorimeters (see Section 5.2.1). PIN photodiodes are also notoriously sensitive to charged particles, a mip may produce a signal equivalent to that 10,000 scintillation photons when it traverses the silicon surface layer. For bright scintillating crystals, such as the ones mentioned above, which produce 10,000 photons or more for every MeV deposited energy, this is not much of a problem. However, for crystals such as PbWO$_4$ which generate three orders of magnitude less light, the passage of a charged particle through the sensor leads to an anomalously large signal when read out with a PIN diode (Figure 3.32).

In Section 5.2.1, we saw that this problem was alleviated when the PIN photodiode was replaced by an Avalanche Photo Diode, which provides an internal signal amplification of the order of 10 to 100. (Figure 5.20). The internal signal amplification, combined with the fact that the active silicon layer, where the detection takes place, is much thinner than in a PIN photodiode, reduces the sensitivity to charged particles substantially, albeit not sufficiently in the case of CMS. The experience of CMS has also shown that the APD performance is extremely sensitive to changes in temperature and in bias voltage (3.1%/V at gain 50) . Thanks to sophisticated temperature control and voltage stability systems, a gain stability at the 0.1% level was achieved [Bar 07].

APDs are successfully applied in several other calorimeters, especially those using PbWO$_4$ crystals (*e.g.*, the ALICE Photon Spectrometer [Aam 08] and PANDA [Kav 11]). Because of the internal gain, the external signal amplification is much less critical than in the case of PIN photodiodes. Despite the excess noise factor typical for the APDs themselves, it is possible to extract much faster signals from the crystals than in case of PIN-readout [Lor 94]. For fast crystals such as PbWO$_4$, this is an important advantage. Avalanche Photon Diodes are also the sensors of choice for the liquid-scintillator based NOVA calorimeters in the Fermilab neutrino beam line [Muf 15].

The concept of APDs has been extended by dividing the silicon surface into very small pixels and operating these in the Geiger mode. Each pixel is thus a binary photon detector, it produces the same signal regardless of the number of photons that hit it simultaneously. However, by making the pixels sufficiently small and packing a large number of them on a small surface area, it becomes possible to obtain approximately the equivalent of an analogue response to a given light pulse, by summing the signals from all pixels. The number of pixels that fired represents the number of photons that constitute the light pulse. These devices have generally become known as silicon photomultipliers (SiPM), although many other names are used as well in the literature: Multi Photon Pixel Counters (MPPC), Digital Pixel Photo Diode (DPPD), Microchannel Avalanche Photo Diode (MAPD), to name just a few.

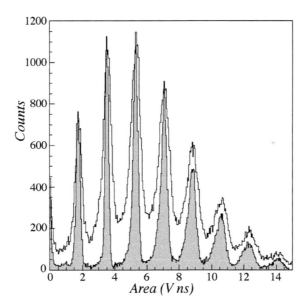

FIG. 5.41. The effect of cooling a SiPM from 25°C (unshaded) to −196°C (shaded) on the single photoelectron peak distribution [Lig 08].

Figure 1.12 shows a picture of a light detector of this type. The improvement in the quality during the past 10 years has been very impressive. Commercially available SiPMs[14] nowadays pack up to 10,000 pixels on a surface area of 1 mm^2. Figure 5.41 shows that SiPMs are excellent single-photon counters. When smaller and smaller pix-

---

[14] http://www.hamamatsu.com/jp/en/4113.html#mppc_pix=2

els are being used, signal linearity and saturation are also less of a concern, since the probability that several photons hit the same pixel is proportional to the surface area of one pixel. On the other hand, since the individual pixels have to be electrically insulated from their neighbors, the effective surface area and thus the efficiency of the SiPM decreases when the pixel density is increased. Figure 5.41 also illustrates that the noise characteristics, which for a long time have been a point of concern, can be improved by cooling the detector [Lig 08].

It is likely that more and more experiments with light-based calorimeters will choose semiconductor based sensors, especially if the time structure of the signals is not an important consideration. Apart from the CMS hadron calorimeter, which will make the change to SiPM readout, these sensors have also been selected for the near detector of the T2K Japanese long-baseline neutrino experiment [Min 07]. And several prototype calorimeters for the experimental program at the proposed International Linear Collider are also equipped with SiPMs (Section 8.3.3)

5.4.1.4 *Signal digitization.* So far, we have limited the discussion to the properties of the light-detecting elements. However, the front-end electronics comprises more than just the generation of electric signals. The signals produced by these detectors need to be digitized. This has to be done fast, within time constraints set by the rates of the interactions in the experiment, and with a resolution commensurate with the dynamic-range requirements, sometimes 15–16 bits. This is by no means a trivial task.

An ingenious solution to meet the dynamic-range and signal-speed requirements in an affordable way was developed at Fermilab, where it was applied in different variations in the readout systems of the calorimeters of the KTeV and CDF experiments. Another variant of this system is applied for reading out some of the calorimeters of the CMS experiment at the Large Hadron Collider.

The front-end electronics of these experiments is based on a pipelined multi-range charge integrator and encoder (QIE) ASIC, with which the detector signals are digitized. The QIE contains five major components: a current splitter, a gated integrator/switch, a comparator, an encoder, and an analog multiplexer [Yar 93].

Since the signals may last longer than the time between events, there may not be enough time to integrate, read out and reset the integrating capacitor before the arrival of the next event. Therefore, the signal processing in the chip is done in a four-stage pipeline (Figure 5.42)

In the first stage, the QIE acquires the signal. It integrates a current pulse (I) from the detector by splitting it into eight binary-weighted current outputs, or "ranges:" $I/2$, $I/4,...I/128$, $I/256$. Each of these signals is applied through a current switch to eight identical integrating capacitors, for example for a period of 132 ns (CDF).

In the second stage, the voltages generated over the eight capacitors are applied to a set of comparators, to determine which of them should be sent to the ADC for digitization. The selected range is encoded as a three-bit number, which is sent to the output, to be used as the exponent of a floating point number. The ADC output forms the mantissa of this number.

In the third stage, the comparator outputs are used to control a multiplexer that

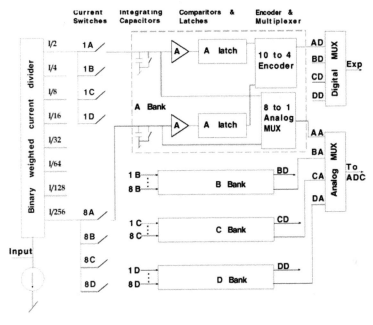

FIG. 5.42. Schematic diagram of the QIE chip used in the front-end electronics of several scintillator-based calorimeter systems. From [Yar 93].

selects one of the capacitor voltages for digitization by the ADC. In the fourth stage, the capacitors are reset.

In this way, an 11-bit ADC (3 bits for the range plus 8 bits for the digitized voltage) delivered information with 16-bit accuracy, and it did so in only 132 ns, the bunch spacing of Fermilab's Tevatron in the Main Injector era.

For operations at the LHC, where the proton bunches pass the interaction area at a rate of 40 MHz, CMS has developed a similar system that can perform the digitization of the calorimeter signals every 25 ns [CMS 08].

5.4.2 *Ionization-chamber calorimeters*

The front-end electronics needed for reading out the signals from liquid-argon and other calorimeters operating in the ionization-chamber mode faces a completely different set of problems than in the case of scintillator-based calorimeter systems.

The signals from ionization-chamber calorimeters are typically considerably slower and also somewhat smaller than the ones discussed above. In Section 5.3.2, we saw that it takes electrons typically of the order of 1 μs to drift to the electrode where they are collected, while the signals from PMTs and other light detectors are typically one to two orders of magnitude faster (see also Section 5.6). It is possible to generate faster signals from LAr calorimeters, but in that case only a fraction of the ionization charge contributes to the signals, which become thus correspondingly smaller.

A typical scintillator-based calorimeter such as the one used in the ZEUS experiment [And 91] produced ~ 100 photoelectrons per GeV of deposited energy. In PMTs with a gain of $\sim 10^5$, these signals were transformed into charge pulses containing $\sim 10^7$ electrons. Taking 70,000 electron–ion pairs produced in 1 cm of liquid argon as a starting point (Section 5.3.2), we find that the signals from 1 GeV showers in a LAr calorimeter with a sampling fraction of 10% are comprised of $\sim 1.6 \cdot 10^6$ electron charges (0.27 pC GeV^{-1}, see also Section 5.6.2.1).

Although the charge produced by ionization-chamber calorimeters may seem in itself comfortably large, reading out the signals from these devices presents some delicate problems, which are not trivial to solve. Large sampling calorimeters consist of a large number of absorber plates, which represent a substantial capacitance. Typical values for the detector capacitance, C_D, range from 1 nF to 100 nF. Charge measurement is performed by observing a fraction of the total ionization charge, which is diverted onto the charge-measuring amplifier. The best low-noise charge amplifiers available are based on junction field-effect transistors (J-FETs) and have an input capacitance, C_A, of ~ 10 pF. If connected directly to the ionization chamber, such a preamplifier would "see" only a fraction of the produced signal charge: C_A/C_D, i.e., only one part in 10,000 for a calorimeter with a capacitance of 100 nF. This is an obvious case of impedance mismatching between the detector and the charge amplifier, which will lead to a very bad signal-to-noise ratio.

A much better charge sharing (and thus a higher signal-to-noise ratio) can be achieved by matching the calorimeter and the charge amplifier by means of a transformer. This is conceptually shown in Figure 5.43a [Wil 74].

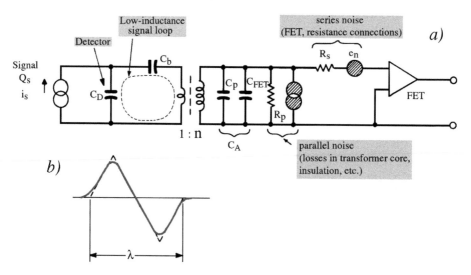

FIG. 5.43. Transformer matching of a high-capacitance ionization-chamber calorimeter to the low-capacitance input of a (FET) charge amplifier (a). The output signal (b) is obtained after triangular (bipolar) shaping of the amplified charge pulses [Wil 74].

The noise level of this type of calorimeter is usually expressed in terms of an "equivalent noise charge" (ENC), that is the energy equivalent of a calorimeter signal that is equally large as the noise. For example, the detector mentioned above, which produced signals of 0.27 pC GeV^{-1}, may have an ENC of $8 \cdot 10^4$ (rms) electrons. This means that the electronic noise translated in terms of energy corresponds to 50 MeV (rms).

Radeka has shown that the noise for non-optimal transformer ratios n scales as [Wil 74]

$$\frac{\text{ENC}}{\text{ENC}_{\text{opt}}} = \frac{1}{2}\left(\frac{n_{\text{opt}}}{n} + \frac{n}{n_{\text{opt}}}\right) \tag{5.4}$$

while the optimal transformer ratio n_{opt} for circuits of the type shown in Figure 5.43a is related to the capacitances as

$$n_{\text{opt}} = \sqrt{\frac{C_{\text{D}}}{C_{\text{A}}[1 + (C_{\text{D}}/C_{\text{b}})]}} \approx \sqrt{C_{\text{D}}/C_{\text{A}}} \tag{5.5}$$

or rather somewhat smaller, since the optimal value for the decoupling capacitor C_{b} is typically about half the detector capacitance.

Therefore, if a detector with a capacitance of 60 nF is coupled to a charge amplifier with a 30 pF input capacitance, optimal noise conditions require a transformer with a ratio of $n \sim 30$. Such a transformer would improve the signal-to-noise ratio by a factor of ~ 15, compared with the case in which no transformer was used.

However, in order to achieve optimal noise conditions, it is essential that the charge be transferred from the detector to the amplifier within a time, t_{c}, that is short compared with the resolving time, λ. This resolving time is determined by the (bipolar) pulse shaping that follows the charge amplification stage (Figure 5.43b), which in turn is chosen commensurate with the fraction of the ionization charge one wants to collect.

This charge transfer time, and hence the detector speed, is also limited by the detector capacitance and by the inductance, L_{C}, of the input signal loop (Figure 5.43a). This loop includes the chamber–plate connections, the transmission line from the chamber to the charge amplifier, the decoupling capacitance C_{b} and the leakage inductance of the transformer. Radeka and Rescia have shown [Rad 88b] that L_{C}, C_{D} and the input impedance of the charge amplifier, R, form a series circuit that is critically damped when:

$$t_{\text{c}} \simeq 4\sqrt{L_{\text{C}}C_{\text{D}}} \simeq 2\sqrt{RC_{\text{D}}} \tag{5.6}$$

Fast ionization-chamber calorimetry thus requires a small detector capacitance and a minimal distance between the liquid gaps and the front-end electronics.

While the use of transformer matching results in a smaller effective detector capacitance and thus in an improved noise performance, it has two practical drawbacks:

1. Transformers are quite bulky, which means that they may jeopardize hermeticity if they are to be installed close to the liquid gaps.
2. Transformers use ferrite cores, and therefore do not work properly in magnetic fields in excess of 10 mT without external shielding.

For the latter reason, the transformer matching technique is in practice not used in experiments where the calorimeter operates in a magnetic field.

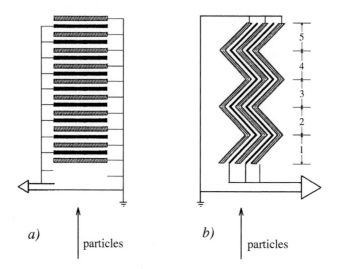

a)   particles   b)   particles

FIG. 5.44. The schematic readout structure of an accordion calorimeter (b), compared to the classical planar structure (a).

An alternative method to reduce the charge transfer time is to decrease the input impedance of the charge amplifiers, by connecting a number of FETs in parallel. This method was used in the D0 uranium/liquid-argon calorimeter and also in the UA1/TMP project [Bac 89].

The capacitance of the detector itself may also be reduced by connecting some of the electrodes in series. This is known as the "Electrostatic Transformer" technique (EST, [Col 90]). This technique is applied in the ATLAS electromagnetic calorimeter, where the traditional planar electrode/absorber configuration is replaced by an "accordion" structure (Figure 5.44b). In this way, the electrodes (pads) that collect the charge from different depth regions inside one calorimeter tower are automatically connected in series, and an EST readout is thus straightforward to implement. The inductance is minimized by mounting the charge preamplifiers on the front and back faces of these accordion towers.

The reduction in capacitance that can be achieved with this technique is considerable. Let us, for example, assume that the capacitance of one readout cell is 200 pF. If 50 such cells are connected in parallel, to form a tower in a planar structure (Figure 5.44a), the total detector capacitance is 10 nF. However, if five such pads are connected in series to form a lateral subtower (with a capacitance of 200/5 = 40 pF) and ten such subtowers are connected in parallel to form one tower (Figure 5.44b), then the total detector capacitance is $10 \times 40 = 400$ pF, a reduction by a factor of 25. Such a readout structure could thus make the calorimeter up to a factor of five faster.

This technique has made it possible to reduce the charge transfer time, and thus the pulse shaping time, to values considerable shorter than the drift time of electrons across the liquid-argon gaps. For example, in the ATLAS calorimeter, a shaping time of 45 ns is used.

5.5 Trigger Processors

The experimental information from calorimeters (energy, position, particle type) is usually available for a very short time, typically $\sim$ 100 ns, after the particle impact. This important feature is extensively used to select or *trigger* on interesting events, based on the characteristics derived from the calorimeter data.

The required trigger selectivity depends on the type of collisions under study. For example, at high-energy electron–positron colliders such as LEP or SLC, the collision rates were very low, at maximum a few Hz. The role of the trigger processor was mainly to distinguish genuine e^+e^- collisions from other types of events, such as beam–gas interactions or cosmic rays.

On the other hand, in modern hadron colliders such as the Tevatron or the LHC, collision rates were/are at the level of 10^7–10^9 Hz. The role of the trigger processors in experiments at such machines is primarily to determine whether individual events are sufficiently interesting to be recorded. Typically, the rate of retained events is of the order of 10–100 Hz, so that the trigger *selectivity* may have to reach levels of the order of $1 : 10^7$.

In the following, we discuss one particular trigger processor in some detail. This processor was developed for the HELIOS experiment, a heavy-ion fixed-target experiment carried out at CERN in the late 1980s. The principles on which the techniques and methods that were used in that experiment were based are still used today in many other experiments. However, thanks to the enormous development in fast electronics and computing (*e.g.*, the availability of Field Programmable Gate Arrays), the trigger menus in modern experiments have become much more elaborate and sophisticated.

5.5.1 *HELIOS' Energy Flow Logic*

The HELIOS experimental setup included several calorimeter systems. A schematic overview of these systems is shown in Figure 5.45 [Ake 87]. The target was surrounded by an almost hermetic "box" of calorimeter modules, covering opening angles up to 96°. In the forward direction, two regions were left open. The first was a 6.3° conical hole centered around the beam axis. When the projectiles interacted in the target, most of the available energy was carried away by particles passing through this hole. This energy was measured in the forward calorimeters, which covered an angular area somewhat larger than the 6.3° cone.

The second open region in the otherwise hermetic calorimeter box consisted of a slit in the horizontal beam plane. The properties of individual particles traversing this slit were measured by a separate spectrometer installed behind it.

All calorimeter towers consisted of an em and a hadronic section, which were read out by PMTs. The signals from the total of 2860 PMTs were used for several purposes, and they were split into two parts to that end. One part, containing two-thirds of the

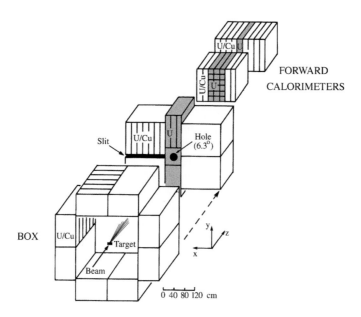

FIG. 5.45. The HELIOS calorimeter layout. See text for details [Ake 87].

total charge, was sent to a charge ADC (CADC), which digitized this signal and determined the total integrated charge. This signal was used for the offline data analyses. The other part, which contained the remaining one-third of the charge, was diverted into the *Energy Flow Logic* (EFL) and was used online for triggering purposes.

The goal of the EFL was to provide a trigger on the total transverse energy (E_T), the total missing energy ($E_{miss} = E_{tot} - E_{beam}$) and the total missing transverse momentum of the events ($\not{p}_T$). This was accomplished by fast analog summing of the physical quantities E_{tot}, E_T and $\not{p}_T^2$, which were obtained from the PMT signals from the various calorimeters in the experiment. These analog sums were digitized with Flash ADCs (FADC) and compared with preset values, corresponding to the trigger thresholds for the physical quantities of interest. If these thresholds were exceeded, the trigger processor generated a signal to record the event in question.

The individual physical quantities were derived from the PMT signals as follows. For E_{tot}, the pulse heights from all individual calorimeter cells were simply added. The missing energy can be derived from E_{tot} by subtracting a fixed number (E_{beam}).

For the quantity $\not{p}_T$, four sums were formed, labelled P_x^+, P_x^-, P_y^+ and P_y^-, where the +/– sign referred to the position of the calorimeter cell with respect to the beam line, in the x, y coordinate system (see Figure 5.45). Each calorimeter cell contributed thus to two of these sums, depending on its position with respect to the beam line. These contributions were weighted by factors of $\sin \theta_i \times \cos \phi_i$ and $\sin \theta_i \times \sin \phi_i$, for the x and y components, respectively, where θ_i and ϕ_i denote the polar and azimuthal angles of tower i.

Digital arithmetic was used to form the quantity p_T^2 from the digitized signals corresponding to these four sums:

$$p_T^2 = (P_x^+ - P_x^-)^2 + (P_y^+ - P_y^-)^2 \tag{5.7}$$

For the quantity E_T, the pulse heights from all calorimeter cells i were added together using weighting factors proportional to $\sin \theta_i$, thus accounting for the geometric position of the calorimeter elements:

$$E_T = \sum_{i=1}^{2860} E_i \,|\sin \theta_i| \tag{5.8}$$

The mentioned weighting factors were applied in hardware, by sending the signals through precision resistors with values inversely proportional to these weighting factors. In other words, to obtain a small weighting factor, the PMT signal was sent through a large resistor, which resulted in a small current.

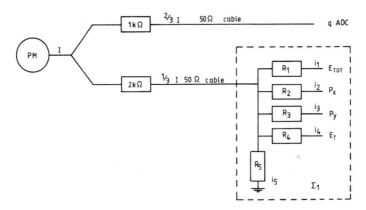

FIG. 5.46. The splitting of the HELIOS PMT signals and the input stage to the summing circuits. See text for details [Ake 87].

This is illustrated in Figure 5.46. First, the PMT signal was split into two signals, representing two-thirds and one-third of the total signal. This was done by means of precision resistors of 1 kΩ and 2 kΩ, respectively. The signal representing one third of the total was further split into four branches, using resistors with values R_1, R_2, R_3 and R_4 inversely proportional to the geometric weighting factors mentioned above. A fifth resistor R_5 was used to match the input impedance of the summing circuit. The currents I_1, I_2, I_3 and I_4 represented the properly weighted contributions of the signals recorded in this particular calorimeter element to the sums E_{tot}, P_x, P_y and E_T, respectively. The rest of the logic consisted of summing these weighted contributions with those from all other calorimeter cells.

The performance of this trigger logic was measured by comparing online values for the mentioned physical quantities, obtained as described above, with the reconstructed offline values derived from analysis of the detailed charge ADC information.

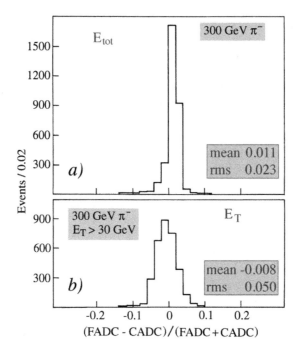

FIG. 5.47. Comparison between flash ADC and charge ADC values for 300 GeV pions incident on a lead target, for the quantities E_{tot} (a) and E_T (b) [Ake 87].

A useful parameter in this comparison is the *asymmetry* between the flash ADC trigger signals and the corresponding charge ADC sum: (FADC-CADC)/(FADC+CADC). Two such asymmetries are shown in Figure 5.47, for a 300 GeV π^- beam incident on a thin lead target. For the data represented in this figure, the threshold on E_T was set at 30 GeV. The figure shows that this asymmetry is centered near zero, which means that the energy scale used at the trigger level was in good agreement with the offline scale. The width of these distributions was determined by fluctuations, which in this case mainly originated from uranium noise.

In practice, the E_T distribution falls very steeply as a function of energy in this type of experiment. The sharpness of the threshold in E_T is therefore of great importance for the physics analysis. A diffuse threshold is likely to result in an event sample that predominantly consists of events that actually did *not* meet the trigger condition. In Figure 5.48, the FADC trigger signal and the corresponding charge ADC sum are shown together for triggered events with $E_T > 30$ GeV recorded in the interactions of 300 GeV π^- mesons with lead. Detailed analysis of the results from this comparison showed that the same fluctuations that were responsible for the width of the asymmetry distribution (Figure 5.47) also caused an inefficiency near the 30 GeV trigger threshold.

In modern experiments, such as those at CERN's Large Hadron Collider, the concept described above has been further developed and become an integral part of the control center of the experiment. This control center is the "trigger menu," which forms the basis

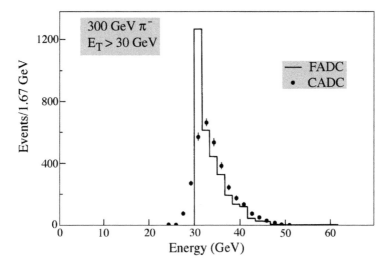

FIG. 5.48. Distributions of the flash ADC trigger signals and of the corresponding charge ADC sums, for a 300 GeV pion beam and a threshold $E_T > 30$ GeV [Ake 87].

for the decisions which of the $\sim 10^9$ events produced every second in the collisions will be retained by the Data Acquisition System for offline analyses. Typically, only $10^{-7} - 10^{-6}$ of the events are selected in this process, which takes place in several steps, called Level 1, Level 2, *etc*. At each level, the decisions made at the previous level are refined and, where necessary, additional selection criteria are applied. Level 1 provides a selectivity of 1/10,000 and reduces the primary event rate accordingly, to $\mathcal{O}(100$ kHz). The calorimeter data are a major tool for this reduction, together with information from the muon counters and the vertex detector (B-physics). In ATLAS and CMS, the L1 trigger searches for high-$p_\perp$ muons, electrons, photons, jets and hadronically decaying τ leptons, as well as large missing and total transverse energy. Using a "sliding window" technique and a limited amount of detector data, the L1 trigger decision is made in a few μs, based on dedicated hardware processors (FPGAs). The higher trigger levels access more detector information, apply dedicated criteria for the identification of electrons and γs, and make use of software-based algorithms. In this process, the final event rate is reduced to $\mathcal{O}(100$ Hz), representing a data rate of $\mathcal{O}(100$ Mbyte/s).

The trigger menu is made up of a substantial number (typically ~ 100) of different trigger selections, which are simultaneously used to examine the data stream. Prescaling of items from the trigger menu is used to make optimal use of the available bandwidth as the luminosity and background conditions change. For detailed reviews of the trigger systems used by ATLAS and CMS, the reader is referred to [Ach 08] and [Ada 05], respectively.

5.6 Time Structure of the Signals

The absorption of highly energetic particles in a calorimeter takes typically of the order of 10 ns or less, except for the thermalization and capture of neutrons produced in this process. However, certain instrumental effects may cause the signals resulting from this absorption process to last for a considerably longer time. This may deteriorate the calorimeter performance, especially in a high rate environment, since signals from the absorption of a certain particle may be contaminated by remnants of previous events. Some of these instrumental effects are described in the following subsections.

5.6.1 *Light-based calorimeters*

The light generated by shower particles in these calorimeters is the result of fluorescent processes or of the Čerenkov effect. Whereas the emission of Čerenkov light by superluminous particles is an instantaneous process, fluorescence is not. Fluorescent processes exhibit one or several characteristic decay constants, ranging from 10^{-9} to 10^{-3} s.

For example, the decay time of BGO, an inorganic scintillator used as an em calorimeter in the L3 experiment [Bak 85, Bak 87], amounts to 300 ns. A considerably faster crystal is CsI, which has two components: 70% of the light is produced with a decay time of 35 ns, the remaining 30% has a decay time of 6 ns. The decay characteristics of these and other scintillating crystals are summarized in Table B.5.

Organic scintillators may even be faster, as is illustrated by the values in Table B.6. Decay times of a few nanoseconds are commonplace in plastic scintillators. The precise decay characteristics often depend on parameters such as the type of matrix material in which the scintillating agents are dissolved and on the concentration of these scintillating agents [Liu 96].

Scintillators with short decay times make it possible to disentangle this instrumental effect from the pulse stretching caused by neutrons contributing to the calorimeter signals. This is illustrated in Figure 3.23, which shows the typical time structure of signals caused by electrons (Figure 3.23a) and pions (Figure 3.23b) showering in the Spaghetti Calorimeter [Aco 91a]. The pion signals exhibit a characteristic tail, with a time constant of about 10 ns (Figure 3.23c), which is absent in the electron signals. The decay time of the scintillator used in this calorimeter was about 3 ns.

A second instrumental effect that stretches the pulses produced by light-based calorimeters originates from the difference between the speed of light in the medium that transports the light signals to the photodetectors and the velocity of the shower particles that generate the light. For example, the scintillating fibers used in SPACAL (Figure 5.49) had an index of refraction $n = 1.59$, so that light generated by the shower particles traveled at a velocity $c/n = 0.63c$ through these fibers. Taking also into account that most of the scintillation light trapped in the fibers underwent many reflections inside the fibers, it traveled typically at a speed of 17 cm ns^{-1} from the location where it was produced to the photodetectors.

Light produced near the front face of this calorimeter therefore took about 13 ns to reach the photodetectors, whereas relativistic particles covered the 2.2 m distance in about 7 ns. This time difference became proportionally smaller as the light was produced deeper inside the calorimeter, *i.e.*, closer to the photodetectors. Even if the absorption process were instantaneous and the response of the light detector were a δ-function, the dispersion caused by the difference between c and c/n would give a time structure to the calorimeter's response function. Details of that time structure would in that case

SPACAL 1989

FIG. 5.49. The Spaghetti fiber calorimeter, and some of the people who were crucial for its realization, from left to right R. DeSalvo, A. Siegrist and G. Iuvino. The detector consisted of a lead absorber structure, into which 176855 scintillating fibers (1 mm diameter) were embedded. These were bunched at the rear end to form 155 towers. The total mass of this calorimeter was about 20 tons.

be determined by the longitudinal shower profile and by fluctuations in the shower's starting point.

The consequences of this effect are nicely illustrated by Figure 5.50, which shows oscilloscope signals from high-energy electron and pion showers in the SPACAL fiber calorimeter. The oscilloscope base was started by the signal from an upstream trigger counter. The figure shows that the hadron signals started a few ns earlier than the electron ones. This is because the light from the hadronic signals was produced substantially deeper inside the calorimeter, *i.e.*, closer to the PMT than the light from the electron showers, which was concentrated near the front face of the calorimeter. Because of the difference between the speed of light in the fibers (c/n) and the speed of the shower particles that generated that light ($\approx c$), the light from the pion showers reached the PMT earlier. How much earlier depended on the starting point of the hadron shower, and the dispersion in that respect is responsible for fluctuations in the starting point of the signals from the hadron showers by several ns, while the signals from the electron showers all started at approximately the same point in time.

An application of these phenomena is shown in Figure 8.35, which illustrates how the time structure of the signals from a fiber calorimeter can be used to determine the depth at which the light was produced, event by event.

Figure 5.50 illustrates one other instrumental effect on the time structure of the signals from scintillating-fiber calorimeters, caused by the aluminized upstream ends of the fibers in the SPACAL detector. The light trapped in the backward direction reflected

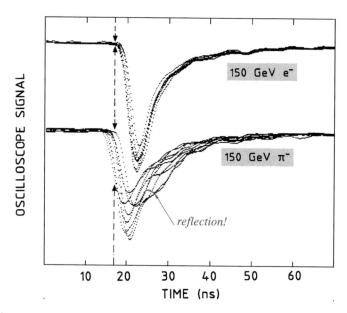

FIG. 5.50. SPACAL signals from 10 different electron and 10 different pion showers at 150
GeV [Des 89]. The time structure of the signals produced by pions interacting at different
depths inside the SPACAL detector show the effect of the mirrored upstream fiber ends.

off those mirrors and arrived later at the photodetectors than light trapped in the forward
direction. If the light were produced at a depth of 50 cm inside the calorimeter, then this
time difference would be equal to the time needed for a 100 cm roundtrip, or 6 ns.

The pulse stretching caused by this effect strongly depends on the depth at which
the first interaction takes place inside the calorimeter. This depth varies strongly from
one event to the next and causes a great variety of different pulse structures. Some
pulses, namely those caused by pions interacting deeply inside the calorimeter, clearly
exhibit a "double-hump" structure. The time interval between the maxima of the two
humps makes it possible to determine the depth at which the shower's center of gravity
is located inside the calorimeter.

Obviously, effects of the type described above also play a role in calorimeters based
on Čerenkov light, albeit that the angular distribution of the Čerenkov light contributing
to the calorimeter signals is less isotropic than for scintillation light (see Section 3.5.1).
Therefore, the "backward component" that leads to the second hump in Figure 5.50 is
considerably less pronounced in this case. On the other hand, the instantaneous nature of
the Čerenkov light makes it very suitable for some of the applications discussed in this
subsection. As an example, Figure 8.36 shows that the time structure of the Čerenkov
signals from a fiber calorimeter can be used to measure the light attenuation length of
these fibers. And Figure 8.18 shows that a comparison between the time structure of the
Čerenkov and the scintillation signals measured for the same showers makes it possible
to determine the contribution of neutrons to the latter signals.

5.6.2 *Charge-collecting calorimeters*

5.6.2.1 *Liquid-argon detectors.* The charge produced by ionizing particles traversing the active calorimeter layers is carried in equal amounts by negative electrons and
positive ions. However, the signals in calorimeters based on liquid argon as the active
medium are, for all practical purposes, exclusively caused by the electrons produced in
the ionization processes. Because of their very low mobility, the positive ions contribute

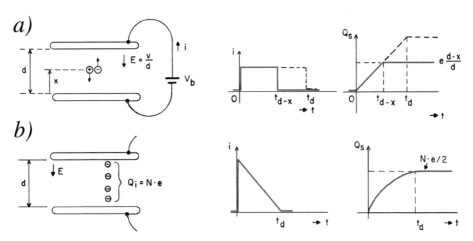

FIG. 5.51. Charge collection in a liquid-argon ionization chamber. Shown are the current and
the collected charge as a function of time, for one electron produced at a distance x from the
negative electrode (a), and for a collection of N electrons, produced uniformly across the gap
(b). From [Wil 74].

very little to the collected charge in the short time it takes the electrons to drift across
the gap.

Consider an ionization chamber with a gap width d, filled with liquid argon. The
voltage across the gap is V, so that the electric field $\vec{E}$ has a strength V/d. The current
resulting from a single electron produced at a distance x from the negative electrode is
then equal to

$$i \;=\; \frac{e}{t_d} \;=\; e\,\frac{v_d}{d} \qquad\qquad (5.9)$$

in which e represents the electron charge, t_d the drift time across the gap and v_d the drift
velocity (see Figure 5.51a). This current lasts as long as it takes the electron to drift to
the positive electrode, *i.e.*, a time $t \;=\; t_d(d-x)/d$. The total collected charge is equal
to the product of this current and the time t:

$$q(x) \;=\; it \;=\; e\,\frac{t}{t_d} \;=\; e\,\frac{d-x}{d} \qquad\qquad (5.10)$$

Each electron thus contributes, on average, $e/2$ to the total collected charge.

Next, consider N electrons, uniformly produced across the liquid-argon gap. This
is the situation that applies, in good approximation, to the ionization charge produced

by most of the charged shower particles that contribute to the signals from liquid-argon calorimeters. The current signal from this collection of N electrons can be derived from Equation 5.9. Each electron contributes a current e/t_d during a time t that depends on the position (x) where the electron was produced: $t/t_d = 1 - (x/d)$. The number of electrons that pass a point located at a distance x from the negative electrode equals $N(x/d)$, so that the current as a function of time reads:

$$I(t) = N\frac{e}{t_d}\left[1 - \frac{t}{t_d}\right] \tag{5.11}$$

The collected charge as a function of time becomes

$$Q(t) = \int_0^t I(t')dt' = Ne\left[\frac{t}{t_d} - 1/2\left(\frac{t}{t_d}\right)^2\right] \tag{5.12}$$

The resulting current and charge waveforms are shown in Figure 5.51. It is interesting to note that for uniform ionization, three-quarters of the observable charge is collected in a time corresponding to half the drift time across the gap. The total observable charge equals half the charge of the electrons produced in the interelectrode gap: $Q_{tot} = Ne/2$.

As we saw in Section 5.3.2, the electrons that form the signals of ionization-chamber calorimeters drift towards the positive electrode with velocities of typically a few km/s, and therefore it takes them of the order of 1 μs to cross the 2–3 mm wide liquid gaps that are used in most calorimeters of this type. To collect all the charge, the pulse duration thus has to be of the order of 1 μs as well. However, in some experiments, individual events may be separated in time by (much) less than 1 μs. In that case, the calorimeter signals have to be generated on the basis of a fraction of the produced ionization charge. This is achieved by pulse-shaping electronics. Bipolar pulse shaping (see Figure 5.52) has developed into the method of choice for this purpose [Rad 88b].

The most important condition for pulse shaping at high count rates is that the positive and negative lobes of the system impulse response have equal total areas. If this condition is fulfilled, then signal pile-up does not produce a net shift in the charge measurement for occasional high-energy events that occur in the presence of high-rate, low-energy background events.

The output waveform of the pulse-shaping system depends on the relation between the electron drift time (t_d, 1 μs in the examples described above) and the impulse response time, t_f. This is illustrated in Figure 5.52, which shows the bipolar impulse response to the characteristic triangular input current (Figure 5.52a), for two situations. In Figure 5.52b, the charge collection takes about as much time as the pulse shaping, $t_d \approx t_f$. In this case, only the ionization charge collected during the period up to $t = t_m$ (the dashed area in Figure 5.52a) contributes to the signals. When the pulse shaping is much faster than the charge collection, the output response takes the form shown in Figure 5.52c. The fraction of the ionization charge that contributes to the signals is much smaller than in Figure 5.52b.

It is thus possible to obtain signals that are considerably faster than the current waveform produced by the calorimeter itself. Obviously, this goes at the expense of the signal-to-noise ratio. As we discussed in Section 5.4.2, the ultimate limit to the speed of response of this type of calorimeter is set by the charge transfer time from the charge-collecting electrodes to the charge amplifiers. The minimum width of the overall system response (t_f) has to be at least five times longer than this charge transfer time [Rad 88b].

An example of a device in which only a fraction of the ionization charge contributes to the signals is the ATLAS LAr calorimeter. In this detector, the electron drift time, t_d, is about 500 ns, and the pulse shaping time, t_p, is 45 ns.

5.6.2.2 Gas calorimeters.

The time structure of the signals from gas-based calorimeters may be very different from that in liquid-argon devices, for the following reasons:

- The signals from gas chambers are formed by induction, as opposed to the signals from LAr detectors, which are derived from the collected charge.

- The signals from gas chambers are caused by the drifting ions, while the signals from LAr detectors are determined by the electrons produced in the ionization processes.

- The signals from gas chambers are caused by charged particles (ions) that are almost all produced in the same area, *i.e.*, the vicinity of the anode wire, where the avalanche develops. On the other hand, the particles responsible for the signals from LAr detectors are created uniformly along the track of the ionizing particle.

- The drift velocities in gases and liquids may be very different.

Typical time structures for pulses from a cylindric proportional counter are shown in Figure 5.53. The dashed line shows the voltage of the anode wire, induced by the

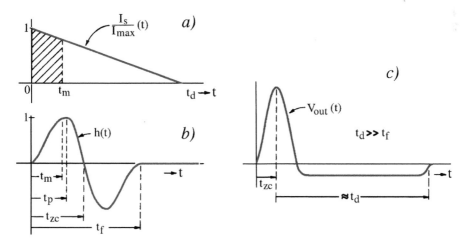

FIG. 5.52. Pulse shaping of signals in ionization-chamber calorimeters. Current waveform induced by electrons traversing the interelectrode liquid gap (*a*). A bipolar impulse response of the detector amplifier system (*b*). The output of the bipolar shaping amplifier driven by the signal in (*a*), for fast pulse shaping, *i.e.*, $t_f \ll t_d$ (*c*). Data from [Rad 88b].

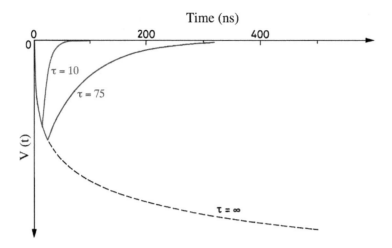

FIG. 5.53. The pulse shape of signals from a proportional wire chamber. The pulse is differentiated by an RC circuit. Results are shown for two values of the time constant (in ns) of this
circuit.

positive ions created in the avalanche as they drift towards the cathode. Because of the
large charge amplification factors achieved in such chambers (10^6, Section 5.3.1), there
is absolutely no need to use all the charge to generate a signal. Usually, the pulse is cut
short by a differentiating RC circuit.

In the example shown in Figure 5.53, the signal rises to about 8% of its ultimate
value in the first 10 ns. The figure shows the pulse shape obtained with differentiating
circuits with time constants of 10 and 75 ns, respectively.

5.6.2.3 *High-rate operation.*   The performance of calorimeters that generate signals
on the basis of the ionization charge produced in the active medium may be affected by
the (sometimes rather long) time it takes to remove all the charged particles produced
in this way from the active calorimeter components.

This ionization charge is carried both by electrons and by ions. The drift velocity
of the ionized atoms is usually several orders of magnitude smaller than for the electrons that constitute the basis for the calorimeter signals, especially in liquids. At very
high event rates, build-up of positive charge that cannot be removed fast enough might
effectively reduce the electric field strength and thus reduce the signals.

This is a serious concern for detectors operating at the LHC, such as the Very Forward Calorimeter for the ATLAS experiment. To minimize this problem, the LAr gaps
in this detector are made much thinner (0.25 mm) than is usual in such calorimeters
(2–3 mm).

In view of these, and other, problems facing experiments at the LHC and other high-
luminosity colliders (*e.g.*, radiation damage), calorimeters using *high-pressure gas* as
the active medium have been proposed [Gio 90] and developed [Dem 93, Bag 95]. If
noble gases such as argon or xenon are highly compressed, to a pressure of 100 atm

or more, their density approaches that of the liquid phase. Such high densities prevent avalanche development, because of the short collision length of the electrons, and it becomes thus conceivable to build a gas calorimeter with unity gain in those conditions. Such a device might offer the same advantages as a liquid-argon detector (stability, radiation hard), without the complications of a cryostat, although a high-pressure vessel poses of course its own challenges in terms of detector hermeticity.

Important for our present discussion is the fact that the drift velocities of the charged ionization products in these high-pressure gases are considerably larger than in the corresponding liquids, thus alleviating the problem mentioned above. Giokaris and coworkers [Gio 90] measured the charge collection time for electrons in 100 atm argon to be about four times faster than in liquid argon, for the same field strength (1 kV mm^{-1}). The latter result was obtained after adding 0.5% of methane to the argon gas, which reduced the signals from mips with $\sim 15\%$ compared with the case of pure argon gas. In the same tests, it was also shown that the purity of high-pressure argon gas is much less of an issue than in liquid argon. At 100 atm, the signal size and speed were practically unaffected by oxygen impurities up to concentrations of 50 ppm, while liquid argon is sensitive to oxygen impurities at the few-ppm level.

Subsequent tests of calorimeter prototypes showed that fast signals could be extracted from such devices. For example, for a hadronic steel/gas calorimeter with a gap width of 1.6 mm, the total width of pion-induced shower signals was measured to be 80 ns, before any pulse shaping was applied [Bag 95]. High-pressure noble gases thus offer an interesting ionization-chamber-mode alternative to liquid dielectrics. The gas detectors are faster, operate at room temperature, and offer comparable performance in terms of energy resolution and radiation hardness.

5.7   Auxiliary Equipment

In most experiments, calorimeters fulfill a variety of different tasks, for example:

- They provide the information for the first-level event selection and triggering.
- They provide crucial information that allows the identification of electrons, muons and neutrinos.
- They provide the information that makes it possible to reconstruct the four-vectors of neutral particles, such as γs, K^0s and neutrons.

Depending on the physics goals of the experiment, one of these tasks may be more important than others. This is usually reflected in the design of the calorimeter. For example, in experiments at electron–positron colliders up to 100 GeV center-of-mass energy, calorimetric detection of hadrons is typically a low-level priority. Also, the event rates at these machines are usually so low that essentially all collisions between particles are recorded and event selection just involves distinguishing interactions between beam particles from beam–gas interactions and cosmic ray events[15]. Therefore, when the calorimeters for these experiments were designed, the main emphasis was on the identification and measurement of leptons produced in the interactions.

---

[15]This statement may no longer be true at the SuperKEKB electron-positron Collider, where a production rate of $\sim 1,000$ $B\bar{B}$ pairs per second is expected at the $\Upsilon(4S)$ resonance.

The performance of a given calorimeter system for what concerns one (or several) of the tasks mentioned above can sometimes be improved by means of certain additional detectors. However, such detectors may also deteriorate other aspects of the calorimeter performance. For example, the additional equipment may improve the identification of electrons, but the associated support structure or readout provision may deteriorate the hermeticity of the system as a whole. In such a case, the choice to install such auxiliary equipment depends on the experimental priorities.

In the following, we briefly mention three detectors that are frequently used in conjunction with a calorimeter system.

5.7.1 *Preshower detectors*

Many calorimeters are equipped with a fine-grained *preshower detector* (PSD), installed directly upstream of the calorimeter itself. Such a PSD is typically a few radiation lengths deep and has a finer granularity than the calorimeter itself. The PSD may be integrated into the calorimeter itself, as in the ATLAS experiment (Figure 9.10), where the presampler is a separate thin liquid-argon layer, which provides shower sampling in front of the active em calorimeter proper, inside the barrel cryostat. It contains interleaved cathode and anode electrodes glued between glass-fibre composite plates, and adds a total thickness of $0.5 - 1 X_0$ to the material upstream of the em calorimeter [Andr 02].

The PSD may also be a complete separate detector, as in CMS, where it is installed just upstream of the ECAL in the kinematic region covered by the $PbWO_4$ endcaps ($1.653 < |\eta| < 2.6$), and is actually an integral part of the tracking system in that region. It is a sampling calorimeter consisting of two layers of lead plates interleaved by layers of silicon strip sensors. The total thickness of this detector is 20 cm, and represents about $3 X_0$ of material [CMS 08].

A PSD may extend the capabilities of a calorimeter system in a variety of ways:

- It may improve the identification of electrons.
- It may improve the distinction between single γs and the much more abundant photon pairs from π^0 decay.
- It may improve the precision of the measured angle of incidence of the particles.

These aspects are discussed in more detail in Chapter 7.

In addition, a PSD may be used to improve the em energy resolution of the calorimeter system, since it may be used to sample energy deposited in dead material upstream of the calorimeter. Especially in liquid-argon detectors such energy deposits, in the cryostat walls, may be very substantial. For this reason, many LAr calorimeters are equipped with a "massless gap," which may be considered a special type of PSD. One popular way to construct a massless gap is to replace the first few absorber plates of the calorimeter with G10. The affected part of the calorimeter, which is located directly behind the cryostat walls, is read out separately and the signals are given an appropriate weight. Hirayama has demonstrated that, in this way, the degrading effects of upstream dead material with a total thickness of as much as $5 X_0$ may be recovered [Hira 91]. A practical example of the benefits of this technique is shown in Figure 5.54, for the

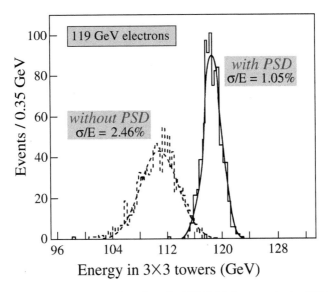

FIG. 5.54. Reconstructed response functions for 119 GeV electrons in the ATLAS em calorimeter, before (dashed line) and after (solid line) adding the weighted signals from the presampler to the measured calorimeter signals [Aub 93b].

measurement of electrons in ATLAS. Also other types of calorimeters may be equipped
with a PSD designed for recovering (the effects of) energy lost upstream of the calorimeter. Hulbert and coworkers describe a dedicated study of the improvement of the em
energy resolution that may be achieved with a PSD *without* fine segmentation, *i.e.*, a
simple sheet of scintillator material [Hul 93].

5.7.2 *Shower maximum detectors*

Advantages similar to some of those listed in the previous subsection may be offered
by a so-called *shower maximum detector* (SMD). A SMD consists of one or several
active layer(s) with a fine-grained readout structure, installed at a depth of ~ 5 radiation lengths inside the absorber structure. Multi-wire proportional chambers [Gaba 78,
Amb 96], silicon strips [Dwu 89], thin scintillating strips [Apo 93, Apo 98b] and scintillating fibers [Aco 95] have been or are being used for this purpose.

As an example of a SMD, we mention the CDF endcap calorimeter, which contained
a layer of scintillating strips installed at a depth of $6X_0$ inside the lead/plastic structure
[Apo 98b]. The strips measured 5×6 mm^2 and varied in length from a few centimeters
to about 1 m. Each strip contained a 0.83 mm thick WLS fiber, which ran longitudinally
in a ball-shaped groove through it (as in Figure 5.25b). This fiber absorbed light produced in the scintillating material and re-emitted it in the green part of the spectrum. In
total, this SMD consisted of 6,400 such strips and the fibers were read out with multi-
anode PMTs. The information provided by this instruments helped CDF improve its jet
energy resolution (Figure 8.40).

5.7.3 Backing calorimeters

In almost all calorimeter systems used in particle physics experiments, shower leakage has to be dealt with at some level. This may be illustrated by the following example. The probability that a high-energy proton penetrates five nuclear interaction lengths or more before initiating a nuclear interaction is 0.7%. Since the nuclear interaction length for a pion is usually considerably longer than for a proton (see Section 2.3.4.1), this means that $\sim 1\%$ of the pions penetrate at least 6–7 λ_{int} before starting to develop a shower. And since few calorimeter systems are deeper than 10 λ_{int}, such events typically exhibit considerable shower leakage.

This shower leakage leads to a systematic mismeasurement of the energy of the particles concerned. Also, the particles leaking out of the back of the calorimeter may traverse detectors downstream of the calorimeters and may thus be misinterpreted as muons, either at the trigger level (affecting the fake-trigger rates) or even in the offline data analysis (leading to flawed physics results).

One could of course eliminate this problem by making the calorimeters sufficiently thick. However, since the volume, and thus the cost, of the detector scales with the radius cubed, this solution is usually prohibitively expensive. It is much cheaper to equip the detector with a *backing calorimeter*, a crudely instrumented block of matter with a thickness of typically 1–2 λ_{int}, installed behind the calorimeter system and intended to detect significant shower leakage. In practice, such instruments are frequently used as *veto counters*: any event in which an anomalously large fraction of the energy is recorded in the backing calorimeter is discarded. However, in the case where the energy fraction deposited in the backing calorimeter is not anomalously large, its signals may be used, in combination with those from the other calorimeter components, to determine the energy of the showering particles.

This technique was, for example, used for the stated purposes in ZEUS [Abr 92]. Also CMS uses a backing calorimeter to complement the measurements of the hadronic shower energy. In their case, the active depth of the calorimeter system is constrained by the fact that it had to fit inside the solenoidal magnetic field. At $\eta = 0$ ($\theta = 90°$), the instrumented depth of the hadronic calorimeter section is therefore only $5.8\lambda_{int}$. The iron return yoke of the magnet is used as absorber material for the backing calorimeter. Large slabs of plastic scintillator are inserted on both sides of the 19.5 cm thick layers of iron. The signals from these scintillators are transported by means of WLS fibers to the SiPM light sensors [CMS 08].

5.8 Operation in a Magnetic Field

5.8.1 Mechanical and electronic effects

In many experiments, the calorimeters operate in a magnetic field that serves to determine the momenta of individual charged particles upstream, and sometimes downstream, of the calorimeter system. This has a variety of practical consequences.

First of all, the use of magnetic materials, such as iron or nickel, should be avoided in the construction of these calorimeters. The forces exerted on these materials when the magnetic field is switched on could cause severe damage to the experimental equipment. But even if everything were so rigidly constructed that nothing physically moves when the field is switched on, the iron could severely distort the magnetic field itself, causing problems for the reconstruction of the trajectories of charged particles.

Operation in a magnetic field also imposes limitations on the readout technology that can be used. If the calorimeter produces light signals, then it is often not possible to use photomultiplier tubes as light detecting elements.

Most PMTs are so sensitive to magnetic fields that already fields as weak as the Earth's magnetic field ($\sim 5 \cdot 10^{-5}$ T) may have a significant effect on the gain. It can be easily checked whether or not this is the case for a certain setup by changing the orientation, since only the non-axial component of the field affects the trajectories of the photoelectrons.

To avoid such effects, PMTs are usually equipped with shields made of material with a high magnetic susceptibility ("μ-metal"), which surround the sensitive areas (mainly the photocathode).

Some PMTs have been specially designed to be able to operate in magnetic fields. This is mainly achieved through a very compact dynode structure, as in *close proximity focusing* tubes [Ori 83]. Also compact structures with only one or two dynodes (known as *vacuum phototriodes* or *phototetrodes*) have been demonstrated to be capable of operating in moderate magnetic fields, up to ~ 1 T. Both OPAL [Akr 90] and DELPHI [Che 89] have used such devices to read out their lead-glass em calorimeters. CMS uses phototriodes for reading out the $PbWO_4$ crystals of the endcap sections of their electromagnetic calorimeter [Baj 00], where the (solenoidal) magnetic field is almost axial. These particular devices have an anode of very fine copper mesh ($10\mu m$ pitch), which allows them to operate with $< 10\%$ gain loss in the 4 T magnetic field of CMS [Bel 03].

Alternative light detectors capable of operating in (strong) magnetic fields include silicon photodiodes, avalanche photodiodes and hybrid photon detectors (see Section 5.4.1). However, each of these alternative solutions may have disadvantages, compared with PMTs. For example, if the light yield is not very large, the electronic noise introduced by silicon photodiodes may cause a substantial deterioration for measurements of small energy deposits. Avalanche photo Diodes may be very unstable under small changes in temperature or bias voltage, while HPDs may be far too expensive for the experiment's budget, and have apparently a limited lifetime (Section 5.4.1.2).

The effects of magnetic fields on light detectors depend strongly on the orientation of the magnetic field with respect to the electric field that accelerates the (photo)electrons towards the collecting electrode. The effects are of course minimal when $\vec{B} \parallel \vec{E}$. They reach a maximum when $\vec{B} \perp \vec{E}$, because the slow (photo)electrons are bent away from the direction they should follow to produce a signal.

This is illustrated by Figure 5.55, which shows the effects of parallel (Figure 5.55a) and perpendicular (Figure 5.55b) magnetic fields on the response of a (proximity-focused) HPD [Arn 94]. Parallel fields had almost no effects, up to fields of about 2 T. The $\pm3\%$ systematic effects observed in these measurements were attributed by the authors to hysteresis and to variations in the position of the light source. However, magnetic fields perpendicular to the direction of the accelerating electric field started to reduce the HPD response significantly already for field strengths in excess of 0.05 T,

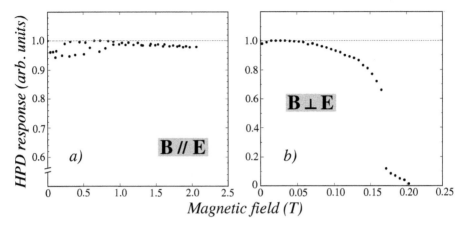

FIG. 5.55. Normalized response of a Hybrid Photon Detector operating in a magnetic field. The magnetic field vector is either oriented parallel to the accelerating electric field (*a*) or perpendicular to it (*b*). Note the blown-up vertical scale in (*a*) and the fact that the horizontal scales differ by a factor of ten. From [Arn 94].

and beyond 0.2 T the device was essentially dead.

One way to deal effectively with the possible problems caused by magnetic fields is to transport the light signals from their source to an area where the magnetic field is so low that it does not prohibit readout with PMTs. Clear plastic fibers allow one to do so with minimal light losses. Tests done in the context of the development of the CDF Plug Upgrade calorimeter by a group at Fermilab have shown that the light losses incurred in coupling fibers in which the detector signals were generated to such clear fibers could be limited to 10–15%, with good reproducibility [Apo 92].

In calorimeters using liquid or gaseous dielectrics, magnetic fields cause limitations for the type of electronic circuits that can be used. For example, voltage or current transformers do not work properly when placed in an external magnetic field. In gas calorimeters based on drift chambers as active elements, the path and the velocity vector of the drifting electrons may be altered by the Lorentz force. In that case, a precise knowledge of the magnetic field map throughout the detector is crucial for determining the modified relationships between drift time and position.

5.8.2 *Effects on the calorimeter signals*

Apart from the effects that magnetic fields have on the functioning of detectors and electronics that handle calorimeter signals, they may also affect these signals themselves. We mention two examples.

5.8.2.1 *Increased light yield.*   The light yield of some plastic scintillators changes when placed in a magnetic field. Typically, the light yield increases, by some 5–10%, for magnetic fields up to 3 T [Bert 87, Cum 90, Blo 92, Man 92, Bert 97]. For larger fields, no further increase was observed by Bertoldi and coworkers [Bert 97], who tested a variety of different scintillators and wavelength shifters for fields up to 20 T. The same authors also reported that the increased light yield was only observed when the scintillators were excited with ionizing particles. In particular, no significant changes in the light yield were observed when the excitation was performed with UV light. This

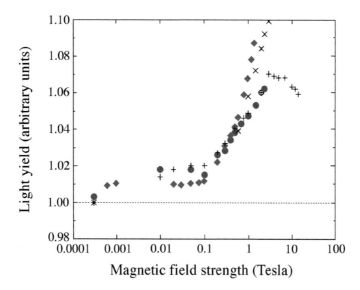

FIG. 5.56. The relative increase in the light yield of plastic scintillators placed in a magnetic field as a function of the field strength, as observed in a variety of experiments [Bert 97].

is a strong indication that the effects are due to an increased production of UV light in the excitation of the polymer base material (polystyrene or polyvinyl toluene), since the excitation of the scintillating fluors with UV light was not affected by the magnetic field.

The light yield is not a simple function of the field strength. This is illustrated in Figure 5.56, which summarizes the results of a variety of experiments. For very small fields, as small as 1 mT, the light yield was already observed to increase by about 1%. A further increase of the field, by two orders of magnitude (to ~ 0.1 T), had little effect of the light yield. Above 0.1 T, the light yield increased rapidly with the field strength, and saturated around 3 T. There is no experimental indication that the described effects depend on the orientation of the magnetic field or on the nature of the ionizing radiation.

5.8.2.2 *Effects on shower profiles.* Magnetic fields may also affect the shower profiles, since the paths of the charged shower particles are subject to the Lorentz force. This may have consequences for the early, non-isotropic shower component. These consequences depend on the strength and the orientation of the magnetic field. If the field is perpendicular to the shower axis, then the lateral shower profile will be broadened. In Section 2.1.4 we saw that a large fraction of the signals from electromagnetic showers comes from soft electrons, produced in Compton scattering and in photoelectric processes. For example, in a copper-based calorimeter, more than half of the shower energy is deposited by electrons softer than 4 MeV (Figure 2.10). The trajectory of these soft electrons is very sensitive to magnetic fields, since they are subject to the Lorentz force. For example, the radius of curvature of an electron with a momentum of 4 MeV/c in a 2 T magnetic field oriented perpendicular to its direction of motion is only about 6 mm.

In Section 3.2.1 we saw that these soft electrons only contribute to the signals from sampling calorimeters if they are produced in a very thin boundary layer. The range of 4 MeV electrons in copper is only 3 mm (Figure 2.47). If a 4 MeV electron escaped from the copper and traversed a 5 mm thick plastic scintillator layer perpendicularly, then it would lose typically 1 MeV. However, if the calorimeter were placed in a strong magnetic field oriented parallel to the sampling layers, then this escaping electron would describe a curved trajectory and lose a considerably larger fraction of its energy in the scintillator. Depending on the gap width between the absorber plates and the strength of the magnetic field, it could even reverse direction and deposit its *entire* energy in the scintillator plate, in a way similar to the "cork-screws" that were the signature of electrons in bubble-chamber pictures (Figure 5.57a).

This effect results in an increased calorimeter response. It increases with the strength of the magnetic field, since the fraction of shower particles trapped in the gaps between the absorber plates increases with the field strength. It would not play a role for magnetic fields oriented perpendicular to the sampling layers.

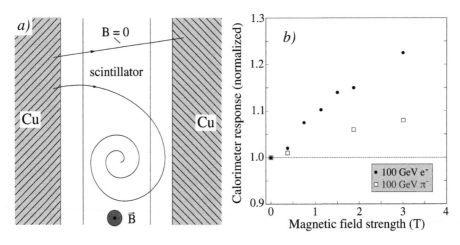

FIG. 5.57. Trajectories of few-MeV electrons contributing to the signals of a sampling calorimeter in the absence and presence of a magnetic field oriented parallel to the sampling layers, pointing into the plane of the figure (*a*). The relative increase in the response of the CMS copper/plastic-scintillator calorimeter, as a function of the strength of a magnetic field oriented parallel to the sampling layers (*b*). The response dependence is given for showers induced by 100 GeV electrons and by 100 GeV pions. In order to eliminate the effects on the specific light yield of the scintillator, all responses have been normalized to those for muons traversing the calorimeter in the same direction as the showering particles [Kun 97].

The described effects were observed by the CMS Collaboration in prototype tests of their hadron calorimeter [Kun 97]. Since the copper/plastic-scintillator CMS calorimeter has to operate in a very strong magnetic field, the effects of this field on the calorimeter performance were studied in great detail. Some results of these studies are shown in Figure 5.57b, where the relative increase in the calorimeter response is plotted as a function of the magnetic field strength, for showers induced by 100 GeV electrons and 100 GeV pions. The field was oriented parallel to the sampling layers in these studies. In order to eliminate the "scintillator brightening" effects discussed in Section 5.8.2.1, the calorimeter responses to these particles were normalized to those for muons that traversed the calorimeter in the same direction. The em response was measured to increase by about 20% when the field strength reached 3T, the maximum value for which measurements were performed in these studies.

The effect of the magnetic field on the hadronic response was clearly smaller, commensurate with what should be expected if only the em shower component was affected by the field. This makes perfect sense, since hadronic shower particles with similar sensitivity to the Lorentz force as the electrons from the example discussed above are extremely non-relativistic and would thus deposit their entire kinetic energy in the scintillator, with or without a magnetic field. Therefore, the field does not affect the response to the non-em shower component. Since only the em shower component is affected, the e/h value of the calorimeter is thus increased when it operates in a magnetic field.

The CMS tests showed no measurable effect on the calorimeter response (over and above the scintillator brightening discussed above), if the magnetic field was oriented *perpendicular* to the sampling layers, *i.e.*, parallel to the shower axes.

5.9   Operation at Very High Luminosity

As the LHC luminosity ramps up towards its design value, and beyond, the experiments will face increasing problems on how to extract the physics data they are looking for from the detector information generated by the enormous event rates. Signal superposition, commonly referred to as "pile-up," features prominently among these problems. For a total pp cross section of 60 mb, and a luminosity of 10^{34} cm^{-2} s^{-1}, the event rate amounts to 600 MHz. At a bunch crossing rate of 40 MHz (bunch spacing 25 ns), this translates into an *average* number of fifteen events per bunch crossing. In other words, every interesting event that one will want to retain is, on average, accompanied by fourteen other events. At the even higher luminosities that are envisaged for the future (HL-LHC), this rate will become correspondingly worse, and one expects eventually having to deal with several hundred events per bunch crossing.

Of course, the overwhelming majority of these "underlying events" are uninteresting, and involve relatively few high-$p_\perp$ particles that contribute to the detector signals. Yet, event pile-up is considered a very serious problem and a large number of studies have been undertaken on how to deal with it and in particular how to mitigate its effects on the detector performance. In the traditional approach, it is considered an additional source of electronic noise. Using a series of signal samples collected at intervals of 25 ns, the properties of the true signal are estimated on the basis of a variance minimization of the noise covariance matrix, a method known as optimal filtering [Cle 94]. The bipolar pulse shaping, introduced by Radeka [Rad 88b], is crucial for the success of this method with the LAr signals from the ATLAS calorimeters. This method works best for Gaussian noise. Pile-up has the tendency to add positive or negative tails to the noise ditribution, which thus becomes non-Gaussian. The extent of these effects depends on the number of underlying events, and thus leads to a luminosity dependence.

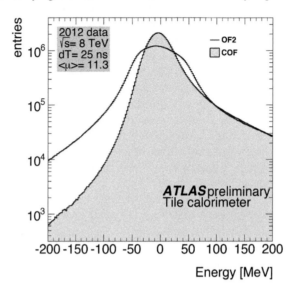

FIG. 5.58. Cell energy distribution reconstructed by the Constrained Optimal Filter (COF) and the Optimal Filtering (OF2) algorithms using 2012 ATLAS pp collision data at $\sqrt{s} = 8$ TeV and 25 ns bunch spacing. The average number of interactions per bunch crossing was 11.3 for this event sample (around 25 millions entries). The COF method is resilient to Out of time signals, therefore, it presents better energy resolution than OF2. Its design is luminosity independent and requires only the information of the pulse shape and pedestal value to compute the 7 amplitudes associated to the 7 samples of the read-out. In this plot only the central sample reconstruction is shown [Fil 15].

Recently, a method has been proposed that is in principle independent of the luminosity. It is based on a deconvolution process of the same type used in digital processing for communication channel equalization, and aims to fully recover the target signal, rather than estimate its amplitude from pulse sampling [Fil 15]. This method has been tested with data from the ATLAS TileCal hadronic calorimeter. Figure 5.58 shows some results from this work.

5.10   Radiation Damage

Increasingly, calorimeter systems developed for particle physics experiments at accelerators have to operate at very high levels of (background) radiation. There are two reasons for this trend:

1. As the collision energy increases, the cross sections for the fundamental processes of interest decrease, in proportion to E^{-2} (Figure 5.59). In order to be able to collect event samples of reasonable size in a reasonable time, the luminosity of the accelerator thus has to *increase* with energy, in proportion to E^2. This is the main reason why the design luminosity of the LHC (where the center-of-mass energy of the collisions is about 14 TeV) is about two orders of magnitude larger than the luminosity of the Tevatron (2 TeV). However, the *total cross section* of the interactions is about the same at these colliders, ~ 0.1 b. This total cross section determines the level of background radiation in which the experiments have to operate.

   In electron–proton colliders and, even more, in electron–positron colliders, the background levels are much smaller than in these hadron colliders. This is because the events of interest represent a much larger fraction of the total number of events. Nevertheless, as the collision energy increases, the same trend applies also for these machines. The background radiation in these colliders is often determined by beam–gas collisions and by synchrotron radiation. And as the luminosities increase to accommodate the smaller cross sections, the radiation levels increase as well.

2. Apart from the reason mentioned above, which links the luminosity requirements of experiments (and thus the levels of the background radiation) to the center-of-mass energy of the particle collisions, high luminosity levels have their own separate physics justification. Increasingly, interesting physics issues require the study of processes with a very small *relative* probability. As examples, we mention the effects of CP violation in the B^0 system, which are expected to manifest themselves at the 10^{-8} level (*i.e.*, samples of 100 million events are needed for these studies), and rare decay modes of particles containing unstable quarks, where some experiments look for signals at the 10^{-13} level [Ada 13b]. Studies of these phenomena are important tools for exploring the boundaries of the Standard Model of Particle Physics. In that respect, they complement the studies that become possible as a result of increased center-of-mass energies.

Progress in our understanding of the fundamental structure of matter thus proceeds on two separate frontiers: the energy frontier and the luminosity frontier. Both frontiers imply increasing levels of background radiation in the experiments. In the past decade, radiation hardness considerations have become a standard design issue for calorimeter systems, at least for the ones operating at accelerator experiments.

There are several, distinctly different, aspects of dealing with radiation in connection with the design, operation and maintenance of a calorimeter system:

- Determining the effects of ionizing radiation and neutrons on calorimeter performance
- Understanding the mechanism through which the effects occur
- Preventing the effects
- Curing the effects if and when they occur
- Coping with the effects that can neither be prevented nor be cured

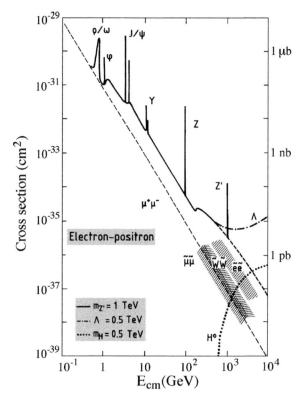

FIG. 5.59. The cross section for scattering processes at the constituent level, as a function of
energy, for electron–positron colliders. The Higgs boson was not yet discovered when this
plot was made. From [Amal 86].

Ideally, detectors should be constructed in such a way that the background radiation
does not affect the performance. However, this is not always possible and in such cases
the last aspect listed above becomes very important.

Historically, the study of radiation damage issues has primarily been developed
in connection with the construction and operation of nuclear power plants. The radiation doses in that type of environment are usually much higher than in particle physics
experiments. On the other hand, the signal-producing equipment used in particle physics
experiments is very delicate, and disastrous effects may occur at much lower dose levels
than those causing the type of structural damage in construction materials that worries
nuclear engineers.

5.10.1 *The effects of ionizing radiation*

The effects of ionizing radiation on particle detectors, and in particular on the active
materials used in calorimeters, depend strongly on the nature of these active materials.

For example, in plastic scintillators, which are widely applied in sampling calorimeter systems, the main effects concern a decrease of the *emission* of scintillation light and

a decrease in the *transmission* of light through the material. The latter effect is usually strongly wavelength dependent, it primarily affects the short-wavelength component. In general, these scintillators become yellow (*i.e.*, opaque for blue light) after receiving a substantial radiation dose.

Another radiation degradation phenomenon that mainly affects polymers and other detector materials based on hydrocarbons (*i.e.*, warm liquids such as TMP and TMS, or liquid scintillators) is *outgassing*. The composition of the gaseous radiolysis products depends upon the molecular composition of the detector material. Most frequently, hydrogen is among these products. This may be problematic, since it may lead to pressure build-up in closed containers [Hol 90].

In liquid-argon calorimeters, outgassing of auxiliary materials such as G10 plates (a mixture of SiO_2 and epoxy) may lead to an increase in the level of (electronegative) impurities in the liquid, in particular oxygen, which will affect the electron lifetime and thus the signals. In chlorinated polymers such as polyvinylchloride, the primary radiolysis product is hydrochloric acid (HCl), which may cause corrosion in the detector.

The exposure of a wire chamber to ionizing radiation may lead to deterioration of the operating characteristics, as a result of the deposition of material on the anode wires and/or the cathode surfaces. Deposits on the wires may reduce the gain and the efficiency of the chamber; those on the cathode may produce discharges. The harmful deposits can originate from the chamber gas, from chamber materials in contact with this gas, or from spurious contaminants. Therefore, the nature and importance of the effects are highly dependent on details of the construction and operation of the detector.

In semiconductor devices, such as silicon detectors, ionizing radiation may lead to a variety of effects, for example increased leakage current, a degradation of the signal-to-noise ratio and even type inversion (p-type Si becomes n-type or vice versa [Pit 92]). These detectors are also particularly sensitive to neutrons (see Section 5.10.4.2).

Many calorimeters contain semiconductor-based electronics located in areas exposed to ionizing radiation. These components may simply stop functioning (properly) as a result of this exposure.

5.10.2 *Dose rate effects*

Many effects of ionizing radiation depend not only on the total dose received by the irradiated object, but also on the *dose rate, i.e.*, on the time during which the object was exposed and received a given total radiation dose.

Dose rate effects are well known to play a role in the irradiation of living organisms. For example, a dose of 200 rem (2 Sv) received in a short period of time by a human being is, if not directly life-threatening, at least considered extremely dangerous and likely to result in serious health problems.

On the other hand, a radiological worker can accumulate the same dose during a working life of 40 years, without exceeding limits that are generally considered safe (5 rem yr^{-1}, *i.e.*, 50 mSv yr^{-1}). Living organisms such as the radiological worker's body have the ability to fully recuperate from the effects induced by relatively low levels of ionizing radiation.

Also, individual cells in a human body have an average life span of about 7 years.

Destroying a few cells faster than foreseen by their schedule does not necessarily have adverse effects on the entire organism. On the other hand, a large dose received in a short time may affect so many cells that certain organs no longer function properly. This is especially dangerous if it occurs in organs that play a role in cell reproduction.

In the above example, dose rate effects make the irradiated object *less vulnerable* to ionizing radiation if the irradiation is spread out over a longer time period. This is the most common form of dose rate effects. The recovery capability of the irradiated object becomes more and more effective as the time allowed for recovery is increased.

This type of dose rate effects is also not uncommon in objects other than living organisms. The observation that many irradiated objects exhibit signs of recovery in the minutes, hours and days following the end of the exposure is already a strong indication of this.

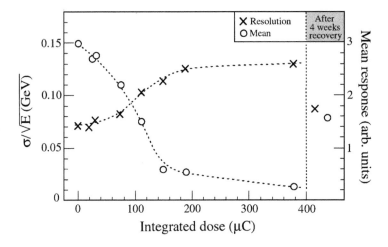

FIG. 5.60. The light output (right ordinate) and the energy resolution (left ordinate) of a scintillating-fiber calorimeter as a function of the integrated radiation dose received by the instrument. The measurements were repeated four weeks after the end of the irradiation runs. From [Joh 92].

Figure 5.60 shows an example of this phenomenon, measured on a scintillating-fiber calorimeter that was irradiated with a beam of 800 MeV protons, during a period of several weeks. The calorimeter response and the energy resolution for em showers are plotted as a function of the integrated dose, expressed in units of absorbed charge (*i.e.*, $1.6 \cdot 10^{-13}$ μC per beam proton). At the end of the irradiations, the response had dropped to only one-tenth of its original value, while the energy resolution had become twice as large. However, in the course of the next four weeks, a substantial recovery occurred. The response increased to about half its original value and the energy resolution recovered almost completely. Similar recovery phenomena have also been observed in a variety of inorganic scintillating crystals [Zhu 98].

In practice, the radiation hardness of materials used in the construction of calorimeters, or nuclear power plants, is often measured in *accelerated tests*. In this procedure, the total dose envisaged during the entire active lifetime of the device (*e.g.*, 10 years) is applied in a much shorter time (*e.g.*, a few days). Since many facilities used for irradiation purposes have limited availability, such accelerated tests are often the only practical possibility.

Clearly, dose rate effects of the type described above will lead to an overestimation of the damage incurred from the given dose, when an accelerated test procedure is followed. In order to come to a more accurate assessment of the problems that may be expected in real life, a series of tests may be conducted in which a given total dose is applied with a variety of different dose rates. For example, a total dose of 1 Mrad (10 kGy) may be applied in irradiations ranging from a few hours to a few days. By studying *differences* in the effects incurred from these different exposures, one may hope to observe trends that may give some confidence in the extrapolations for the real-life case.

Although dose rate effects usually lead to an overestimation of the radiation damage problems when the results from accelerated tests are taken at face value, there is at least one well-documented case in which the opposite situation occurred [Sir 85]. The plastic scintillator used in the uranium/scintillator calorimeter operated by the R807 experiment at CERN's Intersecting Storage Rings was observed to age much *faster* than anticipated on the basis of accelerated tests. This is illustrated in Figure 5.61, which shows the results of tests designed to determine the effects of the ionizing radiation emitted by decaying ^{238}U nuclei on some characteristics of the plastic scintillator, in particular the light yield and the light attenuation length.

Figure 5.61 shows the change in the attenuation length as a function of the applied dose. Several experimental points indicate results obtained under different circumstances (dose rate, environment). Lines have been drawn through experimental points obtained under the same circumstances. The accelerated tests mentioned above were conducted near a beam dump, where the dose rates were ten orders of magnitude higher than those experienced by the scintillator material inside the uranium stack. In these tests, no significant degradation of the scintillator performance was observed for total integrated doses of 1,000 Gy, corresponding to a dose the scintillator would receive as a result of ^{238}U decay during about 100 years of operation. Therefore, it was concluded that the radioactivity of the absorber material was not a problem for the scintillator performance.

However, after a few years of operation, it turned out that the scintillator plates in the uranium calorimeter had become visibly yellow. Measurements revealed that the attenuation length for the scintillation light had decreased by a factor of two to three. It turned out that the observed degradation was *not* correlated with the position of the calorimeter module with respect to the interaction point of the ISR, thus ruling out beam-related radiation damage. On the other hand, the degradation was perfectly correlated with the *age* of the calorimeter module. This clearly indicated that the cause of the observed degradation was intrinsic to the calorimeter itself.

The radiation damage turned out to be caused by the interaction between ultraviolet

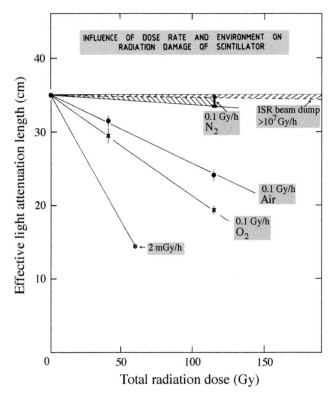

FIG. 5.61. Radiation damage effects observed in the ISR/R807 uranium calorimeter. See text
for details. From [Sir 85].

scintillation light, generated by decaying radioactive uranium nuclei and oxygen that
had diffused into the plastic scintillator. Molecular oxygen was converted into free radicals, such as ozone, in this process. These chemically very active radicals decomposed
the plastic base material. A very similar process occurs in plastics exposed to sunlight,
which turn yellow after a while.

The crucial role of oxygen became clear from a series of systematic experiments,
in which this scintillator material was irradiated with a radioactive source at a dose rate
about 50 times larger than that experienced from uranium decay. The source and the
geometry in which it was placed with respect to the scintillator plates were always the
same in these exposures, only the gas that filled the closed box in which the exposures
took place was varied. Measurements were carried out for air, and for its components,
pure oxygen and pure nitrogen. The results, also shown in Figure 5.61, clearly revealed
that oxygen was responsible for the damage to the scintillator. In the absence of oxygen
(the N_2 case), no significant degradation was observed.

Since diffusion is a very slow process, the effects in the ISR calorimeter were determined and limited by the rate at which fresh molecular oxygen could replenish the
ozonified molecules. In accelerated tests, in which the radiation equivalent of many

years of uranium decay were simulated in exposures of a few hours, the described process did not play a significant role. Once the oxygen that was present at the start of the exposure was converted, the diffusion of new oxygen proceeded much too slowly to be a significant factor in the radiation damage process.

As a result, the accelerated tests dramatically *underestimated* the effects of the uranium-induced radioactivity on the plastic scintillator.

The explanation described above, and in particular the role played by oxygen in this process, was strongly supported by the observation that an identical calorimeter that was built for the HELIOS experiment did *not* suffer from these effects. The HELIOS calorimeter was operated in an inert atmosphere, and after several years of operation, no measurable degradation in the properties of the scintillator had occurred [Ake 87].

5.10.3 *Units*

The radiation dose received by a certain sample is expressed in units called Rads or Grays, which quantify the effects of the processes described above. One Gray (Gy) is defined as 1 J kg^{-1} and equals 100 Rad. These units are in fact *calorimetric* in nature, they express the amount of energy deposited in the form of ionization in an object of a certain mass. The following example may serve to give the reader a feeling for these units.

Imagine a small electromagnetic calorimeter, consisting of lead and plastic scintillator in a ratio 1:1. The detector is $30X_0$ deep and has a diameter of 6 ρ_M, just enough to contain high-energy em showers at the 95% level (see Section 2.1.7). The mass of this detector amounts to ~ 15 kg.

This detector is exposed to a beam of 10 GeV electrons, steered into its center. The radiation dose received by this instrument when it absorbs *one beam particle* can be calculated as follows. The energy of 10 GeV corresponds to $1.6 \cdot 10^{-9}$ J, of which 95% is distributed over 15 kg worth of material. Therefore, the dose received by the detector as a whole equals $\approx 10^{-10}$ Gy. A radiation dose of 1 Mrad (10 kGy) thus requires 10^{14} beam particles to be sent into this detector.

In this example, the received radiation dose is of course not the same for each part of the detector. In the area around the shower maximum, the doses are considerably larger, and in the border regions the doses are considerably smaller. The average shower profiles (see Figures 2.12 and 2.14) are equivalent to the dose profiles in this case, since all particles entered the detector at the same point.

The (average) dose received by the plastic scintillator is *not* the same as the (average) dose received by the whole detector, or by the lead, in this example. Based on the dE/dx values of lead and of plastic scintillator (polystyrene), one finds that a mip deposits 13.6% (2.00/[2.00 + 12.7], see Table B.1) of its energy in the scintillator when traversing the detector. Owing to the fact that $e/mip < 1$, this fraction is smaller for em showers, typically 9–10% (see Figure 3.5).

However, since the plastic scintillator represents 8.3% of the detector mass, the radiation dose received by the scintillator is somewhat larger than the radiation dose received by the lead in this detector.

5.10.4   *Radiation damage mechanisms*

The ISR/HELIOS case described in the previous subsection provided an example of a radiation damage mechanism that was understood in considerable detail and, *as a direct result*, could be prevented. In other cases, the mechanisms are less clear, and sometimes considerably more complex.

5.10.4.1   *Ionizing radiation.*

In organic materials, ionization or excitation of the molecules induced by radiation may lead to dissociation of chemical bonds. In this process, chemically reactive species called free radicals are formed, which initiate further chemical reactions in the matrix. In this sequence of bond breaking and subsequent chemical reactions mediated by free radicals, the molecular structure of the material, and hence its macroscopic properties, are altered.

Polymers such as plastics consist of very long chains of carbon atoms and atoms of other elements (H, O, Cl,...). Individual molecules may consist of tens of thousands of atoms. Side branches and chemical ties between chains (crosslinks) may also be present. The physical properties of the material are determined by the details of this macromolecular structure.

When subjected to ionizing radiation, the most common changes in this structure involve either *scission* or *crosslinking*. In the first process, the bonds between atoms along the backbone of the chain are broken and the macromolecule is chopped into smaller pieces. The other process occurs when free radicals engage in chemical reactions which result in the formation of covalent bonds between adjacent macromolecular chains. Such changes affect many physical properties of the materials. For example, scission makes the polymers usually weaker and softer, while crosslinking makes the material harder and stiffer.

The reasons why some organic materials are considerably more radiation resistant than others is in many cases a mystery. In contrast with intuitive expectation, there is no strong correlation between radiation hardness and chemical or thermal stability. This may be illustrated by the properties of teflon, which is extremely resistant to high temperatures and chemically almost inert, yet it is one of the most vulnerable materials for ionizing radiation.

In general, it is true that polymers that contain aromatic groups (*e.g.*, phenyl rings) are considerably more resistant to radiation than polymers with linear structures. The chemical bonds that keep the ring structure together are much stronger than the ones in the macromolecules discussed above. Therefore, it is much more likely that an excited aromatic ring decays to its ground state than that its internal chemical bonds are dissociated. The aromatic group thus acts as a trap for the excited energy dissipated in the irradiation process. An example of a very radiation resistant aromatic polymer is polyimide, or kapton.

In inorganic crystalline materials, such as scintillating crystals, radiation damage primarily involves creation of defect sites in the structure, resulting from disrupting the chemical bonds between the atoms in the crystal. Such defect sites often appear as color centers in otherwise transparent crystals, which absorb the (scintillation or Čerenkov) light generated in the crystals. There are different types of defects, which are known

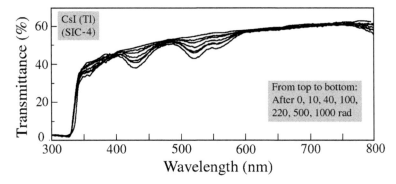

FIG. 5.62. Longitudinal transmittance of a CsI(Tl) crystal after it had received various doses of ionizing radiation [Zhu 98].

as F centers (electrons located in anion vacancies), V centers (holes located in cation vacancies), H centers (interstitial anion atoms) and I centers (interstitial anion ions).

Each type of color center has its own characteristic effect on the light transmittance of the crystal. Typically, this effect consists of well-defined absorption bands. Figure 5.62 shows a number of such absorption bands induced in a CsI(Tl) crystal by ioniz-ing radiation. The absorption bands at wavelengths of 440, 520, 560 and 850 nm are attributed to color centers of the F type [Zhu 98].

Radiation damage of crystals has been and is a very serious concern for CMS. One of the reasons for selecting $PbWO_4$ was the supposed radiation hardness of these crystals, as well as the possibility that eventual damage might be recoverable. However, these crystals face unprecedented radiation levels, especially in the endcap regions. This has been a topic of many detailed studies over the years [Hut 05, Lec 06]. Interestingly, the radiation damage seems to depend on the type of particles that were used in these tests. This is illustrated in Figure 5.63. When γs from a strong ^{60}Co are used to irradi-ate the crystals, the transmission curve is already affected after a dose of $\sim$ 1 kGy (1 hour exposure, Figure 5.63a). However, an exposure of 46 hours did not result in much additional damage (Figure 5.63b), which indicates that there is indeed some recovery going on. On the other hand, irradiation with a beam of high-energy protons clearly resulted in more permanent damage to the light transmission capability of the crystal. The damage is usually expressed in terms of the induced absorption coefficient for light of 420 nm ($\mu_{ind}(420)$), at which wavelength the emission spectrum of $PbWO_4$ peaks. With the ^{60}Co source, $\mu_{ind}(420)$ was measured to be $\sim$ 1 m^{-1}, regardless of the dose, while $\mu_{ind}(420)$ was found to increase about linearly with the integrated dose. Another difference is the shift in the band edge, from $\sim$ 340 nm to $\sim$ 400 nm, observed for protons, while the band edge remains stable in γ irradiations. Similar differences have also been observed for BGO crystals.

These differences are believed to be caused by the extremely dense local ionization caused by nuclear fragments created in nuclear reactions (see Figure 2.49 for an exam-ple). At the moment of this writing, the CMS crystals have received only $\mathcal{O}(1\%)$ of the total integrated luminosity envisaged for the lifetime of the experiment. Yet, radiation

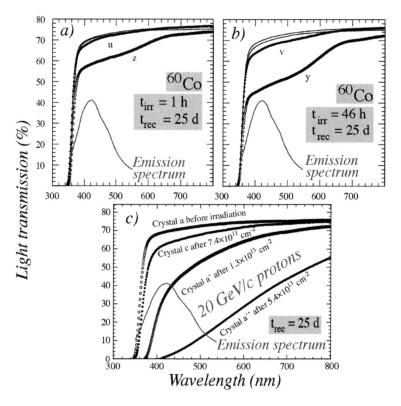

FIG. 5.63. The light transmission in PbWO$_4$ crystals as a function of wavelength, before and after irradiation with ionizing radiation. Results are shown for irradiation with γs from ^{60}Co decay, carried out at a dose rate of 1.1 kGy/hr, for a total period of 1 hour (*a*) and 46 hours (*b*). Also shown are the transmission curves for crystals exposed to beams of 20 GeV/*c* protons, with fluences ranging from $7.4 \cdot 10^{11}$ to $5.4 \cdot 10^{13}$ cm^{-2} (*c*). All measurements were carried out 25 days after the end of the irradiations [Lec 06].

damage effect are already quite significant, especially in the endcap regions. For this reason, it has been decided to replace the endcap calorimeters by some other technology [CMS 15b].

5.10.4.2 *Neutrons.* The radiation damage caused by neutrons often proceeds through the same mechanisms as the ones described above. Neutrons may produce ionizing particles in the form of recoiling nuclei. Protons and αs may be produced in neutron-induced nuclear reactions. Neutrons may also produce γs, in inelastic scattering processes, or when they are captured by a nucleus.

The rate at which these particles are produced depends strongly on the neutron energy and on the nuclear peculiarities (*e.g.*, the cross sections for the various nuclear reactions) of the damaged materials. Therefore, there is no general relationship for the conversion of neutron fluxes into radiation doses.

There is one very specific radiation damage process that is *not* caused by the effects of ionizing particles produced in neutron interactions. This process is the result of *elastic neutron scattering* and it affects semiconductor materials, such as Si, Ge or GaAs. Especially at low energies, below 1 MeV, elastic scattering off nuclei is, in most materials, by far the most likely process to occur, and in many cases it is the *only* process that is energetically possible. For example, in silicon, the lowest-lying excited state in any of the stable isotopes has an excitation energy of 1.27 MeV and inelastic scattering thus requires neutrons of at least that energy to take place.

In elastic scattering, some fraction of the neutron's kinetic energy ($\sim 7\%$, on average, see Section 2.3.3.1), is transferred to the silicon nucleus. This may be sufficient to dislocate this nucleus from its position in the silicon lattice: the neutron thus creates a defect in the lattice structure, consisting of a vacancy and an interstitial silicon atom. If enough kinetic energy is transferred in the collision, the primary Si atom may in turn displace neighboring atoms in the lattice as well.

Such so-called *Frenkel* defects constitute trapping sites for charge carriers. As the number of these defects increases, the silicon structure gradually loses its semiconductor properties, and becomes more and more conducting. The bias voltage applied over the *pn* junction will thus give rise to an increased leakage current.

This type of effect is very specific for low-energy neutrons, which have large cross sections for elastic scattering. Neutrons with kinetic energies down to 160 keV may create lattice defects in a silicon structure [Gro 88]. Such neutrons are abundantly produced in hadronic shower development in calorimeters.

The described process affects all silicon-based electronics, and also the silicon detectors used in particle physics experiments (*e.g.*, the silicon strip detectors used in

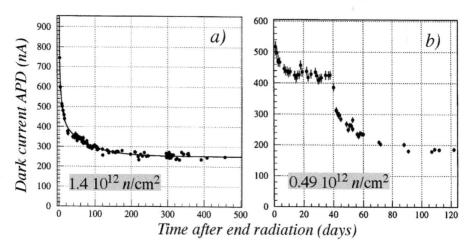

FIG. 5.64. Annealing of the dark current of Hamamatsu APDs, which were irradiated at a nuclear reactor with a fluence of $1.4 \cdot 10^{12}$ n/cm^{-2} (*a*) and $4.9 \cdot 10^{11}$ n/cm^{-2} (*b*), respectively. The annealing took place at room temperature (*a*), or at 0°C for 40 days, followed by room temperature (*b*) [Bac 99].

vertex detectors and in tracking systems). Some of the highest neutron fluences in modern experiments occur inside the calorimeters of experiments at hadron colliders. Figure 2.34, which shows the longitudinal distribution of radioactive nuclei in a calorimeter, may testify to that, since the overwhelming majority of these nuclides were produced in neutron induced reactions. The figure also shows that the maximum fluence occurs at a depth of $\sim 2\lambda_{int}$. This happens to be close to the location of the readout sensors of the crystal ECAL in the CMS detector. In the barrel section of this calorimeter, these sensors are silicon based APDs. Therefore, extensive studies of radiation damage effects in APDs by neutrons have been carried out [Bac 99]. It was found that exposure to large neutron fluences leads to a substantial increase of the dark current, by several orders of magnitude. However, after the radiation was ended, the dark current was observed to drop again (Figure 5.64), albeit only by a factor of two to three, on a time scale of months. This is thus an effect that in practice only will occur during long accelerator shutdown periods. The dark current also seems to drop more and faster when the temperature is increased (Figure 5.64b). It seems that neutrons are not causing any particular radiation damage problems for the vacuum phototriodes, which read out the crystals from the CMS endcaps [Gus 04].

For many years, neutron induced effects in silicon have been a driving force behind the research and development, much of it classified, into radiation-hard electronics.

5.10.5 *Preventing radiation damage effects*

Of course, the best strategy for dealing with radiation damage is to prevent it from happening in the first place. Sometimes, this is possible by installing shielding material, especially if the ionizing radiation comes from a different direction than the particles to be detected with the calorimeter. Readout elements, such as PMTs and the associated electronics, can often be located in a position where the effects of background radiation are negligible.

Another obvious strategy is the use of materials that are sufficiently radiation resistant. Unfortunately, the price of detector components tends to increase strongly with the radiation resistance. For example, quartz fibers with a fluorinated quartz cladding are extremely radiation hard, but also about a factor of five more expensive than more vulnerable plastic-clad quartz fibers. In such cases, the (expected) radiation map of the experimental area is a crucial tool in the design of the detectors. It provides guidance for determining which areas of the detector need to be equipped with the expensive, radiation-hard components, and in which areas cheaper, less resistant components may suffice.

In many experiments, neutrons are the main source of radiation-induced damage to the detectors. These neutrons are produced in hadronic absorption processes. The number of neutrons produced per unit of absorbed energy depends strongly on the Z value of the absorber material. For example, in high-Z materials such as lead, this number is about three times as high as in iron or copper. This is a result of

1. The fact that high-Z nuclei are much more neutron-rich than low-Z ones. For example, the neutron/proton ratio in lead is $\sim 30\%$ larger than in iron or copper.
2. The binding energy per nucleon is larger in iron and copper than in lead, by more

than 10%. Therefore, fewer nucleons are released from the nuclei per unit of absorbed energy.

3. In high-Z materials, the release of protons from the nuclei is much less likely than the release of neutrons, because of the effects of the Coulomb force. In low-Z materials, this effect is much smaller (see Section 2.3.2.3 for more details).

A significant fraction of the neutrons produced in the calorimeters escapes through the front face (albedo, see Section 4.5.3) and may thus affect detectors installed upstream. Therefore, the choice of the absorber material in the calorimeter has direct consequences for the neutron fluences elsewhere in the experimental setup.

In some specific calorimeters, radiation damage may be prevented by choosing appropriate working conditions. For example, the mobility of defects created in crystalline silicon is strongly temperature dependent. It has been demonstrated that the useful life of silicon detectors in a high-radiation environment can be increased considerably by operating these devices at low temperature ($< 5°C$) [Zio 94]

In Section 5.10.2, we saw that the combination of oxygen and UV light was found to be particularly damaging for certain plastic scintillators. By operating this material in an oxygen-free environment, the lifetime of the calorimeter was considerably extended.

5.10.6 *Curing radiation damage effects*

Yet, in modern accelerator-based experiments, which mostly operate at very high luminosities, it is often hard to avoid radiation damage completely. In some materials, the effects of this damage may be cured, for example by thermal annealing or optical bleaching.

Such procedures have been demonstrated to work well for several types of scintillating crystals, such as BGO [Zhu 91], BaF_2 [Zhu 94], and $PbWO_4$ [Zhu 96]. As a matter of fact, the recovery observed in the scintillating-fiber calorimeter discussed in Section 5.10.2 can also be considered as an example of annealing, at room temperature. However, there are also crystals for which such procedures have (almost) no effect, for example CsI where the recovery of the light output proceeds at a rate $\ll$ 1%/day [Zhu 98], while crystals such as BGO may fully recover on a timescale of a few hours.

Both the radiation sensitivity and the recovery speed seem to depend crucially on the presence of certain trace materials in the crystals. This is the reason why these issues have acquired a reputation of black magic, at least among non-experts.

5.10.7 *Coping with radiation damage effects*

It is important to realize that radiation damage is a quantifiable effect. In well-designed experiments, the physics goals will only be compromised if and when the detectors suffer radiation damage *beyond a certain level*. In practice, it is therefore important to monitor the performance of the detectors regularly to keep track of the level of radiation damage. Many experiments have procedures in place precisely for this purpose.

An example of a quantitative study of the effects of radiation damage on the performance of a scintillating fiber calorimeter is given in [Aco 91e]. The authors assumed a certain, parameterized radiation damage profile and calculated the effects of that on important calorimetric properties, such as the energy resolution for em and hadronic

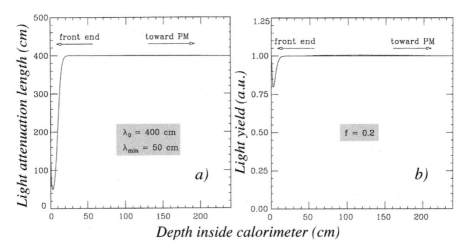

FIG. 5.65. The radiation damage to scintillating fibers in a hadron collider, as parameterized in
Monte Carlo simulations: the impact on the light transmission (*a*) and on the light emission
(*b*) as a function of the position along the fibers. The radiation damage was assumed to be
caused by 1 GeV photons in these simulations. From [Aco 91e].

showers, the $\pi^0/\pi^\pm$ response ratio and the signal linearity.

The radiation damage profile was assumed to be caused by photons with an energy
of 1 GeV. These soft photons are a dominant source of radiation deposited in the calorimeters at high-energy hadron colliders [Gro 88]. The profile was characterized by two
parameters, $\lambda_{\min}$ and f. The parameter $\lambda_{\min}$ describes the light attenuation length at
the maximum of the radiation damage profile, after irradiation. The fractional decrease
of the light emission at that maximum as a result of the irradiation is indicated by f.
The value of these parameters can be measured in radiation damage tests, as a function
of the absorbed dose.

An example of radiation damage profiles used in this study is given in Figure 5.65.
Figure 5.65a shows the light attenuation length as a function of the position along the
scintillating fibers. In the non-irradiated parts of the fibers, this attenuation length was
400 cm. However, at the position where the 1 GeV photons that caused the damage
reached their shower maximum, the local attenuation length has dropped to 50 cm
(*i.e.*, $\lambda_{\min} = 50$ cm). Figure 5.65b shows the light emission as a function of the position along the scintillating fibers, for a reduction of 20% at the shower maximum
($f = 0.2$).

Some results of these simulations are shown in Figure 5.66. In this figure, the ratio
of the responses to neutral and charged pions is given as a function of $\lambda_{\min}$, for three
different values of fractional light emission loss f. The energy of the incoming particles
was fixed at 10 GeV and the radiation damage was assumed to be caused by 1 GeV
photons. The results depend mainly on f. The dependence on $\lambda_{\min}$ is very weak, as
the local transmission loss is only effective across the small (5–10 cm) area where the
photon showers caused damage.

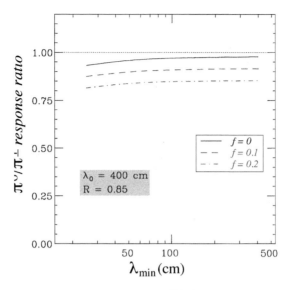

FIG. 5.66. The effect of radiation damage on the π^0/π^+ response ratio of a scintillating-fiber calorimeter. This ratio is shown as a function of $\lambda_{\min}$ for three different values of f. Results of Monte Carlo simulations for 10 GeV particles. The attenuation length before irradiation amounted to 4 m and the radiation damage was caused by 1 GeV γs. From [Aco 91e].

Such results are important for understanding the effect of radiation damage on the jet response. In this example, the π^0/π^+ response ratio decreased by 6.6% for $f = 0.1$. At higher energies, the effects were somewhat smaller, since the π^0 showers deposited a smaller fraction of their energy in the damaged region. At 150 GeV, the decrease of the π^0/π^+ response ratio was only 4.2%.

By combining the results of such Monte Carlo studies with radiation damage tests, which provide experimental values for the parameters $\lambda_{\min}$ and f, one can determine the radiation level that can be tolerated while still maintaining an acceptable calorimeter performance. For example, the authors of [Aco 91e] found in this way that a radiation dose of 6.1 Mrad (61 kGy) would induce a constant term of 0.8% in the em energy resolution of a calorimeter equipped with a certain type of scintillating plastic fibers (SCSN81). To what extent such effects are acceptable depends of course on the physics goals of the experiment in which this calorimeter is used.

5.10.8   *Induced radioactivity*

One aspect of operating in a high-radiation environment that tends to be overlooked is the accumulation of long-lived radioactive nuclides in the detectors. As the collision energy, and thus the luminosity required to do physics, increase, I expect that this will turn out to be an increasingly important issue.

There are two aspects to this problem. First, it affects the measurements, primarily through an increase of the general noise levels. In that sense, all calorimeters become like those using ^{238}U as absorber material. The difference with ^{238}U is that the induced radioactivity may consist of a large variety of different nuclides, each with its own half-life and spectral decay characteristics. As time goes by, the composition of the induced radioactivity changes. The fraction of long-lived nuclides slowly builds up over time, while the short-lived ones are continuously replenished.

The best strategy for dealing with the varying noise levels introduced by this problem is to measure pedestal levels just before and after each bunch crossing. The applicable background level can then be obtained by averaging the integrated charge measured just before and just after the passage of the particles one wants to measure. This procedure is followed, for example, in the ATLAS experiment.

The second aspect of induced radioactivity is the complication of detector maintenance caused by it. As the experiment proceeds, the detector becomes more and more radioactive. Because of this, any modifications/repairs/inspections that need to be carried out become gradually more difficult over time.

In Section 2.3.4, we described shower profile measurements that were carried out by studying the distribution of the induced radioactivity in stacks of metal absorber plates [Ler 86]. Thirty years later, these stacks continue to be stored in a dump site for radioactive waste. Most likely, this will also be the ultimate fate of much of the equipment installed in the LHC interaction areas, once the experiments are finished.

In the profile measurements mentioned above, most of the radioactive nuclides were produced by neutrons. This is in general true for induced radioactivity. This is because neutrons do not have to penetrate the Coulomb barrier to reach a nucleus. Therefore, there is no lower limit to the energy they need to carry in order to be capable of inducing a nuclear reaction. Also, the cross sections for some neutron-induced nuclear reactions are very large, up to kilobarns for some (thermal-neutron) capture reactions.

Because of the dominating role of (thermal) neutrons, the induced-radioactivity distribution in the experimental areas of a collider such as the LHC may be very different from the dose profiles for ionizing radiation. The thermal neutrons behave much like a gas, filling the entire cavity in which the experiment is installed, and they bounce around until one of two things happens: they decay or are captured by a nucleus. And since the half-life for decay is about 15 min, capture is by far the more likely process.

The level of long-lived induced radioactivity depends critically on the choice of materials. For example, small amounts of cobalt (^{59}Co) may turn into very strong, long-lived MeV γ sources of ^{60}Co when exposed to thermal neutrons. Johnson and coworkers, who studied the radiation hardness of a lead/scintillating-fiber calorimeter, found that the long-lived radioactivity induced in this instrument consisted predominantly of the antimony isotopes ^{120m}Sb, ^{122}Sb and ^{124}Sb [Joh 92]. The lead used in the construction of this detector contained a small percentage of antimony (used to improve the mechanical properties), which served as the production target for these nuclides.

As illustrated by this example, it is thus important to consider the propensity of chemical elements to form long-lived, problematic radioactive nuclides as one of the design criteria for experiments in a high-dose environment.

6

CALIBRATING A CALORIMETER SYSTEM

Calorimeters are instruments intended to measure the energy of particles detected by means of absorption. These particles deposit energies that are typically of the order of GeVs. However, the calorimeter produces signals that are typically expressed in pico-Coulombs, or photoelectrons. By calibrating the calorimeter, the relationship between the units in which the calorimeter signals are expressed and the energy of the measured particles is established. The calibration constants therefore have the dimension "GeV pC^{-1}," or its inverse.

These calibration constants can be determined experimentally with particles that deposit a known energy in the calorimeter or in one of its segments. Frequently, test beams of monoenergetic particles, *e.g.*, electrons or pions, are used for that purpose. From the calorimeter signal distribution for such monoenergetic particles, the calibration constants can be determined with (statistically) high precision.

At first sight, the calibration of a calorimeter system may seem to be straightforward and trivial. **It is not.** It is not, to such an extent, that an entire chapter of this book is dedicated to this crucial aspect of working with calorimeters.

In the previous chapters, the various reasons why calibrating a calorimeter system is not a trivial and straightforward job have already been encountered. They are:

1. The calorimeter response depends on the *type* of particle, or the type of jet. The calorimeter response is defined as the *average signal per GeV*. Therefore, the calibration constants also depend on the type of particle, or jet. This also means that calibration constants determined for one particular type of particle, or jet, lead to *systematic mismeasurements* of energy if used for the interpretation of signals caused by other types of particles, or jets.

2. The calorimeter response, and therefore the calibration constants, depend on the *energy* of the particles, or the jets. This means that calibration constants determined for particles or jets of one particular energy lead to systematic mismeasurements of energy if used for particles or jets of other energies.

3. The calorimeter response to showering particles is a function of the *shower age*. In Section 3.2.2, it was shown that the em response of certain calorimeters may change by as much as 30% between the early and late stages of the shower development. Even larger effects may occur in hadronic showers (Section 3.2.8). Ignoring these effects may lead to considerable mismeasurement of energy in individual calorimeter segments, even in compensating calorimeters for which the other effects mentioned above do not play a role.

All these effects may thus lead to a *systematic mismeasurement* of the energy. This fact tends to be ignored. In practice, the precision of the energy measurement with a given calorimeter system is often derived from the width of a signal distribution which is minimized by manipulating the calibration constants for the different calorimeter segments. The fact that different types of particles or jets may give signal distributions with *different mean values* as a direct consequence of such a procedure is not taken into account. In Sections 6.2 and 6.3 these statements are illustrated with examples taken from practice.

Calibration problems are most severe for longitudinally subdivided calorimeter systems, especially if the calorimeter consists of sections with very different e/h values. However, before going into these problems, we first discuss the easier case of a longitudinally unsegmented calorimeter.

6.1   Longitudinally Unsegmented Systems

The best way to calibrate a longitudinally unsegmented calorimeter system is to send a beam of electrons with precisely known energy into the center of each and every calorimeter cell. If the cell size is such that the em showers are completely contained in one cell, the calibration constant of each cell is given by the ratio of the beam energy and the mean value of the signal distribution for that cell.

If the shower energy is only partially contained in one calorimeter cell, the calibration constants C_i have to be determined simultaneously for all cells i, by minimizing the quantity

$$Q = \sum_{j=1}^{N} \left[E - \sum_{i=1}^{n} C_i S_{ij} \right]^2 \tag{6.1}$$

in which E denotes the beam energy and S_{ij} the signal from calorimeter cell i for event j.

In calorimeters where the signals are amplified, it is often convenient to choose the gain factors such that equal energy deposits lead to equal signals, in all calorimeter cells. In that way, signals from different cells may be directly added without applying normalization constants. This is particularly convenient when combinations of calorimeter signals from different cells are used to decide whether an event meets preset selection criteria (triggering).

If the calorimeter is compensating, the calibration constants derived from em shower detection are also valid for hadrons and jets. This is the most ideal situation.

If the calorimeter is non-compensating, energy-dependent correction factors need to be applied to derive the energy of hadrons and jets from the calorimeter signals generated by these objects. These correction factors can be established with test beams of hadrons and electrons, which preferably span the entire energy range of interest in the experiment for which the calorimeter is intended.

6.2   Longitudinally Segmented Systems

Almost all calorimeter systems used in practice consist of several longitudinal segments. Most systems consist of two segments, named the *electromagnetic* and *hadronic* calorimeter sections, but three or even more segments are not unusual either.

There are two main reasons for having at least two longitudinal segments:

1. It allows electrons and photons to be distinguished from hadrons.
2. It makes it possible to adapt important design parameters, such as the sampling fraction and the granularity of the calorimeter, to the particles to be detected, and thus use available resources in the most economic way.

Calibration complications are the price to pay for these advantages. These complications are such that a proper alternative title of this chapter would be:

"(Mis)calibration – The pitfalls of longitudinal segmentation"

As we will see, it is even questionable if there is a correct way to calibrate longitudinally segmented calorimeters at all.

6.2.1 *The basic problem*

We start the description of these complications with a very simple case, involving em showers in a homogeneous calorimeter, which we assume to be based on the detection of Čerenkov light, *e.g.*, a block of lead-glass (Figure 6.1). When electrons are sent into

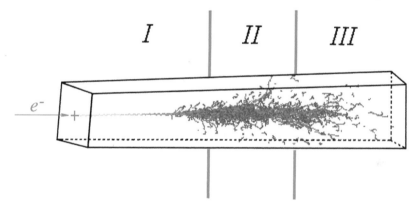

FIG. 6.1. A hypothetical homogeneous lead-glass calorimeter in which an electromagnetic shower develops. This detector is cut in three components, which are read out separately.

this detector, Čerenkov photons are generated in the absorption of the shower, and these photons are converted into photoelectrons in a light sensor, which produces the signals. When we calibrate this device with 100 GeV electrons, the signal from this sensor consists, on average, of 1,000 photoelectrons. The calibration constant for this calorimeter is thus 10 photoelectrons per GeV deposited energy (10 p.e./GeV, or 0.1 GeV/p.e.). Since this calorimeter is linear, a beam of 20 GeV electrons will produce a signal that, on average, consists of 200 photoelectrons (20×10, or $20/0.1$). Since we have defined the calorimeter response as the average signal per GeV deposited energy, we can also say that the *response* of this calorimeter is 10 photoelectrons (per GeV), and the average signal for a 50 GeV electron will thus consist of $50 \times 10 = 500$ photoelectrons.

Next, we are going to cut this detector into three parts, or rather we arrange things in such a way that the signals produced in segments I, II and III are detected separately

(Figure 6.1). This cut is made such that the 100 GeV electrons that were used for the detector calibration deposit, on average, 30% of their energy in segment I, 40% in segment II and 30% in segment III.

Čerenkov light is only produced by the charged shower particles that are sufficiently relativistic, *e.g.*, the electrons and positrons that carry at least 0.3 MeV kinetic energy. Shower particles with energies below this cutoff value do participate in the energy deposition process, but *not* in the signal generation. These soft particles are rather rare in the early stages of the shower development, but they dominate in the late stages. This means that if we now calibrate the three segments of the calorimeter separately, a different relationship will be found between deposited energy (in GeV) and resulting signal (in photoelectrons) for these three segments. For example, we find that in the first segment 15 photoelectrons are produced per GeV deposited energy. In segment II, the signals from the 100 GeV electron showers consist, on average, of 10 photoelectrons per GeV deposited energy, and in segment III the calibration constant is 5 photoelectrons per GeV.

With these new, separate calibration constants, the average total calorimeter signal for 100 GeV electron showers is still the same as before the cut was made. Because of the 30%/40%/30% sharing of the deposited energy, segment I will contribute on average $30 \times 15 = 450$ photoelectrons to the total signal, segment II $40 \times 10 = 400$ and segment III $30 \times 5 = 150$, for a total of $450 + 400 + 150 = 1,000$ photoelectrons.

However, if we now send electrons of another energy into this segmented calorimeter, the energy sharing between these three segments will be different than for the 100 GeV ones, because of differences in the longitudinal shower profile (Figure 2.9). For example, the energy sharing among segments I, II and III for a 20 GeV electron is, on average, 45%/35%/20%, *i.e.*, 9 GeV in segment I, 7 GeV in segment II and 4 GeV in segment III. Based on the calibration constants of these segments established with the 100 GeV electrons, the energy of the 20 GeV electrons will be *underestimated*. This can be easily seen as follows. Imagine that the entire energy of the 20 GeV shower was deposited in segment I. The shower does not know that the calorimeter readout is split into three segments and will still produce 200 photoelectrons, as before. However, these would now be interpreted as an energy deposit of 200/15 = 13.3 GeV, because energy deposit in segment I is converted on the basis of 15 p.e./GeV. Given the mentioned energy sharing for 20 GeV showers, the energy would only be correctly reproduced if the shower produced $9 \times 15 + 7 \times 10 + 4 \times 5 = 225$ photoelectrons, instead of 200. The showers would thus be assigned an energy of $20 \times 200/225 = 17.8$ GeV in this example.

Similarly, the energy of electrons with an energy larger than 100 GeV would be systematically *overestimated* in this example, since they would deposit a relatively large fraction of their energy in segment III, which is more generous in converting photoelectrons into GeVs than the other segments. For example, a 500 GeV electron with energy sharing 20%/30%/50% = 100/150/250 GeV, would be reconstructed at the correct energy if it generated $100 \times 15 + 150 \times 10 + 250 \times 5 = 4,250$ photoelectrons. Since the particles generate, on average, 5,000 photoelectrons, their signals would be interpreted as coming from a 588 GeV electron.

These problems are not limited to electrons with different energies than the one used for calibrating the individual longitudinal segments. Also a γ of 100 GeV would be reconstructed, on average, with a different energy than a 100 GeV electron. This is a consequence of differences between the average longitudinal shower profiles (Section 9.3.4.2). Since γ showers deposit their energy typically deeper inside the calorimeter than electron showers of the same energy, the energy of a 100 GeV γ would thus be interpreted as coming from a higher-energy object, *e.g.*, 105 GeV. Experiments that have developed elaborate calibration schemes for their longitudinally segmented em calorimeter, based on electron detection (*e.g.*, ATLAS, Section 6.2.4) ought to keep this in mind when reconstructing γ showers.

This example illustrates the problems caused by item #3 listed in the introduction of this chapter, namely the fact that the calorimeter response (*i.e.*, the relation between deposited energy and resulting signal) changes as the shower develops, and thus is a function of the depth inside the calorimeter. This implies that the calibration constants, which relate signals to deposited energy, are different for the different segments of a longitudinally segmented calorimeter. And since the shower profiles change with the particle energy, this leads to signal non-linearities. This is a very serious problem, as we will see in the following sections, which deal with real, non-hypothetical calorimeters.

6.2.2 *The HELIOS calorimeter*

The calorimeter system for the HELIOS experiments in the heavy-ion beam at CERN [Ake 87] was a recycled version of the first uranium calorimeter developed for the R807 experiment in the Intersecting Storage Rings [Ake 85]. The calorimeter, based on thin uranium plates interleaved with 3 mm thick plastic scintillator sheets, was longitudinally subdivided into two sections. The first (em) section was 6.4 radiation lengths deep. The second (hadronic) section was $\sim$ 4 nuclear interaction lengths deep. In the transverse plane, the calorimeter consisted of rectangular towers, measuring 20×20 cm^2, read out from two sides by means of wavelength-shifting plates (Figure 6.2).

This calorimeter system was calibrated with electrons of 8, 17, 24, 32 and 45 GeV, and also with cosmic muons. When the electron beam was steered into the center of a calorimeter tower, showers were completely contained in one of the towers, with comparable fractions of the energy deposited in the two longitudinal segments. This is graphically shown in Figure 6.3.

The calibration constants A and B for the em and hadronic calorimeter sections, respectively, were determined by minimizing the width of the total signal distribution, *i.e.*, by minimizing the quantity

$$Q = \sum_{j=1}^{N} \left[E - A \sum_{i=1}^{n} S_{ij}^{\text{em}} - B \sum_{i=1}^{n} S_{ij}^{\text{had}} \right]^2 \qquad (6.2)$$

where E is the electron beam energy and $\sum S^{\text{em}}$ and $\sum S^{\text{had}}$ are the sums of all the signals in the towers i of the em and hadronic calorimeter sections that contributed to the measured signal for event j. With this method, values for A and B and, more importantly, for the intercalibration constant B/A were determined for each calorimeter

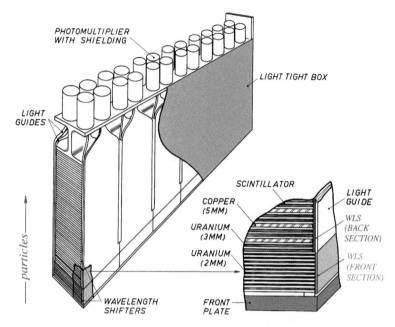

FIG. 6.2. Isometric view of one of the HELIOS calorimeter modules. The enlargement shows the arrangement of the optical readout and the sampling for the em and the hadronic sections. In this particular module, every third uranium plate of the hadronic section was replaced by a 5 mm thick copper plate. Such modules were used in the peripheral regions of the experimental setup. The central calorimeter modules only contained uranium absorber [Ake 87].

tower. However, two fundamental difficulties were encountered when this calibration method was applied:

1. The values of A, B and B/A were found to be energy dependent.
2. The values of B/A differed considerably (on average, more than 20%) from the ones found with muons.

This is illustrated in Figure 6.4, where the fractional widths of various total signal distributions are plotted as a function of the value of B/A (the back/front weighting factor). In the following, we use B/A values that are normalized to the one for muons traversing both calorimeter sections. The value $B/A = 1$ (indicated by the dashed line in the figure) thus represents the calibration result derived from the muon signals, as described below. If $B/A > 1$, then the signals from the hadronic section were given a relatively larger weight for the calculation of the total energy. If $B/A < 1$, then the signals from the em section were given a larger weight.

The B/A value for muons, and thus the normalization factor for all measurements, was experimentally determined as follows by the HELIOS group. Cosmic muons traversing both sections of a given tower were selected by means of a cosmic-ray telescope. These muons generated the characteristic Landau signal distributions in the em and hadronic sections of the tower. The B/A value was determined from the ratio of the

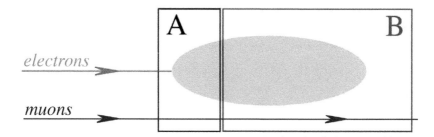

FIG. 6.3. The HELIOS modules were calibrated with beams of electrons. The showers produced by these particles produced signals of comparable strength in both sections of the calorimeter modules . Alternatively, the modules were calibrated with muons, which traversed
the entire module [Ake 87].

most probable signal values in the hadronic and em sections of the tower. This ratio
was compared with its expected value. The latter did not require an experimental measurement, but could be calculated from the composition of the calorimeter and from the
specific ionization of the active and passive materials for mips traversing both sections.
The overall normalization factor was chosen such as to equalize the measured and expected signal ratios. This procedure was repeated for each individual calorimeter tower.

Figure 6.4a shows that the fractional widths of the signal distributions for electrons
reached a minimum for B/A values well below the value expected on the basis of the
muon measurements ($B/A = 1.0$). In addition, the B/A value for which this width
reached a minimum shifted upward with the electron energy, by about 20% over the
energy range for which measurements were done (8–45 GeV). Figure 6.4b shows that
the hadronic energy resolution of the HELIOS calorimeter was much less sensitive to
the front/back weighting factor.

The HELIOS group also found that using B/A values different from 1.0 resulted in
signal non-linearity: the em response became energy dependent. As we saw in Section
3.1.1, signal linearity for em showers is a very fundamental calorimeter property, since
the entire em shower energy is used to ionize and excite the molecules of which the calorimeter consists. Twice as much energy thus leads to twice as many excited and ionized
molecules and should thus lead to calorimeter signals that are twice as large. In the
HELIOS case, it seemed impossible to fulfill this fundamental requirement and to optimize the energy resolution for em showers simultaneously.

The explanation for the described phenomena lies in the fact that the sampling fraction, *i.e.*, the fraction of the deposited shower energy that is converted into a measurable
signal, changes with depth (Section 3.2.2). In the (HELIOS) case of uranium absorber
and plastic scintillator as active material, the sampling fraction *decreases* considerably
as the shower develops, by as much as 25–30% over the volume in which the absorption
takes place. In the early stage of its development, the shower still resembles a collection of mips, but especially beyond the shower maximum the energy is predominantly
deposited by soft (< 1 MeV) γs. The latter are much less efficiently sampled than mips

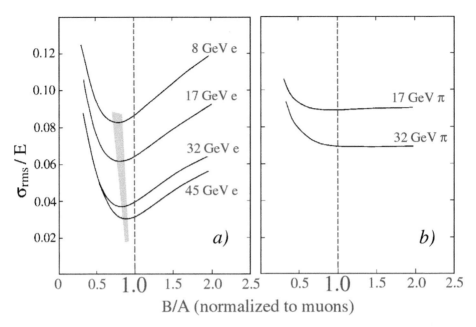

FIG. 6.4. The fractional width σ/E of the signal distributions for electrons (*a*) and pions (*b*) of different energies, as a function of the value of the intercalibration constant B/A of the HELIOS calorimeter system. The dashed line corresponds to the intercalibration constant derived from muon measurements [Ake 87].

in this type of structure, where dominant processes such as photoelectric effect and Compton scattering strongly favor the high-Z absorber material. Therefore, a given energy deposit by the fast (more mip-like) component of the shower in the electromagnetic calorimeter section leads to a considerably larger signal than the same energy deposited in the soft tail in the hadronic calorimeter section (Figure 6.5).

The effect of this on the calibration results is energy dependent. The optimal B/A ratio reflects the difference in (average) sampling fractions between both calorimeter sections. As the electron energy is increased, more energetic shower particles penetrate the hadronic calorimeter section and the response *difference* between the two calorimeter segments becomes smaller, *i.e.*, the optimal value of the B/A weighting factor gets closer to 1 (Figure 6.4).

This phenomenon indicates that there is a fundamental problem. The relationship between the energy deposited by the shower particles and the resulting calorimeter signal (*i.e.*, the calorimeter response) is very different for the two sections of the calorimeter and, moreover, energy dependent. Translating the signals in the individual calorimeter sections into deposited energy is therefore an extremely delicate issue, no matter which calibration constants are being used.

Based on the described phenomena, HELIOS decided to use the B/A values derived from the muon measurements as the basis of their calorimeter calibration. The values of

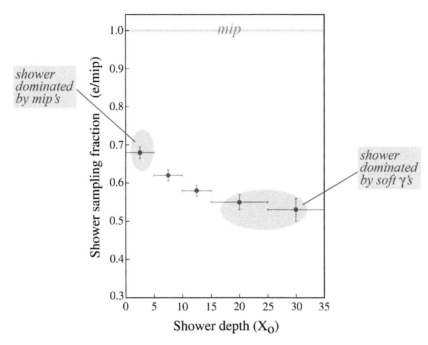

FIG. 6.5. The sampling fraction for 10 GeV electron showers developing in a Pb/plastic-scintillator calorimeter, as a function of depth. This sampling fraction is normalized to that for mips traversing the same calorimeter. Results from EGS4 Monte Carlo simulations [Wig 06b].

A (and thus automatically of B) were fixed such as to reproduce best the electron beam energies over the measured range. In this way, signal linearity was achieved.

In the HELIOS calorimeter, the thickness of the electromagnetic calorimeter section was chosen such as to achieve roughly equal energy sharing for em showers. It turned out that this choice *maximized* the described effects. The hadronic energy resolution was much less sensitive to the value of B/A, because typically only a small fraction of the shower energy was deposited in the front section. Figure 6.4b shows that the hadronic energy resolution was only significantly affected for unrealistically small B/A values, *i.e.*, for a situation in which the small signals in the front section were given an unrealistically large weight.

There are some very important lessons to be learned from this HELIOS experience:

- Showers should *not* be used to intercalibrate sections of a longitudinally segmented calorimeter.

- In a longitudinally segmented calorimeter, no matter how the segments are intercalibrated, only the *total* shower energy derived from the signals is meaningful. The signals from the individual calorimeter segments cannot be interpreted in a straightforward way in terms of deposited energy.

In Chapter 2 we saw that electromagnetic shower development is very well described in the Monte Carlo simulation package EGS4. In the next subsection, we take advantage of this to investigate some of the difficulties described here in a more quantitative way.

6.2.3 *Intercalibration with em showers*

We have used the EGS4 Monte Carlo simulation package for a detailed study of the development of electron showers in longitudinally segmented uranium/plastic-scintillator calorimeters. We had three goals in mind with these simulations:

1. To study the differences between the signals recorded in the two longitudinal segments and the energy deposited in these segments, as well as the energy dependence of these differences.
2. To study the effects of the choice of the intercalibration constant for the signals from the two longitudinal segments on the energy resolution and the response linearity in the specific case of the HELIOS calorimeter.
3. To study the consequences of the non-linearity resulting from the choice of the intercalibration constant in a more general sense.

6.2.3.1 *Energy sharing and signals.* We started by simulating the development of electron showers in a calorimeter for which both segments had exactly the same structure: 3 mm uranium plates, interleaved with 3 mm thick plastic-scintillator plates. This calorimeter was longitudinally subdivided into two segments. The first seven sampling layers $(6.6X_0)$ constituted the first segment, the second segment consisted of the remaining (50) sampling layers.

Electrons of 1, 2, 5, 10, 20, 50 and 100 GeV were sent into this calorimeter. The average fraction of the shower energy deposited in the first section was found to decrease from 62.9% at 1 GeV to 18.6% at 100 GeV. However, these fractions were different if we restricted ourselves to the energy deposited in the scintillator. For example, at 1 GeV, less than 30% of the total signal came from the second calorimeter segment, while 36.7% of the shower energy was deposited in that segment. Therefore, the calorimeter signals did not match the energy sharing between the two sections. This effect is a direct consequence of the decreasing sampling fraction in the developing shower (Section 3.2.2, Figure 3.6).

Table 6.1 lists the sharing of the shower energy and the calorimeter signals between the two sections for all energies. The ratio of the deposited energy and the resulting calorimeter signal is listed as well. This ratio gives an indication of the extent to which the energy in the individual calorimeter segments is mismeasured, if the calorimeter signals are assumed to be a measure of that energy. In Figure 6.6, the fractional mismeasurement of the energy deposited in the individual sections is shown as a function of the energy of the incoming electrons. The energy deposited in the first segment is systematically overestimated; the energy in the second segment is underestimated. Depending on the energy, these effects may exceed 20%.

When this calorimeter was considered an *unsegmented* device, *i.e.*, when the energy deposited in all active layers was added to form the total signal, the detector turned out to

Table 6.1 *The average fractions of the total shower energy (f_E) and of the total calorime-ter signal (f_s) deposited in the first and second segments of a longitudinally segmented ura-nium/plastic-scintillator calorimeter, for different energies of the electrons showering in this de-vice. The ratio of these fractions is given as well.*

Energy	Section 1			Section 2		
(GeV)	f_E	f_s	f_s/f_E	f_E	f_s	f_s/f_E
1	62.9%	70.1%	1.115	36.7%	29.9%	0.815
2	55.1%	62.9%	1.142	44.6%	37.1%	0.832
5	44.9%	53.0%	1.182	54.8%	47.0%	0.857
10	37.9%	45.8%	1.208	61.9%	54.2%	0.875
20	31.1%	38.5%	1.237	68.7%	61.5%	0.895
50	23.1%	29.6%	1.278	76.8%	70.4%	0.918
100	18.6%	24.2%	1.303	81.3%	75.8%	0.932

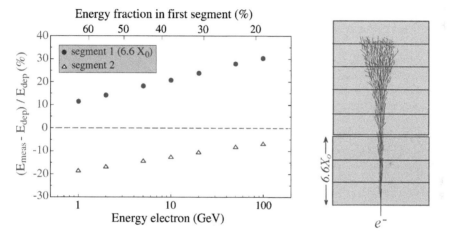

FIG. 6.6. Fractional mismeasurement of the energy deposited in the individual sections of a lon-gitudinally segmented uranium/plastic-scintillator calorimeter, as a function of the energy of the showering electrons (bottom axis) or the energy sharing between the two calorimeter sec-tions (top axis). The energy in the first, $6.6X_0$ deep section is systematically overestimated, the energy in the second segment is systematically underestimated, when the scintillator sig-nals are considered a measure for the deposited energy.

be perfectly linear, the response was constant to within the statistical errors. The energy resolution was found to scale with $E^{-1/2}$ and the signal distributions were perfectly Gaussian. None of these results is surprising in any way.

However, when the signals from one calorimeter segment were given a different weight than the signals from the other segment (the equivalent of $B/A \neq 1$ in the previous subsection), deviations from this ideal picture were observed, especially when the sampling fractions in both segments were chosen to be different, as in the HELIOS calorimeter.

6.2.3.2 *Energy resolution and non-linearity in HELIOS.* Our next set of simulations concerned a structure identical to that used in HELIOS. A 6.3 X_0 deep first segment consisting of ten 2 mm uranium plates, interleaved with 2.5 mm scintillator plates, was followed by a second segment consisting of 3 mm uranium plates, interleaved by 2.5 mm scintillator plates. Figure 6.7a shows the fractional width $(\sigma_{\mathrm{rms}}/E)$ of the signal distributions for electrons of different energies as a function of the relative weight given to the signals from the second calorimeter segment. The phenomena measured with the HELIOS calorimeter (Figure 6.4) were also observed in these Monte Carlo simulations, albeit that the experimental effects seem to be somewhat larger than those found in our Monte Carlo simulations. However, in both cases the width of the total signal distribution was significantly reduced when the signals from the second segment were given a smaller weight than the signals from the first segment.

Since the thickness of the absorber plates in the first section was two-thirds of that in the second section, the sampling fraction in the first section was about one and a

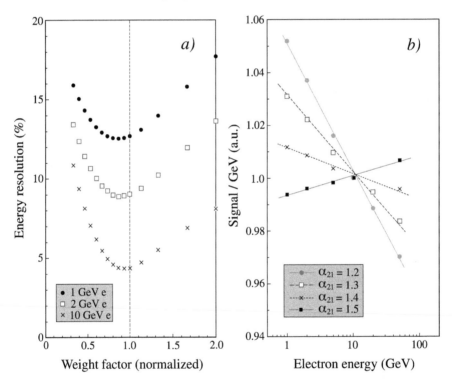

FIG. 6.7. The width σ_{rms}/E of the signal distribution for electrons of 1, 2 and 10 GeV, as a function of the value of the relative weight given to signals from the second section ($> 6.3X_0$) of a uranium/plastic-scintillator calorimeter (*a*). The em response of the same calorimeter as a function of energy, for different values of the intercalibration constant (α_{21}) of the two sections (*b*). Results from EGS4 Monte Carlo calculations. See text for details.

half times larger than in the second section. Giving the signals from both sections the same weight thus meant in this context that the total shower energy was determined as $S_{\text{tot}} = S_1 + 1.5S_2$, where S_1 and S_2 denote the signals from the first and second calorimeter sections, respectively, and $\alpha_{21} = 1.5$ is a factor that represents the difference in sampling fractions between these two sections.

The width of the total signal distribution for 1 GeV electrons reached a minimum when the sum $S_1 + 1.3S_2$ was used to determine the total shower energy, *i.e.*, when the signals from the second segment were given a smaller weight ($1.3/1.5 = 0.87$) than those from the first segment. For 10 GeV electrons, this minimum was reached when a weighting factor of 0.93 was applied. With optimum weighting factors, the fractional width of the total signal distributions was typically a few percent smaller than the width of the distributions corresponding to $S_1 + 1.5S_2$.

The reader will have noticed that we systematically refer to this effect as a "decrease in the fractional width of the signal distribution," or something to that effect, and explicitly avoid the phrase "improvement of the energy resolution." This is because we believe that the observed decrease in the width of the total signal distributions may *not* be interpreted as an improvement in the energy resolution. This can be seen as follows.

Based on the observed improvements in the widths of the signal distributions, one might decide to use a uniform weighting factor of, for example 0.9, for the entire energy range. Signals from the second calorimeter segment are thus multiplied by this factor, while the ones from the first segment are used as such. This will not give optimum results for each and every energy, but it provides a procedure that is easy to use in practice and that promises a significant improvement over the entire energy range. This is precisely the procedure tried (and rejected) by HELIOS (see Section 6.2.1).

Such a procedure leads to a significant signal non-linearity. In Figure 6.7b, the calorimeter response is shown as a function of energy, for different values of the intercalibration constant α_{21} in $S_{\text{tot}} = S_1 + \alpha_{21}S_2$. For a weighting factor of 0.9 (*i.e.*, $\alpha_{21} \approx 1.3$), the em calorimeter response decreases by $\sim 5\%$ over the energy range 1–100 GeV. Different choices for α_{21} lead to different degrees of non-linearity. Linearity is only achieved for one particular value of α_{21}, *i.e.*, when α_{21} equals the ratio of the sampling fractions of the first and second calorimeter segments. In fact, this α_{21} value is not exactly 1.5, but somewhat smaller (1.46), since the sampling fraction is determined by the energy lost in the *combination* of passive and active material and not in the passive material alone. In any case, the α_{21} value needed for signal linearity is different from the one(s) for which the widths of the total signal distributions are minimized and it is therefore not possible to achieve both conditions *simultaneously*. This is illustrated in Figure 6.8.

Our simulations thus clearly confirmed the phenomena that were experimentally observed by HELIOS:

- The width of the total signal distribution can be improved by reducing the weight attributed to signals from the second detector segment.

- The weighting factor for which the width of the total signal distribution is minimized is energy dependent.

- When such reduced weights are applied, the calorimeter becomes non-linear for em shower detection.

6.2.3.3 *Consequences of signal non-linearity.* A calibration procedure in which the width of the total signal distribution of showers that develop in several different calorimeter segments is minimized thus leads *inevitably* to a non-linear response. Now one might argue that there is in principle no reason why a calorimeter that is non-linear for em shower detection, although somewhat inconvenient, should be unacceptable. After all, all non-compensating calorimeters are intrinsically non-linear for hadron and jet detection, and yet these devices are used in many experiments.

Any type of non-linearity could in principle be dealt with by means of a polynomial relationship between the signals S and the corresponding energy E:

$$E = c_0 + c_1 S + c_2 S^2 + c_3 S^3 + \dots \tag{6.3}$$

and the non-zero value of constants other than c_1 might be a small price to pay for improving energy resolution.

This line of reasoning is, however, crucially flawed. The non-linearity introduced by this weighting scheme implies *by definition* that a high-energy π^0, decaying into two unresolved γs produces, on average, a larger signal in this calorimeter than an electron, or one photon, of the same energy. An ω^0 resonance decaying into three unresolved

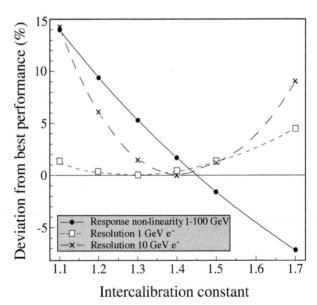

FIG. 6.8. The width of the total signal distribution for 1 and 10 GeV electrons and the em response non-linearity in the energy range 1–100 GeV for a segmented uranium/plastic-scintillator calorimeter with the structure of the HELIOS one [Ake 87], as a function of the intercalibration constant α_{21}. Results from EGS4 Monte Carlo calculations. See text for details.

γs produces an even larger signal, and an energetic K^0 decaying into $\pi^0\pi^0$, or even $\pi^0\pi^0\pi^0$ tops them all.

By introducing a signal non-linearity, the calorimeter response is made dependent on such differences. And since, in practice, the calorimeter information does not always makes it possible to tell whether the signal was caused by one, two, three or even more γs, the systematic differences in the average calorimeter response for those cases are an *integral part of the energy resolution*. Interpreting the width of the signal distribution measured for single electrons from a test beam as the em energy resolution is thus incorrect.

We have used the EGS4 Monte Carlo simulation package for a quantitative study of these effects. We used a generic 3 mm lead/3 mm plastic-scintillator sampling calorimeter for this purpose. This calorimeter was longitudinally subdivided into two segments. The first ten sampling layers $(5.5X_0)$ constituted the first segment, the second segment consisted of the remaining (100) sampling layers. The detector was assumed to be infinitely large in the lateral dimensions. As in HELIOS, calibration constants A and B were used to convert the signals from these segments into energy. Since the structure of both sections is identical, the sampling fractions are the same, and $B/A = 1$ thus refers to the situation where the sampling fraction for mips is the basis for the intercalibration of the signals from both sections. When $B/A < 1$, relatively less weight is assigned to the signals from the second segment, when $B/A > 1$, a larger weighting factor is assigned to these signals.

The non-linearity of this detector is illustrated in Figure 6.9, which shows the calorimeter response as a function of the energy of the showering electrons, for four different values of the intercalibration constant: $B/A = 0.7$, 0.8, 0.9 and 1.1. If we define the non-linearity as the fractional change of the response in the energy interval from 1 to 100 GeV, this nonlinearity ranges from -15% for $B/A = 0.7$ to $+5\%$ for $B/A = 1.1$. Only when $B/A = 1$ is the detector linear for electron detection.

We studied the consequences of this non-linearity described above for the case of 20 GeV em showers, using B/A values of 0.7, 0.8 and 0.9 [Wig 02]. For each B/A value, the detector was calibrated with 20 GeV electrons, *i.e.*, electrons were used to established the values of A (and thus of B). Using the calibration constants found in this way, signal distributions were generated for 20 GeV γs, 20 GeV π^0s (which decay into two γs of 10 GeV each), 20 GeV ω^0s (three γs of 6.67 GeV), 20 GeV K^0_Ss (four γs of 5 GeV) and 20 GeV K^0_Ls (six γs of 3.33 GeV). Each of these signal distributions had its own characteristic mean value. This is illustrated in Figure 6.10, which shows these mean values for the three chosen calibrations ($B/A = 0.7$, 0.8 and 0.9, respectively).

These mean values differed by as much as 1 GeV. When no weighting was applied ($B/A = 1$), all distributions had, within the statistical uncertainty (~ 0.05 GeV), the same mean value.

These effects are not small, given that some of the calorimeters in particle physics experiments are nowadays designed to achieve sub-1% energy resolution. Figure 3.33 shows the signal distributions for 50 GeV γs, 50 GeV π^0s and 50 GeV K^0s (decaying into $\pi^0\pi^0\pi^0$) in our simulated calorimeter, calibrated on the basis of $B/A = 0.8$ with 50

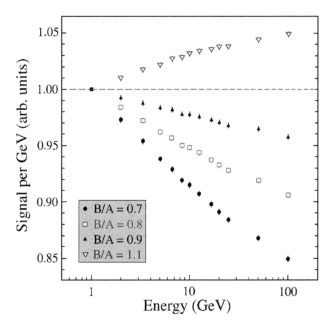

FIG. 6.9. Response nonlinearity for electrons resulting from miscalibration of a longitudinally
    segmented calorimeter. The total calorimeter response is given as a function of energy for four
    different values of the ratio of the calibration constants for the two longitudinal segments,
    B/A. See text for details [Wig 02].

GeV electrons, and assuming an intrinsic energy resolution of 0.5% (*i.e.*, $3.5\%/\sqrt{E}$).
Based on this calibration, single photons were reconstructed with an energy that was, on
average, too low by 0.67 GeV, *i.e.*, 2.7 times the intrinsic energy resolution of the calorimeter. On the other hand, the energy of the kaons was systematically *overestimated*, on
average by 0.85 GeV (3.4 σ). The reconstructed energy of the π^0s was approximately
correct, because the energy sharing between the two calorimeter segments was approximately the same for showers of 25 GeV γs (the decay products of the π^0s) and 50 GeV
electrons (used for the calibration).

   This analysis illustrates a fundamental problem inherent to non-linear calorimeters.
The calorimeter information is intended to determine the energy of particles from the
signals they generate. The precision with which the energy can be measured is determined by the energy resolution. However, in a non-linear calorimeter, *the energy resolution needed in this context does not correspond to the width of the signal distribution
measured for single, monoenergetic particles*. This observation is also particularly important for the evaluation of jet resolutions in a non-compensating calorimeter.

6.2.3.4 *Asymmetric response functions.*   Even though signal linearity can be achieved
by properly choosing the intercalibration constant ($\alpha_{21} = 1.46$ for the case described in
Section 6.2.3.2, $\alpha_{21} = 1.0$ for the other simulated calorimeters), the ideal performance
obtained with longitudinally unsegmented em calorimeters cannot always be restored in

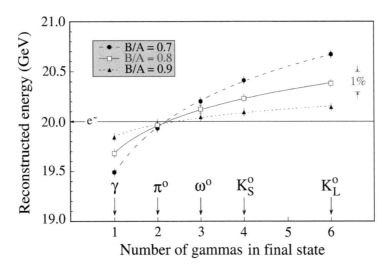

FIG. 6.10. The average reconstructed energy for 20 GeV γs and for 20 GeV particles decaying into multiple γs in a longitudinally segmented Pb/scintillator calorimeter that was calibrated with 20 GeV electrons, for different values of the ratio of the calibration constants for the two longitudinal segments, B/A. See text for details [Wig 02].

every aspect by doing so. For example, when the calorimeter is segmented into sections with different sampling fractions ($\alpha_{21} \neq 1$), the response function inevitably exhibits deviations from a Gaussian shape. This is illustrated in Figure 6.11, which shows the (simulated) response function of the segmented uranium/plastic-scintillator calorimeter from Section 6.2.3.2 for showers induced by 1 GeV photons. In order to see significant details of the tails of this response function (shown here on a logarithmic scale), 100,000 events were simulated.

The response function is clearly asymmetric, which can be understood as follows. The fluctuations in the energy sharing between the two calorimeter segments are non-symmetric. Typically, $\sim 35\%$ of the shower energy is deposited in the second calorimeter segment. However, the probability that more than 45% of the energy is deposited in this segment is 25%, while the probability that less than 25% of the energy is deposited in this segment is only 10%. These asymmetric fluctuations in the energy sharing reflect the asymmetric distribution of the depths at which γs that penetrate this block of matter interact.

As the fraction of the energy deposited in the second, more crudely sampling, calorimeter segment increases, the calorimeter response of this segment to the deposited energy increases as well. This was illustrated in Figure 6.6 and is a consequence of the gradually decreasing sampling fraction as a function of the shower age (Section 3.2.2, Figure 3.6). The simulations showed that the average shower energy deposited in the scintillator layers of our segmented calorimeter amounted to 52.3 MeV, for incident photons of 1 GeV. After application of the factor $\alpha_{21} = 1.46$, this translated into an average signal of 60.1 MeV. In the extreme case of a 1 GeV photon that penetrated the

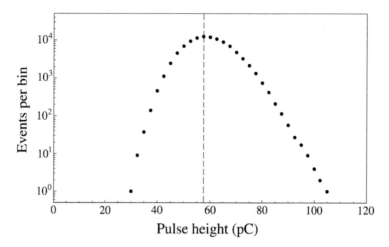

FIG. 6.11. The response function for 1 GeV photons developing in a longitudinally segmented uranium/plastic-scintillator calorimeter with the same structure as the HELIOS calorimeter. The calibration constants were such that the detector is linear for em showers. Results of EGS4 Monte Carlo simulations. See text for details.

entire first segment and deposited its entire energy in the second segment, the energy deposited in the scintillator material amounted to 42.3 MeV which, after application of the factor $\alpha_{21} = 1.46$ would translate into an average signal of 61.8 MeV. On the other hand, photons interacting in the first mm of the calorimeter produced a signal distribution with a mean value of only 59.8 MeV.

These differences, combined with the asymmetric fluctuations in the energy sharing between the two calorimeter segments, led to the asymmetric line shape depicted in Figure 6.11. To restore a Gaussian response function, intercalibration factors $\alpha_{21} \approx 1.0$ would be needed. It should be pointed out that both sides of the distribution shown in Figure 6.11 are well described by parabolas in this logarithmic plot, albeit that the coefficients of the two parabolas are different. It thus seems as if the distribution is characterized by *two different Gaussians*, one for the left-hand side of the plot and another one for the right-hand side.

The asymmetry discussed here is a general feature of longitudinally segmented calorimeters of which the segments have different sampling fractions. The degree of asymmetry depends on a number of factors:

- The effect is larger when the energy fractions deposited in the various calorimeter segments are comparable in size than when most of the energy goes, on average, into one of the calorimeter segments.

- The asymmetry is, for a given calorimeter configuration, energy dependent. This is because the energy sharing for a given configuration is energy dependent.

- The asymmetry is larger for γs than for electrons, since the event-to-event fluctuations in the energy sharing are larger for γ-induced showers [Wig 02].

6.2.3.5 *Summary.* In this subsection we have encountered several undesirable consequences of calibration schemes in which showers that share their energy among several
longitudinal calorimeter segments are used to establish calibration constants by minimizing the width of the total signal distribution (*i.e.*, the quantity Q in Equation 6.2).
Apart from the fact that the calibration constants in such schemes depend on the energy
of the electrons with which the calibration was performed, these procedures also lead inevitably to avoidable signal non-linearities. These linearities are in turn responsible for
differences between the calorimeter responses to showers induced by electrons, photons
and π^0s. We have also seen that the calorimeter response function for em showers may
be asymmetric as well, even if a calibration scheme is chosen that avoids non-linearities.
This asymmetry occurs whenever the longitudinal calorimeter segments have different
sampling fractions for mips.

6.2.4 *Three compartments – The ATLAS LAr calorimeter*

The calibration problems described in the previous subsections become more complicated when the calorimeter consists of more than two longitudinal segments. A recent
example of an experiment that has to deal with this intercalibration issue is ATLAS,
whose Pb/LAr electromagnetic calorimeter consists of three longitudinal segments. At
$\eta = 0$, the depths of these segments are $4.3X_0$, $16X_0$ and $2X_0$, respectively (Figure
9.10). When the particles enter the barrel calorimeter at a non-perpendicular angle, the
total depth of this calorimeter increases (from $22X_0$ at $\eta = 0$ to $30X_0$ at $|\eta| = 0.8$), and
so do the depths of these three segments. The sampling fraction for mips is the same in
all three segments.

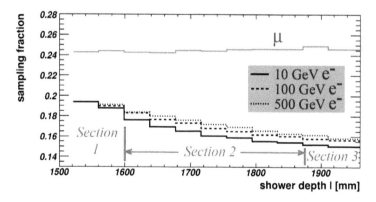

FIG. 6.12. The evolution of the sampling fraction for electron showers of different energies in the three
longitudinal segments of the ATLAS LAr calorimeter, at $\eta = 0$ [Aha 06].

Figure 6.12 shows how the sampling fraction for em showers evolves as a function
of depth, in an energy dependent way. The sampling fraction for muons that traverse
this detector does not change in this process and, therefore, the e/mip value decreases
by 20–25%, depending on the electron energy. In light of our discussions in the earlier
sections of this chapter, these data should by now look very familiar. Not surprisingly,

the problems encountered when calibrating this detector with electron showers were also very similar to the ones experienced by HELIOS, except that there were now three calibration constants to be determined, instead of two. When minimizing Q in

$$Q = \sum_{j=1}^{N} \left[E - A \sum_{i=1}^{n} S_1^{ij} - B \sum_{i=1}^{n} S_2^{ij} - C \sum_{i=1}^{n} S_3^{ij} \right]^2 \qquad (6.4)$$

it turned out that the resulting calibration constants A, B and C not only depended on the electron energy, as in HELIOS, but also on the location of the calorimeter module (the η value). The latter dependence can be understood from the change in the effective depth of the longitudinal segments with the angle of incidence of the particles. And just as in HELIOS, it was found that any choice of the calibration constants resulting from such a minimization procedure introduced a response non-linearity.

Rather than intercalibrating the different longitudinal calorimeter segments with muons (the HELIOS solution), ATLAS decided to approach this problem in a much more complicated way, relying heavily on Monte Carlo simulations in an attempt to achieve the best possible combination of energy resolution and linearity. These elaborate simulations led to a very complicated procedure for determining the energy of a shower detected in the various segments of the calorimeter. This procedure was based on a variety of parameters that depended both on the energy and the η value (Figure 6.13). It was tested in great detail with Monte Carlo events and yielded both excellent signal linearity and good energy resolution [Aha 06].

The energy dependence of the various parameters derives from the change in the longitudinal shower profiles, and thus the energy sharing between the three segments, with the energy of the incoming electron. In Section 4.5, we argued that showers induced by photons are in this respect quite different from showers induced by electrons (see for example Figure 4.31). Particles such as the π^0, ω_0 and K^0 may decay into several γs. If these are unresolved as separate showers by the calorimeter, the longitudinal profile of such objects would be even more different from that of the electrons that were used for this calibration procedure. Figure 6.10 shows the consequences of that for the energy measurement. For these reasons, I believe that the calibration procedure developed by ATLAS on the basis of electron showers (Figure 6.13) would need to be modified for the detection of γs and particles decaying into several γs. No information on this is publicly shared by the ATLAS Collaboration.

6.2.5 Many compartments – The AMS calorimeter

The last example of the pitfalls of calibrating a longitudinally segmented device concerns the calorimeter for the AMS-02 experiment, which detects high-energy electrons, positrons and γs at the International Space Station [Cer 02]. This calorimeter has 18 independent longitudinal depth segments. Each layer consists of a lead absorber structure in which large numbers of plastic scintillating fibers are embedded, and is about $1X_0$ thick. A minimum ionizing particle deposits, on average, 11.7 MeV upon traversing such a layer. The AMS-02 collaboration initially calibrated this calorimeter by sending a beam of muons though it and equalizing the signals from all eighteen longitudinal

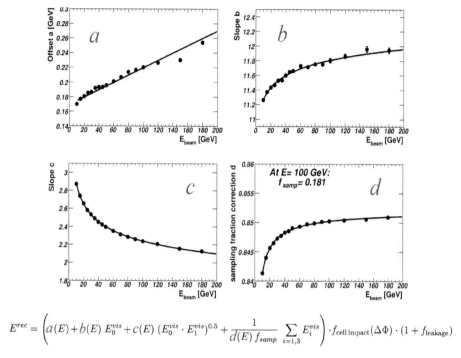

$$E^{rec} = \left(a(E) + b(E) \; E_0^{vis} + c(E) \; (E_0^{vis} \cdot E_1^{vis})^{0.5} + \frac{1}{d(E) \, f_{samp}} \sum_{i=1,3} E_i^{vis} \right) \cdot f_{\text{cell impact}}(\Delta\Phi) \cdot (1 + f_{\text{leakage}})$$

FIG. 6.13. The formula used by ATLAS to determine the energy of a shower developing in the longitudinally segmented ECAL. The energy dependence of the various parameters is shown in graphs $a - d$ [Aha 06].

segments. This seems like a very good thing to do, since all eighteen layers have exactly the same structure. Figure 6.14 shows the average signals from the eighteen longitudinal calorimeter segments, for 20 GeV electron showers developing in this calorimeter. These signals were translated into energy deposits based on the described calibration. Since the calorimeter is only $\sim 18X_0$ deep, the showers were not fully contained in this structure. Had the calorimeter been sufficiently deep, none of the problems subsequently encountered would have occurred.

In order to measure the energy of the detected particles with this calorimeter, an estimate had to be made of the energy leaking out. The authors made this estimate on the basis of a fit of the measured data points to a Γ-function:

$$dE/dt \propto t^\alpha \exp(-\beta t)$$

where t is the layer number and α and β the coefficients to be fitted. The result of this fit is shown for 20 GeV electrons in Figure 6.14. The total shower energy, and thus the energy leakage, was estimated by extrapolating this fit to infinity. As shown in Figure 6.15, this procedure systematically underestimated the leakage fraction, more so as the energy (and thus the leakage) increased. Up to 40 GeV, where the leakage fractions were only a few percent, the estimate was good enough. However, as the energy increased, the

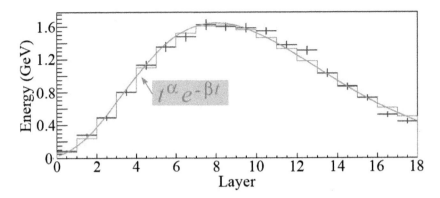

FIG. 6.14. Average signals for 20 GeV electrons in the eighteen longitudinal sections of the AMS-02 lead/scintillating fiber calorimeter. The curve represents the results of a fit to the experimental data. The superimposed histogram is the expected average profile from the Monte Carlo simulations. Data from [Cer 02].

leakage estimate was increasingly too low. At 60 GeV, the measured energy was 13% too low, after the leakage correction it was still 6% too low. At 80 GeV, the measured energy was 21% too low before, and 10% too low after the leakage correction. And at 100 GeV, the measured energy was 33% too low, and the leakage corrections reduced that percentage to 18.

The reason for these discrepancies is the same effect that complicated the energy measurements with the ATLAS ECAL and the HELIOS calorimeter, namely the de-

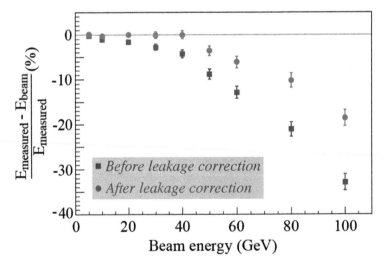

FIG. 6.15. Average difference between the measured energy and the electron beam energy, before and after leakage corrections based on extrapolation of the fitted shower profile (see Figure 6.14) were applied. Data from [Cer 02].

creasing sampling fraction in the developing showers. The fit that was used to determine the longitudinal shower profile and the leakage (Figure 6.14) was based on the measured signals, *not* on the energy deposited in the eighteen shower segments. In the tail of this profile, the sampling fraction was smaller than average. This means that the energy in this tail, and thus also the energy leakage, was systematically underestimated, because the procedure that was used assumed that the relationship between measured signals and the corresponding deposited energy was the same for each segment. The reader is invited to notice the similarity with what happened in the hadronic segment of the HE-LIOS calorimeter, where the energy deposited in that segment was also systematically underestimated (Figure 6.6).

The correct way to estimate the leakage and thus measure the energy of electrons showering in this short calorimeter would have to involve Monte Carlo simulations. After equalizing the gains of the eighteen longitudinal segments with a beam of muons, *e.g.*, at 10 ADC counts for the average signal from such muons in each segment, the average total signal from a muon traversing the calorimeter would consist of 180 ADC counts and correspond to an energy deposit of $18 \times 11.7 = 210$ MeV. The next step should involve the measurement of the signals from low-energy electron showers, preferably fully contained in the calorimeter, *e.g.*, the 5 GeV particles used by AMS-02. The average total signal from these showers would probably be around 2,600 ADC counts (assuming $e/mip \sim 0.6$). The total number of ADC counts for electrons of other energies in an infinitely long calorimeter of this type would follow directly from this result: 5,200 ADC counts for a 10 GeV electron, 26,000 ADC counts for a 50 GeV one, *etc.* Monte Carlo simulations of em shower development in this type of calorimeter would reveal what fraction of the total shower signal would be produced by the modules in the first $18X_0$, as a function of the particle energy. Based on this information, a curve that relates the measured number of ADC counts to the electron energy could be made. This curve could be used to determine the energy of a particle of unknown energy that develops an incompletely contained shower in this calorimeter.

The very complicated issues discussed here will most definitely also affect PFA calorimeters (Section 8.3), which are all based on structures that are highly segmented, both longitudinally and laterally. The underlying problem is that the relationship between deposited energy and resulting signal is not constant throughout a developing shower. As the composition of the shower particles changes, so does the sampling fraction. The examples of the calorimeters discussed here clearly illustrate the problems that may be expected when this is not properly recognized and dealt with. The assumption that the relationship between deposited energy and recorded signal remains the same throughout the developing shower has been the modus operandi for the calorimeters built in the context of the CALICE project [Sef 15].

6.2.6   *Intercalibration with hadronic showers*

In the previous subsections we saw that intercalibration of the different compartments of a longitudinally segmented calorimeter leads to fundamental problems, even for em showers in compensating calorimeters. These problems only get worse when the intercalibration is carried out with *hadronic* showers, in *non-compensating* calorimeters. Yet this technique is in practice frequently used to determine the calibration constants of the em and hadronic compartments of longitudinally segmented calorimeters.

Figure 6.16 shows some results from a detector R&D project carried out in the framework of the preparation for experiments at the SSC[16], known as the "hanging-file calorimeter." [Bere 93] The figure concerns a detector configuration consisting of a lead/scintillator em compartment ($23X_0$, $0.75\lambda_{int}$), followed by an iron/scintillator hadronic compartment ($8\lambda_{int}$). The polystyrene scintillator plates were 3.0 mm thick, the lead plates 3.2 mm and the iron plates 25.4 mm. The sampling fractions for a mip thus amounted to $0.64/(0.64+4.06) = 13.6\%$ in the em compartment and $0.64/(0.64+28.96) = 2.16\%$ in the hadronic compartment of this calorimeter (see Section 3.2 and Table B.1).

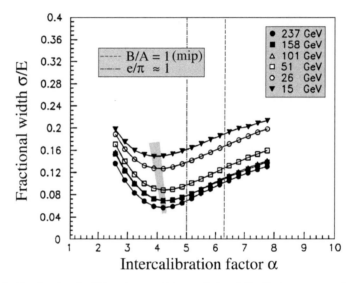

FIG. 6.16. The fractional width, σ/mean, of the total signal distribution, truncated at ± 5 standard deviations from the mean value, as a function of the intercalibration constant α, for pions of different energies detected in a longitudinally segmented scintillation calorimeter. The absorber consisted of 3.2 mm thick lead plates in the em section and 25.4 mm thick iron plates in the hadronic section. Data from [Bere 93].

Pions of different energies were sent into this structure and their total energy was reconstructed as

$$E_{total} = E_{em} + \alpha E_{had}$$

Figure 6.16 shows the fractional width of the E_{total} distribution as a function of the intercalibration parameter α. The smallest values for σ/mean were reached for values of $\alpha \approx 4$, slightly dependent on the pion energy.

---

[16]The Superconducting Super Collider (1987-1993) was an ambitious project to build a 20 + 20 TeV pp colliding-beam machine, near Waxahachie, Texas. After having spent 3 billion dollars on its realization, the US Congress decided to cancel this project.

However, to obtain the same average total signals as for electrons of the same energy, different values for α had to be used, ranging from 5.0 at 237 GeV to 5.6 at 10 GeV. Intercalibration of the two compartments with mips would give $\alpha = 6.3$, since the sampling fraction for mips in the em section was 6.3 (= 13.6/2.16) times as large as in the hadronic section. The value $\alpha = 6.3$, which corresponded to the $B/A = 1$ case discussed before, thus required calibration constants that were $\sim 50\%$ different from those found with the method that minimized the total width of the signal distribution for pions. Calibration constants that would equalize the response to pions and electrons in this calorimeter system would lie somewhere in between these values, and would be energy dependent.

Similar problems were encountered in the calibration of a prototype liquid-argon calorimeter system for the H1 experiment at HERA [Braun 88]. This detector consisted of a 1.06 λ_{int} deep em section based on lead absorber, followed by a hadronic section containing copper absorber (6.12 λ_{int}) and a tail catcher with iron absorber (2.88 λ_{int}). The last section served as a veto counter, it was used to eliminate events if too much energy was deposited in it.

The em and hadronic sections were intercalibrated with beams of pions of 30, 50, 170 and 230 GeV. The total energy deposited by an incident pion was calculated according to $E = C_e Q_e + C_h Q_h$ on the basis of the charges Q_e and Q_h measured in the two compartments. The calibration constants C_e and C_h were determined for each beam energy by the following two conditions:

1. $\langle E \rangle = E_{beam}$, and

2. the width of the total E distribution is minimized

The resulting calibration constants are shown in Figure 6.17. Both C_e and C_h changed with energy. For example, the value of C_e dropped by $\sim 10\%$ over the energy range from 30 to 230 GeV. The ratio C_h/C_e *increased* by $\sim 5\%$ over the same energy range. At the same time, the values of C_e were considerably different from those found with electron beams in the same calorimeter cells.

The calibration constants found in this way thus depended on the type of showering particle and on the energy of this particle. This has a number of very undesirable consequences, for example:

- A 30 GeV photon would be attributed an energy of 35 GeV if the calibration constants derived from the hadronic calibration procedure were used.

- An energetic ρ meson decaying into a $\pi^+\pi^-$ pair would be systematically attributed a larger or smaller energy, depending on whether or not the decay products were recognized as individual particles by the detector.

- For jets, the calorimeter response would become dependent on the multiplicity, *i.e.*, on the average energy of the particles constituting the jet.

The last problem is more severe than it may seem from this example, since jets consist predominantly of particles with much lower energies than the lowest energy considered here, 30 GeV. As the energy drops to lower values, the non-linearities and energy dependencies grow rapidly.

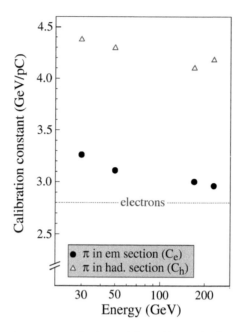

FIG. 6.17. The calibration constants for the em (C_e) and hadronic (C_h) compartments of the H1
liquid-argon calorimeter, obtained from an intercalibration with pion showers, as a function
of energy. The calibration constant of the em compartment obtained with electrons is shown
as well. Data from [Braun 88].

6.2.7   *Each section calibrated with its own particles*

As illustrated by the discussion in the previous subsections, intercalibration of the longitudinal segments of a calorimeter by means of showering particles that deposit part
of their energy in each of the different segments is not a good idea. So what are the
alternatives?

Many calorimeter systems consist of two sections, an electromagnetic section that
is deep enough to contain showers induced by electrons and photons at the 95+% level
(typically $\sim 25X_0$), and a hadronic section in which hadrons typically deposit most
of their energy. One method that is frequently used to calibrate such calorimeters is
described in this subsection, and graphically depicted in Figure 6.18.

In this method, both longitudinal calorimeter segments are *individually* calibrated.
The calibration constants for the calorimeter cells of the em section are determined
with beams of monoenergetic electrons. These electrons deposit their energy entirely
in this section and, therefore, the same method described in Section 6.1 for longitudinally unsegmented devices can be applied to determine the calibration constants for the
individual calorimeter cells that make up the em calorimeter compartment.

The cells of the hadronic calorimeter section are calibrated with beams of hadrons,
but only those events in which the beam particles penetrate the em calorimeter section
without undergoing a strong interaction are selected for this purpose [Bal 88]. Apart

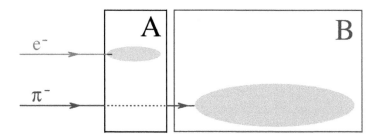

FIG. 6.18. Calibration of the different longitudinal compartments of a calorimeter system with particles that deposit (almost) their entire energy in one of the compartments. Electrons are used to calibrate the em section, while pions penetrating this section without starting a shower are used to calibrate the hadronic calorimeter section.

from a very small fraction of energy lost through electromagnetic interactions in the front section (typically a few hundred MeV), the hadrons dump their entire energy in the hadronic calorimeter section. Calibration constants for the individual hadronic cells are determined in the same way as described in Section 6.1 for longitudinally unsegmented calorimeters.

This method may also be used *in situ*, for example in collider experiments with a central magnetic field (ATLAS and CMS at the LHC, CDF at Fermilab, the LEP experiments ALEPH, DELPHI and OPAL, the HERA experiments ZEUS and H1, *etc.*). Tracks from isolated charged particles that penetrate the em calorimeter section may be selected for that purpose. The momentum of the particles is given by the curvature of the tracks in the magnetic field. For large momenta, the deposited energy is almost independent of the type of particle.

This type of calibration works better when the magnetic fields are strong and extend over a large volume, since the accuracy of the momentum measurement depends on the magnetic field B and the track length ℓ as $(B\ell^2)^{-1}$. A large bending power thus extends the energy range of the method. This energy range is often a weak point that may limit the usefulness of the *in situ* measurements.

At first sight, the idea of calibrating each calorimeter section separately, using the particles that in practice generate the signals from these individual sections, is very appealing. By using calibration constants that correctly reproduce the energy deposited by electrons in the em section and by hadrons in the hadronic section, one hopes to eliminate the problems arising from the different calorimeter response to these particles in non-compensating calorimeters. In almost all calorimeters, the response to hadrons is smaller than that to electrons of the same energy ($e/h > 1$). For such calorimeters, the described method thus corresponds to a choice of $B/A > 1$. In other words, if electrons were to be sent into the hadronic section of a calorimeter calibrated this way, their energy would be overestimated, on average by factor of B/A.

This calibration method works fine for those hadrons that penetrate the em section without interacting. However, this sample usually represents only a small fraction of all hadrons. Most hadrons undergo their first nuclear interaction in the em calorimeter section. They deposit (a sometimes large) part of their energy in the em section and the remainder in the hadronic section. For these events, this calibration method does not produce correct results.

We have studied this problem and its consequences using data from an extremely non-compensating calorimeter, a prototype of the CMS copper/quartz-fiber calorimeter [Akc 97]. This prototype consisted of em and hadronic sections of identical composition: 300 μm thick quartz fibers embedded in a copper matrix, according to a hexagonal pattern, with each fiber located at a distance of 2.2 mm from its six nearest neighbors. Each section was subdivided into 9 towers with lateral dimensions of 52×54 mm^2. The em compartment was about $22X_0$ $(2.1\lambda_{\mathrm{int}})$ deep.

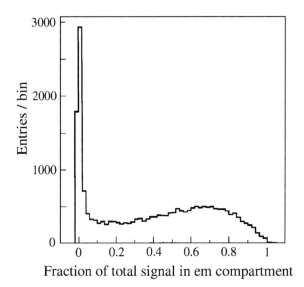

FIG. 6.19. Fraction of the energy deposited in the em section of a longitudinally segmented quartz-fiber calorimeter, for showers initiated by 350 GeV π^- [Gan 98].

Figure 6.19 shows the energy sharing between these two sections for showers initiated by 350 GeV π^-. The peak around $em/total = 0$ represents the fraction ($\sim 15\%$) of the pions that penetrated the em section without interacting. On average, the pions deposited about half of their energy in the em section of this calorimeter.

Both calorimeter sections were separately calibrated, by sending a beam of 80 GeV electrons into the center of each and every tower. Since both sections were calibrated with the same particles, the calibration constants obtained with this procedure thus corresponded, by definition, to $B/A = 1$. However, when these calibration constants were applied to the signals from 350 GeV π^-, the average energy of these pions was found to be only 231 GeV. This was a consequence of the fact that this calorimeter, for all practical purposes, only responded to the em shower core of hadronic showers ($e/h \sim 5$).

When we took the sample of events in which the pions penetrated the em calorimeter section without starting a shower ($em/total = 0$, Figure 6.19), the beam energy

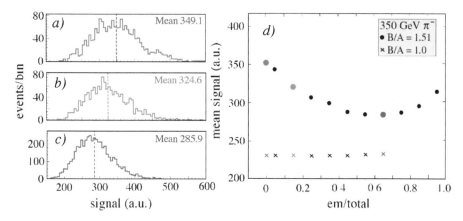

FIG. 6.20. Signal distributions for 350 GeV pion showers in a longitudinally segmented quartz–fiber calorimeter, for events in which different fractions of the (unweighted) shower energy were recorded in the em calorimeter section. Shown are distributions for which this fraction was compatible to zero (*a*), 10–20% (*b*), or 60–80% (*c*). The average calorimeter signal for 350 GeV pions, as a function of this fraction, is shown in diagram (*d*). The calorimeter was calibrated on the basis of $B/A = 1.51$ in all these cases, as required for reconstructing the energy of 350 GeV pions that penetrated the em compartment without undergoing a strong interaction. Diagram (*d*) also contains results (the crosses) obtained for a calorimeter calibration on the basis of $B/A = 1$. From [Gan 98].

could be correctly reproduced, on average, when the amplitudes of the signals from the hadronic section were increased by a factor 1.51 (see Figure 6.20a). In other words, the method in which 80 GeV electrons were used to calibrate both sections separately required a value $B/A = 1.51$ for this calorimeter (and these particles at this energy).

However, when the same weighting factor was applied to events in which the pions started the shower in the em calorimeter section, the energy was systematically underestimated. This is illustrated in Figure 6.20, which shows the signal distributions for events in which 10–20% (Figure 6.20b), or 60–80% (Figure 6.20c) of the unweighted signal was recorded by the em calorimeter section. In the latter case, the event energy was underestimated, on average, by about 20%. Figure 6.20d shows the average total calorimeter signal for these pions as a function of the fraction of the unweighted signal recorded by the em compartment. It varies by $\sim 20\%$. No such effects were observed when no weighting factors were applied ($B/A = 1$).

This analysis shows that the energy of early showering particles was systematically underestimated as a direct result of this calibration method, to an extent that depended on the energy sharing between the two calorimeter compartments. When no weighting factors were applied to the signals ($B/A = 1$), the average total energy was found to be more or less independent of the fraction of the signal recorded in the em calorimeter compartment, at least for events in which this fraction was 70% or less[17]. This is illustrated by the crosses in Figure 6.20d.

These results thus mean that the calibration method discussed in this subsection introduces a strong dependence on the starting point of the showers. Hadrons that happen to interact early in the detector are attributed an energy that is, on average, considerably lower than for hadrons that start showering after penetrating an interaction length or more.

A second undesirable consequence of this calibration method concerns the signal linearity for hadrons. As discussed in Section 3.1.4, all non-compensating calorimeters are intrinsically non-linear for hadron detection, since the average energy fraction carried by the electromagnetically interacting shower component (π^0s) increases with energy. However, this calibration method tends to make the non-linearity worse, as illustrated in Figure 6.21.

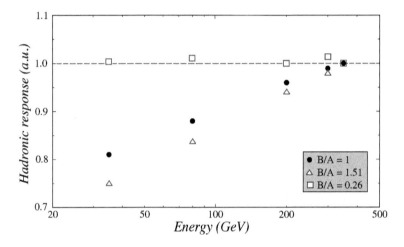

FIG. 6.21. The hadronic calorimeter response as a function of energy for pions detected in a longitudinally segmented Cu/quartz-fiber calorimeter, using calibration schemes with $B/A = 1.51$ and $B/A = 0.26$, as discussed in the text. For comparison, the results obtained with the muon intercalibration method ($B/A = 1$) are shown as well [Gan 98].

This can be understood as follows. As the energy of the incident hadron increases, a larger fraction of the shower energy is deposited in the hadronic calorimeter section. Since the signals from this section are amplified by the applied weighting factor, the hadronic response increases with energy as a direct result of this calibration procedure.

---

[17]For the small fraction of events in which more than 70% of the signal came from the em compartment, an increase in the total signal was observed for this calibration too. However, this increase was due to the fact that the events in this sample were non-typical: π^0s produced in the first nuclear interaction carried an anomalously large fraction of the total energy, and thus developed showers that were almost entirely electromagnetic. These are the same type of events as those populating the high-side tail in hadronic signal distributions from non-compensating calorimeters.

The intrinsic non-linearity, *i.e.*, the increase of the hadronic response because of the increasing fraction of the shower energy carried by π^0s, is thus further enhanced. In Figure 6.21, the intrinsic non-linearity of this quartz-fiber calorimeter is represented by the full circles, while the triangles describe the hadronic non-linearity one would measure if this calibration procedure had been applied.

Although the calibration method discussed in this subsection looked at first sight attractive and logical, it thus turned out to be fundamentally flawed. The experimental data that we used to illustrate the flaws came from a calorimeter that is arguably the most non-compensating one on the market. This feature has the advantage that it strongly amplifies the problems caused by miscalibration and that it makes the origins of these problems very clear. This is because large deviations from $B/A = 1$ had to be made in order to apply this calibration philosophy to this particular detector. However, it should be emphasized that the effects discussed here are an artifact of *all calorimeters with* $e/h \neq 1$. The degree to which the reconstructed hadron energy depends on the starting point of the shower development and the degree to which the non-linearity of the hadronic response are deteriorated as a direct consequence of applying this calibration method depend on the e/h values of both the em and the hadronic calorimeter sections and on the thickness of the em section, expressed in λ_{int}.

Only when $e/h = 1.0$ for both calorimeter sections are the effects described above avoided, because in that case this calibration method is equivalent to using $B/A = 1$.

6.2.8   *Forcing signal linearity for hadron detection*

The third calibration method that we will discuss was inspired by the desire to eliminate the non-linearities in the hadronic response that are characteristic for non-compensating calorimeters. In this method, which is described and applied in [Bag 90], the signals from the em section are given different weights for the detection of electrons and hadrons. The calibration constants for em showers are determined in the same way as described in the previous subsection, *i.e.*, by means of electrons of known energy that deposit all their energy in the em calorimeter section. However, when hadrons are showering in the calorimeter, the signals from the em section are weighted with a factor that is chosen to achieve signal linearity for these particles.

Usually, non-compensating calorimeters have an e/h value larger than 1. This means that the hadronic response of such calorimeters increases with energy (Section 3.1.4). As the energy increases, the fraction of the shower energy deposited in the hadronic calorimeter section increases, on average, as well. Therefore, restoring linearity for hadrons in such calorimeters may be achieved either by suppressing the signals from the hadronic compartment or by boosting the signals from the em compartment. This is equivalent to applying a weighting factor to the intercalibration constant B/A that is smaller than 1.

When hadrons are detected in an under-compensating calorimeter that is calibrated in this way, the signals from the em compartment will thus be attributed a larger energy than for electrons that produce the same signals in the same compartment.

We tested the merits of this calibration method with the same experimental data that were used to investigate the calibration method discussed in the previous subsection [Gan 98]. It turned out that hadronic signal linearity could be achieved in the longitudinally segmented quartz-fiber calorimeter with which these data were obtained, when an intercalibration constant with the value $B/A = 0.26$ was used. The squares in Figure 6.21 represent the resulting hadronic calorimeter response values as a function of energy. The calibration constant A was chosen such as to reproduce the correct mean energy values in the energy domain covered by the pions (35–350 GeV).

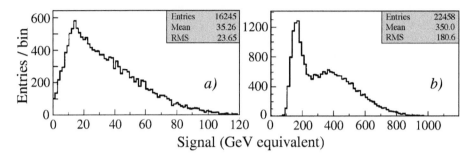

FIG. 6.22. Signal distributions for 35 GeV (*a*) and 350 GeV (*b*) pions detected in a longitudinally segmented quartz-fiber calorimeter that was calibrated with the aim of achieving hadronic signal linearity ($B/A = 0.26$). See text for details [Gan 98].

However, the signal distributions became quite distorted when these calibration constants were applied. This is illustrated in Figure 6.22, which shows signal distributions for pions at 35 GeV and 350 GeV. Especially, the latter distribution looks very unusual. The average signal corresponds indeed to 350 GeV, as it should, but the signal distribution looks as if it were a convolution of several individual distributions with very different average values. That is indeed exactly what happened here.

Figure 6.23 shows the signal distributions for 350 GeV pions, for the same three event selections we used in Section 6.2.7 to demonstrate the problems with the calibration method discussed there: events in which the pions penetrated the em section without interacting in it (Figure 6.23a), and events in which the pions deposited 10–20% (Figure 6.23b) or 60–80% (Figure 6.23c) of their (unweighted) energy in the em calorimeter compartment. These distributions have mean values that differ by more than a factor of three!

When we compare Figures 6.20 and 6.23, we see that the hadronic signal distributions obtained on the basis of the calibration method discussed here depend even more on the starting point of the showers than those based on the calibration scheme from the previous subsection. This is also particularly clear from Figure 6.23d, which shows the average calorimeter signal for 350 GeV pions as a function of the fraction of the (unweighted) shower energy recorded in the em calorimeter compartment. When this fraction is raised from 0 to 0.4, the response changes (increases) by more than 100%, while it changed (decreased) by $\sim 20\%$ in the previous scheme (Figure 6.20d). We repeat and re-emphasize that the response did not depend at all on the starting point of the showers when $B/A = 1$ was used as the basis of the calorimeter calibration.

For reference purposes, the mean signal for single pions averaged over all events

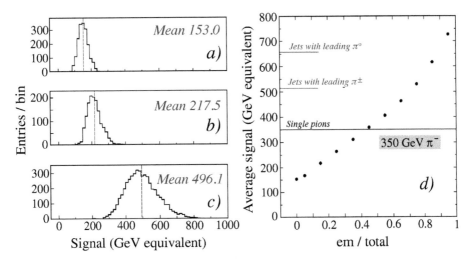

FIG. 6.23. Signal distributions for 350 GeV pion showers in a longitudinally segmented quartz–fiber calorimeter, for events in which different fractions of the (unweighted) shower energy were recorded in the em calorimeter section. Shown are distributions for which this fraction was compatible to zero (a), 10–20% (b), or 60–80% (c). The average calorimeter signal for 350 GeV pions, as a function of this fraction, is shown in diagram (d). The average signals for single pions, and the ones expected for different types of jets of this energy, are also indicated in this diagram. The calorimeter was calibrated on the basis of $B/A = 0.26$, as required for achieving signal linearity for pions. From [Gan 98].

(350 GeV) and the mean signals for 350 GeV jets with leading charged or neutral pions are indicated in Figure 6.23d as well. Showers initiated by such jets have different (average) longitudinal shower characteristics and will thus generate a different response in this calorimeter than single pions, even if the em shower content would be, on average, the same. Especially for jets with a leading π^0, a large fraction of the energy is deposited in the em calorimeter section. As a result, the energy of jets with a leading π^0 is, on average, *overestimated by almost a factor of two* when this calibration method is used.

This calibration method was developed to eliminate the energy dependence of the hadronic calorimeter response. It succeeded in achieving that goal. However, there is a price to be paid, as a direct consequence of this method:

- The hadronic response becomes dependent on the starting point of the showers.
- The response becomes dependent on the longitudinal shower development characteristics. This leads to differences between the calorimeter responses to jets and to single hadrons.

The experimental data used to illustrate these consequences came from an extremely non-compensating calorimeter system. As before, this was done on purpose since it has the advantage of maximizing the effects and clearly indicating their origin. However,

it is important to realize that this calibration scheme will lead to the same effects, albeit of less grotesque proportions, *in any non-compensating calorimeter system.* Only
when $e/h = 1.0$ for both calorimeter sections are the effects described in this subsection avoided, because in that case this calibration method would be equivalent to using
$B/A = 1$.

6.2.9   *No starting point dependence of hadronic response*

The fourth calibration method that we will discuss is, just like the three previous ones,
inspired by a laudable goal. The purpose of this method is to make the hadronic response
independent of the starting point of the shower. It was used by CDF for the calibration of
their "plug upgrade" calorimeter [Alb 02b]. CDF tried also two other calibration meth-

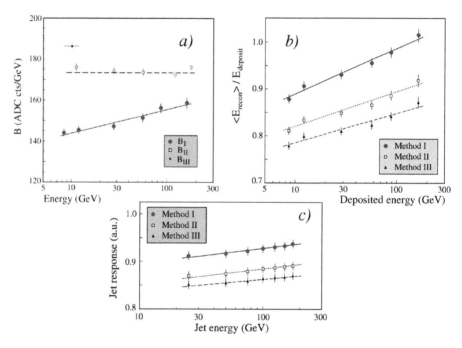

FIG. 6.24.   Results from three different methods investigated by CDF to calibrate their forward
("plug upgrade") calorimeter. The methods are described in the text. Shown are the calibration
constants for the hadronic calorimeter compartment (a) and the response to single pions (b)
and to jets (c) as a function of energy [Alb 02b].

ods and compared the results, some of which are shown in Figure 6.24. The method that
is the topic of this subsection is called "Method III." Method I was based on the calibration procedure described in Section 6.2.7, where the calibration constant ("B_I") of the
hadronic compartment[18] was determined by the response to pions that penetrated the

---

[18]Note that the CDF Collaboration expressed calibration constants in ADC cts/GeV, whereas we systematically use the inverse quantity (GeV/ct) elsewhere in this chapter.

em compartment without starting a shower. As shown in Figure 6.24a, this calibration constant increased with the pion energy, because of the increased average em shower component. This was not the case with Method II, in which the calibration constant of the hadronic compartment was determined in the same way as for the em compartment, namely with electrons. And since both compartments were linear for em shower detection, both A_{II} and B_{II} were independent of the particle energy. This is in essence the method discussed in Section 6.2.11, and which we call the "B/A = 1" method. CDF only tried Method III for pions of one energy, because of concerns that shower leakage might affect the results at higher energies. However, it is likely that also B_{III} would have turned out to be energy dependent if they had used the method at higher energies as well.

Figure 6.24b shows the response to pions as a function of energy, using *all* events, *i.e.*, not only the pions that deposited their entire shower energy in the hadronic compartment. For all three methods, the pion response increased with energy, since $e/h > 1$ in both calorimeter compartments. However, the non-linearity was clearly worst for Method I. As the pion energy increased, a larger fraction of the shower energy was deposited in the hadronic calorimeter compartment. Boosting the signals from that compartment, which is the essence of Method I, thus tended to increase the already existing non-linearity. For Method III, the opposite effect occurred. By giving less weight to the signals from the hadronic compartment, the "natural" response non-linearity was reduced, albeit it not to the extent seen in Section 6.2.8, where the suppression of the signals from the hadronic compartment was so large that it resulted in response linearity.

CDF also studied the effects of these three calibration methods for jets, using a semi-empirical procedure to determine the jet response. This procedure is described in detail in Section 9.3.4.1 for the case of hadronically decaying W bosons. CDF used Equation 2.23 to describe the fragmentation process that generated their jets. Figure 6.24c shows the calorimeter response to these jets, as a function of energy. The same trends are observed as for single pions, but the non-linearities are clearly smaller for all methods. This is due to two factors:

1. Part of the jet signal, on average one-third, comes from γs which develop em showers. The calorimeter is linear for that component of the jet signal. The non-linearity only affects the remaining portion of the jet fragments.

2. The non-linearity is not determined by the jet energy itself, but by *the average energy of the jet fragments*. Since the multiplicity increases with energy, this average energy of the fragments increases more slowly than the jet energy itself.

Yet, the differences between the non-linearities observed for the three calibration methods indicate that the effects described for single pions propagate into the energy measurement of jets.

Even though Method III eliminates the starting point dependence of the hadronic response, it introduces other problems. For example, the response to muons becomes different for the two compartments of the calorimeter system, to an extent determined by the difference between the e/h values of these compartments. If these e/h values would be the same, then this method would actually be indistinguishable from the, in my opinion correct, method discussed in Section 6.2.11. However, in today's experiments, the em calorimeter typically has a larger e/h value than the hadronic compartment, since this is a general consequence of a larger sampling fraction (see, for example, Figures 3.42 and 3.44).

6.2.10   *Dummy compensation*

The calibration methods discussed in Sections 6.2.7 - 6.2.9 tried to deal, in different ways, with the consequences of non-compensation ($e/h \neq 1.0$) for the energy scale of calorimeter systems. One other method has been proposed for that purpose [Fer 97]. The authors proposed to *eliminate* the effects of non-compensation by installing an inactive absorber upstream of the calorimeter. This "dummy section" would absorb, on average, a larger fraction of the energy of incident electrons than for hadrons. For an appropriate thickness of this absorber ($\sim 8X_0$), the average calorimeter signals for hadrons and electrons would then be equal, at least for one energy.

Such an arrangement would correspond to a calibration with $B/A = \infty$, since no signals from the first (dummy) section would be recorded. In Section 4.10 we showed that the main problem introduced by this method is a general deterioration of the calorimeter performance. Only signals corresponding to the shower energy deposited beyond a certain depth in the absorber structure would be recorded in this scheme. By ignoring the fluctuations in the energy deposited in the first $8X_0$ of the absorber structure, the precision of the information that could be obtained on the particles that initiated the showers inevitably deteriorates (Figure 4.67).

With regard to calibration, this method suffers from the same problems as the other calibration schemes based on $B/A \neq 1$:

- The hadronic response becomes dependent on the starting point of the showers.
- The hadronic response is different for jets and single hadrons with the same em shower content.
- The hadronic signal non-linearity is increased for calorimeters with $e/h > 1$.

In addition, the dummy section makes the calorimeter non-linear for electromagnetic showers. This calibration method thus combines the negative aspects of *all* other methods discussed in the previous subsections.

6.2.11   *The right way*

In the previous subsections, we have encountered a number of different methods that are being used to intercalibrate the signals from the different compartments of a longitudinally segmented calorimeter systems. All these methods were based on a specific goal that looked, at least at first sight, quite reasonable and is briefly summarized below:

1. Minimization of the width of the total signal distribution (Section 6.2.6)
2. Correct energy reconstruction of pions penetrating the em compartment without starting a shower (Section 6.2.7)
3. Hadronic signal linearity (Section 6.2.8)
4. Independence of hadron response on starting point shower (Section 6.2.9)
5. Equal response to electrons and pions (Section 6.2.10)

We also saw that each of these approaches introduced specific additional problems. We now conclude this section on the calibration of longitudinally segmented calorimeter systems with a statement that could summarize this entire chapter and therefore deserves to be printed in bold face.

The correct way to intercalibrate the different sections of a longitudinally segmented, non-compensating calorimeter system is by using the same particles for all individual sections. If these particles develop showers, then they can only be used to calibrate sections in which these showers are completely contained.

Only in this way is the relationship between the deposited shower energy (in GeV) and the charge (in picoCoulombs) generated as a result established unambiguously. We have referred to this as the $B/A = 1$ method. The use of a beam of muons to intercalibrate the eighteen segments of the AMS-02 electromagnetic calorimeter (Section 6.2.5) definitely qualifies as a viable method in this respect. The mistake made in that case did not concern the calibration method itself, but the interpretation of the results.

An example of a good $B/A = 1$ method for calibrating calorimeters consisting of separate em and hadronic compartments is to determine the calibration constant of the em compartment with electrons of known energy and to intercalibrate the different longitudinal segments with a beam of muons. In non-compensating calorimeters, the energy of showering hadrons is not correctly reproduced in this way. However, that problem can be handled by applying an overall correction factor, which can be measured independently. An example of a calorimeter that was calibrated in this way is the CCFR Target calorimeter that operated in Fermilab's neutrino beam [Sak 90].

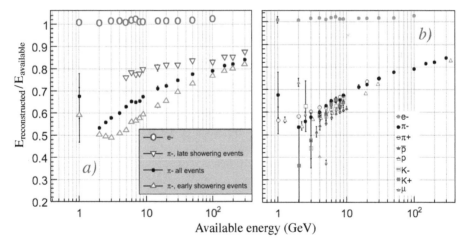

FIG. 6.25. The response to electrons and pions as a function of energy, for the CMS barrel calorimeter. The pion events are subdivided into two samples according to the starting point of the shower, and the pion response is also shown separately for these two samples (a). The ratio of the total reconstructed energy in the CMS barrel calorimeter and the available beam energy (b) for a variety of beam particles at different energies [CMS 07].

Another correct ($B/A = 1$) calibration method is to expose the individual longitudinal segments separately to an electron beam and derive the calibration constants from the signals recorded in that way. In many cases, constructional details prohibit such a procedure from being applied. Yet, this method was used for the calibration of the CMS calorimeter system.

The CMS calorimeter system suffers from a particularly nasty complicating factor, namely the fact that the em and hadronic compartments have very different e/h values. The em section of the barrel calorimeter is made of $PbWO_4$ crystals ($e/h \approx 2.5$), the hadronic section consists of brass plates interleaved with plastic scintillator ($e/h \approx 1.4$). For a systematic study of the hadronic performance of this calorimeter system, both compartments were calibrated with a 50 GeV electron beam. Figure 6.25a shows that the response to pions, represented by the black dots, is very non-linear. This non-linearity is especially evident below 10 GeV, which is important since pions in this energy range carry a large fraction of the energy of jets at the LHC. More troublesome is that the response depends on the starting point of the showers. The figure shows results for two event samples, selected on that basis: showers starting in the em section ($\triangle$) or in the hadronic section ($\triangledown$). At low energies, the response was measured to be more than 50% larger for the latter (penetrating) events. In practice, in an experiment, it is often hard/impossible to determine where the shower starts, especially if these pions are traveling in close proximity to other jet fragments (*e.g.*, γs from π^0 decay) which develop showers in the em section.

The signal distributions for 100 GeV/c π^- beam particles are displayed in Figure 6.26. A sizeable fraction of the pions experienced their first nuclear interaction in the em section. This can be concluded from the signal distribution in Figure 6.26a, which exhibits a clear mip peak caused by particles that penetrated the ECAL without starting a shower, as well as a broad distribution of larger signals caused by pions that did start a shower. The signals in the HCAL (Figure 6.26b) show a complementary distribution, *i.e.*, small signals for the early showering particles and larger signals for the ones that penetrated the ECAL. The total signal, shown in Figure 6.26c, is reconstructed with an average signal equivalent to 79% of the signal from an electron of the same energy. The hadronic energy resolution, σ_{rms}/E, is relatively large (15%), mainly as a result of fluctuations in the starting point of the showers.

CMS has performed a detailed study of the response of their calorimeter system to particles with momenta in the range from 1.5 - 350 GeV/c. The results are shown as a function of the *available energy* in Figure 6.25b. The available energy is the energy that results in a contribution to the calorimeter signal. For protons this is the kinetic energy, for pions the total energy and for anti-protons the total energy plus the energy equivalent of a proton mass. The response turned out to be dependent of the type of particle, because of details such as the need to conserve baryon number (in proton induced showers) and strangeness (kaons), and the characteristics of charge exchange reactions (differences between $\pi^+/\pi-$) in the shower development [Akc 12a]. In CMS, the energy dependence of the response shown in Figure 6.25c is the basis of the correction for the observed non-linearity; measured signals are simply multiplied with the inverse of the response value for that particular energy value. The use of this energy dependent

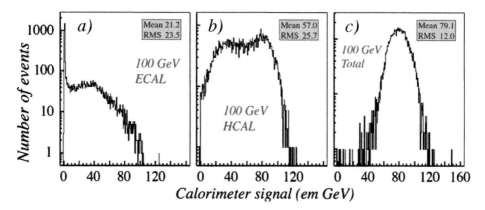

FIG. 6.26. The signal distributions for 100 GeV/c π^- are shown for the CMS em calorimeter section (a), the hadronic calorimeter section (b), and the combined system (c). The experimental data also included the measured longitudinal energy leakage [Akc 12a].

overall correction factor restores hadronic signal linearity on average, even though the starting point dependence remains. If the starting point of the hadron shower can be determined, then the inverse of the $\triangle$ or $\triangledown$ curves from Figure 6.25a could be used as the correction factor instead.

Just like in CMS, the em and hadronic sections of the ATLAS calorimeter system also have different e/h values, but the differences are much smaller in that case. The consequences of these differences are qualitatively the same, but less dramatic than indicated in Figure 6.26. More information about the performance of the ATLAS and CMS calorimeters in given in Chapter 7.

As discussed in Sections 6.2.7 and 6.2.8, compensating calorimeter systems offer more calibration possibilities than ATLAS and CMS, because several other methods are equivalent to $B/A = 1$ in that case.

6.2.12   *Validation*

When all is said and done, it is important to check the correctness of the chosen calibration scheme with experimental data. For this, one needs to have sources of known experimental energy deposits. One such source is a particle of precisely known mass, whose decay products are detected by the calorimeter. This method offers excellent possibilities to validate the calibration of em calorimeters, because of the availability of a variety of particles that are abundantly produced in today's accelerator experiments and which cover a large mass (*i.e.*, energy) range: π^0 (mass 135.0 MeV/c^2), η (547.9 MeV/c^2), J/ψ (3.097 GeV/c^2), Υ (9.460 GeV/c^2) and Z^0 (91.19 GeV/c^2) all decay into particles (e^+e^- or γ pairs) that develop electromagnetic showers. Examples of mass peaks reconstructed from these decay products are shown at various places in this book (*e.g.*, Figures 6.34, 7.27, 7.53, 7.61).

Unfortunately, there are no such clearcut calibration sources that can be used for hadron calorimeters. Yet, there are certainly possibilities. At low energies, the decay

$K_S^0 \to \pi^+\pi^-$ could be used for this purpose and at high energies the hadronic decay modes of the intermediate vector bosons W and Z. In the latter case, the problem is the QCD background, which makes it hard to extract a sample of boson decay events from the di-jet invariant mass distributions. The only experiment that has managed to do so was UA2 (Figure 7.70). However, at the LHC one could take advantage of the high production rate of $t\bar{t}$ events, and select a much cleaner sample of hadronically decaying Ws from the dominant decay mode $t \to Wb$.

Alternative methods to validate the hadronic calorimeter calibration include:

- A comparison between the momenta of isolated charged tracks and the corresponding energy deposits in the calorimeter.
- The use of events in colliders for which the energy deposited in the calorimeters is precisely known. Hadronically decaying Z bosons produced at LEP and neutral current events in the ep collisions at HERA, in which the hadronic four-vector was completely constrained, can be mentioned as examples.
- A method called "γ + jet p_T balancing," in which events are selected with one clear jet, plus another object that develops an em shower with no associated charged-particle tracks. The transverse momentum of the jet is assumed to be equal to that of the "γ." This method has, for example, been used by CDF [Boc 01], and is also part of the CMS calibration procedure [CMS 09].

However, none of these alternative methods provides a check with the same level of rigorosity as the ones based on the reconstruction of a particle with a precisely known mass from its decay products.

6.3   Consequences of Miscalibration

As we have seen in the previous section, the calibration of a longitudinally segmented calorimeter system is far from trivial. It would, therefore, not be surprising at all if many of the calorimeter systems in current and past experiments were calibrated incorrectly. However, it is in practice not easy to find experimental proof of that. One example of a case in which miscalibration manifests itself beyond a shadow of a doubt is when particles of precisely known mass (*e.g.*, the J/ψ meson or the Z^0 boson) are reconstructed with a significantly different mass value on the basis of their calorimetrically measured decay products (*e.g.*, e^+e^- for the mentioned particles).

Sometimes, there may also be indirect indications of problems caused by miscalibration. In this section we discuss examples of this, as well as some practical consequences of miscalibration.

6.3.1   *Jets at 90 GeV*

An example of an indication that something went wrong with the calibration is shown in Figure 6.27. This figure contains data from pions developing showers in the ALEPH [Bag 90] and ZEUS [Beh 90] calorimeters. The ALEPH calorimeter consisted of iron absorber, read out with wire chambers; ZEUS operated a uranium/plastic-scintillator device. In Figure 6.27a, the average fraction of the pion energy deposited in the em calorimeter compartment is plotted as a function of the pion energy for both calorimeter systems.

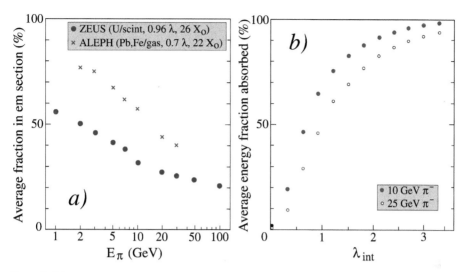

FIG. 6.27. Energy sharing between the electromagnetic and hadronic calorimeter sections for pions detected with the ALEPH and ZEUS calorimeters. Diagram (a) shows the fraction of the energy deposited in the em calorimeter section, as a function of energy. Experimental data from [Bag 90, Beh 90]. Diagram (b) shows the energy fraction deposited by 10 and 25 GeV pions in iron absorber as a function of the absorber thickness. Experimental data from [Cat 90].

The pions deposited a considerable fraction of their energy in the em calorimeter section. At 10 GeV, the average fraction was measured to be $\sim$ 30% for ZEUS and $\sim$ 60% for ALEPH. The differences between these two sets of results were very large. Moreover, they had the wrong sign. The ALEPH em calorimeter compartment was *thinner* than the ZEUS one (0.7 *vs.* 0.96 nuclear interaction lengths). Based on measurements of the hadronic shower containment as a function of the absorber thickness, also done by ALEPH [Cat 90], one would expect the energy fraction deposited by pions in ALEPH's em calorimeter section to be 40% *smaller* than for ZEUS, instead of a factor two larger (Figure 6.27b).

Given this large discrepancy (a factor of 2.5 at 10 GeV), the conclusion that one of the two experiments (or maybe both) mismeasured the energy deposited by pions in the electromagnetic calorimeter compartment by a considerable factor is hard to avoid.

The above-mentioned discrepancy can be understood by analyzing the calibration methods that were used to obtain these data. Contrary to ZEUS, the ALEPH calorimeter was strongly non-compensating, with e/π ratios ranging from 1.7 at 2 GeV to 1.3 at 30 GeV. This means that pions produced signals that were considerably smaller than those from electrons at the same energy. To get to the correct energy, the calibration constants ALEPH used for pions were therefore considerably larger than for electrons.

Usually, showering pions deposit some fraction of their energy (f_{em}) in the form of electromagnetic showers, generated by π^0s. Such π^0s are mainly produced in the initial stages of the shower development, *i.e.*, in the em calorimeter section. However, the sampling fraction for this shower component is the same as for electrons. As a result, the energy deposited by showering pions in the em calorimeter section was *overestimated* by ALEPH.

This effect was further amplified as a result of their method to restore *signal linearity* for hadrons. Since $\langle f_{em} \rangle$ is energy dependent, non-compensating calorimeters are intrinsically non-linear for hadrons. In the case of ALEPH, the hadronic response increased with energy. As the energy increases, so does the fraction of the energy deposited in the hadronic calorimeter section. In an attempt to restore linearity, the signals of the em calorimeter section were deliberately given a larger weight than the signals from the hadronic section. This is essentially the $(B/A < 1)$ calibration method we discussed in Section 6.2.8.

As a result of this, the energy deposited in the em calorimeter section was overestimated even more. The combination of both effects led to a discrepancy of more than a factor of two between the energy deposited in the em section of the ALEPH calorimeter and the em section of the compensating ZEUS calorimeter.

The philosophy of the ALEPH calibration method was to achieve a situation in which

1. pions and electrons of the same energy would produce, *on average*, the same signals, and

2. the *average* pion signal was proportional to the energy.

The price paid for reaching that goal was a considerable mismeasurement of the energy of particles or jets whose signals did *not* resemble those of the average pion. By overestimating the energy deposited in the em calorimeter section by more than a factor of two, the energy of particles or jets that happened to deposit a considerable fraction of their energy in this calorimeter section was similarly overestimated. And the energy of particles that penetrated the em section without undergoing a strong interaction was systematically underestimated by a considerable factor.

The most serious problems resulting from this calibration strategy may be expected for jets. First, since jets contain some fraction of π^0s, which *always* deposit their *entire* energy in the em calorimeter section, the *average* energy sharing between the two sections may be quite different for charged pions and for jets of the same energy. The latter deposit, on average, a larger fraction of their energy in the em calorimeter section. As a result, the jet energy is, on average, overestimated. However, this effect is (partly) offset by the fact that the jet may contain particles which do not deposit their (entire) energy in the calorimeter (muons, neutrinos) or which travel outside the cone through which the jet is defined.

Second, the jet energy measurement is strongly correlated with the jet topology. This may be illustrated with the following example, involving a high-energy jet produced by a fragmenting u quark. Let us assume that the leading particle (which contains the original u quark) carries half of the jet energy. This particle has (approximately) equal probabilities of being a π^+ or a π^0.

If the leading particle were a π^0, then all its energy would be deposited in the em calorimeter section. If it were charged, then most of its energy would go to the hadronic section. The ALEPH detector would measure a very different value for the energy of these two jets. Assuming that the other jet particles would generate the same signals in both cases, the difference between the energies measured for these two jets could be as much as 35% [Wig 91]. The signal distribution for mono-energetic jets detected with the ALEPH calorimeter would thus be considerably broader than the signal distribution for mono-energetic pions of the same energy, and possibly shifted to higher values, as a result of the chosen calibration scheme.

Possible evidence for the problems that this calibration scheme may have caused for jet detection may be derived from Figure 6.28. This figure shows the total energy distribution for hadronically decaying Z^0 particles measured with the ALEPH calorimeter system, and may thus be considered a jet measurement at 90 GeV.

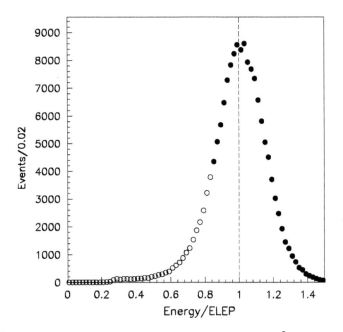

FIG. 6.28. The total energy distribution of hadronically decaying Z^0 particles, measured with the calorimeters of the ALEPH experiment at LEP. The black data points were used for the fit described in the text [Wig 91].

Some of the jets initiated by b and c quarks may have contained neutrinos and muons which were not (completely) absorbed by the calorimeters. Also, some energy may have leaked out since the detector did not cover the complete 4π solid angle. Such phenomena cause a low-energy tail in the signal distribution, which can actually be observed in Figure 6.28.

When this tail was ignored and the distribution in the area $E/E_{\mathrm{LEP}} = 0.85 - 1.5$ (the black experimental data points in Figure 6.28) was fitted to a Gaussian function, a fractional width σ/mean of 14.3% was obtained. This may be compared with the width of 8.8% that could have been expected on the basis of the energy resolution quoted for single hadrons ($\sigma/E = 84\%/\sqrt{E}$ [Bag 90]).

The fact that not one, but a multitude of hadrons (an unknown and fluctuating mixture of π^0s and other particles) were measured simultaneously by the calorimeter, introduced thus an extra contribution of $\sim 11\%$ (*i.e.*, $\sqrt{14.3^2 - 8.8^2}$) to the energy resolution in the case of 90 GeV jets.

It should be emphasized that this discussion is by no means intended to criticize ALEPH. This example was chosen because of its educational value. It clearly illustrates the problematic consequences of an incorrect calibration strategy.

In the LEP studies of Z^0 decay, hadronic calorimetry played only a very modest role. The ALEPH experiment was equipped with very powerful tracking capabilities and offered, therefore, excellent alternatives to calorimetry for studying the Z^0 physics. The ALEPH Collaboration demonstrated that the distribution shown in Figure 6.28 could be considerably improved when the calorimetric information on the charged hadrons was replaced by the measured track momenta [Bus 95]. If, in addition, the event sample was limited to Z^0s that produced jets in the central detector region, then the resolution improved from 14.3% to 6.9%.

However, it should also be emphasized that the very clean nature of the $e^+e^- \rightarrow Z^0$ events, which makes such improvements possible, is by no means representative for today's experiments in particle physics, which increasingly rely on excellent and correctly calibrated calorimeter systems for their physics analyses.

6.3.2   *Never intercalibrate with showers!*

In Section 6.2.1 we saw how HELIOS ran into trouble when they intercalibrated the two longitudinal sections of their calorimeter with high-energy electrons which deposited part of their shower energy in each of these sections. When they treated the calibration constants for these two sections as parameters whose values were determined by minimizing the width of the total signal distribution (*e.g.*, Equations 6.2 or 6.4), they found that the results were energy dependent. They also found that calibration constants determined in this way made their calorimeter non-linear for electrons. It was thus impossible to minimize the total width of the signal distribution and achieve signal linearity simultaneously.

The HELIOS calorimeter was almost perfectly compensating. This example thus illustrates that intercalibration problems of longitudinal calorimeter segments are by no means a unique feature of non-compensating calorimeters. They are not limited to hadron showers either, since this example concerned electrons. The problem was caused by the fact that the relationship between the deposited energy and the resulting signal (*i.e.*, the calibration constant C_z) in a sampling calorimeter varies with the shower age (or the depth, z). The two calorimeter sections thus sampled different regions of the developing showers, with different C_z values, even if the sampling structure of the calorimeter was exactly the same in the two sections. In fact, the C_z values were unique for each individual event, depending on the energy sharing between the two calorimeter sections. Had the opportunity existed to determine the starting point of individual γ-induced showers in HELIOS, and had Equation 6.2 been applied for different subsets of events, distinguished on the basis of the starting point of these showers, then different sets of calibration constants would have been found for each of these subsets. It is only because the *average* energy sharing between the two calorimeter sections depended on the shower energy that the calibration constants produced by Equation 6.2 were energy dependent.

The width of the total signal distribution that results from a minimization procedure as described above is only equivalent to an energy resolution for a very limited subset of events (*e.g.*, electrons of one particular energy that emit an energetic bremsstrahlung γ between $0.5X_0$ and $1.0X_0$). For all other cases (electrons of other energies, electrons of the same energy with different shower characteristics, γs of the same energy, π^0s of the same energy) this signal distribution has the wrong mean value and its width cannot be interpreted as the applicable energy resolution.

Intercalibrating different sections of a calorimeter system with showers is thus a bad idea. Determining calibration constants of individual calorimeter sections through a procedure in which the width of the total signal distribution is minimized may also lead to wrong results for a *longitudinally unsegmented* calorimeter.

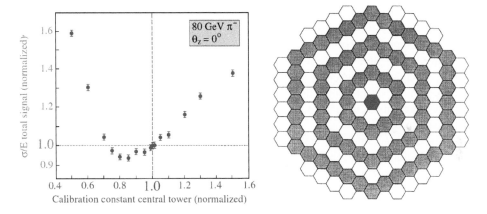

FIG. 6.29. The fractional width, σ/E, of the signal distribution for 80 GeV pions in the SPACAL detector as a function of the weighting factor applied to the signals from the central calorimeter tower, into which the pion beam was steered. The calorimeter towers were individually calibrated with high-energy electrons [Aco 91c].

Figure 6.29 shows an example, taken from SPACAL [Aco 91c]. The towers of this longitudinally unsegmented calorimeter were all calibrated individually with a beam of high-energy electrons. However, it turned out that the total width of the signal distribution for high-energy pions could be made narrower by applying a weighting factor to the signals from the central tower, which recorded typically about half of the total signal, and almost all of the em shower component. At 80 GeV, the width could be reduced by about 5% by reducing the signals from this tower by 15%. However, at 10 GeV, the effects of such manipulations were almost zero.

Also in this case, the width of the total signal distribution could thus be improved by deliberately changing the calibration constants. Also in this case, this is a result of differences between the sampling fractions for the energy deposited in different parts of the detector (*i.e.*, the central tower and the peripheral towers, respectively). And also in this case, manipulating the calibration constants with the goal of reducing the total width of the signal distribution would have led to a number of undesirable effects, such as response non-linearity and systematic mismeasurement of energy. For example, had this calorimeter been calibrated with a beam of 80 GeV pions, sent into each individual tower, and had the calibration constants been determined by minimizing the width of the total signal distribution, then the energy of 80 GeV electrons would have been reconstructed as 68 GeV (85% of the true value, see Figure 6.29).

6.3.3   *Meaningless e/π ratios*

One calorimetric figure of merit that often completely loses its meaning as a result of calibration procedures is the e/π signal ratio. In Section 3.1.4, the e/π signal ratio was introduced. We explained its relationship with the e/h value, its energy dependence and the resulting response non-linearity for pion detection. However, it was always implicitly assumed that the relationship between deposited shower energy and the charge generated as a result by the calorimeter was the same for electrons and pions, *i.e.*, that the same calibration constants (in units of GeV pC^{-1}) were applied for all particles.

However, if calibration constants are used that depend on the type of particle, on its energy, or on the place where it starts showering, as in some of the calibration methods discussed in this chapter, then the e/π signal ratio is turned into an arbitrary number that says more about the manipulations with the calibration constants than about the calorimeter properties.

Therefore, e/π values of longitudinally segmented calorimeter systems **only** contain meaningful information if the relationship between deposited energy and generated charge has been established in exactly the same way for all calorimeter segments, *i.e.*, if the segments were intercalibrated with a $B/A = 1$ method.

6.3.4   *Calibration and hadronic signal linearity*

Hadronic signal non-linearity is a general characteristic of non-compensating calorimeter systems. It has been demonstrated that one of the common consequences of calibration schemes that aim for minimization of the width of the total signal distribution ($B/A > 1$) is an increase in the hadronic signal non-linearity, over and above the intrinsic non-linearity that occurs even in a correct calibration scheme (Figure 6.21). Hadronic signal non-linearity is bad, especially when it comes to jet detection. It is bad because it makes the reconstructed energy dependent on the jet topology. A jet with a leading particle that carries a large fraction of the energy of the fragmenting object will be reconstructed with a different energy than a jet whose energy is more evenly divided among the different constituent particles (Figure 6.23d). This phenomenon will, for example, lead to different calorimeter responses to gluon and quark jets of the same energy, because of the different composition of such jets. Calibration schemes that increase such differences should be avoided. Unfortunately, in practice this is not the case since the focus is usually on the energy resolution. However, one should keep in mind that the width of a signal distribution is equivalent to a measure of the energy resolution, *i.e.*, the precision with which the energy of a showering object can be determined, *if and only if* the central value of the distribution has the correct energy. The effects of signal non-linearity invalidate this presumption.

6.4   Offline Compensation

In the previous sections, several examples have been shown of problems that may arise when a calorimeter system consisting of two longitudinal segments is incorrectly calibrated. Similar problems occur when the calorimeter consists of three, four, five or even more longitudinal segments. Also for such systems, the simple rule formulated in Section 6.2.11 holds: the only correct way to intercalibrate the various segments of such a system is with particles that behave, on average, identically in each and every segment.

Only in this way is the energy deposited in the active calorimeter material measured unambiguously. An overall scale factor can be used to account for the differences between the visible energy in showers induced by electrons, hadrons and jets. As we have seen in Sections 6.2 and 6.3, alternative calibration methods, designed to correct the effects of these differences through weighting factors, optimized for an "average shower," tend to amplify these effects for non-average events.

Sometimes, a longitudinal subdivision into many segments is applied in the hope of achieving the advantages offered by an intrinsically compensating calorimeter, for example the superior hadronic energy resolution [Braun 88, Braun 89]. In Section 4.10, we argued that this cannot be achieved through general weighting factors, since resolutions are determined by event-to-event fluctuations, and not by mean values. If and only if the em shower content is measured *event by event* can the effects of event-to-event fluctuations in the em shower content be reduced. Such fluctuations tend to dominate the hadronic energy resolution of non-compensating calorimeters.

The WA1 Collaboration [Abr 81] managed to achieve this for high-energy single hadrons of known energy. Their method and the obtained results are discussed in Section 4.10. The H1 Collaboration [Braun 89] applied a calibration method in which large local energy deposits were selectively suppressed. This also eliminated some of the consequences of event-to-event fluctuations in the em content of hadronic showers.

The H1 LAr calorimeter [Braun 89] consisted of an em compartment subdivided into five longitudinal segments and a hadronic compartment subdivided into six such segments. The total energy was calculated as

$$E \;=\; C_{\mathrm{e}}^{w} \sum_{i=1}^{5} Q_i(1 - \eta_{\mathrm{e}}Q_i) + C_{\mathrm{h}}^{w} \sum_{i=1}^{6} Q_i(1 - \eta_{\mathrm{h}}Q_i) \tag{6.5}$$

where Q_i represented the charges in the $(5 + 6)$ individual longitudinal segments, C_{e}^{w} and C_{h}^{w} set the overall scale, and the factors $(1 - \eta_{\mathrm{e}}Q_i)$ and $(1 - \eta_{\mathrm{h}}Q_i)$ were required to exceed a minimum value (δ). The latter requirement suppressed large energy deposits in individual cells.

This method thus involved five parameters ($C_e^w, C_h^w, \eta_e, \eta_h, \delta$). With the exception of δ, these parameters were found to be energy dependent. The problems introduced because of that have already been discussed in Section 6.2.3.

Both the WA1 and the H1 calibration methods led to improvements in the resolution characteristics of single pions from beams with known energy, at least for energies in excess of 30 GeV. Also, the line shape became clearly more symmetric as a result of the efforts to recognize and correct for the anomalously large concentrations of energy deposit responsible for the high-end tails of pionic signal distributions in a non-compensating calorimeter (see Figure 4.69a).

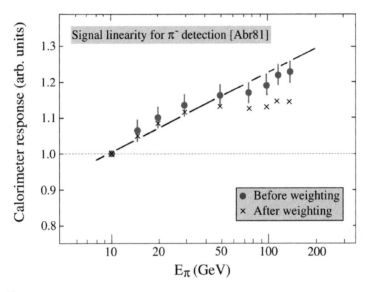

FIG. 6.30. The calorimeter response to pions as a function of energy for the WA1 detector, before and after application of an offline weighting procedure, that was intended to correct for the difference between the em and non-em responses event by event. The response was normalized to one at 10 GeV. Experimental data from [Abr 81].

The experimental data showed that procedures of this type worked especially well at high energies, where π^0s and non-em forms of energy deposit led to distinctly different signals in individual longitudinal segments of the detectors. At low energies, where these differences were much less spectacular, the improvements in energy resolution were found to be marginal, at best. Also the hadronic signal non-linearity did not disappear at energies below 30 GeV (Figure 6.30).

Neither WA1 nor H1 have demonstrated any beneficial effects of these "offline compensation" methods for jets, or more generally for a situation in which energy is deposited in the calorimeter system by a collection of particles with unknown composition and energies. Experimentally, this is not easy to achieve, since accelerators typically do not produce test beams of jets that can be used for that purpose. However, in the *ep* collisions at HERA, certain types of neutral-current event samples, in which the hadronic four-vectors were kinematically completely constrained, could well have been used to check the merits of the applied calibration procedures.

The ATLAS experiment has been inspired by the technique developed by H1 to correct their hadronic energy measurements for the effects of non-compensation. The ATLAS barrel calorimeter consists of a LAr ECAL plus an iron/plastic-scintillator HCAL, each of which is subdivided into three longitudinal compartments. In addition, there is a presampler (Section 5.7.1) intended to recover energy lost in the material upstream of the calorimeter system. In the beam tests on the basis of which their "cell weighting method" was developed, the hadronic calorimeter consisted of four longitudinal compartments. In addition, a so-called "midsampler" was installed to account for energy lost in the cryostat walls separating both calorimeter systems [Akh 00].

Also in their approach, the reconstruction of the energy of the showering hadron relies on upwards corrections of relatively small signals. The signals from cells that are relatively small compared to those from cells in which em shower components deposit energy are given a weight intended to equalize the response of those two types of cells. The total energy is determined with a formula similar to 6.5.

$$E = \sum_{\text{em cells}} W_{\text{em}}(E_{\text{cell}}, E_{\text{beam}})E_{\text{cell}} + \sum_{\text{had cells}} W_{\text{had}}(E_{\text{cell}}, E_{\text{beam}})E_{\text{cell}} + E_{\text{cryo}}$$

(6.6)

The weight factors applied to the signals from each individual calorimeter cell are a function of the compartment in which the calorimeter cell is located and of the beam energy:

$$W_{\text{em}} = A_E + B_E/E_{\text{cell}} \tag{6.7}$$
$$W_{\text{had}} = A_H + B_H/E_{\text{cell}} \tag{6.8}$$

The parameters A_E, B_E, A_H and B_H were taken from a fit and all depend on the hadron energy. Since finding the hadron energy was the purpose of the entire exercise, the parameter values had to be determined in an iterative process. The term E_{cryo} is intended to account for the loss of energy in the cryostat and is proportional to the geometric mean of the energy lost in the last em compartment and the first hadronic one. It turned out that E_{cryo} defined in this way was nicely correlated with the energy detected in the "midsampler," so that there was no need to include the latter device in the design of the final detector.

As stated before, procedures of this type may be applied with some success in beam tests, where the properties of the particle that generates the signals are precisely known. However, the benefits in the messy environment of a high-luminosity hadron collider are not clear, and in any case remain to be demonstrated.

6.5 Calibration of Calorimeters with Many Channels

Calorimeter systems often consist of thousands of individual cells. This large number prohibits establishing the calibration constants for each and every cell in a test beam at an accelerator. In that case, one sometimes relies on a calibration of *all calorimeter cells* with cosmic rays. A subset of the cells is exposed to showering particles in a test beam. By comparing the signal distributions for cosmic rays and the distributions for showering particles in these cells

1. The conversion factors needed to establish the calibration constants may be determined on the basis of the cosmic-ray measurements alone.

2. The precision of the calibration constants may be established on the basis of cosmic-ray measurements alone.

This procedure was followed for the calorimeters used in the R807 [Ake 85], ZEUS [Amb 92] and CMS [CMS 13] experiments. Its accuracy depends primarily on the statistical uncertainty in the mean values of the muon signals, and thus on the counting time. All mentioned experiments reported accuracies at the few percent level.

Calibration of the electromagnetic calorimeter channels was a particularly challenging task for CMS. This calorimeter consists of a barrel containing 61,200 $PbWO_4$ crystals, supplemented by two endcaps made of 7,324 crystals each. All these crystals had to be individually calibrated with a precision that would make it possible to detect high-energy γs with an energy resolution better than 1%. Cosmic rays played an important role in this process, which is briefly described below. For more details, the reader is referred to [CMS 13], and references therein.

The calibration constants of the ECAL channels were determined using relative and absolute calibrations. Relative callibrations are intercalibrations between one channel and another. For the absolute calibrations, which relate the intercalibrations to the mass scale, $Z \to e^+e^-$ decays are a crucial benchmark. The first step in the calibration process was an initial set of laboratory measurements, in which the light yield and the attenuation characteristics of *each individual crystal* were determined with radioactive sources. Crystal-to-crystal light yield variations (rms) were measured to be $\approx$ 15% for the barrel crystals and $\approx$ 25% for the endcap ones. Based on the results of these measurements, crystals with similar characteristics were grouped together in so-called supermodules. This reduced the rms of the variations within one supermodule to $\approx$ 8% in the barrel region.

Each supermodule was placed on a cosmic-ray stand for a period of about one week. A muon traversing the full length of a crystal deposited typically an energy of 250 MeV. In this way, intercalibration information was obtained for all crystals. A subset of the supermodules was exposed to a beam of high-energy electrons (90 and 120 GeV). A comparison of the experimental data taken with these electrons and with the cosmic rays revealed that the precision of the intercalibration coefficients obtained with the cosmic-ray calibration was about 1.5% (Figure 6.31).

CMS also uses collision data to check and possibly improve the intercalibration coefficients, especially for the crystals that have not been exposed to electron test beams. Three different methods are being used in this context:

1. The ϕ *symmetry method* uses a large sample of minimum bias events. It is based on the expectation that the total deposited transverse energy in this event sample should be, on average, the same for all crystals at the same pseudorapidity. This method provides a fast intercalibration of all crystals located in a particular η ring.

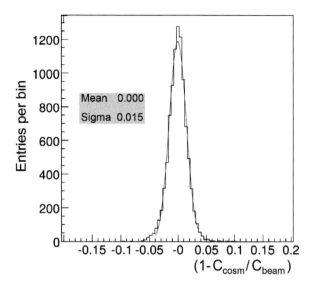

FIG. 6.31. Distribution of the relative differences between the intercalibration coefficients measured with high-energy electrons and those obtained from cosmic-ray muons [CMS 08].

2. The π^0 *and η-meson calibrations* use the invariant mass of γ pairs from the decay of these mesons for intercalibration purposes. A special data stream is used to profit from the copious production of these particles at the LHC. These events are collected at a rate of about 10 kHz, with minimal impact on the CMS readout bandwidth.

3. *Intercalibrations with isolated electrons from W- and Z-boson decays* are based on a comparison of the energy measured in the em calorimeter to the track momentum measured in the silicon tracker. In 2011, 7.5 million isolated electrons of this type were collected, *i.e.*, ~ 100 electrons per crystal.

Figure 6.32 summarizes the results of these efforts. The combined intercalibration precision is 0.4% for the central barrel crystals ($|\eta| < 1$), and 0.7–0.8% for the rest of the barrel. In the endcaps, the precision is better than 2% up to the limit of the electron and photon acceptance ($|\eta| \sim 2.5$). The variation of the precision with pseudorapidity arises partly from the size of the data sample, and partly from the amount of material in front of the ECAL.

The diameter of one PbWO$_4$ crystal is about $1\rho_M$. As a result, a showering electron or photon deposits only a fraction of its energy in one crystal, even when its trajectory coincides with the crystal axis. For that reason, the intercalibration methods 2 and 3, mentioned above, are based on (3×3) clusters of crystals, rather than on single ones. The response of the PbWO$_4$ crystals depends in a complicated way on their history in the colliding-beam environment at the LHC. The radiation to which they are exposed causes a loss in transparency, which is (partially) recovered during beam-off periods. A sophisticated laser system continuously monitors these effects in all crystals during the

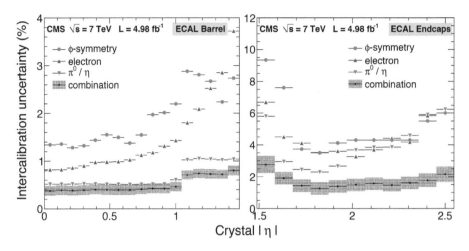

FIG. 6.32. The intercalibration precision obtained in 2011 using different intercalibration methods and the resulting precision, with its uncertainty, for the combination of the methods, in
the CMS barrel (left) and endcap (right) calorimeters [CMS 13].

runs, and provides the information to correct the signals. Details of this procedure are
described in Section 6.6.1.

Figure 6.33 shows the invariant mass distributions for events in which the electrons
and positrons from the decay $Z \rightarrow e^+e^-$ ended up in the barrel (Figure 6.33a) or in
the endcap (Figure 6.33b) region of the calorimeter. The figure shows the effects of
including the information from the collision data and from the laser monitoring system
on the distributions. The latter corrections were much more important in the endcap
than in the barrel, because of the much higher radiation levels.

The absolute calibration of the energy scale was initially based on the results of
the exposure of a subset of the crystals to electron beams, with energies in the energy
range relevant for the envisaged physics studies (vector boson decay, Higgs production), *i.e.*, 90 and 120 GeV. The calibration constants were established by equalizing the
signals in a 5×5 cluster of crystals to the beam energy.

An important test of the correctness of these calibration constants is how well the
mass of a known particle that decays electromagnetically is reproduced when these
constants are used to convert the measured signals. The process $Z \rightarrow e^+e^-$ is ideally
suited for this purpose, since the events have a very clear signature and are produced at
sufficiently high rates for this purpose. However, for determining the invariant mass of a
Z boson decaying into an e^+e^- pair, a number of effects have to be taken into account
that do not play a role in beam tests, for example:

- The electrons and positrons lose energy through synchrotron radiation on their
  way through the magnetic field to the calorimeter. These γs may or may not contribute to the signals produced by the crystals in the cluster selected for determining the electron energy

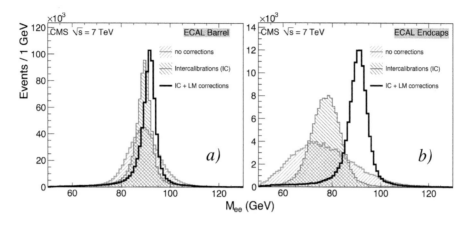

FIG. 6.33. Reconstructed invariant mass from $Z \rightarrow e^+e^-$ decays, for single-channel corrections set to unity, for final intercalibration, and for both final intercalibration and laser monitoring corrections, in the barrel (a) and endcap (b) CMS calorimeters [CMS 13].

- The electrons and positrons lose energy in the material upstream of the calorimeter. The amount of material varies with η, but is in some cases about $2X_0$.

- In the beam tests, the electrons hit the crystals in their geometric center. In the LHC, the Z decay products may hit the crystals anywhere, $e.g.$, close to the boundary between crystals. Energy losses in the cracks between crystals are thus not the same.

CMS estimated these effects with elaborate GEANT4 based Monte Carlo simulations. Figure 6.34 shows the measured e^+e^- invariant mass distribution for two scenarios: both the electron and the positron were detected in the barrel calorimeter (Figure 6.34a) and both particles were detected in the endcap (Figure 6.34b). The experimental data were fitted with a curve that is the convolution of a Breit-Wigner line shape and a Crystal Ball (CB) function. The BW parameters of the Z^0 are very precisely known, thanks to the LEP experiments. Therefore, the parameters of the CB function provide information about the displacement and the energy resolution associated with this measurement.

The results show that the Z^0 mass was well reproduced in both parts of the ECAL ($\Delta m/m < 1\%$). The relative energy resolution, represented by the Gaussian term of the CB function, amounted to 1.1% in the barrel and 2.6% in the endcaps. In studies performed with $Z \rightarrow \mu^+\mu^-\gamma$ events, the value of the Z mass was also reproduced within 1.5%, in all regions of the calorimeter.

A much less elaborate effort was expended to calibrate the 10,000 channels of the ATLAS hadron calorimeter, which is also based on optical readout. Also in this case beams of electrons were used to determine the absolute energy scale for a limited set of calorimeter modules. The calibration constants of the other channels were and are determined with a radioactive source [Sta 02]. This source, ^{137}Cs with a strength of about 10 mCi, is moved hydraulically throughout the entire calorimeter body. The signals cre-

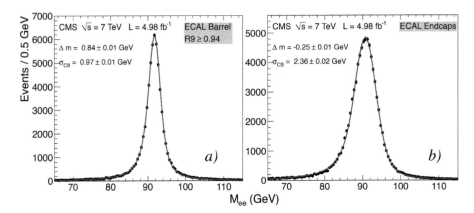

FIG. 6.34. The dielectron invariant mass distribution for Z decays with both electrons in the
CMS barrel calorimeter (a) or both electrons in the endcap (b). The parameters listed in each
panel represent the difference between the mean value and the true boson mass (Δm) and the
width of the Gaussian term of the (Crystal Ball) function used to fit the distribution (σ_{CB})
[CMS 13].

ated by the 662 keV γs and the β electrons it emits are generated in the scintillators
and follow the same optical chain as the light produced by showering particles. This
system was used for the intercalibration of all the channels of this calorimeter and for
tuning the high voltage setting in each and every individual channel. During dedicated
calibration runs, this source traverses each of the 463,000 tiles of this detector, and the
current generated in each of these tiles is extracted. This system has made it possible to
equalize the gains of all PMTs to within $\pm 3\%$. The absolute calibration constants were
also in this procedure derived from the relationship between the current generated by
the source and the energy deposited in the beam-tested ones.

Calibrating a large LAr calorimeter system is in some respects easier than a system
based on optical readout. The lack of internal amplification of the ionization charge,
as well as the absence of radiation damage affecting the signals (only the noise), are
important factors in that context. The aim of the calibration of this type of calorimeter
is to determine the conversion factor between the measured signal, in ADC counts, and
the signal current, in μA, for each channel. This is typically done by injecting a voltage
pulse via a precision capacitor into each preamplifier. This procedure can be repeated on
a regular basis, thus allowing monitoring of the stability of the calibration results. The
conversion from a signal expressed in μA to a signal in GeV is done on the basis of the
results of testbeam measurements. Figure 6.35 shows these conversion factors for the
ATLAS ECAL, as a function of pseudorapidity. At this stage, ATLAS also had to deal
with a separate issue, namely the complications caused by the longitudinal segmentation
of the ECAL (Section 6.2.4). As in CMS, reconstruction of the Z^0 boson through its
decay into an electron and a positron serves as the benchmark for verification of the
absolute energy scale.

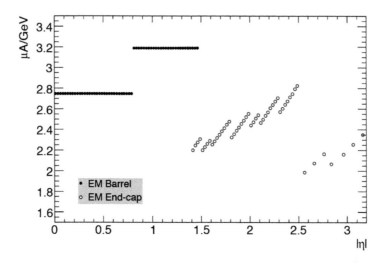

FIG. 6.35. Ionization current per unit deposited energy from em showers in the barrel and end-cap electromagnetic calorimeters of ATLAS. These conversion factors depend on the sampling fraction and the drift velocity, and are therefore sensitive to small temperature variations (Section 6.6.2.1). The initial values have been derived from electron testbeam data [Aad 08].

Figure 6.36 shows an example of the electronics chain for the treatment of charge pulses from a LAr calorimeter, taken from the H1 experiment [Braun 88]. Each electronic channel consisted of the following modules: a charge-sensitive preamplifier, a twisted-pair line driver, a differential amplifier (receiver), a pulse shaper and a charge-sensitive ADC. In order to limit the capacitance, and thus the noise level, as much as possible, the preamplifiers were located as close as possible to the feedthroughs of the signal cables out of the cryostat, $\sim$ 9 m from the LAr cells. The line driver differentiated the signals and transmitted them over 60 m to the differential amplifier. The shaping time was 2 μs.

This chain was calibrated by feeding a voltage pulse via the capacitor C_C into each preamplifier. The charge input could be attenuated in a computer-controlled way, so that the whole dynamic range of the electronics could be scanned in fine steps. This method made it possible to determine the pedestal value, the conversion factor from charge to ADC channel and the linearity of the ADC for each electronic channel. Also the noise levels and the electronic cross-talk between individual channels could be measured in this way. In H1, this calibration procedure had been fully automated and was typically repeated on a daily basis.

The main difference between the H1 and ATLAS LAr calorimeters is that in the latter only part of the ionization charge contributes to the signals. The fast shaping of the readout (Figure 5.52) requires that the distribution of the calibration signals be done via injection resistors placed at the input of the detector cell, *i.e.*, inside the liquid, and not at the input of the preamplifier located outside the cryostat.

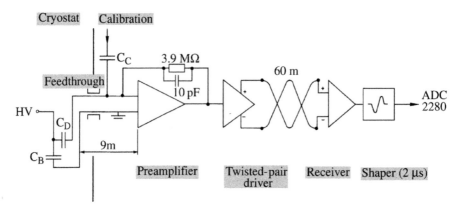

FIG. 6.36. The electronics chain for the treatment of the LAr pulses in the H1 experiment at HERA [Braun 88].

6.6 Checking and Maintaining the Calibration

Calorimeter systems in particle physics experiments are nowadays typically intended to operate for at least a decade. On a timescale like that, a variety of processes may occur that critically affect the calorimeter response. This is particularly true for, but not limited to, calorimeters with optical readout systems.

The light yield, the light attenuation, the light collection efficiency and the quantum efficiency of the light detectors may deteriorate over time as a result of natural processes (often summarized in the term *ageing*), or as a result of induced radiation.

In readout systems in which the primary signals are amplified (*e.g.*, PMTs in optical systems, electronic amplifiers in calorimeters based on the detection of ionization charge), the gain of the amplifiers may be dependent on parameters such as the temperature, the humidity, or the stability of power supplies.

In calorimeters with gaseous active media, the gas gain is often extremely sensitive to the precise composition of the gas mixture, and in calorimeters based on liquids operating in the ionization-chamber mode, the signals are extremely sensitive to very small contaminations of electronegative substances (in particular oxygen), and to temperature variations.

Once the calorimeter has been calibrated, a major effort is therefore needed to monitor the system for eventual changes in the signal-to-energy conversion factors. In most experiments, elaborate systems have been developed for this purpose. The details of these systems depend, obviously, on the specific characteristics and needs of the calorimeter for which they are used. We first discuss the needs of calorimeters based on optical signals.

6.6.1 *Calorimeters with optical readout*

The stability of the calibration in calorimeters based on scintillation or Čerenkov light has several aspects:

- The amount of light generated in the active calorimeter layers may change with time.
- If separate wavelength-shifting elements are applied, then the light collection of these elements, or the efficiency of the conversion may change with time.
- The light attenuation in the active layers, and/or in the wavelength-shifting elements (if applied) may change with time.
- The quantum efficiency of the light detectors may change with time.
- The gain of the light detectors may change with time.

The last aspect is usually the easiest to monitor. For this purpose, a sufficiently stable (typically better than 1%) light source is needed. Often, a pulsed laser that generates light with a wavelength and a signal shape roughly equal to those of the light generated in the calorimeter forms the basis of such a system. The laser beam is defocused and the light is distributed over a large number of optical fibers in a reproducible way. These fibers are guided to each and every light detector that has to be monitored. Pulse-to-pulse fluctuations in the amount of light produced by the laser are separately monitored, for example with a high-quality photodiode, which receives a small, but fixed fraction of the light. Systems of this type are in use in most current, light-based calorimeters. (MINOS, BELLE 2, T2K, ATLAS, CMS, ALICE *etc.*).

The main purpose of such a system is to keep track of the gain of the light detectors, so that corrections can be made, either online (*e.g.*, by changing the voltage) or offline (*e.g.*, by correcting the calibration constants), if and when needed. However, in practice such changes are cumbersome and one would like to avoid them as much as possible. Ideally, the role of this monitoring system is therefore a *preventive* one. The optical system itself is usually designed in such a way that the gain is intrinsically stable at the required level. Especially in crystal calorimeters, which aim for energy measurements with a precision of 1% or better, this is not trivial to achieve. For example, in CMS the response variations must be kept within 0.4% to retain the excellent intrinsic energy resolution of the electromagnetic calorimeter (Figure 6.32).

Light detectors such as PMTs are devices with an exponential gain–voltage characteristic. Typically, a change of about 100 V in the applied voltage between the photocathode and the anode corresponds to a gain change by a factor of two. Therefore, stability at the fraction of 1 V level is required to limit gain changes to the 1% level. This level of gain stability usually also requires temperature stability at better than $1°C$. Other types of light detectors, *e.g.*, Avalanche Photo Diodes, are even more sensitive to changes in the applied (bias) voltage and in temperature (-2% K^{-1}). CMS has built an impressive cooling system, that stabilizes the temperature in the calorimeter and limits deviations from the operating temperature of $18°C$ to less than $0.03°C$ in the barrel section and $0.08°C$ in the endcaps. The APD bias voltage of 380 V is kept stable to better than 65 mV.

After having discussed monitoring of the light detectors, we now turn to the light itself. As indicated above, the amount of light that constitutes the (average) calorimeter signal for a given energy deposit may change over time as a result of several independent effects, *e.g.*, chemical ageing and/or radiation damage. Therefore, monitoring the stability of the light yield has several aspects, which may or may not be addressed in the chosen method.

The best opportunities to monitor the (stability of the) generation and transportation of light are available in calorimeters with ^{238}U absorber. This uranium is radioactive, it disintegrates through a sequence of α and β decay processes to end up as ^{206}Pb. The αs, βs and γs produced in this process generate light in the active elements, through the same processes as the particles produced in shower development generate light. Therefore, such calorimeters are equipped with a "natural" light source, which is extremely stable, since the half-life of ^{238}U amounts to $4.5 \cdot 10^9$ yr.

Both the HELIOS [Ake 85] and ZEUS [And 91] experiments, whose calorimeters were based on this type of absorber, exploited this feature to monitor the stability of the light generation and transportation in their detectors. The HELIOS measurements indicated that the radioactivity-induced light, integrated over 10 μs, corresponded, on average, to the light resulting from an energy deposit of 0.4 GeV and 3.0 GeV per tower in the em and hadronic calorimeter sections, respectively.

The rms widths of the distributions of these uranium-induced signals were found to be 11% and 5.6%, respectively. Therefore, a dedicated measurement of 1,000 events in which the light produced by the calorimeter was integrated over periods of 10 μs, yielded a measurement of the light yield with a relative precision of a fraction of 1%. ZEUS used the uranium "noise" in a similar way.

The only drawback of this method is that the spatial profile of the light production in particle showers is different from the profile of the light production by the natural radioactivity of the calorimeter. The latter profile is uniform in all dimensions, whereas the light from showers is predominantly produced in the upstream part of the calorimeter.

This only has consequences if the light yield is actually observed to decrease over time. In that case, the implications of such an effect for the calibration constants of the calorimeter are not straightforward to determine, since these consequences depend on the precise cause of the effect. For example, if the decrease in the signals was caused by a general decrease in the light production of the scintillators, due to some chemical ageing effect, the observed change in the radioactivity signals would be the same as the change in the calibration constants to be applied for shower signals.

If, on the other hand, the decrease was caused by light attenuation, e.g., in the wavelength-shifting plates used in these calorimeters, then the effects on the calibration constants for shower signals would be much larger than the observed effect on the uranium signals. Since the uranium light was, on average, produced much closer to the PMTs, it was much less affected by light attenuation than the shower light.

The uranium signals are therefore ideally suited to monitor the calorimeter *stability*. However, in the case of observed instabilities, other methods are needed to establish the cause and consequences of these instabilities. These other methods usually involve moveable radioactive sources, which can be inserted in the calorimeter to record the light production profile, e.g., along the wavelength-shifting plates. ZEUS had incorporated a system of this type in its uranium calorimeters. In calorimeters with non-radioactive absorber material, such moveable source systems are also frequently used

[Bee 84, Bon 87, Hah 88].

A modern version of such a system is used to monitor the stability of the ATLAS hadronic calorimeter [Sta 02]. In this case, the source is ^{137}Cs. The electrons and (662 keV) γs emitted in β decay of this nuclide generate signals in the scintillators, and these signals are used to monitor the performance of the optical part of the calorimeter, including the front-end electronics and the PMTs. During dedicated runs, this source traverses the 463,000 tiles of the detector, and the current generated in each of these tiles is extracted. Because of the limited penetration of the βs or γs from such a source, this method does not provide a complete test of all aspects of the light generation and transportation in the calorimeter. Therefore, this method is primarily intended as a diagnostic tool. As long as the results from source scans are stable[19], it is assumed that this stability also applies to the light generated in the shower development. If instabilities are observed, then more work is needed to establish the consequences.

When instabilities in the light generation and/or transportation are observed, typically several affected calorimeter modules are remeasured in test beams with particles of precisely known energies, in order to re-establish the calibration constants. In some experiments, such a remeasurement is part of the standard operating procedure, whether or not the monitoring system revealed instabilities. An example was set by the UA2 Collaboration [Jen 88].

The UA2 lead/iron/scintillator calorimeter was not equipped with a moveable source system. However, during each shutdown period the signals from a strong ^{60}Co source placed in front of each module were measured to check the stability of the light yield. Changes in calibration constants were made assuming that the source signals were representative for shower signals. This assumption was checked regularly by remeasuring a subset of calorimeter modules in a test beam.

Results of these measurements are shown in Figure 6.37. After several years of operation, the calibration constants turned out to be correct to within about 2%, thanks to this method.

Instabilities in the light signals are a major point of concern for the CMS electromagnetic calorimeter (ECAL). The response of the PbWO$_4$ crystals varies under irradiation, due to the formation of color centres that reduce the transparency. The crystal transparency recovers, at least partially, through spontaneous annealing [CMS 05]. A monitoring system, based on the injection of laser light into each crystal, is used to track and correct for response changes during LHC operation [Anf 08]. This laser light has a wavelength close to that of the emission peak of the PbWO$_4$ scintillation light (440 nm). Additional light sources provide separate information on the stability of the light detectors.

The evolution of the ECAL response to the laser light is shown in Figure 6.38, as a function of time during the 2011 LHC operations. An average value is shown for each of six pseudorapidity ranges. The data are normalized to the response measured at the start of 2011. The corresponding instantaneous luminosity is also shown. The

---

[19]Because of the radioactive decay of this nuclide, the currents it creates decrease every year by 3.3%. This has to be taken into account when determining the (changes in) the calibration constants.

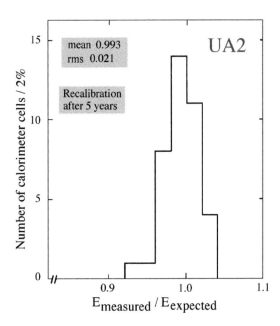

FIG. 6.37. Long-term stability of a number of UA2 calorimeter modules. After several years of operation, remeasurement in a test beam demonstrated the validity of the energy scale within $\sigma \sim 2\%$ [Jen 88].

response dropped during periods of LHC operation, and recovered partially during off-beam periods. The magnitude of the effects clearly increased with the pseudorapidity, as a result of the increasing dose rate [Zhu 15].

The results of the laser monitoring system are used to change the calibration constants of the crystals. In doing so, the fact that the spatial distribution of the scintillation light generated in the crystal is not the same as that of the laser light should be taken into account. The conversion factors turned out to depend, among other things, on the producer of the crystals. The effect of these changes is shown in Figure 6.39, which shows the E/p ratio for electrons from $W \rightarrow e\nu_e$ decay as a function of time, before and after the corrections were made. The mean value of the reconstructed energy of these electrons remained constant to within 0.1% for the barrel calorimeter, and 0.3% in the endcap [Ghe 15].

The decrease in the transparency of the PbWO$_4$ crystals continued in 2012. At the end of the 2012 LHC run, the transparency was reduced to 94% of its initial value for $|\eta| < 1.4$, to 70% of that value for $2.1 < |\eta| < 2.4$, while the transparency was reduced to only one third of its initial value for the crystals in the high-η region. Based on these observations, CMS has decided to replace the crystals in the endcap by another system [CMS 15b].

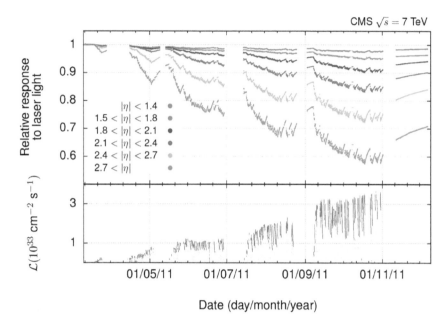

FIG. 6.38. Relative response of the CMS PbWO$_4$ crystals to laser light during 2011, normalized to the response measured at the start of 2011. An average is shown for each pseudorapidity range. The bottom plot shows the corresponding instantaneous luminosity. After the last LHC technical stop, a recovery of crystal transparency was observed during the low luminosity heavy-ion data-taking at the end of 2011[CMS 13].

6.6.2 Calorimeters based on direct charge collection

Calibration systems for calorimeters based on proportional-gain wire-chamber readout have to cope with both the inherent, comparatively large gain variations between different wires, and with the long-term global variations resulting from changes in gas composition, temperature and pressure.

The uniformity in gas-gain calorimetry can be optimized by tightly controlling the mechanical tolerances of the wire-chamber elements. In very carefully constructed devices, the gain variations ($\sigma_{\rm rms}$) can be limited to 5–10% [Are 89, Cam 88, Cih 88].

The mentioned global gain variations can be monitored with a representative set of reference chambers [Dew 89]. The signals from such reference chambers are sometimes used in a feedback circuit, adjusting the chamber high-voltage to maintain constant gain.

For charge-collecting calorimeters of the ionization-chamber type (*e.g.*, liquid argon), uniformity of the response also requires tight mechanical tolerances in the construction. For example, a difference between absorber plates of 2.0 mm and 2.1 mm sandwiching the same liquid gap corresponds to a 5% difference in the sampling fraction. Since LAr operates without internal charge amplification, the relationship between the ionization current measured from a calorimeter cell and the energy deposited in that cell should, once it is established, in principle not change with time, provided that the

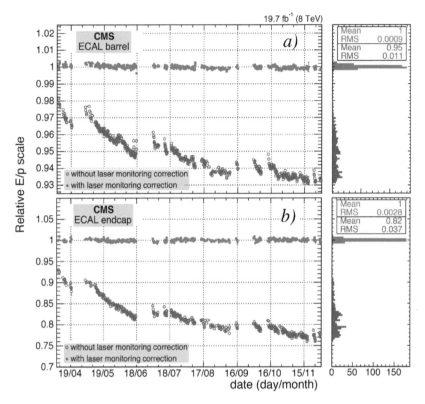

FIG. 6.39. Ratio of the energy measured by the CMS ECAL over the momentum measured by the tracker, for electrons selected from $W \to e\nu_e$ decays, as a function of the date at which these were recorded. The ratio is shown both before and after the application of transparency corrections obtained from the laser monitoring system, both for a supermodule from the barrel (a) and one from the endcap (b) calorimeters. Histograms of the values of the measured points, together with their mean and RMS values are shown next to the main plots [CMS 15a].

purity of the liquid and its operating temperature are rigorously controlled, and rate and radiation effects are properly taken into account.

6.6.2.1 *Temperature effects.* In a LAr calorimeter in which all the ionization charge carried by electrons is collected, the signals depend very weakly on the temperature. Only changes in the density of the liquid affect the signals, at the level of $\sim 0.5\% K^{-1}$. However, if only a fraction of the ionization charge is being collected, as in the AT-LAS calorimeters, the signals are also affected by the temperature dependence of the electron mobility. The effect of this on the signals is considerably larger, and the total temperature dependence of the signals is about $2\% K^{-1}$ in these calorimeters. This rather strong dependence calls for a uniform temperature throughout the cryostat. AT-LAS aims for a temperature difference smaller than 0.3 K between any two points of the calorimeter, which would limit the energy-independent contribution to the energy

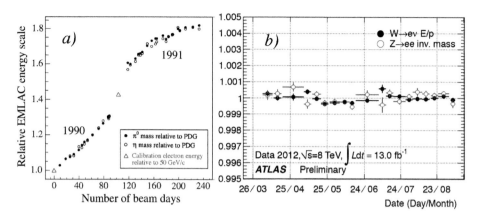

FIG. 6.40. Energy scale of the E-706 LAr calorimeter as a function of time, during a period of two years. The points are normalized to a calibration run taken at the beginning of operations in the year 1990 [Apa 98]. The closed and open circles indicate the change of the apparent mass of reconstructed π^0s and ηs, respectively (*a*). The energy scale of the ATLAS LAr calorimeter during the five months of the 2012 LHC run [Nik 13]. The mass of the reconstructed vector boson decays into electrons is shown as a function of time (*b*).

resolution resulting from this effect to 0.2%. Figure 6.40b shows that this goal was indeed achieved during the five months of the 2012 LHC Run. A comparison with Figure 6.40a illustrates the progress that has been made in this respect since the early days of liquid-argon calorimetry. The E-706 Collaboration built and operated a large LAr calorimeter for an experiment at Fermilab to study the production of photons and jets with high transverse momentum. Figure 6.40a shows the response of this calorimeter during two years of operation. The overall energy scale changed by as much as 80% over this period [Apa 98].

6.6.2.2 *Rate effects.* The signals of LAr sampling calorimeters are formed by electrons liberated in ionization processes. However, for each electron, there is also an ion. The mobility of these ions is much smaller than the mobility of the electrons, by some five orders of magnitude at the argon boiling point. In the presence of a very high flux of incident particles, this low mobility might lead to positive-charge build-up in the liquid gaps, which in turn could distort the electric field and thus reduce the measured signals. This effect was clearly observed during dedicated tests of a prototype module of the ATLAS barrel ECAL [Atl 96b]. It was found to depend strongly on the electric field strength.

Figure 6.41 shows the calorimeter response as a function of the beam intensity, for different settings of the high voltage across the LAr gaps, *i.e.*, for different values of the ion mobility. At the highest η value covered by the endcap calorimeters (3.2), the energy flux is estimated to reach values of a few million GeV cm^{-2} s^{-1}. According to this figure, the reduction in the signals will be limited to less than 1% for a field strength of 10 kV cm^{-1}. On the other hand, in the forward calorimeter, positive-ion

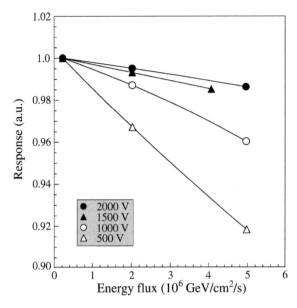

FIG. 6.41. The response of the 2 m ATLAS barrel calorimeter prototype as a function of the incident energy flux, for various settings of the high voltage across the LAr gaps [Atl 96b].

build-up might well lead to significant signal losses at high luminosity. The very small gap width chosen for this component of the calorimeter system (0.25 mm instead of 2 mm) will certainly limit the effects, since the ion drift time is reduced proportionally. On the other hand, the energy fluxes in this part of the detector reach very large values, two orders of magnitude larger than in the endcap compartment.

6.6.2.3 *Radiation effects.* Liquid-argon calorimeters have a reputation for being radiation hard. That is certainly true for the active medium itself, which could eventually also be easily replaced. However, there is more inside the cryostat than the liquid alone. Given the sensitivity of Si based electronics, ATLAS decided to move the preamplifiers outside the cryostat in some vulnerable parts of the LAr calorimeter. Also the materials contained in the electrode structure are a reason for concern. A dedicated exposure of a small ATLAS test calorimeter to high neutron fluences showed a significant reduction in the signals due to pollution of the argon by electronegative trace materials released from materials such as kapton, epoxy, *etc*. These tests also revealed measurable effects of a build-up of the radioactive isotope ^{41}Ar, which is produced by neutron capture in ^{40}Ar and decays with a half-life of 1.8 hr to ^{41}Ar [Atl 96b].

6.7 Conclusions

Calibration is an incredibly important aspect of working with calorimeters. Physicists want to measure particle energies with their calorimeters. The calorimeters produce electric signals. The calibration gives the recipe for converting one into the other.

While usually enormous efforts and ingenuity are invested in the design and the construction of calorimeters, their calibration is often a stepchild, left to be taken care of by someone who has nothing better to do. This rather general disregard for this aspect of calorimetry probably stems from the perception that the job is trivial and straightforward. In this chapter, we have attempted to show that this is by no means the case.

We have distinguished longitudinally unsegmented and segmented calorimeter systems. The first category took less than one page to discuss, the rest of the chapter was almost entirely spent on the second category.

Calibrating longitudinally segmented calorimeter systems is very tricky indeed. This is because the calorimeter response depends on the type of particle, on the particle's energy and on the age of the shower it developed. Especially in jets, the various effects that play a role interfere in ways that are impossible to disentangle.

In this chapter, we have discussed four examples of strategies used in practice to calibrate a calorimeter system consisting of two longitudinal compartments:

1. In the first strategy, the calibration constants were chosen such as to optimize the energy resolution for particles that deposited their energy in several calorimeter compartments (Sections 6.2.1–6.2.4, 6.2.6).
2. In the second strategy, the energy scale was set by particles that deposited their entire energy in one of the two compartments. Electrons were used to determine the calibration constants for the em compartment, the calibration constants for the hadronic compartment were derived from the signals generated by pions that penetrated the em compartment and showered in the hadronic compartment (Section 6.2.7).
3. The third strategy aimed for hadronic signal linearity in combination with reconstruction of the beam energy for electrons and pions simultaneously. This was done by increasing the signals from the em calorimeter compartment for hadron showers by a weighting factor (Section 6.2.8).
4. The fourth strategy aimed to achieve equal responses for electrons and pions by multiplying the signals from the ($8X_0$ thick) first calorimeter segment with zero (Section 6.2.9).

All four strategies looked, at first sight, reasonable and logical. They were all designed to achieve a specific beneficial result: optimization of the energy resolution (method 1), signal linearity for hadrons (method 3), equalization of the response to electrons and pions (method 4). Yet all four methods turned out to be fundamentally flawed at more or less severe levels. In each case, the benefit for which the calibration method was designed was offset by the introduction of new, sometimes very severe, problems. Among the undesirable side effects that were encountered, we mention:

- Signal non-linearity for em showers (1,4)
- Systematic mismeasurement of the energy of early (2) or late (3,4) showering particles
- Systematic mismeasurement of the energy of hadrons that penetrate the em calorimeter compartment without starting a shower (3,4)
- Systematic mismeasurement of the energy of jets (2,3,4)

- A general degradation of the energy resolution and the line shape (3,4)

- Systematic mismeasurement of the energy of γs and π^0s (1,4)

The underlying reason for all these problems is that the mentioned calibration strategies only work *on average*. The envisaged goals are only achieved for a particular subset of events, for example the pions that penetrate the em compartment without interacting (method 2), or electrons of a given energy (method 1), or pions that deposit $44 \pm 1\%$ of their energy in the em calorimeter compartment (method 3). At the same time, new problems are introduced for events that do not resemble the average for which the calibration method was designed. Manipulating calibration constants may thus seem to work out fine for one particular subset of events, but may have very negative consequences for other, possibly more relevant, subsets.

The only way to avoid such problems is to calibrate the individual sections of a longitudinally segmented calorimeter system in exactly the same way $(B/A = 1)$. If the individual compartments can be separated and are deep enough to contain electron showers, then an electron beam may be used for this purpose. If that is not possible, then these sections may be intercalibrated with muons traversing the entire depth of the calorimeter. Only in this way is the relationship between the deposited shower energy and the resulting signal established unambiguously, identically for all calorimeter sections.

In doing so, the conditions that exist automatically in longitudinally *unsegmented* calorimeters and that make these devices trivial to calibrate are reproduced. This leads us to the conclusion that longitudinal segmentation, in general, *does not serve any purpose* with regards to the precision with which the energy of showering particles can be measured in a calorimeter.

Only if the segmentation makes it possible to determine the em shower content **event by event** could some benefit be expected. However, even the feasibility of this is questionable in environments other than a test beam where particles of precisely known energy are delivered to the detector.

This conclusion may be illustrated by considering a perfectly compensating calorimeter, *i.e.*, a calorimeter in which event-to-event fluctuations in the em shower content do not contribute to the energy resolution. If such a calorimeter were longitudinally segmented, then the extra information on the shower development would not allow one to improve the energy resolution. In fact, one could only deteriorate the excellent resolution of the unsegmented device by calibrating the segments in a way that differs from the only correct one: $B/A = 1$.

Given all the potential pitfalls that result from longitudinally segmenting a calorimeter, the question arises why this is done in the first place. The answer to that question has typically three components:

1. It makes it possible to use available resources in the most economical way, by constructing a fine-sampling, high-granularity electromagnetic section, followed by a more crudely designed hadronic compartment

2. It allows easy identification of electrons and γs

3. It helps to improve the hadronic performance

Whereas the first argument certainly addresses a valid point, it also calls for limiting the total number of segments to two. More segments means that more money has to be spend on readout. Concerning the second argument, it has been demonstrated many times that there are several alternative methods to identify electrons and γs with comparable success in longitudinally unsegmented calorimeters. Some examples of these alternative methods are discussed in Section 8.2.6.2. Also for this argument, there is no benefit from having more than two segments. The third argument is the one that is typically used to defend a segmentation into more than two longitudinal segments. AT-LAS has six segments and H1 had eleven for that reason. However, as we have seen in this chapter, segmentations of this type lead to very complicated and possibly unsolvable calibration problems, while the advantages are questionable at best and, in my opinion, non-existent. If one was really concerned about hadronic performance, then I think it would make much more sense to design the hadronic section of the calorimeter as a monolithic *compensating* device. The crudely sampling lead/plastic-scintillator calorimeter built by the ZEUS group in their R&D phase could be taken as an excellent example [Bern 87]. It had $e/h = 1.0$ and detected hadrons with an energy resolution of $43\%/\sqrt{E}$.

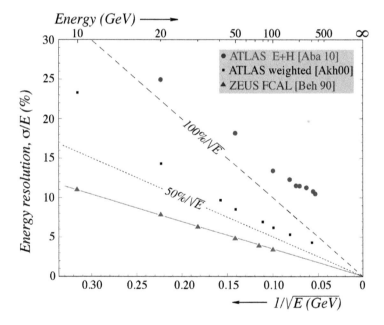

FIG. 6.42. The energy resolution measured for pions as a function of energy in the ATLAS and ZEUS calorimeters. The measured data, based on the standard calibration constants are shown as full dots for ATLAS [Aba 10] and as triangles for ZEUS [Beh 90]. Also shown are the results of an "offline compensation" procedure using the signals in the six longitudinal segments of the ATLAS calorimeter system [Akh 00].

I would like to illustrate this point by comparing the published hadronic performance results of the ATLAS and ZEUS calorimeters. Figure 6.42 shows the energy resolution measured for pions as a function of energy in the ATLAS and ZEUS calorimeters. The measured data, based on the standard calibration constants are shown as full dots for ATLAS [Aba 10] and as triangles for ZEUS [Beh 90]. Also shown are the results of an "offline compensation" procedure using the signals in the six longitudinal segments of the ATLAS calorimeter system [Akh 00]. To set the scale, the figure also depicts what energy resolutions of $50\%/\sqrt{E}$ and $100\%/\sqrt{E}$ would look like. Even after applying the "offline compensation" method, the ATLAS resolution is substantially worse than $50\%/\sqrt{E}$, at all energies. This, in my opinion very unrealistic, resolution is also about a factor of two worse than that reported by ZEUS.

I have noticed that there is a general belief that collecting more information about the absorption process inside calorimeters, which is the goal of having multiple longitudinal segments, must necessarily lead to achieving better performance characteristics. I think that this is an illusion, for the same reason as it is an illusion to think that measuring the four-vectors of all the individual molecules of the gas in a certain container would make it possible to determine the temperature inside that container more precisely. It would just have the effect of complicating the issues to a degree that would make it almost impossible to avoid making mistakes. Something very similar is happening here. Simple is better, and money for readout is much better spent on lateral segmentation (granularity!) than on an increased number of longitudinal compartments.

7

PERFORMANCE OF CALORIMETER SYSTEMS

Calorimeters exist in a wide variety and are used in very different types of experiments. The design of a particular calorimeter system is usually driven by requirements stemming from the physics goals of the experiment, and by the available budget. These factors lead to large differences in performance between the various calorimeter systems used in particle physics experiments.

In this chapter we review the performance of calorimeter systems. The factors that determine and limit the energy, position and time resolution of calorimeters are discussed, and results from representative calorimeters are presented. The possibilities for using calorimeter information for particle identification and for particle–particle separation are also discussed. We finish this chapter with a brief description of the various tasks of calorimeter systems in modern particle physics experiments.

7.1 Energy Resolution

7.1.1 *Caveats*

Calorimeters are instruments to measure energy. The precision with which this is done is the energy resolution. Not surprisingly, the energy resolution is often considered the most important parameter for describing the calorimeter performance. When a new experiment is designed, the energy resolution is often a very important criterion for comparing the different calorimeter options. Before reviewing the energy resolution achieved with different types of calorimeters, several important caveats have to be kept in mind, especially when comparing results.

1. The energy resolution is typically determined as the relative width (σ/E) of a signal distribution obtained in a test beam where particles of precisely known energy enter the detector at a fixed point, usually the geometric center of a calorimeter cell. One should keep in mind that this signal distribution is, strictly speaking, only valid for these particles in these conditions.

2. If the central value of the signal distribution does *not* correspond to the true value of the energy one tries to measure, then the width of the signal distribution does *not* represent the energy resolution. An example of this was shown in Figure 3.33, where the mean value of the signal distribution depended on the type of particle absorbed by the calorimeter. This effect was caused by a response non-linearity that resulted from the calibration procedure that was used. One should be very cautious with energy resolutions quoted for non-linear calorimeters, especially when this non-linearity is the result of the calibration procedure.
   Response non-linearity is also characteristic for almost all calorimeters intended

for hadronic shower detection. The consequences of this for the jet energy resolution are discussed at the end of Section 7.1.4.

3. In some longitudinally segmented calorimeters (*e.g.*, H1 and ATLAS), complicated "cell weighting" procedures are being used in an attempt to reduce the effects of non-compensation on the hadronic performance. While such procedures
may be successful in the clean, well defined conditions of a test beam, the merits
in the real conditions of a colliding-beam experiment are less clear, especially for
jets.

4. The width of a signal distribution is typically determined with a Gaussian fit to the
data. However, this result is only indicative for the precision with which energy
can be measured if the distribution is indeed Gaussian. If that is not the case, $\sigma_{\rm rms}$
should be used.

5. Energy resolutions quoted for calorimeters in which signal saturation occurs, and
in particular for *digital* calorimeters, are meaningless. This issue is discussed in
detail in Section 9.3.3

Last, but not least, it should be mentioned that there may be important differences
between

1. the energy resolution determined from the width of the measured signal distribution for mono-energetic particles from a test beam, and

2. the precision with which the energy of a given particle can be determined from
the signal(s) produced in a detector (system).

In practice, it is of course the latter precision that matters. However, measurements of
the precision are invariably based on the mentioned method. These differences may
be especially significant when energy dependent correction factors are applied to the
signals. In Section 9.2, we elaborate on this issue with some practical examples.

7.1.2 *Electromagnetic showers*

7.1.2.1 *Homogeneous calorimeters.*   The best resolutions for em shower detection
are obtained with homogeneously sensitive detectors. Such detectors are the instruments
of choice when ultimate performance is needed. Their em energy resolution is usually
limited by fluctuations in the light-collection efficiency (see Section 4.2.4).

Detectors based on NaI(Tl) have been around for decades and have consistently delivered excellent energy resolutions, for example in the Crystal Ball experiment, initially
at SPEAR (SLAC), later at DESY [Blo 83]

$$\sigma/E = 0.026/\sqrt[4]{E} \qquad (7.1)$$

However, NaI(Tl) crystals are not ideal for all applications that require ultimate
energy resolution. They are hygroscopic, their density is not very high (3.67 g cm^{-3},
see Table B.5), the fluorescent decay is slow and, last but not least, the crystals are
relatively expensive, *i.e.*, the cost per cubic radiation length, the relevant unit in this
context, is high.

For these reasons, a number of alternative crystals have been developed, and in modern experiments equipped with a crystal calorimeter, NaI(Tl) is no longer the material

of choice. It has been abandoned in favor of CsI or CsI(Tl), BGO, BaF$_2$ or PbWO$_4$. Currently running and planned experiments that are using crystal calorimeters include CMS [CMS 08] and ALICE [Aam 08] at the LHC (PbWO$_4$), BELLE-II [Abe 10] at SuperKEKB (CsI), BES-III [Don 08] at BEPC (CsI(Tl)), Mu2e [Pez 15] at Fermilab (BaF$_2$), PANDA [Kav 11] at FAIR (PbWO$_4$), the rugby ball BGO-OD [Ban 15] at ELSA (BGO) and A2-TAPS [Nei 15] at MAMI (BaF$_2$). The Fermi γ-ray space telescope [Sad 01] studies cosmic phenomena with a CsI(Tl) calorimeter, while the CALET experiment [Mar 12] does the same at the ISS with PbWO$_4$.

Apart from these, some other types of scintillating crystals, not yet applied in large-scale experiments, have potentially very interesting properties. Examples are CeF$_3$ (fast) [Auf 96a], BSO (the cheap alternative for BGO) [Shi 05, Akc 11b] and the radiation hard, but extremely expensive lutetium-based LSO and LYSO [Zhu 11] crystals.

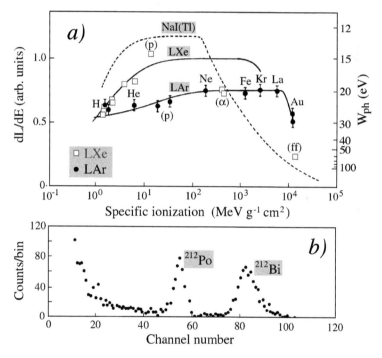

FIG. 7.1. Saturation properties of liquid argon and liquid xenon for densely ionizing particles. Diagram a) shows the scintillation light yield (number of photons per unit deposited energy) for a variety of ionizing particles as a function of the specific ionization of these particles. The ionizing particles include a variety of relativistic ions; (p), (α) and (ff) represent 20 and 40 MeV protons, α particles and uranium fission fragments, respectively. The righthand scale shows the inverse quantity W_{ph}, the energy needed to produce one photon. Diagram b) shows an α spectrum measured with a LXe detector. The peak labelled ^{212}Po corresponds to an energy of 6.1 MeV, the ^{212}Bi peak contains a collection of α transitions with energies ranging from 9.9 to 10.5 MeV [Dok 99].

Liquified noble gases are also known to be very bright scintillators [Dok 90]. For example, Séguinot and coworkers reported having measured ~ 400 photoelectrons per MeV of deposited energy in liquid xenon [Seg 92]. One interesting aspect of this type of scintillator is that the saturation characteristics are different from those in scintillating crystals. In these crystals, the specific light yield (*i.e.*, the number of photons per MeV) for densely ionizing particles, such as αs, is typically an order of magnitude smaller than for mips, while in the liquids no significant saturation occurs for these particles (Figure 7.1a). For this reason, these scintillators can be successfully used for the detection of α particles, as illustrated in Figure 7.1b. Relevant scintillation properties of the liquified noble gases are summarized in Table B.7.

A major experiment that uses the scintillating capabilities of liquified noble gases is MEG, which searches for the extremely rare decay $\mu^+ \rightarrow e^+\gamma$ at PSI in Villigen (Switzerland) and uses a 900 liter liquid xenon detector for this purpose [Saw07, Mih 11]. Figure 7.2 shows the signal distribution for 55 MeV γs in this detector, which is of course the relevant energy for detecting this decay mode for stopped muons (mass 106 MeV/c^2).

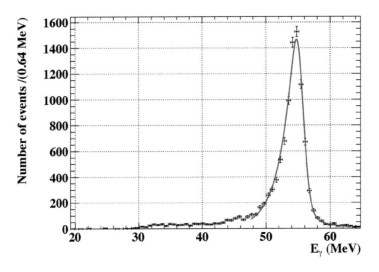

FIG. 7.2. The signal distribution for 54.9 MeV γs in the LXe calorimeter of the MEG experiment [Mih 11].

One problem with these scintillators is the wavelength of the emitted light, which is in the vacuum ultraviolet region of the spectrum, ranging from 128 nm for LAr to 174 nm for LXe. This causes self-absorption in the liquid (which might be responsible for the low-energy tail in the spectrum) and makes light detection non-trivial. MEG detects this light with 846 PMTs, which are immersed in the liquid and observe the scintillation photons without any transmission window. These specially developed PMTs are equipped with a quartz window which is transparent to 80% of the light, and are capable to operate at cryogenic conditions with minimum heat generation from the base circuit.

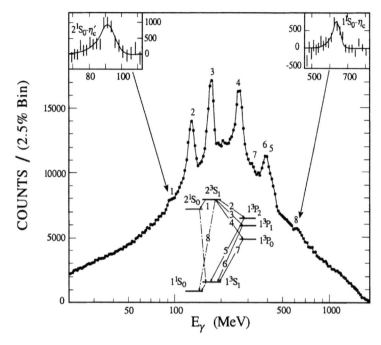

FIG. 7.3. Signal distribution for γs emitted in the radiative decay of the $\psi(2S)$, measured with the Crystal Ball NaI(Tl) calorimeter [Blo 83].

The excellent energy resolution that can be achieved with crystal calorimeters is illustrated in Figure 7.3, which shows a γ-ray spectrum in the energy range from 50 to 1000 MeV. This spectrum was recorded with the NaI(Tl) calorimeter of the Crystal Ball experiment, which studied charmonium states from 1978 - 1981 at the SPEAR e^+e^- storage ring (SLAC), and was later used for measurements of Υ resonances at DESY [Blo 83]. It concerns the radiative decay of charmonium in the $\psi(2S)$ state. The various decay modes are indicated in the figure. Most of the associated γ-rays were well resolved by the calorimeter, including the weak transitions feeding the η_c (~ 600 MeV) and η_c' (~ 90 MeV) states of the $c\bar{c}$ system. This is illustrated by the blown-up sections of the relevant portions of the γ-ray spectrum, shown in the inserts.

The two examples described above illustrate the performance of homogeneous scintillation calorimeters at energies (far) below 1 GeV. In CMS, a crystal calorimeter is used to detect electrons and γs at much higher energies. Figure 7.4 shows the signal distribution for 120 GeV electrons. The relative energy resolution is quoted as 0.44%, but the figure shows that this result was only obtained by combining the signals in an array of 25 crystals, and even then additional corrections had to be applied to account for energy leakage outside this array.

The excellent resolution achievable with crystal calorimeters is mainly important at *low energies*, *e.g.*, below 10 GeV. At high energies, the performance of all calorimeters, including these, is limited by *instrumental effects*. Regardless of the type of calorime-

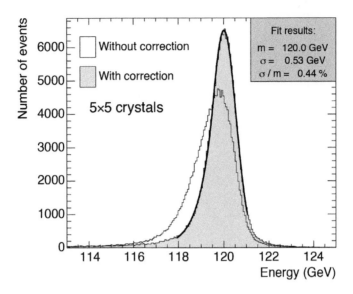

FIG. 7.4. Signal distribution for 120 GeV electrons, reconstructed in a 5×5 matrix of PbWO$_4$ crystals of the CMS barrel em calorimeter. Results are given before and after correcting for shower leakage outside this matrix [Adz 07].

ter, it requires a major effort to keep these instrumental effects under control to such an extent that resolutions better than 1% can be achieved in practice. At energies above 100 GeV, em energy resolutions better than 1% are by no means a monopoly of crystal calorimeters. As we shall see in Section 7.1.2.2, such resolutions can also be achieved with sampling calorimeters of various types, namely those with a stochastic resolution term smaller than $0.1/\sqrt{E}$.

However, at energies below 10 GeV, the small energy-dependent term of crystal calorimeters really makes a crucial difference for the energy resolution. Therefore, it is understandable that all experiments studying the spectroscopy of the $b\bar{b}$ system at the $\Upsilon(4S)$ resonance, in which γs of the order of 1 GeV and below play a crucial role, use crystal calorimeters. As an example, Figure 7.5b shows that γs in the energy range of 50 MeV - 1 GeV are measured with a relative precision of $\sim 2\%$ with the CsI(Tl) calorimeter of the BELLE-II experiment [Ike 00]. In Figure 7.5a, the energy resolution is given for the photon spectrometer of the ALICE experiment at the LHC [Ale 05]. The γs with energies in the 1 - 10 GeV range, which are important in the context of the ALICE study of the formation of the Quark-Gluon Plasma, are measured with energy resolutions better than $\approx 3\%$ in the PbWO$_4$ crystals that constitute this instrument.

A completely different type of homogeneous calorimeter for the detection of em showers is based on the signals produced by Čerenkov light: *lead-glass*. A major advantage of these devices are the blazingly fast signals they generate, a consequence of the instantaneous nature of the Čerenkov mechanism. As discussed in Section 4.2.3, the resolution of these calorimeters is limited by signal quantum fluctuations. Lead-glass

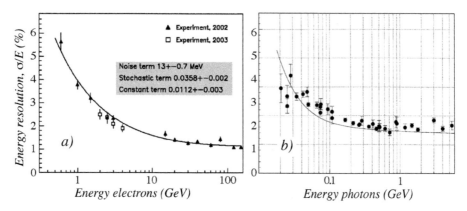

FIG. 7.5. Electromagnetic energy resolution of the PHOS calorimeter of ALICE, as a function of energy. Results are given for electron signals in a 3×3 matrix of PbWO$_4$ crystals (a) [Ale 05]. Electromagnetic energy resolution of the Belle-II calorimeter, as a function of energy. Results are given for γ signals in a 5×5 matrix of CsI(Tl) crystals (b) [Ike 00].

calorimeters have a long tradition in experimental particle physics. However, because of their susceptibility to radiation damage, they are nowadays only used in experiments where the dose rates are small, e.g., at e^+e^- colliders or at fixed-target experiments in electron beams [Mkr 13]. One of the largest arrays of lead-glass blocks ever assembled was used in the OPAL experiment at LEP. The em energy resolution achieved with this detector amounted to $\sim 5\%/\sqrt{E}$ [Akr 90]. Other experiments that have operated large lead-glass em calorimeters include VENUS at KEK [Oga 86], NOMAD at CERN [Aut 96], HERMES at DESY [Ava 96], E-852 at BNL [Cri 97] and the Fermilab experiments E-760 [Bart 90] and E-781 [Bal 05].

A material that may be used as an alternative for lead-glass is lead fluoride, PbF$_2$ [Ande 90]. It has a considerably higher density than lead-glass, 7.77 g cm^{-3} (vs. typically $4 - 5$ g cm^{-3} for lead-glass with a high lead content). Its radiation length of only 9.3 mm is almost as short as that of the densest known scintillating crystal (PbWO$_4$, 8.9 mm). Irradiation with γs and neutrons does affect the transmission at short wavelengths, but these effects can be cured completely when the irradiated crystals are exposed to daylight [Ande 90, App 94]. Only for dose levels of the order of 10 kGy or more does a small permanent reduction of the light transmission remain. It has been reported that doses of ~ 4 kGy actually *improved* the light transmission of certain PbF$_2$ crystals [Ande 90]. Measurements with a small prototype showed that high-energy electron showers can be detected with an energy resolution of $\sim 6\%\sqrt{E}$ in calorimeters made of this material [App 94]. Experiments that are using em calorimeters consisting of PbF$_2$ crystals include A4 at MAMI [Bau 11] and the g-2 experiment at Fermilab [Fie 15].

Another material that has been proposed as a potentially much more radiation-hard alternative for lead-glass is BaYb$_2$F$_8$ [Ase 92]. However, until now tests of this material have been limited to small crystal samples.

Liquified noble gases are also used as homogeneous em calorimeters detecting the *ionization charge* generated by the shower particles. In Novosibirsk, a 70-ton liquid-krypton detector (KEDR) operates at the electron–positron collider VEPP-4M [Pel 09]. The energy resolution of this device was measured with positrons on a 400 kg proto-type. The measured σ/E values ranged from $\sim 5.7\%$ at 0.13 GeV to 1.7% at 1.2 GeV [Aul 90].

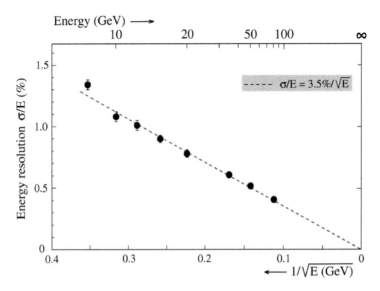

FIG. 7.6. The em energy resolution of the NA62 liquid-krypton calorimeter [Barr 96].

The NA62 experiment operates a $27X_0$ (1.25 m) deep LKr calorimeter at CERN's SPS, which earlier served the predecessor experiment NA48 [Fan 07]. Its em energy resolution was measured with electrons at high energies (10–80 GeV) [Barr 96]. The results are shown in Figure 7.6.

Also LAr and LXe are used in this mode. The CMD-3 Collaboration operates a 400-liter LXe detector at the e^+e^- Collider in Novosibirsk [Pel 09]. For a 40-liter LXe device, a resolution was reported of $3.4\%/\sqrt{E}$, for electrons in the energy range 1–6 GeV [Bara 90]. Because of the small size of this detector, shower leakage probably contributed significantly to this result. The latter problem does not play a role for the multi-ton LAr detectors ICARUS, MicroBooNE and DUNE. Despite the very long radiation length (14 cm), these detectors are sufficiently large to fully contain em showers (Figure 5.33). A first attempt to measure the energy of em showers developing in liquid argon was performed by members of the ICARUS collaboration, who used signals from such showers in their 600-ton detector to select $\pi^0 \to \gamma\gamma$ events [Ank 08]. Using a restricted sample of "clean" events with an average energy of 700 MeV, they measured the π^0 mass with a resolution of 16%.

7.1.2.2 *Sampling calorimeters.* Whereas the em energy resolution of crystal calorimeters is usually limited by instrumental effects, such as fluctuations in the light collection efficiency (Section 4.4.4), the resolution of sampling calorimeters is determined and limited by fluctuations that directly result from the very fact that the shower energy is sampled: *sampling fluctuations.*

In some very specific cases, other fluctuations may also contribute significantly to the em energy resolution, for example:

- In calorimeters based on liquid argon, or similar media operating in the ionization-chamber mode, electronic noise dominates and limits the energy resolution at low energy (see Section 4.4.1).
- In calorimeters based on the detection of Čerenkov light, fluctuations in the number of photoelectrons tend to dominate the em energy resolution (see Section 4.2.3).

Fluctuations in the number of photoelectrons may also contribute to the em energy resolution of scintillation calorimeters. However, if these fluctuations are comparable or even larger than the sampling fluctuations, then the calorimeter design is clearly flawed. In a well-designed sampling calorimeter, the energy resolution is dominated by sampling fluctuations, at least in the energy range most relevant for the physics goals of the experiment. If this is not the case, then money is wasted.

In Section 4.3, it was shown that in calorimeters with dense active media, sampling fluctuations are dominated by fluctuations in the number of different shower particles contributing to the signals. In order to understand the origin of these fluctuations and their effects on the em energy resolution, it is important to realize that the energy of a showering photon or electron is primarily deposited in the absorbing medium through a very large number of very soft electrons (Section 2.4). These electrons, with energies far below the critical energy, have a range that is typically much smaller than the distance between consecutive active elements. Especially when the Z of the absorber material is much larger than the Z of the active layers, the overwhelming majority of these electrons are produced in the absorber, because of the favorable cross sections for Compton scattering and for the photoelectric effect (Section 2.1.4).

In general, most of the shower electrons therefore do not contribute to the calorimeter signal at all and sampling fluctuations may be interpreted as the statistical fluctuations in the *number* of different shower electrons that *do* contribute.

This number can be increased by increasing the total surface of the boundary between the active and passive material in the calorimeter volume. This can be achieved either by incorporating more active layers of a given type, say with thickness d, in the calorimeter volume (increased sampling fraction f_{samp}), or by reducing the thickness d of the individual active elements for a given total amount of active material (increased sampling frequency).

In Section 4.3.1, it was shown that the contribution of sampling fluctuations to the em energy resolution can be described with the following equation:

$$\frac{\sigma}{\sqrt{E}} = c\sqrt{d/f_{\text{samp}}} \qquad (7.2)$$

Evidence for the validity of this relationship was presented in Figures 4.10 and 4.11.

There are thus two different and independent ways to reduce the sampling fluctuations, and hence to improve the em energy resolution of sampling calorimeters:

1. By increasing the sampling *fraction*
2. By increasing the sampling *frequency*

In the first case, the amount of active material in the volume in which the showers develop is increased. In the second case, the number of independent active elements sampling the showers is increased for a fixed sampling fraction.

Traditionally, attempts to construct high-resolution em calorimeters have focused on the first method. Most of the calorimeters with a resolution better than $10\%/\sqrt{E}$ have a large sampling fraction. Examples are the LAr calorimeters built for high-resolution experiments, *e.g.*, ATLAS [Aha 06] and NA31 [Bur 88a]. The sampling fraction of these calorimeters exceeded 20%.

One of the disadvantageous consequences of this approach is the relatively low detector density that may result. For example, the LAr calorimeter used in the NA31 CP-violation experiment at CERN had an active depth of 58 cm, for 25 X_0. The sampling fraction of this device was 28%, and its em energy resolution was $\sim 7.5\%\sqrt{E}$.

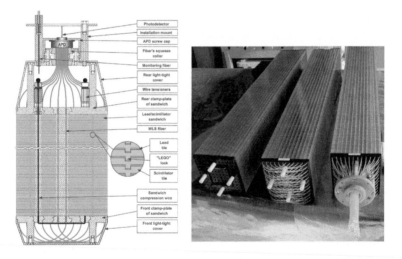

FIG. 7.7. The structure of a shashlyk calorimeter module (*left*) and shashlyk modules at different stages of assembly [Ato 08].

For experiments in which the longitudinal size of the em calorimeter is not an important concern, *shashlyk* structures offer a cheap option to achieve very good em energy resolution. This type of calorimeter, first used about 25 years ago [Ato 92], consists of very thin lead sheets, alternated with plastic-scintillator plates that are up to an order of magnitude thicker. The scintillation light is transported to the rear end by means of wavelength-shifting (WLS) fibers, which run longitudinally through the modules, of which each plate has been perforated with holes for that purpose. Each WLS fiber collects light from two regions of the module, since it loops at the upstream end and then returns through another path of holes (Figure 7.7). The WLS fibers from one module are bunched, polished and connected to a PMT or another type of light detector. The best detectors of this type have achieved energy resolutions of 5–6% for γs from 0.2 – 0.4 GeV [Ato 08], i.e., a factor of two to three larger than those obtained with state-of-the-art crystal calorimeters, such as the CsI(Tl) calorimeter of the Belle-II experiment (Figure 7.5b). These shashlyk modules consist of lead foils with a thickness of 0.275 mm, alternated by 1.5 mm thick polystyrene scintillator plates, that were mass produced by means of injection molding. They were designed to be used in the KOPIO experiment at BNL[20], for the study of the rare decay process $K_L^0 \rightarrow \pi^0 \nu \bar{\nu}$. The sampling fraction for mips in this detector is close to 50%, and the effective radiation length is almost 4 cm. Calorimeters of this type are also used in the LHCb [Mac 09] and COMPASS-II [Anf 13] experiments at CERN. The latter detector is read out by means of silicon photomultipliers with a very high pixel density.

Even though such calorimeters are relatively cheap, their longitudinal size makes them a less favorable choice for 4π collider experiments, since radial space in such experiments is very expensive. Apart from that, a low effective calorimeter density also has physics implications. In a low-density calorimeter, both the effective radiation length and the effective Molière radius are relatively large, so that the showers require a lot of space to develop in *all* directions. Therefore, overlapping showers and the resulting difficulties for applying the isolation criteria needed to select certain physics processes are a larger problem for such detectors than for high-density calorimeters.

In calorimeters based on plastic-scintillator readout, another consequence of a large sampling fraction is the relatively large e/h value of the detector.. This is due to two effects:

1. The e/mip response ratio increases as the thickness of the absorber layers decreases (Figure 3.7).

2. The n/mip response ratio decreases as the sampling fraction for mips increases (Figure 3.41a).

Compensation based on boosting the response to the shower neutrons and/or reducing the relative response to the em shower component requires a small sampling fraction (Section 3.4) and is, therefore, hard to combine with excellent em energy resolution,

In the second method, a high sampling frequency is applied to achieve very good em energy resolution. It has been demonstrated by several groups that this approach may lead to relatively high-density instruments with excellent em energy resolution characteristics.

Fiber calorimeters are among the detectors that explore this second road. In the framework of the RD1 project at CERN, a calorimeter consisting of scintillating plastic fibers with a diameter of 0.5 mm, embedded in a lead matrix at a 20% volume ratio,

---

[20]This experiment was approved, but subsequently canceled (2005) because of a lack of funds. It is now carried out at the Japan Proton Accelerator Research Complex (J-PARC). The calorimeter detecting the γs in this alternative experiment is based on CsI crystals [Dor 05].

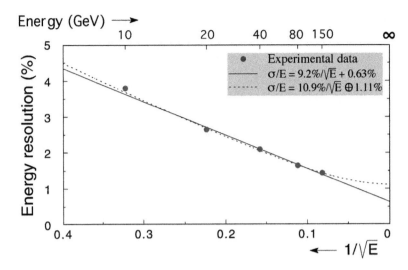

FIG. 7.8. The energy resolution for electrons of the RD1 fiber calorimeter [Bad 94a].

was built and tested. The em energy resolution of this device is shown in Figure 7.8 [Bad 94a].

The sampling fraction of this calorimeter for showers was only 2.3% (3.8% for mips), and the effective radiation length was 7.2 mm. This means that this detector was more than three times as compact as the NA31 calorimeter mentioned above ($X_0 = 23.2$ mm), for a resolution that was only slightly worse. Even better em energy resolutions were obtained with the following calorimeters, all based on high sampling frequency:

- The JETSET lead/fiber calorimeter [Her 90], used in an experiment at CERN's Low-Energy Anti-proton Ring. The em energy resolution of this detector was $6.3\%/\sqrt{E}$, for an effective radiation length of 16 mm. The sampling fraction of this device was $\sim 17\%$ for mips.

- The KLOE lead/fiber calorimeter [Ant 95, Adi 02], used at the ϕ-factory DAPHNE in Frascati. The em energy resolution of this detector, mainly intended for photons with energies in the 20–300 MeV range, was measured to be $5.7\%/\sqrt{E}$, for an effective radiation length of 16 mm. This detector had a sampling fraction of 15% for mips.

- The GlueX lead/fiber calorimeter at the Thomas Jefferson Lab [Lev 08] was modeled after the KLOE one (effective radiation length 14.5 mm), and obtained comparable performance. Figure 7.9 shows the energy resolution as a function of energy for γs from a tagged-photon beam. These resolutions are about a factor of two larger than those obtained for γs in the same energy range with the KOPIO shashlyk calorimeter [Ato 08]. However, the latter has an effective radiation length that is almost three times larger.

- The SPAKEBAB lead/plastic-scintillator calorimeter [Dub 96]. In contrast with the other detectors mentioned in this context, this one consisted of a large number of very thin plates, read out with wavelength-shifting fibers. The resolution for electrons was measured to be $5.5\%/\sqrt{E}$ for an effective radiation length of 15 mm. The sampling fraction for mips was 20%.

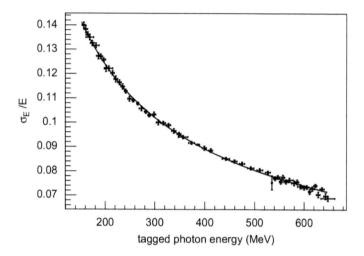

FIG. 7.9. The energy resolution of the GlueX lead/fiber calorimeter as a function of energy for γs from a tagged-photon beam [Lev 08].

It should be pointed out that improvement of the em energy resolution through an increased sampling frequency has its limits. Some of these limits are purely practical. Working with increasingly thinner fibers becomes rapidly very cumbersome. Moreover, the resolution improves only very slowly. For example, the 0.5 mm RD1 calorimeter contained four times as many fibers per unit volume as the SPACAL detector, which had the same sampling fraction, but was equipped with 1 mm fibers. Yet the energy resolution improved only by a factor $\sqrt{2}$. In other words, the resolution scales with the fourth root of the number of components that have to be handled.

When the sampling frequency is increased while the sampling fraction is kept constant, signal quantum fluctuations (photoelectron statistics) will start to dominate the em energy resolution at some point, since the number of signal quanta does not change in this process. In the mentioned high-resolution scintillation calorimeters, this point had not yet been reached. For example, the light yield of the SPAKEBAB calorimeter was measured to be $1,300 \pm 90$ photoelectrons/GeV, so that fluctuations in the number of photoelectrons contributed $2.8\%/\sqrt{E}$ to the measured energy resolution of $5.5\%/\sqrt{E}$ [Dub 96]. In the case of KLOE (resolution $4.8\%/\sqrt{E}$), photoelectron statistics accounted for $2\%/\sqrt{E}$ [Bab 93].

In fiber calorimeters that operate at an angle near $0°$ (*i.e.*, the particles enter the detector approximately parallel to the fibers), increasing the sampling frequency has one additional beneficial effect for the energy resolution. It turns out that the em resolution of such calorimeters contains an energy-independent contribution, which depends sensitively on the angle between the incident particles and the scintillating fibers (and on the angle with the fiber planes). This constant term originates from the variation of the sampling fraction, and thus the calorimeter response, with the impact point of the particles (Section 4.4.2).

By increasing the sampling frequency for a given sampling fraction, this effect is reduced. When 0.5 mm fibers are used instead of 1.0 mm ones, a given area contains four times as many fibers and the distance between the positions of two neighboring fibers is reduced by a factor of two. Therefore, the amplitude of the oscillations observed in Figure 4.21, and thus the constant term in the em energy resolution, are expected to be reduced by a factor of two as well.

This was confirmed by experimental observations. The constant term in the em energy resolution of the 0.5 mm fiber calorimeter built by RD1 was measured to be 0.63%, at 2° (Figure 7.8 [Bad 94a]), while SPACAL (1.0 mm fibers) found 1.23%, at 3° [Aco 91c].

Dozens of different em calorimeters have been used in completed experiments, are being used in current experiments, or are being developed for future experiments. Their main characteristics are summarized in the tables of Appendix C. Here, we just mention a few representative devices.

- Lead/plastic-scintillator calorimeters are a popular choice, because it allows for cheap, compact and fast em shower detectors with good performance character-istics. Examples of completed experiments in which such calorimeters were used are ARGUS ($e^+e^- \to \psi(4S)$ at DESY [Hof 79]), UA2 at CERN's $p\bar{p}$ collider [Bee 84], CDF at Fermilab's Tevatron collider [Bal 88] and the neutrino oscilla-tion experiment CHORUS at CERN [Dic 96]. The ZEUS experiment at DESY [And 91] used ^{238}U as absorber material, and the WA70 experiment at CERN op-erated a large detector based on *liquid* scintillator as active material. The liquid, mineral oil doped with scintillating agents, was contained in teflon tubes. The absorber consisted of 4.2 mm thick lead sheets.
  Current experiments using Pb/plastic-scintillator em calorimeters include the BNL heavy-ion experiments PHENIX [Aph 03] and STAR [Bed 03, All 03], the LHCb [Mac 09] and COMPASS-II [Anf 13] experiments at CERN, the muon $g - 2$ experiment at Fermilab [McN 09] and the AMS-02 experiment at the Interna-tional Space Station [Adl 13c]. The LHCf experiment at CERN [Mas 12] uses tungsten instead of lead as absorber material in a calorimeter of this type. The re-ported em energy resolutions for the mentioned calorimeters vary between 8 and $18\%/\sqrt{E}$. Details are given in Appendix C.

- Calorimeters read out with wire chambers of one type or another are typically used in experiments that require a large instrumented volume, and do not aim for ultimate performance. Examples are the ALEPH em calorimeter at LEP [Cat 86, Dep 89]), the CHARM II calorimeter used for the study of (anti-)neutrino–electron scattering at the CERN SPS [Dew 89] and the calorimeter used for the MAC experiment at SLAC [All 89]. Currently, the MIPP experiment at Fermilab [Nig 09] uses an em calorimeter based on proportional wire chambers. The reported em energy resolutions for the mentioned calorimeters, all of which used lead as absorber material, except CHARM II (which used marble), vary between $18–28\%/\sqrt{E}$. A somewhat better resolution was reported in the context of an R&D project, in which a novel type of gas (C_3F_8) was tested in a $21X_0$ deep lead based calorimeter module [Bez 02].

- Liquid-argon calorimeters were originally mainly used in fixed-target experiments, where the needs of cryogenic operation did not jeopardize the hermeticity of the experimental setup, and the bunch-spacing did not preclude the use of long signal integration times. Examples are the calorimeters used in the E-706 experiment at Fermilab [Lob 85] and the NA31 CP-violation experiment at CERN [Bur 88a]. However, starting in the 1980s, this detector technique has also become increasingly popular in high-luminosity collider experiments. Ingenious tricks that made it possible to build hermetic 4π structures and to extract fast, low-noise signals from these devices have played an important role in this development. The experiments D0 (Fermilab, [Aba 94]), SLD (SLAC, [Dub 86], and H1 (DESY, [Braun 89]) were all built around large LAr calorimeters. Currently, the ATLAS experiment at the LHC [Aad 08] operates a large LAr-based em calorimeter. With the exception of D0 (^{238}U), all mentioned calorimeters use(d) lead as absorber material. The reported em energy resolutions of the mentioned detectors vary between $7–16\%/\sqrt{E}$. Details are given in Appendix C. Stable operation and radiation hardness are generally considered major advantages of this detector technique.

7.1.3   *Pion showers*

In Section 4.7, we saw that the calorimetric energy resolution for the detection of hadrons and jets tends to be dominated by fluctuations in the em shower content, except when the calorimeter is compensating. In that case, the responses to the em and non-em shower components are equal ($e/h = 1$), so that the large event-to-event fluctuations in the energy sharing between these components do not affect the energy resolution.

These fluctuations are not governed by the laws of Poisson statistics and, therefore, the hadronic energy resolution of calorimeters in which these fluctuations dominate, does not scale with $E^{-1/2}$. This is illustrated with some representative examples in Figure 7.10.

In this figure, the energy resolution for single-pion detection is given as a function of the particle energy, which is, as usual in this book, plotted on a scale linear in $1/\sqrt{E}$. In this way, $1/\sqrt{E}$ scaling implies that the experimental points follow a straight line through the bottom right corner of the graph. For reference purposes, the lines representing resolutions $\sigma/E = 40\%/\sqrt{E}$, $50\%/\sqrt{E}$, $60\%/\sqrt{E}$ and $70\%/\sqrt{E}$ are also drawn in Figure 7.10.

Only the resolutions for the (almost) compensating ZEUS and SPACAL calorimeters (Figure 7.10d) fulfill this condition approximately. These resolutions scale as $30–35\%/\sqrt{E}$, apart from a small additional term for the SPACAL device.

The data in Figures 7.10a,b,c exhibit clear deviations from such scaling behavior. The resolution for the undercompensating ($e/h \approx 1.5$) iron/plastic-scintillator calorimeter of the WA1 Collaboration (Figure 7.10a) amounted to $55\%/\sqrt{E}$ at low energies. However, for energies beyond ~ 50 GeV the resolution barely improved and levelled off at $\sim 8\%$. Similar results were obtained with the overcompensating ($e/h \approx 0.7$) uranium/plastic-scintillator calorimeter of the WA78 experiment (Figure 7.10b).

The term describing the deviation from $E^{-1/2}$ scaling, which dominates the resolution at high energies, is clearly smaller for the uranium/liquid-argon calorimeter

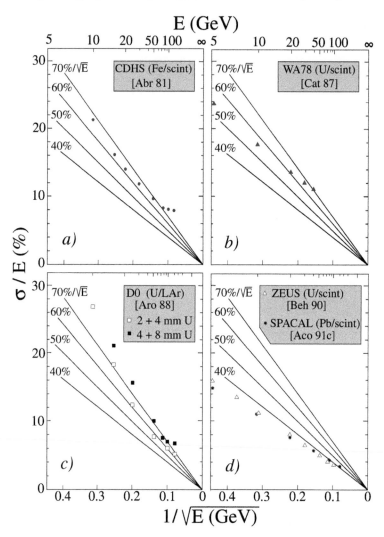

FIG. 7.10. The energy resolution for single pions, for some representative calorimeters. The energy resolution is shown as a function of the pion energy, which is plotted on a scale linear in $E^{-1/2}$. The straight lines represent $E^{-1/2}$ scaling. Results from WA1 [Abr 81] (a), WA78 [Cat 87] (b), D0 [Aro 88] (c), ZEUS [Beh 90] and SPACAL [Aco 91c] (d).

operated by the D0 Collaboration (Figure 7.10c). This should be expected since this detector is more compensating than the WA1 and WA78 ones ($e/h \approx 1.2$). However, this instrument exhibits another feature, typical for LAr calorimeters, namely the rapid deterioration of the energy resolution at low energies.

This effect is caused by electronic noise, which contributes a term proportional to $1/E$ to the energy resolution (Section 4.4.1). The contribution of this noise to the energy resolution is worse for hadronic showers than for em ones, since the latter require a much smaller detector volume to develop. In first approximation, at a given energy the signal/noise ratio is inversely proportional to the square root of the volume over which the signals are integrated for the energy measurement.

The results shown in Figure 7.10 are representative for a great variety of hadron calorimeters that have been built and tested over the past decades. The resolutions reported for individual devices can be found in the tables of Appendix C. When these resolutions are compared with each other, it should be kept in mind that such a comparison has its limitations. We mention a few reasons, in addition those listed in Section 7.1.1:

1. The response functions for pions are typically non-Gaussian. This does not prevent one from making Gaussian fits. Some authors report the results of such fits as the measured resolution, others give the rms values, and yet others fit the signal distributions to some *ad hoc* function and derive the widths of the distributions from such fits [Sak 90].

2. Some authors have tested their calorimeters with beams that contained a large fraction of protons. As was shown on various occasions (*e.g.*, Section 4.7.2), non-compensating calorimeters may respond differently to protons and pions. Sometimes, the fraction of protons in a test beam is energy dependent, which may further complicate the assessment of the results.

3. The chosen calibration procedure may have had a large effect on the obtained energy resolutions. In some cases, minimization of the width of the total signal distribution was the procedure of choice in a longitudinally subdivided calorimeter system. In other cases, different criteria have been applied in the calibration (*e.g.*, hadronic response linearity, response equality for electrons and pions). As we have seen in Chapter 6, such differences in the calibration philosophy may lead to substantial differences in the measured energy resolutions *for the same calorimeter system.*

4. Instrumental effects may have been treated in different ways. Shower leakage is a good example. Typically, tested prototype calorimeters are too small to fully contain the hadronic showers. In some cases, this problem was dealt with by selecting only events in which the showers were fully contained [Fab 77]. In other cases, the effects of shower leakage were studied with some type of Monte Carlo simulation and the measured resolutions were corrected (improved) on this basis. In yet other cases, the experimental results were presented as measured.

However, in spite of the difficulties associated with detailed comparisons between different sets of experimental results, the overwhelmingly dominating role of fluctuations in the em shower content (and the related fluctuations in invisible energy) on the

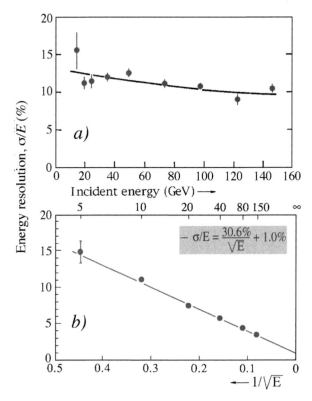

FIG. 7.11. The hadronic energy resolution as a function of energy for a homogeneous calorimeter consisting of 60 tons of liquid scintillator (*a*) [Benv 75] and for the SPACAL lead/fiber calorimeter (*b*), which had a sampling fraction of only 2.3% for showers [Aco 91c].

hadronic energy resolution of calorimeters is crystal clear. This is perhaps most dramatically illustrated with Figure 7.11a, which shows the energy resolution for pions in a *fully homogeneous* calorimeter consisting of 60 tons of mineral oil doped with scintillating agents [Benv 75]. All other sources of fluctuations, such as sampling fluctuations, have been completely eliminated in this device. The energy resolution turned out to be only weakly dependent on the energy of the pions and did not drop below $\sim 10\%$, even at the highest energies (150 GeV) at which this detector was tested. For comparison, the resolution of SPACAL, a calorimeter in which only 2.3% of the pion energy was sampled, was about three times better at this energy (Figure 7.11b).

Fluctuations in em shower content also dominate the hadronic performance of the two large general purpose experiments at the LHC, ATLAS and CMS. The calorimeter systems in these experiments were optimized for good em energy resolution, and as a result the em calorimeters have large e/h values. I estimate these to be ~ 1.5 for the ATLAS Pb/LAr calorimeter and ~ 2.4 for the CMS crystals (*cf.* Table 7.1). In any case, the e/h values are larger than those of the plastic-scintillator based hadronic sections,

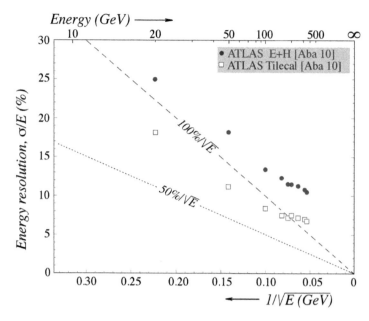

FIG. 7.12. The energy resolution measured for pions as a function of energy in the ATLAS calorimeters. The measured data, based on the standard calibration constants are shown as full dots. The squares give the resolution measured for pions that penetrated the em compartment without starting a shower and deposited their energy in the hadronic calorimeter section (TileCal) [Aba 10]. The dashed and dotted lines, representing resolutions of $100\%/\sqrt{E}$ and $50\%/\sqrt{E}$, are drawn for reference purposes.

which are ~ 1.3 in both cases.

Some consequences of the large discrepancy between the e/h values of the two calorimeter compartments of CMS were already discussed in Section 6.2.11 (Figures 6.25, 6.26). However, ATLAS is not immune to these effects either. Figure 7.12 shows that the energy resolution is actually significantly better for pions that deposit their entire energy in the hadronic section than for events in which the showers develop in the combination of this detector and the fine-sampling LAr em compartment [Aba 10]. Yet, in all cases the resolution is considerably worse than the $52\%/\sqrt{E}$ that is often quoted [Aad 08].

However, the ATLAS resolutions are still better than the CMS ones. This is illustrated in Figure 7.13, where both are shown together. Good hadronic performance was clearly not a priority in the design of these experiments, as evidenced by the results. The performance of CMS is even so poor that better jet resolutions are obtained when the momenta of the charged jet fragments are used rather than the calorimeter information. This issue is further discussed in Section 8.3.2.

Poor hadronic energy resolution is not the only consequence of non-compensation. One of the other effects is the difference in the response to different types of hadrons. For example, the average signal per GeV is smaller for protons and kaons than for pions

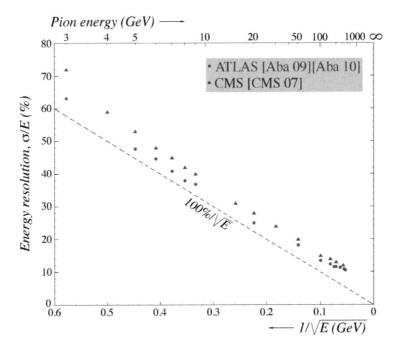

FIG. 7.13. The energy resolution measured for pions as a function of energy in the ATLAS
[Aba 09, Aba 10] and CMS [CMS 07] calorimeters. The dashed line, representing a resolution of $100\%/\sqrt{E}$, is intended for reference purposes.

in a calorimeter with $e/h > 1$. These, and other consequences of $e/h \neq 1$ are further
discussed in Section 7.5.

We want to re-emphasize, once again, that the elimination of the devastating effects
of fluctuations in the em shower fraction on the hadronic energy resolution can *only*
be achieved if the calorimeter is either intrinsically compensating or if it allows a determination of the em shower content event by event. In particular, it is *not* possible
to achieve any meaningful improvement in the energy resolution by means of overall
weight factors applied to the different longitudinal sections into which the calorimeter
system is subdivided (see Chapter 6 for details).

7.1.4   *Jets and multi-particle events*

7.1.4.1   *Jets vs. single hadrons.*   The real challenge in modern experiments is not so
much the measurement of the properties of single hadrons, but rather of jets, the signatures of fragmenting quarks, diquarks and gluons. Especially at high energies, these jets
are characterized by a rather collimated bunch of particles: hadrons, photons (from π^0
decay), and sometimes also leptons (from fragmenting b or c quarks). A typical LHC
jet, generated by a 1 TeV fragmenting quark, consists of tens of particles. The leading
particle, which contains the original fragmenting object, usually carries a considerable
fraction ($\sim 30\%$) of the energy.

From a calorimetric perspective, jets are not very different from single hadrons. Both generate showers that consist of a mixture of em and non-em components, and the energy carried by each of these components may fluctuate strongly from one event to the next. Also for jets, these fluctuations dominate the energy resolution, especially at high energy.

Differences between the energy resolutions for jets and for single hadrons can be expected as a result of the following effects:

- The multiplicity of jets is typically larger than the number of secondaries produced in the first collision of a single hadron of the same energy as the jet. Therefore, the event-to-event fluctuations in the em shower content might be somewhat smaller for jets. This effect *improves* the jet energy resolution compared with the single-hadron one.

- The jet fragments are spread out over a certain region. Usually, a jet is experimentally defined by means of a cone emerging from the interaction vertex. For example, in hadron collider experiments, this cone, defined in units of the pseudorapidity η and the polar angle ϕ, is often chosen as $R = \sqrt{(\Delta\eta)^2 + (\Delta\phi)^2} \approx 0.3 - 0.4$. However, some of the particles generated in the fragmentation process fall outside this cone. This effect is equivalent to lateral shower leakage (Section 4.5.2) and thus *deteriorates* the jet resolution compared with the single-hadron one.

In practice, the jet energy resolution is much harder to establish than the resolution for single hadrons, since test beams of jets with precisely known energy and composition are not available. Nevertheless, there are ways to measure the jet energy resolution.

- In LEP experiments, the hadronic decay $Z^0 \to q\bar{q}$ provided two jets with known energy (45 GeV). In Section 6.3.1 we argued therefore that the calorimetric measurement of hadronically decaying Z^0s may be considered a good experimental approximation for 90 GeV jets (Figure 6.28).

- In the *ep* scattering experiments at HERA, neutral-current events provided jets with known four-vectors. A measurement of the direction and the energy of the scattered electron constrained the momentum and the energy of the struck quark, assuming that the other two quarks are spectators in this process.

- Semi-empirical jet signals may be constructed on the basis of a chosen fragmentation function and experimental signal distributions for the jet fragments. This method and its applications were discussed in Section 4.7.3.

7.1.4.2 *Multi-particle events.* Apart from these methods, studying so-called *multi-particle events* in a test beam may also provide an impression of the calorimeter performance for jet detection. This is done as follows.

Mono-energetic beam particles are sent onto a target, mounted somewhere upstream of the calorimeter. Interactions in this target are selected with the help of scintillation counters sandwiching it. Typically, a useful interaction is characterized by a mip signal in the upstream scintillator and a signal of at least three times the mip value in the

downstream scintillator. In fact, the signals in the downstream scintillator may be used
to select interactions according to their multiplicity.

In this way, several (but not all) important characteristics of a jet are approximated
in a controlled way. These characteristics include:

- A number of particles (photons and hadrons) enter the calorimeter simultaneously.

- The energies of the individual particles, and their nature, are unknown. However,
  the sum of their energies *is* known, which constitutes an important controlled
  aspect of this type of measurement.

- The energies of the individual particles, and in particular the energy sharing between the em and non-em reaction products, fluctuates strongly from event to
  event.

- The spatial distribution of the reaction products (the "jet fragments") can be varied through the distance between the target and the calorimeter. This is another
  important controlled aspect of this type of measurement.

These multi-particle events differ from jets in their (average) multiplicity and in the
energy distribution of the particles of which they are composed. Therefore, the results
obtained with this type of test have to be interpreted with caution.

An example of a multi-particle signal distribution, measured in the way described
above, is shown in Figure 7.14. This distribution concerns the reaction products from
150 GeV π^- impinging on a paraffin target installed 44 cm in front of the SPACAL
detector, which was used to measure these reaction products [Aco 91c].

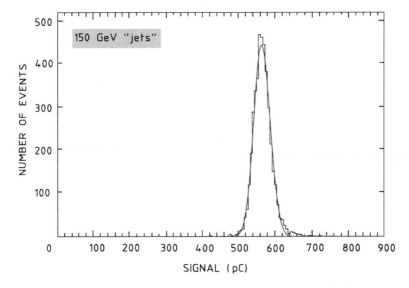

FIG. 7.14. Signal distribution for the secondary particles produced by 150 GeV π^- in an upstream target, measured with the Spaghetti Calorimeter [Aco 91c].

This figure combines all interactions in the target for which a multiplicity of at least three was measured in the downstream scintillator. It turned out that both the energy resolution and the quality of the Gaussian fit improved significantly as the multiplicity requirement was raised. The reasons for this have to do with instrumental peculiarities of the SPACAL detector, namely the contributions of light attenuation in the scintillating fibers to the response function. In the case of the "pseudo-jet," the showers were initiated by a number of reaction products, which entered the calorimeter *simultaneously*. As a result, the overall longitudinal fluctuations in the light production were considerably smaller than for single pions. In spite of additional fluctuations caused by energy losses in the target and by side leakage out of the calorimeter, the signal distribution measured for the multi-particle events was considerably narrower and more symmetric than the one measured for single pions at the same energy.

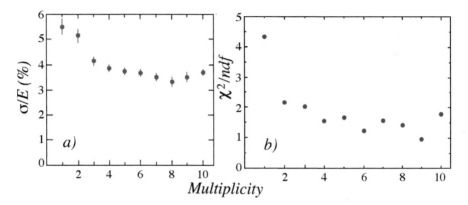

FIG. 7.15. The multi-particle energy resolution (a) and the χ^2 of a Gaussian fit to the multi-particle signal distribution (b) as a function of the multiplicity of the interactions. Data are for interactions induced by 150 GeV π^- in a paraffin target 44 cm upstream of the SPACAL calorimeter [Aco 91c].

Figure 7.15 shows that the improvement in the hadronic calorimeter performance was indeed clearly related to the *multiplicity* of the interactions in the upstream target. The larger the number of charged secondaries, the better the measured energy resolution of the "pseudo-jet" (Figure 7.15a). Figure 7.15b shows that the improvement in the energy resolution went hand in hand with a change in the line shape. As the multiplicity increased, the response function became more and more Gaussian. This illustrates the reduced effects of light attenuation, which caused the response function to be skewed to the high-energy side (Figure 4.30a,c,e).

It should be emphasized that the effects described above were only measurable because of the excellent hadronic performance of the SPACAL calorimeter. In almost any other calorimeter the described effects would have gone totally unnoticed, given contributions of other energy-independent effects to the hadronic energy resolution, and natural asymmetries in the hadronic response function.

7.1.4.3 *Heavy-ion showers.*   The best hadronic energy resolutions have been reported
for the detection of high-energy heavy ions (Figure 7.71). The HELIOS Collaboration
measured a resolution $\sigma/E = 1.9\%$ for 3.2 TeV ^{16}O ions [Ake 87], and the WA80 Collaboration, which also operated a uranium/scintillator calorimeter, found a resolution of
1.7% for 6.4 TeV ^{32}S ions [You 89].

One should realize, however, that in these cases a convolution of either 16 or 32
independent 200 GeV nucleon showers was measured. Hence, strictly speaking, these
results only say something about the precision of the energy measurement for a 200 GeV
nucleon shower. If sixteen signals from such showers are convolved, then the resulting
signal has a resolution σ/E that is four ($= \sqrt{16}$) times smaller than the resolution for
the individual signals from 200 GeV nucleons. In other words, if the resolution for 200
GeV protons (or neutrons) was 7.6%, then a resolution of 1.9% should be expected for
^{16}O ions with an energy of 3.2 TeV. The measured resolution for heavy ions at multi-
TeV energies is thus by no means indicative for the resolution that may be expected for
the detection of single hadrons or jets carrying such energies.

A similar statement applies to the "determination" of the energy resolution for high-
energy em shower detection in liquid xenon, based on convolving the signals from large
numbers of low-energy electrons (100 keV) recorded in a small cell [Seg 92]. Also in
this case, the measurements only revealed something about the energy resolution for the
detection of these low-energy electrons. In a high-energy em shower, a variety of new
effects, absent or negligible in the case of these electrons, affect the signals and their
fluctuations. As an example of such effects, we mention that the (174 nm) shower light
is produced in a large detector volume. Light attenuation, *e.g.*, through self-absorption
and shower leakage, are the likely consequences of this.

These examples illustrate that, in general, measurements made for low-energy particles cannot be used to determine the high-energy calorimeter performance.

7.1.4.4 *Effects of non-linearity.*   We finish this section on energy resolution with
some comments on the effects of hadronic signal non-linearity, which is a general
characteristic of non-compensating calorimeter systems. The energy dependence of the
hadronic calorimeter response can of course be measured with hadron beams, and the
calibration constants should reflect the non-linearity. However, it is important to realize
that these constants represent averages and may be systematically different for different
types of hadrons, *e.g.*, protons and pions (Section 7.5). The non-linearity is especially
bad for jet detection, since it causes the reconstructed energy to be dependent on the
jet topology. A jet with a leading particle that carries a large fraction of the energy of
the fragmenting object will be reconstructed with a different energy than a jet whose
energy is more evenly divided among the different constituent particles (Figure 6.23d).
This phenomenon will, for example, lead to different calorimeter responses to gluon
and quark jets of the same energy, because of the different composition of such jets.

The energy resolution is the precision with which the energy of a showering object
can be determined. The width of the signal distribution recorded for a beam of mono-
energetic particles is only a measure of the energy resolution if the central value of the
distribution has the correct energy. The above considerations illustrate that this is not

the case for hadron detection in a non-compensating calorimeter. The non-linearity is thus responsible for a separate, additional contribution to the energy resolution. This contribution is not governed by Poisson fluctuations and will tend to dominate the over-all resolution at high energies. The same conclusion holds for non-linearities that are introduced by the chosen calibration scheme of a longitudinally segmented calorimeter system (Section 6.3.4).

7.2 Position and Angular Resolution

7.2.1 Electromagnetic showers

The most frequently used method to determine the position of a particle that showers in a calorimeter is by reconstructing the center of gravity $(\bar{x}, \bar{y})$ of the energies E_i deposited in the various detector cells (with coordinates x_i, y_i) that contribute to the signal:

$$\bar{x} = \frac{\sum_i x_i E_i}{\sum_i E_i} \tag{7.3}$$

and a similar expression for the $\bar{y}$ coordinate.

However, impact points calculated in this way tend to be systematically shifted towards the center of the cell hit by the showering particle [Ako 77, Car 79, Ori 83, Aco 91b]. This is a consequence of the steeply falling lateral em shower profile. Typically, the energy density decreases by an order of magnitude over radial distances less than 5 cm (see, for example, Figure 2.16). As a result, in most events in practical calorimeters, at least half of the shower energy is deposited in one individual calorimeter cell, and the exact percentage is only weakly dependent on the location of the impact point.

This effect is illustrated in Figure 7.16, which shows the signals generated by 80 GeV showering electrons in neighboring cells of the SPACAL calorimeter [Aco 91b]. Only in events in which the impact point was located close to the boundary between different calorimeter cells was the shower energy really shared between different cells. In these cases, the particle's position was most precisely and most accurately measured. In other situations, one cell received a lion's share, and the remaining "halo" was shared among the surrounding cells.

Consider an event (E_1) in which the showering particle entered the calorimeter halfway between the center of a cell (say cell #1) and the boundary with another cell (cell #2), e.g., at the position $y \approx 20$ mm in Figure 7.16. For this event, the fraction of the shower energy deposited in cell #1 would not be very different from the fraction recorded for another event (E_2), in which the particle entered in the center of cell #1 ($y \approx 40$ mm). In both cases, the contribution of cell #1 to the sum in Equation 7.3 would be about the same. In both cases, the coordinates of the particle's impact point would be calculated (with Equation 7.3) under the assumption that *the entire energy* recorded in cell #1 was deposited in its center (point C).

Small differences between the signals recorded in the various surrounding cells would in events of the type E_1 thus be the only basis for finding an impact point deviating from the center (C) of the hit calorimeter cell. However, the reconstructed impact point would always be located too close to C, because the bulk of the signal was mis-attributed as originating from this point.

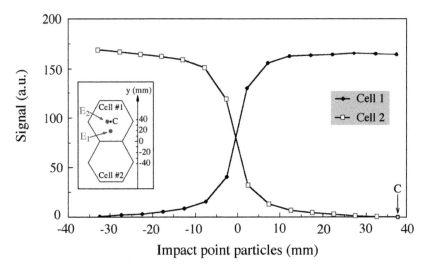

FIG. 7.16. Signals from 80 GeV electrons, recorded in neighboring calorimeter cells, as a function of the (y coordinate of the) impact point of the particles [Aco 91b].

The effects described above are responsible for the peculiar patterns in Figure 7.17. This figure shows scatter plots for 80 GeV electrons detected with SPACAL, in which the impact point of the particles, reconstructed from the calorimeter data, is plotted versus the "true" impact point, measured with wire chambers installed upstream of the calorimeter [Aco 91b].

Figure 7.17a, in which this information is given for the x' coordinate (see Figure 7.18c for the definition of the coordinate system in these hexagonal cells), shows that the impact point was only correctly reconstructed when the particles entered either in the center of a calorimeter cell ($x' = 0$), or near the point where three hexagonal cells joined ($x' \approx 40$ mm). In these cases, the points in the scatter plot cluster around the dashed line, which represents the equality of the x' values reconstructed on the basis of the calorimeter data on the one hand and the upstream wire chambers on the other.

However, if the electron entered the calorimeter anywhere else than in the mentioned areas, then the impact point was always reconstructed too close to the cell's center. Figure 7.17a shows that practically all events with (true) impact points in a cylinder with a radius of about 25 mm around the cell's center were reconstructed as entering very close to the calorimeter center ($x'_{CAL} = 0$). These events constitute the horizontal band in this figure. The systematic mismeasurement of the particle position could thus be as large as 30 mm when the calorimeter data were used at face value.

Figure 7.17b shows similar data for the y' coordinate. Also here, the experimental points cluster in a horizontal band (around the calorimeter center, $y'_{CAL} = 0$), and only near the boundaries with neighboring modules were the impact points correctly reconstructed. In the area between the center and these boundaries, the y' coordinate was also systematically mismeasured when the calorimeter data were used at face value.

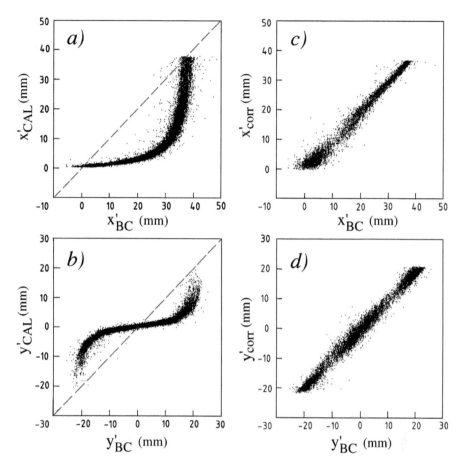

FIG. 7.17. Scatter plots for 80 GeV electrons detected with the SPACAL calorimeter, showing the relations between the coordinates of the particle's impact point, measured with wire
chambers upstream of the calorimeter (horizontal), and determined from the calorimeter data
on the basis of the center-of-gravity method (vertical). The data shown in a) and c) concern
the x' coordinate, b) and d) refer to the y' coordinate. The plots in c) and d) were obtained
after applying the corrections from Equation 7.4 to the calorimeter data. See text for details
[Aco 91b].

This problem can be solved in several ways. In one method that is sometimes applied, a larger weight is given to cells in which only a small fraction of the shower
energy is deposited. In another method, the position of the center of gravity found with
Equation 7.3 is shifted, using an empirical algorithm. For example, an algorithm of the
type

$$x'_{corr} = A\arctan(Bx') \tag{7.4}$$

applied to the data from Figure 7.17a, reproduced the x coordinate of the impact points,
on average, very well (Figure 7.17c). A similar algorithm changed the picture for the y'

coordinate (Figure 7.17b) to the pattern shown in Figure 7.17d.

The position resolution is given by the widths of the bands in Figures 7.17c (for the x' coordinate) and 7.17d (y'). Not surprisingly, this width was found to depend on the impact point of the particles. Figure 7.18 shows how the position resolution varied with the coordinates x' and y'. The smallest resolutions were obtained in the boundary areas between different calorimeter cells. Averaged over one cell, the position resolutions $\sigma_{x'}$ and $\sigma_{y'}$ were found to be 1.8 mm and 1.6 mm, respectively, for 80 GeV electrons.

The position resolution may be expected to scale with $1/\sqrt{E}$ on the basis of the following arguments. The energy deposit E_i in each cell i has a relative precision σ_i/E_i. This relative precision improves as $1/\sqrt{E}$ with the total shower energy E, *provided that the (average) shower profile stays the same*. In that case, the energy sharing between the various calorimeter cells i is, on average, independent of the shower energy. If all the terms in Equation 7.3 have a relative precision that scales with $1/\sqrt{E}$, and if the relative contributions of the individual terms to the sum are energy independent, then the relative precision of the final result (*i.e.*, the sum of all the terms, or the value of the position coordinate) must also scale with $1/\sqrt{E}$.

This is indeed in agreement with experimental observations. For electrons entering the detector in the center of a cell, SPACAL measured a position resolution

$$\sigma_{y'} = \frac{17.1 \text{ mm}}{\sqrt{E \text{ (GeV)}}} \qquad (7.5)$$

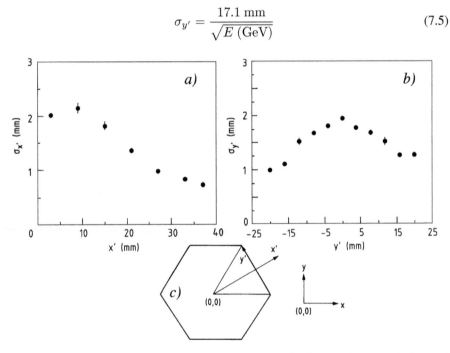

FIG. 7.18. The position resolution for 80 GeV electrons in SPACAL as a function of x' (*a*) and y' (*b*) [Aco 91b]. The coordinates (x', y') used for the hexagonal SPACAL geometry (*c*), and their relationship to the cartesian (x, y) coordinates.

and a similar result was obtained for the x' coordinate. The cell center represents the worst possible case for what concerns the position resolution (see Figure 7.18a). Averaged over the entire surface, the position resolution was found to be about 20% better than indicated in Equation 7.5 [Aco 91b].

The position resolution is not only determined by the energy resolution, but also by the cell size. The smaller the cell size (measured in the relevant units of the Molière radius, ρ_M), the more cells contribute to the signals, and the more accurately the shower's center of gravity can be determined.

The cell size of SPACAL, which had an effective radius of $1.9\rho_M$, was by no means optimized for electron impact-point determination, since an electron hitting a cell in its central region deposited typically $\sim 95\%$ of its shower energy in this one cell. The position resolution of the RD1 projective prototype fiber calorimeter, which had a cell size with an effective radius of $1.1\rho_M$, and an em energy resolution similar to SPACAL, was found to be smaller by more than a factor of two: $\sigma_{x,y} = 7.5$ mm/$\sqrt{E}$ [Bad 94a].

Even better position resolutions for em showers were reported for the SPAKEBAB calorimeter [Dub 96], a very-fine-sampling lead/scintillator calorimeter of the "shish-kebab" type (see Section 4.2.5 for details). This calorimeter had square cells with an effective radius of $0.7\rho_M$. When an electron entered the calorimeter in the center of a cell, typically half of the shower energy was deposited in this one cell, 40% in the six surrounding cells, and the remaining 10% was distributed over the next "ring" of 12 calorimeter cells (see Figure 4.5 for the geometry of this detector).

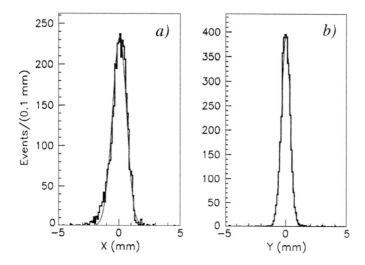

FIG. 7.19. Distribution of the differences between the impact point's x coordinates, measured with wire chambers placed upstream in the beam line, and those measured with the SPAKE-BAB calorimeter, for 100 GeV electrons entering the calorimeter in the center of a cell (a). A similar distribution for the y coordinate (b). The calorimeter was oriented in such a way that the electrons entered it at a $3°$ angle with the y, z plane [Dub 96].

Figure 7.19a shows the distribution of the differences between the impact point's x coordinates, measured with wire chambers placed upstream in the beam line, and those measured with SPAKEBAB, for 100 GeV electrons entering the calorimeter in the center of a cell. A similar distribution is shown in Figure 7.19b, for the y coordinate.

In these measurements, the calorimeter was rotated over a small angle ($3°$) with the vertical (y, z) plane. This is the reason why the x and y distributions are slightly different. Event-to-event fluctuations in the *longitudinal* shower development translated in a *lateral* fluctuation in the (x) position of the shower's center of gravity. This fluctuation is of the order of $X_0 \sin 3°$ (~ 0.6 mm). The contribution of this effect to the position resolution showed up as an asymmetric tail in the distribution of the x coordinate of the reconstructed impact points (Figure 7.19a).

The data depicted in Figure 7.19 concern electrons that entered the calorimeter in the center of a cell, *i.e.*, the point where the position resolution reaches its largest value. Nevertheless, this position resolution (*i.e.*, the rms width of the distributions shown in Figure 7.19) was found to be excellent: 0.58 mm for the x coordinate and 0.34 mm for the y coordinate.

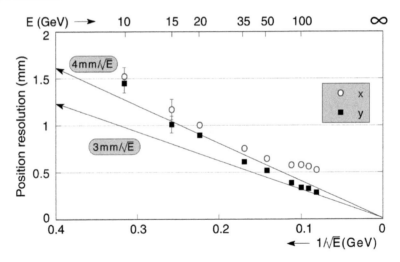

FIG. 7.20. The position resolution for electrons in the SPAKEBAB calorimeter as a function of energy [Dub 96].

Figure 7.20 shows the position resolution as a function of energy for electrons entering into the center of a cell at an angle of $3°$, as before. The difference between the resolutions for the x and y coordinates was found to become smaller at lower energies. This may be understood from the energy-independent term (0.6 mm) that the effect responsible for this difference contributed to the position resolution. At high energies, this term was large compared with the stochastic contributions. This resulted in an asymmetric distribution of the x coordinate (Figure 7.19a) and in considerable differences between the resolutions of the x and y coordinates of the reconstructed position of the impact point.

However, at low energies, the stochastic fluctuations dominated. These were independent of the fluctuations in longitudinal shower development, which caused the 0.6 mm constant term. Therefore, the σs from both effects may be added in quadrature. This resulted in much more symmetric x position distributions and vanishing differences between σ_x and σ_y.

Another energy-independent contribution to the measured position resolution came from the wire chambers, whose accuracy was limited to about 0.2 mm. At low energies, contributions from pedestal fluctuations and PMT noise affected the results as well. This type of fluctuation contributed a term proportional to $1/E$ to the position resolution of this calorimeter (cf. Section 4.4.1).

For these reasons, the authors of [Dub 96] parameterized their experimental results as follows

$$\sigma_{x,y} = c_1 \oplus \frac{c_2}{\sqrt{E}} \oplus \frac{c_3}{E} \tag{7.6}$$

The values for c_2 and c_3 were found to be 3 mm and 10 mm, respectively, for both σ_x and σ_y. The energy-independent term, c_1, was different for both coordinates, as discussed above: 0.6 mm for σ_x and 0.2 mm for σ_y.

The wire chamber resolution ultimately limited the position resolution of this calorimeter to $\sim 200~\mu$m at high energies. This result illustrates the very precise reconstruction of the four-vectors of electrons (and photons) that can be achieved with high-resolution calorimeters that measure the showers developed by such particles.

In Figure 7.21, a compilation is made of the position resolutions of a wide variety of em calorimeters, for 10 GeV electron showers. The figure shows that the cell size, which is expressed in units of the Molière radius, is indeed a dominating factor for the precision with which the impact point of the particles can be reconstructed from calorimetric data.

In Figure 7.21a, the position resolution is given in mm, in Figure 7.21b in Molière radii. The dotted curve gives a good description of most experimental points. This means that the position resolution for em showers can be reasonably estimated with the following formula

$$\sigma_{x,y} \approx \frac{0.1 R_{\text{eff}}}{\sqrt{0.1E}} \tag{7.7}$$

with the energy E expressed in units of GeV.

For example, the OPAL em calorimeter consisted of tapered lead-glass blocks with a lateral cross section at the front face of 92×92 mm^2. The effective radius of these blocks was thus $\sqrt{92 \times 92/\pi} = 52$ mm. At 50 GeV, one may thus expect a position resolution of $5.2/\sqrt{5} = 2.3$ mm. This is in good agreement with the experimental observations [Akr 90].

7.2.2 Hadron showers

Hadron showers can be localized with the same techniques as described above for em showers. Especially at high energies, hadron showers consist of a narrow (em) shower core, surrounded by a halo of particles extending to several times the core diameter

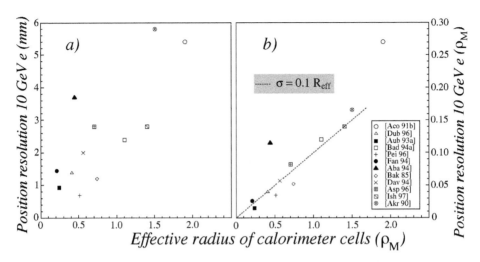

FIG. 7.21. The position resolution for 10 GeV em showers in a variety of calorimeters as a function of the effective Molière radius of one calorimeter cell. The position resolution σ is given in mm (a) or in Molière radii (b).

[Ses 79, Aco 92b]. The shower core is especially dominant in the early stages of the shower development. Therefore, showering hadrons are, in principle, best localized in these early stages. However, in practice, the calorimeter depth at which these early stages occur fluctuates wildly from event to event. In a calorimeter system consisting of an em and a hadronic section, typically 20–30% of the hadrons penetrate the em section without interacting and thus leave only the signals from the hadronic section for measuring their position.

The hadronic position resolution is subject to the same general rules as the position resolution for em showers:

- The better the energy resolution, the better the position resolution.
- The position resolution improves with increasing energy. As for the energy resolution, this improvement usually deviates from simple $1/\sqrt{E}$ scaling, because of a variety of factors which may include non-compensation, noise, *etc.* In addition, the lateral shower profile, and thus the (average) energy sharing between different calorimeter cells, is energy dependent. Such details cause differences between the precise energy dependences of the em and hadronic position resolutions of the same instrument.
- The position resolution improves when the cell size is reduced.

The latter effect was evaluated by Binon and coworkers [Bin 81]. They concluded that the hadronic position resolution improves as

$$\sigma_x^h \sim \exp(2d) \tag{7.8}$$

where d represents the linear transverse dimension of a calorimeter cell, expressed in units of interaction lengths.

Let us consider, as an example, a calorimeter with a cell size measuring 0.5×0.5 λ_{int}^2. If this cell size were reduced by a factor of four, to $0.25 \times 0.25\ \lambda_{int}^2$, then the hadronic position resolution would thus improve by a factor $\exp(1 - 0.5) = 1.65$, according to this formula. Equation 7.8 also suggests a limit on the useful transverse segmentation, of $d \approx 0.1\lambda_{int}$. If the segmentation was reduced from $0.2 \times 0.2\ \lambda_{int}^2$ to $0.1 \times 0.1\ \lambda_{int}^2$, the fourfold increase in the number of channels would only result in a 22% improvement in the position resolution.

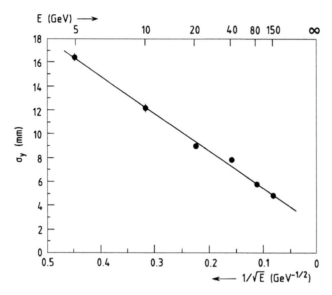

FIG. 7.22. The position resolution for pions in the SPACAL calorimeter as a function of energy, for particles that entered the detector into the center of a calorimeter cell [Aco 91b].

The conditions of excellent hadronic energy resolution and a fine transverse granularity ($0.12\ \lambda_{int}^2$) were combined in SPACAL. This resulted in a very good position resolution for hadronic showers (Figure 7.22):

$$\sigma_y\,(\pi^-) = \left(\frac{31.4}{\sqrt{E}} + 2.4 \right) \mathrm{mm} \qquad (7.9)$$

For the highest energies at which this detector was tested (150 GeV), position resolutions smaller than 5 mm were obtained. Also for hadronic showers, the position resolution was found to depend on the impact point of the particles. As for electrons, the position was most accurately measured for particles that entered the calorimeter near the boundaries between different cells, while the areas near cell centers allowed the least precise particle localization.

For 80 GeV pions, the position resolution varied between 4.5 mm near cell boundaries and 5.8 mm in the cell centers. Averaged over the entire cell surface, the resolution amounted to 5.1 mm for these particles. The results shown in Figure 7.22 are for pions that entered the calorimeter in the center of a cell.

The SPACAL Collaboration also studied the position resolution as a function of the number of calorimeter cells included in the calculation of the center of gravity of the energy deposition profile (Equation 7.3). They found that the best resolutions were obtained if the procedure was limited to those (19) cells located within a cylinder with a radius of 17 cm (slightly less than 1 λ_{int}) around the shower axis (see Figure 6.29 for the geometry of this calorimeter). The signals from the more remote calorimeter cells were, on average, so small that the main effect of including them in the center-of-gravity calculation was an increase of the overall noise level, resulting in a deterioration of the position resolution.

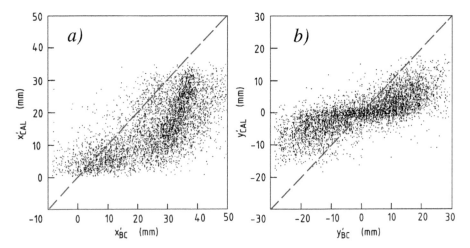

FIG. 7.23. Scatter plots for 80 GeV pions detected in SPACAL, showing the relations between the coordinates of the particle's impact point, measured with wire chambers upstream of the calorimeter (horizontal), and determined from the calorimeter data on the basis of the center-of-gravity method (vertical), for the x' (a) and y' (b) coordinates [Aco 91b].

The calorimeter with which these measurements were performed was relatively fine-grained, its cell size was much smaller than for typical hadronic calorimeters used in particle physics experiments. Still, the hadronic position resolution could clearly benefit from a further reduction of this cell size. In Section 7.2.1, we saw that the center of gravity measured from the energy deposit profile of *electron showers* is systematically displaced when the granularity of the calorimeter is too coarse compared with the gradient of the shower profile (Figure 7.17). Similar displacement effects, albeit much less pronounced than for electrons, were observed for hadron showers in SPACAL (Figure 7.23). Also in this case, the systematic mismeasurement of the shower position caused by these effects could be eliminated with a correction function of the type described in Equation 7.4.

7.2.3 Angular resolution

Determination of the angle of incidence of a showering particle requires a measurement of the particle's position at several (≥ 2) points along its path. When calorimeters are involved in the measurement of this angle, the particle position may be determined with the center-of-gravity method discussed in the previous subsections.

In experiments in which the calorimeter system is preceded by a particle tracking system (*e.g.*, experiments in a high-energy colliding-beam geometry), the angle of incidence of charged particles is usually best determined with the information from this tracking system at a number of points along the particle track. Such a system typically offers position resolutions at the fraction-of-one-millimeter level. In such experiments, the calorimeter data are mainly important for measuring the angle of incidence of *neutral particles*[21], such as γs, neutrons and K^0s.

A calorimetric measurement of the angle of incidence requires a longitudinally segmented detector, since the position of the shower has to be determined at different depths. This position is most accurately measured *early* in the shower development, when the shower is still very narrow.

Many calorimeters are equipped with a fine-grained *preshower detector*, which can be beneficially used for this purpose. Such a PSD is up to a few radiation lengths deep and has a smaller granularity than the calorimeter itself. As an example, we mention the CMS experiment, where the PbWO$_4$ em calorimeter is preceded in the endcap region ($1.653 < |\eta| < 2.6$) by a sampling device consisting of two layers of lead plates interleaved by layers of silicon strip sensors. The total thickness of this detector is 20 cm, and it represents about $3X_0$ of material [CMS 08]. In prototype tests, the position of electrons showering in such a detector could be determined with a precision of [Auf 98]

$$\sigma = \frac{1.3 \text{ mm}}{\sqrt{E}} \oplus 0.26 \text{ mm}$$

In other calorimeters, Shower Maximum Detectors are used to determine the position of electrons and photons with high precision (see Section 5.7.2). This gives a measurement point at a depth of $\sim 5X_0$ inside the calorimeter which, in combination with measurements at other depths, can be used to measure the direction of the showering particles.

Apart from precise shower localization, both SMDs and PSDs are also useful (and intended) for other purposes, such as particle identification (Section 7.6) and particle–particle separation (Section 7.6). In addition, a PSD is often used to determine the energy deposited in "dead" material installed upstream of the calorimeter. This is particularly true for ATLAS, where the em calorimeter is preceded by an amount of material that varies between $\sim 2X_0$ at $\eta = 0$ and $\sim 4X_0$ at the maximum η (1.8) covered by the PSD. Figure 5.54 illustrates the crucial role of the PSD for achieving the necessary precision with which the energy of electrons and γs can be reconstructed [Aub 93b].

Whereas the auxiliary equipment mentioned above determines the particle position at a very early stage of the shower development, the electromagnetic calorimeter itself

---

[21]When a γ or K^0_S is produced close to the primary vertex, the position of this vertex is of course also an important asset for the measurement of the angle of incidence in the calorimeter.

provides a measurement of the shower position at greater depth. The longitudinal center of gravity of em showers with energies in the 10–100 GeV range is located at a depth of ∼ 10–15 X_0 (see Figure 2.9). By combining the results of position measurements at different depths, the angle of incidence of the showering γs can be measured.

In some experiments, the em calorimeter is longitudinally segmented with the main purpose to be able to measure the trajectory of incoming γs. An example is the ALEPH em calorimeter [Dec 90], which consisted of three longitudinal sections, with depths of $4X_0$, $9X_0$ and $9X_0$, respectively.

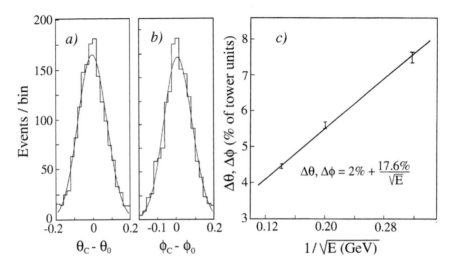

FIG. 7.24. The angular resolution for em showers in the ALEPH calorimeter. Distributions of the difference between the angle at which 10 GeV electrons entered the detector and the angle reconstructed from the calorimeter data are given both for the polar (a) and the azimuthal (b) angles. In diagram c), the angular resolution is plotted as a function of energy. The angles are expressed as a fraction of the angular acceptance of one projective calorimeter tower. This angular acceptance varied from 17×17 mrad^2 at $\theta = 90°$ to 9×10 mrad^2 at $\theta = 40°$ in this calorimeter [Dec 90].

The angular resolution of this calorimeter is shown in Figure 7.24. A 10 GeV electron beam was sent into a calorimeter tower and the shower position was reconstructed in the three different sections, using the center-of-gravity techniques described above. By fitting a line through these three points, the direction of the showering particle was measured. This direction could be compared with the beam direction. The distribution of the difference between the measured and the known direction is shown in two projections, in Figures 7.24a and 7.24b. The angular resolution, *i.e.*, the width of distributions of this type, is given as a function of energy in Figure 7.24c.

The projective ALEPH towers each covered an angular area ($\Delta\theta \times \Delta\phi$) with respect to the interaction vertex that varied from 17×17 mrad^2 at $\theta = 90°$ to 9×10 mrad^2 at $\theta = 40°$. The resolutions in Figure 7.24c are expressed as a fraction of the angular coverage of one such tower. For example, at 10 GeV, the resolution was measured to be $\sim 8\%$, which corresponds to ~ 1.3 mrad in both θ and ϕ. Such accurate measurements of the angle of incidence of γs are, among other reasons, needed to determine whether they resulted from the decay of a specific particle produced in the interactions. For example, a 50 GeV π^0 decays in two γs with an opening angle that is typically only ~ 6 mrad.

Good angular resolution for γ detection was also an important goal of the LHC experiments ATLAS and CMS, which are both designed to be able to detect the Higgs boson through its decay mode $H^0 \to \gamma\gamma$. A preshower detector is an important asset in that context. Figure 7.25 shows the angular resolution for em showers measured with

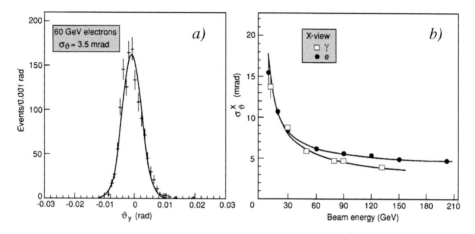

FIG. 7.25. The angular resolution for em showers in the ATLAS PSD/calorimeter system. Diagram a) shows a distribution of the difference between the angle at which 60 GeV electrons entered the detector and the angle reconstructed from the detector data. In diagram b) the angular resolution is shown as a function of energy for γs and electrons [Aub 93b].

a prototype of the ATLAS PSD/calorimeter system [Aub 93b]. Apart from accurate measurements of the angle of incidence of high-energy γs, this fine-grained system was also crucial for distinguishing the interesting single photons from the much more abundant π^0s. This aspect is discussed in Section 7.7.2.

The angular resolution achieved with calorimetric measurements of *charged particles* is important for experiments in which the calorimeter not only serves as a detector, but also as a target. This is, for example, the case in neutrino scattering experiments, which usually require a multi-ton target/detector.

In these experiments, measurements of important kinematic variables such as Bjorken-x and y, and the Q^2 transferred in the reaction, hinge on a precise determination of the four-vectors (and thus of the angles with respect to the direction of the incoming neutrino) of the leptonic and hadronic final-state components.

The angular resolution for em and hadronic showers has been measured for several calorimeters used in this type of experiment [Did 80, Bog 82, Dew 86]. For the same reasons as for the position resolution, the angular resolution also improves (at least for em showers) as $1/\sqrt{E}$ with increasing energy. For em showers, the CHARM Collaboration (CERN) measured [Dew 86]

$$\sigma_\theta^e (\text{mrad}) \approx \frac{20}{\sqrt{E}} \tag{7.10}$$

The angular resolution of this calorimeter for hadron showers was measured to be [Abt 83]

$$\sigma_\theta^h (\text{mrad}) \approx \frac{160}{\sqrt{E}} + \frac{560}{E} \tag{7.11}$$

in which the energy E is expressed, as usual, in units of GeV.

7.2.4 Localization through timing

A completely different way of determining the position of the showering particles was used in the KLOE experiment at the ϕ factory DAPHNE (Frascati, Italy), which studied CP violation in K^0 decays and several other physics topics [Fra 06]. When the ϕs, produced at rest in $e^+ e^-$ collisions at $\sqrt{s} = 1.020$ GeV, decay into two K^0s these carry a momentum of 110 MeV/c each.

This experiment operated a 4π electromagnetic calorimeter [Ant 95, Adi 02] built to perform three major tasks:

- Determine the vertex for the decays $K^0 \to \pi^0 \pi^0$.
- Reject the $K^0 \to \pi^0 \pi^0 \pi^0$ events with good efficiency.
- Provide the trigger for the experiment.

In order to perform these tasks, the calorimeter needed to measure the γs from the decays $\pi^0 \to \gamma\gamma$ with good precision. These γs typically had energies in the range from 50 to 300 MeV.

The calorimeter consisted of scintillating fibers embedded in lead. The barrel section of the calorimeter had a length of 3.75 m and an outer diameter of 4 m, with the fibers running parallel to the beam line. Because of the very fast signals generated by the fibers, the time resolution was extremely good: 34 ps/$\sqrt{E}$ [Ant 95].

Light travels through these fibers at a speed of ~ 17 cm ns^{-1}. By comparing the arrival times of the scintillation light at both ends of the fibers, the shower position along the fiber direction could thus be determined with a precision of only 1–3 cm, for the γs of interest (Figure 7.26, righthand scale).

In this experiment, a good localization of the γs was very important to identify the parent particle. The invariant mass of a particle decaying into two γs is given by

$$M = \sqrt{2E_1 E_2 (1 - \cos\theta_{12})} \tag{7.12}$$

The precision with which the mass can be measured is thus not only determined by the energy resolution, $i.e.$, the measurement uncertainty on the γ energies E_1 and E_2, but also by the relative uncertainty on the angle (θ_{12}) between the directions of these γs.

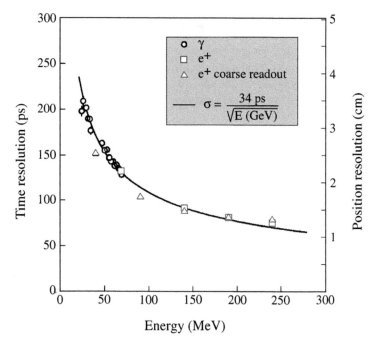

FIG. 7.26. Time resolution (left-hand scale) of the KLOE calorimeter for signals generated by γs as a function of energy, and the position resolution (right-hand scale) in the direction along the fiber that can be achieved by comparing the arrival times of the shower light at both ends of the fiber [Ant 95].

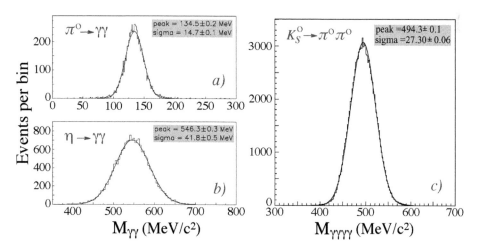

FIG. 7.27. Invariant mass distributions of two γs and four γs detected in the same event by the KLOE em calorimeter. The mass peaks of the π^0 (a), η (b) and K_S^0 (c) are located within 1% of their established values [Adi 02].

The latter uncertainly is of course directly affected by the precision of the mentioned localization procedure.

Figure 7.27 shows the invariant mass distributions of $\gamma\gamma$ and $\gamma\gamma\gamma\gamma$ combinations recorded in the e^+e^- collisions at the ϕ resonance in DAPHNE. The masses of the π^0, η and K_S^0, which decay in these exclusive modes, were reconstructed, on average, to within 1% of their established values, with good resolutions [Adi 02].

7.3 Time Characteristics

The time characteristics of the calorimeter signals are especially important when the event rates are high, for example in hadron colliders, where these event rates are measured in MHz (ISR, Tevatron), or even in GHz (LHC).

When discussing these time characteristics, two aspects should be distinguished:

1. The accuracy with which a given calorimeter signal can be attributed to a certain event or bunch crossing. We will refer to this aspect as the *timing* aspect.
2. The signal *duration*, *i.e.*, the time it takes before all the charge or light constituting the calorimeter signal is collected. At high rates, remnants from previous events may affect the precision of the measured signals.

In this section, we first concentrate on this second aspect.

The duration of the calorimeter signals is determined by the physics of the shower development, and by the mechanism(s) through which the signals are generated. In Chapters 2 and 3 we saw that a considerable fraction of the energy carried by showering hadrons is deposited by slow neutrons, mainly through elastic scattering. In materials with a large cross section for thermal-neutron capture, and in particular in ^{238}U, the energy released in that process may represent some 10% of the total shower energy. These processes are very slow. The energy deposition through elastic neutron–proton scattering has a time constant of the order of 10 ns, while thermal neutron capture occurs on a timescale of ~ 1 μs (Section 3.2.9).

The time structure of the calorimeter response function thus depends on the relative contribution of these processes to the signals. This contribution may vary from absolutely negligible to some 20%, depending on the calorimeter composition and on the mechanism through which the signals are generated. The examples in the following subsections may serve to illustrate this point.

7.3.1 *Čerenkov calorimeters*

Calorimeters based on the detection of Čerenkov light emitted by the shower particles are practically insensitive to the peculiarities of the neutron component of the shower development. Only the high-energy γ-rays produced in inelastic neutron scattering or in thermal neutron capture might create electrons or e^+e^- pairs above the Čerenkov threshold. However, by choosing an absorber material in which relatively few neutrons are produced (low-Z) and/or which has a small cross section for thermal neutron capture (*e.g.*, lead), this slow component of the signals can be made negligibly small.

The very short duration of the signals from Čerenkov calorimeters is illustrated in Figure 7.28, which shows the time structure of signals measured with the RD52 dual-readout fiber calorimeter [Lee 16a]. This calorimeter was equipped with two types of

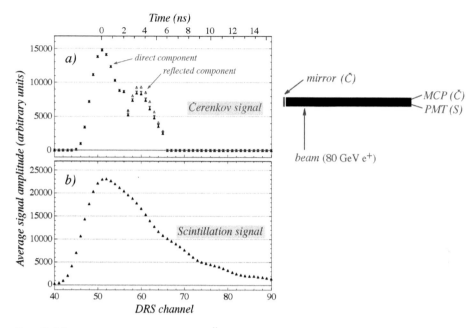

FIG. 7.28. Typical time structure of the Čerenkov (*a*) and scintillation (*b*) signals measured for 80 GeV positrons entering the RD52 calorimeter perpendicular to the fibers. The impact point was located at 29.5 cm from the mirrored front end of the fibers. The direct and reflected light components of the Čerenkov signals are clearly separated [Lee 16a].

fibers, which detected the Čerenkov and the scintillation light produced in the showers, respectively. The upstream end of the Čerenkov fibers was aluminized. The figure shows the time structure of the signals produced by 80 GeV positrons that were send into this calorimeter perpendicular to the fiber direction. The impact point was located 29.5 cm from the front face. The Čerenkov signals (Figure 7.28a) clearly show the direct and the reflected signal components, which are clearly distinguishable, even though the difference in arrival time was only slightly more than 3 ns. The scintillation signal (Figure 7.28b) was clearly less fast than the Čerenkov ones, because of the (10 ns) decay time of the scintillating dye. Since the upstream end of the scintillating fibers was not aluminized, no reflected signals were observed in this case.

The emission of Čerenkov light is an instantaneous process. A time structure in the signals produced by this type of calorimeter may be caused by the fact that the Čerenkov light is produced by particles traveling (approximately) at a speed c, but propagates at a speed c/n, where n is the index of refraction. For example, in the quartz-fiber calorimeter ($n = 1.46$) used in the very-forward region of the CMS experiment, which has a total depth of 1.5 m, the difference between the arrival times of light generated near the upstream and downstream end faces of the detector amounts to 2.3 ns as a result of this effect [Akc 97].

In large-area underwater neutrino telescopes, with PMTs installed at a distance of

50 m from each other, this time difference may become as large as 50 ns. In such detectors, the precise time structure of the signals may be used as a tool to determine the energy and the direction of multi-TeV particles. See Chapter 10 for more details on this point.

The other main effect contributing to the time resolution of the signals from Čerenkov calorimeters is the jitter in the transit time of the photomultiplier tube that converts the light into electric signals. Different types of PMTs clearly yielded different time resolutions for the CMS quartz-fiber calorimeter [Akc 97].

7.3.2 *Compensating calorimeters*

Compensating calorimeters, based on plastic-scintillator readout, do rely on the contribution of neutrons to the hadronic calorimeter signals. Therefore, the time structure of the processes through which these neutrons lose their kinetic energy is an intrinsic aspect of the time structure of the signals from such calorimeters.

Figure 3.23 illustrates this feature, for signals from a compensating lead/plastic-scintillator calorimeter (SPACAL). The pion signal clearly exhibits an exponential tail, which is absent for the electron signal. This tail is the result of the elastic neutron–proton scattering process. A similar tail was used to measure the em fraction of hadron showers event by event by the RD52 Collaboration (Section 8.2.5).

The compensation condition cannot be realized if one would want to extract very fast signals from a scintillation calorimeter such as SPACAL, comparable in duration to the Čerenkov ones shown in Figure 7.28a, since the neutrons do not contribute adequately to the signals when the allowable charge collection time is very short. This is illustrated in Figure 7.29, which shows the e/π signal ratio measured as a function of the charge

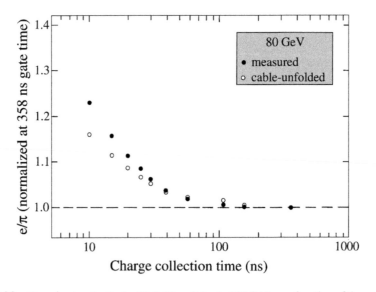

FIG. 7.29. The e/π signal ratio for 80 GeV particles in SPACAL as a function of the gate width [Aco 92a].

integration time. This ratio is normalized to the one obtained for the standard gate width of 358 ns: e/π (80 GeV) = 1.03. Only if the signals were integrated over at least 50 ns did the measured e/π signal ratio get to within a few percent of this value. For shorter gate widths, the ratio rapidly increased. For a gate width of 10 ns, which would be long enough to collect all the charge from a Čerenkov fiber calorimeter, the 80 GeV pion signals in SPACAL would be, on average, $\sim 20\%$ too small to reach the compensation condition.

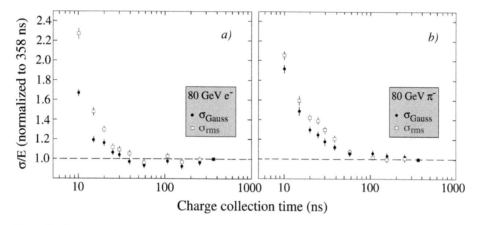

FIG. 7.30. The energy resolutions for 80 GeV electrons (a) and pions (b) in SPACAL as a function of the charge collection time. The resolutions are normalized to the one obtained for a gate width of 358 ns [Aco 92a].

A related effect is shown in Figure 7.30. Since the hadronic energy resolution depends strongly on the e/h ratio, the energy resolution deteriorated rapidly when the gate width was reduced. A similar effect was also observed for the electron signals, as a result of the very long cables through which the signals were transported from the calorimeter to the counting room in these experiments. These cables affected the charge collection considerably. However, in spite of that, the figure shows a significant difference between the deterioration of the em and hadronic energy resolutions. This difference can be attributed to the effects of "incomplete compensation."

The timescale of thermal-neutron capture is, in general, too long to be of practical use for calorimetry (Section 3.2.9). Also, the neutron capture may take place at large distances from the shower axis (see Figure 3.20). This aspect limits the usefulness of including the signals from such processes in the shower response as well.

7.3.3   *Ionization calorimeters*

Calorimeters with liquid argon as active medium are insensitive to the elastic scattering process through which the soft neutrons lose their kinetic energy. Tails caused by the physics of the shower development, as in Figures 3.23 or 8.18, are therefore absent in the signals from such calorimeters.

However, owing to the mechanism through which the signals are generated in these devices, the duration of the signals is very long. The same can be said about *all* calorimeters that rely on charge drifting to an electrode, where it is collected to form a signal. It takes typically several microseconds before all the electrons liberated by the ionizing shower particles from an event are collected. The removal of the positive ions from the active gaps takes a multiple of that time.

This does not mean that no fast signals can be obtained from these calorimeters. Radeka has shown that, with bipolar pulse-shaping techniques, representative signals with a duration of about 100 ns can be extracted from LAr calorimeters [Rad 88a]. Even in the demanding LHC environment, such signals allow unambiguous identification of the bunch crossing in which they were generated (Section 5.6.2.1). In other words, they can be used for fast and accurate timing purposes.

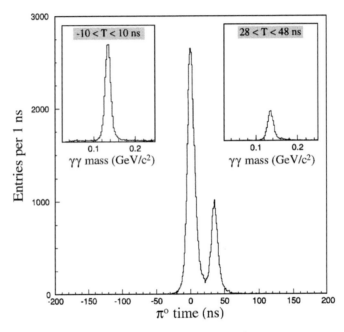

FIG. 7.31. The time spectrum of events in which high-p_T π^0s were detected in the LAr calorimeter of the E-706 experiment at Fermilab. The inserts show the $\gamma\gamma$ invariant-mass spectra for two time intervals, one in-time (0 ns) and one out-of-time (38 ns later) [Apa 98].

The value of this has been demonstrated in practice by the E-706 Collaboration at Fermilab, who operated a large fine-grained liquid-argon calorimeter in an experiment aimed at studying the production of high-p_T photons and jets [Apa 98]. The electronics designed and implemented for the readout of this calorimeter included timing circuitry to provide information on the arrival time of detected showers. In this experiment, some rare events were recorded in which a second interaction took place, $\approx$ 38 ns after the nominal interaction time. The occurrence of such events was signalled by scintillation counters installed near the interaction vertex. Figure 7.31 shows the distribution of times assigned to high-p_T π^0s produced in these rare events. This time spectrum shows two clear peaks, separated by 38 ns. The inserts, which show the $\gamma\gamma$ invariant-mass distributions for two time intervals associated with these two different interactions, prove that both peaks in the time spectrum concern π^0 production. The excellent time resolution of this calorimeter and its associated electronics are clearly illustrated by this figure.

Although the timing capability of ionization calorimeters is excellent, the *charge collection* in these devices continues for a relatively long time after the event. During that time, new showers developing in the same region of the calorimeter may be inaccurately measured, due to "old charge" contributions to the new signals, an effect commonly known as *(out-of-time) pile-up*. This effect limits the tolerable event rates and thus the acceptable luminosity level at which the calorimeter can operate.

The effects of this type of pile-up can be limited by narrowing the active gaps and/or by increasing the electric field strength, thus reducing the "dead time" associated with the detection of a shower.

It should be emphasized that out-of-time pile-up is not an exclusive problem of calorimeters measuring direct ionization. Some scintillators or wavelength shifters have a fluorescent component with a very long decay time. For example, in the HELIOS experiment, it was noted that about 5% of the total light production in em showers occurred more than 1 μs after the absorption of the particle [Ake 87]. Also some scintillating crystals have light components with a very long decay constant, *e.g.*, CsI(Tl) (1.3 μs), or BaF$_2$ (0.63 μs). Complete information about the intrinsic time structure of the signals generated by the scintillators used in calorimeters can be found in Tables B.5 – B.7.

7.3.4   *Timing and pile-up*

The luminosity at modern particle colliders is such that during each bunch crossing multiple interactions occur. This is especially true at hadron colliders, but also at SuperKEKB, an e^+e^- collider designed to operate at $\mathcal{L} = 10^{36}$ cm^{-2}s^{-1}, this may become a problem. The LHC currently operates at $\mathcal{L} \approx 10^{34}$ cm^{-2}s^{-1} and at this luminosity, on average, about 15 events occur during each bunch crossing. The bunch crossings are spaced by 25 ns. This event rate is expected to increase by an order of magnitude in the high-luminosity era. Of course, the overwhelming majority of these "underlying events" are non-interesting minimum-bias events, and the main effect of these is a general increase in the "noise" level of the calorimeter channels. This effect is usually called *in-time pile-up*, and should be distinguished from *out-of-time pile-up* (OOT), which is due to events that took place during bunch crossings other than the one in which the interesting event under study occurred. The latter form of pile-up comes in two varieties: "early" OOT pile-up, which refers to energy left in the calorimeters from events that occurred in previous bunch crossings, and "late" OOT pile-up, which comes from events in later bunch crossings that is integrated along with the trailing portion of the pulse from the buch crossing of interest.

This is illustrated schematically in Figure 7.32, which shows the calorimeter signal in a certain channel as a function of time. The "triggered" or "signal" bunch crossing is located at "0." The next crossing, also indicated by a dashed vertical line, occurs 25 ns

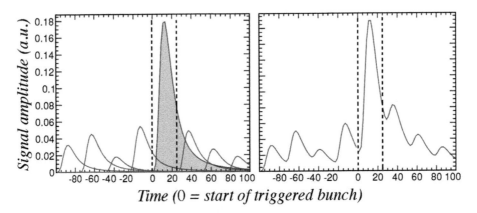

Time (0 = start of triggered bunch)

FIG. 7.32. Illustration of out-of-time pile-up. The left frame shows individual pulses from a
detector as a function of time. The "triggered" or "signal" bunch crossing is located at "0" and
is bounded by the vertical dashed lines. The "signal" pulse is shaded and extends into the next
two 25 ns "buckets," which also contain some pulse height from the later pile-up collisions.
Some energy from the preceding "bucket" also falls into the "signal" bunch crossing. The
right frame shows the sum of all pulse heights as a function of time.

later. The "signal" pulse is shaded and extends into the next two 25 ns "buckets," which
also contain some pulse height from the later pile-up collisions. Some energy from the
preceding "bucket" also falls into the "signal" bunch crossing. The right frame of Figure
7.32 shows the sum of all pulse heights as a function of time. The challenge is to extract
the height or the integrated charge of the shaded pulse from this pattern. Timing and
pulse shape information may be very important tools to achieve this goal. In Section
5.9, an example is shown of the results ATLAS achieved in dealing with the effects of
OOT pile-up, for runs in which the average number of pile-up events contributing to the
(TileCal) signals was 11.3.

In-time pile-up poses challenges of a different order. Most events occurring in the
same bunch crossing as the one of interest contain only low-P_T particles and deposit
little energy in the calorimeters. A relatively small fraction of the events are "hard" collisions that contain high-transverse-momentum particles. The presence of some tracks
and energy from such additional hard collisions in the same bunch crossing can confuse
or degrade the triggers and the offline reconstruction of the event of interest. One way
of dealing with this problem is to use available pointing capability to associate the calorimeter signals with a particular interaction vertex along the beam line. For example,
the ATLAS presampler has proven to be a valuable tool in this context. Also tracker
information may be used for this purpose. Alternatively, the (small) time differences
between the events occurring in the same bunch crossing may be exploited. However,
timing information at the 10 ps level is needed to be successful in that respect.

7.4   The e/h Ratio for Different Types of Calorimeters

We have seen in this and in previous chapters that the hadronic performance of calorimeters is crucially affected by the e/h ratio. When a calorimeter system consists of longitudinal compartments with different e/h ratios, additional complications arise as a consequence of that fact alone. In this section, we briefly review the expected e/h values for different types of calorimeters. The basis for the values given below is Equation 3.9

$$\frac{e}{h} = \frac{e/mip}{f_{\text{rel}} \cdot rel/mip + f_p \cdot p/mip + f_n \cdot n/mip}$$

In Chapter 2, the values of the various parameters in this equation and their dependence on the absorber and active material are discussed in detail. We use the numbers from the Monte Carlo simulations by Gabriel (Table 2.6) as a guideline for evaluating the consequences for the e/h values of different calorimeter types.

- Čerenkov calorimeters. These are relatively easy to evaluate, since none of the products from the nuclear reactions contributes to the signals. Assuming that the relativistic pions are similar to mips (i.e., rel/mip =1), we find for fully active calorimeters (i.e., $e/mip = 1$) such as lead-glass, an e/h value of $1/0.14 \approx 7$. For dual-readout calorimeters based on copper absorber, such as DREAM (Section 8.2), $e/mip \sim 0.8$, which leads to an estimated e/h ratio of $0.8/0.14 \approx 5 - 6$. When lead absorber is used ($e/mip \sim 0.6$), e/h values of $0.6/0.14 \approx 4$ should be expected.

- Homogeneous crystal calorimeters. The e/h value of such calorimeters is smaller than for those from the previous category, since we can now also expect a contribution to the signals from spallation protons. However, the signals from these densely ionizing particles are affected by saturation effects (light quenching). This makes $p/mip < 1$. No significant contribution from neutrons to the signals should be expected. The effective Z values of most crystals are in the 55 (e.g., CsI, BaF$_2$) to 75 (e.g., PbWO$_4$) range. Using a value $f_p = 0.36$, as suggested by Table 2.6, we find $e/h = 1/(0.14 + 0.36 \, p/mip)$. This means that $e/h > 2.0$. Assuming that $p/mip = 0.8 - 0.9$, as suggested by Figure 3.24, then $e/h \sim 2.2 - 2.4$. In any case, it seems unlikely that a crystal calorimeter with $e/h < 2$ can be made.

- Imaging LAr calorimeters. The difference with the previous category is that signal quenching for densely ionizing particles (spallation protons) is not a problem in this case (Figure 7.1). Since the Z value of argon is not very different from that of iron, we use the iron value for f_p from Table 2.6. The expected result is thus $e/h \approx 1/(0.14 + 0.42) = 1.8$.

- LAr sampling calorimeters. The difference with the homogeneous LAr detectors concerns the e/mip ratio. According to Figure 3.5, this value is ~ 0.9 when iron or copper is used as the absorber material, and ~ 0.7 for lead or uranium. The expected e/h value for Fe/LAr or Cu/LAr calorimeters is thus $\approx 0.9/(0.14 + 0.42) = 1.6$, while for Pb/Lar $e/h \approx 0.7/(0.14 + 0.33) = 1.5$. If ^{238}U is used as absorber material, the additional energy produced in nuclear fission may bring the latter value further down, to $1.1 - 1.2$. However, also in these calorimeters, neutrons do not contribute significantly to the signals.

- *Plastic-scintillator sampling calorimeters.* The difference with all previous categories concerns the neutrons which are abundantly produced in the shower development. These neutrons may be sampled very efficiently in such calorimeters, through their interactions with protons in the active material. The precise contribution to the signals depends on the sampling fraction for the charged shower particles. The relationship between n/mip and the sampling fraction for mips is discussed in detail in Section 3.4.3.2, and depicted in Figure 3.41. In addition, the e/mip values are somewhat smaller than for the calorimeters with liquid-argon readout. Figure 3.5 suggests e/mip values of 0.85 for iron or copper absorber, and 0.65 for lead or uranium. Thus, $e/h \approx 0.85/(0.14 + 0.33\, p/mip + 0.05\, n/mip)$ for plastic-scintillator sampling calorimeters with Fe or Cu absorber, while $e/h \approx 0.65/(0.14 + 0.42\, p/mip + 0.09\, n/mip)$ when Pb or ^{238}U is used as absorber material.

This information is summarized in Table 7.1

Table 7.1 *The e/h values expected for different types of calorimeters.*

Calorimeter type	Expected e/h	Value (range)
Čerenkov (lead-glass)	$1/0.14$	≈ 7
Čerenkov (Cu/quartz fiber)	$0.8/0.14$	5 - 6
Čerenkov (Pb/quartz fiber)	$0.6/0.14$	≈ 4
Scintillating crystals	$1/(0.14 + 0.36\, p/mip)$	> 2
Homogeneous LAr	$1(/0.14 + 0.42)$	≈ 1.8
Fe/LAr sampling	$0.9/(0.14 + 0.42)$	≈ 1.6
Pb/LAr sampling	$0.7/(0.14 + 0.33)$	≈ 1.5
^{238}U/Lar sampling	$0.65/(0.14 + 0.33 \times 1.2)$	1.1 - 1.2
Fe/plastic-scintillator	$0.85/(0.14 + 0.42\, p/mip + 0.05\, n/mip)$	≤ 1.5
Pb/plastic-scintillator	$0.65/(0.14 + 0.33\, p/mip + 0.09\, n/mip)$	≤ 1.3
^{238}U/plastic-scintillator	$0.6/(0.14 + 0.33\, p/mip + 0.12\, n/mip)$	≤ 1.2

7.5 Aspects of (Non-)Compensation

The hadronic performance of non-compensating calorimeters is usually dominated by effects that are a direct consequence of this non-compensation. In Section 7.1.2, this was illustrated for the hadronic energy resolution. Fluctuations in the energy sharing between the em and non-em shower components of hadrons and jets tend to dominate all other types of fluctuations in calorimeters that have a different response to em and non-em types of energy deposit.

However, non-compensation also has other consequences for the (hadronic) calori-meter performance. Among these, we will discuss

- the signal non-linearity,
- the non-Gaussian response function, and
- differences between the response functions for different types of hadrons.

7.5.1   The response to pions

The hadronic signal non-linearity is a direct consequence of the energy dependence of the average fraction of the shower energy carried by the em component, $\langle f_{em} \rangle$ (Sec-tion 2.3.1). The degree of non-linearity depends on the e/h value: the more this value deviates from 1.0, the stronger the non-linearity.

This effect has been observed both for undercompensating ($e/h > 1$) and over-compensating ($e/h < 1$) calorimeters. Figure 3.12 shows the calorimeter response to pion as a function of energy, for three different calorimeters: the undercompensating iron/plastic-scintillator calorimeter used in the WA1 (also known as CDHS) experiment [Abr 81], the compensating HELIOS calorimeter [Ake 87] and the overcompensating WA78 calorimeter [Dev 86, Cat 87]. Both the HELIOS and the WA78 calorimeters were composed of uranium plates, interleaved with plastic scintillator. These calorimeters differed in their sampling fraction, 9% for HELIOS *vs.* 5% for WA78, for minimum ionizing particles.

The hadronic response of the CDHS calorimeter ($e/h \sim 1.5$) increased by $\sim 22\%$ between 10 and 138 GeV. When the pion energy was changed by a factor of ten inside this energy range, the response of this calorimeter changed by $16 \pm 3\%$. The response of the WA78 calorimeter ($e/h \sim 0.8$) *decreased* by about 8% between 10 and 40 GeV [Cat 87]. The response of the HELIOS calorimeter ($e/h \sim 0.98$) was found to be con-stant, within experimental errors, for pions with energies from 17 to 200 GeV.

Another measurement of the hadronic signal non-linearity, made with the extremely non-compensating QFCAL detector (Figure 3.49), gave a response increase of $20 \pm 1\%$, for pions with energies between 35 and 350 GeV [Akc 97].

The curves in Figure 7.33 show the expected signal non-linearity as a function of the e/h value, for calorimeters that use lead or iron as absorber material. The non-linearity is defined here as the response ratio to pions of 100 GeV and 10 GeV. It was calculated on the basis of Equations 3.5 and 2.19. The latter equation, which gives the value of $\langle f_{em} \rangle$ as a function of the pion energy, was used with the "standard" parameter values of $k = 0.82$ and $E_0 = 0.7$ GeV for iron and 1.3 GeV for lead.

The experimental results mentioned above are included as data points in Figure 7.33. This was not always directly possible, since the response ratio between 100 GeV and 10 GeV was only measured by WA1 and HELIOS. In other cases, the energy dependence of $\langle f_{em} \rangle$ was used to derive that response ratio from the experimental data that were available. For example, using Equations 3.5 and 2.19, the response ratio between pions of 350 GeV and 35 GeV in an iron calorimeter was found to be $\sim 2/3$ of the response ratio between pions of 100 GeV and 10 GeV, for e/h values around 5. Therefore, the experimental QFCAL result was increased by a factor 1.5 in Figure 7.33, in order to make a comparison with the other experimental data possible.

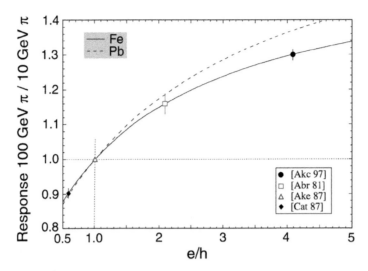

FIG. 7.33. The change in the response to pions between energies of 10 GeV and 100 GeV, as a function of the e/h value of the calorimeter. The curves were calculated on the basis of Equations 3.5 and 2.19. See text for details.

This figure thus gives a quantitative impression of the hadronic non-linearity that may be expected for non-compensating calorimeters. Conversely, the measured hadronic signal non-linearity may be used to make a rough estimate of the degree of non-compensation, *i.e.*, the e/h value of the instrument.

In Figure 6.30, the response of the CDHS detector is shown before and after application of an offline weighting procedure that was intended to correct for the difference between the em and non-em response functions event by event. This procedure is described in detail in Section 4.10. It allowed for a considerable improvement of the hadronic energy resolution, especially at high energy. It was demonstrated that application of this procedure restored $E^{-1/2}$ scaling of the hadronic energy resolution (Figure 4.69b). However, at low energies the event-by-event determination of the em shower content became increasingly uncertain, and for energies below ~ 30 GeV, the resolution improvement resulting from this weighting procedure was found to be marginal.

The results of this weighting procedure for the hadronic signal non-linearity were in complete agreement with those on the energy resolution. After applying the procedure, the calorimeter response became approximately constant for energies above 30 GeV. However, between 10 and 30 GeV, nothing changed. The response increased by about 10% over this energy range, both before and after the weighting procedure.

7.5.2 *The hadronic line shape*

The hadronic response function (also known as the *line shape*) of non-compensating calorimeters is non-Gaussian. This reflects the fact that the event-to-event fluctuations in f_{em} are not governed by the laws of Poisson statistics.

In hadronic shower development, there is no symmetry between the production of charged and neutral pions. Owing to the irreversibility of the π^0 production, the probability that the em shower component carries an anomalously large fraction of the total energy is not equal to the probability that a similarly anomalous fraction of the total energy is carried by the non-em component (Section 4.7.1).

As a result, the f_{em} distribution tends to be asymmetric, skewed to the high-end side (see Figure 4.52). The hadronic response function of an undercompensating calorimeter $(e/h > 1)$ is characterized by the same shape, since showers with an anomalously large f_{em} value will produce an anomalously large signal. On the other hand, in overcompensating calorimeters $(e/h < 1)$, such showers will lead to an *anomalously small* signal. The hadronic line shape of such calorimeters may thus be expected to be skewed to the low-energy side.

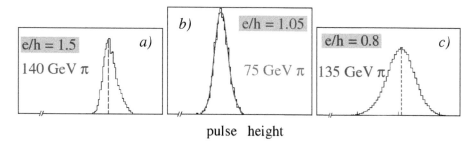

pulse height

FIG. 7.34. Signal distributions for mono-energetic pions in calorimeters with different e/h values. Data from WA1 [Abr 81], ZEUS [Beh 90] and WA78 [Dev 86].

Figure 7.34 shows typical signal distributions for mono-energetic pions in three calorimeters with e/h values larger, (approximately) equal and smaller than 1, respectively. The response function of the compensating calorimeter is well described by a Gaussian distribution, whereas the other two distributions are clearly asymmetric in the sense described above.

The non-Gaussian response function may cause problems in experiments that require a high trigger selectivity. For example, if one wants to trigger on transverse energy, then it is very problematic to unfold a steeply falling $E_\perp$ distribution and a non-Gaussian response function, and it is impossible to do so without introducing biases. If $e/h > 1$, then such a trigger is likely to select predominantly events in which a large fraction of the calorimeter signal was electromagnetic in nature. In an overcompensating calorimeter, events in which little or no energy was deposited in em form are likely to be overrepresented in the selected data sample.

7.5.3 *The response to different types of hadrons*

The production of π^0s in hadronic shower development may depend on the type of showering hadron. As a result, the hadronic response of non-compensating calorimeters may depend on the type of hadron as well.

In Section 2.3.1 it was argued that the requirement of baryon number conservation in the shower development process leads to f_{em} values that are, on average, smaller for proton-induced showers than for pion showers of the same energy. Experimental data taken with a quartz-fiber calorimeter, which was practically only sensitive to the em component of hadron showers, showed indeed a significant difference between the response to pions and protons of the same energy (Figure 2.26).

The QFCAL Collaboration investigated this effect in more detail and found that the response of this calorimeter to protons and pions differed by $\sim 10\%$ in the energy range 200–375 GeV. In addition, it turned out that the energy resolution for protons was significantly better than for pions, and that the line shape for proton showers was much better described by a Gaussian function than the response function for pions [Akc 98].

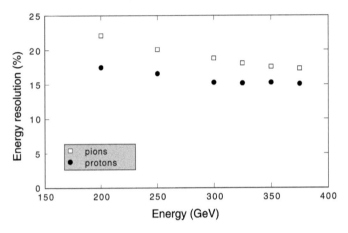

FIG. 7.35. The energy resolution of a quartz-fiber calorimeter for protons and pions as a function of energy [Akc 98].

Figure 7.35 shows the energy resolution for protons and pions, measured with this detector, as a function of energy. Apparently, the constraints imposed by the requirement of baryon number conservation had a considerable impact on the level of fluctuations in the fraction of energy carried by the π^0s produced in the shower development. The energy resolution for protons was typically about 20% better than for pions in this energy range. When the contributions of fluctuations in the number of photoelectrons to the energy resolution were eliminated from the measured values, the differences between the (intrinsic) resolutions for protons and pions were found to be even larger.

Also the line shapes for proton- and pion-induced showers in this type of calorimeter were different. This is because the production of leading π^0s, responsible for the skewed signal distributions for mono-energetic pion showers, was absent for protons, where the leading particle was most likely a baryon. It turned out that the signal distributions for mono-energetic protons were very well described by Gaussian functions. Figure 7.36 shows the response functions for 300 GeV pions and protons in the QFCAL detector. The χ^2 value for a Gaussian fit to the pion distribution was found to be a factor of five larger than for the proton distribution. The latter fit is also shown in Figure 7.36c.

Since the differences between the response functions to protons and pions are caused

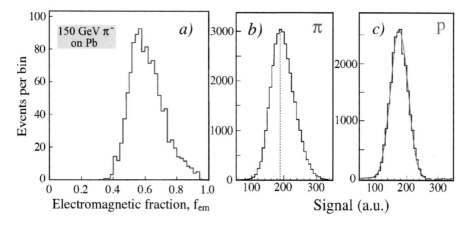

FIG. 7.36. Event-to-event fluctuations in the em fraction of 150 GeV π^- showers in lead [Aco 92b]. Signal distributions for 300 GeV π^- (b) and 300 GeV protons (c), measured with the copper-based QFCAL detector [Akc 98]. The curve in diagram c) represents the result of a Gaussian fit to the signal distribution.

by the properties of the em shower component, these differences are most pronounced in calorimeters with the largest e/h values, such as the quartz-fiber calorimeter in which these effects were first observed. However, the effects have also been observed in calorimeters with an e/h value much closer to 1.0, such as the ATLAS TileCal, for which an e/h value of 1.36 was reported [Adr 09]. Also in this calorimeter the response function for protons was measured to be narrower and more symmetric than that for pions at the same energy, as illustrated in Figure 7.37b. Figure 7.37a shows that the response to protons in this calorimeter was significantly smaller than that to pions, by 4.0% at 50 GeV, 3.1% at 100 GeV and 2.4% at 180 GeV. Note that this energy range is comple-

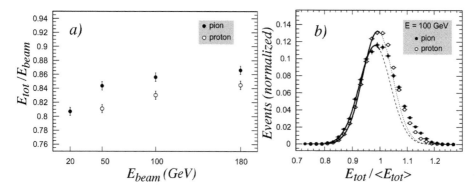

FIG. 7.37. The response of the ATLAS TileCal to pions and protons as a function of energy, normalized to the electron response (a). The response functions of this calorimeter for 100 GeV pions and protons (b) [Adr 10].

mentary to that for which the quartz-fiber calorimeter reported results. The observation that the differences between protons and pions become smaller as the energy increases is consistent with the results obtained with the latter detector (Figure 7.35). It may indicate that the differences in the π^0 production characteristics become less pronounced when a larger number of generations of particle production are needed for a complete absorption of the shower.

Similar effects may also be expected for other types of particles whose interactions are restricted by conservation laws governing the workings of the strong force, and in particular *strange particles*. Due to the requirements of strangeness conservation, less π^0 production should be expected in showers induced by kaons than in pion showers of the same energy. In addition, the fluctuations in f_{em} should be smaller and more symmetric in kaon-induced showers. Experimental data obtained by CMS in beam tests of their calorimeters seem to confirm that the response to kaons is indeed somewhat smaller compared to that of pions [Abd 09]. This is shown in Figure 7.38, which also confirms the reduced response to baryons.

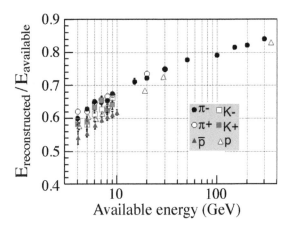

FIG. 7.38. The ratio of the reconstructed energy in the CMS calorimeter and the available energy for a variety of beam particles at different energies [Abd 09]. The vertical scale is normalized to 1.0 for electrons.

7.5.4 $E^{-1/2}$ scaling

Compensation has been achieved in several types of calorimeters:

- *Uranium/plastic-scintillator.* Several compensating calorimeters of this type have been built and operated in experiments, most notably in HELIOS [Ake 87], ZEUS [And 91] and WA80 [You 89]. The sampling fraction required to achieve compensation is about 9% for mips and the hadronic energy resolutions achieved with these calorimeters were typically in the 35–40%/$\sqrt{E}$ range.

- *Uranium/gas.* This technique was exploited in the hadronic section of the L3 calorimeter system [Gal 86]. It was experimentally demonstrated that the e/h

ratio could be tuned through the hydrogen content of the gas mixture used in the wire chambers (Figure 3.40). With a compensating gas mixture, the hadronic energy resolution was found to scale as $\sim 70\%/\sqrt{E}$.

- *Lead/plastic-scintillator.* This combination turned out to be compensating for a sampling fraction of $\sim 4\%$ for mips. For a device consisting of 10 mm thick lead plates, interleaved with 2.5 mm plastic scintillator, a hadronic energy resolution of $44\%/\sqrt{E}$ was measured [Bern 87]. A similar detector that operated at the E-814 experiment at Brookhaven National Laboratory, had a resolution of $(43 \pm 3\%)/\sqrt{E}$ [Fox 92]. Sampling fluctuations played a dominant role in these rather crudely sampling detectors. With much finer-sampling calorimeters, in which 1 mm scintillating fibers were used as active elements, the hadronic resolution improved to ~ 32–$35\%/\sqrt{E}$ [Aco 91c, Arm 98].

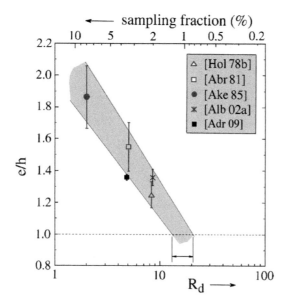

FIG. 7.39. The e/h value for plastic-scintillator calorimeters with iron or copper absorber, as a function of the sampling fraction for mips (top horizontal scale), or the volume ratio of the amounts of passive and active material (bottom horizontal scale).

In principle, compensation could also be achieved with relatively low-Z absorber materials, such as copper or iron. Plastic-scintillator calorimeters that use(d) iron or copper as absorber materials and have reported e/h values include:

- The ISR-807 copper calorimeter, which had an e/h value of ~ 1.9, for a sampling fraction (for mips) of $\sim 8\%$ [Ake 85].
- The WA1 iron calorimeter ($e/h \sim 1.5$, sampling fraction for mips $\sim 3\%$) [Abr 81].
- A prototype version of the latter detector, based on somewhat thinner scintillator plates [Hol 78b]. This calorimeter had a sampling fraction for mips of $\sim 2\%$ and

an e/h value of ~ 1.3.

- The CDF Plug Upgrade calorimeter [Alb 02a]. The hadronic compartment of this detector had a steel-to-plastic ratio of 8.5:1 (sampling fraction for mips 2%). The e/h value was reported to be 1.36 ± 0.05 [Alb 02b].
- The ATLAS TileCal consists of steel and plastic at a ratio 4.7:1. The e/h value reported for this calorimeter, which has a sampling fraction of 3.6%for mips, varies between 1.36 [Adr 09] and 1.44 [Sim 08].

Figure 7.39, in which these e/h values are plotted as a function of the volume ratio of the amounts of passive and active material (bottom axis), and as a function of the sampling fraction (top axis), suggests that compensation could indeed be achieved for this combination of passive and active materials. However, the sampling fraction would have to be very small, $\sim 1\%$ for mips, for e/h to become 1. This corresponds to a structure in which the scintillator would occupy only 4–8% of the calorimeter volume ($R_d = 12 - 21$). Obviously, sampling fluctuations would necessarily be very large in such a device, both for em and for hadronic showers.

The WA1 Collaboration operated an iron/plastic-scintillator calorimeter for its neutrino experiments at the CERN Super Proton Synchrotron. This calorimeter consisted of 2.5 cm thick iron slabs interleaved with 5 mm thick plastic scintillator sheets. In their prototype studies for this experiment, WA1 varied the thickness of the iron slabs over a large range and measured the effect of that on the hadronic energy resolution. The results are shown in Figure 7.40, where the hadronic energy resolution, expressed here as $\sigma/\sqrt{E}$, is plotted as a function of the thickness of the iron plates.

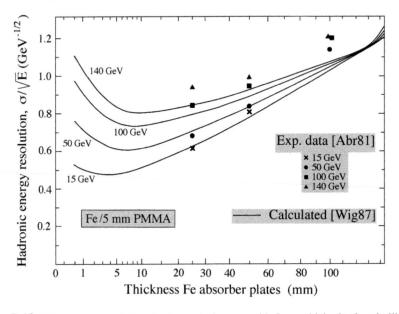

FIG. 7.40. The energy resolution for iron calorimeters with 5 mm thick plastic scintillator as active material, as a function of the thickness of the absorber plates [Abr 81, Wig 87].

In Section 4.7.1 we showed that the hadronic energy resolution of non-compensating calorimeters is characterized by a deviation from $E^{-1/2}$ scaling in the high-energy domain, where the (non-Gaussian) fluctuations in the energy sharing between the em and non-em shower components become the dominant factor. The more the value of e/h deviates from 1.0, the more the hadronic energy resolution deviates from $E^{-1/2}$ scaling (Equation 4.28):

$$\frac{\sigma}{E} = \frac{c}{\sqrt{E}} + \mathcal{F}(e/h) \tag{7.13}$$

This means that if the energy resolution is expressed in terms of

$$\frac{\sigma}{E} = \frac{c'}{\sqrt{E}} \tag{7.14}$$

the value of c' is energy dependent to a degree determined by the e/h value. Only for $e/h = 1$ is the value of c' independent of the pion energy.

Figure 7.40 shows the values of c' for various energies used in these tests (15 – 140 GeV). For the iron thickness that was eventually chosen for the WA1 calorimeter structure, 2.5 cm, the value of c' increased by 38% between 50 and 140 GeV (see also Figure 4.69b).

However, as the plate thickness increased, the c' value became less energy dependent, even though its absolute value increased as a result of the cruder shower sampling. This decreased energy dependence of c' reflects the decrease of the e/h value with the sampling fraction (Figure 7.39). For 10 cm thick iron plates, the c' values for all energies in the range 50–140 GeV were found to be equal to within 7%. For 15 cm thick plates, the c' values were measured to be more energy dependent again.

These results indicate that the e/h value indeed crossed the 1.0 value for an iron plate thickness near 10 cm (i.e., $R_d = 20$, thus confirming the prediction derived from Figure 3.44), and that the calorimeter with the 15 cm thick iron plates actually operated in an *overcompensated* mode.

This analysis illustrates that iron absorber, in combination with plastic scintillator as active material, allows for a compensating structure, provided that these materials are used in the right proportion. The authors of [Abr 81] measured an energy resolution of $\sim 10\%$ for 140 GeV pions for a calorimeter consisting of 10 cm thick iron plates interleaved with 5 mm plastic-scintillator sheets. There is no fundamental reason why they should not have been able to achieve 1% at 14 TeV with this crude instrument.

7.6   Particle Identification

Owing to differences in the characteristic energy deposit profiles, and to other features of shower development, calorimeter signals may be used to identify the particles that caused the signals. In the following, we discuss various methods that are being used for this purpose, and show some representative results.

7.6.1   *Electron/pion distinction*

Electron/pion discrimination can be achieved with calorimeters by exploiting the very different longitudinal and lateral energy deposition profiles of em and hadronic showers.

As we have seen in Chapter 2, the development of em and hadronic showers scales (approximately) with the radiation length and the nuclear interaction length, respectively. The distinction between electrons and hadrons thus works best in materials with very different radiation and nuclear interaction lengths.

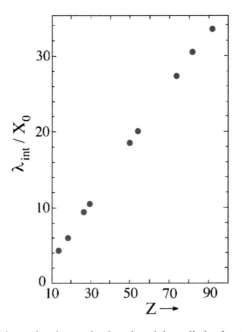

FIG. 7.41. Ratio of the nuclear interaction length and the radiation length as a function of Z.

Figure 7.41 shows the ratio of these two parameters as a function of Z. This ratio is almost proportional to Z. Therefore, the best separation between em and hadronic showers is achieved with high-Z absorber materials, such as uranium, lead or tungsten. These elements are indeed most frequently used as absorber material in the em section of calorimeter systems.

7.6.1.1 *Using the longitudinal shower information.* In almost all experiments that use calorimeter information for electron identification, differences between the *longitudinal* development of em and hadron showers are exploited. This is in practice often one of the most important reasons for separating calorimeter systems in em and hadronic sections.

The potential for electron/pion separation based on longitudinal shower information is illustrated in Figure 7.42. This figure shows distributions of the ratio of the energy deposited in the hadronic and electromagnetic sections of the lead based RD1 calorimeter [Bad 94b], for 80 GeV electron and pion events. These two distributions are very different (note the logarithmic vertical scale).

The electron/pion separation that was achieved on the basis of these differences is given in Figure 7.43, where the fractions of electrons and pions passing a cut of the

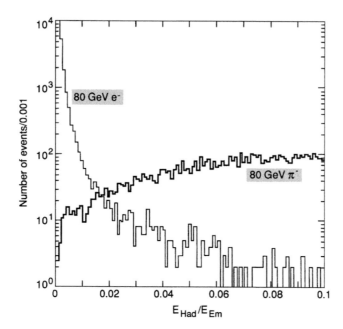

FIG. 7.42. Distribution of the ratio of the energies deposited in the hadronic and electromagnetic sections of the RD1 calorimeter, for electrons and pions of 80 GeV [Bad 94b].

type $E_{had}/E_{em} < f_{cut}$ are shown, as a function of f_{cut}, for two different energies. For example, if one required that less than 1% of the particle energy be deposited in the hadronic calorimeter section ($f_{cut} < 0.01$), a pion rejection factor of 330 was obtained, for 95% electron efficiency, at 80 GeV. At lower energy, the quality of the e/π separation degraded. For example, at 20 GeV a pion rejection factor of 25 was obtained, for 92% electron efficiency.

The calorimeter with which these studies were performed is schematically shown in Figure 5.24. It was not segmented into em and hadronic sections in the classical longitudinal sense. The hadronic sections consisted of the wedges surrounding the square "rods" that formed the em sections of the projective towers. In this structure, the ratio of the volumes occupied by the em and hadronic sections thus decreased *gradually* as a function of depth, while in classical longitudinally segmented devices, this ratio exhibits a step function, going from 1 to 0 at the boundary between the two sections.

Nevertheless, the results shown above are very similar to those found in these classical structures. Studies of this type have been performed for many longitudinally segmented calorimeter systems. Typical values for the pion rejection factor achieved on the basis of longitudinal shower information are of the order of 10 to 100 in the energy range from 1 GeV to 10 GeV [Hit 76, Cob 79, Eng 83, Mic 84]. At high energies, values in excess of 1000 have been achieved [App 75, Bau 88].

Especially in LAr-based detectors, the em calorimeter is easily segmented into several longitudinal sections. In that case, more detailed longitudinal shower information

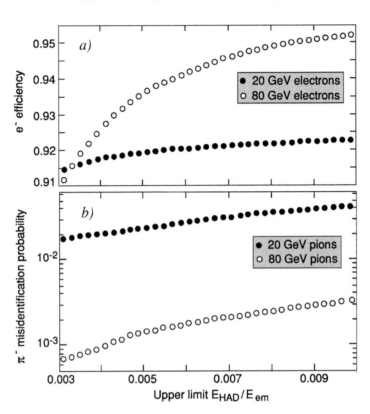

FIG. 7.43. The fraction of electrons (a) and pions (b) that passed a cut $E_{had}/E_{em} < f_{cut}$ as a function of f_{cut}, for the projective RD1 calorimeter [Bad 94b].

may be used to distinguish electrons from hadrons. An example of this was given by Cobb and coworkers [Cob 79], who built a finely segmented Pb/LAr calorimeter for an experiment at CERN's Intersecting Storage Rings. This em calorimeter consisted of five longitudinal sections, the second of which covered the depth region from $2.7X_0 - 5.2X_0$. This section was equipped with two sets of readout electrodes that were inclined at angles of $\pm 20°$ with the beam direction. These signal electrodes, called u and v, were read out separately.

Figure 7.44 shows scatterplots in which the *difference* between the signals recorded in these u and v strips is plotted versus the total energy recorded in this segment of the calorimeter, *i.e.*, the *sum* of the signals measured on these strips. This figure thus gives an impression of the shower fluctuations in this calorimeter segment. These fluctuations were clearly larger in the case of 4 GeV/c pions (Figure 7.44b) than for 4 GeV/c electrons (Figure 7.44a). This is no surprise given that the hadron signals are dominated by protons released in nuclear reactions of the type shown in Figure 2.29. Differences between correlation patterns of this type may also be used to increase the electron/hadron distinction capability of the calorimeter.

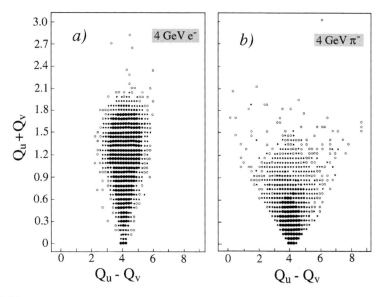

FIG. 7.44. Fluctuations in the longitudinal energy deposit ($Q_u - Q_v$, arbitrary units) *vs.* total energy deposit ($Q_u + Q_v$), for 4 GeV/c electrons (a) and 4 GeV/c pions (b) showering in a longitudinally segmented Pb/LAr calorimeter [Cob 79]. See text for details.

The principal limitation to the level of separation that can be achieved with the techniques described in this subsection comes from physics. The charge exchange reactions, $\pi^- + p \to \pi^0 + n$ and $\pi^+ + n \to \pi^0 + p$, generate purely em showers if the entire momentum carried by the charged pions is transferred to the neutral ones. When such reactions take place in the first few radiation lengths of the calorimeter, the resulting showers are practically indistinguishable from those generated by electrons. At energies of a few GeV, the cross sections for these reactions amount to about 1% of the total inelastic cross sections. This fraction decreases logarithmically with increasing energy [Bar 70].

7.6.1.2 *Using the lateral shower information.*   Differences between em and hadronic shower development do not only manifest themselves in the longitudinal energy deposit profile, but also *laterally*. These lateral differences in the shower profiles may equally well be used for the purpose of electron/pion separation, and with fine-grained detectors excellent results can be obtained.

The SPACAL Collaboration has extensively studied e/π separation methods based on the lateral energy deposit profiles in their longitudinally unsegmented calorimeter. One of the algorithms that was found to be very useful for electron identification in the offline analysis, was defined through a quantity

$$R_{\mathrm{p}} = \frac{\sum_i r_i E_i^{0.4}}{\sum_i E_i^{0.4}} \qquad (7.15)$$

where E_i represented the energies deposited in individual calorimeter towers i and r_i

indicated the distance between the center of tower i and the center of gravity of the energy deposit profile. All 155 SPACAL towers were included in the calculation of the R_p value for individual showers.

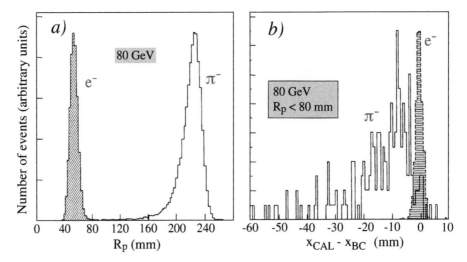

FIG. 7.45. Distribution of the effective width R_p (see text) for 80 GeV electron and pion showers in SPACAL, oriented at an angle of $2°$ between the fibers and the incoming particles (*a*). Distribution of the displacement of the shower's center of gravity with respect to the particle's impact point, for 80 GeV electrons and for 80 GeV pions that produced showers that were laterally indistinguishable from these electron showers (*b*) [Aco 91b].

It turned out that electron and pion showers were very cleanly separated through this parameter, which may be considered a measure of the effective shower width (Figure 7.45a). At 80 GeV, the R_p values clustered around 50 mm for electron showers, while the average value for pion showers at the same energy was about 220 mm.

However, a small fraction ($\sim 0.1\%$) of the pion events could not be distinguished from electrons solely on the basis of the R_p value. This sample contained, for example, events in which (almost) all the pion energy was transferred to one or several π^0s in the first nuclear interaction. In such events, the showers looked laterally very similar to em ones. However, unlike the showers initiated by electrons, the starting point of such hadron showers was located at a certain depth (z) inside the calorimeter, with a probability distributed as $\exp(-z/\lambda_{int})$. Therefore, if this starting point could be determined, one might distinguish these hadron showers from the ones initiated by electrons.

The SPACAL Collaboration used the lateral displacement of the center of gravity of the energy deposit profile with respect to the impact point of the showering particles to determine z, event by event. All pion events that could not be distinguished unambiguously from electrons on the basis of their lateral shower profile ($R_p < 100$ mm, see Figure 7.45a) were used for this analysis.

In these beam tests, the calorimeter was oriented in such a way that the particles entered it at an angle of only $2°$ with respect to the fibers. A (typical) event in which the center of gravity of the energy deposit profile was located at a depth of 50 cm inside the calorimeter thus exhibited a lateral displacement of this center of gravity with respect to the particle's impact point equal to $50 \times \tan 2° = 1.7$ cm. A good position resolution was therefore imperative for the success of this method.

The distribution of the lateral displacement of the shower's center of gravity with respect to the pion's impact point ($x_{CAL} - x_{BC}$) for this event sample is shown in Figure 7.45b. For comparison, the distribution of this variable is also shown for a random sample of electron events. Thanks to the good position resolution of this calorimeter, all pions with a shower starting point located more than a few centimeters inside the detector could be distinguished from electrons with this method.

With the longitudinally unsegmented SPACAL detector, pion rejection factors of several hundred were obtained, with electron efficiencies better than 95%, in the energy range 40–150 GeV, on the basis of lateral shower information *alone* [Aco 91a]. When the longitudinal shower information was incorporated, in the way described above, the rejection factor improved to more than 6,000, at 80 GeV.

The improvement in the quality of the e/π separation that can be obtained by including lateral shower information in a longitudinally segmented calorimeter system depends strongly on the granularity of the detector, and on the energy of the particles. Cobb and coworkers used both longitudinal and lateral shower information to distinguish electrons from hadrons in the energy range from 1 to 4 GeV (Figure 7.44, [Cob 79]). Their detector was a fine-sampling lead/liquid-argon calorimeter that was longitudinally subdivided into five sections. The first two sections (which covered about $5X_0$ in depth) had readout strips that were 2 cm wide, while the signal electrodes were 10 cm wide in the deeper sections of the detector. The e/π separation was found to improve by a factor of three to five when they included lateral shower information in the selection criteria.

A systematic study of the effects of detector granularity on the hadron rejection capability was performed by Engelmann *et al.* [Eng 83]. They used an array of 60 extruded lead-glass bars for this purpose. Each bar had a cross section of 6.5×6.5 cm^2 and a length of 140 cm. The long axes of these bars were oriented vertically and an array of 12 layers consisting of 5 bars each was constructed. Alternate layers were displaced transversely by 10 mm to avoid cracks. Beam particles traversed the 12 layers (total depth $24.2X_0$) at normal incidence. The standard segmentation was thus $5 \times 2X_0$ laterally and $12 \times 2X_0$ longitudinally. The authors used beams of 15 GeV/c particles to study the effects of grouping layers and/or rows on the hadron rejection factor. This factor was determined from cuts in the hadronic energy deposition profiles that were passed by 90% of the electrons.

They concluded from these studies that fine transverse segmentation was essential. The rejection factor decreased by a factor of 30 to 100 when only longitudinal shower information was taken into account, depending on the longitudinal segmentation. Four longitudinal segments turned out to be adequate. As long as the segment containing the shower maximum was not too large, increasing the number of longitudinal segments did not improve the rejection significantly.

7.6.1.3 *Using the time structure of the calorimeter signals.*   The SPACAL Collaboration also developed a completely different method of electron identification. This method is based on small, but very significant differences between the time structures of em and hadronic calorimeter signals (Section 3.2.9). These differences are illustrated in Figure 5.50, which shows 10 electron and 10 hadron signals, chosen at random [Des 89].

The electron signals all look rather identical, while the pion signals exhibit a variety of shapes, which can be attributed to fluctuations in the starting point of the showers, in combination with the effects of upstream mirrors (see Section 5.6.1), and to the contribution of neutrons to the signals (Section 3.2.9).

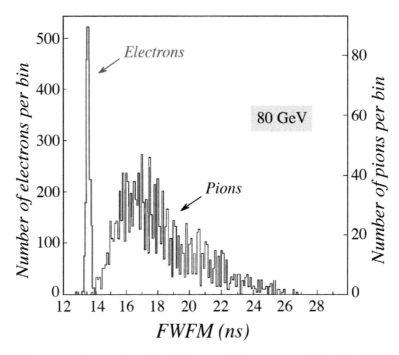

FIG. 7.46. The distribution of the full width at one-fifth maximum (FWFM) for 80 GeV electron and pion signals in SPACAL [Aco 91a]. The left-hand scale applies to the electron signals, the right-hand scale to the pion signals.

SPACAL developed a method in which these differences were exploited to provide a very fast (25 ns) electron trigger [Aco 91a]. This method discriminated between electrons and pions on the basis of the signal width, measured at a fixed fraction of the amplitude. An appropriate fraction was found to be 20% and the corresponding signal width was therefore called the Full Width at one-Fifth Maximum (FWFM).

Distributions of the FWFM for 80 GeV electron and pion signals, measured with an electronic circuit that was developed to measure this quantity for trigger purposes, are shown in Figure 7.46. As before, the electron signals were found to be all nearly identical, with a standard deviation in their FWFM of only 0.13 ns. On the other hand, the distribution of the pion signal widths reflected the large event-to-event fluctuations also observed in Figure 5.50. Almost all pion signals were found to be wider than the electron ones, and a pion rejection factor of $\sim 1,000$ (at 80 GeV) was obtained with this method.

The availability of detailed information on the time structure of the signals provides several other options for distinguishing between electron and pion induced showers. Some of these options have been exploited by the RD52 (DREAM) Collaboration, in their studies of dual-readout calorimeters [Akc 14a]:

- The ratio of the signal amplitude and the integrated charge is typically quite different for these two types of showers (Figure 8.25d).

- The starting time of the signal, measured with respect to an upstream trigger signal, can be used to distinguish between electron and pions (Figure 8.25c).

Both effects are based on the fact that the light produced in the fibers that form the active material of these calorimeters travels at a lower speed (c/n) than the shower particles that are responsible for the production of that light, which typically travel at speeds close to c. As a result, the deeper the light is produced inside the calorimeter, the earlier it arrives at the PMT that converts it into an electric signal. Since the light from hadron showers is typically produced much deeper inside the calorimeter, the PMT signals start earlier than for em showers, which produce light close to the front face of the calorimeter. These, and other methods used by RD52, are further discussed in Section 8.2.6.2.

7.6.1.4 The E/p method.

Often, experiments in which calorimetric methods are used for e/π separation, also have alternative methods available for the identification of electrons. One such method, the so-called E/p method, is based on a comparison of the energy, measured calorimetrically, and the momentum of the (charged) particle, measured from the curvature of its track in a magnetic field upstream of the calorimeter. For electrons depositing their entire energy in (the em section of) the calorimeter, this ratio is approximately 1.0. Pions deposit typically only a fraction of their energy in the em calorimeter section and, therefore, yield an E/p value less than 1.0 when the energy measurement is based on signals from this section.

An example of results obtained with this method is given in Figure 7.47. This figure shows E/p distributions for electron candidates produced in a Fermilab experiment (E-70) that combined a powerful magnetic spectrometer with a lead-glass calorimeter (see Section 7.6.1.8 for more details). Electrons that deposited their entire energy in this calorimeter showed up as a clear peak centered around 1.0 in these distributions. Events in which significant leakage occurred, *i.e.*, electrons that entered the calorimeter near its boundaries, or pion events that were not removed by various cuts that were applied, were characterized by E/p values less than 1.0.

The width of the peak, and thus the rejection power of this method, is determined both by the energy resolution of the calorimeter and by the momentum resolution of the spectrometer. The energy resolution improves with energy, while the momentum resolution degrades ($\Delta p/p$ is proportional to p). Therefore, the latter dominated the

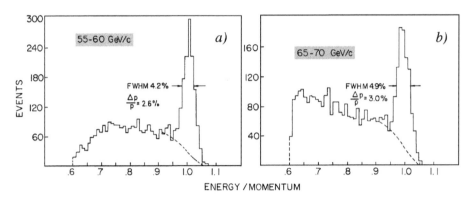

FIG. 7.47. Distribution of the E/p values of a sample of candidate electrons recorded in the E-70 experiment at Fermilab, for particles with momenta ranging from 55 to 60 GeV/c (a) and from 65 to 70 GeV/c (b) [App 75].

rejection power of this method at high energies. This is illustrated by the fact that the peak in Figure 7.47b (for the momentum bin 65–70 GeV/c) is broader than that in Figure 7.47a (55–60 GeV/c).

This example illustrates that the additional information provided by the magnetic spectrometer may further improve the particle identification beyond the level achievable on the basis of calorimetric information alone. In Section 7.6.1.8, this is worked out in some detail.

7.6.1.5 *Preshower detectors.* In Section 7.6.1.1, we saw that the large difference between the radiation length and the nuclear interaction length in high-Z materials can be successfully exploited for electron/pion separation. In the data shown in Figure 7.42, this difference manifested itself in the form of a very different energy sharing between the longitudinal calorimeter sections, for showers induced by electrons and pions, respectively.

However, the difference between (the very early phases of) em and hadronic shower development may also offer excellent opportunities for e/π separation in structures that are very much thinner than a shower-absorbing calorimeter system. This feature forms the basis of many *preshower detectors*, which we encountered earlier in this chapter in the context of measurements of the angle of incidence of the showering particles (Section 7.2.3).

An extremely simple preshower detector may consist of a plate of lead, 1 cm (1.9 X_0, 0.06 λ_{int}) thick, followed by a sheet of plastic scintillator. When a beam consisting of a mixture of high-energy electrons and pions is sent through this device, almost all pions (96%) traverse it without strongly interacting. These pions produce a minimum ionizing peak in the scintillator. On the other hand, the electrons lose a considerable fraction of their energy by radiating large numbers of bremsstrahlung photons. Some of these photons convert into e^+e^- pairs in the PSD and thus contribute to the scintillation signals produced by this device.

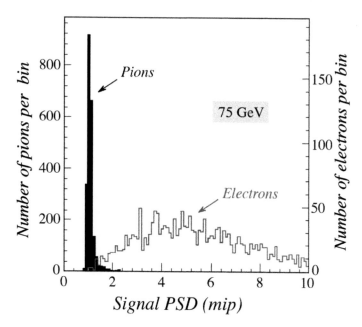

FIG. 7.48. Signal distributions for 75 GeV pions and electrons in a preshower detector used in beam tests of CDF calorimeters.

The result is a very clear separation between electrons and pions. Figure 7.48 shows the signal distributions for 75 GeV electrons and pions in the described device, used in beam tests of the CDF Plug Upgrade calorimeter [Apo 98a, Alb 99]. Even with such simple devices, pion rejection factors of the order of hundred are readily achieved.

Such preshower detectors are particularly helpful during beam tests, for example for obtaining a pure sample of pion events. Especially at low energy, pion beams at accelerators are often contaminated with electrons. By placing a device of the type described above in front of the calorimeter to be tested, this electron contamination can be effectively eliminated, through a cut on the PSD signals. Since pions are defined as events that cause a mip signal in the PSD, the pion sample is not biased by this procedure.

Although the electron signals from the described PSD are substantial (compared with the signal from a mip), the fraction of the energy lost in this very early shower development phase is small. Provided that the PSD is placed directly in front of the calorimeter, the effect on the energy resolution of the latter is even smaller.

SPACAL made a study of these effects, for a 1.7 X_0 thick PSD placed 12 cm in front of the calorimeter [Aco 91c]. Electrons with energies of 20 GeV and above lost, on average, less than 1% of their energy in this PSD. The effect of event-to-event fluctuations in this energy loss on the energy resolution of the calorimeter was barely significant: at 20 GeV, the energy resolution for electrons was found to be $3.99\% \pm 0.08\%$ without the PSD and $4.06\% \pm 0.08\%$ with the PSD installed in front of the calorimeter.

More sophisticated PSDs than the ones described above are sometimes used to distinguish between electrons and photons. This application is further discussed in Section 7.6.3.

7.6.1.6   *Other methods to identify electrons.*   Dual-readout calorimeters, which provide Čerenkov signals in addition to signals that measure the energy deposited by *all* charged shower particles, offer additional possibilities to distinguish between electrons and pions. This was demonstrated by the RD52 (DREAM) Collaboration [Akc 14a], who used the ratio of both signals for this purpose (Figure 8.25b). They employed several other methods in combination with this one to determine the capability to identify electrons in their longitudinally unsegmented fiber calorimeter. A multivariate neural network analysis showed that the best e/π separation achievable for 60 GeV beams was 99.8% electron identification, with 0.2% pion misidentification.

7.6.1.7   *Triggering on electrons.*   High-energy electrons are among the most powerful indicators of interesting physics processes in particle physics. Such electrons played an important role in the discovery of several heavy quarks, of the τ lepton and of the intermediate vector bosons.

Especially in hadron colliders, the interesting electrons are outnumbered by hadrons and photons in the same energy range, by several orders of magnitude. The success of experiments at hadron colliders depends often, in no small measure, on the electron detection efficiency in these difficult conditions.

All the techniques previously described in this section, as well as other techniques not depending on calorimetry, may be exploited for the purpose of building a fast and efficient electron trigger. Among the non-calorimetric techniques, we mention the detection of Čerenkov radiation [Seg 94], transition radiation [Cob 77] or synchrotron radiation [Mel 92], characteristically emitted by high-energy electrons. This problem has two aspects:

1. Establishing that the signal is caused by an em shower and not by a hadronic one.
2. Establishing that the signal is caused by an electron and not by one or several photons.

The latter aspect requires the association between a charged track and the calorimeter signal(s). A similar association is needed for the application of the E/p method (Section 7.6.1.4). The establishment of such an association requires elaborate computations and is in practice only achieved at the highest trigger level. In practice, this requirement is often the main factor determining the (low-energy) limit on the achievable energy range for electron detection.

The distinction between em and hadronic energy deposit on the basis of the available calorimeter information can be achieved much faster. The FWFM module developed by SPACAL (Section 7.6.1.3) delivered signals from electron candidates within 25 ns to the trigger logic.

Signatures derived from the energy deposit profile require some processing, in which signals from different (groups of) calorimeter cells have to be compared, but can usually be applied at the second trigger level. An example of a simple, fast algorithm that

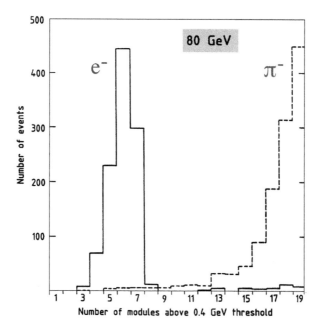

FIG. 7.49. Distributions of the number of calorimeter cells recording an energy exceeding a preset limit (0.4 GeV), for electron and pion showers at 80 GeV [Aco 91a].

may effectively distinguish between electrons and pions in a fine-grained calorimeter is given in Figure 7.49. This algorithm is based on counting the *number of calorimeter cells* in the shower region that exhibit signals exceeding a preset threshold. Since em showers are much narrower than hadronic ones, a clear separation can be achieved in this way.

When defining electrons as particles for which the number of different cells with a signal is at maximum seven, the electron efficiency was about 99% in the given example, while 1% of the pions passed such a cut.

7.6.1.8   Electron identification in practice.   One of the first experiments in which the calorimetric tools described in this subsection were fully exploited for a physics analysis was carried out by a Columbia/FNAL group at Fermilab in the early 1970s [App 75]. In this experiment (E-70), high-energy electron production in hadron–hadron interactions was studied. This led eventually to the discovery of the fifth (*b*) quark [Her 77].

Their detector consisted of 45 identical lead-glass blocks, with dimensions $15 \times 15 \times 35$ cm^3 each, read out with PMTs. The blocks were arranged as shown in Figure 7.50, to provide three longitudinal layers of $6.2X_0$, $6.2X_0$ and $14.8X_0$, respectively. The detector thus covered a cross-sectional area of 75×75 cm^2, but in practice a somewhat smaller fiducial area (60×60 cm^2) was used to exclude particles entering near the edges, thus avoiding unnecessary leakage effects. Between the first and second layers a curtain of scintillation counters was installed (T2). This provided useful triggering information. Upstream of the calorimeter, a 1.3 cm ($2.3X_0$) thick lead sheet, sandwiched between

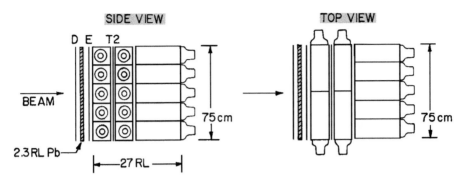

FIG. 7.50. Diagram of the lead-glass em calorimeter of the Columbia/FNAL experiment E-70 [App 75].

two scintillation counters (D and E), served as a preshower counter.

This calorimeter was part of a much larger experimental setup (Figure 7.51), which included a powerful magnetic spectrometer, with a momentum resolution $\Delta p/p = 4.5 \cdot 10^{-4} p$ (GeV/c). The em energy resolution of the calorimeter was reported as $10\%/\sqrt{E} + 1.5\%$ (FWHM), which corresponds to $\sigma/E = 4\%/\sqrt{E} + 0.6\%$. For particles with energies below 30 GeV, the measurement error on the E/p value was thus dominated by the calorimeter. For higher energies, the error in the momentum measurement was larger.

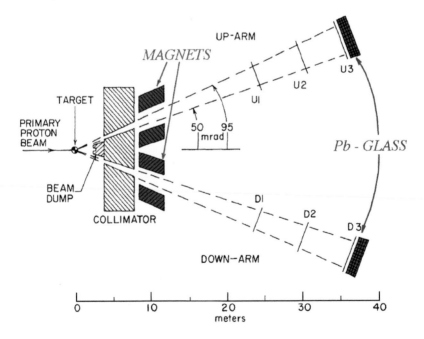

FIG. 7.51. The magnetic spectrometer of the Columbia/FNAL experiment E-70 [App 75].

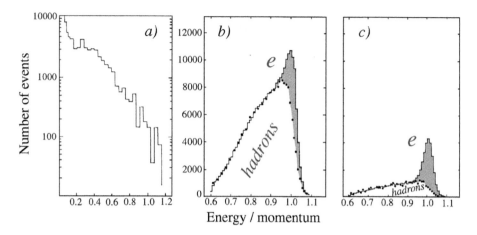

FIG. 7.52. Results of an experiment searching for high-energy electron production in hadron–hadron collisions. Shown are the E/p distributions for incident charged particles with momenta in the 40–45 GeV/c range before (a) and after application of cuts described in detail in the text (b and c). The dots in diagrams b) and c) represent the hadron background, established in measurements in which electrons were prevented from reaching the detector [App 75].

Figure 7.52 shows E/p distributions for charged particles with a momentum in the range 40–45 GeV/c. The raw data are displayed in Figure 7.52a. The distribution falls very steeply with the E/p value. Only a small fraction of 1% of the particles had E/p values in the vicinity of 1.0, the expected value for electrons.

Part of the hadron rejection was accomplished on the basis of the signals from the scintillation counters T2 (see Figure 7.50). These counters were installed near the shower maximum for electrons in this energy range. A cut on the T2 signal turned out to eliminate a large fraction ($\sim 99.8\%$) of the pions, while retaining almost all of the electrons. This is illustrated in Figure 7.52b, which suggests the presence of an electron peak near 1.0 in the E/p plot.

Additional cuts further enhanced the significance of this peak, as illustrated in Figure 7.52c. These cuts involved:

1. The signals in the preshower counter. By requiring a signal of at least seven times the signal from a mip in the E counter (see Figure 7.50), 70% of the remaining hadrons were eliminated, while less than 5% of the electrons were cut in this way.

2. The energy sharing between the three longitudinal calorimeter sections. Fluctuations in this energy sharing were significantly smaller for electrons than for hadrons, and cuts in the energy fractions deposited in these three sections provided additional hadron rejection.

The result of all these cuts, combined with the requirement that $0.95 < E/p < 1.05$ led to a hadron rejection factor of more than 10^4, i.e., less than one hadron out of 10,000 passed all electron cuts and ended up in the mentioned E/p bin.

Figures 7.52b,c also show the hadron background. This background was obtained from additional measurements, in the following way. By placing two inches of lead in the secondary beam, downstream of the target and upstream of the magnets (Figure 7.51), any electrons contained in the beam were for all practical purposes absorbed and an effectively pure hadron beam was created. The dots in Figure 7.52b and Figure 7.52c were measured in this configuration and thus represent the E/p distributions of the hadrons that passed the electron cuts. The agreement between these distributions and those obtained without the lead brick in the E/p region $0.6 - 0.9$ is evidence that the electron cuts did not create an artificial "electron" peak. Subtraction of the hadron background measured in this way further increased the hadron rejection factor to $\sim 10^5$.

An example of a modern experiment that uses similar techniques is LHCb, which studies CP violation and rare decays of particles containing b quarks at the LHC [Alv 08]. Identification of electrons is extremely important in these studies. Apart from a magnetic spectrometer, this experiment also uses a system of powerful Ring Imaging Cherenkov counters to identify electrons buried in the huge background of heavier particles [Yps 95]. Figure 7.53a shows the E/p signal ratio for electrons and hadrons. The low-energy tail in the electron peak is the result of bremsstrahlung losses in the material upstream of the calorimeter. The importance of correctly identifying electrons (and positrons) in this experiment is illustrated in Figure 7.53b, which shows the invariant mass distribution of e^+e^- pairs in a sample of $B_s^0 \rightarrow J/\psi\phi$ events, with the J/ψ decaying into an electron–positron pair. The background in this event sample is mainly due to pion tracks with low transverse momentum, and could be efficiently removed with a p_T cut that barely affected the electrons. It turns out that the average efficiency for identifying electrons from $J/\psi \rightarrow e^+e^-$ decays in such events is 95%, with a pion mis-identification fraction of 0.7%.

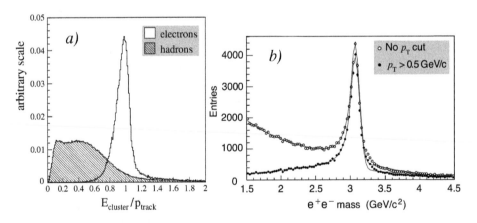

FIG. 7.53. The ratio of the energy measured by the ECAL to the momentum measured by the magnetic spectrometer in the LHCb experiment (a). Invariant mass plots for the e^+e^- pairs in a sample of $B_s^0 \rightarrow J/\psi\phi$ signal events, showing the effect of a cut $p_T > 0.5$ GeV/c for the $e^\pm$ candidates (b) on the reconstructed J/ψ peak [Alv 08].

7.6.2 *Muon identification*

There are several different possibilities to discriminate between muons and hadrons with the help of calorimetric methods.

1. In very deep calorimeters with fine longitudinal segmentation, high-energy muons are clearly recognized as isolated, minimum ionizing tracks, ranging far beyond the tracks produced in hadronic shower development (Figure 7.54). This is the technique used in experiments with incident neutrinos.

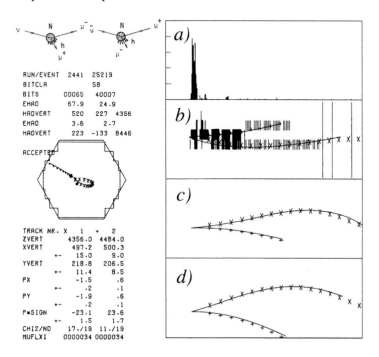

FIG. 7.54. Event display for a neutrino interaction in the WA1 detector. In this interaction, two muons were produced. These are represented by the crosses in diagrams *c*) and *d*), which show projections on two different planes of hits recorded in the wire chambers. The corresponding signals in the plastic scintillators are shown, in two projections, in diagrams *a*) and *b*). Courtesy of the WA1 Collaboration, CERN.

2. In calorimeters used in 4π experiments at colliders, which for practical reasons are barely deep enough to contain hadron showers, the lateral energy deposition profile may be used to discriminate between muons and hadrons (Figures 7.55 and 7.56). The quality of π/μ separation with this method depends strongly on the lateral granularity of the calorimeter.

3. Calorimeters may simply absorb all hadrons, so that charged tracks recorded beyond the calorimeter are by definition caused by muons.

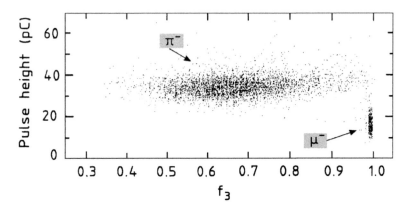

FIG. 7.55. Calorimetric π/μ separation through differences in the lateral energy deposit profile. The scatter plot shows the fraction of the total energy recorded in the three calorimeter towers with the largest signals, versus the total calorimeter signal, for a mixed beam of 10 GeV pions and muons [Aco 91d].

4. In dual-readout fiber calorimeters, the signals from muons in the scintillating and Čerenkov fibers differ by a precisely known amount, independent of the muon energy (Figure 3.48). This difference corresponds to the energy deposited by ionization of the absorber structure, since the Čerenkov signals only register the radiative component of the energy loss, and can be calculated on the basis of the material composition of the calorimeter [Akc 04].

It is important to realize that a calorimeter that absorbs hadrons is, at the same time, also a source of muons. In hadron showers, many charged pions and kaons are produced. Each of these secondary particles has a small, but finite, decay probability, in which process typically a muon is produced.

Kaons are a relatively more important source of such muons than are pions. This is because kaons of a given energy are about ten times more likely to decay when traveling a certain distance than pions of the same energy that travel the same distance. For example, in a detector in which the interaction length for mesons amounts to 20 cm, the probability that a 10 GeV kaon decays in flight is about 0.26%, versus 0.036% for 10 GeV pions.

Since K^+s are more abundantly produced in nuclear interactions than K^-s (because in the associate production of $s\bar{s}$ pairs, the s quark has the option to end up in the target baryon, contrary to the $\bar{s}$ quark), this type of background is characterized by an excess of positive muons over negative ones.

Muons generated in hadronic shower development can be easily distinguished from penetrating beam particles (e.g., pions that decay in flight before reaching the calorimeter), provided that the calorimeter has a sufficiently fine transverse granularity. Figure 7.56 shows energy deposit profiles in SPACAL and in a leakage calorimeter of the same structure that was installed directly behind it, with the fibers (and thus the calorimeter towers) oriented *perpendicular* to the direction of the beam particles, which enter

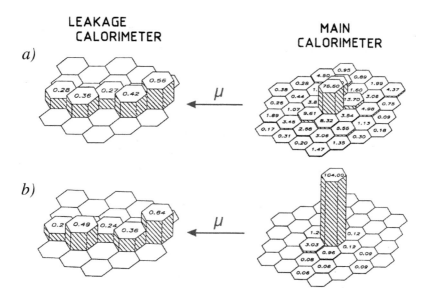

FIG. 7.56. Energy deposit profiles in the SPACAL calorimeter and in the leakage calorimeter installed behind it. Typical examples of a pion shower in which a muon is produced (*a*) and of a muon contaminating the beam and penetrating the calorimeters (*b*) [Aco 91d]. The numbers in the calorimeter cells are a measure for the deposited energy.

from the right side in this figure. Muons escaping from the main calorimeter left a track in this leakage calorimeter, characterized by approximately equal signals in all towers traversed. The origin of these muons could be deduced from the corresponding energy deposit pattern in the main calorimeter. If the muon was generated in the decay of a particle produced in the hadronic shower development, then the energy deposit pattern in this calorimeter was typical for that of hadronic showers, with many towers recording a significant fraction of the shower energy (Figure 7.56a). If, on the other hand, the muon was a beam particle that traversed the main calorimeter as well as the leakage calorimeter, then the pattern was very different: practically all energy deposited in the main calorimeter was concentrated in the tower traversed by the muon (Figure 7.56b).

The probability for the production of an escaping muon in hadronic shower development was measured to be about $10^{-4} E_\pi$ (GeV) for SPACAL (Figure 7.57). For 80 GeV π^-, the muons generated in the showers and escaping this 9.5 λ_{int} deep calorimeter had an average momentum of $\sim$ 3 GeV/c (see Figure 4.48).

Considerably more energetic leakage muons may be produced in pairs as decay products from ρ^0 decay. The ρ^0 resonance is abundantly produced in hadronic shower development and has a branching ratio of $4.6 \cdot 10^{-5}$ for decay into a $\mu^+ \mu^-$ pair. The RD1 Collaboration measured dimuon production in hadronic shower development at $\sim$ 6% of the single-muon rate (see Figure 4.49).

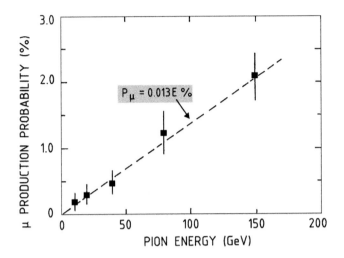

F$_{IG.}$ 7.57. The probability that an escaping muon is created in hadronic shower development in the SPACAL detector, as a function of the energy of the incoming pion [Aco 91d].

7.6.3 *Electron/γ/π⁰ distinction*

Usually, the distinction between electrons and photons can be made on the basis of the presence of a charged-particle track pointing to the calorimeter region in which an em shower developed. However, there are situations in which an energetic photon, accompanied by a nearby charged hadron or muon, may be misinterpreted as an electron. This is especially a problem for the recognition of electrons that belong to jets, which may be a powerful tool for selecting jets that result from the fragmentation of heavy quarks (see Section 7.7.3).

The showers developed by γs are only different from electron showers in the very early stage [Wig 02]. Therefore, the distinction between electrons and photons can only be made in the very first part of the absorber if the calorimeter data are needed for this purpose.

Most high-energy γs produced in particle–particle collisions are decay products of π^0s, which are abundantly produced in such collisions. These photons are thus part of a *doublet*. If the π^0 is extremely relativistic ($E \gg m$), then the angle ($\Delta\theta$) between the two members of this γ-doublet is approximately equal to m/E. When the two γs enter the calorimeter, located at a distance D from the production vertex, they are thus separated by a distance

$$d \approx \frac{Dm}{E} \qquad (7.16)$$

For example, if the calorimeter is located at a distance of 1.5 m from the interaction vertex, the γs from a 10 GeV π^0 are typically separated by about 2 cm upon entering this detector.

In Section 7.6.1 we saw that many experiments use a preshower detector, installed right upstream of the calorimeter system. This PSD can fulfill many different useful

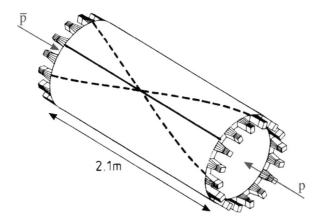

FIG. 7.58. Geometry of the scintillating-fiber detector that served as a PSD in the $\bar{p}p$ UA2 experiment at CERN. The orientation of the various fiber layers with respect to the beam line and the arrangement of the readout elements are visible in this figure [Alit 89].

tasks. One of these is the distinction between electrons, photons and photon pairs.

The UA2 Collaboration used their PSD in an elegant way to distinguish electrons from photons, and also managed to achieve this when the photon entered the detector close to a charged pion [Alit 89]. Their PSD consisted of a $1.5X_0$ thick lead absorber, combined with eight active planes. Six of these planes were installed upstream of the absorber and served as a tracking system for charged particles. The other two active planes were located downstream of the absorber and detected the early phases of developing em showers. Each active layer consisted of three layers of 1 mm thick scintillating plastic fibers, arranged as stereo triplets (Figure 7.58). In each stereo triplet, one layer of fibers was oriented parallel to the beam line (the *axial* layer), the two other layers were oriented at angles $\pm\alpha$ (the *stereo* layers), with $\alpha = 15.75°$ for the upstream section of the detector and $\alpha = 21°$ for the downstream section. The entire detector formed a cylinder around the beam pipe, with a thickness of 6 cm, an average radius of 41 cm and a total length of 2.1 m, thus covering polar angles from 20° to 160° in the experimental setup. The scintillating fibers were read out individually by means of an array of charge coupled devices (CCDs).

The pattern recognition was based on an algorithm that tried to reconstruct spatial points in each stereo triplet layer, and tracks connecting such points in the different active layers. When a charged particle crossed the detector, it produced light in the fibers it traversed. These fibers formed a pattern of three segments in the $z = 0$ plane, *i.e.*, the plane perpendicular to the beam line through the center of the interaction region. In this projection, the stereo fibers that produced a signal were located at equal distances on either side of the traversed axial fibers. If the track originated from the center of the interaction region, then all three track segments were radial. In other cases, the stereo projections were slightly tilted with respect to the axial projection.

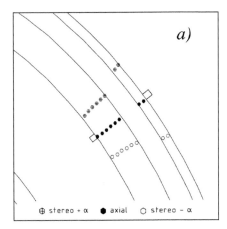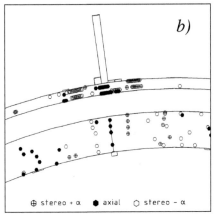

FIG. 7.59. Representation of an ideal track in the UA2 PSD (*a*), and the hit pattern recorded for a high-energy electron produced in the $\bar{p}p$ collisions (*b*). The hit fibers are drawn in the $z = 0$ plane [Alit 89].

Figure 7.59a shows the reconstructed hit pattern for an ideal track, *i.e.*, a track created by a particle that produced signals in all fiber planes it traversed. In practice, that usually did not happen, because of detection inefficiencies. The single-fiber detection efficiency was measured to be $\sim 84\%$, on average. It was found to be dominated by (downward) fluctuations in the number of photoelectrons produced by a crossing particle. This also caused a small position-dependent effect, because of light attenuation effects in the fibers. In practice, charged-particle tracks were required to have produced at least four hits out of six in the upstream tracking planes.

Em showers produced very large signals in the downstream fiber planes. For example, 40 GeV electrons produced signals in the downstream planes that were, on average, ~ 20 times larger than those from 40 GeV pions (see also Figure 7.48). A typical example of the hit pattern created in this detector by a high-energy electron produced in the $\bar{p}p$ collisions studied by the UA2 Collaboration is shown in Figure 7.59b.

The hit pattern in the various active planes was different for the four situations that needed to be distinguished:

1. A single photon would produce no hits in the upstream detector planes, but it would convert in the absorber and produce a signal in the downstream planes that was, on average, equivalent to several mips.

2. A π^0 (photon pair) would produce no hits in the upstream planes, and two multi-mip signal clusters in the downstream planes, separated by a distance commensurate with Equation 7.16.

3. An electron would produce mip signals in the upstream planes (*i.e.*, it would be recognized as a charged track), and multi-mip signals in the downstream planes.

4. A pion accompanied by a photon would produce mip signals in the upstream planes, and a mip signal plus a multi-mip signal nearby in the downstream planes.

Thanks to the excellent spatial resolution of the fiber planes (0.39 mm rms per active layer in the tracking part) and the very fine granularity, this PSD was extremely useful for the identification of electrons in UA2. In practice, the positions of the em shower and the extrapolated track matched within ~ 1 mm in the (R, ϕ) plane and ~ 3 mm in the z coordinate (along the beam line). This excellent performance was crucial for virtually eliminating the contamination of the genuine electron sample by events in which the much more numerously produced photons or photon pairs, in combination with nearby pion tracks, mimicked the electron signature.

The two-track resolution of this detector was ~ 3 mm. This made it possible to distinguish nearby charged tracks separated by at least that distance. It also allowed the distinction between π^0s and single γs for energies up to ~ 15 GeV.

An excellent capability to distinguish between high-energy electrons, γs and π^0s turned out to be an absolutely crucial tool for discovering the Higgs boson through its $H^0 \rightarrow \gamma\gamma$ decay mode. Both the ATLAS and CMS experiments were designed to be able to do just that. This was no easy task, since the ratios of inclusive photons

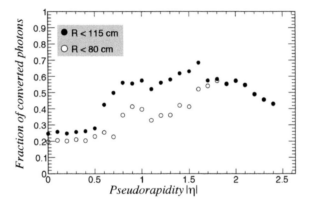

FIG. 7.60. The fraction of high-energy γs that convert into electron–positron pairs in the AT-LAS detector before reaching the presampler [Aad 08].

and electrons to jets from QCD processes are even one to two orders of magnitude smaller than at the Tevatron. For example, at a transverse momentum of 40 GeV/c, the electron-to-jet ratio is $\sim 10^{-5}$. An additional complication for identifying high-energy γs derives from the conversion of a large fraction of the photons into e^+e^- pairs before reaching the presampler (Figure 7.60). This is a consequence of the substantial amounts of material installed upstream of the calorimeter system.

CMS use their high-granularity multi-layer silicon tracking system to distinguish between converted photons and electrons [CMS 15a]. Photon identification requires that there be no charged-particle track with a hit in the inner layer of the pixel detector, pointing to the photon cluster in the ECAL. Photon conversion candidates can be distinguished from electrons from b, c quark decays by exploiting the fact that the momenta of the conversion electrons are approximately parallel, since the photon is massless. These criteria reduced the remaining photon inefficiency almost entirely to photons converting in the beam pipe. The validity of the identification procedure of converted photons was checked with a sample of $Z \rightarrow \mu^{+}\mu^{-}\gamma$ events, in which the photons were exclusively associated with a conversion track pair. The $\mu\mu\gamma$ invariant mass distribution of these events is shown in Figure 7.61a, together with the simulation result, which is in very good agreement with the experimental data.

However, in physics analyses based on photon signals, the main background does not come from electrons, but from π^{0}s produced in (QCD) jets. In the transverse momentum range of interest, the γs from π^{0} decay are collimated and reconstructed as a single photon. The background tends to be dominated by π^{0}s that carry a substantial fraction of the total jet p_{T} and are thus relatively isolated from jet activity in the detector. Single-γ identification is based on shower-shape and isolation variables. The different distributions of one such variable for signal and background event samples are shown in Figure 7.61b.

Information from this and many other variables were combined in a multivariate analysis (TMVA [Hoc 07]), which employed a boosted decision tree (BDT). This technique defines a single discriminating variable characterizing each photon (the BDT score), which results from the combination of many variables discriminating prompt photons from background candidates. The separation of signal and background can-

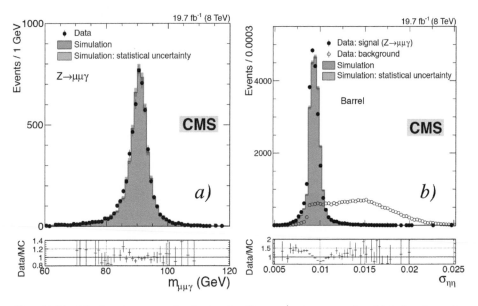

FIG. 7.61. The invariant mass distribution for $Z \rightarrow \mu^{+}\mu^{-}\gamma$ events, in which the photon is associated with a conversion track pair (a). Distribution of the shower shape variable $\sigma_{\eta\eta}$ for γs from a sample of such decays and for background-dominated γ candidates, in the CMS barrel calorimeter (b). The signal and background distributions are normalized to the same total number of photons [CMS 15a].

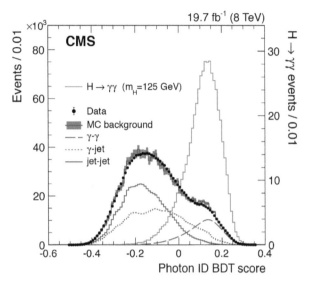

FIG. 7.62. The photon identification BDT score of the lower-scoring photon of di-photon pairs with an invariant mass between 100 and 180 GeV/c^2, for events passing the $H \rightarrow \gamma\gamma$ preselection in the 8 TeV dataset of CMS, and for simulated background events. The solid line histogram on the right (righthand vertical axis) is for simulated Higgs boson signal events [CMS 15a].

didates can be seen in Figure 7.62. The photons from $H \rightarrow \gamma\gamma$ decay typically had a BDT score around 0.15. The BDT scores of background events peaked near values of ~ -0.2, but the distribution of these events extended well into the region covered by genuine signal events. Using this technique, a $H \rightarrow \gamma\gamma$ signal could thus only be established statistically, not on an event-by-event basis.

Unlike CMS, ATLAS has a fine-grained presampler installed directly upstream of the barrel calorimeter. This instrument plays an important role in distinguishing single high-energy γs from π^0s. An example of the results is shown in Section 7.7.1.

7.6.4 *Meson/baryon distinction*

In the previous subsections we discussed a number of methods with which particles can be identified on a case-by-case basis, using the calorimeter information. However, the distinction between mesons and baryons can only be made *statistically*, on the basis of available calorimeter data, just as candidate γs from Higgs decay could only be statistically separated from the much more abundant unresolved $\pi^0 \rightarrow \gamma\gamma$ events.

In Section 7.5.3 it was shown that non-compensating calorimeters in general, and the QFCAL in particular, respond quite differently to proton- and pion-induced showers. The differences may be summarized as follows:

- The calorimeter response to protons is systematically smaller in undercompensating calorimeters. For particles in the energy range 200–375 GeV, QFCAL measured the proton response to be $\sim 12\%$ smaller than the response to pions.

- The energy resolution for protons is considerably better for protons than for pions. For particles in the 200–375 GeV range, QFCAL measured this difference to be about 20%.

- The line shape for protons is well described by a Gaussian function, while the signal distribution for pions is skewed to the high-energy side.

- The interaction length for pions is considerably longer than for protons. For copper absorber, QFCAL measured this difference to be $\sim 20\%$. This leads to considerable differences in the energy sharing between the different sections of a longitudinally segmented system.

The last feature is independent of the calorimeter's e/h value. The other features do depend on the e/h value. They are most pronounced in extremely non-compensating devices such as QFCAL, but were also clearly observed in the ATLAS TileCal ($e/h = 1.36$).

The QFCAL Collaboration exploited these differences to determine the composition of mixed proton/pion event samples, with a precision of a few percent. The following independent event characteristics were used for this purpose:

1. The fraction of the total energy deposited in the em calorimeter section. For example, at 200 GeV, the mean fraction of the signal deposited in this $22.6X_0$ deep section was found to be $47.8\% \pm 1.3\%$ for protons and $42.9\% \pm 0.2\%$ for pions.

2. The fraction of the energy detected in the tower in which the particle entered the calorimeter. For example, at 300 GeV, the average fraction of the signal recorded in this central tower was measured to be $70.3\% \pm 0.4\%$ for protons and $75.5\% \pm 0.1\%$ for pions.

Other event characteristics that could be used for this purpose include quantities measuring the asymmetry of the line shape, such as the ratio of the numbers of events with energies above and below the average value. Also the signal itself could be utilized, provided that an independent measurement of the particle momenta was available. In that case, the fact that the calorimeter response to protons is significantly smaller than that to pions would make the calorimeter signal itself a powerful tool for distinguishing these particles.

The mentioned differences could also be exploited to enrich a given event sample. By applying proper cuts, *e.g.*, by eliminating events in which the energy fraction deposited in the hit tower, or the calorimeter response itself, exceeds a certain lower limit, and/or in which the energy fraction deposited in the em calorimeter section falls below a certain upper limit, an event sample that originally consisted of a 50/50 mixture of protons and pions could be enriched to contain 95% protons, with a reasonable efficiency (*i.e.*, without losing too many protons in the selection process).

The enrichment of event samples achieved in this way is comparable to particle identification methods based on differences in specific ionization [Leh 78]. Also in that case, the dE/dx value measured for an individual particle does not allow its identification with great certainty, because of the large event-to-event fluctuations in that number. However, if the average dE/dx values are sufficiently different for different types of particles, then the measured values for individual particles may be used to generate enriched event samples. Similar comments apply to particle identification methods in Ring Imaging Čerenkov Counters [Seg 94].

Enriched event samples obtained with such methods may be very helpful in a variety of physics analyses, and in particular for spectroscopic studies. For example, baryons containing a c or or b quark will contain a baryon among their decay products, and frequently this final state baryon is a proton. An event sample enriched in protons may considerably improve the signal-to-background ratio in the invariant mass plots through which the properties of these heavy baryons are studied.

The calorimeter response to protons is smaller than that to pions in undercompensating calorimeters, because of baryon number conservation in the shower development. In Section 7.5.3 we argued that a similar phenomenon might occur for showers induced by kaons, where the requirement of strangeness conservation in the shower development would also limit the energy available for π^0 production. The methods discussed above might therefore, at some level, also be applicable for π/K separation.

7.6.5 *Neutrinos and LSPs*

Neutrinos only interact extremely weakly with matter, their interaction lengths are sometimes measured in units of light years. The signature of neutrinos in calorimetric measurements is, therefore, the *absence* of a signal, rather than a signal.

Apparent missing (transverse) energy or missing (transverse) momentum has become a powerful means of inferring the presence of neutrinos among the collision products. The first and most famous application of this measurement technique led in the early 1980s to the discovery of the W boson [Arn 83a], through the processes $W \to e + \nu_e$ and $W \to \mu + \nu_\mu$. The details of this success story are discussed in Section 11.1.

Missing energy relies on a measurement of the total energy, using 4π calorimetric coverage for all particles produced in the collisions. This can in practice be achieved for experiments at e^+e^- colliders or in fixed-target experiments. Neutrino production is inferred whenever the measured energy is *significantly* lower than the total available energy, *i.e.*, incommensurate with the energy resolution function of the detector.

A total-energy measurement is not practical at hadron colliders, since a considerable fraction of the total energy is produced at angles too close to the beam line to be accessible for calorimetric measurements. In this case, neutrino production may be signaled through an imbalance in the *transverse* energy: $E_{\mathrm{T}}^{\mathrm{miss}} = |-\sum_i \vec{p}_{\perp,i}| \neq 0$, incommensurate with detector resolution. In the large 4π experiments at the Tevatron (CDF and D0) and the Large Hadron Collider (ATLAS and CMS), neutrinos from W decay were/are identified in this way.

Typical values for the intrinsic quality of such measurements are as follows [Arn 84, Bag 84, Jen 88]

$$\sigma\left(\frac{E_{\mathrm{T}}^{\mathrm{miss}}}{E_{\mathrm{total}}}\right) \approx \frac{0.7}{\sqrt{E_{\mathrm{total}}}} \tag{7.17}$$

Another particle whose signature would consist of the absence of a calorimeter signal, if it existed and were produced, is the *Lightest Supersymmetric Particle*. In the theory of Supersymmetry (SUSY), our world of matter, built of fermionic quarks and leptons which interact by exchanging bosons, is assumed to be complemented by a SUSY world consisting of bosonic squarks and sleptons which interact by exchanging fermionic field quanta (photinos, gluinos, *etc.*). If SUSY particles were produced in high-energy collisions, then they would instantaneously decay to the lightest member of this family (the LSP) and this particle would not measurably interact with regular matter.

In order to be sensitive to LSPs in the entire accessible energy range, the LHC experiments have emphasized the sensitivity and resolution for missing transverse energy in the design of their calorimeter systems. The most important calorimeter parameter relevant for this issue is hermetic coverage. Both the ATLAS and CMS calorimeters cover the phase space up to about five units of pseudorapidity.

7.7 Particle–Particle Separation

In Section 7.2 it was shown that the lateral granularity of the calorimeter is a crucial factor for the precision with which showering particles can be localized in the detector. This granularity also determines to what extent the calorimeter data can be used to distinguish particles entering the calorimeter simultaneously at a small distance from each other as separate entities.

The possibilities in this respect depend, apart from the granularity, also on the characteristic widths of the showers generated by the particles. In a given calorimeter, it is easier to recognize two showers as such when they are generated by two electrons than when it concerns an electron–pion or pion–pion doublet.

7.7.1 *Electromagnetic shower doublets*

The capability of recognizing close shower doublets is very relevant for some important physics analyses. Consider, for example, the process $H^0 \rightarrow \gamma\gamma$. In the energy region of interest, from 50 to100 GeV, photons from π^0 decay outnumber these Higgs γs by several orders of magnitude. The two photons from π^0 decay are separated by only a few mm (Equation 7.16). It is therefore crucially important to be able to distinguish these "trivial doublets" from single photons that might originate from Higgs decay, at an adequate level.

The detection of leptons (electrons, muons, neutrinos) produced in the collisions has, more than once, turned out to be a crucial ingredient for new discoveries in particle physics. In collider experiments without a central magnetic field (UA2, D0), the measurement of the energy of electrons relied entirely upon the calorimeter system. However, if an electron is accompanied by one or more energetic γs, which do not leave traces in the upstream tracking system, then the electron energy may be severely overestimated on the basis of the calorimetric measurements. To prevent this type of problem, a good particle-particle separation capability is essential as well.

These are examples of tasks for which many experiments rely on the preshower detector, installed just upstream of the calorimeter, or a shower maximum detector, installed at a depth of a few radiation lengths. In the very early phase of their development, em showers are extremely narrow. With a sufficiently fine-grained PSD or SMD, em showers separated by only a few mm may be recognized as doublets [Alit 89, Dwu 89, Apo 93, Hul 93, Aco 95, Aki 95, Gru 95, Auf 98]. An example of the application of such a detector for this purpose is discussed below.

The ATLAS presampler plays an important role in distinguishing single energetic photons from the much more abundant doublets from π^0 decay. The high granularity of the presampler ($\Delta\eta = 0.025, \Delta\phi = 0.1$) has turned out to be indispensable for this purpose. The γ/γ separation power of the presampler/calorimeter prototype was studied in the prototype phase with data obtained with a tagged photon beam (consisting of hard bremsstrahlung photons emitted by electrons traversing a radiator) [Aub 93b]. The decay of a 50 GeV π^0 was simulated by superimposing two single photons with a total energy of 50 GeV, using applicable kinematics. Assuming that the presampler was located at a distance of 140 cm from the beam line, the distribution of the distance between the impact points of the two γs peaked at the minimum value of 7.5 mm, while 96% of the pairs were separated by less than 30 mm.

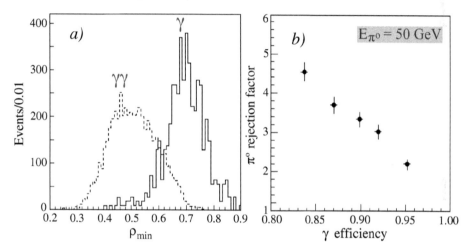

FIG. 7.63. The capability of the ATLAS presampler to distinguish single γs from doublets caused by π^0 decay, at 50 GeV. Shown are the distribution of the shape variable ρ (a) and the rejection against π^0s provided by the combined presampler/calorimeter system versus the efficiency fo single γs (b). See text for details [Aub 93b].

A variety of criteria were used in an attempt to distinguish these overlapped events from single 50 GeV γ showers. One such criterion was based on the width of the showers in the presampler. The authors defined a shape variable ρ, defined as the maximum fraction of the total presampler signal detected in any three-cell cluster. Figure 7.63a shows the distribution of this variable for single 50 GeV γs and overlapped $\gamma\gamma$ events as described above. As expected, the ρ distribution peaked at higher values for the single photons. A cut $\rho > 0.6$ gave a π^0 rejection of a factor of five, while 91% of the single γs passed the cut.

Of course, this method only worked for events in which both γs from the decaying π^0 converted in the presampler. For other situations, alternative criteria were developed and tested. The final result of these tests is shown graphically in Figure 7.63b, where the π^0 rejection factor is plotted *vs.* the fraction of single γs that were retained [Aub 93b]. It thus turned out that at this energy, the needed rejection factor of three against isolated π^0s could be safely achieved with an efficiency of 90% for single photons.

In many experiments, PSDs and SMDs have turned out to be multi-functional devices, which may play an important role in a variety of physics analyses. In this chapter, we have encountered the following applications:

- Angular measurements for photons (Section 7.2.3)
- Electron/pion distinction (Section 7.6.1.5)
- Electron/photon distinction (Section 7.6.3)
- Photon/π^0 distinction (Section 7.6.3, this section)
- Clean-up of electron sample, by identifying $e + \gamma$ events (this section)

In addition, PSDs may play an important role in improving the em energy resolution of the calorimeter, by recovering the (effects of) energy lost in material upstream of the calorimeter (Section 5.7.1).

7.7.2   Doublets involving hadrons

Since hadron showers have strongly fluctuating starting points, PSDs are not very useful for recognizing doublets involving hadrons. Signals from mips (charged hadrons) tend to be completely insignificant near a developing shower, and neutral hadrons (K_L^0, n) are likely not to leave any signal at all in a PSD.

Therefore, particle–particle separation relies in this case on the calorimeter system proper. Because of the much larger cell size, and because hadron showers are considerably wider than em ones in their very early stages, the separation results are much less impressive when hadrons are involved.

The SPACAL Collaboration investigated the $\pi - \pi$ separation capability of their detector, which had a hexagonal cell structure (86 mm apex-to-apex), on the basis of the energy deposit patterns of the particle showers. They performed this study by merging, in pseudo-events built offline, the showers from two different pions entering the calorimeter at a relative distance d. The resulting energy deposit profile was compared with the *average* shower profile for one pion entering the detector at an intermediate position. This comparison was done by defining a separation parameter

$$K = \sum_i r_i \left[E_i - f(r_i) \right]^2 \tag{7.18}$$

where E_i is the fraction of the total energy deposited in tower i, located at a distance r_i from the shower's center of gravity, and $f(r_i)$ describes the fractional energy deposit in this cell for an average shower generated by a single pion with the same center of gravity.

The value of K therefore made it possible to determine how well individual showers were described by the average profile, $f(r)$, for showers generated by single pions.

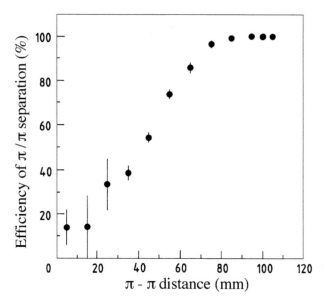

FIG. 7.64. The probability of recognizing a 80 GeV + 80 GeV π^- pair as a function of their relative distance, in the SPACAL detector [Aco 91b]. See text for details.

In order to identify pion pairs, a cut on the value of K was applied. All events with $K < K_{cut}$ were considered to be single pions; all other events were considered to be pion pairs. The value of K_{cut} was chosen to keep the probability for misidentifying a single pion as a pair below 5%.

Some results of this study are shown in Figure 7.64, which shows the probability of recognizing an 80 GeV + 80 GeV π^- pair as such, as a function of the relative distance (d) between the impact points of the two particles. For distances $d > 8$ cm, the separation efficiency was measured to be larger than 95%.

The distance between the centers of two neighboring cells in this calorimeter was 7.4 cm, and it is therefore not a surprise that the detector's capability to recognize two particles as a pair started to deteriorate quickly for distances d smaller than that. Had the cell size been smaller, pion–pion separation would certainly have been possible for even smaller values of d.

This remarkably good pion–pion separation capability was made possible by the narrow em shower components, caused by π^0s and ηs generated in the absorption process. As a result, typically $\sim 50\%$ of the shower energy was deposited in a single calorimeter cell. In Section 7.2.2, we saw that the narrow profile of this central core caused "displacement effects" for the localization of the showering particle. Here, we see that it also allows for the separation of two pions separated by a small distance.

From this perspective, a very fine lateral granularity is certainly not an overkill in hadronic calorimeters. Such a granularity provides a degree of detail that might be quite useful, for example in certain jet analyses. Figure 7.65 shows an event display of a

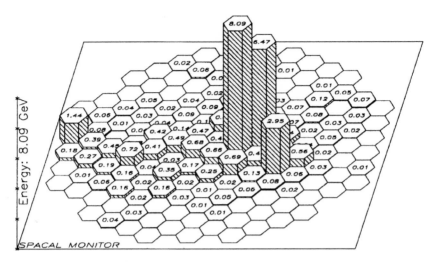

FIG. 7.65. A SPACAL event display of the reaction products from a pion interaction in an upstream target. The numbers denote the energy (in GeV) deposited in the individual calorimeter cells [Aco 91b].

"jet" detected by SPACAL. Several individual particles can be clearly recognized in this event. To put this figure in perspective, it should be mentioned that the typical cell size of hadron calorimeters currently in use for particle physics experiments is equivalent to some 25 cells of the size displayed in this figure.

7.7.3   *Multiplets involving electrons or photons*

In some physics analyses, the main challenge is the identification of electrons in the vicinity of a number of other particles, predominantly pions and photons (from π^0 decays). The search for top quark production can be mentioned as an example of such an analysis. Heavy quarks have significant leptonic decay probabilities. Therefore, it is quite likely to find an electron among the final-state products of the decay chain $t \to b \to c \to u$. By selecting jets that contain energetic electrons, one may hope to create an event sample that is substantially enriched in events in which t-quark production actually took place (a technique known as "soft lepton tagging").

One of the largest problems in identifying electrons inside jets is the background caused by overlapping pion–photon pairs. Especially in cases where the γ is much more energetic than the pion, the combination of an em shower-like energy deposit pattern in the calorimeter plus a charged track pointing to the calorimeter region where the shower is recorded provides exactly the type of signature characteristic for electrons. Since pions and γs are abundantly present in *any* jet, dealing with this background is the main challenge in analyses of this type.

A dedicated study of this problem was performed by Engelmann and coworkers [Eng 83], for a calorimeter system consisting of 60 lead-glass bars with a cross section of 6.5×6.5 cm^2 and a length of 140 cm. These bars formed an array with 12 layers of five bars each. Their method to identify electrons in a complex environment was based upon the covariance matrix describing energy deposits and correlations in this 60-element detector array.

For a sample of N electrons, they constructed the 60×60 matrix $\mathbf{M}$, whose elements were given by

$$M_{ij} = \frac{1}{N} \sum_{n=1}^{N} (E_i^{(n)} - \bar{E}_i)(E_j^{(n)} - \bar{E}_j) \tag{7.19}$$

where $E_i^{(n)}$ represented the energy observed in the ith element of the detector for the nth electron event, and $\bar{E}_i$ was the average energy deposited in that detector bar for the entire event sample. The inverse matrix $\mathbf{H} \equiv \mathbf{M}^{-1}$ was used to define a variable ξ for test events:

$$\xi = \sum_{i,j=1}^{60} (E_i - \bar{E}_i) H_{ij} (E_j - \bar{E}_j) \tag{7.20}$$

The value of ξ was a measure of how much "electron-like" a particular test event was.

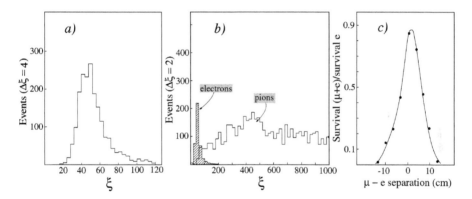

FIG. 7.66. Identification of electrons in a jet, using a covariance method. Shown are the ξ distributions for samples of mono-energetic electrons (a) and same-energy electrons and pions (b) in a 60-element detector array. Diagram c) shows the probability that the combination of an electron shower plus a minimum ionizing track is identified as a single electron, as a function of the distance between the shower and the track [Eng 83].

Figure 7.66a shows the ξ distribution for the same sample of 31.4 GeV electron events that was used to define the $\mathbf{H}$ matrix. This distribution closely resembles the standard χ^2 distribution, with a mean value of about $\xi = 60$ (i.e., the number of independent calorimeter cells). When showers induced by pions were confronted with the covariance matrix for electrons (normalized to the proper energy), the resulting ξ values were usually considerably larger. This is illustrated by the pion distribution in Figure 7.66b, which has a broad maximum around $\xi = 500$, with very few events at $\xi < 100$, where 90% of the electrons were found.

The probability of recognizing an em shower in the vicinity of other particles was studied with the same covariance method. The authors built event samples by superimposing the energy deposit patterns of muons traversing their detector at various impact positions upon electron events. The resulting energy deposit patterns were subjected to the ξ test based on the covariance matrix $\mathbf{H}$ derived for the electrons alone. Figure 7.66c shows the probability that such $(\mu + e)$ events were identified as single electrons, as a function of the distance between the shower and the track. The ξ value needed for identification was chosen such that 90% of the single electrons passed this test. The figure shows that for separations in excess of 10 cm, the probability of accepting a $(\mu + e)$ event as a single electron was less than 10%, even though the ratio of the energies deposited by the two particles was only 1/50. When this energy ratio changed in favor of the "contaminating" particle, it became easier to recognize and reject the overlapping events with this method.

Another frequently used method to distinguish overlapping showers is based on yet another measure of the effective shower width, namely the second moment of the lateral energy distribution, also known as the *dispersion, D*:

$$D = \frac{\sum A_i x_i^2}{\sum A_i} - \left(\frac{\sum A_i x_i}{\sum A_i} \right)^2 = \overline{x^2} - \bar{x}^2 \tag{7.21}$$

where A_i is the pulse height in the ith detector module and x_i the module position. In a more general description, the one-dimensional Equation 7.21 has to be extended to include the y coordinate as well.

The dispersion is, on average, much smaller for em showers than for hadronic ones. Typical values in lead-glass are ~ 3 cm^2 for single em showers and two to three times that value for hadronic ones. The dispersion also depends on the cell size, on the impact point and on the angle of incidence of the particles. The smallest values are found for normal incidence near the center of a detector cell.

Overlapping showers are recognized on the basis of the fact that the dispersion of the energy deposit pattern is larger than that of a typical single shower with the same impact point. Usually, such a dispersion analysis tests the hypothesis that a certain energy deposit profile is caused by overlapping showers. The success of this testing method depends, apart from the factors mentioned above, also on the distance between the particles causing the overlapping showers and on their relative energies. The larger this distance and the smaller the energy difference, the easier it is to recognize two showering particles as a doublet.

Berger and coworkers used a method of this type to study π^0 production in high-energy collisions (200 GeV per nucleon) between ^{32}S ions and gold nuclei [Berg 92]. They used experimental data collected with the SAPHIR calorimeter [Bau 90], a 48×28 array of lead-glass modules (length 46 cm, cross section 3.5×3.5 cm^2 each), for this purpose. Figure 7.67a shows a typical event pattern recorded in this calorimeter for a single reaction of this type.

In Figure 7.67b, invariant-mass spectra are plotted for $\gamma\gamma$ pairs found from the dispersion analysis. In order to make the dispersion parameter less sensitive to geometric

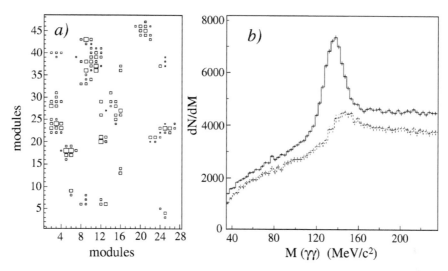

FIG. 7.67. ADC pulse height distribution in the SAPHIR calorimeter for a single 6.4 TeV ^{32}S + Au reaction. Each square is a measure for the logarithm of the signal in an individual lead-glass module (*a*). The invariant-mass spectrum for $\gamma\gamma$ combinations produced in such reactions. The solid histogram required both candidate γs to have D_{xy}^{corr} values < 3 cm^2, the dashed histogram contains pairs where at least one D_{xy}^{corr} value was in the range from 3 to 6 cm^2 (*b*) [Berg 92].

factors (impact point, angle of incidence, cell size), the authors introduced a *corrected dispersion*, D_x^{corr}, defined as $D_x^{\text{corr}} = D_x - D_x^{\text{min}}$, where the minimum possible value D_x^{min} depended on the mentioned factors. The same was done for the y coordinate. The larger of the two values D_x^{corr} and D_y^{corr} was used as the two-dimensional corrected dispersion D_{xy}^{corr}.

Single γs were defined as em showers with $D_{xy}^{\text{corr}} < D^{\text{lim}}$. The smaller the value of D^{lim} was chosen, the more likely it became that the energy cluster was indeed caused by a single photon. The solid histogram in Figure 7.67b was obtained for photon pairs in which clusters met the criterion $D_{xy}^{\text{corr}} < 3$ cm^2. This spectrum shows a very clear peak around the π^0 mass (~ 135 MeV/c^2). When the requirements on D_{xy}^{corr} were relaxed, *i.e.*, when the value of D^{lim} was increased, this π^0 peak became rapidly less significant, as illustrated by the dashed histogram in Figure 7.67b. In that case, at least one of the two γs had a D_{xy}^{corr} value in the range from 3 to 6 cm^2.

In the LHC experiments ATLAS and CMS, recognizing the presence of (soft) electrons inside jets is an important tool in the context of their physics studies involving b quarks, such as the production of Higgs bosons decaying into $b\bar{b}$. Even though at most 21% of b-jets contain a soft lepton of a given flavor, tagging such events may lead to samples of relatively high purity. More importantly, these event samples are only very weakly correlated to those based on other b-tagging algorithms (which concentrate, for example, on the occurrence of a displaced secondary vertex). This is important

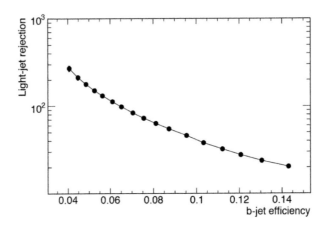

FIG. 7.68. Rejection of light jets versus b-tagging efficiency in $pp \rightarrow WH$ events in ATLAS, using the soft-electron b-tagging algorithm [Aad 08].

for checking and cross-calibration purposes.

Reconstructing soft electrons inside jets in the em calorimeter is difficult because of the overlap with showers from other jet fragments. The techniques employed for this purpose are all based on matching a track in the inner detector to a cluster in the em calorimeter. The success rate of this procedure is highly dependent on the track multiplicity in the jet, and also on the amount of material upstream of the calorimeter (photon conversions).

Figure 7.68 shows the rejection rate of events in ATLAS in which Higgs bosons produced in association with a W boson in pp collisions decay into a light $q\bar{q}$ pair, versus the efficiency of the soft-electron b-tagging algorithm. For example, a light-jet rejection of a factor of 90 can be achieved in combination with a b-tagging efficiency of 7%. It turns out that a substantial fraction of the surviving light-jet events is tagged by electrons from γ conversions, illustrating the limitations on the detector performance resulting from the upstream material in ATLAS (see also Figure 7.60).

7.8   (Multi-)Jet Spectroscopy

As the energies at which the properties of matter were probed increased, the experimental emphasis has gradually shifted from measurements aimed at reconstructing the four-vectors of all individual particles produced in the interactions to more global event characteristics, indicative of interesting processes at the constituent level. Prominent among those characteristics is the production of jets, the bunches of particles that are produced in the fragmentation process of a quark, diquark or gluon.

In the 1960s, many new particles and resonances with masses in the few-GeV range were discovered through a reconstruction of their decay products. As the energy increased, more and more massive new objects became accessible, culminating in the discovery of the W and Z intermediate vector bosons (1982) and of the top quark (1995).

The phase space available for the decay products of these heavy objects is so large that the hadronic decay products frequently manifest themselves in the form of collimated jets of particles. Figure 7.69 shows an example of a Z^0, produced at LEP and detected by the ALEPH Collaboration. The Z^0 displayed in this event decayed hadronically into a $q\bar{q}$ pair. The quark and the anti-quark, produced back-to-back, each with an energy of 45 GeV, fragmented into two bunches of particles that can be clearly recognized.

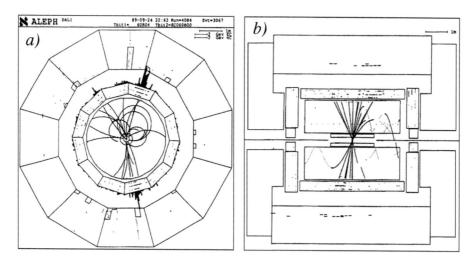

FIG. 7.69. Event display of a hadronically decaying Z^0 boson produced at LEP and detected by the ALEPH Collaboration. Diagram $a)$ shows a projection of the event on a (x, y) plane perpendicular to the beam line. A projection on a (r, z) plane containing the beam line is shown in diagram $b)$ [Dec 90].

When the first jets were discovered, at the e^+e^- collider PETRA in the late 1970s [Sod 81], one needed imagination, and mathematical savvy, to demonstrate the existence of preferential directions among the particles produced in the interactions, and new terms were invented to quantify the "thrustiness" of the reaction products.

However, as the energy increased, jets became more and more collimated, as illustrated by the event in Figure 7.69. At the LHC, where many intermediate vector bosons are produced with large longitudinal momenta, the Lorentz-boosted fragmenting quarks from their decay appear as highly collimated bunches of photons and pions heading toward the forward calorimeters.

One of the attractive features of calorimeters is that the resolution (in energy, position, and time) improves as the energy increases. This feature, combined with the increasingly collimated nature of high-energy jets and the increased importance of jets as decay products of heavy particles, has created a new task for calorimeter systems: *multi-jet spectroscopy*.

The first example of use of a calorimeter as a jet spectrometer was given by the

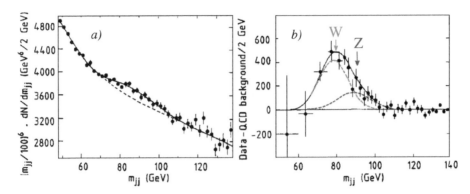

FIG. 7.70. Two-jet invariant mass distributions from the UA2 experiment [Alit 91]. Diagram
  a) shows the measured data points, together with the results of the best fits to the QCD
  background alone (*dashed curve*), or including the sum of two Gaussian functions describing
  $W, Z \rightarrow q\bar{q}$ decays. Diagram b) shows the same data after subtracting the QCD background.
  The data are compatible with peaks at $m_W = 80$ GeV and $m_Z = 90$ GeV. The measured width
  of the bump, or rather the standard deviation of the mass distribution, was 8 GeV, of which 5
  GeV could be attributed to non-ideal calorimeter performance [Jen 88].

UA2 Collaboration, who capitalized on the fine granularity and good energy resolution
of their Pb/Cu/plastic-scintillator calorimeter system to study the hadronic decay of the
intermediate vector bosons W and Z through the jet–jet decay mode (Figure 7.70).

The figure shows a clear and significant bump above the QCD background in the jet–
jet invariant mass distribution, in the region from 70 to 100 GeV. When this background
is subtracted from the experimental data (Figure 7.70b), one might even see evidence
for a double-peak structure, corresponding to hadronic W and Z decay.

Another example that illustrates the advantages of high-resolution calorimetry for
spectroscopic measurements at high energy was given by the heavy-ion experiments
WA80 and HELIOS at CERN. Figure 7.71 shows the total energy distribution measured
by the WA80 Collaboration when the 6.4 TeV ^{32}S beam from the CERN SPS was
dumped onto their uranium/plastic-scintillator calorimeter [You 89].

It turned out that ^{32}S nuclei were not the only particles present in this beam. A few
percent of the nuclei apparently dissociated under way and those with the proper $A/Z =$
2 ratio made it through the acceleration process. The good energy resolution of the
calorimeter allowed a detailed study of these contaminating lower-mass nuclei. Similar
results were reported by HELIOS [Ake 87]. As an aside, we mention that the mass
resolution obtained in these measurements would be adequate to separate hadronically
decaying W and Z bosons.

For the reasons mentioned above, this type of calorimeter application is likely to
gain in importance as the collision energies further increase. At the LHC and future
colliders, multi-jet spectroscopy might turn out to be a major and powerful technique
for many physics analyses.

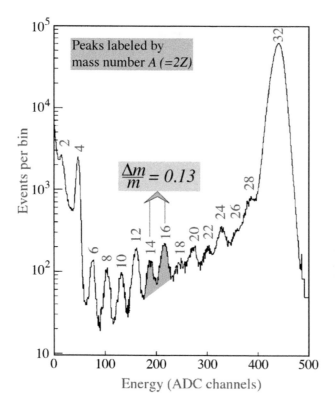

FIG. 7.71. The WA80 calorimeter as a high-resolution spectrometer. Total energy measured with the calorimeter for minimum-bias events revealed the composition of the momentum-selected CERN heavy-ion beam [You 89].

7.9 Calorimeter Tasks in Experiments

The calorimeter tasks in modern particle physics experiments can be subdivided as follows:

- Event selection (triggering)
- Lepton identification
- Energy measurement (jets, leptons, energy flow)
- Hadron absorption
- Vertex reconstruction
- Mitigation of pile-up effects

Event selection is a prime task in all experiments, but especially in those in which the event rate is orders of magnitude beyond the rate at which events may be recorded for offline analysis. The latter is typically limited to 10–1,000 Hz, depending on the complexity of the experimental setup.

Experiments in which processes with a very small cross section are studied, such as ν scattering or high-energy e^+e^- collisions (LEP, SLC), mainly use the calorimeter signals to distinguish the events of interest from background (cosmic rays, beam–gas scattering, *etc.*), for example through the total deposited energy.

On the other hand, experiments studying strong-interaction processes often have to deal with primary event rates in the MHz–GHz range. In such experiments, the trigger not only has to determine that an interaction occurred, but also whether it was a potentially interesting event. In Section 5.5, several examples of trigger schemes that are being used in practice are discussed.

Some of the methods used for the identification of leptons were discussed in Section 7.6. Usually, calorimeter information is combined with data from other detectors, for example a downstream muon spectrometer, or an upstream tracking system or preshower detector. The latter systems are also essential for recognizing and eliminating events in which an electron is accompanied by a nearby γ or π^0 (Section 7.7).

Accurate measurement of the electron and photon energy is emphasized in almost all experiments. This goes, inevitably, at the expense of the energy resolution for hadrons and jets, since the requirements for excellent em resolution and compensation are orthogonal. Excellent em resolution is easiest to achieve with detectors that have a large sampling fraction (*e.g.*, 100%, as in crystal calorimeters), while compensation can only be achieved in calorimeters that have a small sampling fraction (Section 3.4).

In many experiments, this emphasis on electromagnetic performance is fully justified. For example, in most e^+e^- colliders, the energies are low and so are the event rates. Each of the charged hadrons can be accurately measured with a tracking system, and the production of K_L^0 mesons and neutrons, for which excellent hadron calorimetry could possibly make a difference, is relatively unimportant.

At high-energy hadron colliders, and in particular also in the *ep* experiments at DESY, good hadron calorimetry is given higher priority. In these experiments, the hadronic energy resolution is often directly determining or limiting the measurement precision of important physical quantities. For example, the experimental uncertainty in the top mass is still dominated by the (imprecision of the) measured energy of its hadronic decay products [Aal 12b, Aba 14, Aab 16, Kha 16]. At HERA, the hadronic energy resolution was the main factor limiting the precision of the measurement of the structure function at high values of Q^2. And the experimental possibilities to discover Supersymmetry, should it exist, are largely determined and limited by the quality of the available calorimetry.

Hadron absorbers have long been used for studying muon physics. The use of instrumented absorbers that allow π/μ discrimination, and recognition of muons generated in hadronic showers (Section 7.6.2) was pioneered in neutrino scattering experiments, but offers also great possibilities in 4π experiments. As the energy increases, energy losses by muons on their way through the hadron absorber become rapidly more important (Section 2.2). Also in this respect, an instrumented hadron absorber may serve an important purpose, since it makes it possible to measure these losses, event by event. The dual-readout calorimeters discussed in Chapter 8 make it even possible to separate the ionization and radiative energy losses by muons on their way through the absorber

(Figure 3.48).

The need for vertex reconstruction on the basis of calorimeter information is specific for experiments in which the calorimeter not only serves as detector, but also as target (*e.g.*, in ν scattering experiments) and/or as source (nucleon decay experiments). Therefore, such experiments usually emphasize angular resolution in their design.

Other examples of physics phenomena for which vertex reconstruction is crucial include the study of atmospheric neutrinos (Chapter 10), since it determines the detector's capability to distinguish between particles that came from the atmosphere above the detector, or from below (traversing the entire Earth). Also the study of the process $H^0 \to \gamma\gamma$ at LHC requires reconstruction of the vertex position, which is located somewhere in a 30 cm long region along the beam line, together with several dozen other vertices of *pp* interactions occurring in the same bunch crossing. A precise measurement of this vertex position is as essential for detecting the Higgs boson as the energy resolution of the calorimeter.

The LHC experiments have to deal with significant pile-up effects. In Run I, on average, 15 events occurred every bunch crossing. In the high-luminosity era, that rate is expected to increase to more than 100 events per bunch crossing. Excellent timing might make it possible to locate the vertex of an interesting hard collision, both in space (along the beam line) and in time, and select the detector information associated with that particular event, while (to some extent) excluding the unrelated information generated in the same bunch crossing. The difficulty scale of this problem is set by the rms time of 170 ps for *in-time pile-up* events, for 25-ns bunch crossings. This means that time resolutions of the order of 10 ps will be needed to achieve substantial mitigation of the problems caused by this phenomenon. Nevertheless, calorimeters based on instantaneously produced Čerenkov light, readout by ultrafast light detectors may provide the key to the solution [Akc 16].

8

NEW CALORIMETER TECHNIQUES

Progress in our understanding of the structure of matter and its fundamental properties has always been driven by the availability of new, more powerful particle accelerators. In the past fifteen years, discussions about accelerator projects in the post-LHC era have mainly concentrated on a high-energy electron–positron collider, with a center-of-mass energy that would allow this machine to become a factory for $t\bar{t}$ and Higgs boson production. Both linear colliders (ILC, CLIC) and circular ones (FCCee, CEPC) are being considered in this context. A sufficiently large circular collider could subsequently be used to further push the energy frontier for hadron collisions beyond the LHC limits.

In order to take full advantage of the experimental opportunities created by such colliders, adequate particle detectors will be needed, since the quality of the scientific information that can be obtained will, to a very large extent, be determined (and limited) by the quality of the detectors with which experiments at these machines will be performed. In these experiments, that quality primarily concerns the precision with which the four-vectors of the scattered objects produced in the collisions can be measured. At the TeV scale, these objects are leptons, photons and fragmenting quarks, di-quarks and gluons. The fragmenting objects are commonly referred to as jets. Achieving the best possible precision for the momentum and energy measurements of these objects is usually a very (if not the most) important design goal of the proposed experiments.

These considerations have determined the directions in which calorimeter R&D has evolved in the past fifteen years. In this chapter, we discuss the prospects for calorimetry at colliders that will open up the TeV domain for detailed studies. The discussion focuses on hadron calorimetry, where the challenges will be greatest, although the detection of electromagnetic showers also faces problems, especially in finely segmented instruments.

8.1 Calorimetry in the TeV regime

An often mentioned design criterion for calorimeters at a future high-energy linear e^+e^- collider is the need to distinguish between hadronically decaying W and Z bosons[22]. The requirement that the di-jet masses of $W \rightarrow q\bar{q}$ and $Z \rightarrow q\bar{q}$ events are separable by at least one Rayleigh criterion implies that 80–90 GeV jets should be detected with a resolution of 3–3.5 GeV . This goal can be, and has been achieved with compensating calorimeters for single hadrons [Beh 90, Aco 91c], but not for jets.

---

[22] An important reaction to be studied is $e^+e^- \rightarrow H^0 Z^0$. By using the hadronic decay modes of the Z^0 (in addition to e^+e^- and $\mu^+\mu^-$ decay), an important gain in event rates can be obtained. However, more abundant processes such as $e^+e^- \rightarrow W^+W^-$ will obscure the signal unless the calorimeter is able to distinguish efficiently between hadronic decays of W and Z bosons.

However, because of the small sampling fraction required for compensation, the em energy resolution is somewhat limited in such devices. And because of the crucial role of neutrons produced in the shower development, the signals would have to be integrated over relatively large volumes and time intervals to achieve this resolution, which is not always possible in practice. In this chapter, we discuss two methods that are currently being pursued to circumvent these limitations. However, first, we briefly review the factors that determine and limit the hadronic calorimeter resolution.

8.1.1 *Hadronic energy resolution*

In the TeV domain, it is incorrect to express calorimetric energy resolutions in terms of $a/\sqrt{E}$, as is often done. Deviations from $E^{-1/2}$ scaling are the result of non-Poissonian fluctuations. These manifest themselves typically predominantly at high energies, where the contribution of the Poissonian component becomes very small. It is often assumed that the effect of non-compensation on the energy resolution is energy independent ("constant term"). Figure 4.55 illustrates that this is incorrect. It has been demonstrated (Section 4.7) that the effects of fluctuations in f_{em} are more correctly described by a term that is very similar to the one used to describe the energy dependence of $\langle f_{em} \rangle$ (see Figure 2.25). This term should be added in quadrature to the $E^{-1/2}$ scaling term which accounts for the Poissonian fluctuations:

$$\frac{\sigma}{E} = \frac{a_1}{\sqrt{E}} \oplus a_2 \left[\left(\frac{E}{E_0} \right)^{l-1} \right] \qquad (8.1)$$

Here, E_0 is a material dependent constant related to the average multiplicity in the hadronic interactions and l (which has a value of 0.72 in copper) is determined by the energy dependence of $\langle f_{em} \rangle$ [Gab 94]. The parameter a_2 is determined by the degree of non-compensation. It varies between 0 (for compensating calorimeters) and 1 (for extremely non-compensating calorimeters). Following Groom [Gro 07], we assume a linear relationship for intermediate e/h values:

$$a_2 = |1 - h/e| \qquad (8.2)$$

In Chapter 4, we showed that several sets of experimental hadronic energy resolution data, *e.g.*, [Ari 94], are well described by a linear sum of a stochastic and constant term

$$\frac{\sigma}{E} = \frac{c_1}{\sqrt{E}} + c_2 \qquad (8.3)$$

which is, in the energy range covered by the current generation of test beams, *i.e.*, up to 400 GeV, indistinguishable from Equation 8.1, albeit that the stochastic parameters differ ($c_1 > a_1$).

Interestingly, the similarity between the two expressions disappears when the energy range is extended into the TeV domain. This is illustrated in Figure 8.1. The curves in these graphs represent Equation 8.1 for the energy range from 0.2 - 10 TeV. Figure 8.1a shows the contributions of the stochastic and the non-compensation term as a function of energy, as well as the total energy resolution, for a calorimeter with $e/h = 1.3$

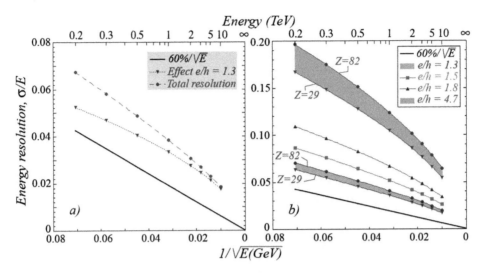

FIG. 8.1. Hadronic energy resolution in the TeV domain, calculated with Equation 8.1.

and a stochastic term of $60\%/\sqrt{E}$. It is clear that, even for an e/h value that is usually considered quite good, the effects of fluctuations in f_{em} dominate the hadronic energy resolution in the TeV regime. Figure 8.1b shows the total energy resolution for calorimeters with different e/h values. Especially for large e/h values, the energy dependence of the resolution is no longer well described by a straight line in this plot (thus invalidating Equation 8.3). Figure 8.1b also shows the effects that may be expected as a result of material dependence. These derive from the value of E_0 in Equation 8.1, which is almost a factor of two larger in high-Z absorber materials such as lead, compared to copper or iron.

8.2 Dual-Readout Calorimetry

As stated above, the main drawbacks of compensating calorimeters derive from the need for a high-Z absorber material, such as lead or uranium. This absorber material both reduces the em response and generates a large number of neutrons, the two ingredients that are crucial for achieving the compensation condition, $e/h = 1.0$, and thus for eliminating the contribution of fluctuations in the em shower fraction, f_{em}. However, the small e/mip value, typically ~ 0.6 in these absorber materials, leads to large response non-linearities for low-energy hadrons, which lose their kinetic energy predominantly through ionization of the absorber medium, rather than through shower development (Figure 8.2). Such particles account for a significant fraction of the energy of high-energy jets, such as the ones produced in the hadronic decay of the W and Z intermediate vector bosons. Figure 8.3 shows the distribution of the energy released by Z^0s (decaying through the process $Z^0 \rightarrow u\bar{u}$) and Higgs bosons (decaying into a pair of gluons) at rest, that is carried by charged final-state particles with a momentum less than 5 GeV/c. These distributions were provided by Prof. Bryan Webber, a uni-

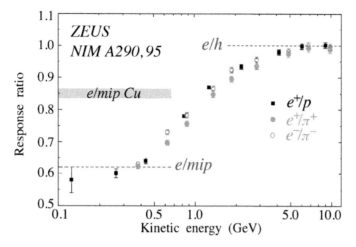

FIG. 8.2. The response of the uranium based ZEUS calorimeter to low-energy hadrons. Experimental data points from [And 90]. For comparison, the e/mip ratio for a copper-based calorimeter is shown as well.

versally recognized expert in this area[23]. The figure shows that, on average, 21% of the energy equivalence of the Z^0 mass is carried by such particles, and the rms event-to-event fluctuations are such that this fraction varies between 13% and 35%. For Higgs bosons decaying into a pair of gluons, the average fraction is even larger, 34%, with rms variations between 23% and 45%. Because of the absence of a leading-particle effect, a gluon jet has a higher average multiplicity than a jet resulting from a fragmenting quark or anti-quark.

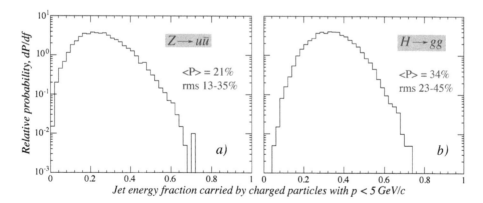

FIG. 8.3. Distribution of the fraction of the energy released by hadronically decaying Z^0 (a) and H^0 (b) bosons at rest that is carried by charged final-state particles with a momentum less than 5 GeV/c [Web 15].

[23] I thank Dr. Webber for carrying out these calculations at my request.

As a result of the important contribution from soft jet fragments, and the large event-by-event fluctuations in this contribution, the energy resolution for intermediate vector bosons measured with the compensating ZEUS uranium calorimeter was worse than expected on the basis of the single-pion resolution. Also, the small sampling fraction required to achieve compensation limited the em energy resolution, to $18\%/\sqrt{E}$ in ZEUS. And the (properly amplified) contributions of neutrons, which are equally essential for this purpose, made it necessary to integrate the hadronic signals over a rather large time (~ 50 ns) and calorimeter volume (~ 1 m^3).

An alternative approach to eliminate the effects of the fluctuations in the em shower fraction, which dominate the hadronic energy resolution of non-compensating calorimeters, is to *measure* f_{em} for each event. It turns out that the Čerenkov mechanism provides unique opportunities to achieve this.

Calorimeters that use Čerenkov light as signal source are, for all practical purposes, only responding to the em fraction of hadronic showers (Section 3.5, [Akc 97]). This is because the electrons/positrons through which the energy is deposited in the em shower component are relativistic down to energies of only 200 keV. On the other hand, most of the non-em energy in hadron showers is deposited by non-relativistic protons generated in nuclear reactions. Such protons do generate signals in active media such as plastic scintillator or liquid argon. By comparing the relative strengths of the signals representing the visible deposited energy (dE/dx) and the Čerenkov light produced in the shower absorption process, the em shower fraction can be determined and the total shower energy can be reconstructed using the known e/h value(s) of the calorimeter. This is the essence of what has become known as *dual-readout* calorimetry.

The Dual-REAdout Method (DREAM) allows the elimination of the mentioned drawbacks of intrinsically compensating calorimeters:

1. There is no reason to use high-Z absorber material. An absorber such as copper has an e/mip value of 0.85, which strongly mitigates the effects of non-showering hadrons on the jet energy resolution. Also, by using copper instead of lead or uranium, a calorimeter with a given depth (expressed in nuclear interaction lengths) will need to be much less massive.

2. The sampling fraction of detectors based on this method can be chosen as desired. As a result, excellent em energy resolution is by no means precluded.

3. The method does not rely on detecting neutrons (although these may offer some additional advantages, as shown in the following). Therefore, there is no need to integrate the signals over large times and detector volumes.

8.2.1   *Initial attempts: ACCESS*

The idea to use the complementary information from scintillation and Čerenkov light was first applied in a prototype study for ACCESS, a high-energy cosmic-ray experiment proposed for the International Space Station [Nag 01][24]. Because of the very severe restrictions on the mass of the instruments, the ACCESS calorimeter had to be very

---

[24]The ACCESS project was canceled after the 2003 accident with the Columbia space shuttle.

thin, less than 2 λ_{int}. It was therefore imperative to maximize the amount of information obtained per unit detector mass.

When high-energy hadrons develop showers in such a thin calorimeter, the response function is completely determined by leakage fluctuations. These fluctuations are very likely correlated with the fraction of energy spent on π^0 production inside the detector. In general, π^0s produced in the first nuclear interaction develop em showers that are contained in the detector, while charged pions typically escape. Therefore, events in which a large fraction of the initial energy is converted into π^0s in the first interaction will exhibit little leakage (a large detector signal), while events in which a small fraction of the energy has been transferred to π^0s will be characterized by large leakage (small detector signals). A dual-readout calorimeter that measures both the ionization losses (dE/dx) and the production of Čerenkov light might distinguish between events with relatively small and large shower leakage, since the ratio of the two signals would be different in these two cases: a relatively large Čerenkov signal would indicate relatively little shower leakage, while a small Čerenkov signal (compared to the dE/dx signal) would suggest that a large fraction of the shower energy escaped from the detector.

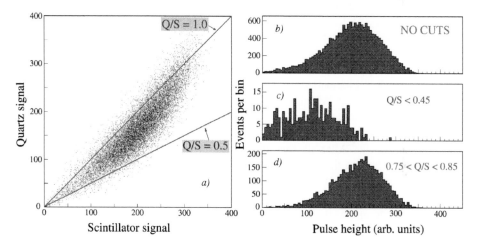

FIG. 8.4. Results of tests of the dual-readout ACCESS calorimeter with 375 GeV pions. Scatter plot of the signals recorded in the quartz fibers *vs.* those in the scintillating fibers (*a*). The signal distributions from the scintillating fibers for all events (*b*) and for subsets of events with a small (*c*) or average (*d*) fraction of Čerenkov light [Nag 01].

The dual-readout calorimeter prototype built for ACCESS consisted of a 1.4 λ_{int} deep lead absorber structure, in which alternating ribbons of two types of optical fibers were embedded. The signals from the scintillating fibers provided a measure for the total energy deposited by the showers, while quartz fibers recorded the Čerenkov light produced in the absorption process. Figure 8.4 shows some results of the tests of this instrument. These tests were carried out at CERN with a beam of 375 GeV pions. In Figure 8.4a, the signals recorded by the quartz fibers are plotted versus those from the scintillating fibers. The non-linear correlation between these signals indicates that they indeed measure different characteristics of the showers.

The scintillator signal distribution, *i.e.*, the projection of the scatter plot on the horizontal axis, is shown in Figure 8.4b. This distribution is skewed to the low-energy side, which may be expected as a result of shower leakage. The *ratio of the signals from the quartz fibers and from the scintillating fibers (Q/S)* corresponds to the slope of a line through the bottom left corner of Figure 8.4a. The two lines drawn in this figure represent $Q/S = 1$ and $Q/S = 0.5$, respectively.

In Figure 8.4c, the signal distribution is given for events with a small Q/S value $(Q/S < 0.45)$. These events indeed populate the left-side tail of the calorimeter's response function (Figure 8.4b). This distribution is very different from the one obtained for events with Q/S ratios near the most probable value, shown in Figure 8.4d. The average values of the scintillator signal distributions in Figures 8.4c and 8.4d differ by about a factor of two.

These results demonstrate that events from the tails of the Q/S distribution correspond to events from the tails of the (dE/dx) response function. Therefore, the ratio of the signals from the quartz and the scintillating fibers does indeed provide information on the energy containment and may thus be used to reduce the fluctuations that dominate the response function of this very thin calorimeter. The authors showed that the resolution could be improved by 10–15% using the Q/S information and that this improvement was primarily limited by the small light yield of the quartz fibers, 0.5 photoelectrons per GeV. Fluctuations in the number of Čerenkov photoelectrons determined the width of the "banana" in Figure 8.4a and thus the selectivity of Q/S cuts. Therefore, the relative improvement in the energy resolution also increased with the hadron energy.

It is remarkable that the dual-readout technique already worked so well in this very thin calorimeter. After all, this instrument only detects the very first generation of shower particles and the non-em shower component has barely had a chance to develop. The overwhelming majority of the non-relativistic shower particles, in particular the spallation and recoil protons, are produced in later stages of the hadronic shower development. The signals from these non-relativistic shower particles are crucial for the success of the method, since they are the ones that do produce scintillation light and no Čerenkov light. The fact that the technique already appeared to work so well in this very thin calorimeter therefore held the promise that excellent results might be expected for detectors that fully contain the showers.

8.2.2   *The DREAM project*

Inspired by the results obtained with the ACCESS calorimeter, the authors embarked on a follow-up project intended to contain hadron showers in a much more complete way. The instrument they built became known as the DREAM calorimeter. As before, the two active media were scintillating fibers which measured the visible energy, while clear, undoped fibers measured the generated Čerenkov light. Copper was chosen as the absorber material. The basic element of this detector (see Figure 8.5) was an extruded copper rod, 2 m long and 4×4 mm^2 in cross section. This rod was hollow, the central cylinder had a diameter of 2.5 mm. In this hole were inserted seven optical fibers. Three

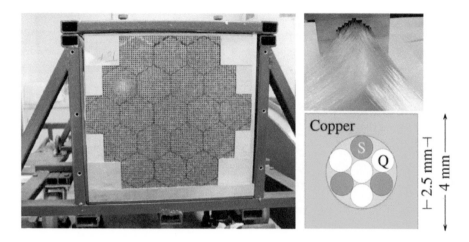

FIG. 8.5. The DREAM detector. The basic building block is an extruded hollow copper rod. Seven optical fibers (four Čerenkov and three scintillating fibers) are inserted in the central hole. The two types of fibers are split into separate bunches as they exit the downstream detector end. The hexagonal readout structure is indicated. The Čerenkov fibers of the central tower and the six towers of the Inner Ring were filled with quartz fibers, in the twelve towers of the Outer Ring clear PMMA fibers were used for this purpose.

of these were plastic scintillating fibers, the other four fibers were undoped. All fibers had an outer diameter of 0.8 mm and a length of 2.50 m. The fiber pattern was the same for all rods, and is shown in Figure 8.5.

The DREAM detector consisted of 5580 such rods, 5130 of these were equipped with fibers. The empty rods were used as fillers, on the periphery of the detector. The instrumented volume thus had a length of 2.0 m, an effective radius of $\sqrt{5130 \times 0.16/\pi} = 16.2$ cm, and a mass of 1030 kg. The effective radiation length (X_0) of the calorimeter was 20.1 mm, the Molière radius (ρ_M) was 20.4 mm and the nuclear interaction length (λ_{int}) 200 mm. The composition of the instrumented part of the calorimeter was as follows: 69.3% of the detector volume consisted of copper absorber, while the scintillating and Čerenkov fibers occupied 9.4% and 12.6%, respectively. Air accounted for the remaining 8.7%. Given the specific energy loss of a minimum-ionizing particle in copper (12.6 MeV/cm) and polystyrene (2.00 MeV/cm), the sampling fraction of the copper/scintillating-fiber structure for mips was thus 2.1%.

The fibers were grouped to form 19 towers. Each tower consisted of 270 rods and had an approximately hexagonal shape (80 mm apex to apex). The layout is schematically shown in Figure 8.5: a central tower, surrounded by two hexagonal rings, the Inner Ring (six towers) and the Outer Ring (twelve towers). The towers were longitudinally unsegmented.

The depth of the copper structure was 200 cm, or 10.0 λ_{int}. The fibers leaving the rear end of this structure were separated into bunches: one bunch of scintillating fibers and one bunch of Čerenkov fibers for each tower, 38 bunches in total. In this way, the

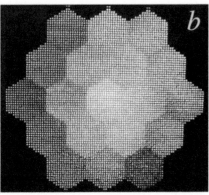

FIG. 8.6. The DREAM calorimeter. Shown are the fiber bunches exiting from the rear face of the detector (a) and a picture taken from the front face while the rear end was illuminated (b).

readout structure was established (see Figure 8.5). Each bunch was coupled through a 2 mm air gap to a photomultiplier tube.

Figure 8.6 shows photographs of the assembled detector. In Figure 8.6a, the fiber bunches exiting the downstream end of the calorimeter and the 38 ferrules that hold and position the fibers for the PMTs that detect their signals are shown. In total, this detector contained about 90 km of optical fibers. Figure 8.6b shows the front face of the calorimeter, when the fibers were illuminated with a bright lamp located behind the detector. The hexagonal readout structure is clearly visible.

This calorimeter thus produced two types of signals for the showers developing in it, a scintillation signal (S) and a Čerenkov signal (C). Both signals can be calibrated with electrons of known energy E, so that $\langle S \rangle = \langle C \rangle = E$ for em showers, and the calorimeter response to em showers, $R_{em} = \langle S \rangle / E = \langle C \rangle / E = 1$. For a given event, the hadronic response of this calorimeter can then be written as

$$R_S = f_{em} + \frac{1}{(e/h)_S}(1 - f_{em})$$

$$R_C = f_{em} + \frac{1}{(e/h)_C}(1 - f_{em}) \tag{8.4}$$

i.e., as the sum of an em shower component (f_{em}) and a non-em shower component $(1 - f_{em})$. The contribution of the latter component to the reconstructed energy is weighted by a factor h/e. When $f_{em} = 1$ or $e/h = 1$, the hadronic shower response is thus the same as for electrons: $R = 1$. However, in general $f_{em} < 1$ and $e/h \neq 1$, and therefore the hadronic response is different from 1. The reconstructed energy will thus be different (typically smaller) than E:

$$S = E\left[f_{em} + \frac{1}{(e/h)_S}(1 - f_{em}) \right]$$

$$C = E\left[f_{em} + \frac{1}{(e/h)_C}(1 - f_{em})\right] \qquad (8.5)$$

The dual-readout method works because $(e/h)_S \neq (e/h)_C$. The larger the difference between both values, the better. The em shower fraction f_{em} and the shower energy E can be found by solving Equations 8.5, using the measured values of the scintillation and Čerenkov signals and the **known** e/h ratios of the Čerenkov and scintillator calorimeter structures. We will describe later in this subsection *how* the latter ratios can be determined.

A crucial implication of Equations 8.5 is that the ratio of the two measured signals S and C is *independent of the shower energy E*. There is thus a one-to-one correspondence between this measured signal ratio and the value of the em shower fraction, f_{em}. This fraction can thus be determined event-by-event, since

$$\frac{C}{S} = \frac{f_{em} + 0.21\,(1 - f_{em})}{f_{em} + 0.77\,(1 - f_{em})} \qquad (8.6)$$

where 0.21 and 0.77 represent the h/e ratios of the Čerenkov and scintillation calorimeter structures, respectively.

Figure 8.7 shows the signal distributions for 100 GeV π^- detected with this calorimeter. The energy scale was determined with electrons, and the average hadronic response was thus 0.8166 for the scintillating fiber structure and 0.6404 for the Čerenkov

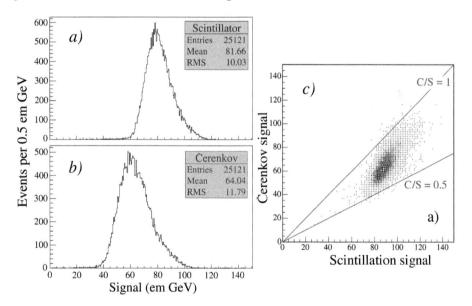

FIG. 8.7. Signal distributions for 100 GeV π^- recorded by the scintillating (*a*) and Čerenkov (*b*) fibers of the DREAM calorimeter, and a scatter plot showing the correlation between both types of signals (*c*). The signals are expressed in the same units as those for em showers, which were used to calibrate the calorimeter (em GeV). Data from [Akc 05b].

one. The response functions exbibit the asymmetric shape that is characteristic for hadrons in a non-compensating calorimeter. The correlation between both types of signals is shown in Figure 8.7c. This scatter plot may be compared with the one obtained for the ACCESS calorimeter (Figure 8.4a). The events are now concentrated in a smaller area of the scatter plot, as a result of the better shower containment. However, the fact that the events are, as before, not concentrated along the diagonal, illustrates the complementary information provided by both signals [Akc 05b].

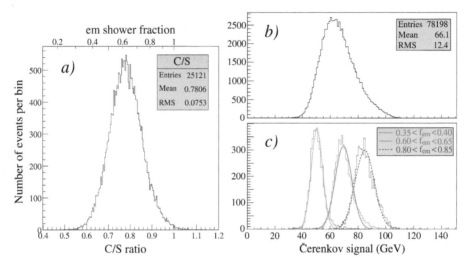

FIG. 8.8. The relationship between the ratio of the Čerenkov and scitillation signals and the electromagnetic shower fraction, derived for the 100 GeV π^- events on the basis of Equation 8.6 (a). The total Čerenkov signal distribution for these events (b) and distributions for subsamples of events selected on the basis of the measured f_{em} value (c). Data from [Akc 05b].

The merits of the dual-readout method are clearly illustrated by Figure 8.8 [Akc 05b]. The distribution of the event-by-event signal ratio, and thus of the em shower fraction derived on the basis of Equation 8.6, is shown in Figure 8.8a. The value of f_{em} (top scale) varies from 0.3 to 1, with a maximum around 0.6. The f_{em} value, which can thus directly be derived from the Čerenkov/signal ratio for each individual event, can be used to dissect the overall signal distributions, as illustrated in Figure 8.8b/c, which show the overall Čerenkov signal distribution for the 100 GeV π^- events, as well as distributions for three subsamples selected on the basis of their f_{em} value. Each f_{em} bin probes a certain region of the overall signal distribution, and the average value of the subsample distribution increases with f_{em}. The overall signal distribution is thus a superposition of many such (Gaussian) subsample signal distributions, and the shape of the overall signal distribution reflects the (asymmetric) distribution of the f_{em} values.

Instead of three f_{em} bins, one could also use a much larger number, and plot the average calorimeter signal as a function of f_{em}. The results are shown in Figure 8.9 for 200 GeV "jets," separately for the Čerenkov (Figure 8.9a) and scintillation (Figure

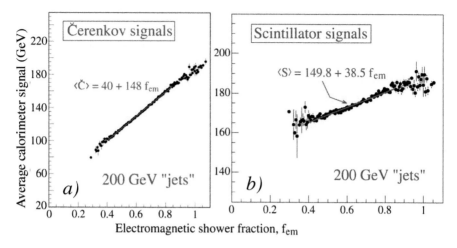

FIG. 8.9. The average Čerenkov (a) and scintillation (b) signals for 200 GeV "jets," as a function of the em shower fraction, f_{em} [Akc 05b].

8.9b) signals[25]. The figure shows linear relationships between these signals and the em shower fraction, as is assumed in Equations 8.5. These relationships makes it possible to determine the e/h values of the calorimeter for the two types of signals. According to Equation 8.4, the response should vary between $R = h/e$ for $f_{em} = 0$ and $R = 1$ for $f_{em} = 1$. The value $R = 1$ is obtained if we assume that the detected energy was 188 GeV instead of 200, which is reasonable, since some fraction of the particles produced in the upstream pion interactions have not, or only partially been detected by the calorimeter. Under that assumption, the fits from Figure 8.9 lead to $h/e = 40/188$ for the Čerenkov calorimeter and $h/e = 149.8/188$ for the scintillation calorimeter. Had we instead assumed that the entire 200 GeV was deposited in the calorimeter, e/h values of 200/40 = 5.0 and 200/149.8 = 1.34 would have been found for the Čerenkov and scintillation calorimeter structures, respectively. These values change to 188/40 = 4.7 and 188/149.8 = 1.26, respectively, under the stated leakage assumption.

These results may serve to provide a feeling for the experimental uncertainties in the em shower fraction (Equation 8.6), as well as the energy of the showering hadrons. The latter can be found by solving the two Equations 8.5 for the parameter E (instead of f_{em}):

$$E = \frac{S - \chi C}{1 - \chi}, \quad \text{with } \chi = \frac{1 - (h/e)_S}{1 - (h/e)_C} \sim 0.3 \text{ for this calorimeter} \qquad (8.7)$$

This expression essentially determines the shower energy by calculating what the calorimeter response would have been for $f_{em} = 1$, based on the actually measured f_{em} value.

---

[25]These "jets" were in fact not fragmenting quarks or gluons, but high-multiplicity multi-particle events created by pions interacting in a target placed upstream of the calorimeter.

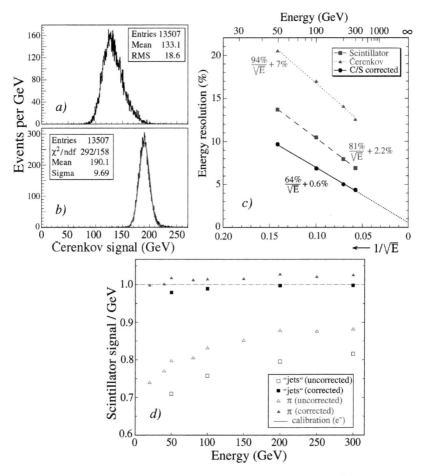

FIG. 8.10. Effects of the corrections applied on the basis of the observed Čerenkov/scintillator signal ratio. Čerenkov signal distributions for high-multiplicity 200 GeV "jets" in the DREAM before (*a*) and after (*b*) the corrections. The energy resolution for multi-particle "jets," measured separately with the scintillation and Čerenkov signals, and after applying the mentioned corrections (*c*). The calorimeter response before and after the corrections, both for single pions and for multi-particle "jets." The electron response, which was the basis for calibrating the calorimeter signals, is shown as well (*d*). Data from [Akc 05b].

A comparison of the scintillation and Čerenkov signals thus made it possible to correct the experimental data in a straightforward way for the effects of non-compensation. In this process, the energy resolution improved, the signal distribution became much more Gaussian and, most importantly, the hadronic energy was correctly reproduced, both for single hadrons and for jets. The results for 200 GeV "jets" are shown in Figure 8.10. Using only the *ratio* of the two signals produced by this calorimeter, the resolution for these "jets" improved from 14% to 5%, in the Čerenkov channel (Figure 8.10a,b). It was shown that this 5% resolution was in fact dominated by fluctuations in side leakage in this (small, only 1030 kg instrumented volume) detector. Interestingly, the energy resolution turned out to scale almost perfectly with $E^{-1/2}$ after this C/S correction (Figure 8.10c), while the energy resolution measured for each of the two signals separately showed large deviations from such scaling. Fluctuations in (side) leakage contributed to the value of the scaling term.

Also the jet energy was well reconstructed as a result of this procedure. Whereas the raw data gave a mean value of 133.1 GeV for these 200 GeV "jets," the described procedure led to hadronic energies that were within a few percent the correct ones, *in an instrument calibrated with electrons*. In the process, hadronic signal linearity (a notorious problem for non-compensating calorimeters, see Section 3.2.6) was more or less restored as well (Figure 8.10d). Any remaining effects can be ascribed to side leakage and would most likely be absent in a larger detector.

Simultaneous detection of the scintillation and Čerenkov light produced in the shower development turned out to have other, unforeseen beneficial aspects as well. One such effect was discussed in Section 3.5.2, where Figure 3.48 shows the signals from muons traversing the DREAM calorimeter along the fiber direction [Akc 04]. The gradual increase of the response with the muon energy is a result of the increased contribution of radiative energy loss (bremsstrahlung) to the signals. The Čerenkov fibers are *only* sensitive to this energy loss component, since the primary Čerenkov radiation emitted by the muons falls outside the numerical aperture of the fibers. The constant (energy-independent) difference between the total signals observed in the scintillating and Čerenkov fibers thus represents the non-radiative component of the muon's energy loss. Since both types of fibers were calibrated with em showers, their response to the radiative component was equal. This is a unique example of a detector that separates the energy loss by muons into radiative and non-radiative components.

Following the successes of the DREAM project, a new collaboration was formed to explore the possibilities opened up by this new calorimeter technique. This became known as the RD52 Collaboration, and its activities were part of the officially supported CERN detector R&D program. All experimental activities were concentrated in the H8 beam of CERN's Super Proton Synchrotron. In the following subsections, highlights of the achievements of this project are presented.

8.2.3 *Crystals for dual-readout calorimetry*

Once the effects of the dominant source of fluctuations are eliminated, the resolution is determined and limited by other types of fluctuations. In the case of the DREAM detector, these fluctuations included, apart from fluctuations in side leakage which can be eliminated by making the detector sufficiently large, *sampling fluctuations* and fluctuations in the *Čerenkov light yield*. The latter effect alone contributed $35\%/\sqrt{E}$ to the measured resolution, since the quartz fibers generated only eight Čerenkov photoelectrons per GeV deposited energy. Both effects could in principle be greatly reduced by using crystals for dual-readout purposes. Certain dense high-Z crystals (PbWO$_4$, BGO) produce significant amounts of Čerenkov light. The challenge is of course to separate this light effectively from the (overwhelmingly) dominant scintillation light. Precisely for that reason, the idea to use such crystals as dual-readout calorimeters met initially with considerable doubt. Yet, the RD52 Collaboration demonstrated that it could be done.

For the proof-of-principle measurements, lead tungstate (PbWO)$_4$) crystals were used. This material has the advantage of producing very little scintillation light, while the large refractive index promised a substantial Čerenkov light yield. Čerenkov light is emitted by charged particles traveling faster than c/n, the speed of light in the medium with refractive index n in which this process takes place. The light is emitted at a characteristic angle, θ_C, defined by $\cos\theta_C = 1/\beta n$. In the case of sufficienty relativistic particles (*i.e.*, $\beta \sim 1$) traversing PbWO$_4$ crystals ($n = 2.2$), $\theta_C \sim 63°$ [26].

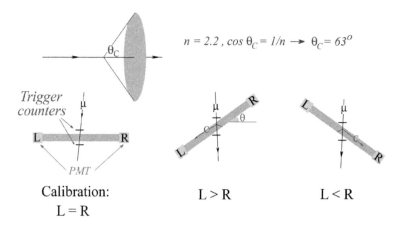

FIG. 8.11. Principle of the asymmetry measurement that was used to establish the contribution of Čerenkov light to the signals from the PbWO$_4$ crystals. Depending on the orientation, this directionally emitted light contributed differently to the signals from the left and right photomultiplier tubes [Akc 07b].

In order to detect the contribution of Čerenkov light to the signals from a PbWO$_4$ crystal, both ends of the crystal were equipped with a photomultiplier tube. By varying the detector *orientation* with respect to the direction of the incoming particles, a contribution of Čerenkov light would then manifest itself as an angle-dependent asymmetry. This is illustrated in Figure 8.11, which shows the setup of the initial measurements that were performed with a cosmic-ray telescope to test this principle [Akc 07b]. The PMT gains were equalized for the leftmost geometry, in which the crystal was oriented horizontally. By tilting the crystal through an angle (θ) such that the axis of the crystal is at the Čerenkov angle θ_C with respect to the particle direction, Čerenkov light produced by the cosmic rays traversing the trigger counters would be preferably detected in either the L (central geometry) or R (rightmost geometry) PMT. By measuring the

---

[26]The reality may be somewhat more complicated, because of the anisotropic optical properties of lead tungstate crystals [Bac 97, Chi 00], which might affect some aspects of Čerenkov light emission [Del 98].

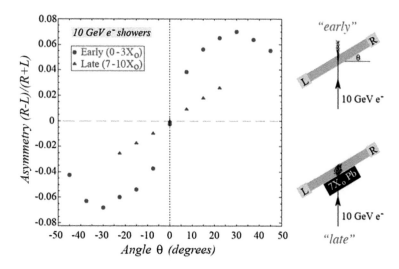

FIG. 8.12. Left-right response asymmetry measured for 10 GeV electrons showering in a $2.5X_0$ thick $PbWO_4$ crystal, as a function of the orientation of the crystal (the angle θ). Results are shown for the early and the late components of the showers. The latter measurements were obtained by placing 4 cm of lead upstream of the crystal [Akc 07b].

response asymmetry $(R - L)/R + L)$ as a function of the tilt angle θ, the contribution of Čerenkov light to the detector signals could be determined.

The initial cosmic-ray measurements indicated that the contribution of Čerenkov light was at the level of 15–20% [Wig 07]. Because of the extremely low event rates and the tiny signals (typically 20–30 MeV), systematic follow-up studies were carried out with particle beams at CERN's SPS. Figure 8.12 shows some of the results of this work, and in particular the characteristic "S" shape which indicates that the Čerenkov component of the light produced in the developing showers was most efficiently detected when the crystal axis was oriented at the Čerenkov angle with the shower axis. The figure also shows that placing a lead brick upstream of the crystal had the effect of making the angular distribution of the light produced in the crystal more isotropic, thus reducing the left-right asymmetry [Akc 07b].

It turned out that the scintillation light yield, and thus the fraction of Čerenkov light in the overall signal, depends very sensitively on the temperature of the $PbWO_4$ crystals: $\sim -3\%$ per degree Celsius [Akc 08b]. For this reason, the large em calorimeters that are based on these crystals (CMS, ALICE, PANDA) all operate at very low temperatures, and maintaining the temperature constant at the level of $\pm 0.1°C$ is an essential requirement for obtaining excellent energy resolution. There is also another temperature dependent phenomenon that affects the efficiency at which the Čerenkov and scintillation components of the light produced by these crystals can be separated: the decay time of the scintillation signals. RD52 found this decay time to decrease from ~ 9 ns at $13°C$ to ~ 6 ns at $45°C$ [Akc 08b].

The difference in the time structure of the signals is another important characteristic

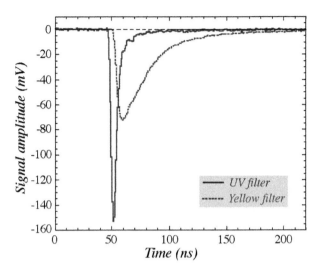

FIG. 8.13. Average time structure of the signals from a PbWO$_4$ crystal doped with 1% Mo, generated by 50 GeV electrons. The angle θ was 30° in these measurements. Shown are the results obtained with UV and yellow filters, respectively [Akc 09b].

that can be used to distinguish between the scintillation and Čerenkov components of the light produced by high-energy particles in crystals. And of course, the larger the difference in the time structure, the better the separation works. The RD52 collaboration managed to improve the applicability of PbWO$_4$ crystals for dual-readout calorimetry by doping them with small amounts, $\mathcal{O}(1\%)$, of molybdenum [Akc 09b]. This had two beneficial effects: it increased the decay time of the scintillation light and it shifted the spectrum of the emitted scintillation light to larger wavelengths. The effects of that are illustrated in Figure 8.13, which shows the calorimeter signals generated by 50 GeV electrons traversing a crystal of this type. This crystal was oriented such as to maximize the relative fraction of Čerenkov light in the detected signals. By selecting the UV light by means of an optical filter, almost the entire detected signal was due to (prompt) Čerenkov light, while a yellow transmission filter predominantly selected scintillation light, which had a decay time of ~ 26 ns as a result of the Mo-doping. Whereas the differences in angular dependence were very suitable to demonstrate that some of the light generated in these crystals is actually the result of the Čerenkov mechanism, the combination of time structure and spectral characteristics provides powerful tools to separate the two types of light in real time. One does not even have to equip the calorimeter with two different light detectors for that. This was demonstrated with a calorimeter consisting of bismuth germanate (Bi$_4$Ge$_3$O$_{12}$, or BGO) crystals [Akc 09c]. Even though Čerenkov radiation represents a very tiny fraction of the light produced by these crystals, it is relatively easy to separate and extract it from the signals. The much longer scintillation decay time (300 ns) and the spectral difference are responsible for that[27].

---

[27]The BGO scintillation spectrum peaks at 480 nm, while Čerenkov light exhibits a λ^{-2} spectrum.

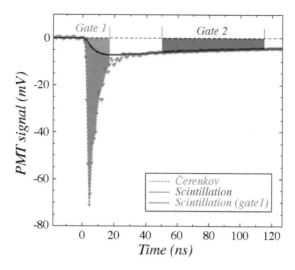

FIG. 8.14. The time structure of a typical shower signal measured in the BGO em calorimeter
equipped with a UV filter. These signals were measured with a sampling oscilloscope, which
took a sample every 0.8 ns [Akc 09c]. The UV BGO signals were used to measure the relative
contributions of scintillation light (gate 2) and Čerenkov light (gate 1).

Figure 8.14 shows the time structures of signals from a BGO calorimeter recorded
with a UV filter. The "prompt" component observed in the ultraviolet signal is due
to Čerenkov light. A small fraction of the scintillation light also passes through the
UV filter. This offers the possibility to obtain all needed information from only one
signal. An external trigger opens two gates: one narrow (10 ns) gate covers the prompt
component, the second gate (delayed by 30 ns and 50 ns wide) only contains scintillation
light. The latter signal can also be used to determine the contribution of scintillation to
the light collected in the narrow gate. In this way, the Čerenkov/scintillation ratio can
be measured event-by-event on the basis of one signal only [Akc 09c].

The same possibility was offered by BSO crystals. These have a similar chemical
composition as BGO, with the germanium atoms replaced by silicon ones. Both the
(scintillation) light yield and the decay time of this crystal are about a factor of three
smaller than for BGO. Tests with BSO crystals showed that this made the separation of
Čerenkov and scintillation light somewhat more efficient, while maintaining the possibility to obtain all necessary information from one calorimeter signal. This makes BSO a
potentially interesting candidate for a crystal-based dual-readout calorimeter, especially
also because the rare and expensive germanium component is not needed, [Akc 11b].

Apart from the time structure and the spectral differences, there is one other characteristic feature of Čerenkov light that can be used to distinguish it from scintillation
light, namely the fact that it is *polarized* [Akc 11a]. The polarization vector is oriented
perpendicular to the cone of the emitted Čerenkov light. RD52 used a BSO crystal to
demonstrate this possibility (Figure 8.15). This crystal was placed in a particle beam

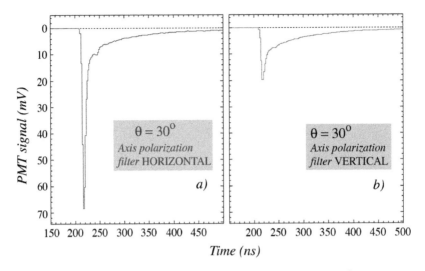

FIG. 8.15. Average time structure of the signals generated by 180 GeV π^+ traversing a BSO crystal in its center at $\theta = 30°$ and passing through a U330 optical transmission filter, followed by a polarization filter. The transmission axis of the latter filter was either oriented horizontally (*a*) or vertically (*b*). The time scale describes the time passed since the start of the time base of the oscilloscope [Akc 11a].

and oriented such as to maximize the fraction of Čerenkov light that reached the PMT (as in Figure 8.12). A UV filter absorbed most of the scintillation light, and the time structure of the transmitted signals showed a clear prompt Čerenkov signal, as well as a 100 ns tail due to the transmitted component of the scintillation light. In addition, a polarization filter was placed directly in front of the PMT. Rotating this filter over 90° had a major effect on the prompt Čerenkov component, while the scintillation component was not affected at all [Akc 11a].

8.2.4   *Tests of crystal-based dual-readout calorimeters*

The RD52 collaboration also performed tests of calorimeter systems in which the em section consisted of high-Z crystals, while the original DREAM fiber calorimeter served as the hadronic section [Akc 09c]. Two matrices of crystals were assembled for this purpose. The first one consisted of 19 PbWO$_4$ crystals borrowed from the CMS Collaboration (total mass $\sim$ 20 kg). The second matrix consisted of 100 BGO crystals that were previously used in the em calorimeter of the L3 experiment (total mass $\sim$ 150 kg) [Sum 88]. The Čerenkov and scintillation components of the light produced in these crystals were separated as described above, exploiting the differences in time structure and spectral composition. Figure 8.16 shows results from the measurements with the BGO matrix, obtained for high-multiplicity multi-particle events ("jets") generated by 200 GeV π^+ in an upstream target. The overall Čerenkov signal distribution is shown, together with subsets of events selected on the basis of the measured Čerenkov/scintillation signal ratio, *i.e.*, on the basis of the em shower content of the

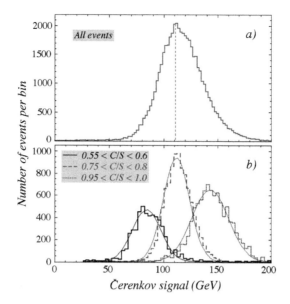

FIG. 8.16. The Čerenkov signal distribution for 200 GeV "jet" events detected in the BGO+fiber calorimeter system (a), together with the distributions for subsets of events selected on the basis of the ratio of the total Čerenkov and scintillation signals in this detector combination (b). Data from [Akc 09c].

events. A comparison with Figure 8.8 indicates that the dual-readout method also worked for this detector combination [Akc 09c].

Yet, after elaborate studies of many crystals, and dedicated efforts to tailor the crystal properties to the specific requirements for dual-readout calorimetry, the RD52 Collaboration decided that this was not the most promising avenue for improving the performance obtained with the original DREAM calorimeter. I recall that the main motivation for examining the option of using crystals for dual-readout calorimetry was the possibility to eliminate the effects of sampling fluctuations, and the potentially higher Čerenkov light yield, thus reducing the main sources of fluctuations that limited the performance of the DREAM fiber calorimeter. However, it turned out that the use of crystals introduced new, worse sources of fluctuations.

The main problem is that the (short wavelength) light that constitutes the Čerenkov signals is strongly attenuated, because of the absorption characteristics of the crystals. The attenuation length was in some cases so short that it led to $\mathcal{O}(10\%)$ response nonlinearities for electron showers[Akc 12b]. Since the depth at which the light is produced increases only logarithmically with the electron energy (Figure 2.9), this indicates that the attenuation length is of the order of a few radiation lengths. Such a short attenuation length affects several aspects of the calorimeter performance in major ways, since it causes the signal to depend sensitively on the location where the light is produced. For comparison, we mention that the attenuation lengths of the fibers used in the dual-

readout fiber calorimeters were orders of magnitude longer. In some cases, λ_{att} was
measured to be more than 20 m. Another problem is that a large fraction of the potentially available Čerenkov photons needs to be sacrificed in order to extract a sufficiently
pure Čerenkov signal from the light produced by the crystals.

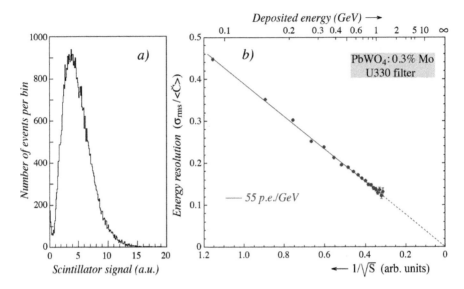

FIG. 8.17. The scintillation signal distribution for 50 GeV electrons traversing a PbWO$_4$ crystal
at $\theta = 30°$ (*a*) and the fractional width of the Čerenkov signal distribution as a function of
the amount of energy deposited in the crystal, as derived from the scintillation signal (*b*). The
crystal was doped with 0.3% Mo [Akc 10]. See the text for more details.

Figure 8.17 illustrates how this Čerenkov light yield can be measured in practice
[Akc 10]. It concerns measurements on a PbWO$_4$ crystal doped with 0.3% of molybdenum. This crystal was placed at an angle $\theta = 30°$ with the beam line (as in Figure
8.12). One PMT (R) was equipped with a UV filter, in order to select the Čerenkov
light, for which the detection efficiency is largest at this angle. At the other side of the
crystal only scintillation light was detected. EGS4 calculations indicated that the beam
particles (50 GeV electrons) deposited on average 0.578 GeV in this crystal, which was
slightly thicker than $2X_0$ in this geometry. This made it possible to calibrate the scintillation signals, the distribution of which is shown in Figure 8.17a. This distribution was
subdivided into 20 bins. For each bin, the signal distribution on the opposite side of the
crystal, *i.e.*, the Čerenkov side, was measured. The fractional width of this distribution,
$\sigma_{\mathrm{rms}}/C_{\mathrm{mean}}$, is plotted in Figure 8.17b versus the average scintillator signal in this bin,
or rather versus the inverse square root of this signal ($S^{-1/2}$). It turned out that this fractional width scaled perfectly with this variable, *i.e.*, with $E^{-1/2}$. Since the relationship
between the energy E and the scintillation signal S is given by the calibration described
above, it was also possible to indicate the energy scale in Figure 8.17b. This is done on
the top horizontal axis. The observed scaling of $\sigma_{\mathrm{rms}}/C_{\mathrm{mean}}$ with $E^{-1/2}$ means that the

energy resolution is completely determined by stochastic processes that obey Poisson statistics. In this case, fluctuations in the Čerenkov light yield were the only stochastic processes that played a role, and therefore the average light yield could be directly determined from this result: 55 photoelectrons per GeV deposited energy. For an energy deposit of 1 GeV, this led to a fractional width of 13.5%, and therefore the contribution of Čerenkov photoelectron statistics amounts to $13.5\%/\sqrt{E}$. This is not much better than what could be achieved in a dedicated fiber sampling calorimeter.

Other considerations that led to the decision to pursue other alternatives for improving the DREAM results were the high cost of the crystals, as well as the problem that the short-wavelength light needed to extract the Čerenkov signals made the crystals very prone to radiation damage effects. The alternative chosen by the RD52 Collaboration was the improvement of dual-readout fiber calorimetry, by constructing a very-fine-sampling device, that became known as the SuperDREAM calorimeter. However, before describing this device in detail, I want to discuss one other aspect of high-precision calorimetry: neutron detection.

8.2.5 *Benefits of neutron detection*

Should one succeed to eliminate, or at least greatly reduce the contributions of sampling fluctuations and photoelectron statistics to the hadronic energy resolution, then the last hurdle toward ultimate performance is formed by the fluctuations in invisible energy, *i.e.*, fluctuations in the energy fraction used to break up atomic nuclei. In Chapter 3, it was shown that the elimination of fluctuations in f_{em} takes care of the effects of the *average* contribution of invisible energy. However, for a given value of f_{em}, the invisible energy fluctuates around this average. The kinetic energy carried by the neutrons produced in the shower development process is correlated to this invisible energy loss. Efficient neutron detection, a key ingredient for compensating calorimeters, not only brings e/h to 1.0, but also greatly reduces the contribution of fluctuations in invisible energy to the hadronic energy resolution. It has been demonstrated that this reduces the ultimate limit on this resolution to $\sim 13\%/\sqrt{E}$ [Dre 90], in compensating lead/plastic-scintillator calorimeters (see Section 4.8).

Detailed measurements of the time structure of the calorimeter signals, examples of which are given in Figures 8.13 and 8.14, make it also possible to measure the contribution of neutrons to the shower signals. Figure 8.18a illustrates this with data taken with the original DREAM fiber calorimeter [Akc 09a]. The figure shows the average time structure of Čerenkov and scintillation signals measured with a sampling oscilloscope for showers from 200 GeV multi-particle events developing in this calorimeter. The scintillator signals exhibit an exponential tail with a time constant of ~ 20 ns. This tail has all the characteristics expected of a (non-relativistic) neutron signal (Section 3.2.9) and is thus absent in the time structure of the Čerenkov signals. It was also not observed in the scintillation signals for em showers [Akc 07a]. The distribution of the contribution of this tail to the hadronic scintillation signals (f_n) is plotted in Figure 8.18b. By measuring the contribution of this tail *event by event*, the hadronic energy resolution could be further improved [Akc 09a].

It was found that the fraction of the scintillation signal that could be attributed to

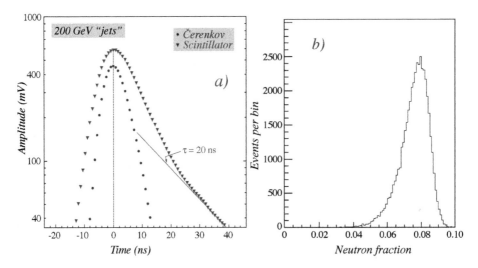

FIG. 8.18. The average time structure of the Čerenkov and scintillator signals measured for the showers from 200 GeV "jets" in the DREAM tower located on the shower axis. The measured (oscilloscope) signals have been inverted (*a*). Event-by-event distribution of the fraction of this scintillation signal caused by neutrons (*b*). Data from [Akc 09a].

neutrons (f_n) was anti-correlated with the Čerenkov/scintillation signal ratio, and thus with f_{em} (Figure 8.19a). This is no surprise, since a large em shower fraction implies that a relatively small fraction of the shower energy has been used for the processes in which atomic nuclei are broken up. This anti-correlation means that the essential advantages of the dual-readout method, which derived from the possibility to measure f_{em} event by event, could also be achieved with *one readout medium*, provided that the time structure of the (scintillation) signals is measured in such a way that the contribution of neutrons can be determined event by event.

This is illustrated in Figure 8.19b, which shows the total Čerenkov signal distribution for all 200 GeV "jet" events, as well as the distributions for subsamples of events with $0.06 < f_n < 0.065$ (the blue downward pointing triangles), $0.07 < f_n < 0.075$ (red squares) and $0.08 < f_n < 0.085$ (green upward pointing triangles). Clearly, the different subsamples each probe a different region of the total signal distribution for all events. This total Čerenkov signal distribution for all events is thus a superposition of many distributions such as the ones for the subsamples shown in this figure. Each of these distributions for the subsamples has a different mean value, and a resolution that is substantially better than that of the overall signal distribution. The signal distributions for the subsamples are also much more Gaussian than the overall signal distribution, whose shape is simply determined by the extent to which different f_n values occurred in practice. And since the f_n distribution is skewed to the low side (Figure 8.18b), the overall Čerenkov signal distribution is skewed to the high side.

A measurement of the relative contribution of neutrons to the hadronic scintillator signals thus offers similar possibilities for correcting the effects of non-compensation as

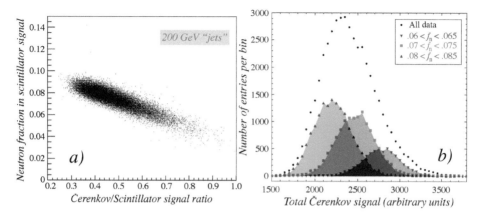

FIG. 8.19. Scatter plot for 200 GeV "jets." For each event, the combination of the total
Čerenkov/scintillator signal ratio and the fractional contribution of neutrons to the total scintillator signal is represented by a dot (*a*). Distribution of the total Cherenkov signal for 200
GeV "jets" and the distributions for three subsets of events selected on the basis of the fractional contribution of neutrons to the scintillator signal (*b*). Data from [Akc 09a].

an event-by-event measurement of the em shower fraction (Figure 8.8). However, when
both f_{em} and f_n are being measured, even better results may be expected. By selecting
a subsample of hadronic events, all with the same f_{em} value, there would still be event-
by-event differences in the share of invisible energy. The nuclear reactions taking place
in the non-em shower development process vary from event to event, and so does the
nuclear binding energy lost in these processes. For this reason, in calorimeters such as
the DREAM fiber calorimeter, measurements of f_n provide information *complementary*
to that obtained from the C/S signal ratio [Akc 09a].

This is illustrated in Figure 8.20a, which shows that the energy resolution of a sample of events with the same em shower fraction is clearly affected by the relative contribution of neutrons to the signals. As f_n increases, so does the fractional width of the
Čerenkov signal distribution. A larger f_n value means that the average invisible energy
fraction is larger. This in turn implies that the event-to-event fluctuations in the invisible energy are larger, which translates into a worse energy resolution, even for signals
to which the neutrons themselves do not contribute. Figure 8.20b shows the response
function obtained with the combined information on the em shower fraction and the contribution of neutrons to the signals. This Čerenkov signal distribution concerns 200 GeV
"jet" events with a C/S value between 0.70 and 0.75 and a fractional neutron contribution to the scintillator signals between 0.045 and 0.065. The distribution is very well
described by a Gaussian fit, with an energy resolution of 4.7%[28]. The resolution was
further reduced, to 4.4%, when the relative neutron contribution was narrowed down to
the interval 0.05–0.055. It is important to notice that these results were achieved for a

---

[28]When only information on the em shower fraction was used, the resolution for 200 GeV multi-particle
events was found to be 5.1% (Figure 8.10b).

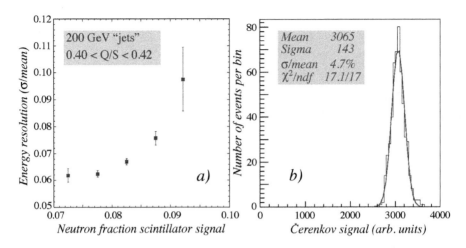

FIG. 8.20. The energy resolution for 200 GeV "jets" with the same em shower fraction, as a function of the fractional neutron contribution to the scintillator signals (a). Čerenkov signal distribution for 200 GeV "jets" with $0.70 < C/S < 0.75$ and $0.45 < f_n < 0.65$, together with the results of a Gaussian fit (b). Data from [Akc 09a].

calorimeter with an instrumented mass of only about one ton.

8.2.6 The RD52 fiber calorimeter

The design of the new dual-readout fiber calorimeter was driven by the desire to reduce the factors that limited the hadronic energy resolution of the original DREAM fiber calorimeter as much as possible. These factors concerned side leakage, the Čerenkov light yield and sampling fluctuations. The fluctuations in side leakage could be reduced in a trivial way, *i.e.*, by making the calorimeter sufficiently large. It was estimated that the instrumented mass has to be about 5,000 kg to contain hadronic showers at the 99% level, and thus limit the effects of leakage fluctuations on the hadronic energy resolution to $\sim 1\%$.

Fluctuations in the number of Čerenkov photons would be limited by maximizing the Čerenkov light yield. To that end,

- The numerical aperture of the Čerenkov fibers is increased,
- The upstream end of the Čerenkov fibers is aluminized,
- The quantum efficiency of the PMT photocathodes is increased,

while sampling fluctuations are limited in the following way:

- Fibers are individually embedded in the absorber structure, instead of in groups of seven,
- The packing fraction of the fibers is maximized, *i.e.*, roughly doubled compared to the DREAM calorimeter.

The fiber structure of the RD52 calorimeter is schematically shown in Figure 8.21. On the same scale, the structures of the DREAM and SPACAL calorimeters are shown as

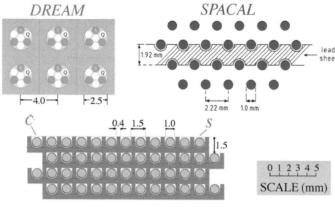

Figer pattern RD52

FIG. 8.21. The structure of the RD52 fiber calorimeter (copper-based modules), compared to that of two other fiber calorimeters: DREAM [Akc 05b] and SPACAL [Aco 91c].

well. Compared with DREAM, the number of fibers per unit volume, and thus the sampling fraction, is approximately twice as large in the RD52 calorimeter. And since each fiber is now separately embedded in the absorber structure, the sampling *frequency* has also considerably increased. Since both factors determine the electromagnetic energy resolution, a substantial improvement of that resolution should thus be expected.

Figure 8.22 shows pictures of the front face and the back end of a calorimeter module. Each module consists of four towers, and each tower produces a scintillation and a Čerenkov signal. The transverse dimension of the module was chosen such that the eight PMTs would fit within its perimeter, and the maximum fiber density was determined by the total photocathode surface of these PMTs (which corresponds to more than half of the module's lateral cross section).

The Čerenkov light yield was increased by using clear plastic fibers instead of the quartz ones used in the DREAM calorimeter. The numerical aperture of these plastic fibers[29] is larger (0.50 *vs.* 0.33). Also, the Čerenkov fiber density was increased, by $\approx 65\%$. In addition, the new PMTs have a higher quantum efficiency, thanks to a Super Bi-alkali photocathode. As a result, the number of Čerenkov photoelectrons measured for em showers increased by about a factor of four, from 8 to 33 Cpe/GeV [Akc 14b].

Another important difference between the RD52 and DREAM fiber calorimeters concerns the readout, which in the RD52 one is based on a Domino Ring Sampler (DRS) circuit [Rit 10] that allows time structure measurements of each signal with a sampling rate of 5 GHz (*i.e.*, 0.2 ns time bins). In the previous subsections it was shown that detailed measurements of the time structure are an invaluable source of information, not only for separating the Čerenkov and scintillation signals from crystals, but also to identify and measure the contribution of neutrons to the scintillation signals [Akc 09a].

---

[29] The light yield is proportional to the numerical aperture *squared*.

9.3 x 9.3 x 250 cm
150 kg
4 towers, 8 PMTs
2 x 2048 fibers

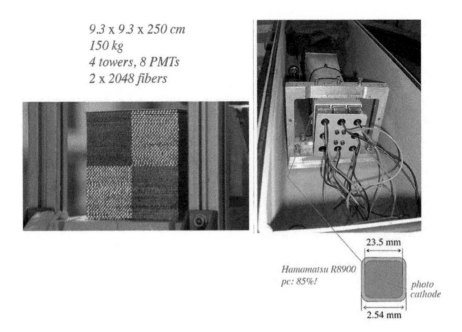

Hamamatsu R8900
pc: 85%!

23.5 mm

photo
cathode

2.54 mm

FIG. 8.22. Front (left) and rear (right) view of one of the RD52 fiber calorimeter modules. The tower structure is made visible by shining light on two of the eight fiber bunches sticking out at the back end. See text for more details.

Another important goal of the time structure measurements is to determine the depth at which the light is produced in this longitudinally unsegmented calorimeter. As is shown in Section 8.2.7, this can be achieved by making use of the fact that the light signals travel at a slower speed in the fibers ($\sim$ 17 cm/ns) than the particles producing this light (30 cm/ns).

It turned out to be very difficult to produce copper plates with the required specifications for this very-fine-sampling calorimeter structure. Therefore, the collaboration initially built nine modules using lead, which is relatively easy to extrude, as the absorber material. At a later stage, also several copper modules were built.

8.2.6.1 *Electromagnetic performance.* The RD52 calorimeter modules were extensively tested with beams of electrons, with energies ranging from 6 to 80 GeV. For reasons discussed in Section 8.2.7, the scintillation resolution turned out to be very sensitive to the angle of incidence of the particles, when these angles were very small ($\lesssim$ 3° between the beam line and the direction of the fibers) and the electron energy was high. Figure 8.23 shows the response functions for 40 GeV electrons, separately for the scintillation and the Čerenkov signals [Akc 14b]. For comparison, the response functions measured with the original DREAM fiber calorimeter are shown as well. The energy resolution was considerably better, and the response functions were also better described by a Gaussian function, especially in the case of the scintillation signals, even though the RD52 measurements were performed at a much smaller angle of incidence:

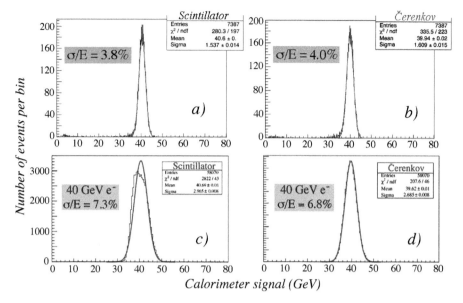

FIG. 8.23. Comparison of the em response functions measured with the RD52 copper-fiber calorimeter [Akc 14b] and the original DREAM copper-fiber calorimeter [Akc 05a], for 40 GeV electrons. Results are given separately for the scintillation and Čerenkov signals.

$(\theta, \phi) = (1.5°, 1.0°)$ *vs.* $(3°, 2°)$ for DREAM [Akc 05a].

One advantage of the new fiber pattern used in the RD52 calorimeters is that the scintillation and Čerenkov readout represent completely independent sampling structures. Therefore, by combining the signals from the two types of fibers, a significant improvement in the energy resolution is obtained. This was not the case for the original DREAM calorimeter [Akc 05a], where the two types of fibers essentially sampled the showers in the same way. Figure 8.24 shows that the energy resolution of the combined signal deviates slightly from $E^{-1/2}$ scaling. The straight line fit through the data points suggests a constant term of less than 1% [Akc 14b]. In any case, the energy resolution is substantially better than for either of the two individual signals, over the entire energy range covered by these measurements. It is also better than the performance reported for other integrated em+hadronic fiber calorimeters, such as SPACAL and DREAM. Careful analysis of the measured data showed that the contribution of sampling fluctuations to the total signal was $8.9\%/\sqrt{E}$ and that fluctuations in the number of Čerenkov photoelectrons about the average yield of 33/GeV increased the total stochastic term to $13.9\%/\sqrt{E}$. The small deviation from $E^{-1/2}$ scaling is due to the dependence of the scintillation response on the impact point.

This impact point dependence was of no consequence for the linearity of the calorimeter response. The average signals were measured to be proportional to the electron energy to within 1%, regardless of the angle of incidence of the electrons [Akc 14b].

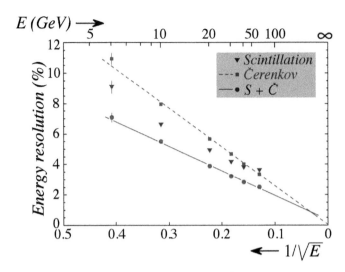

FIG. 8.24. The em energy resolution measured with the Čerenkov fibers (*a*), the scintillating fibers (*b*) and the sum of all fibers (*c*) in the copper-fiber calorimeter [Akc 14b].

8.2.6.2 *Particle identification.*   Traditionally, the calorimeter systems in high-energy physics experiments have been separated into two sections: the electromagnetic (em) and the hadronic section. This arrangement offers a certain number of advantages, especially for the identification of electrons and photons, which deposit all their energy in the em section and can thus be identified as such based on this characteristic.

The RD52 fiber calorimeter is longitudinally unsegmented, it does not consist of separate electromagnetic and hadronic sections. It is calibrated with electrons, and the calibration constants established in this way also provide the correct energy for hadronic showers developing in it. This eliminates one of the main disadvantages of longitudinal segmentation, *i.e.*, the problems associated with the intercalibration of the signals from different longitudinal sections (Chapter 6). Another advantage derives from absence of the necessity to transport signals from the upstream part of the calorimeter to the outside world. This allows for a much more homogeneous and hermetic detector structure in a 4π experiment, with fewer "dead areas."

Despite the absence of longitudinal segmentation, the signals provided by the RD52 fiber calorimeter offer several excellent possibilities to distinguish between different types of particles, and especially between electrons and hadrons. Identification of isolated electrons, pions and muons would be of particular importance for the study of the decay of Higgs bosons into pairs of τ leptons, if a calorimeter of this type were to be used in an experiment at a future Higgs factory.

Figure 8.25 illustrates the effects of the different identification methods [Akc 14a]:

1. There are large differences in *lateral* shower size, which can be used to distinguish between em and hadron showers. One advantage of the RD52 calorimeter structure is that the lateral granularity can be made arbitrarily small, one can make the tower size (defined by the number of fibers connected to one readout element)

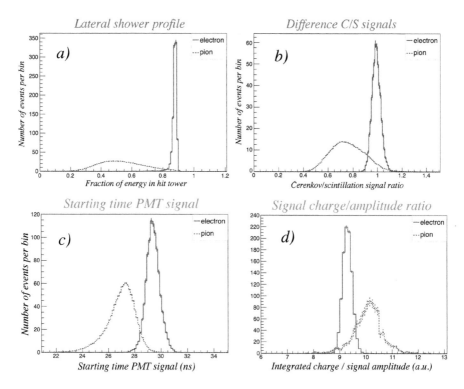

FIG. 8.25. Effects of four different shower characteristics that may be used to distinguish between electron and hadron showers in the longitudinally unsegmented RD52 lead-fiber calorimeter. Shown are the fraction of the total signal recorded by the tower in which the particle entered (*a*), the ratio of the Čerenkov and scintillation signals of the event (*b*), the starting time of the signal in the PMT, measured with respect to an upstream trigger signal (*c*), and the ratio of the total integrated charge and the amplitude of the signal (*d*). Data obtained with 60 GeV particle beams [Akc 14a].

as large or small as desired. Figure 8.25a shows the distributions of the fraction of the shower energy deposited in a RD52 tower located on the shower axis, for 60 GeV electrons and pions.

2. The availability of both scintillation and Čerenkov signals from the same events offers opportunities to distinguish between em and hadronic showers. For example, the ratio between the two signals is 1.0 for electrons (which are used to calibrate the signals!) and smaller than 1.0 for hadrons, to an extent determined by f_{em} and e/h. Figure 8.25b shows the distributions of the Čerenkov/scintillation signal ratio for 60 GeV electrons and pions.

3. The next two methods are based on the fact that the light produced in the fibers travels at a lower speed (c/n) than the particles responsible for the production of that light, which typically travel at c (see Section 8.2.7). As a result, the deeper

inside the calorimeter the light is produced, the earlier it arrives at the PMT. Since the light from hadron showers is typically produced much deeper inside the calorimeter, the PMT signals start earlier than for em showers, which all produce light close to the front face of the calorimeter. Figure 8.25c shows distributions of the starting time of the PMT signals for 60 GeV electron and pion showers.

4. The same phenomenon also leads to a larger width of the hadron signals, since the light is produced over a much larger region in depth than for electrons. Therefore, the ratio of the integrated charge and the signal amplitude is typically larger for hadron showers. Figure 8.25d shows distributions of that ratio for showers induced by 60 GeV electrons and pions[30].

One may wonder to what extent the different methods mentioned above are correlated, in other words to what extent the mis-identified particles are either the same or different ones for each method. It turned out that by combining different e/π separation methods, important improvements could be achieved in the capability of the longitudinally unsegmented calorimeter to identify electrons with minimal contamination of mis-identified particles. A multivariate neural network analysis showed that the best e/π separation achievable with the variables used for the 60 GeV beams was 99.8% electron identification with 0.2% pion misidentification. Further improvements may be expected by including the full time structure information of the pulses, especially if the upstream ends of the fibers are made reflective [Aco 91a].

The longitudinally unsegmented RD52 fiber calorimeter can thus be used to identify electrons with a very high degree of accuracy. Elimination of longitudinal segmentation offers the possibility to make a finer lateral segmentation with the same number of electronic readout channels. This has many potential benefits. A fine lateral segmentation is crucial for recognizing closely spaced particles as separate entities. Because of the extremely collimated nature of em showers (Section 8.2.7), it is also a crucial tool for recognizing electrons in the vicinity of other showering particles, as well as for the identification of electrons in general. Unlike the vast majority of other calorimeter structures used in practice, the RD52 fiber calorimeter offers almost limitless possibilities for lateral segmentation. If so desired, one could read out every individual fiber separately. Modern silicon PM technology certainly makes that a realistic possibility.

8.2.6.3 *Hadronic performance.* The hadronic performance of the RD52 fiber calorimeter has until now only been measured with a detector that, just as its DREAM predecessor, was too small to fully contain hadronic showers. Moreover, because of problems encountered with the large-scale production of the required copper absorber structure, only data obtained with a 1.5 ton lead module are available at this time. The (9-module) calorimeter was subdivided into $9 \times 4 = 36$ towers, and thus produced 72 signals for each event. In order to get a handle on the shower leakage, the detector was surrounded

---

[30]This result may be compared with Figure 7.46, which shows the difference between the widths of electron and pion signals in the SPACAL calorimeter. However, in that case, the differences were greatly enhanced by the aluminized upstream ends of the fibers. This had a much larger effect on the signal structure of the hadron showers than for the electron ones.

FIG. 8.26. The RD52 fiber calorimeter, installed in the H8C beam area at CERN. The system of trigger counters and beam defining elements is visible in the left bottom part of the figure. The calorimeter is surrounded on four sides by "leakage counters." The insert shows the front face of the (lead-)fiber calorimeter.

by an array of 20 plastic scintillation counters (measuring $50 \times 50 \times 10$ cm^3 each). Figure 9.4 shows a picture of the setup in which this detector combination was tested at CERN. As usual, all 72 calorimeter signals were calibrated with electrons. The leakage counters were calibrated with a muon beam, the muons deposited on average 100 MeV in each module they traversed. Next, pion beams were sent into the central region of the calorimeter. Figure 8.27 shows the scintillation and Čerenkov signal distributions for 20, 60 and 100 GeV π^- showers, as well as the ones corrected on the basis of the measured em shower fraction, using Equation 8.7 [DRE 13]. The latter distributions exhibit the familiar benefits of the dual-readout method: a relatively narrow, Gaussian signal distribution centered around the correct mean value, $i.e.$, the energy of the electrons that were used to calibrate the channels. The energy resolution is not very different from the one obtained with the original DREAM calorimeter (Figure 8.10), which is no surprise since in both cases leakage fluctuations were a dominant contribution to the hadronic energy resolution.

8.2.6.4 *The rotation method.* Recently, the RD52 Collaboration developed a new method to analyze the hadronic data taken with their dual-readout fiber calorimeters [Lee 17]. This method is described below. Rewriting Equations 8.5, the ratios of the calorimeter signals and the particle energy can be expressed as follows:

$$S/E = (h/e)_S + f_{\mathrm{em}}\left[1 - (h/e)_S\right]$$
$$C/E = (h/e)_C + f_{\mathrm{em}}\left[1 - (h/e)_C\right] \qquad (8.8)$$

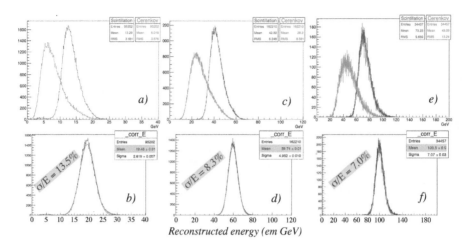

Reconstructed energy (em GeV)

FIG. 8.27. Signal distributions for π^- beam particles of 20, 60 and 100 GeV showering in the lead-fiber calorimeter. The top row (a, c, e) shows the signal distributions measured for the scintillation and Čerenkov signals. The bottom row (b, d, f) shows the signal distributions that were obtained after combining the S and C distributions according to Equation 8.7, with $\chi = 0.45$ [DRE 13].

As shown in Figure 8.28, the experimental data points for hadron showers detected with a dual-readout calorimeter are thus located on a straight (red) line in the C/E vs. S/E diagram. This line links the point $[(h/e)_S, (h/e)_C]$, for which $f_{em} = 0$, with the point (1,1), for which $f_{em} = 1$. The experimental data points for electron showers are concentrated around the latter point. The S and C signal distributions measured with the calorimeter correspond to the projections of the scatter plot on the horizontal and vertical axes, respectively. The asymmetric shape of these distributions reflects the distribution of the data points around the straight line.

The slope of the line around which the hadronic data points are clustered, *i.e.*, the angle θ, only depends on the two e/h values, and is thus *independent of the hadron energy*. The parameter χ, introduced in Equation 8.7, is related to this angle as

$$\cot \theta = \frac{1 - (h/e)_S}{1 - (h/e)_C} = \chi \tag{8.9}$$

and is thus also independent of the energy. Because of this feature, the scintillation and Čerenkov signals measured for a particular hadron shower can be used to reconstruct its energy in an unambiguous way (Equation 8.7). This is graphically illustrated in Figure 8.28, since Equation 8.7 implies that the data point (S, C) is moved up along the straight (red) line until it intersects the (dashed) line defined by $C = S$. If this is done for all hadronic data points, the result is a collection of data points that cluster around the point (1,1), just like the data points for electron showers. Since this feature is independent of the particle energy, this calorimeter will thus be as linear for hadrons as for electrons.

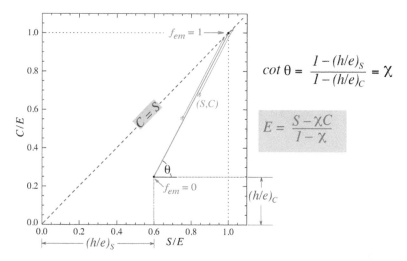

$$cot\,\theta = \frac{1 - (h/e)_S}{1 - (h/e)_C} = \chi$$

$$E = \frac{S - \chi C}{1 - \chi}$$

FIG. 8.28. The $S - C$ diagram of the signals from a dual-readout calorimeter. The hadron events are clustered around the straight red line, the electron events around the point (1,1).

Interestingly, a dual-readout calorimeter produces signal distributions with the same average value for event samples of pions, protons and kaons of the same energy. The f_{em} distributions are quite different for showers produced by these different types of hadrons, as a result of conservation of baryon number and strangeness in the shower development (Section 4.7.2). For the dual-readout calorimeter, this difference translates in a different clustering of data points around the straight (red) line defined by Equations 8.8. However, Equation 8.7 will reproduce the beam energy regardless of the f_{em} distribution. This property distinguishes this type of calorimeter clearly from calorimeters for which the energy reconstruction is based on "offline compensation" techniques, where the mentioned differences translate into a systematic uncertainty in the hadronic energy measurement, unless one knows what type of hadron caused the shower (Section 7.5.3). For dual-readout calorimeters, the relationship (8.7) is universally valid for all types of hadrons, and also for jets.

The independence of θ and χ on the energy and on the particle type offers an interesting possibility to measure the hadronic energy with unprecedented precision, at least for an ensemble of particles with the same energy. In practice, the energy resolution is usually determined in that way, *i.e.*, as the fractional width (σ/E) of the signal distribution for a beam of mono-energetic particles produced by an accelerator. The so-called "rotation method" works as follows (see Figure 8.29). First, the experimental hadronic data points are fitted with a straight line. This line intersects the $C = S$ line at point P. If the calorimeter is calibrated with electrons, then P is the point around which all electron showers that carry the same energy as the hadron beam are clustered. In other words, the coordinates of point P reveal the energy of the hadron beam. Next, the measured distribution of the hadronic data points is rotated around point P, over an angle $[90° - \theta]$. This procedure corresponds to a coordinate transformation of the type

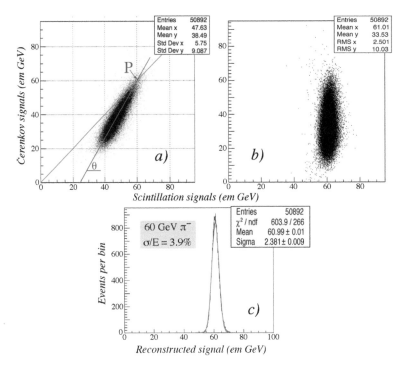

FIG. 8.29.   Signal distributions of the RD52 dual-readout lead/fiber calorimeter for 60 GeV pions [Lee 17]. Scatter plot of the two types of signals as recorded for these particles (*a*) and rotated over an angle of 30° around the point where the two lines from diagram *a* intersect (*b*). Projection of the latter scatter plot on the *x*-axis (*c*).

$$\begin{pmatrix} S' \\ C' \end{pmatrix} = \begin{pmatrix} \sin\theta & -\cos\theta \\ \cos\theta & \sin\theta \end{pmatrix} \begin{pmatrix} S \\ C \end{pmatrix} \qquad (8.10)$$

The projection of the rotated scatter plot on the *x*-axis is a narrow signal distribution centered around the correct energy value.

Figure 8.29 shows an example of results obtained in practice with a procedure of this type [Lee 17], for a beam of 60 GeV π^-. The resulting signal distribution is well described by a Gaussian function with a central value of 61.0 GeV and a relative width, σ/E, of 3.9%. The narrowness of this distribution reflects the clustering of the data points around the axis of the locus in Figure 8.29a. Since this clustering is affected by fluctuations in lateral shower leakage, even more narrow signal distributions may be expected when these fluctuations are reduced, *i.e.*, in a larger calorimeter.

The RD52 Collaboration applied exactly the same procedure for beams of other energies, and also compared the results for beams of protons and pions. Results for 20 GeV π^+ and 125 GeV protons are shown in Figures 8.30a and 8.30b, respectively. As expected, the beam energy is well reproduced for these different energies and particle types, using the same value of θ in all cases. Figure 8.30c shows that the calorimeter is very linear, and that the responses to protons and pions are equal within 1–2%,

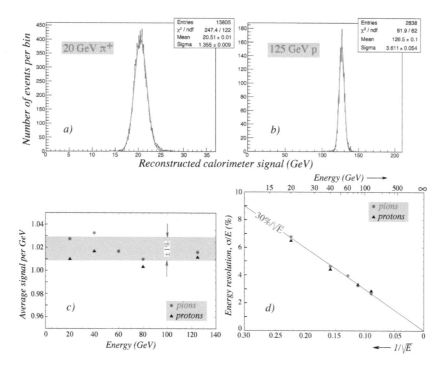

FIG. 8.30. Signal distributions measured with the RD52 dual-readout lead/fiber calorimeter for 20 GeV pions (a) and 125 GeV pions (b). The average signal per unit energy (c) and the fractional width of the signal distribution (d) are given separately for pions and protons. Data obtained with the rotation method [Lee 17].

despite response differences of $\sim 10\%$ in the measured Čerenkov signal distribitions. The fractional width of the signal distributions (σ/E) scales with $30\%/\sqrt{E}$, without any significant additional contributions (Figure 8.30d). It should be pointed out that the energy of the beam particles was *not* used to obtain these signal distributions. In all cases, the straight line that was used to fit the experimental data points in the scatter plot intersected the $C = S$ line at approximately the correct energy.

One may (correctly) argue that the width of signal distributions such as those shown in Figures 8.29 and 8.30 does not really constitute an energy resolution. Strictly speaking, the energy resolution denotes the precision with which the energy of one arbitrary particle (of unknown energy) absorbed in this calorimeter may be determined. And even though no additional information about the particles in the test beam, such as the energy or the composition, was used, the availability of an *ensemble* of events was essential for the success of the method described above. This issue is further discussed in Section 9.2.

8.2.6.5 *Monte Carlo simulations.*   In order to estimate the improvement that may be expected in a calorimeter that is large enough to contain the showers at a sufficient level,

elaborate GEANT4 based Monte Carlo simulations were performed. The reliability of
these simulations was assessed by comparing the results with the experimental data obtained with the DREAM calorimeter[Akc 14c]. It turned out that the Čerenkov response
function (Figure 8.7b) was well described by these simulations. On the other hand, the
scintillation distribution (Figure 8.7a) was more narrow, less asymmetric and peaked at
a lower value than for the experimental data. This is due to the rather poor description of
the non-relativistic component of the shower development, which is completely dominated by processes at the nuclear level, by the standard FTFP_BERT hadronic shower
development package used in GEANT4. Both the average size of this component, as
well as its event-to-event fluctuations, are at variance with the experimental data. This
non-relativistic shower component only plays a role for the scintillation signals, *not* for
the Čerenkov ones.

Yet, some aspects of hadronic shower development that are important for the dual-
readout application were found to be in good agreement with the experimental data,
e.g., the shape of the Čerenkov response function and the radial shower profiles. Attempts to use the dual-readout technique on simulated shower data reasonably reproduced some of the essential characteristics and advantages of this method: a Gaussian
response function, hadronic signal linearity and improved hadronic energy resolution.
The problem that the reconstructed beam energy was systematically too low may be
ascribed to inaccuracies in the description of the non-relativistic shower component.

As stated above, the main purpose of these very time consuming simulations was to
see if and to what extent the hadronic performance would improve as the detector size
is increased. Figure 8.31a shows the signal distribution obtained for 100 GeV π^- in a
copper-based RD52 calorimeter with a lateral cross section of 65×65 cm^2. The mass of
such a ($10\lambda_{int}$ deep) device would be ~ 6 tonnes. According to these simulations, the
average calorimeter signal, reconstructed with the dual-readout method, would be 90.2
GeV, and the energy resolution would be 4.6%.

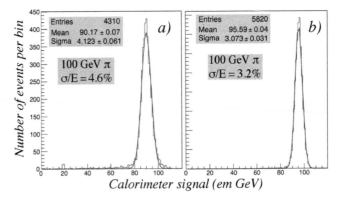

FIG. 8.31. GEANT4 simulations of the response function to 100 GeV π^- particles of a
dual-readout fiber calorimeter with the RD52 structure, and lateral dimensions of 65×65
cm^2 [Akc 14c]. Results are shown for the standard FTFP_BERT hadronic shower simulation
package (*a*), and with the high-precision version of this package, FTFP_BERT_HP (*b*).

In order to see to what extend these simulations depend on the choice of the hadronic shower development package, the simulations were repeated using the high precision version of the hadronic shower simulation package (FTFP_BERT_HP), which seems to provide a much more elaborate treatment of the numerous neutrons produced in the shower process. Indeed, the results of this work (Figure 8.31b) show a clear improvement: the average calorimeter signal has increased to 95.6 GeV, and is thus within a few percent equal to that of an em shower developing in the same calorimeter structure (one of the crucial advantages of calorimeters based on the DREAM principle). Also the energy resolution improved significantly, from 4.6% to 3.2%. Simulations for 200 GeV hadron showers with the FTFP_BERT_HP package yielded an average signal of 191 GeV and an energy resolution of 2.4% [Akc 14c].

These simulations thus suggest that resolutions of a few percent are feasible, and that the hadronic performance of a sufficiently large copper-based RD52 calorimeter would be at the same level as that of the compensating SPACAL and ZEUS calorimeters, or even better. It should also be emphasized that the results shown in Figure 8.31 are for *single hadrons*. There is an important reason why the *jet* energy resolution of copper-based dual-readout fiber calorimeters may also be expected to be much better than that of the high-Z compensating calorimeters [Wig 13]. A sizeable component of the jet consists of soft hadrons, which range out rather than develop showers (Figure 8.3). The response of calorimeters such as ZEUS to these soft particles was considerably larger than the response to the showering γs and high-energy hadrons. The scale for the difference between these responses is set by the e/mip value, which was measured to be 0.62 in ZEUS [Dre 90] and 0.72 in SPACAL [Aco 92c]. The advantage of an absorber material with much lower Z is an e/mip value that is much closer to 1.0 (the value at which this effect ceases to play a role). For the copper-based DREAM calorimeter, an e/mip value of 0.84 was measured [Akc 04]. The possibility to measure jets with superior resolution compared to previously built high-Z compensating calorimeters was one of the main reasons why the dual-readout project was started.

8.2.7 *Other RD52 results*

Detailed studies with the fine-grained RD52 fiber calorimeter have revealed important information about the showering particles that are of interest for other calorimeters as well. In this subsection, we address the em shower profiles and the time structure of the showers.

8.2.7.1 *The electromagnetic shower profiles.*

The fine-grained RD52 fiber calorimeters are ideally suited for precision measurements of the lateral shower profiles. This was done by moving a very narrow electron beam across the boundary between neighboring towers and measuring the energy fraction deposited in each of these towers. The narrow beam was obtained by selecting beam particles based on the coordinates of the points where they traversed upstream wire chambers. Figure 8.32 shows the profile measured for 100 GeV electrons in the lead-based RD52 calorimeter. Since the calorimeter is longitudinally unsegmented, the profile is integrated over the full depth. It exhibits a very pronounced central core, which is presumably caused by the extremely collimated nature of the showers in the early stage of the shower development, before the shower

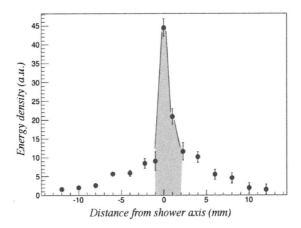

FIG. 8.32. The lateral profile of 100 GeV electron showers in the RD52 lead-fiber calorimeter, measured with the scintillation signals [Akc 14b].

maximum is reached. In this stage, the shower mainly consists of energetic bremsstrahlung photons, which convert into energetic e^+e^- pairs that travel in the same direction as the beam particles. According to Figure 8.32, a considerable fraction of the shower energy ($\sim 20\%$) is deposited in a cylinder with a radius of 1 mm about the shower axis [Akc 14b].

This feature has important consequences for this type of calorimeter, where the distance between neighboring fibers of the same type is 2-3 mm (see Figure 8.21). The calorimeter signal (from this early shower component) depends crucially on the impact point of the beam particles, if these enter the calorimeter parallel to the fibers. This dependence is quickly reduced when the electrons enter the calorimeter at a small angle with the fibers. As the angle increases, this early collimated shower component is sampled more and more in the same way as the rest of the shower. However, at angles where this is not the case, this effect adds an additional component to the em energy resolution. This is clearly observed in Figure 8.33, which shows the energy resolution for 20 GeV electrons as a function of the angle of incidence [Car 16]. This effect is, in first approximation, energy independent and thus results in a constant term in the em energy resolution. The measured em energy resolution of the scintillation signals of the RD52 copper-fiber calorimeter (Figure 8.24) exhibits indeed a clear deviation from $E^{-1/2}$ scaling. Because of the extreme dependence on the angle of incidence, one should be careful when comparing the em performance measured with different fiber calorimeters. For example, the improvement in the em scintillation resolution of the RD52 calorimeter with respect to the DREAM one is much larger than suggested by the comparison in Figure 8.23, because the angles at which the DREAM measurements were performed were twice as large as in case of the RD52 calorimeter. The distance separating neighboring fiber clusters in DREAM was such that the position dependence of the scintillation signal in these measurements even led to a non-Gaussian response function (Figure 8.23c).

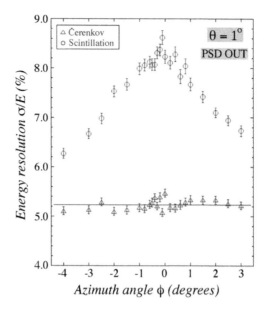

FIG. 8.33. The energy resolution measured for 20 GeV electrons in the scintillation and the Čerenkov channels of the copper-fiber RD52 calorimeter, as a function of the azimuth angle of incidence (ϕ) of the beam particles. The tilt angle θ was 1° [Car 16].

Now, why does this position dependence of the response function only affect the resolution measured with the scintillation signals? The reason is that the collimated early shower component does *not* contribute to the Čerenkov signals, since the Čerenkov light produced by shower particles traveling in the same direction as the fibers falls outside the numerical aperture of the fibers. For the 20 GeV electrons, the Čerenkov fibers thus only register shower particles that travel at relatively large angles with the shower axis ($20 - 60°$), and such particles are for all practical purposes only found beyond the shower maximum, where the shower is wide compared to the typical distance separating neighboring fibers of the same type. The "constant" term that affects the scintillation resolution is thus practically absent for the Čerenkov signals, as illustrated by Figure 8.24.

In Section 8.2.2, another consequence of this difference between the two types of signals from dual-readout calorimeters was shown (Figure 3.48). When muons traverse this calorimeter parallel to the fibers, the Čerenkov fibers only register the radiative component of their energy loss, because the Čerenkov light emitted in the non-radiative (ionizing) component falls outside the numerical aperture of the fibers.

These results also have consequences for other types of calorimeters. Typically, the lateral granularity is chosen on the basis of the Molière radius of the calorimeter structure, with the argument that this parameter determines the radial shower development. However, the results shown here indicate that em showers have a very pronounced, extremely collimated core, whose radial dimensions are very small compared to the

Molière radius. A much finer granularity would make it possible to resolve doublets, or recognize electromagnetic components of jets much better than in a calorimeter designed on the basis of the Molière radius. Fibers offer this possibility, since the lateral granularity of a calorimeter of the RD52 type could be made arbitrarily small.

8.2.7.2   *Time structure of the showers.*   Earlier in this chapter, it was shown how the time structure of the calorimeter signals could provide crucial information. It could, for example, be used

- to distinguish and separate the scintillation and Čerenkov components of the light signals from crystals (Figures 8.13, 8.14),

- to identify showers initiated by electrons and photons (Figure 8.25), and

- to recognize and measure the contribution of neutrons to the calorimeter signals (Figure 8.18).

The timing information is particularly important for fiber calorimeters such as the ones discussed here. Even though light attenuation is not a big effect in the optical fibers used as active media in these calorimeters, it may have significant consequences for the hadronic performance. The depth at which the light is produced in these showers fluctuates at the level of a nuclear interaction length, *i.e.*, ~ 25 cm in the RD52 fiber calorimeters. The light attenuation length in the scintillating fibers amounts to ~ 8 m in the scintillating fibers while in the Čerenkov ones values up to 20 m have been measured. But even for an attenuation length of 20 m, the mentioned depth fluctuations introduce a constant term of $\sim 1\%$ in the hadronic energy resolution, and this term increases correspondingly for shorter attenuation lengths.

If the depth at which the light is produced was known, the signals could be corrected event by event for the effects of attenuation. The timing information of the calorimeter signals provides this information, thanks to the fact that light in the optical fibers travels at a lower speed than the particles that generate this light. The effective speed of the light generated in the fibers is c/n, with n the index of refraction. For an index of 1.59, typical for polystyrene based fibers, this translates into a speed of 17 cm/ns. On the other hand, the shower particles that are responsible for the generation of light in the fibers typically travel at a speed close to c.

The effects of this difference are illustrated in Figure 8.34, which shows how the starting time of the PMT signal varies with the (average) depth at which the light is produced inside the calorimeter. The deeper inside the calorimeter the light is produced, the earlier the PMT signal. For the polystyrene fibers, the effect amounts to 2.55 ns/m. For the RD52 lead-fiber calorimeter, which has an effective nuclear interaction length (λ_{int}) of ~ 27 cm, this corresponds to ~ 0.6 ns/λ_{int}.

This was measured experimentally with 60 GeV electron and pion event samples, using a Time-to-Digital Converter (TDC) [Akc 14a]. The TDC was started by the signal produced by an upstream trigger, and stopped by the signal from the central calorimeter tower. Figure 8.35a shows the TDC signal distribution for the electron showers. In these showers, the light was, on average, produced at a depth of ~ 12 cm inside the calorimeter ($10X_0$), with event-to-event variations at the level of a few cm. The width of this

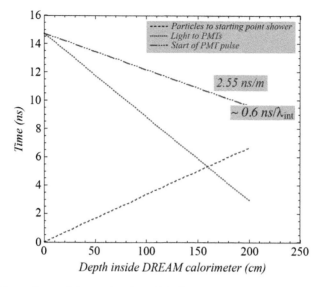

FIG. 8.34. Dependence of the starting time of the PMT signals on the average depth (z) inside the calorimeter where the light is produced (the dash-dotted line). This time is measured with respect to the moment the particles entered the calorimeter. Also shown are the time it takes the particles to travel to z (the dashed line) and the time it takes the light to travel from z to the PMT (the dotted line) [Akc 14a].

distribution, ~ 0.5 ns, is thus a good measure for the precision with which the depth of the light production can be determined for individual events, ~ 20 cm.

Figure 8.35b shows the measured TDC distribution for 60 GeV π^-. This distribution peaked ~ 1.5 ns earlier than that of the electrons, which means that the light was, on average, produced 60 cm deeper inside the calorimeter. The distribution is also asymmetric, it has an exponential tail towards early starting times, *i.e.*, light production deep inside the calorimeter. This measured TDC signal distribution could be used to reconstruct the average depth at which the light was produced for individual pion showers. The result, shown in Figure 8.35c, essentially represents the longitudinal shower profile of the 60 GeV pion showers in this calorimeter [Akc 14a].

In earlier studies of longitudinally unsegmented calorimeters, the depth of the light production was measured from the displacement of the lateral center-of-gravity with respect to the entrance point of the beam particles. To use this method, it was necessary to rotate the calorimeter over a small angle with respect to the beam line [Akc 05b]. The study described here does not require such a rotation. And unlike the displacement method, it also works for jets and neutral particles.

Figure 8.36 shows results of measurements performed to assess the effects of light attenuation in the fibers. The scatter plot in Figure 8.36a shows the calorimeter signal for the Čerenkov light from 80 GeV π^- versus the average depth at which that light was produced inside the calorimeter. As the light is produced deeper inside, the signal tends to be, on average, somewhat larger. This effect is quantified in Figure 8.36b, which

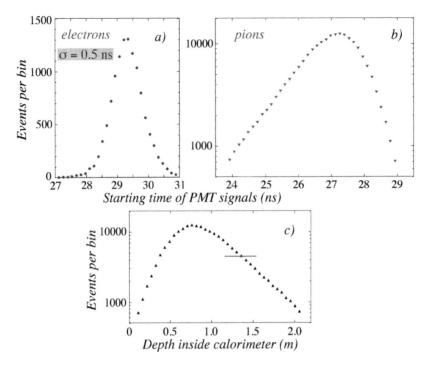

FIG. 8.35. The measured distribution of the starting time of the calorimeter's scintillation signals produced by 60 GeV electrons (*a*) and 60 GeV pions (*b*). This time is measured with
respect to the moment the beam particle traversed trigger counter T1, installed upstream of
the calorimeter (see Figure 9.4). These data were used to determine the distribution of the
average depth at which the light was produced in the hadron showers (*c*) [Akc 14a].

shows the average signal as a function of the depth at which the light was produced. The
data points are well described with an exponential curve with a slope of 8.9 m, which
thus represents the attenuation length of these fibers. This means that the signal changes
by 2–3%/λ_{int} as a result of light attenuation. And since this calorimeter is intended for
hadronic energy measurements at the level of 1%, elimination of the energy independent
term caused by light attenuation effects is important.

Until a few years ago, detailed measurements of the time structure of the calorimeter signals required a high-quality digital sampling oscilloscope[31]. In recent years, the
developments in microelectronics have made it possible to obtain this type of capability for a fraction of the cost. For example, CAEN is now offering a 36-channel VME
module (V1742), based on the DRS4 chip [Rit 10], which provides 5 GSample/s sampling. The RD52 Collaboration used such a module to measure the time structure of 30

---

[31] The results shown in Figures 8.13, 8.14 and 8.18 were all performed with a Tektronix TDS 7254B digital
oscilloscope, which provided a sampling capability of 5 GSample/s, at an analog bandwidth of 2.5 GHz,
i.e., the signals were sampled every 0.4 ns.

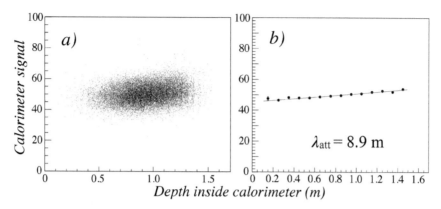

FIG. 8.36. Light attenuation in the Čerenkov fibers. The scatter plot (*a*) shows the calorimeter signal for the Čerenkov light from 80 GeV π^- versus the average depth at which that light was produced inside the calorimeter. The projection of this scatter plot on the vertical axis provides the effective light attenuation curve of the fibers (*b*) [Akc 14a].

different calorimeter signals simultaneously [Wig 16].

The results shown in Figure 8.37 concern a 40 GeV positive beam, which consisted of a mixture of electrons, pions and muons. The different particles were identified with external counters, *i.e.*, a preshower detector, a tail catcher and a muon counter. The figure shows the average time structure of the Čerenkov signals from the calorimeter tower into which the beam particles were steered. The muons produced light over the full 2.5 m length of the calorimeter module, and therefore the signals started earlier than for the electrons and pions, which produced most of the light close to the front face of the calorimeter. The shower maximum for the pions was located about 20 cm deeper inside the calorimeter than for electrons, and therefore the pion signals also started a bit

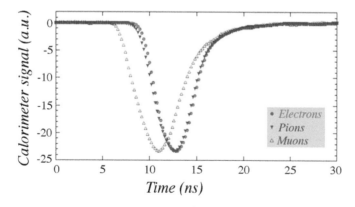

FIG. 8.37. Average time structure of the Čerenkov signals produced by the RD52 lead-fiber calorimeter, for 40 GeV electrons, pions and muons, measured with the DRS4 chip. The time base was started by the signal in the upstream trigger counters [Wig 16].

earlier than the electron ones.

The scintillation signals for the same events had a longer duration. This is due to the time constant that is characteristic for the scintillation process, plus the fact that the non-relativistic particles which contributed to the scintillation signals (but not to the Čerenkov ones) were distributed over a greater depth in the calorimeter volume than the relativistic ones. The pion component of the beam was also completely responsible for any signals recorded by the leakage counters. Figure 8.38 shows the average signals

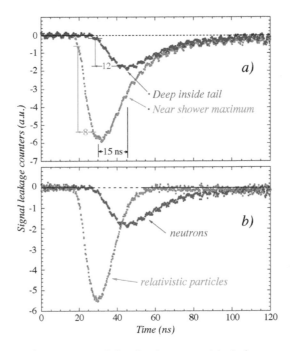

FIG. 8.38. Average time structure of the signals measured in leakage counters surrounding the RD52 lead-fiber calorimeter, for 40 GeV π^- steered into the center of this calorimeter. Diagram a shows the signals measured in a counter located close to the shower maximum (not far from the front face of the calorimeter) and in a counter located near the shower tail, *i.e.*, about 2 m from the front face of the calorimeter. In diagram b, the signal from the upstream counter is unfolded into a "neutron" and a "prompt" component [Wig 16].

recorded in two different leakage counters. These counters (see Figure 8.26) were located close to the shower maximum (the early signal), and near the end of the calorimeter module (the late signal). The latter signal consisted very likely exclusively of recoil protons produced by elastic neutron scattering, while the early signal may also contain a contribution from relativistic particles produced in the shower development and escaping the calorimeter. In the hadronic shower development, typically a few thousand neutrons are released from the nuclei in which they were bound. They typically carry a few MeV kinetic energy and lose that energy predominantly by means of elastic scattering off protons in the plastic components of the detectors, with a time constant of ~ 10 ns (Figure 3.23). The time difference between the two signals shown in Fig. 8.38 and the difference in rise time are consistent with the above assessment.

These are only a few examples of the information that can be obtained on the basis of time information about the calorimeter signals. Most likely, more applications will be developed, especially if even faster light detectors become available.

8.2.8 *Challenges*

Many years of experience have shown that detectors based on light as the source of experimental information have their own characteristic problems. Among these, I mention the effects of light attenuation, short-term instabilities arising from temperature and other environmental effects, and long-term effects of radiation damage and other aging phenomena. Perhaps the most daunting challenges should be expected from light attenuation. In the previous subsection, we saw that to limit the effects of spatial shower fluctuations on the signals from a longitudinally unsegmented fiber calorimeter to 1%, the light attenuation length of the readout elements has to be ~ 20 m, unless the possibility exists to determine the depth at which the light is produced, event by event[32]. The requirements in this context will be extremely hard to achieve with active media other than optical fibers, and in particular with crystals. On the other hand, it is very non-trivial to make a 4π detector structure with longitudinal optical fibers, although some useful ideas have been pursued with that purpose in mind [Anz 95b].

A second set of problems for light-based calorimeters arises from the need to operate in a magnetic field. Not only does this field affect the light production characteristics of some media, it also restricts the choice of light detectors. New types of light detectors developed to deal with these problems (APD, SiPM) exhibit encouraging, but certainly by no means ideal characteristics. Silicon photomultipliers offer the additional possibility to obtain longitudinal segmentation in fiber calorimeters, without sacrificing much in terms of compactness [Ber 16].

8.3 Particle Flow Analysis

A completely different method that has been proposed to meet the challenging requirements of recognizing, reconstructing and separating hadronically decaying W and Z bosons in future particle physics experiments is the so-called *Particle Flow Analysis* (PFA). This method is based on the combined use of a precision tracker and a highly-granular calorimeter. The idea is that the charged jet particles can be precisely measured with the tracker, while the energy of the neutral particles is measured with the calorimeter. Such methods have indeed successfully been used to improve the mass resolution of hadronically decaying Z^0s at LEP [Bus 95], and the jet energy resolution using γ-jet p_T balancing events at CDF [Boc 01] and at CMS [CMS 09]. Two detector concepts studied in the context of the experimental program for the proposed International Linear Collider (ILC) are based on this method as well [Beh 13].

---

[32]This number (20 m) applies to single hadrons. The requirements are much less stringent for the detection of jets, since the depth fluctuations in light production decrease considerably as the number of particles that simultaneously develop showers in the same structure increases [Aco 91c, Akc 05b]

The problem that limits the success of this method is that the calorimeter does not know or care whether the particles it absorbs are electrically charged. Therefore, the detected calorimeter signals will have to be corrected for the contributions of the charged jet particles. Proponents of this method have advocated a fine granularity as the key to the solution of this "double-counting" problem [Sef 15]. However, it has been argued by others that this, for practical geometries, is an illusion [Lob 02]. Especially in jets with leading charged particles, the overlap between the showers from individual jet particles makes the fine granularity largely irrelevant.

In order to increase the spatial separation between showers induced by the various jet particles, and thus alleviate the double-counting problem, the concept detectors for the ILC that are based on the PFA principle count on strong solenoidal magnetic fields (4 - 5T). Such fields may indeed improve the validity of PFA algorithms, especially at large distances from the vertex, since they open up a collimated beam of particles. It is important to be quantitative in these matters. After having traveled a typical distance of 1 m in a 4 T magnetic field, the trajectory of a 10 GeV pion deviates by 6 cm from that of a straight line, *i.e.*, less than one third of a nuclear interaction length (the characteristic length scale for lateral hadronic shower development) in typical calorimeters. The field is not always beneficial, since it may also have the effect of bending jet particles with a relatively large transverse momentum with respect to the jet axis *into* the jet core.

Of course, in the absence of reliable Monte Carlo simulations[33] the only way to prove or disprove the advocated merits of the proposed PFA methods is by means of dedicated experiments in realistic prototype studies.

8.3.1   *The importance of calorimetry for PFA*

The first statement in any talk about PFA mentions that 2/3 of the final-state particles constituting a jet are electrically charged, and that the momenta of these particles can be measured extremely precisely. This is of course true, but the implication that the calorimeters of PFA based detector systems don't have to be very good, since they only have to measure one third of the jet energy, is incorrect. In the absence of calorimeter information, based on tracker information alone, the jet energy resolution would be determined by the *fluctuations* in the fraction of the total jet energy that is carried by the charged fragments. This issue was studied by Lobban *et al.* [Lob 02], who found that these event-to-event fluctuations are very large. Depending on the jet fragmentation algorithm, the $\sigma_{\rm rms}$ of the energy fraction carried by charged particles was found to be 25–30% of the average value, *independent of the jet energy*.

One may wonder why these fluctuations do not become smaller at higher energies, given that the *number* of jet fragments increases. The reason for this is that the observed increase in multiplicity is uniquely caused by the addition of more *soft* particles. The bulk of the jet energy is invariably carried by a small number of the most energetic particles. This means that the fraction of the jet energy carried by charged particles is strongly dependent on the extent to which these particles participate in the "leading"

---

[33]Concern about the absence of reliable simulations for hadronic shower development was the main reason for a special workshop held at Fermilab in 2006 [HSS 06]. To my knowledge, the fundamental problems addressed at this workshop, *e.g.*, with regard to the hadronic shower widths that are crucial for PFA, still exist.

component of the jet. Therefore, the event-to-event fluctuations in this fraction are large and do *not* become significantly smaller as the jet energy increases.

As an aside, we mention that the same argument thus necessarily also applies for the event-to-event fluctuations in the fraction of the neutral particles (mainly π^0s). These fluctuations are responsible for the poor jet energy resolution of non-compensating calorimeters, especially at high energy, since the response of such calorimeters is usually considerably larger for em showers than for non-em ones.

In the absence of a calorimeter, measurements of jet energy resolutions better than 25–30% can therefore not be expected on the basis of tracker information alone, *at any energy*. And since the contributions of showering charged particles to the calorimeter signals have to be discounted properly for the PFA method to work, the quality of the calorimeter information is in practice very important to achieve the performance needed to separate hadronically decaying W and Z bosons.

8.3.2 *PFA at LEP, the Tevatron and the LHC*

One of the conclusions of the analysis described in [Lob 02] was that the PFA approach may result in improving the jet energy resolution of "poor" calorimeter systems to become "mediocre," but that it will do little for the performance of calorimeters with "mediocre" or "good" resolution.

Figure 8.39a shows the calorimetric jet resolution of a generic LEP detector (the dashed line), as well as the jet resolution one might expect when applying the PFA (originally known as the Energy Flow Method) to jets measured with this detector, assuming that the momenta of the charged jet fragments are precisely known. The results are shown as a function of the jet energy. At low energies, the method is seen to improve the jet resolution by $\sim 35\%$. As the energy increases, the relative improvement slowly decreases, to $\sim 18\%$ at 1,000 GeV. The main reason for this energy-dependent difference is that the jets become increasingly collimated at higher energies. Therefore, the problems due to overlapping showers increase and the slow hadronic fragments that are swept away by the magnetic field represent a decreasing fraction of the jet energy. The error bars in Figure 8.39a indicate the effect of choosing different parameters for the jet fragmentation[34].

Figure 8.39b shows the improvement of the energy resolution that can be achieved for calorimeters with different energy resolutions. The benefits have a tendency to decrease when a better calorimeter system is used. This can be concluded from simulations in which the calorimeter was replaced by one with a hadronic resolution of $43\%/\sqrt{E} + 3\%$ and $e/h = 1.3$. On the other hand, they increase when a more inferior calorimeter system is used (resolution $100\%/\sqrt{E} + 8\%$ and $e/h = 2.0$). These tendencies can be understood by considering the extreme cases: for a perfect calorimeter, there is nothing left for a tracker to improve upon, while for no calorimeter at

---

[34]The authors used a fragmentation function of the type $D(z) = (\alpha + 1)(1 - z)^\alpha / z$ in which $D(z)$ denotes the probability that a jet fragment carries a fraction z of the energy of the fragmenting object. The parameter α can be chosen as desired. It has been demonstrated that a function of this type gives a reasonable description of the fragmentation processes measured at LEP and at the Tevatron, for parameter values $\alpha = 3$ and $\alpha = 6$, respectively [Gre 90].

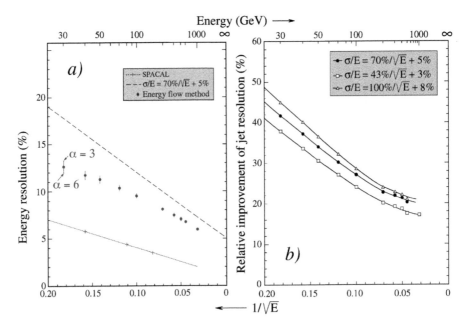

FIG. 8.39. In (a), the jet energy resolution is given as a function of energy, obtained after applying the PFA (originally known as the Energy Flow Method, represented by the full circles), using simulated data from a calorimeter with a jet resolution given by the dashed curve. For comparison, the hadronic resolution of the SPACAL calorimeter is given as well ([Aco 91c], the dotted curve). In (b), the relative improvement of the jet resolution that can be obtained by using PFA is given for various hypothetical calorimeters as a function of the jet energy. Results of Monte Carlo simulations [Lob 02].

all, the tracker still gives a 30% resolution for the jets (Section 7.1.3.1). However, the differences are relatively small for the three simulated systems.

Practical experience so far seems to confirm this assessment. The first experiment in which PFA was elaborately applied was ALEPH [Bus 95], one of the LEP experiments. The hadron calorimeter was not considered a very important component of the LEP detectors, which had, on the other hand, excellent tracking systems. Using a specific (biased) subsample of hadronically decaying Z^0s at rest[35], the authors exploited the properties of this tracking system to the fullest extent and achieved an energy resolution of 6.2 GeV, an improvement of about 25% with respect to the reported hadronic energy resolution of the stand-alone calorimeter system.

The CDF experiment at the Tevatron also used PFA techniques to improve their jet energy resolution. Figure 8.40 shows the effects of including information from the tracking system and the shower max detectors on the measured jet energy, for jets produced at central rapidities, *i.e.*, fragments entering the calorimeter in the barrel region

---

[35] Any event in which energy was deposited within 12° of the beam line, as well as any event in which more than 10% of the total energy was deposited within 30° of the beam line, was removed from the event sample used for this analysis.

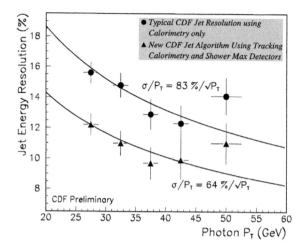

FIG. 8.40.   The effect of including information from the tracking system and the shower max detectors on the jet energy resolution measured with the CDF detector, for jets in the central rapidity (barrel) region [Boc 01].

[Boc 01].

The CMS experiment took advantage of their all-silicon tracking system, plus a fine-grained ECAL, to improve their mediocre jet energy resolution. Figure 8.41a shows the expected improvement of the jet energy resolution if PFA techniques would be used, which ranged from ~ 55% at 20 GeV to ~ 30% at 50 GeV, ~ 20% at 100 GeV and ~ 10% at 500 GeV. The experimental data (Figure 8.41b) show a somewhat smaller improvement, especially at the lowest and highest energies measured for this purpose. However, the improvement that resulted from the use of PFA was also here certainly significant.

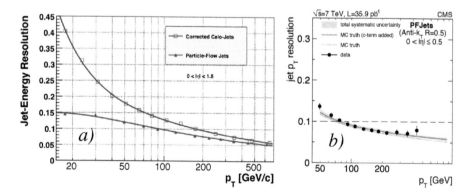

FIG. 8.41.   Simulated jet energy resolution in CMS as a function of transverse momentum, effect of PFA techniques (a). Measured jet energy resolution using PFA techniques as a function of energy (b) [CMS 09].

On the other hand, ATLAS did not observe any benefits, despite extensive efforts. This might be interpreted as a confirmation of the prediction of Lobban *et al.* mentioned at the beginning of this subsection, since the ATLAS calorimeter system measures jets with much greater precision than CMS.

Encouraged by the observed improvements in the jet performance, CMS has recently decided to replace its entire endcap calorimeter system with a dedicated PFA detector [CMS 15b]. This system, which is designed to comprise about six million electronic channels, is scheduled to replace the current endcap calorimeters in the forward region around 2025. It is intended to mitigate the problems of radiation damage and event pile-up, which are expected to have rendered the current system (consisting of $PbWO_4$ crystals, backed up by a brass/plastic-scintillator hadronic section) ineffective by then. The new system will consist of $5\lambda_{int}$ deep fine-grained calorimetry, 40 sampling layers with 1 to 1.5 cm^2 silicon pads as active material, backed up by another $5\lambda_{int}$ of "conventional" calorimetry. Also the tracking system upstream of this calorimeter will be replaced, with upgrades foreseen both in granularity and in η-coverage.

8.3.3 *PFA calorimeter R&D*

A large collaboration, called CALICE, has set out to test the viability of the PFA ideas. In the past 10-15 years, they have constructed a variety of calorimeters, both for the detection of em showers as well as hadronic ones. The replacement of the CMS endcap calorimeters, mentioned in the previous subsection, is based on and inspired by the work of this collaboration [Sef 15].

The calorimeters constructed by CALICE have one thing in common: a very high granularity. The calorimeter modules have a very large number of independent electronic readout channels, $\mathcal{O}(10^4)$ in most modules, up to half a million in one specific case. The active elements are either:

1. Silicon pads, typically with dimensions of 1×1 cm^2,

2. Small scintillator strips, read out by SiPMs,

3. Small Resistive Plate Chambers (RPCs), operating in the saturated avalanche mode.

4. As an alternative, micromegas and GEMs are being tested.

These readout elements are interspersed between layers of absorber material. Typically, tungsten is used for the detection of em showers. Its Molière radius (9.3 mm) is the smallest of all practical absorber materials (platinum is better!), so that the lateral development of the em showers is limited as much as possible. This is important for separating showers from several particles that enter the calorimeter in close proximity. For the deeper sections of the calorimeter, typically stainless steel is being used. Figure 8.42 is often shown to illustrate the advantage of using tungsten. The nuclear interaction length and especially the Molière radius, which determine the extent of the lateral shower development for hadronic and electromagnetic showers, respectively, are considerably smaller than for steel. Of course, what really matters here is the *effective* value of these parameters in the calorimeter, which also includes low-Z materials such as plastic, silicon and air. A second thing to keep in mind is that the particles in reality

$X_0 = 1.8cm, \lambda_I = 17cm$

$X_0 = 0.35cm, \lambda_I = 9.6cm$

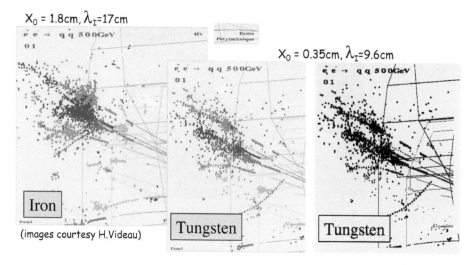

(images courtesy H.Videau)

FIG. 8.42. Simulated shower development of jet fragments in a calorimeter based on iron (left) or tungsten (center, right) as absorber material.

are not colored. The difference between the colored and bitmap versions of the tungsten image illustrates that the task to assign calorimeter hits to individual jet fragments may in practice be quite daunting indeed, even in the densest possible absorber structures.

The largest calorimeter that was specifically designed for em shower detection is a tungsten/silicon device [Rep 08]. It has an active surface area of 18×18 cm^2 and is 20 cm deep, subdivided longitudinally into 30 layers. The first 10 layers are $0.4X_0$ thick, followed by 10 layers of $0.8X_0$ and finally another 10 layers of $1.2X_0$, for a total absorption thickness of $24X_0$. The active layers consist of a matrix of PIN diode sensors on a silicon wafer substrate. The individual diodes have an active surface area of 1×1 cm^2, and there are thus $18 \times 18 = 324$ calorimeter cells per layer, 9,720 in total. These are read out by means of a specially developed Application-Specific Integrated Circuit (ASIC).

CALICE also built and tested a large hadron calorimeter, a sandwich structure based on 38 layers of 5 mm thick plastic scintillator, interleaved with absorber plates [Adl 10]. For this instrument, they either used 17 mm thick steel or 10 mm thick tungsten plates. This absorber material thus represents a total thickness of about $4\lambda_{int}$ in both cases. The active layers are housed in steel cassettes with 2 mm cover plates on both sides. This increased the total depth of the instrumented volume to $\sim 5.3\lambda_{int}$. The transverse dimensions of the active layers are 90×90 cm^2. Figure 8.43 shows a picture of one of the active layers. The layer is subdivided into tiles, small ones in the central region and larger ones in the outer regions (and also in the rear of the calorimeter module). The smallest tiles measure 3×3 cm^2. Each tile has a circular groove in which a wavelength shifting fiber is embedded. This fiber collects the scintillation light produced in the tile, re-emits the absorbed light at a longer wavelength and transports it to a SiPM, which converts it into an electric pulse. In total, this calorimeter contains 7,608 tiles

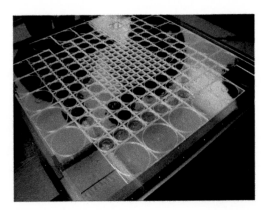

FIG. 8.43. An active plastic-scintillator plane, used to detect the signals in the scintillator based CALICE hadron calorimeter [Adl 10].

(*i.e.*, electronic channels). This was the first large-scale application of SiPMs in a particle detector.

Another CALICE module has a lateral cross section of ≈ 1 m^2 and a similar depth as the previously mentioned one. The effective depth can be varied through the choice of the absorber material and the thickness of the absorber plates. In between each two plates an array of RPC cells with dimensions of 1×1 cm^2 is inserted, *i.e.*, about 10,000 per plane [Dra 07]. In total, there are 54 independent longitudinal segments, so that the total number of active elements is about half a million. These RPCs operate in the saturated avalanche mode, and thus provide a "yes" or "no" signal when a particle develops a shower in this device. This is thus a "digital" calorimeter. An event display in this detector consists of a pattern of RPC cells that fired when the particle that created it was absorbed. These patterns may be very detailed, as illustrated by the example shown in Figure 8.44.

The performance of the mentioned devices as stand-alone calorimeters is not particularly impressive. For example, the em energy resolution of the W/Si em calorimeter was reported as $16.5\%/\sqrt{E} \oplus 1.1\%$ [Adl 09], which is almost twice as large as that of the ATLAS lead/liquid-argon ECAL [Aha 06]. The proposed new endcap calorimeter for CMS, which is based on this CALICE design has an envisaged em energy resolution that is even worse (20–24%/$\sqrt{E}$ [CMS 15b], considerably worse than the resolution provided by the crystals it will replace.

In stand-alone mode, the hadronic energy resolution of the iron/plastic-scintillator calorimeter was reported as $57.6\%/\sqrt{E} \oplus 1.6\%$, for a heavily biased event sample and after applying corrections that will not be applicable in a collider experiment [Adl 12]. Since this calorimeter was not deep enough to fully contain high-energy hadron showers, the event sample used to obtain this result was limited to showers that started developing in the first few layers of the calorimeter module. When this device was combined with the high-granularity W/Si ECAL, the resolution deteriorated, as illustrated in Figure 8.45 [Sef 15]. This figure shows the signal distribution for 80 GeV pion showers

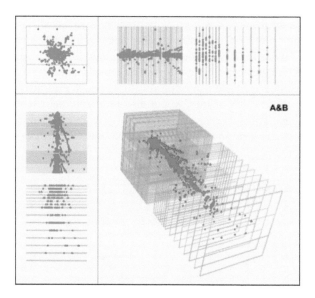

FIG. 8.44. Event display for a 120 GeV π^- showering in the CALICE digital hadron calorimeter [Rep 13, Sef 15].

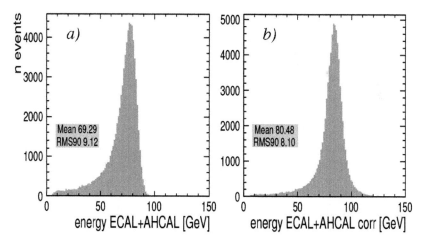

FIG. 8.45. The line shape of the CALICE W/Si + Fe/plastic combination for 80 GeV pion showers, before (a) and after (b) correction procedures were applied based on the starting point of the showers and the estimated leakage [Sef 15].

before (a) and after (b) a variety of corrections were applied. The resulting energy resolution is more than twice as large as the value reported by ZEUS [Beh 90].

More cumbersome than the unremarkable energy resolutions reported for these calorimeters is the response non-linearity. Figure 8.46 shows the average signal of the

Fe/plastic detector for positrons, as a function of energy [Adl 11a]. The measured data points exhibit a significant response non-linearity, namely a $\sim 10\%$ decrease in the energy range from 10 to 50 GeV. According to the authors, this is due to saturation of the SiPM signals, and they expect that this may be remedied when SiPMs with a larger dynamic range become available. Unfortunately, not enough information is supplied to verify this explanation which, if true, would also invalidate the energy resolution reported by the authors. This is because, as a matter of principle, signal saturation implies that the fluctuations that determine the energy resolution are partially suppressed (see Section 9.3.3 for more on this subject).

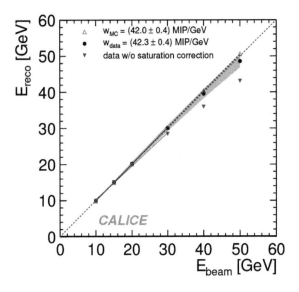

FIG. 8.46. The average signal for positrons in the CALICE analog hadron calorimeter, as a function of the beam energy. Shown are the measured data points, before and after corrections for saturation in the SiPM readout were applied, and the Monte Carlo prediction. The shaded area represents the systematic uncertainty in the corrections [Adl 11a].

The signal saturation phenomenon reaches very substantial proportions in the "digital" calorimeter built by CALICE. The resulting response non-linearity (Figure 9.20) is even so large that it leads to apparent *overcompensation* [Rep 12, Sef 15]. Because of the large suppression of fluctuations in the shower development process, the quoted energy resolutions are not very meaningful. The CALICE Collaboration has apparently also realized these problems and has embarked on equipping the RPCs with a 2-bit readout system. This provides the possibility to subdivide the signals into three categories, on the basis of different threshold levels. This is called the "semi-digital" option [Ste 14]. However, the RPCs still operate in avalanche mode, and the relationship between the different thresholds (corresponding to bit settings 1/0, 0/1 and 1/1, respectively) and the deposited energy is not a priori clear. At the moment of this writing, this

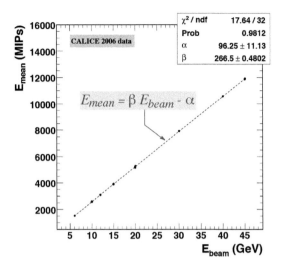

FIG. 8.47. The average signal measured for electrons in the CALICE W/Si calorimeter, as a function of the beam energy, together with a fit to the experimental data [Adl 09].

is still a work in progress.

An effect with a different origin than signal saturation may be responsible for the response non-linearity observed in the silicon based em calorimeter (Figure 8.47, [Adl 09]). The authors showed that the experimental data could be fit with a straight line and concluded, incorrectly, that this is evidence for linearity (see Section 9.3.5). The straight line does not extrapolate to zero for zero deposited energy. The average signal measured per unit deposited energy gradually increases, by $\sim 5\%$ over the energy range from 6 to 45 GeV (Figure 9.26). This phenomenon may well be a consequence of the calibration procedure, which is the topic of the next subsection, and have the same origin as the AMS problem (Figure 9.11) described in Section 9.1.3.

The large signal-nonlinearity observed for small signals is important since high-energy jets, such as the ones from the hadronic decay of intermediate vector bosons and the Higgs boson, consist of a considerable number of low-energy final-state particles, which together represent a significant fraction of the jet energy. Quantitative information on this point is given in Section 8.2, Figure 8.3.

8.3.4 Calibration

Calibration of these megachannel PFA devices must be a daunting task. The basis for the calibration is what is referred to as the MIP scale, which is obtained by exposing each active element to enough muons to determine the average signal, in ADC counts. These averages are equalized for all channels. Figure 8.48 shows a distribution of 6471 calibration constants (expressed in ADC counts per mip) derived from an exposure of the em CALICE calorimeter to muons [Adl 07] . The MIP scale is determined on the basis of these constants and the local sampling fraction for mips in the vicinity of the position where the readout element in question is located. For example, in the W/Si em

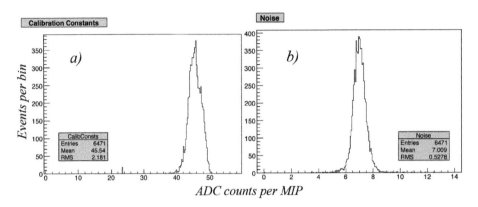

FIG. 8.48. Distributions of the calibration constants (*a*) and the rms noise (*b*) for 6471 channels of the em section of a CALICE calorimeter [Adl 07].

calorimeter the tungsten plates in the first section have a thickness of 1.4 mm, and the average energy deposited by a mip traversing such a plate is 3.1 MeV (see Table B1)[36]. The energy deposited in a readout element located in this section is thus calculated by multiplying the observed signal (in terms of that recorded for a muon) times 3.1 MeV. For example, a signal measured to be 15.8 times the average value of that of a muon is attributed an energy of $15.8 \times 3.1 = 49$ MeV. The second section of the W/Si em calorimeter consists of tungsten plates that are twice as thick (2.8 mm). A mip would deposit twice as much energy in such a plate and, therefore, a signal measured to be 15.8 times the average value of a muon recorded in a readout element located in that section would be assigned an energy of $15.8 \times 6.2 = 98$ MeV. This is the MIP scale.

Until this point, the calibration is rather straightforward, albeit laborious with so many channels to take care of. However, these calorimeters are intended to measure the energy of em and hadronic *showers*, and that is not the same as measuring the energy deposited by a muon traversing the detector. Not only is the conversion factor from MIPs into deposited energy different for these different particles (electrons, hadrons and muons, due to things such as e/h and e/mip ratios), it also varies *within a developing shower*.

Consider, for example, em showers [Wig 06b]. The sampling fraction of a given calorimeter structure depends on the stage of the developing showers. In calorimeters consisting of high-Z absorber material (*e.g.*, lead) and low-Z active material (plastic, liquid argon), the local sampling fraction decreases by as much as 25 to 30% as the shower develops. In the early stage, the shower still resembles a collection of mips, but as it develops, soft (< 1 MeV)γs increasingly dominate the energy deposition pattern. These γs are much less efficiently sampled than mips in this type of structure, where dominant processes such as the photo-electric effect and Compton scattering strongly favor the high-Z absorber material (Figure 8.49).

---

[36]The CALICE authors use a slightly different number, 3.4 MeV.

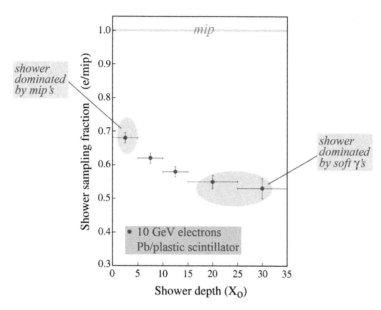

FIG. 8.49.   The sampling fraction changes in a developing shower. The e/mip ratio is shown as a function of depth, for 10 GeV electrons in a Pb/scintillating-plastic calorimeter. Results of EGS4 calculations [Wig 06b].

Similar phenomena also play a role in non-em shower components, where different particles (pions, non-relativistic protons, neutrons) may be sampled very differently. And since the composition of these showers changes with depth (age), the relationship between deposited energy and resulting signal also becomes a function of depth.

In order to correctly determine the deposited energy that corresponds to a given signal measured in a PFA calorimeter cell, both the type of particle responsible for this signal and the stage of its shower development when this signal was produced must be known. This strikes me as an impossible task. It needs to be stressed that this problem cannot be solved with an overall depth dependence of the calibration constants (*i.e.*, the conversion factors from the MIP scale to the energy scale). One reason for this is that hadron showers may contain em components that penetrate very deep into the absorber structure. This is illustrated in Figure 8.50, which shows longitudinal shower profiles for twelve randomly selected events in which 270 GeV pions were absorbed in a sampling calorimeter of which the signals from each individual sampling layer were read out separately [Gre 94]. Showers initiated by energetic π^0s produced in the developing shower manifest themselves as narrow peaks in these profiles. Such peaks were frequently observed at depths of $3\lambda_{int}$ or deeper, *i.e.*, way beyond the region traditionally covered by the em section of a calorimeter system.

The papers in which the CALICE results are described talk about these problems in general terms: overall corrections to set the "em" and "hadronic" energy scales, weighting schemes, χ^2 minimizations, corrections for boundary effects between em

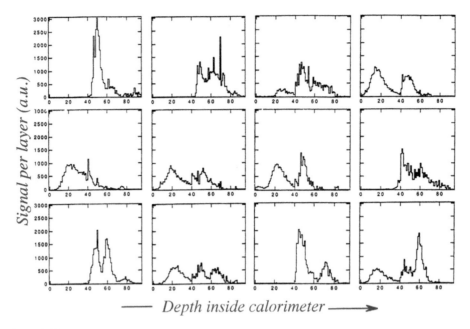

FIG. 8.50.   Longitudinal energy deposit profiles for twelve randomly selected 270 GeV π^- showers in a
lead/iron/plastic-scintillator calorimeter. This calorimeter test module had 100 layers, with a total depth
of 6λ. Experimental data from [Gre 94].

and hadronic calorimeter sections and between different calorimeter modules. And everything is being "validated" with Monte Carlo simulations, which lack sufficient precision. How difficult these problems may be in practice is illustrated by the experience of
ATLAS, whose em calorimeter consists of only three longitudinal segments [Aha 06]. It
took the Collaboration several years to figure out how to intercalibrate these segments.
See Section 9.1.4 for details on this.

Let us now look again at Figure 8.49, which shows how the sampling fraction
evolves with depth. The shower gradually changes from a collection of mips into a
collection of soft γs, and the latter are less efficiently sampled in this high-Z calorimeter. The figure concerns 10 GeV electrons. If the MIP scale would be used to calculate
the energy of these electrons, the result would be that the energy would be underestimated, to be something like 6.5 GeV. In order to convert the MIP scale to the em scale,
CALICE would multiply all signals by a factor 10/6.5.

However, the relative underestimation of the energy deposited in this calorimeter
would probably be larger for 1 GeV electrons (*e.g.*, 40% instead of 35%), since these
showers would resemble more a collection of soft γs than the 10 GeV ones. On the
other hand, the underestimation of the energy would be smaller for 100 GeV showers
(*e.g.*, 30% instead of 35%). Such showers differ from the low-energy ones because they
possess an early, strongly collimated component consisting predominantly of mip-like
particles. In a sense, 1 GeV showers may be considered to represent the late stages

of a 100 GeV shower, and the early stages of the latter are more mip-like than the 1 GeV showers[37]. If CALICE applied the same overall correction factor to account for the average difference between the em scale and the MIP scale at all energies, then the above considerations might explain the response non-linearity observed for positrons in the W/Si calorimeter.

The measurements of the response linearity (Figure 8.47) are restricted to the energy range 6 to 45 GeV. It would be interesting to extend these measurements to lower energies and see how the observed non-linearity behaves in the range 0 to 6 GeV. As pointed out in the previous subsection, the latter energy range is also very important for the jet energy measurements envisaged by CALICE.

8.3.5 *PFA algorithms*

Experimental verification of the merits of the envisaged PFA method involves much more than checking the validity of the applied calibration procedures. The calibration issues only address the problems associated with measuring the energy deposited in the calorimeter. However, the fundamental problem faced in the application of the PFA method is how to avoid double counting, *i.e.*, how to eliminate from the measured calorimetric energy deposit pattern the contributions from charged hadrons, whose momenta have been measured by the tracking system. In that sense, a correct calibration is somewhat less important in these PFA calorimeters, because a major purpose of these fine-grained devices is *to throw away the information they provide*.

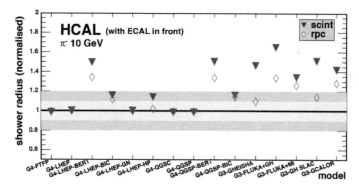

FIG. 8.51. The radius of 10 GeV π^- showers in a sampling calorimeter read out with scintillator or with gas (RPCs), as calculated in fifteen different hadronic shower simulation models [Adl 07].

The fundamental problem here is that there is no such thing as a "typical hadronic shower profile" that could be used for this purpose. Figure 8.50 shows the longitudinal energy deposit profiles of twelve randomly selected events of 270 GeV π^- showering in a Pb/Fe/plastic-scintillator calorimeter [Gre 94]. The lateral shower profiles exhibit similarly large event-to-event variations.

---

[37] Support for this explanation can be derived from Figure 6.12.

Since the main problem for resolving the confusion caused by overlapping energy deposit profiles of neighboring particles is the *lateral* extent of the developing showers, it is important to note that there is no consensus about this aspect of hadron showers among the different available Monte Carlo packages that are used in particle physics. Figure 8.51 compares the effective radius of 10 GeV π^- showers in the CALICE calorimeters (W/Si ECAL + Fe HCAL), as calculated in fifteen different hadronic shower simulation programs [Adl 07]. Differences of more than 50% are observed among the different packages represented in this plot.

It should be mentioned that the wealth of experimental data obtained with the various CALICE calorimeter modules has been an important inspiration for the developers of these Monte Carlo simulations to "fine-tune" their code. Unfortunately, despite many efforts, the important radial width of hadron showers remains systematically underestimated (Figure 8.52). However, even if there would be consensus and agreement on this radial width issue, it would not be sufficient for an adequate PFA algorithm. The resolution of calorimeters is determined by *fluctuations*. Therefore, successful elimination of double counting depends on how this can be achieved *event by event*, not on average. In that sense it is encouraging to notice that in Figure 8.52, the relative width and the shape of the various distributions, which are a measure for the event-by-event fluctuations, seem to be better reproduced than the mean value.

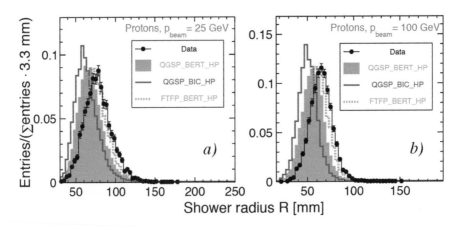

FIG. 8.52. Distribution of the energy weighted mean radius of proton showers of 25 (*a*) and 100 (*b*) GeV in the W/scintillator CALICE hadron calorimeter, together with the predictions of the most commonly used hadronic shower Monte Carlo simulations [Sic 13].

CALICE also uses detailed experimental information on the sub-structure of hadron showers to compare measured event-by-event fluctuations with simulated ones. An example of this is given in Figure 8.53 [Adl 13a], which shows results from an attempt to measure track multiplicities in developing hadron showers. To that end, they use an algorithm searching for secondary tracks in the developing shower, which point to em sub-showers (caused by π^0s produced by a particle produced in an upstream interaction) or other centers of dense activity. The left frame of Figure 8.53 shows the multiplicity distribution of such tracks in showers induced by 25 GeV pions in their Fe/scintillator hadron calorimeter; the right frame shows the average multiplicity as a function of the energy of the showering particle. The lower panels show the differences between the experimental results and several Monte Carlo simulation packages. Disregarding the apparent discrepancy between some of the results shown in the top and bottom parts of the right panel, it seems that this aspect of the shower development is best reproduced by the QGSP_BERT physics list,

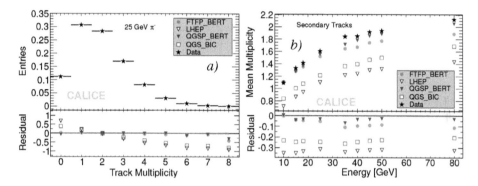

FIG. 8.53. Track multiplicity distribution for 25 GeV pion showers (a) and the average number of secondary tracks as a function of energy (b) in the CALICE Fe/scintillator hadron calorimeter. The lower panels show the residuals of the comparisons with several hadronic Monte Carlo simulations [Adl 13a].

The algorithms used for PFA make use of this sub-structure. The name of one of the algorithms was even inspired by the tree-like shower sub-structure: ARBOR [Rua 14]. It attempts to tag the trajectory of charged particles produced in the shower development and uses this information to reconstruct showers on an event-by-event basis. The terminology used in this algorithm distinguishes "seeds, branches and leafs" in the event topology.

Another algorithm generally used for this purpose is called Pandora [Tho 09]. It was specifically developed in the context of design studies for experiments at the ILC. It is a very complex algorithm, containing more than 10,000 lines of C++ code. It is beyond the scope of this text to discuss in detail how it works. However, the following observations are in order:

1. The author makes unrealistic assumptions about the performance of the calorimeters used for evaluating the performance of his algorithm. Both the assumption of an energy resolution of $15\%/\sqrt{E}$ for photons and of $55\%/\sqrt{E}$ for neutral hadrons have not been met by any of the prototypes built by the CALICE Collaboration. The effects of the response non-linearities that were evident in the performance of these devices were not taken into account at all.

2. The author deliberately ignores the contributions of neutrons to the signals, with the (understandable) argument that these are impossible to assign to a particular

shower. However, neutron contributions are essential for achieving a good energy resolution for calorimeters with hydrogenous readout, such as plastic scintillator. By ignoring these contributions, the assumptions made about the energy resolution for detecting the neutral hadron component of the jet are even more unrealistic.

3. The jet energy distributions obtained with this algorithm are extremely non-Gaussian. Normally speaking, the energy resolution should thus be quoted in terms of σ_{rms}. However, in order to reduce the effects of the tails of these distributions, the author introduces a new variable to characterize the precision with which the jet energy resolution can be determined, which he calls rms_{90}. For the record, it should be mentioned that, even for a perfectly Gaussian signal distribution, this variable gives a result that is 21% better (smaller) than the σ that would result from a Gaussian fit to this distribution.

Keeping this in mind, the results presented by the author for jets produced by fragmenting (anti-)quarks from the decay of Z^0 bosons in the ILC and detected by the ILD experiment are listed in Table 8.1.

Table 8.1 *Results obtained with the Pandora algorithm for the precision with which the energy can be reconstructed for jets produced by fragmenting (anti-)quarks from the decay of Z^0 bosons in the ILC and detected by the ILD experiment. Shown are for each jet energy the values of σ_{rms} and rms_{90}, in GeV [Tho 09].*

Jet energy (GeV)	σ_{rms} (GeV)	rms_{90} (GeV)
45	3.4	2.4
100	5.8	4.1
180	11.6	7.6
250	16.4	11.0
375	29.1	19.2
500	43.3	28.6

The relative precision with which the energy of these jets can be reproduced, expressed in terms of σ_{rms}/E and rms_{90}/E, is also plotted as a function of energy in Figure 8.54. For comparison, the hadronic energy resolution of the SPACAL calorimeter (the σ from the fit to a Gaussian signal distribution) is shown as well. It turns out that the latter is equally good at low energy, and better at high energy, compared to the Pandora rms_{90} value, which, it should be stressed, is incorrectly referred to as the energy resolution by the author of [Tho 09], as well as in numerous CALICE publications.

The author also provides information about the different contributions to the precision with which the jet energy can be determined. Figure 8.55 shows that the increase in rms_{90} with energy is due to the "confusion" term. This is because it becomes increasingly difficult to disentangle a jet into its various components as the energy increases and the jet becomes more and more collimated. Interestingly, the plot also suggests that the author believes that the highly granular calorimeters used for PFA provide a better energy resolution in stand-alone mode than "traditional" calorimeters. I am not aware

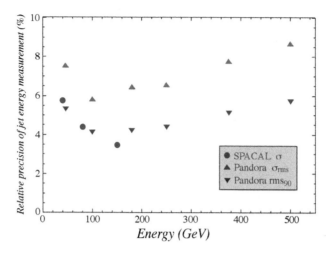

FIG. 8.54. The relative precision with which the energy can be determined for jets produced by fragmenting (anti-)quarks from the decay of Z^0 bosons in the ILC and detected by the ILD experiment, as a function of energy. Shown are the σ_{rms} and rms_{90} values obtained with the Pandora algorithm [Tho 09]. For comparison, the published hadronic energy resolution of the SPACAL calorimeter is shown as well [Aco 91c].

of any experimental information that would support this belief.

One annoying aspect of this entire business is the tendency of PFA authors to equate their rms_{90} values to the energy resolution. This unfortunate habit is reflected in Figure 8.55, where the PFA data are expressed in terms of rms_{90}, as indicated by the vertical axis, and where the genuine energy resolution of the "traditional" calorimeter is plotted in a way that explicitly assumes that $\sigma_{Gauss} = rms_{90}$. The suggestion that these two variables represent the same thing, which is implied by this figure, is extremely misleading.

The main goal of all these algorithmic exercises was of course to investigate to what extent hadronically decaying W and Z bosons could be separated by means of a PFA based detector system. Figure 8.56 shows the answer to that question, for WW and ZZ pairs produced in e^+e^- collisions at a center-of-mass energy of 250 GeV. According to Table 8.1, this result was obtained for a total energy resolution (σ_{rms}) of 16.4 GeV (6.6%). Figure 8.54 suggests that a good stand-alone calorimeter should be able to do as well, if not better.

I want to conclude the discussion about the Pandora algorithm by quoting the author:

"The performance of PFlow calorimetry depends strongly on the reconstruction software. For the results obtained to be meaningful, it is essential that both the detector simulation and the reconstruction chain are as realistic as possible. The use of Monte Carlo information at any stage is likely to lead to an overly-optimistic evaluation of the potential performance of PFlow calorimetry."

It is indeed very good to keep this in mind.

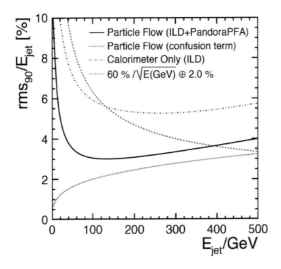

FIG. 8.55.   The empirical functional form of the jet energy resolution obtained from PFlow calorimetry
(PandoraPFA and the ILD concept). The estimated contribution from the confusion term only is shown
separately (the dotted curve). The dot-dashed curve shows a parameterisation of the jet energy resolution obtained from the total calorimetric energy deposition in the ILD detector. In addition, the dashed
curve, $60\%/\sqrt{E} \oplus 2.0\%$, is shown to give an indication of the resolution achievable using a traditional
calorimetric approach [Tho 09].

8.3.6   Experimental checks of the PFA approach

Given the complications associated with the calibration of the calorimeters and the limitations of the PFA algorithms, it is *extremely important* to check the merits of this
approach *experimentally*, using as little simulation as possible. One method that has
been repeatedly suggested, and which has actually been applied to assess the jet performance of the CMS calorimeter system, is described below. The suggested method
would exploit the fact that the CALICE Collaboration has accumulated, over the years,
a database containing about 500 million events recorded in their various calorimeters.
Events induced by beams of electrons, positrons, $\pi^{\pm}$, $\mu^{\pm}$ or protons spanning an energy
range from 1 to 100 GeV.

The goal of the described procedure is to determine the signal distributions for
hadronically decaying W and Z bosons. First, a decay mode is selected, say $W^{+} \rightarrow u\bar{d}$.
Then, a fragmentation function is used to determine the final state particles into which
each of the two quarks hadronizes. The four-vectors of the listed particles are the basis for determining the calorimeter signals. For example, if the list contains a π^{-} with
a momentum of 11.8 GeV/c, a random event from a beam test with 10 GeV pions is
chosen and the measured signals in all contributing calorimeter cells are multiplied by
a factor 1.18 (11.8/10). The energy deposit profile from this particle has to be shifted in
the transverse plane from the center of the detector (which is taken to be the intersection
between the direction of the fragmenting quark and the calorimeter's front surface) to
the point where the particle in question hits the calorimeter front surface. This distance

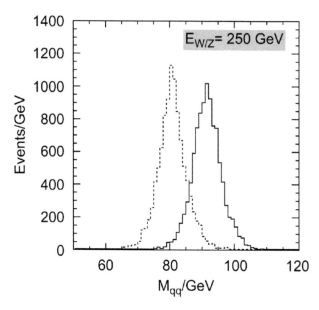

FIG. 8.56. The reconstructed invariant mass distribution for the hadronic system in the processes $ZZ \rightarrow d\bar{d}\nu\bar{\nu}$ and $WW \rightarrow u\bar{d}\mu^-\bar{\nu}_\mu$, simulated in one of the detectors proposed for the International Linear Collider, at a total collision energy of 250 GeV [Tho 09].

is determined by the x, y components of the momentum vector, by the magnetic field vector and by the distance between the production vertex and the calorimeter front surface. The latter two properties have to be fixed for this analysis. For conditions such as in CMS (a 4 Tesla field inside a cylinder with a radius of about 1 meter), the displacement of this particular jet fragment (assuming a typical p_T of 0.3 GeV/c with respect to the jet axis) would amount to 2.5 cm as a result of transverse momentum, combined with 5.5 cm as a result of bending in the magnetic field.

If the next particle in the list is a proton with a momentum of 4.2 GeV/c, a random event from a 5 GeV/c proton run is selected and the signal from each contributing calorimeter cell multiplied by a factor 0.84 (4.2/5). The displacement of this particle, and its generated energy deposit profile, with respect to the jet axis as a result of the transverse momentum component and bending in the magnetic field, are to be calculated as described above. If the next particle in the list is a 18.2 GeV γ (e.g., from the decay of an energetic π^0), a random event from a 20 GeV electron run is chosen and the measured signals multiplied by a factor 0.91 (18.2/20). In this case, the displacement of the energy deposit profile with respect to the jet axis is only determined by the transverse momentum vector, since the magnetic field does not affect the path traveled by the γ.

And so on, until calorimeter signals have been assigned for each final-state fragment of the jet. For neutral hadrons, such as K^0s or neutrons, one should use random events from the pion or proton runs, leaving out the (small) ionization energy deposits upstream of the starting points of the showers.

The combined energy deposit profiles of all the jet fragments represent the energy deposit profile for this particular jet. This procedure may be repeated a large number of times for the same jet, by choosing different random events from the vast database to compose a total calorimeter signal for the listed collection of jet fragments. The procedure may also be repeated for a completely different jet, as well as for different decay modes of the intermediate vector boson.

A procedure of this type was carried out to assess the jet response functions for hadronically decaying W and Z bosons in CMS. The resulting signal distributions obtained in this way are shown in Figure 9.22. The described procedure would also be ideally suited to test claims about the performance of PFA based systems experimentally. To that end, one should apply one's favorite PFA algorithm to eliminate the contributions from charged hadrons, of which the impact points and four-vectors are precisely known, from the corresponding jet energy deposit profiles. When that is done, the remaining calorimeter energy can be determined from the remnants of this profile. This energy should then be added to the (precisely known) energy of the charged hadrons, and the result is the (PFA) jet energy.

For each collection of final-state particles a very large number of different jet energy deposit profiles could be constructed, by randomly choosing different events from the vast database. Also, the collection of final-state particles could be changed by letting the quarks fragment in a different way. A procedure of this type would result in jet response functions that are based as closely as possible on the experimentally measured information.

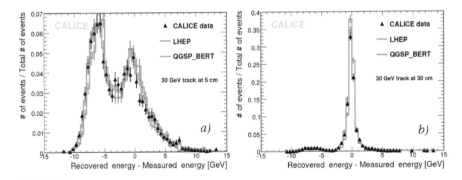

FIG. 8.57. Difference between the recovered energy and the real energy for a neutral 10 GeV hadron hitting the calorimeter at 5 cm (*a*) or 30 cm (*b*) from the shower axis of a 30 GeV charged hadron. The experimental data, taken with a combination of the W/Si ECAL and the Fe/scintillator HCAL are compared to Monte Carlo predictions [Adl 11b].

Until now, experimental verification of the validity and claims of the PFA methods, as suggested above, has been lacking. CALICE has performed a specific study of what they call the "confusion effect" (Figures 8.57 and 8.58.). The energy deposit profiles of two hadron showers developing in the W/Si ECAL + Fe/scintillator HCAL calorimeter system are combined together. The distance between the two impact points, *i.e.*, shower axes, is varied between 5 cm and 30 cm. Figure 8.57 shows how well the energy of a 10 GeV hadron can be recovered by the PFA algorithm, in the vicinity of a 30 GeV hadron, for a distance of 5 cm (*a*) or 30 cm (*b*). The peak around -6 GeV in Figure 8.57a indicates that the most likely energy to be assigned to this 10 GeV particle is 4 GeV when the showers are separated by only 5 cm. Not surprisingly, proper recognition of the 10 GeV is much better when the 30 GeV shower is 30 cm away.

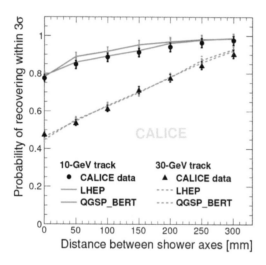

FIG. 8.58. Probability of recovering the energy of a 10 GeV neutral hadron within 3 standard deviations from its real energy in the vicinity of showers induced by charged 10 GeV (circles and continuous lines) or 30 GeV (triangles and dashed lines) hadrons. This probability is shown as a function of the distance between the shower axes, for beam data taken with a combination of the W/Si ECAL and the Fe/scintillator HCAL, and for Monte Carlo simulated data [Adl 11b].

In Figure 8.58, the probability that a 10 GeV hadron is reconstructed with its proper energy ($\pm$ 3 standard deviations) when it hits the calorimeter in the vicinity of another 10 GeV (dots) or 30 GeV (triangles) hadron is plotted as a function of the distance between the two shower axes. Since the relative hadronic energy resolution is reported to be about 20% at 10 GeV, the 3σ criterion means that everything reconstructed with an energy between 4 and 16 GeV counts as "recovered" in this plot. These results give some insight in the selectivity of the PFA algorithm, since even when two 10 GeV showers are completely merged (0 cm distance between the shower axes), it apparently recognizes an excess energy between 4 and 16 GeV (over and above the 10 GeV that was removed since a 10 GeV/*c* track pointed to the shower axis) 80% of the time. Such close proximity of showering jet fragments is probably not very frequent, but jet fragments with a momentum of 10 GeV/*c*, or less, are very common in the hadronic decay of intermediate vector bosons.

Unfortunately, the paper describing these studies is not very clear on specifics. Are the selected events always the same, or are they chosen at random, or is one of them the same and the other one chosen at random, *etc.*? Yet, this is an important and valuable study to assess the practical value of the PFA concept, and the Collaboration should be encouraged to do much more in this respect with their half-billion recorded events.

8.3.7   *Future applications of PFA in particle physics*

Despite the many remaining question marks and the lack of reliable experimental verification of the advocated merits, several experiments at proposed and/or planned accelerator facilities have adopted the PFA approach for detecting jets in their detector design. These include the experiments at the proposed International Linear Collider (ILC, [Beh 13]) and at the CERN Linear Collider (CLIC, [Lin 12]). As mentioned in Section 8.3.2, the CMS experiment at CERN's Large Hadron Collider has decided to replace their endcap calorimeter by a PFA based structure [CMS 15b].

One of the early ILC concept detectors that planned to use Particle Flow Analysis did not gamble entirely on the advocated PFA merits, but chose a compensating lead/plastic-scintillator hadronic calorimeter section. Measurements of hadronic showers developing in prototypes of this detector (called GLD) gave a resolution of ~ 4.5 GeV in the most relevant 80–90 GeV region [Uoz 02], in good agreement with earlier measurements for a similar structure [Beh 90].

8.4   Concluding remarks

High-quality energy measurements will be an important tool for accelerator-based experiments at the TeV scale. There are no fundamental reasons why the four-vectors of all elementary particles could not be measured with a precision of 1% or better at these energies. However, reaching this goal is far from trivial, especially for the hadronic constituents of matter. Unfortunately, little guidance is provided by hadronic Monte Carlo shower simulations in this respect. In the past thirty years, progress in this domain has, therefore, primarily been achieved through dedicated R&D projects, and this is still the way to go today.

Several major R&D efforts are underway to further improve the quality of hadronic energy measurements. These efforts are characterized by very different styles. R&D in the framework of the Particle Flow Analysis concept is to a large extent concentrated on the technicalities of detector design, whereas very fundamental questions concerning crucial aspects of the applicability of this concept (*e.g.*, calibration) tend to be ignored. On the other hand, the dual-readout R&D project concentrates strongly on experimental tests of the validity of the principles on which improvement of the hadronic calorimeter performance is based, and tends to ignore issues concerning the incorporation of this type of detector into a 4π experiment, and simulations in general.

Until now, the dual-readout approach has been remarkably successful. It combines the advantages of compensating calorimetry with a reasonable amount of design flexibility. Since there is no limitation on the sampling fraction, the factors that limited the energy resolution of compensating calorimeters (SPACAL, ZEUS) to $\sim 30\%/\sqrt{E}$ can be eliminated, and the theoretical resolution limit of $\sim 15\%/\sqrt{E}$ seems to be within reach. Dual-readout detectors thus hold the promise of high-quality calorimetry for *all* types of particles, with an instrument that can be calibrated with electrons.

The potential of the PFA concept for obtaining excellent jet performance may be real, but will need to be *experimentally* demonstrated. Given the cost and the time scales involved, the development of the complex multi-million channel detectors needed for this purpose will require solid experimental proof of the merits of this approach.

9

ANALYSIS AND INTERPRETATION OF TEST BEAM DATA

The physics on which calorimetry is based is quite complicated. The calorimetric performance of a particular instrument is affected in intricate and sometimes counter-intuitive ways by the usual design parameters, such as the dimensions of the active and passive materials. As a result, the development of calorimeters from relatively crude devices for very specific applications (*e.g.*, neutrino scattering) into the precision instruments that form the prime experimental tool in many modern experiments, did not happen overnight.

Much of what is known about calorimetry and about the factors that determine and limit the performance of calorimeters has been learned from dedicated R&D projects, not linked to any particular particle physics experiment. In the previous chapter, two such projects are described in some detail. But also the prototype work for several major experiments has provided very useful information in this context. A shining example: the ZEUS experiment.

Typically, the experimental information is obtained by exposing the calorimeter module(s) to beams of particles provided by an accelerator. Many accelerator centers provide dedicated beams for this purpose. Whereas some facilities provide photon beams, in general, charged particles are used to test calorimeters. The detectors are exposed to beams of protons, pions, muons, electrons and sometimes kaons. A system of magnetic dipoles and quadrupoles, as well as collimators makes it possible to select particles within a certain momentum bite and focus these onto a small beam spot in the detector test area.

Note that the conditions in these "beam tests" may be quite different from those encountered in the experimental circumstances for which the detector is intended. In "real life," it is typically not known what type of particle is entering the calorimeter, where it is entering and what its momentum is. The detected signals may also very well be the result of several particles entering simultaneously. This should be kept in mind when interpreting the results of the beam tests.

The analysis and the interpretation of calorimeter data obtained in beam tests is the topic of this chapter. Several common mistakes that are made in practice are discussed, using examples from the available literature. This is a very tricky subject, almost anyone who has ever been involved in (the analysis of data from) calorimeter beam tests has fallen into one of the many traps, including myself. The examples I have chosen for this chapter are not intended to criticize anyone in particular, but rather to illustrate the issues as clearly as possible.

9.1 Analysis of the Measured Test Beam Data

One of the most common mistakes made when analyzing the performance of a calorimeter derives from the selections that are made to define the experimental data sample. This selection process may easily lead to biases, which distort the performance characteristics one would like to measure. In some extreme cases, this may lead to very wrong conclusions, as in the example discussed below.

9.1.1 *Biased event samples*

In 1974, Radeka and Willis built the first calorimeter based on direct detection of the non-amplified ionization charge produced in the shower development process. Their calorimeter consisted of thin iron absorber plates interspersed by liquid argon [Wil 74], and thus operated in cryogenic conditions (87 K). The operating principle is described in Section 5.3.2. A large calorimeter module, based on this principle and schematically depicted in Figure 9.1, consisted of seven hexagonal containers. These containers were longitudinally segmented into six segments each, labeled A,B,C,D,E,F.

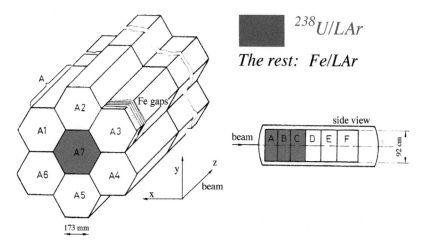

FIG. 9.1. Schematic structure of the first uranium/liquid-argon calorimeter built at CERN in the 1970s by the Fabjan/Willis/Radeka group [Fab 77]. See text for details.

In six of the seven containers, thin iron plates served as the absorber medium. In the seventh container, located in the center of the detector, these iron plates were replaced by plates made of depleted uranium (^{238}U), in the first three depth segments (A - C). This setup was used to make comparative measurements between calorimeters based on iron and ^{238}U absorber. The results were the first ones obtained with a calorimeter based on ^{238}U as absorber material and generated a lot of interest because of the seemingly equal response to em and hadronic showers, and the resulting myth that contributions from nuclear fission were responsible for this phenomenon [Fab 77].

In the paper in which these results are described, the authors also reported an energy resolution for 10 GeV pions of 9.6%, *i.e.*, about a factor of two better than achieved with any other calorimeter available at that time. This apparent benefit of *compensation* triggered a great demand for depleted uranium among designers of new experiments in particle physics. D0, ZEUS and L3 can be mentioned as examples. However, neither the mentioned excellent hadronic energy resolution, nor the reported e/π signal ratios have ever been reproduced for a calorimeter with this structure. What was the reason for that?

In order to analyze the uranium test beam data, and to distinguish the results from those obtained with an all-iron configuration, the authors of [Fab 77] required that the events that were used to determine the performance of the uranium calorimeter *only* produced signals in the uranium part of the calorimeter structure. Events that contained signal contributions from the surrounding iron-based detector segments were discarded. At first sight, this requirement does not look unreasonable. However, it led to a severely biased event sample, that was mainly populated by events in which the energy carried by the incoming beam particle was in large part converted into π^0s which subsequently developed em showers in the calorimeter. As a result, it should not be a surprise that the average signals for electrons and pions of the same energy were almost equal and that the energy resolution for pion showers was not much worse than that for electron showers of the same energy.

The cited paper does not provide specific information about the fraction of pion events that were discarded in the analysis, but it must have been quite substantial, given that the effective radius of the uranium section amounted to less than one nuclear interaction length.

In order to provide an independent check of this phenomenon, I used data from recent beam tests of a dual-readout fiber calorimeter, the 1.3 ton lead-based detector shown in the insert of Figure 8.26. The signals from the plastic leakage counters that surrounded this calorimeter were added to the scintillation signals from the fiber calorimeter for this purpose. The energy equivalence of the leakage signals was determined on the basis of the measured lateral shower profile, which indicated that on average 6.4% of the shower energy was carried by shower particles that escaped sideways.

Figure 9.2 shows the signal distribution for 60 GeV π^- in this calorimeter [Lee 17]. The top diagram was obtained for *all* events, after applying the standard dual-readout procedure based on the measured scintillation and Čerenkov signals (Equation 8.7). The bottom diagram shows the signal distribution for events in which the showers were fully contained in this calorimeter. Even though the leakage counters, which were used to select these events provided only partial coverage, the energy resolution was found to be more than 50% better for this event sample. These results illustrate that the hadronic energy resolution of the dual-readout calorimeters discussed in Section 8.2 was dominated by (lateral) shower leakage. Light attenuation in the (scintillating) fibers also contributed, as evidenced by the slightly asymmetric response function.

This type of event selection could lead to spectacular performance in any calorimeter. For example, one could take a BGO crystal, which serves as an em calorimeter in several experiments, and expose it to a test beam of pions. By requiring that the pion showers are fully contained in this crystal, basically only charge exchange events are selected, in which the pion converts its entire energy to a π^0 in the first interaction.

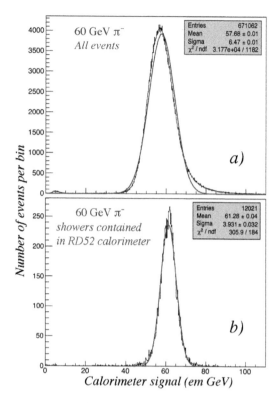

FIG. 9.2.   The signal distribution for 60 GeV π^- in the RD52 dual-readout lead/fiber calorimeter, for all events (a) and for events where the showers were contained in this detector (b). See text for details.

The resulting signal distribution will look very much like that for electrons of the same energy as the pions. Of course, the selected event sample represents only a very small, non-representative, fraction of the total.

These examples illustrate the importance of *unbiased* event samples in determining the performance of a calorimeter. Almost every analysis of test beam data I know of suffers from bias problems, the question is only *to what extent* the obtained results are affected by this. Obviously, it may also make a comparison between results obtained with different calorimeters problematic, and less meaningful.

An important practical reason for the need to select event samples for the data analysis is the fact that the test beams provided by accelerators are typically not (sufficiently) pure. I illustrate this with an example from one of my own test beam campaigns in the H8 beam of the CERN Super Proton Synchrotron. The primary 400 GeV proton beam provided by the SPS is sent onto a target, from which secondary beams can be derived. The charge sign and the momentum of these secondary particles are determined by means of a series of bending magnets. Beams of lower energy particles can be obtained by sending the secondary beam onto a target and selecting the reaction products from the interactions in that target by means of a properly tuned set of downstream bending

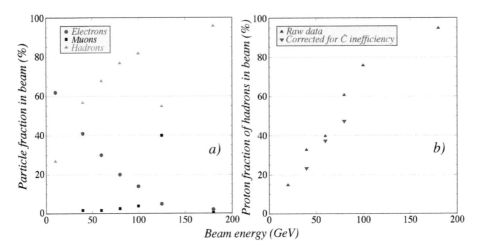

FIG. 9.3.   Measured composition of the H8 beam at the CERN Super Proton Synchrotron.

magnets, and transporting the selected particles to the area where the detectors to be tested are installed.

The beams can be enriched in certain types of particles by different methods. For example, a muon beam can be easily produced by installing a thick (*e.g.*, 10 λ_{int}) absorber in the beam line. Charged particles penetrating this absorber without noticeably changing momentum are by definition muons. The secondary beam can be enriched in electrons by sending it through a thin high-Z radiator, in which electrons are much more likely to lose energy than the hadronic beam particles. Downstream of that radiator, bending magnets optimized for the degraded energy transport the remnants of the beam to the experimental area.

The final composition of the beam arriving at the detector depends on the particle energy, but also on the widths of collimators placed at various positions in the beam line, as well as the cross sectional area of the beam selected by trigger counters placed directly upstream of the calorimeter to be tested. This is because the transverse size is typically different for the different beam components.

Figure 9.3a shows the composition of the H8 beam measured for hadron beams derived from a 180 GeV secondary beam. It shows that at low energies, the beam actually consists predominantly of electrons, even though no specific efforts were made to enrich the beam in these particles. Muons typically make up a few percent of the beam particles, except in the energy range from 120 to 170 GeV, where muons produced in the decay in flight of the 180 GeV pions and kaons are found. Some 1.3% of these pions and 9.8% of the kaons decay in the 130 m separating the primary and secondary targets.

The measurements of the beam composition were carried out with a number of sensors specifically installed in the beam line for this purpose. Threshold Čerenkov counters only provided a signal for beam particles whose velocity exceeded a threshold determined by the gas pressure in these counters. At a given momentum, these counters thus provided information that could be used to distinguish high-mass particles (no signal) from the low-mass ones. A thin $(1X_0)$ preshower detector (PSD) installed directly upstream of the calorimeter was very helpful for identifying electrons, which produced a substantially larger signal in this device than the other beam components (Figure 7.48). A scintillation counter installed 20 m behind the calorimeter and 1.6 m of concrete made it possible to identify beam muons unambiguously.

Figure 9.3b shows the measured fraction of protons in the hadron beams, which had positive polarity in these measurements. The distinction between the pion and proton components of the hadron beam was made with the Čerenkov counters, of which the gas pressure was chosen such that the protons did *not* produce a signal. The inefficiency of these counters was measured by repeating the measurements for negative polarity. Since anti-protons were virtually absent in that case, the absence of a Čerenkov signal provided information on the basis of which this inefficiency could be determined. Another tool for that purpose was the fraction of the pion events in which only one of the two Čerenkov counters produced a signal.

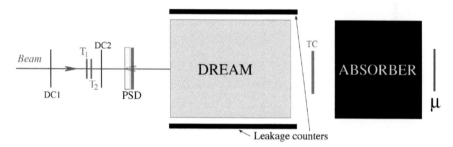

FIG. 9.4.   Schematic overview of the arrangement of the auxiliary detectors that were used to identify the individual beam particles in the H8 beam of the CERN SPS [Akc 14a].

The main reason for elaborating on this issue is to illustrate the methods that can (and should) be employed to obtain a pure sample of hadron events, namely on the basis of information from *external* detectors, which makes it possible to eliminate events that are caused by contaminating particles of other types. The event selection should **not** be based on the measured calorimeter data. Violation of this important principle will inevitably lead to a biased event sample. Yet, in practice violation of this principle is almost unavoidable, and in any case tempting. The following example illustrates how one can deal with the non-purity of test beams in practice. The setup schematically shown in Figure 9.4 was used to identify each event that produced a signal in the calorimeter either as an e^-, a μ^- or a π^-. To that end, the following cuts were applied, for negative-polarity beams of particles [Akc 14a]:

1. *Electrons* were identified as particles that produced a signal in the PSD that was larger than the combined average signals produced by two minimum-ionizing particles traversing this detector simultaneously. Additional requirements were that no signals incompatible with electronic noise were produced in the tail catcher (TC) and the muon counter. To avoid the effects of electronic noise and cosmic rays that produced spurious trigger signals, the total scintillation signal in the

calorimeter was required to be larger than a certain (small) minimum value.

2. *Pions* were identified as particles that produced a signal in the PSD that was compatible with a minimum-ionizing particle traversing it, and no signal incompatible with noise in the muon counter.

3. *Muons* were identified as particles that produced signals in the PSD, the tail catcher and the muon counter that were compatible with minimum-ionizing particles traversing these detectors.

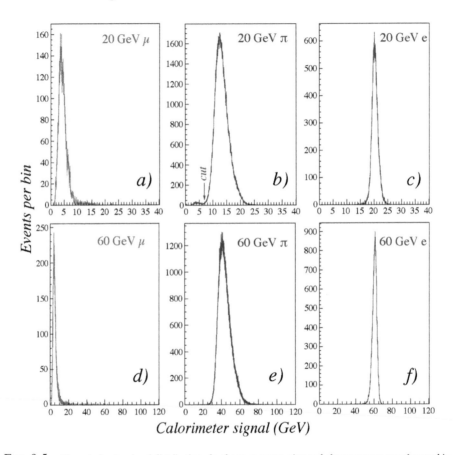

FIG. 9.5.   The calorimeter signal distributions for the pure muon, pion and electron event samples used in the analyses [Akc 14a]. See text for details.

Figure 9.5 shows the results of these procedures for the 20 and 60 GeV beams [Akc 14a]. Shown are the total calorimeter (scintillation) signal distributions for events that were classified as muons, pions and electrons based on the information from the preshower detector, the tail catcher and the muon counter. The 20 GeV pion sample contained a small contamination of muons that did not produce a signal in the muon counter, because of multiple scattering or inefficiencies of this counter. This contamination was removed by the cut indicated in the figure.

The latter cut constitutes a violation of the principle that only signals from auxiliary detectors should be used for event selection, since the calorimeter signals themselves are used to eliminate a very small fraction of the events. In this particularly case, the effect that necessitated this procedure was quite obvious and did not lead to a significant bias. However, in the next subsection a much less obvious example of the pitfalls that may be encountered when violating this fundamental principle is discussed. This example also comes from my own beam tests of the RD52 calorimeter (see Section 8.2).

9.1.2 Using the calorimeter data for event selection

As illustrated in Figure 9.3, the H8 beams at the CERN SPS typically had a mixed composition. While measuring the electromagnetic performance of the RD52 dual-readout fiber calorimeter for very small angles of incidence, we wanted to determine the effects of the preshower detector, which absorbed a small fraction of the beam particle's energy, on that performance. This preshower detector, which consisted of a 5 mm thick plate of lead followed by a plastic scintillator, installed upstream of the calorimeter, was very important for removing the significant residual pion contamination from the electron event samples. The question was how to deal with this contamination when the preshower detector was removed. This turned out to be a challenging problem.

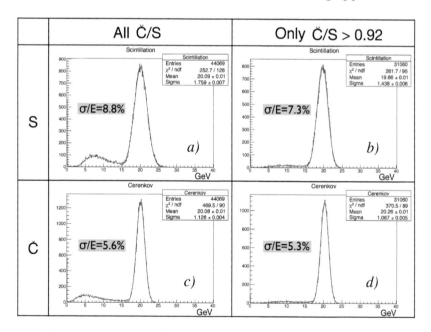

FIG. 9.6. Signal distributions for the Čerenkov and scintillation signals with and without a cut on the ratio of these two signals. The angle of incidence of the 20 GeV electrons was ($\phi = 0 \pm 0.5°$, $\theta = 1°$). The preshower detector was out of the beam line [Car 16]. See text for more details.

In a previous study [Akc 14a], we had found that the Čerenkov/scintillation calorimeter signal ratio was an efficient tool to discriminate between electrons and pions. Therefore, we tried to clean up our electron event samples by means of a cut on that signal ratio [Car 16]. Figure 9.6 shows that this worked quite well. The pion contamination, visible as a broad bump below 15 GeV, was considerably reduced by retaining only events with $C/S > 0.92$, while the contents of the electron peak were barely affected. However, this figure shows one other amazing effect, namely the improvement in the energy resolution for the electrons, from 8.8% to 7.3% in the scintillation signal. These data concern the measurements at a tilt angle of 1° and an azimuth angle $\phi = 0 \pm 0.5°$, *i.e.*, the beam particles entered the calorimeter almost parallel to the direction of the fibers. As explained in Section 8.2.7.1, the scintillation energy resolution is in this configuration affected by a position dependence of the signals (see Figure 8.33). The contribution of the early, highly collimated shower component to the calorimeter signal depends in that case sensitively on the distance between the impact point of the particles and the nearest scintillating fiber. For reasons explained in Section 8.2.7.1, this effect does not play a significant role for the Čerenkov signal. By retaining events with a large C/S signal ratio, events with a large scintillation signal are selectively eliminated. These are predominantly events in which the beam particles entered the calorimeter close to a scintillating fiber. In this procedure, the effect of the impact point dependence on the energy resolution is diminished, and therefore the measured energy resolution improves, especially in the scintillation channel, where this dependence plays a key role.

This experience constitutes another example of the importance of one of the fundamental rules of testing calorimeters: to measure the performance of a calorimeter, only *external* detectors should be used to select events (in this case to eliminate pions from the mix), not the calorimeter signals themselves, since this might lead to biased event samples.

9.1.3 *Separate tests of em/hadronic calorimeter sections*

In most experiments in High Energy Physics, especially the ones at the large particle accelerators, the calorimeter systems consist of separate electromagnetic and hadronic sections. Typically, these sections are designed, built and tested by separate groups of people, without much consideration for the other components of the detector system. This may lead to some very specific problems, an example of which is described below.

This example concerns the calorimeter system of the CMS experiment at CERN's LHC. When this experiment was designed, much emphasis was placed on excellent performance for the detection of high-energy photons, in view of the envisaged discovery of the Higgs boson through its $H^0 \to \gamma\gamma$ decay mode. To that end, the Collaboration decided to use $PbWO_4$ crystals for the em section of the calorimeter, since these would provide $\mathcal{O}(1\%)$ energy resolution for the γs produced in this process. Since the crystals would have to operate in a strong magnetic field, Avalanche Photo Diodes were chosen to convert the light produced by these crystals into electric signals. Given the available sizes of APDs at that time, and in order to take full advantage of the light yield, each crystal was equipped with two APDs (Figure 9.7a). However, in order to save some money, these APDs were ganged together and treated as one device in the data acquisition system.

The hadronic section of the CMS calorimeter consists of a sampling structure, based on brass absorber and plastic scintillator plates as active material. Both sections were developed completely independently, and tested separately in different beam lines at CERN. For the tests of the em section, high-energy electron beams were used, the hadronic section was exposed to beams of all available types of particles (electrons, pions, kaons, protons, muons). The performance of both sections was documented in detail, and found to be in agreement with expectations [Adz 07, Abd 08].

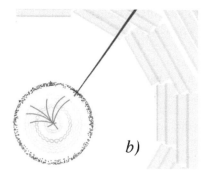

FIG. 9.7. Photograph of two CMS APDs (active area 5×5 mm^2) mounted in a capsule (a). CMS event display of a pp collision event, showing an isolated ECAL spike (top-right) simulating a 690 GeV transverse energy deposit (b) [Pet 12].

Yet, when the entire calorimeter system was assembled and exposed to high-energy hadrons, an unexpected surprise occurred [Pet 12]. In some fraction of the events, anomalously large signals were observed. An example of this phenomenon is shown in Figure 9.7b. What was going on?

The APDs that convert the light produced in the crystals are also extremely sensitive to ionizing particles. In fact, a mip traversing such an APD may create a signal equivalent to several thousand light quanta [Hau 94]. The signal produced by densely ionizing charged particles is correspondingly larger. In Figures 2.29 and 2.49, examples are given of interactions that are typical when a high-energy hadron strikes an atomic nucleus. Many densely ionizing nuclear fragments are visible in these pictures, with dE/dx values that are 100 or more times that of a mip. If such a nuclear interaction would take place in the vicinity of an APD, a very large signal could result as a consequence of the nuclear fragments traversing the active detector surface area. Since the light yield of these crystals is such that one GeV deposited energy corresponds to $\mathcal{O}(10^4)$ detected photons, the occurrence of a nuclear reaction of the type shown in these figures close to an APD could thus easily mimic an energy deposit of 100 GeV or more.

This phenomenon was only discovered when the em and hadronic calorimeter sections were assembled together and exposed to high-energy hadrons. Since the em section corresponds to about one nuclear interaction length, a substantial fraction of hadrons entering the calorimeter start the shower development process in the em section, and therefore the process that generates the "spikes" described above becomes possible.

As stated above, the nuclear reactions have to take place "close to" the sensor surface of the APD in order to produce this effect. The scale on Figure 2.49 clarifies what "close" means in this context, *i.e.*, within 100 μm or so. This also provides the answer to the question how this problem could have been avoided. Since the two APDs that read out each crystal are separated by several mm, a nuclear interaction of the type discussed here would never affect both APDs, but only one of them. Therefore, if the two APDs had been read out separately, instead of being treated as one detector unit in the data acquisition system, "spike" events would be easily recognized since the signal in only one of the APDs would be anomalously large, while the signal from the other one would still provide the useful information one would like to obtain from the event in question.

However, since the two groups that were responsible for the two calorimeter sections worked in their own individual, separate universes, this problem was only discovered when it was too late to make the corrections needed to avoid it. I include this example here because it illustrates that it is important to realize that a calorimeter built for a given experiment is not a stand-alone device, but is part of an integrated system of detectors. The different components of this detector system may affect the performance of each other and it is important to realize and test this in the earliest possible stage of the experiment.

9.1.4 *Miscalibration*

As explained in Chapter 6, calibration is probably the most important, as well as the most underrated aspect of working with calorimeters. Beam tests are the best option to study the problems (and possible solutions) in this respect. In this subsection, three examples of complications encountered in the process of calibrating longitudinally segmented calorimeter systems are described.

The first example concerns the CMS calorimeter system, which was also mentioned in the previous subsection. The main complicating factor for the calibration of this system are the very different e/h values of the em and hadronic sections. The crystal em section has a value of 2.4, while $e/h = 1.3$ for the hadronic section.

The hadronic performance of this calorimeter system was systematically studied with various types of particles ($e, \pi, K, p, \bar{p}$), covering a momentum range from 1–300 GeV/c [Abd 09]. Figure 9.8 shows some results from this study. Both sections were calibrated in stand-alone mode with 50 GeV electrons. The figure shows that the hadronic section of the calorimeter is very linear for em showers, just like the em section. However, the response to pions, represented by the black dots, indicates that the combined calorimeter system is extremely non-linear for these particles. This non-linearity is especially evident below 10 GeV, which is important since pions in this energy range carry a large fraction of the energy of typical LHC jets, even in the TeV domain. More troublesome is that the response strongly depends on the starting point of the showers. The figure shows results for two event samples, selected on that basis: showers starting in the em section ($\triangle$) or in the hadronic section ($\triangledown$). At low energies, the response is more than 50% larger for the latter (penetrating) events. In practice in an experiment, it

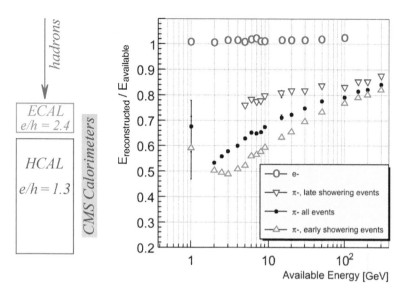

FIG. 9.8. The response to electrons and pions as a function of energy, for the CMS barrel calorimeter. The pion events are subdivided into two samples according to the starting point of the shower, and the pion response is also shown separately for these two samples [Akc 12a].

is often hard/impossible to determine where the shower starts, especially if these pions are traveling in close proximity to other jet fragments (*e.g.*, photons from π^0 decay) which develop showers in the em section.

Response non-uniformity is not only a problem for hadron showers, it also affects electrons and photons that develop showers in a longitudinally segmented em (sampling) calorimeter. The basic reason is that the sampling fraction of a given calorimeter structure depends on the stage of the developing showers. For example, in calorimeters consisting of high-Z absorber material (*e.g.*, lead) and low-Z active material (plastic, liquid argon), the sampling fraction may vary by as much as 25 - 30% over the volume in which the absorption takes place [Wig 06b].

In the early stage of its development, the shower still resembles a collection of mips, but especially beyond the shower maximum, the energy is predominantly deposited by soft (< 1 MeV)γs. The latter are much less efficiently sampled than mips in this type of structure, where dominant processes such as photo-electric effect and Compton scattering strongly favor reactions in the high-Z absorber material (Figure 9.9).

The decrease of the sampling fraction as the em shower develops may have very important practical consequences. These include:

- *Systematic mismeasurement* of energy [Cer 02],

- Electromagnetic signal *non-linearity* [Ake 87], and

- *Differences in response* to showers induced by electrons, photons and π^0s [Wig 02].

These issues are especially relevant in longitudinally segmented calorimeters, where

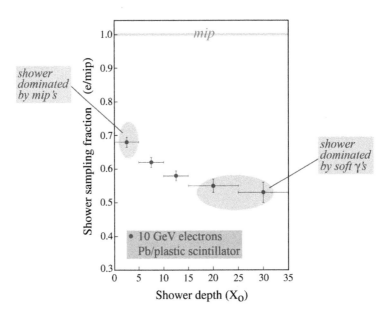

FIG. 9.9.   The sampling fraction changes in a developing shower. The e/mip ratio is shown as a function of depth, for 10 GeV electrons in a Pb/scintillating-plastic calorimeter. Results of EGS4 calculations [Wig 06b].

one has to decide which calibration constants to assign to the different segments. This *intercalibration* is more often than not done incorrectly, resulting in the consequences mentioned above.

A recent example of an experiment that has to deal with this intercalibration issue is ATLAS, whose Pb/LAr ECAL consists of three longitudinal segments (Figure 9.10). Figure 6.12 shows how the sampling fraction for em showers evolves in this detector as a function of depth, in an energy dependent way. Elaborate Monte Carlo simulations played a crucial role in the ATLAS solution of the intercalibration problems, for which they developed a very sophisticated procedure, based on a variety of energy-dependent parameters (Figure 6.13). Typically, in more empirical approaches to this problem, only one of the performance characteristics is pursued in isolation, and the results are far from optimal (or even incorrect) for other aspects [Wig 02, Wig 06a]. Even though the ATLAS approach led to a combination of good energy resolution and signal linearity for electrons, there are reasons to believe that the performance would be different for γs and for particles decaying into several unresolved γs ($\pi^0, \eta, \omega_0, ...$). See Section 6.2.4 for more details on this.

The third example of the pitfalls of calibrating a longitudinally segmented device concerns the calorimeter for the AMS-02 experiment at the International Space Station [Cer 02]. This calorimeter has eighteen independent longitudinal depth segments. Each layer consists of a lead absorber structure in which large numbers of plastic scintillating

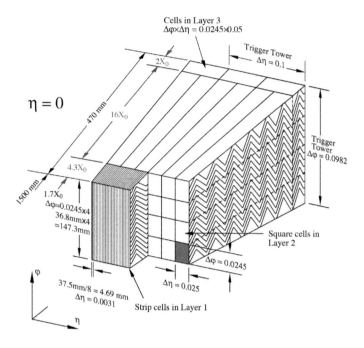

FIG. 9.10. Schematic view of the ATLAS em calorimeter [Aad 08].

fibers are embedded, and is about $1X_0$ thick. A minimum-ionizing particle deposits 11.7 MeV upon traversing such a layer. The AMS-02 collaboration initially calibrated this calorimeter by sending muons through it and equalizing the signals from all eighteen longitudinal segments. This seems like a very good method to calibrate this detector, since all layers have exactly the same structure.

However, when this calorimeter module was exposed to beams of high-energy electrons, it turned out to be highly non-trivial how to reconstruct the energy of these electrons. Figure 9.11a shows the average signals from 20 GeV electron showers developing in this calorimeter. These signals were translated into energy deposits based on the described calibration. The measured data were then fitted to a Γ-function and since the showers were not fully contained, the average leakage was estimated by extrapolating this fit to infinity. As shown in Figure 9.11b, this procedure systematically underestimated this leakage fraction, more so as the energy (and thus the leakage) increased. The reason for this is the same effect that complicates the energy measurements with the AT-LAS ECAL. Since the sampling fraction decreases as the shower develops, a procedure in which the relationship between measured signals and the corresponding deposited energy is assumed to be the same for each depth segment will cause the energy leakage to be systematically underestimated, more so if that leakage increases. More details on this are given in Section 6.2.5.

This very complicated problem will most definitely also affect PFA calorimeters, which are all based on structures that are highly segmented, both longitudinally and

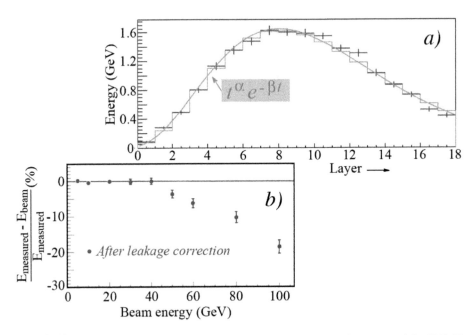

FIG. 9.11. Average signals measured for 20 GeV electrons in the 18 depth segments of the AMS-02 lead/scintillating-fiber calorimeter (*a*). Average relative difference between the measured energy and the beam energy, after leakage corrections based on extrapolation of the fitted shower profile (*b*). Data from [Cer 02].

laterally. The underlying problem is that the relationship between deposited energy and resulting signal is not constant throughout a developing shower. As the composition of the shower changes, so does the sampling fraction. Figure 9.11 provides a clear example of the problems that this may cause.

9.1.5 Calibration and hadronic signal linearity

Hadronic signal non-linearities and starting point dependence of the response are general characteristics of non-compensating calorimeter systems, although the effects are usually not as spectacularly large as shown in Figure 9.8. Many calibration schemes used in practice set the energy scale of the em calorimeter section with electrons, while the hadronic section is calibrated with pions that penetrate the em section without undergoing a nuclear interaction and deposit their (almost) entire energy in the hadronic section (Figure 9.12a). Other schemes choose the calibration constants for the hadronic section on the basis of a minimization of the width of the total signal distribution (Figure 9.12b). It has been shown repeatedly that both types of schemes are **fundamentally flawed** [Alb 02b, Wig 02, Wig 06a, Wig 06b]. It has been demonstrated experimentally that one of the consequences of these schemes is an *increase* in the hadronic signal non-linearity, over and above the non-linearity that would result from a correct calibration scheme [Alb 02b].

Hadronic signal non-linearity is bad, especially when it comes to jet detection. It is

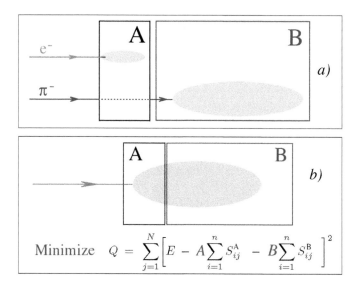

$$\text{Minimize} \quad Q = \sum_{j=1}^{N}\left[E - A\sum_{i=1}^{n} S_{ij}^{A} - B\sum_{i=1}^{n} S_{ij}^{B}\right]^{2}$$

FIG. 9.12. Two often applied (inter)calibration schemes for longitudinally segmented calorimeters.

bad because it makes the reconstructed energy dependent on the jet topology: a jet with a leading particle that carries a large fraction of the energy of the fragmenting object will be reconstructed with a different energy than a jet whose energy is more evenly divided among the different constituent particles. This phenomenon will, for example, lead to different calorimeter responses to gluon and quark jets of the same energy, because of systematic differences in the composition of these two types of jets. Calibration schemes that increase such differences *should be avoided*, but in practice this is not the case since one is usually completely focused on "resolution." However, it should be realized that the width of a distribution is only an indication of the resolution (*i.e.*, the precision with which the energy of an event can be determined) if the central value of the distribution has the correct energy. The effects of signal non-linearity invalidate this presumption.

9.1.6 *Concluding statements*

I would like to end this section about calibration issues with some very personal statements, based on 35 years of experience with these detectors.

1. All the problems encountered in calibrating longitudinally segmented calorimeters can be avoided by using a longitudinally unsegmented calorimeter. The least that can be said is that, in that case, the temptation to make mistakes is absent.
2. There is **nothing** that can be achieved as a result of longitudinal calorimeter segmentation that cannot be achieved (better) by other means.

I realize that these statements may be counter-intuitive and, therefore, controversial. However, I have never seen any experimental evidence to the contrary.

9.2 Interpretation of the Results of Beam Tests

One of the most important tasks of a calorimeter system is to measure the energy of particles or jets that are absorbed in it. The energy resolution is a measure of the calorimeter quality in this respect. As described in the previous section, the energy resolution is typically determined from the measured signal distribution for a beam of mono-energetic particles that enter the calorimeter in (approximately) the same impact point, usually the center of a module. However, in this section I will argue that one has to be careful interpreting the results of such measurements. Experimental data obtained with the RD52 dual-readout lead/fiber calorimeter (Section 8.2.6) are used to illustrate this.

In Section 8.6.2.4, the so-called rotation method was used to analyze the hadron data obtained with this calorimeter (Figure 8.29). In this method, the data points in the $S - C$ scatter plot are fitted with a straight line. The intersection of this line with the diagonal, the $S = C$ line, provides the particle energy. By rotating the data points over a fixed angle (30° for this calorimeter), the points cluster around the vertical axis. The projection of these data points on the horizontal axis gives a very narrow, Gaussian signal distribution. This method provided excellent results, both for different types of hadrons (pions, protons) as well as for multi-particle "jets". The calorimeter turned out to be very linear over the entire energy range studied, the response was the same for all particles, and the fractional width (σ/E) of the signal distribution was measured to scale with $30\%/\sqrt{E}$, both for protons and pions (Figure 8.30).

Yet, while the very narrow signal distributions for the beam particles were exclusively based on the measured calorimeter data, and *not* on any knowledge about the energy of the beam particles or the composition of the beam, I don't think it is correct to interpret these results as the energy resolution of the calorimeter. The energy resolution of a calorimeter represents the precision with which the energy of an arbitrary particle with unknown energy, absorbed in this calorimeter, may be determined. In the described method, the determination of the coordinates of the rotation point, and thus the energy scale of the signals, relies on the availability of an *ensemble of particles of the same energy*. In practice, however, one is only dealing with *one* event and the described procedure can thus not be used in that case.

It is possible to determine the energy of an unknown particle showering in the dual-readout calorimeter that is *not* affected by this problem. In this procedure, described in Section 8.2.2, the em shower fraction of the hadronic shower is derived from the ratio of the Čerenkov and scintillation signals. Using the known e/h values of the two calorimeter structures, the measured signals can then be converted to the em energy scale ($f_{em} = 1$). The energy resolutions obtained with this method are not as good as the ones given above. Figure 9.13 graphically illustrates the difference between the values obtained with the two methods discussed here. The precision of the energy measurement is represented by the arrows in the two diagrams.

The message I want to convey in this section is that one should not confuse the precision of the energy determination of a given event based on calorimeter signals alone with the width of a signal distribution obtained in a test beam, because the latter is typically based on additional information that is not available in practice. In the example described above, this additional information derived from the availability of a

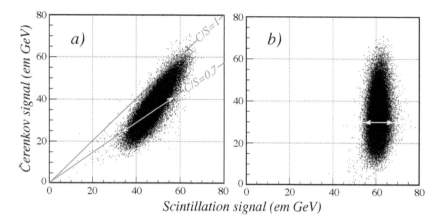

FIG. 9.13. Scatter plots of the Čerenkov *vs.* the scintillation signals from showers induced by mono-energetic hadrons (*a*). The arrow indicates the precision with which the em shower fraction, and thus the energy, of an individual particle can be determined on the basis of the measured ratio of the Čerenkov and scintillation signals, 0.7 in this example. The rotation procedure for an ensemble of mono-energetic pions leads to the scatter plot shown in diagram *b*. The precision of the measurement of the width of that distribution is indicated by a white arrow as well [Lee 17].

large number of events generated by particles of the same energy. In other cases, additional information may be derived from knowledge of the particle energy and/or the particle type. This is especially true for calorimeters whose energy scale depends on "offline compensation" (Section 6.4), or other techniques intended to minimize the total width of the signal distribution from a detector system consisting of several longitudinal segments (Section 6.2.4). Such techniques make use of calibration constants whose values vary with energy (Figure 6.13), and also depend on the type of showering particle (Section 9.3.4).

9.3 Reporting the Results of Beam Tests

As discussed in the previous sections, testing the performance of calorimeters in particle beams, and interpreting the results of such measurements, requires a very careful approach, and rigorous procedures with respect to data selection and calibration. Reporting the results of such tests is an entirely separate issue, with its own idiosyncrasies. These may either be based on incorrect preconceived ideas on *how* calorimeters should perform, or inspired by an honest desire to make the results look as good as possible (and, therefore, in fact better than they really are). One may also conveniently disregard some important contributions to the performance, which may be studied in isolation (and thus separated out) in beam tests, but not in "real life." The results may be incorrectly interpreted as being valid for conditions other than the ones in which they were obtained. And finally, there are unfortunately examples in which deliberate attempts are being made to mislead the audience to which the results are communicated. In this section, examples of all of this are presented.

9.3.1 *Quoting energy resolution in terms of $x\%/\sqrt{E}$*

One of the attractive aspects of calorimeters is the improvement of the relative precision with which the energy of showering particles can be measured at increasing energy. This is because fluctuations governed by Poisson statistics, such as fluctuations in the number of signal quanta or sampling fluctuations, contribute a term that scales with $E^{-1/2}$ to the relative energy resolution, σ/E. However, typically there are also other factors that contribute to the energy resolution, and these contributions may (and usually do) have a completely different energy dependence than the stochastic ones. Such other contributions may even dominate the energy resolution in certain energy regimes, *e.g.*, at very high energy. By quoting the measured relative energy resolution just in terms of $x\%/\sqrt{E}$, the experimental result is then either oversimplified and/or presented in a misleading way. This is actually rather common practice. If some calorimeter measures the relative energy resolution for 100 GeV pions as 5%, this is often immediately interpreted as "the calorimeter has a hadronic energy resolution of $50\%/\sqrt{E}$."

In order to get a good impression of the calorimeter performance, it is necessary to have experimental data over as large an energy range as possible. Only then will it be possible to assess the energy dependence of the calorimeter, and the various terms contributing to it. But even then, this is far less trivial than it may seem. As an example,

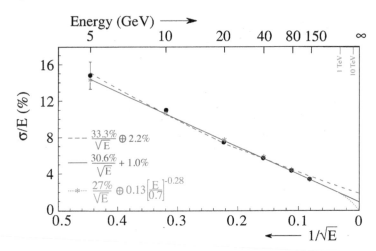

FIG. 9.14. The experimental data points measured for hadrons in the SPACAL lead/fiber calorimeter, together with three fits through these data points [Aco 91c]. The asterisks represent the values of the dotted line fit at the energies of the experimental data points.

Figure 9.14 shows the hadronic energy resolution measured with the SPACAL calorimeter [Aco 91c]. The experimental data, represented by the black dots, are well described by a straight line in this graph, where the energy is plotted on a scale linear in $E^{-1/2}$. If the energy resolution was only affected by stochastic fluctuations, the data points should be located on a straigth line *through the bottom right corner*, *i.e.*, extrapolating to zero resolution for an infinite energy deposit. This is clearly not the case here, the intercept for $E = \infty$ occurs for $\sigma/E \approx 1\%$. It is therefore no surprise that the authors described the fit with the sum of a stochastic and a constant term: $\sigma/E = 30.6\%/\sqrt{E} + 1.0\%$, represented by the solid line in Figure 9.14.

At the time when these measurements were performed, the prevailing view was that the effect of non-compensation ($e/h \neq 1$) on the hadronic energy resolution was represented by a constant term. However, since the effects of fluctuations in f_{em} were completely uncorrelated to the stochastic fluctuations that determined the $E^{-1/2}$ scaling term, such a constant term had to be added in quadrature, rather than linearly, as in the fit mentioned above. For that reason, the authors of [Aco 91c] also performed a fit to an expression in which both terms were added in quadrature. This fit, which described the experimental data points less well gave the following result: $\sigma/E = 33.3\%/\sqrt{E} \oplus 2.2\%$, and is represented by the dashed line in Figure 9.14.

However, as explained in Section 4.7, it is completely incorrect to describe the effects of fluctuations in f_{em} by means of an energy independent term, since there would otherwise be very little difference between the hadronic energy resolutions measured with compensating and strongly non-compensating calorimeters at low energy (e.g., $E \sim 10$ GeV), which is not true at all. A more correct way to describe the (energy dependent) contributions of these effects is presented in Equation 4.29. A fit to an expression of this type is also shown in Figure 9.14: $\sigma/E = 27\%/\sqrt{E} \oplus 0.13\left[(E/0.7)^{-0.28}\right]$. This fit is represented by the dotted line and is near the experimental data points practically indistinguishable from the straight line. It is only beyond the highest-energy data point, in the TeV range, that the differences between the results of these different fits become substantial.

It will be clear from this example that it does not make much sense to claim the measured hadronic energy resolution of the SPACAL calorimeter in terms of $x\%/\sqrt{E}$. In comparisons with the results obtained with other calorimeters, the only meaningful thing to do is to compare the resolutions measured at (a list of) specific energies. And, in order to be meaningful, such a comparison also requires that similarly rigorous procedures have been applied in the collection of the data samples. Finally, it almost goes without saying, it is also important that the same definitions for the performance characteristics (such as signal linearity and energy resolution) are being used. However, as pointed out in Section 9.3.5, even this is not a trivial matter.

9.3.2 *Elimination of important resolution contributions*

The energy resolution of a calorimeter typically contains contributions from a variety of sources (see Chapter 4). Instrumental effects often play an important role. Sometimes, these effects and their contribution to the resolution can be estimated in isolation. This is especially the case in beam tests, where a collection of particles, all of the same precisely known energy, enter the calorimeter at approximately the same position. As an example, I mention beam tests of the ATLAS Forward Calorimeter. Figure 9.15 shows the published results of these tests [Hee 09], as black dots. The dashed line represents the results of the published fit to these data points. However, in obtaining these data points, the authors subtracted the contribution of electronic noise to the measured energy resolution of this liquid-argon calorimeter. As shown in the insert, the energy

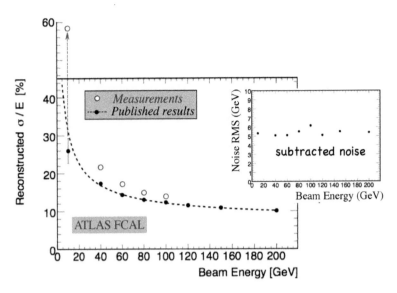

FIG. 9.15.   The hadronic energy resolution of the ATLAS Forward Calorimeter [Hee 09]. Shown are the measured data points and the actually measured resolution. See text for details.

equivalent of the rms fluctuations in these noise signals was typically of the order of 5 GeV. Therefore, the contribution of such fluctuations to the measured relative energy resolution, σ/E, was typically of the order of 5 GeV/E, to be added in quadrature to other contributions. This noise term thus contributed most strongly at low energy. Based on this information, I have derived the actually *measured* energy resolution, and this is represented by the open circles in Figure 9.15, which deviate quite considerably from the black dots, especially at the lowest energy.

Of course, the possibility to eliminate the contributions of electronic noise is limited to beam tests. It is completely impossible during the regular operations of this detector in the LHC environment. By presenting the black dots and the dashed curve as the results of the measurements of the calorimeter resolution (which is done both in the abstract and the summary of the cited paper), the authors gave a misleading account of the performance of their detector.

A second example of an instrumental effect that is easily recognized in beam tests, but not in real life, is the so-called "Texas tower effect." As illustrated and explained in Section 4.8.2, this effect is a consequence of the very small sampling fraction of some calorimeters, and in particular those with a gaseous active medium that contains hydrogen. The sampling fraction of such calorimeters is typically $\mathcal{O}(10^{-5})$, which means that the energy deposited by a showering 100 GeV hadron in the active material amounts to ~ 1 MeV. However, when one of the numerous MeV-type neutrons produced in this process scatters off a hydrogen nucleus, a comparable amount of energy may be deposited in the gas layer *by that one recoil proton!*.

In beam tests, where a beam of monoenergetic particles is sent into the calorimeter, such anomalous events are easily recognized, and can be removed from the event samples. However, during the operation of such a calorimeter in an accelerator environment, this is not (always) possible, especially if no additional information about the event in question, *e.g.*, from a tracker system, is available. This difficulty led the CDF Collaboration to the decision to scrap their gas-based forward calorimeter, and replace it with a device that had a sampling fraction that was three orders of magnitude larger [Alb 02a].

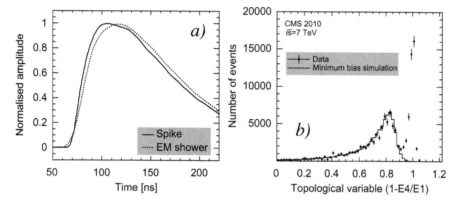

FIG. 9.16. Distinguishing characteristics of CMS "spike" events. Shown are the average pulse shape (*a*) and the distribution of the socalled *Swiss Cross* topological variable for the highest energy deposit in each event (*b*). See text for details [Pet 12].

This phenomenon is very similar, at least in its consequences, to the "spikes" observed for hadron detection in the CMS calorimeter system (Section 9.1.3). Also there, one low-energy shower particle may cause an event in which an anomalously large amount of energy seems to be deposited in the detector (see Figure 9.7). Contrary to CDF, CMS has made an effort to deal with the real-life consequences of this phenomenon, and has succeeded in recognizing, and eliminating, affected events to a reasonable extent. This is illustrated in Figure 9.16, which shows two characteristics that distinguish these "spike" events from the "normal" ones. One effective method is a cut on the so-called *Swiss Cross* variable, in which the signal in each tower of the em calorimeter is compared to the sum of the signals in the four neighboring towers. Monte Carlo simulations showed that a cut of events in which $E4/E1 < 0.05$ is an effective tool for eliminating the "spike" events [Pet 12].

Also the time structure of the spike events is a differentiating characteristic. Since the signals in this case are caused by shower particles traversing the APDs, they are faster than the signals based on detection of the scintillation light generated by the shower particles in the crystals. The latter are delayed because the molecules excited in the scintillation process take some time to decay ($\sim$10 ns). Figure 9.17 shows the timing distribution of the hits in the ECAL barrel with a reconstructed energy above 1

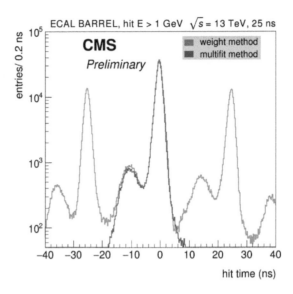

FIG. 9.17. CMS "spike" events may be distinguished from regular (scintillation) events by means of the time structure [CMS 10]. See text for details.

GeV, for an 80 ns time slice of the LHC operations. The 25 ns bunch structure is clearly visible. The events caused by the scintillation light are characterized by the peaks at -25 ns, 0 ns and + 25 ns in this plot. These peaks are preceded by smaller peaks that occur about 10 ns earlier. These are the spike events, and the figure shows that these represent about 3% of the total. It should be emphasized that this plot concerns the time characteristics of the events that *survived* the Swiss Cross cuts.

Based on the understanding of the underlying cause of this phenomenon, it is expected that the frequency of these events will increase both with the luminosity and the center-of-mass energy of the pp collisions in the LHC. The reason why this example is included in this context is that it is an inherent feature of the CMS calorimeter system. Even though attempts to deal with the problem may seem successful, it is good to keep in mind that any data selection based on the calorimeter information alone may lead to a biased event sample (see Section 9.1.1). Also, it is inevitable that some fraction of the events in which the process that causes a "spike" occurs will *not* be eliminated by the cuts devised to deal with the problem.

The third and final example of an instrumental problem that is easily recognizable in beam tests, but may well escape the attention in the normal running conditions of an experiment is *signal saturation*. I use data from one of my own experiments to illustrate this. The SPACAL calorimeter (Figure 9.18) consisted of 155 hexagonal towers. Each of these towers was calibrated by sending a beam of 40 GeV electrons in its geometric center. Typically, 95% of the shower energy was deposited in that tower, the remaining 5% was shared among the six neighbors. The high-voltage settings were chosen such that the maximum energy deposited in each tower during the envisaged beam tests

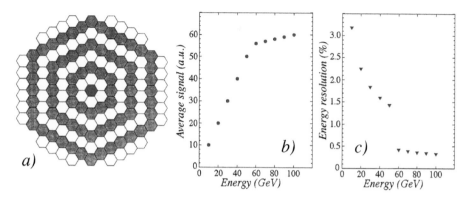

FIG. 9.18. Saturation effects in one of the towers of the SPACAL calorimeter (*a*). Shown are the average signal (*b*) and the energy resolution (*c*) as a function of energy, measured when a beam of electrons is sent into this tower.

would be well within the dynamic range of that tower. For most of the towers (except the central septet), the dynamic range was chosen to be 60 GeV. When we did an energy scan with electrons in one of these non-central towers, the results shown in Figures 9.18b and 9.18c were obtained. Up to 60 GeV, the average calorimeter signal increased proportionally with the beam energy, but above 60 GeV, a non-linearity became immediately apparent (Figure 9.18b). The signal in the targeted tower had reached its maximum value, and would from that point on produce the same value for every event. Any increase in the total signal was due to the tails of the shower, which developed in the neighboring towers. A similar trend occurred for the energy resolution (Figure 9.18c). Beyond 60 GeV, the energy resolution suddenly improved dramatically. Again, this is because the signal in the targeted tower was the same for all events at these higher energies. The energy resolution was thus completely determined by event-to-event fluctuations in the energy deposited in the neighboring towers by the shower tails.

The spectacular energy resolution shown in Figure 9.18c might give rise to great enthusiasm among the researchers who carry out these tests, until they discover the true cause of this phenomenon. As with the other examples mentioned in this subsection, this is a phenomenon that can be perfectly studied in the completely controlled conditions of a beam test, but is much harder to recognize (if at all) when running a particle physics experiment.

9.3.3   *The importance of signal linearity*

As already indicated in the previous subsection, signal saturation is **bad**. It creates a completely wrong impression of the precision with which energy can be measured. This was already realized by Muzaffer Atač, who around 1980 built the first *digital* calorimeter [Ata 83], a sampling structure based on readout with wire chambers operating in the saturated avalanche mode. This device was longitudinally subdivided into five sections, each representing about five radiation lengths.

This detector turned out to be quite non-linear for the detection of electron showers.

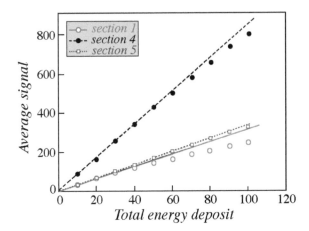

FIG. 9.19. The average signal as a function of energy for electrons in a calorimeter based on digital readout (wire chambers operating in the saturated avalanche mode). The calorimeter was longitudinally subdivided into five sections, the first section covered the first $5X_0$ (*i.e.*, before the shower maximum), sections 4 and 5 were located at depths of about 15 and $20X_0$, respectively [Ata 83].

The response dropped by more than 10% over an order of magnitude in energy deposit (see Figure 3.29a). Interestingly, most of this non-linearity derived from the signal contributions from the first two sections. Figure 9.19 shows the average signal recorded in several sections of this calorimeter, as a function of the total energy deposit. The non-linearity was clearly much larger in section 1 (which covered the first five radiation lengths) than in sections 4 and 5, which were located far beyond the shower maximum, even though the average signals (especially in section 4) were much larger than in section 1. This difference is due to the fact that the em shower is much more collimated in the early phase of its development (see Figures 2.16b, 8.32). Therefore, the probability that multiple shower particles traversed the same calorimeter cell simultaneously for a given event was much larger in this early phase than beyond the shower maximum, where the profile had broadened considerably. And since the signal from one cell of this calorimeter was the same, regardless of whether it was caused by 1, 4, 9 or 36 shower particles, the first sections were the main contributors to the observed signal non-linearity of this calorimeter.

Digital calorimeters thus inevitably lead to signal non-linearity. This is not only true for em showers, but also for hadronic ones. Electromagnetic shower components, *e.g.*, from the decay of the π^0s abundantly produced in hadronic shower development, are also extremely collimated. Therefore, many shower particles from such a π^0 shower may traverse an individual calorimeter cell. If this cell is "digital" (*i.e.*, producing "yes" or "no"), then the signal will be the same, regardless whether it is caused by 1, 3 or 29 shower particles. As a result, an important source of fluctuations is suppressed.

The response and the energy resolution of digital calorimeters are thus meaningless. As explained above, this was already discovered around 1980 and the further development of such devices was thus abandoned, for very good reasons. Yet, in recent years, the proponents of Particle Flow Analysis have rediscovered this option. In the context of the CALICE project, a digital hadron calorimeter was built, with no fewer than 500,000 readout channels. Each readout element was a 1×1 cm^2 RPC cell. Iron or tungsten was used as the absorber material.

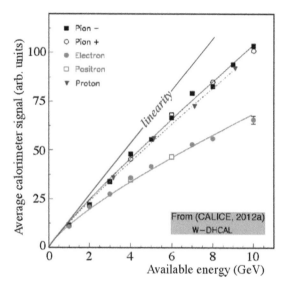

FIG. 9.20.   The average signal as a function of energy for electrons and hadrons in a large calorimeter based on digital readout. About 500,000 RPC cells were embedded in a structure consisting of tungsten absorber plates [Sef 15].

Some results of the beam tests of this device are shown in Figure 9.20 [Sef 15]. The average calorimeter signal, which in this case means the average *number* of RPC cells that produced a "yes" signal, is shown as a function of energy, for electrons and hadrons. It turned out that this detector was strongly non-linear, both for em and hadron showers. To illustrate this point, I have added a line that corresponds to linearity, *i.e.*, a signal that is proportional to the energy. Startlingly, this detector produced signals for hadron showers that were even larger than those for the em showers of the same energy. In other words, the detector seemed to be strongly *overcompensating* ($e/h < 1$). Of course, this feature just reflects the fact that the em showers are more collimated than the hadron ones and, therefore, it happened more frequently that several shower particles traversed the same RPC cell.

At the conference where these data were first presented, the author suggested that non-linearity might not be an important issue for calorimetry, and that linearity could be achieved by means of correction factors [Rep 13]. This is an argument in the same category as the one that was used to achieve "dummy" compensation, by installing passive absorber material in front of the calorimeter, which would affect the em response more than the hadronic one (Section 4.10, Figure 4.67).

The performance of any calorimeter is determined by fluctuations, and in this type of calorimeter fluctuations are suppressed and eliminated in a major way. Therefore, any conclusion about the performance of this device, especially in terms of energy resolution, is completely *meaningless*. To those who think that the intrinsic non-linearity of digital calorimeters is not a problem, I suggest a Geiger counter as the ideal calorimeter. It is perfectly compensating (the same signal for electrons and pions) and has an energy resolution of $0\%/\sqrt{E}$! It is, of course, also perfectly non-linear.

Calorimeter signal non-linearity may also be caused by factors other than signal saturation, in particular

1. *Miscalibration.* This issue was discussed at length in Section 6.2. Especially calibration schemes based on showers which deposit part of their energy in different longitudinal detector sections are prone to this phenomenon (see Figure 6.7). In the case of em showers, a consequence may be that different particles producing the same total signal (γ, e, π^0) may be reconstructed with slightly different energies [Wig 02].

2. *Non-compensation.* Calorimeters with $e/h \neq 1.0$ are intrinsically non-linear for hadron detection, as a result of the energy dependence of the average value of the em shower fraction (see Figure 3.12). In Chapter 6, procedures are described to reconstruct the energy of individual hadron events correctly in non-compensating calorimeters.

However, the non-linearity resulting from signal saturation is fundamentally different from these cases. In the CALICE digital calorimeter, the non-linearity depends on the size of the RPC cells, it is thus an instrumental effect. In the other cases, the non-linearity has nothing to do with the granularity of the readout, and the fluctuations in the shower development are not related to or affected by the non-linearity in any way, shape or form.

9.3.4   *Single hadrons vs. jets, electrons vs. γs*

Typically, in beam tests of calorimeters for experiments in high-energy physics, beams of charged particles are used, since these are readily supplied by accelerators. In some cases, efforts have been made to use beams of γ-rays, but these are typically much less monoenergetic and collimated than beams of electrons or pions, and thus less suited for high-precision measurements.

In modern particle physics experiments, one is more interested in detecting fragmenting quarks or gluons (which produce jets of particles) rather than single hadrons. Yet, the accelerators do not provide jet test beams and, therefore, beams of single hadrons are used instead to determine the calorimeter performance. However, it is important to realize that the performance for single hadrons is not necessarily representative of that for jets.

Something similar happens for electromagnetic showers. Sometimes, the experiment is primarily focused on detecting high-energy γs with high precision, as in the case of the experiments searching for the Higgs boson through its $H^0 \rightarrow \gamma\gamma$ decay mode. In the absence of high-energy photon beams, the calorimeters are tested with electrons or positrons. Also here, it is important to realize that the performance is not necessarily the same for electrons and photons.

9.3.4.1 *Single hadrons vs. jets.* There are significant differences between the detection of hadrons and jets in calorimeters. Jets involve some additional sources of fluctuations, *e.g.*, fluctuations in the energy sharing between the em and non-em jet fragments, and fluctuations in the energy sharing between the different non-em jet fragments. Because of this, response non-linearities are also important for the jet response function. In calorimeters with separate em and hadronic sections, the energy sharing between these sections and the fluctuations in this energy sharing are also different for single hadrons and jets. For this reason, offline compensation techniques (Section 4.10) based on the results of beam tests with hadron beams are not applicable for jets.

The ZEUS experiment found that the compensating ^{238}U/plastic-scintillator calorimeter, which had achieved record-setting energy resolutions for the detection of single hadrons, did not perform nearly as well for jets. One reason was that the jets in this experiment contained many soft hadrons, for which the response was non-linear. Figure 8.2 shows that the response ratio for electrons and hadrons gradually drops for energies below 5 GeV, from 1.0 at high energy to an asymptotic value of 0.62 at the point where the hadrons predominantly lose their kinetic energy by ionization, *i.e.*, the e/mip value. This is an important effect. For example, it was shown [Web 15] that in the hadronic decay of Z^0 bosons at LEP, on average 25% ($\pm 10\%$, σ_{rms}) of the total energy was actually carried by hadrons with momenta < 5 GeV/c (Figure 8.3).

So how could the jet response function of a calorimeter (system) be devised on the basis of the measured performance for electrons and hadrons? It is important to realize that, from the perspective of the calorimeter, a jet is simply a collection of particles, mainly pions and γs, that enter the detector simultaneously. If one has a database of events generated by beam particles of different types and energies, then these experimental data could be used to reconstruct the energy deposit profile for a given jet in many different ways.

This procedure was used to determine the response functions for W and Z bosons in the CMS calorimeter system [Gum 08]. Figure 9.21 shows relevant characteristics, covering a wide dynamic range, for different beam particles whose signal distributions were measured in this calorimeter system [Abd 09]. The average reconstructed energy is shown as a function of momentum (Figure 9.21a), and as a function of *available energy* (Figure 9.21b), which is the energy that contributes to the calorimeter signals. For protons this is the kinetic energy, for pions the kinetic energy plus the mass, and for anti-protons the kinetic energy plus twice the mass.

The procedure to determine the signal distributions for hadronically decaying W and Z bosons worked as follows [Gum 08]. First, a decay mode was selected, say $W^+ \rightarrow u\bar{d}$. Then, a fragmentation function was used to determine the final state particles into which each of the two jets hadronizes. This list of particles was the basis for determining the calorimeter signals. For example, if the list contained a π^+ with a momentum of 12.4 GeV/c, a random event from a beam test with 10 GeV pions was chosen and the measured signal in all participating calorimeter cells was multiplied by a factor 1.24 (12.4/10). In order to take into account the effects of non-compensation, the

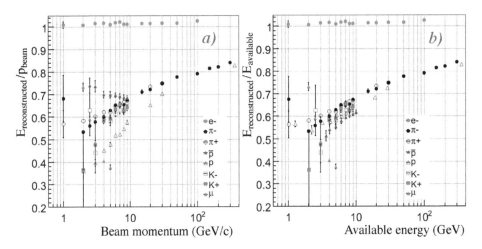

FIG. 9.21. The ratio of the reconstructed energy in the CMS calorimeter and either the momentum (*a*) or the available energy (*b*) for a variety of beam particles at different energies [Abd 09].

signal was also multiplied by a factor 1/0.7, where 0.7 represents the reduction of the calorimeter response for pions of this energy (see Figure 9.21b), compared to electrons (which were used to set the absolute energy scale in this calorimeter).

If the next particle in the list was a proton with a momentum of 4.8 GeV/*c*, the available (kinetic) energy was determined to be 3.95 GeV, a random event from a 5 GeV/*c* proton run was selected and its signal multiplied by a factor 0.96 (4.8/5). To account for the suppression of the signal from a proton with a kinetic energy of 3.95 GeV, a multiplication factor of 1/0.54 was applied as well. If the next particle in the list was a 18.3 GeV γ (*e.g.*, from the decay of an energetic π^0), a random event from a 20 GeV electron run was chosen and the measured signal multiplied by a factor 0.915 (18.3/20). In this case, no non-compensation effects had to be taken into account. And so on, until each final-state fragment of the jet was assigned a calorimeter signal. The sum of all these signals would then represent the calorimeter signal for this particular jet. And in the case of a boson produced at rest, the sum of the signals from both jets (u and $\bar{d}$) would constitute the boson mass.

In the described CMS analysis, this procedure was repeated many times for the same two jets, by choosing different random events from the vast database to describe the calorimeter signals for the jet fragments. The procedure was also repeated for two completely different jets, as well as for different decay modes of the intermediate vector bosons. The resulting signal distributions for the hadronically decaying W and Z bosons obtained in this way are shown in Figure 9.22. The total numbers of W and Z events were chosen such as to represent the production cross section ratio in high-energy pp collisions. Clearly, the energy resolution of the CMS calorimeters is not good enough to identify a hadronically decaying vector boson as a W or a Z.

The described procedure would also be ideally suited to test claims about the performance of PFA-based systems experimentally. For that, it would also be necessary to

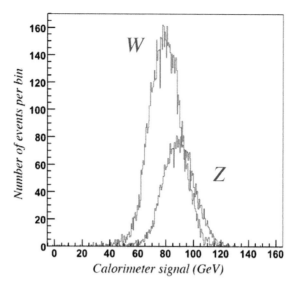

FIG. 9.22. The signal distributions for hadronically decaying W and Z bosons in the CMS calorimeter, reconstructed on the basis of the jet fragmentation functions and the response functions for individual jet fragments. The total numbers of W and Z events were chosen such as to represent the production cross section ratio in high-energy pp collisions [Gum 08].

use the transverse momentum of the jet fragments with respect to the direction of the fragmenting quark or gluon. This transverse momentum, together with the effects of the magnetic field, is needed to determine the distance over which the measured shower profile has to be shifted inside the calorimeter. Using the procedure described above for CMS, the hit profile of the jet in the calorimeter could be determined for a given choice of final-state particles. Next, one's favorite PFA algorithm could be applied to eliminate the contributions from charged hadrons and determine the remaining calorimeter energy. This energy could then be added to the (precisely known) energy of the charged hadrons to give the jet energy. As before, for each collection of final-state particles a large number of different jet hit profiles could be constructed, by randomly choosing different events from the vast database. Also, the collection of final-state particles could be changed by letting the quarks fragment in different ways. A procedure of this type would result in jet response functions that are based as closely as possible on the experimentally measured information.

The PFA proponents have chosen not to follow this procedure. Instead, they prefer to run Monte Carlo simulations, which are known to be wrong in reproducing crucial features such as the width of hadronic showers, to generate a jet response function. Then they use highly unconventional (unrelated to Gaussian) statistics to derive from this the precision of the jet energy measurement [Tho 09]. More on this in Section 9.3.5.

9.3.4.2 *Electrons vs. γs.* In [Wig 02], a systematic study is described of the differences between the detection of electrons and γs in calorimeter systems. Some of the re-

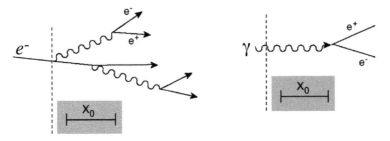

FIG. 9.23. The different starts of em showers initiated by high-energy electrons (*left*) and γs (*right*) entering an absorbing medium.

sults are discussed in Section 5.1.4. Electron and γ induced showers are different in the way they start developing (Figure 9.23). Upon entering an absorber medium, electrons start to radiate immediately. In the first radiation length, they lose, on average, 63.2% of their kinetic energy to bremsstrahlung. On the other hand, high-energy γs travel, on average, 9/7 radiation lengths in the absorbing medium before their first interaction.

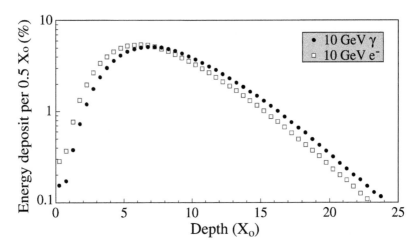

FIG. 9.24. The average longitudinal shower profiles for 10 GeV electrons and γs developing in a calorimeter made of $Z = 50$ absorber material. Results of EGS4 simulations [Wig 02].

This difference between the interaction mechanisms has two types of consequences for the em showers initiated by photons and by electrons/positrons:

1. Photon-induced showers deposit their energy, on average, deeper inside the absorbing structure than do em showers induced by charged particles of the same energy (Figure 9.24).

2. The fluctuations in the amount of energy deposited in a given slab of material are larger for showers induced by photons than for showers induced by e^+ or e^-.

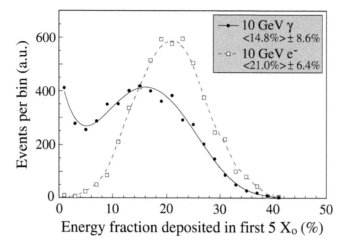

FIG. 9.25. Distribution of the energy fraction deposited in the first five radiation lengths by 10 GeV electrons and γs showering in lead. Results of EGS4 simulations [Wig 02].

The first effect results from the fact that the photons travel a certain distance in the absorbing structure before they start losing energy, while electrons and positrons start losing energy immediately upon their entry. Moreover, the starting point of the photon-induced showers fluctuates from event to event, which leads to the second effect.

These effects are illustrated in Figure 9.25, which shows the distribution of the energy deposited by 10 GeV electrons and 10 GeV photons in a $5X_0$ (2.8 cm) thick slab of lead. On average, the electrons deposit more energy in this material than the photons (2.10 GeV $vs.$ 1.48 GeV). However, the *fluctuations* in the energy deposited by the photons are clearly larger than those in the energy deposited by the electrons (0.86 GeV $vs.$ 0.64 GeV). The distribution for the photon showers exhibits an excess near zero, which is the result of photons penetrating (almost) the entire slab without interacting. The "punch-thru" probability for a high-energy γ is in this example $\exp(-35/9) \approx 2\%$.

The different effects of dead material installed in front of the calorimeter on electrons/positrons and γs is particularly relevant for the ATLAS experiment, where the electromagnetic calorimeter is "hidden" in a cryostat, although the effects are less dramatic than suggested in Figure 9.25 because this this cryostat is made of aluminium. However, another consequence of the differences between electron and γ induced showers for ATLAS is that the very complicated calibration scheme that was developed for electrons showering in the three longitudinal segments of the ECAL (Figure 6.13) is *not* necessarily the optimal solution for γ detection in this calorimeter.

9.3.5 *Misleading presentation of results*

In any field of science, it is important that the participants have a common understanding of what is meant by certain terms, and follow generally agreed-upon procedures to determine the value of quantities of interest in that field. In the absence of this, confusion and inefficiency will reign, and outsiders will quickly cease to pay attention to the achievements.

In calorimetry, detector performance is defined in terms of (non-)linearity, energy resolution and response function (line shape). These terms have a very precise meaning.

- A calorimeter is said to be *linear* if the signal it produces is proportional to the deposited energy. This means that the average signal per unit deposited energy is constant.

- The *energy resolution* of a calorimeter is a measure for the precision with which the energy of a showering object can be determined. This precision is expressed in terms of a *standard deviation*, which is the σ_{rms} of the signal distribution for repeated deposits of the same energy. If this distribution is dominated by Poisson fluctuations, then the standard deviation becomes equal to the σ from a Gaussian fit.

- The *response function* or line shape is the signal distribution for monoenergetic particles of a certain type. The response function is typically different for different types of particles (electrons, pions, protons), in terms of shape, width and central value[38].

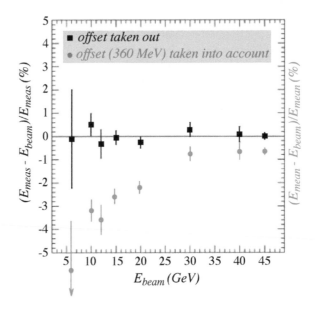

FIG. 9.26. Residual signals in the CALICE W/Si calorimeter, before and after taking out a 360 MeV offset. Data from [Adl 09].

---

[38] In this text, we use the term *response* to indicate the *average* signal per unit deposited energy (GeV), but this is not universal practice.

It is very unfortunate that some people feel the need to redefine these terms for their own purpose, in order to make the performance of their detector look better than it really is. Let us first consider the linearity of the calorimeter response. A calorimeter is *not* necessarily linear when the measured signals plotted versus the deposited energy can be described with a straight line. This straight line has to extrapolate through the origin of this plot, in other words the signal has to be zero when no energy is deposited. In [Adl 09], the authors fit the measured signals of their calorimeter with the following expression (see Figure 8.47):

$$E_{\text{mean}} = \beta \, E_{\text{beam}} - 360 \text{ MeV} \tag{9.1}$$

Then, they define

$$E_{\text{meas}} = E_{\text{mean}} + 360 \text{ MeV} \tag{9.2}$$

and plot

$$(E_{\text{meas}} - E_{\text{beam}})/E_{\text{meas}}$$

as a function of the beam energy. The result is represented by the squares in Figure 9.26. They conclude that "*the calorimeter response is linear to within approximately 1%.*" This is highly misleading. When the calorimeter signals they actually *measured* are used to check the linearity, *i.e.*, when

$$(E_{\text{mean}} - E_{\text{beam}})/E_{\text{mean}}$$

is plotted as a function of the beam energy, the results, represented by the full circles in Figure 9.26, look quite different. I conclude from this that the authors measured a response non-linearity of 5% over one decade in energy.

We now turn to the energy resolution. Often, the measured signal distributions exhibit non-Gaussian tails. In that case, it is common practice to quote the σ_{rms} value as the energy resolution. However, the author of [Tho 09] defined another variable, in order to make the results less dependent on the tails of his signal distributions (see Figure 8.45 for an example), and thus look better. This variable, called rms_{90}, was defined as the root-mean-square of the energies located in the smallest range of reconstructed energies which contained 90% of the total event sample. For the record, it should be pointed out that for a perfectly Gaussian distribution, this variable gives a 21% smaller value than the true σ_{rms} (*i.e.*, σ_{fit}).

Of course, everyone is free to define variables that suit his/her needs. However, this author quotes the results of his simulations, both in the abstract and the summary of his paper, in terms of the "jet energy resolution," and compares results obtained by him in terms of rms_{90} with genuine energy resolutions from calorimeters with Gaussian response functions (see Figure 8.55). This practice is also generally followed by the proponents of PFA.

9.3.6   *Concluding remarks*

This chapter is mainly intended to warn the reader to be extremely careful in comparing different calorimeters. As shown in Section 9.1, the use of biased event samples may make the results look better than they really are, by a considerable factor. And in any comparison between results obtained with PFA calorimeters and traditional ones with Gaussian response functions, the energy resolution of the latter should be multiplied by a factor 0.79. Unfortunately, I have had to notice on several occasions that many people who are actively involved in this business are not or insufficiently aware of this.

Understandably, the developers of new instruments tend to make their devices look as good as possible. However, a misleading presentation of the results of performance studies does, in my opinion, a great disservice to the scientific community which depends on these detectors for its continued success.

10

CALORIMETERS FOR MEASURING NATURAL PHENOMENA

All calorimeters discussed in the previous chapters were intended to detect events created by particles produced by powerful accelerators. This was even true for experiments that took place at a long distance from the accelerators, such as OPERA [Aga 15] and ICARUS [Alm 04] in the Gran Sasso tunnel, which detected neutrinos sent there from CERN, 730 km away. Or the MINOS [Mic 08] and NOVA [Pat 13] experiments, located at the Soudan mine in Minnesota, which study interactions by neutrinos produced at Fermilab, 810 km away. In a later stage, these experiments will be joined by DUNE in South Dakota, 1,300 km from the neutrino source at Fermilab. In East Asia KAMLAND [Gan 11] and Daya Bay [An 12] have detected interactions by neutrinos produced at a variety of nuclear reactors throughout Japan, China and even Korea.

In this chapter, we describe calorimeters that are primarily intended to study natural phenomena, by detecting the particles resulting from these processes. These phenomena include:

- The production of neutrinos in the core of the Sun
- The production of neutrinos in Supernova explosions
- The production of neutrinos by cosmic rays in the Earth's atmosphere
- The absorption of extremely-high-energy cosmic rays in the Earth's atmosphere
- Interactions by extremely-high-energy extragalactic neutrinos
- Proton decay

For some of these studies, the natural environment itself is instrumented to act as a calrimeter. We start this chapter, however, with a manmade instrument that has played a pioneering role in what is nowadays called *Neutrino Astronomy*, or *Astroparticle Physics*.

10.1 SuperKamiokande

10.1.1 *Physics goals*

SuperKamiokande (also known as SuperK) is a second-generation water-Čerenkov detector. It succeeded Kamiokande, which has been operating at the same location from 1983 to 1995. That location is the Kamioka mine, inside Mt. Ikenoyama, about 400 km northwest of Tokyo. Both Kamiokande and SuperK operate(d) $\sim 1,000$ m below the top of the mountain, which serves as a shield for cosmic rays with a thickness equivalent to 2,700 m of water.

Kamiokande was originally primarily intended as a detector to study proton decay, and has been one of the major players that have pushed the limits on the proton lifetime to levels beyond $5 \cdot 10^{32}$ years, for decay modes such as $p \to e^+\pi^0$.

However, as time went by (and the protons refused to decay), gradually more emphasis was placed on the study of cosmic phenomena, and in particular on processes in which neutrinos are involved. This required several experimental modifications, such as a complete overhaul of the water purification system, since the detector had to become sensitive to much smaller energy deposits for this purpose. The detection threshold was pushed down, initially to ~ 7 MeV, which allowed for studies of solar neutrinos [Hir 91]. In later years, the threshold was gradually further lowered to below 5 MeV [Fuk 03].

Kamiokande's first moment of glory came on February 23, 1987, when it detected 11 neutrinos in a time span of 13 seconds, with energies ranging from 7 to 36 MeV. It turned out that these neutrinos were released in the explosion of Supernova 1987a [Hir 88]. This marked the first time that neutrinos from this type of source were observed. Not only did these experimental results allow for invaluable tests of models describing stellar collapse, they also provided a completely new, tight upper limit on the possible rest mass of the (electron) neutrino. On the basis of the measured dispersion after a journey of 160,000 years, it was concluded that this mass had to be smaller than 5.7 eV/c^2 [Lor 02]. This limit is not much larger than that derived from endpoint studies of the ^{3}H β-spectra. In 2002, M. Koshiba was awarded the Nobel prize in phyics for his pioneering contributions to astrophysics, which included this discovery [Hir 87], as well as the detection of solar neutrinos in real time by the Kamiokande team [Hir 90].

SuperK is a larger version of Kamiokande, its total active mass is about 16 times larger (50 kiloton *vs.* 3 kiloton), while its fiducial mass has increased by twice that factor (22 kiloton *vs.* 0.68 kiloton). The scientific goals of the SuperK experiment are very similar to those of Kamiokande, but can be pursued with considerably increased statistical sensitivity. This is especially important for the study of ν-induced processes, which are strongly limited by the interaction rates. The experimental program that is currently carried out includes:

- The study of solar neutrinos. These are produced in the nuclear reactions through which the Sun generates energy. This production takes place in the core of the Sun, at a rate of $\sim 10^{45}$ per second. Much of what is experimentally known about these neutrinos comes from radiochemical experiments, pioneered by Ray Davis and his team in the Homestake mine in South Dakota [Dav 68]. In that experiment, the solar neutrinos initiated the reaction $\nu_e + ^{37}$Cl $\to ^{37}$Ar $+ e^-$ in an enormous tank filled with cleaning liquid. The number of argon atoms produced in this way was counted and the solar ν_e flux was calculated on the basis of the results.

  Experiments of the same type, but sensitive to lower-energy neutrinos and thus to the major processes through which the Sun generates energy (*i.e.*, the pp cycle: $4p \to ^4$He $+ 2e^+ + 2\nu_e + 26.7$ MeV), were carried out in the 1990s, using large quantities of gallium as a target [Abd 96, Ham 96]. The neutrinos revealed themselves through the reaction $\nu_e + ^{71}$Ga $\to ^{71}$Ge $+ e^-$ in these experiments.

  All experiments of this type showed a *deficit* of neutrinos. The neutrino flux derived from these measurements was smaller than expected on the basis of solar model calculations, by a factor of typically two to three. This deficit is the origin of the "solar neutrino problem." [Bah 95]

In contrast to the mentioned experiments, SuperK has the capability of detecting solar neutrinos *in real time*, through the reaction $\nu_e + e^- \rightarrow \nu_e + e^-$. The Čerenkov light from the recoil electrons is the source of experimental information. This has three major advantages compared with the radiochemical experiments. First, the *direction* of the incident neutrinos may be determined. This makes it possible to identify the Sun unambiguously as the source of the detected signal. Second, the *time* at which the interactions occur is known. Time-dependent phenomena, such as a possible day/night effect, can therefore be studied. And third, the neutrino *spectrum* may be measured. This offers additional possibilities to compare the results with theoretical predictions.

The threshold for detecting the $\nu_e e$ reaction in SuperK is $\sim 4 - 5$ MeV. This means that SuperK, just like the Homestake radiochemical experiment, is only sensitive to neutrinos produced in the more exotic branches of the nuclear reaction chain, and in particular to the neutrinos produced in the decay of ^{8}B. This process is only responsible for $\sim 10^{-4}$ of the solar energy production and the rates of these neutrinos are considerably more model dependent than the neutrino rates from the dominant processes. The maximum energy of the solar neutrinos is about 15 MeV.

- The study of atmospheric neutrinos. These are generated in the decay of pions, kaons and muons produced by cosmic rays interacting in the Earth's atmosphere. They are considerably more energetic than the solar neutrinos. Most of the atmospheric neutrinos detected in SuperK carry energies in the range from 0.1 to 10 GeV.

In contrast with the solar neutrinos, which are only observed through $\nu_e e$ scattering, atmospheric neutrinos generate several types of observable reactions in the SuperK detector. Among those reactions, the most interesting ones are the charged-current reactions induced by $\nu_e, \bar{\nu}_e, \nu_\mu$ and $\bar{\nu}_\mu$. In these reactions, usually a large fraction of the available energy is transferred to a relativistic electron or muon. This lepton travels typically in approximately the same direction as the incoming (anti-)neutrino, so that the direction of the latter may be determined by measuring the direction of the charged lepton.

The study of atmospheric neutrinos is a very important tool in the search for *neutrino oscillations*, a phenomenon that is likely to occur if neutrinos are massive particles. One sign of this phenomenon is the apparent change in composition (as measured through charged-current interactions) of a neutrino beam as it moves away from the production source. Since the source of the atmospheric neutrinos surrounds the entire Earth, and since neutrino absorption in the Earth itself is an insignificant effect, SuperK has the opportunity to study these particles for a wide variety of distances traveled, ranging from a few km to $\sim 13,000$ km (the diameter of the Earth).

The second moment of glory for SuperKamiokande came in 1998, when they announced the discovery of neutrino oscillations and thereby implicitly proved that neutrinos are massive particles [Fuk 98]. In 2015, T. Kajita received the Nobel Prize in physics for this discovery. I believe that this was only the second time in history that two Nobel Prizes have been awarded for different discoveries to the same group of experimentalists[39]. The neutrino oscillation phenomenon, and the way it was discovered in SuperK, are discussed in detail in Section 11.2.

- Since 2000, the SuperKamiokande detector has also been used to study inter-actions by neutrinos that are produced at the Japanese Center for High Energy Physics KEK, about 300 km to the southeast, and then beamed to the Kamioka mine. This project was initially known under the acronym K2K (KEK-to-Kamioka [Ahn 06]), when the neutrinos were produced by the 12 GeV proton synchrotron in Tsukuba. In a second stage, which started in 2005, the neutrinos which have a spectrum that peaks at $\sim$ 0.6 GeV, are provided by the new, more powerful J-PARC accelerator, located at the Tokai campus of KEK. The project is now known as T2K (Tokai-to-Kamiokande) and has already obtained several interesting re-sults [Abe 16]. The experiments are intended to provide complementary infor-mation for the atmospheric neutrino studies. The direction, the spectrum (which peaks at $\sim$ 0.6 GeV) and the composition of this neutrino beam are precisely known. This is important for studying systematic effects that may play a role in the analysis of the atmospheric neutrino data. In addition, a "long-baseline" experiment of this type has its own place in the search for neutrino oscillations. In that context, the T2K Collaboration also operates a "near detector" (ND280), located at 280 m from the neutrino source.

The SuperK detector is not only used for carrying out studies like those mentioned above, but it is of course also on permanent alert to detect the neutrino burst from an-other Supernova explosion, should one occur in the "vicinity," *i.e.*, within a distance of the order of one million lightyears. Also the proton decay studies are likely to be updated from time to time.

In the meantime, a start has been made with the upgrade of the detector by an-other factor of twenty. HyperKamiokande, as the future detector is known, will consist of two cylindrical tanks, each with a diameter of 48 m and 250 m long. These will contain one million metric tons of ultrapure water. Hamamatsu has built a special fac-tory for the production of the 99,000 light sensors that will detect the feeble Čerenkov light pulses produced in this detector. These sensors have a diameter of half a meter, a 50% larger quantum efficiency and better timing characteristics than the SuperK ones. The design of HyperKamiokande was obviously inspired by the Andromeda Galaxy. When completed, its more than 800 times larger fiducial volume will make the Hy-perKamiokande detector substantially more sensitive to supernova explosions in An-dromeda than Kamiokande was to SN1987a, which occurred a factor fifteen closer to Earth, in the Large Magellanic Cloud. Its strongly improved pointing capability would make it possible to locate the source of the neutrinos from such explosions unambigu-ously. And the fact that HyperKamiokande consists of two independent detectors would eliminate the need for "dead time."

---

[39]The honor was also bestowed on members of the MARK II experiment at SLAC, in 1976 and 1995.

10.1.2 *Experimental requirements*

The experimental requirements needed to carry out the experimental program outlined in the previous subsection include

- A detection threshold that is pushed to the lowest possible limit. This threshold level is closely related to the purity of the water used as detector material, and in particular to its natural-radioactivity content. Radon gas is the most serious background for the solar neutrino analysis.

- A full-proof capability to recognize and reject signals that are caused by reaction products of interactions taking place outside the fiducial detector volume. The interaction cross sections for the neutrinos of interest are extremely small. To set the scale, the mean free path of 1 GeV neutrinos in water is of the order of the distance between Earth and the outermost planets of the solar system. In spite of the impressive flux of neutrinos traversing the huge volume of the SuperK detector, the event rates are thus very small. For example, solar neutrino interactions of the type $\nu_e + e^- \rightarrow \nu_e + e^-$ are recorded at a rate of about 15 per day. That is about twice the rate of interactions induced by atmospheric neutrinos.

  These rates are some five orders of magnitude smaller than the rates at which relativistic charged particles originating from other sources produce Čerenkov light inside the fiducial detector volume. Even with a veto system that rejected 99.999% of these background events, the remaining background would still be of the same order of magnitude as the signal itself.

- The capability to distinguish between the signals generated by muons and by electrons with energies in the region of interest (0.1–10 GeV). This is crucial for the atmospheric-neutrino analysis. At energies in the GeV range, electrons develop showers. This means that the energy is deposited by a large number of relativistic particles. Some fraction of these particles, in particular the shower electrons produced by Compton scattering of bremsstrahlung photons, may travel at large angles with respect to the direction of the showering electron. At low energies, multiple scattering has a larger effect on the trajectory of electrons than on that of muons. In both cases, electrons produce Čerenkov light that is spread out over a larger angular range with respect to the direction of the incoming particle than do muons. The particle identification techniques employed in SuperK are based on this effect.

- The capability to reconstruct the trajectories of the muons and electrons produced in neutrino interactions with adequate angular precision. This is important for the reconstruction of the trajectories of the neutrinos that produced the charged leptons. In the case of solar neutrinos, there is a very strong correlation between the direction of the neutrino and that of the recoiling electron. This is illustrated in Figure 10.1, which shows the angular distribution of the recoil electrons in $\nu_e e$ reactions induced by solar neutrinos, measured with respect to the direction of the incoming neutrino. These distributions are given for neutrinos from ^{8}B decay, for different values of the low-energy cutoff energy. For the high-energy tail of this spectrum ($E > 10$ MeV), the angle between the neutrino and the recoil elec-

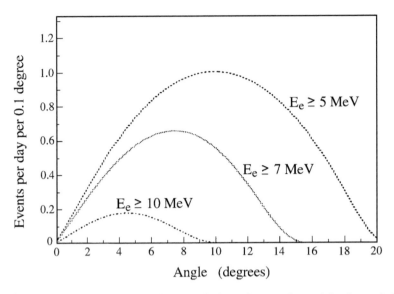

FIG. 10.1. Distributions of the angle between the incoming neutrino and the detected electron in $\nu_e + e^- \rightarrow \nu_e + e^-$ events induced by solar neutrinos, for various energy regions [Kos 98].

tron is very small, typically $\sim$ 4–5°. If the cutoff energy is lowered, the angles increase.

For charged-current interactions induced by atmospheric neutrinos, the correlation between the direction of the incoming neutrino and the scattered charged lepton is typically less strong, since part of the energy is transferred to the struck nucleon in this case. For example, for charged-current events induced by 1 GeV ν_μs, the average angle between the neutrino and the muon is 25°, and in 68% of the cases this angle has a value between 20° and 40°. For lower energies, the correlation becomes rapidly weaker.

These numbers give an impression about the desirable angular resolution for the various analyses. A resolution of a few degrees would be very useful for the solar neutrino studies, but anything better than 20° would not make much difference for the atmospheric neutrinos.

10.1.3 The detector

10.1.3.1 *The SuperK tank and its contents.*   The SuperK detector [Fuk 03] is a huge cylindrical vessel made of stainless steel and filled with 50,000 tons of pure water. The height of the tank is 41.4 m, and its diameter 39.3 m. Inside this tank an elaborate stainless steel frame holds the PMTs that record the Čerenkov light produced in the water. This support structure is also used to separate the detector volume into two parts, called the Inner Detector and the Outer Detector, which are optically completely isolated from each other. This is achieved with black polyethylene terephthalate sheets. The Inner Detector is also a cylinder, with a height of 36.2 m and a diameter of 33.8 m. It is completely surrounded by the Outer Detector and contains 32,000 tons of water.

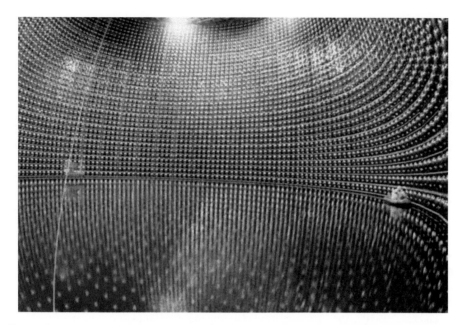

F<small>IG</small>. 10.2. Picture taken inside the SuperK detector, when it was partially filled with water. Courtesy of the SuperKamiokande Collaboration.

The Inner Detector is viewed by 11,146 20-in. PMTs. Of these, 7,650 are attached to the barrel parts of the support frame, the rest are equally distributed between the top and bottom walls. The PMTs form a grid with a spacing of 70 cm. In total, more than 40% of the wall surface of this water-filled cathedral is covered with photocathode material! Figure 10.2 shows a picture taken inside the detector while it was being filled.

The Outer Detector completely surrounds the Inner Detector. It has a thickness of 2.7 m in the barrel part and 2.6 m on top and bottom. This detector is viewed by 1,885 8-in. PMTs. These PMTs are mounted on the outward side of the support frame and thus "look" in the opposite direction, compared with the PMTs that view the Inner Detector. There are two 8-in. Outer PMTs for each group of twelve 20-in. Inner PMTs. Therefore, the photocathode surface of the PMTs viewing the Outer Detector represents less than 1% of the wall surface of this detector. In order to maximize the light collection efficiency, the rest of the wall surface is covered with Tyvek, a material that has a very high light reflectivity.

The Outer Detector plays a crucial role in reducing the background, since it acts as a veto counter for the Inner Detector. It identifies cosmic muons, which enter the apparatus at a rate of about 2 Hz. Since it represents an absorber with a thickness of more than $7X_0$ and $4\lambda_{\mathrm{int}}$, it also shields the Inner Detector against the γ and neutron background produced in the surrounding rock.

In practice, many analyses were carried out for a fiducial volume with a mass of about 22 kiloton. The boundaries of this fiducial volume were located 2 m inside the boundaries of the Inner Detector. The reconstructed vertex position of the interactions is subject to measurement uncertainties and the fiducial volume was chosen so as to make sure that event vertices were located inside the Inner Detector.

On November 12, 2001, this remarkable facility suffered a terrible accident in which a chain reaction of failures destroyed more than 6,000 photomultiplier tubes. The tank was being refilled with water after some of its burned-out tubes had been replaced. The conclusion of the investigating teams was that workmen standing on styrofoam pads on top of some of the bottom PMTs must have caused microfractures in the neck of one of the tubes, which led to an implosion of that tube. Owing to Pascal's principle (which states that pressure is transmitted undiminished in an enclosed static fluid), the pressure pulse of that implosion caused a chain reaction of implosions throughout the water-filled portion of the tank. It took almost five years and an estimated $25 million to bring the detector back into its pre-accident state, with each PMT now housed in a protective casing of fiber-reinforced acrylic.

10.1.3.2 *The PMTs.* The 20-in. PMTs used in the Inner Detector were custom-made for this experiment. A schematic view of one of these PMTs is shown in Figure 10.3. The PMT has a bi-alkali photocathode, with a spectral sensitivity that matches the Čerenkov spectrum. The quantum efficiency peaks at 22% near a wavelength of 360 nm, and reaches 10% for wavelengths of 310 nm and of 510 nm.

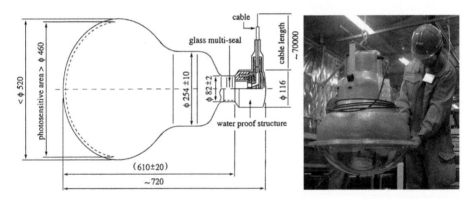

FIG. 10.3. Schematic structure of the PMTs used to detect the Čerenkov signals in SuperK. All dimensions are expressed in mm [Suz 93], together with a picture of the real thing, including the protection added after the 2001 accident. Picture courtesy of the SuperKamiokande Collaboration.

In order to ensure a good collection efficiency, a Venetian-blind type of dynode structure is used in this tube. Because of the large photosensitive area of this structure, the transit time and the event-to-event fluctuations in this transit time are rather long. These timing characteristics determine directly the resolution of the reconstructed vertex, and therefore it is very important that they be optimized. This was achieved through the design of the voltage divider. An 11-stage voltage divider, with 40% of the total voltage applied between the cathode and the first dynode, gave the best results: a collection efficiency of 70%, combined with an average transit time of 100 ns and a standard deviation of 2.2 ns in the transit time for signals at the one-photoelectron level. Each of these PMTs has a dynamic range from 1 to 300 photoelectrons. SuperKamiokande can detect events in which as little energy as 4.5 MeV is deposited in the inner Detector. The upper limit is about 1 TeV.

In order to ensure a uniform response over the entire photocathode surface, the residual magnetic field inside the PMTs has to be kept at a level below 10^{-5} T. This is achieved with 26 Helmholtz coils, which surround the entire detector and compensate for the geomagnetic field. As a result, the residual geomagnetic field is kept below the mentioned upper limit *over the entire detector volume*.

The 8-in. PMTs used in the Outer Detector are each connected to a 1.3 cm thick WLS plate with dimensions of 50×50 cm^2. These plates absorb the UV Čerenkov light and re-emit it in the blue part of the spectrum, where the PMTs are most sensitive. As a result, the light collection efficiency of the Outer Detector is increased by $\sim 60\%$. The PMTs used in SuperK are described in considerably more detail in [Suz 93].

10.1.3.3 *Electronics and data acquisition.*

The PMT signals are sent to front-end electronics modules, where they are digitized and compared with preset thresholds. If the signal exceeds the threshold value, set at $\sim 1/4$ p.e., then a rectangular pulse with a width of 200 ns and a height of 15 mV is generated. This pulse is called a "HITSUM" signal. It is added to all other HITSUM signals recorded in the same 200 ns time interval. If the total HITSUM signal exceeds a preset threshold, then a trigger is generated and the hit pattern as well as the timing information of all PMTs that participated in the event are recorded for offline analysis. The signals from the Outer Detector are used to veto against charged particles coming from the outside world. A veto signal is formed if 19 or more HITSUM signals from PMTs viewing the Outer Detector are observed in the same 200 ns time interval.

In practice, several threshold levels are used. For the solar neutrino analysis, the threshold was initially set at 320 mV. After subtracting the average dark background rate, this was equivalent to 29 hits, which means that a global trigger was generated when at least 29 PMTs viewing the Inner Detector were hit in any 200 ns time window. This is equivalent to the signal expected when 50% of the Čerenkov photons generated by a 5.7 MeV electron are detected. With these conditions, the trigger rate was ~ 11 Hz. For the atmospheric neutrinos, proton decay studies and cosmic muon measurements, the trigger threshold was set higher, while for some dedicated solar-neutrino studies it was set at a lower level. Of course, the trigger rate increased sharply when the threshold was lowered. For example, for a threshold equivalent to the average signals generated by 3.6 MeV neutrinos, the trigger rate was 2.1 kHz. This increase was the result of γs from the rock surrounding the detector, of radioactive decay in the PMT glass itself and of radon dissolved in the water.

Not surprisingly, the vast majority of the background events turned out to originate from the perimeter of the Inner Detector. By reconstructing the vertex in real time and rejecting events with a vertex outside the 22.5 kton fiducial volume, the trigger rate was drastically reduced. Intelligent triggering like this has allowed SuperK to lower the threshold at which solar neutrinos can be studied gradually to ∼ 4.5 MeV. This is important since the neutrino spectrum from ^{8}B decay peaks at ∼ 5 MeV.

Inspired by the SN1987A experience, SuperK is employing a GPS system that makes is possible to synchronize the absolute times of events with other sites, with an accuracy of ∼ 100 ns. This is not only important in case another detectable supernova explosion occurs, but also for the long-baseline neutrino beam experiment.

10.1.3.4 *Water and air purification.*   There is a rich supply of natural water near the detector site. This water is used as detector material. A highly sophisticated water purification system circulates and purifies 30 m^3 of water every hour to maintain its quality. The entire contents of the tank is thus treated by this system once every ten weeks. The water is pumped from the top of the tank, treated by the purification system and returned to the bottom of the tank. A schematic overview of this system, which has been systematically improved over the years, is shown in Figure 10.4.

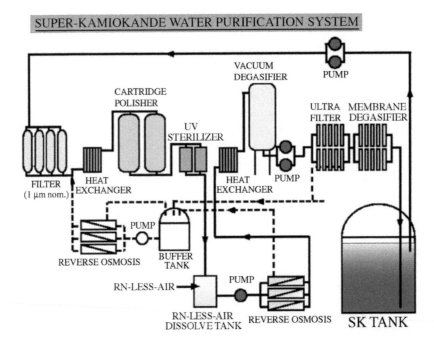

FIG. 10.4. Schematic overview of the water purification system used in SuperK [Fuk 03]. The figure shows the status of this system in 2002.

The purpose of the water purification system is to remove bacteria and radioactive material, and to maintain the highest possible water transparency. With the help of heat exchangers, the temperature of the water is kept constant at 13° C to suppress bacteria growth. The cartridge polisher removes heavy ions, which reduce water transparency and include radioactive species. An ion-exchanger, that was originally included in the water purification system, was removed when it was discovered that the resin it used was a significant source of radon. Radon gas dissolved in water is a major and consistent source of background. A tank to dissolve radon-reduced air in the water increases the radon removal efficiency of the vacuum degasifier, which removes dissolved gases from the water. Its efficiency for removing radon gas is estimated at 96%. It also removes 99% of the oxygen, and is thus another weapon to discourage the growth of bacteria. As a final step in the battle against radon, a membrane degasifier removes 83% of the remaining radon. It uses the radon-reduced air from the air purification system as a purge gas. The ultra filter removes almost all dust particles with a diameter down to 10 nm. Together with the reverse osmosis system, it reduces the concentration of particles of size lager than 200 nm from $\sim$ 1,000 particles per cm^3 before purification to 6 particles per cm^3 after purification.

After the water purification, the specific radioactivity, which amounts to 10 kBq m^{-3} in the natural mine water (mainly as a result of ^{222}Rn), has over time been reduced to levels below $\sim 10^{-3}$ Bq m^{-3}, a reduction of seven orders of magnitude! To achieve these low levels of specific radioactivity in the water, it was also crucial to take proper care of the air in the mine, which may be heavily contaminated with radon gas. It turned out that the radon concentration in the air changes by two orders of magnitude with the seasons, as a result of the changing direction of the air flow. During the summer, radon accumulates and typically reaches levels of 2,000–3,000 Bq m^{-3}. During the winter, fresh air enters the mine from outside and the concentration drops to 100–300 Bq m^{-3}.

The very first step in the battle against radon consisted of covering the walls of the cave with a polyurethane material intended to prevent the emanation of radon gas from the rock. In order to prohibit the radon that does enter into the cave from entering the detector, the water tank was tightly sealed. There is a 60 cm space between the water surface and the top of the tank. The air let into this region is sent through an elaborate "radon-free air" system, which reduces the contamination level to $\sim 10^{-2}$ Bq m^{-3}. The air inside the tank is kept at a slight overpressure with respect to the outside, to prevent non-purified air from leaking in.

Another crucial goal of the water and air purification systems is to maintain the transparency of the water. This is important for two reasons:

1. The signals are small, especially those from solar neutrinos. Light attenuation makes them smaller, and thus effectively raises the threshold for solar neutrino detection.

2. A change in the light attenuation characteristics of the water leads to a change in the energy scale of the detector, and thus requires recalibration.

The attenuation length is wavelength dependent. The best transparency occurs for $\lambda \approx 400$ nm, where in good-quality water attenuation lengths in excess of 100 m are routinely measured. For longer wavelengths, the transparency rapidly decreases ($\lambda_{\text{att}} \approx 30$ m at 500 nm). Near 330 nm, attenuation lengths of ~ 60 m are typical for SuperK. What matters in practice is the attenuation of the "effective" light spectrum that results from a convolution of the Čerenkov spectrum and the wavelength-dependent quantum efficiency of the PMTs. This information is provided by cosmic muons that traverse

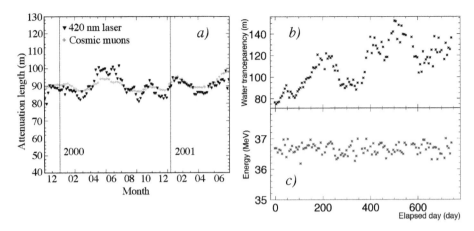

FIG. 10.5. The light attenuation length of the water, as a function of time. Separate results are shown for measurements with a 420 nm laser and with the entire Čerenkov spectrum emitted by muons traversing the tank from top to bottom (*a*) [Fuk 03]. After the accident, the transparency has further improved (*b*). After correcting for the effects of light attenuation, the reconstructed energy remains constant (*c*) [Abe 11].

the water tank from top to bottom. The attenuation length for this effective spectrum is typically ~ 90 m and has gradually improved over time. Figure 10.5a shows the results of measurements of the light attenuation length made with a dedicated laser system (420 nm), and with the Čerenkov light emitted by cosmic muons. The latter method has the additional advantage that it does not require dedicated running time, since the data sample for this analysis is automatically collected during normal data taking. The results of both methods are in good agreement. When the detector was rebuilt after the 2001 accident, the water transparency further improved, as can be seen in Figure 10.5b, which shows results of measurements with cosmic muons over a period of about two years in 2006–2008 [Abe 11].

The typical distance Čerenkov photons travel in the tank before reaching a PMT is about 20 m. The probability that they are absorbed on their way is therefore $1 - \exp(-20/90) \sim 20\%$ when the attenuation length is 90 m, and $\sim 15\%$ for an attenuation length of 120 m. All SuperK measurements are corrected event-by-event for the effects of light attenuation, taking into account the vertex position at which the light is produced and the attenuation length at the moment of the event. Figure 10.5c shows that the reconstructed calibration signal, in this case the average energy of electrons produced by decaying muons stopped in the water tank, remained very constant over time, thanks to this procedure [Abe 11].

10.1.4   *Calibration and monitoring*

As for every other calorimeter, a precise energy calibration in the region of interest is essential for the quality of the scientific results. Since the detector is located 300 km away from the nearest particle accelerator, traditional methods based on particle beams with precisely known characteristics cannot (easily) be employed in this case. There-fore, the SuperK Collaboration developed a variety of ingenious alternatives, which are briefly described below.

One characteristic difference between the SuperK detector and the other calorimeter systems described in this chapter is that in SuperK the Čerenkov light produced in the in-teractions is always shared among a large number of PMTs. In accelerator experiments, it is not uncommon that the majority of the energy carried away by the particles pro-duced in an interaction is deposited in a single calorimeter tower. This does not happen in SuperK. In typical interactions induced by atmospheric neutrinos, more than 1,000 PMTs record a signal above threshold, and the largest contribution to the total signal by a single PMT rarely exceeds a few percent. This feature is illustrated by the event displays shown in Figure 10.13. It is also illustrated by the fact that any event that does not generate coincident signals in at least 29 different PMTs is considered insignificant (Section 10.1.3.3)

As a consequence, it is much less important that the gains of all individual SuperK PMTs are precisely calibrated than it is in typical accelerator experiments. It is important that the relationship between deposited energy and the resulting *total sig-nal* generated in the detector be established as precisely as possible. However, random variations of individual PMT gains of the order of 10% are inconsequential for the de-tector performance.

10.1.4.1   *Gain equalization.*   Before the PMTs were installed in the detector tank, their gain-voltage characteristics were established with standard techniques. Each PMT has its own high-voltage supply, and after the PMTs were installed in the detector, the individual high-voltage values were set so as to equalize the PMT gains, *i.e.*, to equalize the number of pico-Coulombs per photoelectron.

The quality of this gain equalization was tested *in situ* with a uniform, isotropically emitting light source that could be installed at various positions inside the tank (Figure 10.6a). This light source was an acrylic ball filled with a wavelength shifter and magne-sium oxide powder that served to diffuse the light. This object received UV light from a xenon lamp through an optical fiber. The wavelength shifter absorbed this light and re-emitted is at wavelengths typical for the Čerenkov spectrum. This light was recorded in all PMTs installed in the Inner Detector. After correcting the observed signals for various instrumental effects, such as differences in acceptance and light attenuation, the relative gains of all PMTs could be compared.

The result of this test is shown in Figure 10.6b. The distribution of the relative PMT gains measured in this way is reasonably described by a Gaussian function, with a stan-dard deviation of $\sim 7\%$. When these measurements were repeated after 2.5 years of operations, the standard deviation had increased to 7.7% [Abe 11]. After the accident that necessitated the reconstruction of the SuperK detector, these calibrations were re-peated with a somewhat different system, and a standard deviation of 5.9% was obtained [Abe 14].

10.1.4.2   *Absolute energy calibration methods in the 5–100 MeV region.*   The most precise absolute calibration results were obtained with a linear electron accelerator

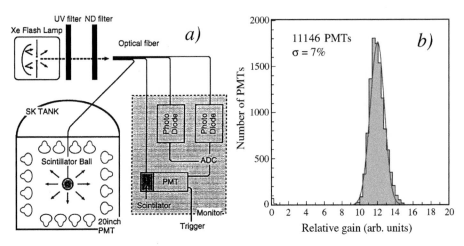

FIG. 10.6. Schematic overview of the experimental setup that was used to test the gain equalization of the PMTs in the Inner Detector (*a*), and the distribution of the gains measured in
this way (*b*) [Kos 98].

(LINAC) which could be installed at various positions inside the water tank. This LINAC,
which was originally built for medical purposes, produced a mono-energetic electron
beam with an energy that could be tuned between 5 and 17 MeV. The beam pipe could
be inserted into any of a number of holes, spaced by about 2 m along one axis in the
ceiling of the water tank. This setup is schematically shown in Figure 10.7. The LINAC
produced electrons at a rate of typically 1–3 Hz. The LINAC data were used to set the
absolute energy scale and to determine the resolution of the energy, position and angle
measurements as a function of energy, between 5 and 17 MeV. An example of the total
signal distribution recorded for 17 MeV electrons from the LINAC is shown in Figure 10.8a. On average, the total signals produced by these particles consisted of ~ 80
photoelectrons. The width of the signal distribution is approximately what one would
expect on the basis of event-to-event fluctuations in that number: $\sigma/\text{mean} \sim 13\%$. Figure 10.8a also shows that the experimental signal distribution was well matched by the
MC simulations, which took account of all experimental effects (light attenuation in the
water, quantum efficiency of the PMTs, *etc.*).

Unfortunately, the LINAC could only be installed in a limited number of positions.
Also, the electron beam could only be sent downward into the tank. Only a limited number of PMTs could thus be calibrated in this way. Therefore, more universal calibration
methods were developed in addition.

The first method, known as the "nickel calibration," was based on the use of nuclear
γ-rays produced in thermal-neutron capture by nickel. The neutrons originated from
nuclear fission, for which a ^{252}Cf source was used. In each fission process, on average
3.8 neutrons with an energy of typically 2 MeV are produced. In order to be captured
by a nickel nucleus, these neutrons need to be thermalized. This thermalization process,
in the surrounding moderator (water), and the subsequent emission of a γ-ray by the

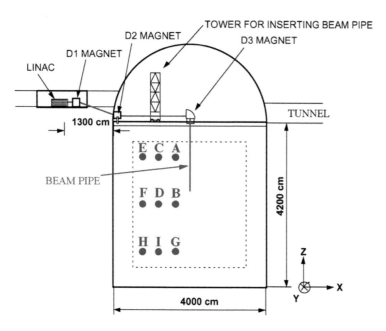

FIG. 10.7. Schematic layout of the SuperK tank and the LINAC that was used for calibrating the PMTs. The lettered positions indicate the locations where the end of the beam pipe is positioned in the tank for LINAC data taking [Fuk 03].

excited compound nucleus, takes place on a timescale of $\sim 100~\mu$s. The energy of the most abundant γ-ray emitted in this process is 9.0 MeV. All other prominent transitions have energies in the energy range from 6 to 9 MeV. This calibration system consisted of a 60 kBq ^{252}Cf source located in the center of a cylindrical polyethylene container (with a height and a diameter of 20 cm), which for the rest was filled with pure water and 2.8 kg of nickel wire. A proportional chamber located in the center of the container was used to tag the fissions. Capture γs emitted in the time interval between 10 and 200 μs after a fission trigger were used for calibration purposes.

A major advantage of this system was the isotropic emission of the calibration particles. In addition, it could be installed essentially anywhere in the detector. Therefore, all PMTs could be calibrated with this method. This source was used extensively during the first few years of SuperK operations, but was subsequently replaced, because of the inherent difficulty in understanding the transport of the complicated γ spectrum through the massive source, as well as its shadowing effects on the Čerenkov photons, with the required precision. After extensive cross calibration, it was replaced by a ^{16}N source. The decay of this nuclide has a Q value of 10.4 MeV, and is dominated by a β branch of 4.3 MeV, coincident with a 6.1 MeV γ ray. The detectable energy from this decay thus peaks near 8.3 MeV. The ^{16}N source, which has a half-life of 7.13 s, was produced in situ with a deuterium-tritium neutron generator, through the reaction $n + ^{16}$O $\rightarrow ^{16}$N$ + p$. This generator could be installed anywhere in the detector by

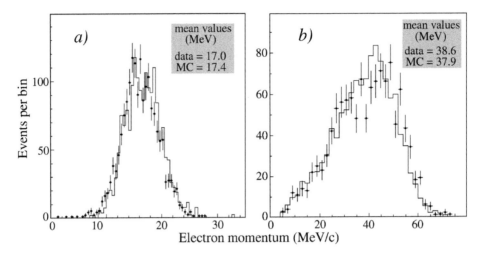

FIG. 10.8. Calibration results in the low-energy region, using electrons from the LINAC (*a*) or from the decay of stopping muons (*b*). The dots represent the experimental data, the histograms give the Monte Carlo predictions [Oku 99].

means of a computer controlled crane. In 2009, SuperK went back to the Ni-Cf calibration method, using a newly built spherical source [Abe 14].

At somewhat higher energies, electrons from the decay of stopping muons provide a useful source of absolute calibration. As in the case of the sources, the calibration particles are not mono-energetic, but are distributed according to a precisely known spectrum. The maximum energy of electrons from this source is 53 MeV, and their mean energy is 37 MeV. The mean energy of these decay electrons is used to check the stability of the calibration in Figure 10.5c. In this case, one has to make sure that the events used for the calibration are indeed caused by the decay of muons stopping in the fiducial volume of the tank. This means that the incoming muon has to be positively identified by the Outer Detector and that its decay vertex has to be located inside the fiducial detector volume. To eliminate Čerenkov light from the stopping muon itself from the signals, only light recorded in a time window from 1.5 to 8 μs after the muon came to rest is taken into account (the muon's lifetime is 2.2 μs). The recorded spectrum of these decay electrons is shown in Figure 10.8b. Also this spectrum is in good agreement with the Monte Carlo simulations.

10.1.4.3   *Calibration in the 0.1–1 GeV region.*   In the energy range of interest for the atmospheric neutrino studies, a number of other calibration techniques, all based on the experimental data themselves, have been employed at SuperK. Figure 10.9 shows results obtained with two of these techniques.

In the energy range up to $\sim$ 0.4 GeV, muons are not yet ultra-relativistic. For example, a muon with a momentum of 200 MeV/c has a β of 0.88. This means that the characteristic Čerenkov angle, $\arccos(\beta n)^{-1}$, is not 41° ($\beta \approx 1$), but only 31°. By measuring the opening angle of the Čerenkov cone, one therefore knows the momen-

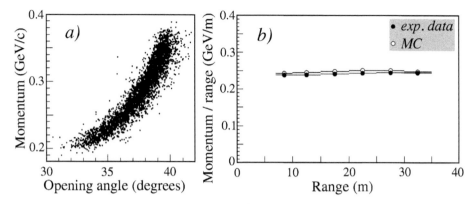

FIG. 10.9. Calibration results in the 0.1-1 GeV region, using the opening angle (*a*) or the range
(*b*) of stopping muons [Oku 99].

tum of the muon. Figure 10.9a shows a scatterplot of the measured opening angle versus
the calculated muon momentum (obtained by subtracting the muon mass in quadrature
from the measured energy). There is a very clear correlation between these quantities.

The muons used for this purpose were produced in the fiducial detector volume
by atmospheric neutrinos. That was also true for the muons used for the second method
applied in this energy range. This method was used for the more energetic muons, which
produced a track with measurable length in the detector. In that case, the well-known
range/energy relationship could be used to determine the energy (or the momentum) of
the muons event by event. This method was used for muons with energies in the range
from $\sim$ 1.5 to 8 GeV, which left tracks with a length ranging from $\sim$ 8 to 35 m.

In all calibration methods, the experimental results were compared in detail with
Monte Carlo simulations and the precision of the particular method was derived from
that comparison. For example, in the case of the data displayed in Figure 10.9a, one
could determine the momentum distribution for muons with opening angles between
$32°$ and $33°$. This distribution could then be compared with the momentum distribution
for MC events with opening angles in the same bin. The ratio between the average
momenta of the experimental data and the MC events was then considered a measure of
the quality of this calibration method.

All calibration methods discussed here (and several others which we skipped) led
the investigators to the conclusion that the absolute energy scale of the SuperK detector
in the energy range from 5 MeV to 8 GeV was known to within $\pm 2.5\%$.

10.1.4.4 *Time calibration.* The time calibration of the individual PMTs is very important, since the position of an interaction vertex in SuperK is reconstructed on the
basis of the arrival times of the signals in all the PMTs that recorded Čerenkov light
from the event. The arrival time of a PMT signal at the front-end electronics, as defined
by the moment a HITSUM signal is generated (Section 10.1.3.3), depends on the cable
length, on the PMT's transit time and also on the pulse height of the signal. The latter dependence is due to the slewing effect in the discriminator and necessitates timing

measurements as a function of pulse height for each PMT.

Figure 10.10a shows the experimental setup used for this calibration. Short (~ 0.4 ns FWHM) light pulses are produced by a nitrogen laser. The wavelength of this light is shifted from 337 nm to 384 nm (a fairly typical value for the detected Čerenkov light) by means of a dye laser module. This light is split into two parts. One (small) part is sent to a fast-response (monitor) PMT, which generates a signal that serves as the trigger that starts the TDCs for all individual SuperK PMTs.

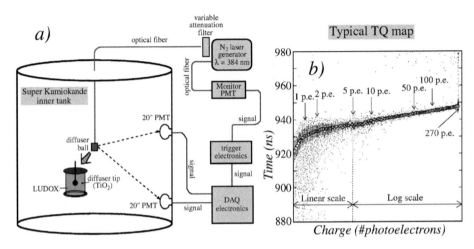

FIG. 10.10. Overview of the setup for the time calibration of the PMT signals (a) and a typical map of the relationship between the pulse height Q and the signal arrival time T (b). The circles correspond to central values, and the error bars represent the 1σ fluctuation levels [Oku 99].

The intensity of the remaining light can be changed with a filter, to accommodate measurements at various pulse heights. This light is sent to a diffuser ball located in the center of the tank, which was designed to diffuse the light in all directions without affecting its time structure. The diffuser ball consists of a titanium oxide tip surrounded by LUDOX, a gel made of very small (20 nm) glass fragments.

A typical result of measurements made with this system is shown in Figure 10.10b. This figure, called a *TQ map*, shows how the timing of the PMT signals changes as a function of the pulse height. This *time-walk* effect is due to the fact that the rise time of a large pulse is relatively shorter than that of a small pulse. Especially for small signals, the event-to-event fluctuations are also substantial. These fluctuations are important, since they translate directly into the uncertainty in the reconstructed vertex position. Figure 10.11 provides quantitative information about this aspect. Figure 10.11a shows the distribution of the timing for signals corresponding to about one photoelectron, for all PMTs of the SuperK detector. This distribution is asymmetric, because of the different time characteristics of the contributions of direct and indirect light (from reflection and scattering) to the signals. This asymmetry is also evident in Figure 10.10b, which

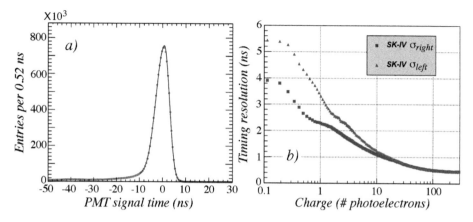

FIG. 10.11. Timing distributions added over all readout channels for signals produced by ~ 1 photoelectron, together with the result of a fit with an asymmetric Gaussian (a). The time resolution, *i.e.*, the width of Gaussian fits to the left and right portions of the timing distribution, as a function of the integrated charge of the PMT signals (b) [Abe 14].

shows that there are more signals arriving relatively late than relatively early. The distribution has been fitted to the sum of two Gaussians with different widths, the early direct light has a smaller σ than the late indirect light. Figure 10.11b shows the timing resolution, *i.e.*, the values of both σs, as a function of the pulse height of the signal. For signals of the order of one photoelectron, which are typical for solar neutrino studies, this timing resolution is 2–3 ns. Since Čerenkov light travels at a speed of 22.6 cm ns^{-1} in water, this translates into an uncertainty in the vertex reconstruction for solar neutrinos of less than 1 m.

10.1.4.5 *Monitoring.* The PMT gains and the light attenuation length of the water are continuously monitored in this experiment. Cosmic-ray muons, which enter the detector from the top and exit through the bottom, are used for this purpose. For a given event of this type, the charge Q in each PMT can be calculated. It is a function of the distance l the light has to travel to the PMT, of the orientation of the PMT with respect to this light (since the effective surface area seen by the Čerenkov photons depends on the angle θ between the PMT axis and the direction of the flight path) and of the attenuation length (λ_{att}) of the water:

$$Q = a\frac{f(\theta)}{l} \exp\left(-\frac{l}{\lambda_{\text{att}}}\right) \tag{10.1}$$

where a is a normalization constant. A large sample of cosmic-ray muons gives a large sample of possible distances l_i, angles θ_i and associated signals Q_i. These data can be grouped in bins and plotted as in Figure 10.12, which shows the product $Ql/f(\theta)$ as a function of the flight path l. An exponential fit through these points gives the effective attenuation length of the Čerenkov light (see Section 10.1.3.4). The gain of the PMT can be derived from the intercept of this curve with the vertical axis ($l = 0$).

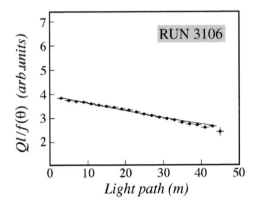

FIG. 10.12. Typical result of the attenuation length/PMT gain measurement on the basis of cosmic-muon signals [Oku 99].

Figures 10.5a,b show the variation in the effective light attenuation length as a function of time, obtained with this method. The PMT gains drift typically less than 5% per year.

10.1.5 *Particle identification and event reconstruction*

One of the most important aspects of the atmospheric neutrino analysis is the distinction between charged-current events induced by muon neutrinos and by electron neutrinos. This distinction is made on the basis of the different characteristics of the Čerenkov rings produced by the charged leptons in both reactions. These differences are illustrated in Figure 10.13, which shows event displays for typical charged-current interactions by atmospheric electron neutrinos (Figure 10.13a) and muon neutrinos (Figure 10.13b).

These event displays show the hit patterns of the PMTs in the Inner Detector and (in the upper left corner) the Outer Detector. For both detectors, the hit patterns are shown separately for the top and bottom planes and for the (unfolded) barrel region in between these planes. Each PMT with a signal above the trigger threshold is indicated with a dot at its proper position in this display; the size of the dot is a measure for the amplitude of the signal in that PMT. The absence of significant activity in the Outer Detector indicates that the light detected in the Inner Detector resulted from neutrino interactions in the latter.

Both event displays clearly exhibit a Čerenkov ring. The Čerenkov ring produced by the muon (which had an energy of ~ 800 MeV in this event) is characterized by its sharp edges; the ring produced by the electron ($E \sim 1$ GeV) is considerably more fuzzy. This difference is a result of the fact that the electron develops a shower. Many of the shower particles, and in particular the Compton electrons produced by bremsstrahlung photons, travel in a different direction than the original incoming electron. Their Čerenkov light thus hits the detector walls in a different area than the light produced by that electron.

SuperK developed an algorithm to quantify the differences between these hit patterns. This algorithm is based on a comparison of the measured hit pattern with those expected (on the basis of Monte Carlo simulations) for typical electrons or muons of

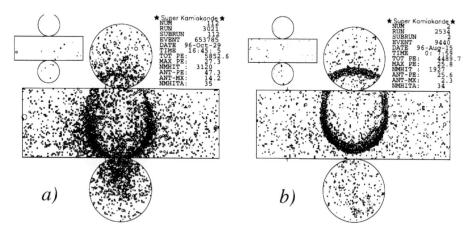

FIG. 10.13. Event displays for charged-current interactions of an atmospheric ν_e (a) and an atmospheric ν_μ (b) in the fiducial volume of SuperK [Oku 99].

the measured energy produced in the reconstructed vertex position and traveling in the measured direction. This leads to probabilities $P_{\text{pattern}}(e)$ and $P_{\text{pattern}}(\mu)$.

At energies below ~ 400 MeV, an additional measured quantity that can be used to discriminate between muons and electrons is the reconstructed opening angle of the Čerenkov cone. At these energies, muons become measurably non-relativistic, which means that their Čerenkov cones have a smaller opening angle (see Figure 10.9a). For electrons, the opening angle remains constant down to much smaller energies. The measured value of the opening angle leads to probabilities $P_{\text{angle}}(e)$ and $P_{\text{angle}}(\mu)$ in the same way as described above.

The total probability P that the event is caused by an electron (muon) is given by the product $P_{\text{pattern}} \times P_{\text{angle}}$. Finally, a particle identification (PID) estimator is defined as follows:

$$\text{PID} = A\left[\sqrt{-\log P(\mu)} - \sqrt{-\log P(e)}\right] \tag{10.2}$$

where A is a constant. If this estimator is positive, then the event is classified as e-like. If the estimator is negative, then the event is classified as μ-like.

The PID distribution for a sample of atmospheric neutrino interactions is shown in Figure 10.14. Electron-like and muon-like events are clearly separated. It is estimated that the misidentification probability for charged-current atmospheric neutrino interactions is of the order of 1%.

For the reconstruction of the events, all available experimental information is used. Apart from the hit patterns in the Inner and Outer Detectors, this also includes the time and the pulse height information.

The simplest situation occurs when a muon is produced that travels downward, parallel to the axis of the tank. In that case, the Čerenkov light projects as a circular ring on the bottom of the tank. The outer radius of this ring (R_{out}) is related to the height (h) at which the muon was produced:

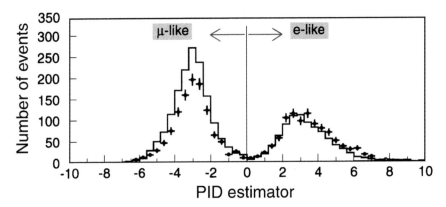

FIG. 10.14. Distribution of the "PID estimator" for sub-GeV atmospheric neutrino interactions in the SuperK fiducial volume. The points represent the experimental data, and the histogram MC-simulated events. Events with a positive (negative) PID value are identified as electron–like (muon-like). See text for details [Oku 99].

$$h = R_{\text{out}} \tan \theta_{\text{C}} \tag{10.3}$$

where θ_{C} is the Čerenkov angle. The inner radius (R_{in}) makes it possible to determine the track length. Ignoring the changes in θ_{C} during the last, non-relativistic phase of the muon's journey, this length (l) can be written as:

$$l = (R_{\text{out}} - R_{\text{in}}) \tan \theta_{\text{C}} \tag{10.4}$$

The applicable value of θ_{C} can be derived from the total signal, *i.e.*, from the total energy of the muon. If this energy is larger than ~ 0.4 GeV, then $\theta_{\text{C}} \simeq \arccos(1/n)$; for lower energies the fact that $\beta < 1$ has to be taken into account (see Figure 10.9a).

If the muon leaves the Inner Detector, then the Čerenkov ring turns into a *disk*. The PMTs located near the center of this disk see considerable light intensity, since it is produced nearby. The signals from the central PMT also arrive earlier than those from the peripheral areas of the disk, since the Čerenkov light travels at a slower speed (c/n) than the particle that created it ($\approx c$). In this case, the escaping muon also produces a signal in the Outer Detector.

Some of these features are observed in the event display shown in Figure 10.15, in particular the disk, the increasing light intensity near the center of this disk and the corresponding signal in the Outer Detector. The muon in this event did not travel parallel to the detector axis, but escaped at an angle through the barrel part of the detector. This causes the shape of the ring to deviate from a simple circle. These deviations can be quite complicated.

A relatively simple complication occurs when a muon travels horizontally, through the center of the tank. In that case, the projection of the Čerenkov cone on the barrel wall of the Inner Detector becomes an ellipse. The light emitted in the horizontal plane arrives earlier at the PMTs than the light from the other parts of the cone, which has to travel a longer distance to reach a PMT.

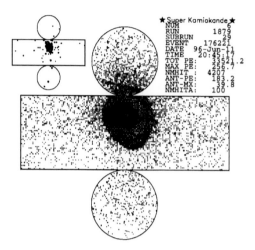

FIG. 10.15. Event display for an interaction induced by a ν_μ in the Inner Detector. The muon produced in this interaction escaped and left a signal in the Outer Detector (upper left corner) [Oku 99].

When the muons travel at other angles, the symmetry in the hit pattern (*i.e.*, the shape of the ring) is usually lost. However, details of the shape of the ring, of the intensity distribution over the PMTs and of the time dispersion between different parts of the hit pattern are all powerful tools for making a reasonably accurate educated guess about the vertex position.

The final determination of the vertex position and the direction of the charged lepton is made with a likelihood method similar to the one used for the particle identification: the experimental information is compared with that from Monte Carlo events with vertices and directions in the region of interest. This procedure also makes it possible to estimate the uncertainty in the vertex position and the direction of the lepton. For the atmospheric neutrino studies, these uncertainties were typically found to be ~ 30 cm for the vertex position and $\sim 2°$ for the charged-lepton direction.

10.1.6 *Selected results*

In Section 11.2 we elaborate on the results from the atmospheric neutrino studies and their implications for the important issues of neutrino oscillations and neutrino mass. In this section, some results from the solar neutrino analysis are described.

Figure 10.16 shows an event display for a 20 MeV ν_e interaction in the fiducial volume. This is a relatively convincing case, also because of the Čerenkov ring that has been drawn to guide the eye. Most ν_e candidate events are much less clear, especially when the energy gets lower. In the displayed event, 94 PMTs produced a signal above the trigger threshold. Events such as this one make it clear why at least 29 PMT signals above threshold are required to consider an event at all as a candidate for a solar neutrino interaction.

The techniques discussed in the previous subsections for event reconstruction and

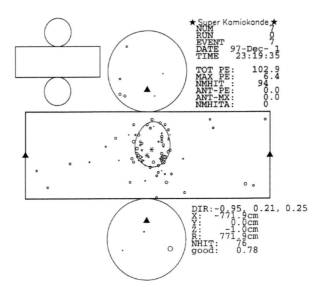

FIG. 10.16. Event display for a 20 MeV ν_e interaction in the fiducial volume of SuperK
[Kos 98].

energy measurement were also applied in this analysis. However, because of the much
lower energy, the experimental uncertainties in the position of the interaction vertex,
the direction of the electron and its energy were considerably worse than for the atmospheric neutrinos.

The detector performance at these low energies was determined in great detail with
experimental data from the LINAC and the Ni-Cf and ^{16}N sources (see Section 10.1.4.2).
The results are shown in Figure 10.17, where the performance data obtained before the
accident (the dotted curves) are compared with the ones as of 2011, obtained after the
detector was refurbished (the solid lines). The differences between these two sets of
results are mainly due to improvements in the vertex reconstruction [Abe 11].

The uncertainty in the reconstructed vertex position is shown in Figure 10.17a. The
position resolution decreases from about 1 m at the lowest energy (5 MeV) to ~ 50 cm
at 16 MeV. The latter value is actually only a factor of two worse than the resolutions
typically achieved in the atmospheric neutrino studies, at energies that are two orders of
magnitude larger.

The angular resolution of the electron measurements is shown in Figure 10.17b.
Here, an increase in energy clearly pays off in much better performance. Whereas the
direction of the electrons and muons produced in atmospheric neutrino interactions can
be typically determined to within a few degrees, the direction of the solar neutrinos
varies from ~ 20° at 16 MeV to almost double that value near the low end of the energy
range.

The energy resolution in the solar-ν energy range is limited by photoelectron statistics. The dashed curve in Figure 10.17c represents the contribution of this source for
a light yield of 7 photoelectrons per MeV, which is approximately the experimental

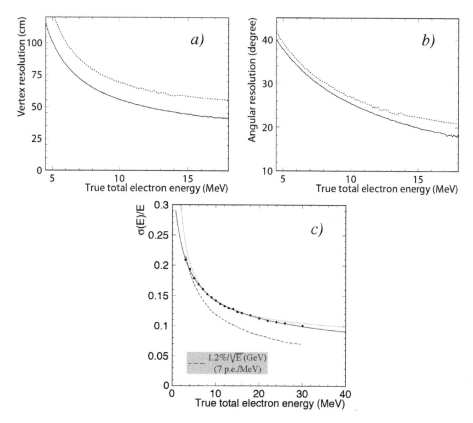

FIG. 10.17. The resolutions in the reconstructed vertex position (*a*), the angle of incidence (*b*)
and the energy (*c*) of the solar neutrinos, as a function of energy. The dotted curves represent
the results obtained before the detector accident in 2001, the solid curves were obtained after the detector was refurbished [Abe 11]. The dashed line shows the energy resolution that
should be expected if only stochastic fluctuations in the number of photoelectrons played a
role (for a light yield of 7 p.e./MeV.

value. SuperKamiokande describes the experimental data points with the following expression:

$$\frac{\sigma(E)}{E} = \frac{0.376}{\sqrt{E}} + 0.0349 - \frac{0.123}{E}, \quad \text{with } E \text{ in units of MeV}$$

which is represented by the solid curve in Figure 10.17c [Abe 11].

At higher energies, in the range from 0.1 to 10 GeV, the energy resolution is well
described by the following parameterization [Oku 99]:

$$\frac{\sigma}{E}(e) = \frac{2.6\%}{\sqrt{p\,(\text{GeV}/c)}} + 0.6\%, \qquad \frac{\sigma}{E}(\mu) = \frac{0.7\%}{\sqrt{p\,(\text{GeV}/c)}} + 1.7\%$$

These resolutions are considerably better than those obtained with other Čerenkov calorimeters, and in particular lead-glass detectors. There are two main reasons for that.

1. The density of lead-glass is typically four times larger than that of water. The specific ionization, $\langle dE/dx \rangle$, and thus the total track length, is almost three times as large. Since the emission of Čerenkov light is proportional to the effective track length, the Čerenkov light yield (photons per MeV) is thus three times larger in water than in lead-glass.

2. The collection efficiency in SuperK is larger than in typical lead-glass calorimeters. In total, $\sim 40\%$ of the Inner Detector surface is covered by photocathode material. The light collection efficiency of lead-glass calorimeters depends somewhat on the geometry, but rarely exceeds 20% for em showers (in Section 4.2.3, we estimated a maximum of 25%).

Taking also into account that water is much more transparent than lead-glass to the Čerenkov photons that contribute to the signals, it is thus reasonable to estimate that the number of photoelectrons per MeV in SuperK is some five to ten times larger than in a typical lead-glass calorimeter. The energy resolution is correspondingly better.

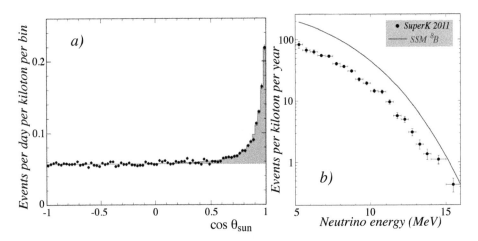

FIG. 10.18. Angular distribution of the electron direction of the $\nu_e + e^- \rightarrow \nu_e + e^-$ candidate events, measured with respect to the direction of the Sun (a), and the measured spectrum of the solar neutrino component of this sample (b). For comparison, the predicted spectrum of the ^{8}B component of the solar neutrinos is shown as well [Abe 11].

Because of these high light yields, the recoil electrons from the $\nu_e + e^- \rightarrow \nu_e + e^-$ reactions induced by solar neutrinos produce significant signals. The limiting factor in the analysis is not so much the size of the signals, but rather the extremely small event rates (15 events per day) and thus the sensitivity to background processes that produce similarly significant signals. Even though its levels have been reduced by some seven orders of magnitude (Section 10.1.3.4), natural radioactivity, and in particular ^{222}Rn, has remained the main source of this background.

The crucial factor for distinguishing this background from the genuine solar neutrino interactions turned out to be SuperK's capability to reconstruct the direction of the electrons that caused the signals. Even though the angular resolution was only modest in the energy range of interest ($20°$–$40°$, see Figure 10.17b), this made an enormous difference, as illustrated by Figure 10.18a.

For each candidate solar-ν event, the angle (ϑ_{sun}) between the reconstructed track and the line pointing from the Sun to the SuperK detector at the time of the event was calculated. Figure 10.18a shows a distribution of candidate events recorded after the detector was refurbished (2006–2008) as a function of this angle. This distribution is mainly flat in $\cos \vartheta_{sun}$, as expected for background events, for which the position of the Sun is irrelevant. However, the distribution also shows a very clear excess of events near $\cos \vartheta_{sun} = 1$, i.e., events in which the electron seemed to come from the direction of the Sun. Out of a total of some 30,000 candidate events, recorded over a period of 548 days, 8,148 could be attributed to the Sun on this basis, with an uncertainty of 2.7%, dominated by systematic effects.

Compared with detailed model predictions of the solar neutrino flux, the measured rate is a factor of two to three too small (Figure 10.18b). This experimental result, earlier observed in radiochemical measurements and known as the "solar neutrino puzzle," was historically the first indication of the occurrence of solar neutrinos. Figure 10.18b shows the calculated spectrum of neutrinos produced in β decay of ^{8}B, the dominant process through which the Sun produces neutrinos in this energy range [Bah 04]. The figure shows that the discrepancy tends to become a bit smaller near the high-energy endpoint. This is believed to be evidence for the contribution of an even more rare process to the solar neutrino production [Bah 98]:

$$^3\text{He} + p \rightarrow \ ^4\text{He} + e^+ + \nu_e$$

commonly referred to as the *hep* process. The expected *hep* neutrino flux is three orders of magnitude smaller than the ^{8}B solar neutrino flux. However, since the end point of the *hep* neutrino spectrum is about 18.8 MeV, compared to about 16 MeV for the ^{8}B neutrinos, the high-energy end of the measured SuperK spectrum should be relatively enriched with *hep* neutrinos.

During the initial period of solar neutrino measurements (1,496 days), before the 2001 accident, in total 22,404 solar neutrinos were detected, with an uncertainty of 3.6%. This high-statistics run made it even possible to observe a small, but significant, seasonal variation in the solar neutrino flux, caused by the elliptic nature of the Earth's orbit around the Sun (Figure 10.19). The average neutrino flux measured by SuperK is reported as $2.35 \pm 0.02(stat) \pm 0.08(syst)$ cm^{-2} s^{-1} for the pre-2001 data [Hos 06] and as $2.40 \pm 0.04(stat) \pm 0.05(syst)$ cm^{-2} s^{-1} for the post-2006 data [Abe 11].

The data shown in this subsection illustrate the enormous progress that has been made since the days that Ray Davis (Nobel laureate 2002) was counting ^{37}Ar atoms in his cleaning liquid, and how far we have come in understanding the details of the mechanism through which our star generates the energy that makes life on our planet possible.

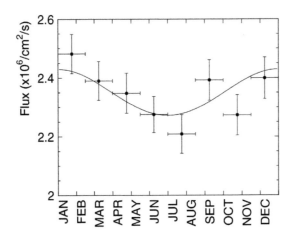

FIG. 10.19. Seasonal variation in the measured solar neutrino flux. The solid line is the prediction based on the eccentricity of the Earth's orbit [Hos 06].

10.2 Natural Water Based Telescopes

10.2.1 *Introduction*

In the next sections we look into efforts to use our natural environment as a calorimeter. The driving force behind all these efforts is the opportunity to create a very large instrument in this way. Typical detector volumes are measured in units of km^3, *i.e.*, three orders of magnitude larger than HyperKamiokande, which will have by far the largest instrumented volume of any man-made calorimeter. Such large volumes are needed to achieve the scientific goals of the experiments, which focus on the study of very rare natural phenomena. Examples of such phenomena include the absorption of extremely-high-energy protons, αs or heavier atomic nuclei of extraterrestrial origin in the Earth's atmosphere and interactions of extra-galactic neutrinos in the Earth itself.

With few exceptions, all natural calorimeters are based on light as the source of experimental information. Usually, the Čerenkov mechanism is the source of this light, for example in detectors where sea water or Arctic ice serve as the absorber medium. Čerenkov light is usually also an important source of experimental information in detectors using the Earth's atmosphere as a calorimeter. In some experiments of the latter type, scintillation light is used as well.

We start the discussion about natural calorimeters with instruments based on the same principles as the one from the previous section: water Čerenkov counters.

The idea to use the water from a sea, an ocean or a lake as a natural calorimeter has many attractive aspects. Superluminal charged particles produce Čerenkov light, and water is very transparent to a large fraction of this light. Water is relatively dense, its radiation length and nuclear interaction length are 0.36 m and 0.85 m, respectively. Therefore, even the highest-energy showers are absorbed in a volume of manageable size (< 100 m^3), with dimensions that are much smaller than the effective light attenuation length.

When water at great depths is used as a detector, it is automatically covered by an effective shield against background sources of light, such as the Sun and cosmic muons. Minimum ionizing particles lose ~ 0.2 GeV m^{-1} in water. Because of additional losses due to radiative processes, a detector operating at a depth of 3,000 m is in practice shielded against all cosmic muons produced in the atmosphere with energies up to ~ 6 TeV. This is important if the detector is intended as a *neutrino telescope*, which is the primary goal of all projects in which a sizeable volume of natural water at great depth is instrumented.

The term "neutrino telescope" implies that detectors of this type are not only designed to measure the *energy* of the neutrinos interacting within their boundaries, but also their *trajectory*, and thus (hopefully) their point of origin in the Universe. In practice, the neutrino's properties are measured through the particles it creates in interactions with matter, and in particular through the charged lepton created in charged-current processes. Fortunately, as a result of Lorentz boosting, the angle between the directions of the incoming neutrino and this lepton decreases rapidly with increasing energy. This is illustrated in Figure 10.20a, which shows the average value of this angle as a function of the neutrino energy. For energies in excess of 1 TeV, the direction of the charged lepton (a μ in this case) is typically to within 1° the same as that of the incoming neutrino that created it.

In practice, neutrino telescopes focus primarily on detecting *muon neutrinos*, for the following reasons:

- High-energy muons travel a long distance in a straight line. It is therefore easier to measure their direction accurately than the direction of the em showers created by high-energy electrons.
- Since high-energy muons may travel several kilometers, the detector would also be sensitive to neutrino interactions occurring at a large distance from the instrumented volume, provided that the neutrino traveled in the right direction. As a result, the effective detector volume available for studying interactions by high-energy ν_μs is considerably larger than for other neutrino species.

Figure 10.20b shows the range–energy relationship for muons in water. The energy loss can be parameterized as:

$$\left\langle \frac{dE}{dx} \right\rangle = \alpha(E) + \beta(E) \cdot E \tag{10.5}$$

where the first term describes the energy loss by ionization and the second one the energy loss by radiation (bremsstrahlung). In water, the values of the parameters α and β are 0.2 GeV m^{-1} and $3.4 \cdot 10^{-4}$ m^{-1}, respectively. The critical energy, *i.e.*, the energy at which the average losses by ionization and radiation are equal, thus amounts to α/β, which is approximately 600 GeV. In rock, this critical energy is somewhat smaller, due to the larger (average) Z value: ~ 500 GeV. For energies up to several hundred GeV, the muon range is approximately proportional to the energy. However, at higher energies, bremsstrahlung takes over as the dominant energy loss mechanism and the range gradually starts to increase logarithmically with the muon energy.

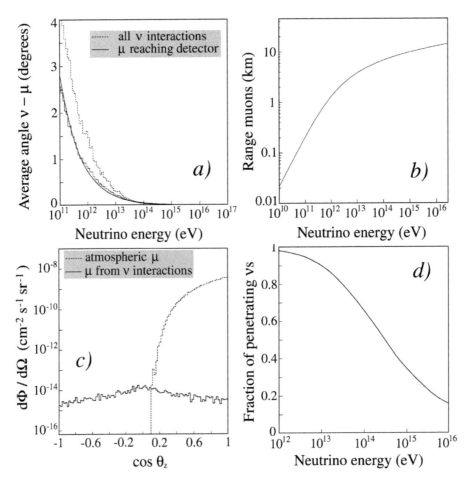

FIG. 10.20. The average angle between the ν_μ and the muon it creates in a charged-current interaction, as a function of the neutrino energy (a). The relationship between the energy and the range of muons in water (b). Distribution of muons originating in the Earth's atmosphere and muons produced by ν_μ interactions in the sea as a function of the zenith angle, measured at a depth of 2,300 m (c). The fraction of neutrinos traversing the entire Earth without interacting, as a function of energy (d).

This feature has important consequences for the background levels caused by muons produced (from π and K decay) in the Earth's atmosphere, in deep-sea experiments. The effect of adding an extra kilometer of water on top of the detector is determined by the logarithmic part of the curve in Figure 10.20b, not by the linear part. For example, 1 km of water stops all muons with energies up to 0.8 TeV, but if the water thickness is increased from 4 km to 5 km, the stopping power increases from ~ 15 TeV to ~ 30 TeV. Similar advantages apply to experiments carried out in deep mines, in tunnels traversing high mountains, or in any other setup where the detector is shielded by a large amount of material against "cosmic rays."

As a result, the muons measured at great depth in the oceans appear to come predominantly from the direction of the zenith, more so than when measured close to the water surface. This is illustrated in Figure 10.20c, which shows the distribution of muons measured at a depth of 2,300 m, as a function of the zenith angle. Muons that originated in the atmosphere were ten times less likely to reach that depth if they entered the water surface at an angle of 60°, compared with normal incidence. And if they entered the water surface at an angle larger than 80°, then they did not reach the mentioned depth at all.

On the other hand, muons generated in the water itself, by charged-current ν_μ interactions, are more or less isotropically distributed at this depth, or at any other depth for that matter. Even though these muons are less abundant than the atmospheric ones, by six orders of magnitude at this depth, an angular cut is an extremely powerful way to select a clean sample of them. This leads to an important conclusion: *upward traveling muons are almost exclusively produced by neutrino interactions nearby.* For this reason, deep-underwater neutrino telescopes typically have a large fraction of their PMTs oriented *downward*, facing the ocean floor in search of muons traveling upward. Muons observed in this way must have been produced by neutrinos that have traversed the entire Earth and interacted less than $\sim$ 10 km away from the instrumented sea volume.

Although the cross sections for neutrino interactions are notoriously small, they are proportional to the neutrino energy. Therefore, at the very high energies neutrino telescopes are aiming for, the probability that the neutrinos are absorbed on their way through the Earth is no longer negligible. This is illustrated in Figure 10.20d. For example, at 10 PeV, this probability has already reached 85%. For energies higher than that, the best possibilities for the neutrino telescopes are located *sideways*, *i.e.*, near $\cos\theta_z = 0$ (see Figure 10.20c). Depending on details of the location, the neutrinos travel in that case only of the order of 100–200 km through the Earth before they reach the detector, *i.e.*, only $\sim$ 1–2% of the distance they would have to travel through the Earth in the case of $\cos\theta_z = -1$.

10.2.2 *Performance characteristics*

To get an idea of the performance characteristics of deep-underwater neutrino telescopes, it is useful to make a comparison with SuperKamiokande. This detector is based on the same principles and has been operating for a number of years.

10.2.2.1 *Light yield.* It is generally believed that an instrument intended for high-energy neutrino astrophysics studies should have an instrumented volume of $\sim$ 1 km^3 to be able to fulfill its envisaged tasks. That is 20,000 times the instrumented volume of SuperK and 50,000 times the fiducial volume of that detector. The total photocathode surface of the PMTs recording the light in SuperK's Inner Detector is $\sim$ 2,300 m^2, which represents $\sim$ 40% of the wall surface. As a result of this almost hermetic coverage, the recorded light yield is very high: $\sim$ 7 p.e. MeV^{-1}.

If the same total photocathode area was used to instrument the boundary surface of a cube with a volume of 1 km^3, then it would cover only 0.038% of that surface, three orders of magnitude less than in SuperK. In that case, the light yield could thus not be expected to be higher than 7 p.e. GeV^{-1}. In fact, it would be even *lower* than that, since the effects of light attenuation have to be taken into account as well. Because of the absence of sophisticated water purification systems, the light attenuation length in natural calorimeters is considerably shorter than in SuperK. Measurements *in situ* have yielded values that are typically a factor of two shorter than in SuperK, at comparable wavelengths. For example, the ANTARES group has reported an attenuation length varying between 46 and 60 m for blue light (473 nm) and between 22 and 26 m for UV light (375 nm), measured at a depth of 2,300 m in the Mediterranean Sea, off the coast near Toulon, France [Agu 05]. The NESTOR group has measured a light attenuation length of 55±10 m (at 460 nm) at a depth of 3,500 m, off the coast near Pylos (Greece) [Ana 94]. And NEMO reports to have measured 67 m (at 440 nm) at a depth of 3,000 m in the Capo Passero basin, southeast of Catania (Sicily) [Ric 07].

With light attenuation at that level, a 1 km^3 detector instrumented as described above, would effectively only "see" light from the peripheral $\sim$ 20% of the detector volume. In order to make it also sensitive to events taking place in the interior regions of the cube, PMTs would have to be deployed at various layers inside the cube, spaced by a distance of the order of the effective light attenuation length. If the 2,300 m^2 of photocathode surface considered here were redistributed to achieve this, the light yield would thus decrease to a level of $\sim$ 1 p.e. GeV^{-1}.

Light yields at this level make it of course completely impossible to use such neutrino telescopes for studying solar neutrinos, Supernova explosions and other phenomena that require sensitivity at the MeV level. Also studies of atmospheric neutrinos will have to be limited to the high-energy tails of the spectra, if at all possible.

10.2.2.2 *Noise.* One important difference with SuperK concerns the noise. In Section 10.1.3 we saw that SuperK made great efforts to reduce the noise caused by radioactive impurities (mainly radon gas) dissolved in the water. As a result, signals as low as $\sim$ 5 MeV ($\sim$ 35 p.e.) became significant enough to be recorded.

Sea water contains enormous quantities of salt, mainly chlorides of sodium and potassium. Potassium has a long-lived radioactive isotope, ^{40}K, which is one of the most important sources of natural radioactivity. This nuclide emits γs of 1,460 keV, which in water are mainly absorbed through Compton scattering. The Compton electrons are typically sufficiently relativistic to emit Čerenkov light. The radioactive salt in the sea water thus manifests itself by producing a continuous stream of signals in the PMTs of the neutrino telescope. At the ANTARES test site, the rate of these signals in a given PMT was measured to vary between 17 and 47 kHz. This rate is represented by the baseline level in Figure 10.21a.

This figure shows the count rate in an individual PMT as a function of time, during a 10-min period [Amr 00]. Apart from the continuous level caused by the radioactive salt, this plot also shows numerous spikes. During some of those spikes, which typically lasted several seconds, the integrated light intensity was several orders of magnitude larger than that of the light produced by ^{40}K decay. These spikes were caused by living organisms, mainly light-emitting bacteria. Their activity varied considerably with time. It showed both seasonal effects, as well as a dependence on the velocity of underwater

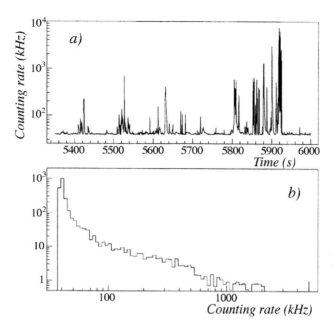

FIG. 10.21. Short-term environmental effects on the performance of the PMTs in the ANTARES underwater neutrino telescope. Shown are the count rate as a function of time in one of the PMTs (a) and the distribution of the counting rates for this particular 10-min time stream (b) [Amr 00]. See text for details.

currents. Figure 10.21b shows the distribution of the counting rates for this particular time interval. The sharp cutoff at 35 kHz corresponds to the background light continuum from ^{40}K decay.

Figure 10.22 shows another environmental effect on the performance of the PMTs in an underwater neutrino telescope. As time goes by, opaque sediments accumulate on the surface of the PMTs, thus reducing the light sensitivity. In three months, the glass cover of upward-looking PMTs installed at the ANTARES test site became $\sim 20\%$ less transparent to light coming from above. As might be expected, these sediments did not affect the surface of the PMTs uniformly. For example, PMTs looking sideways recorded only a reduction of $\sim 3\%$ in the light intensity [Amr 03].

Despite the much higher noise levels, the impact of this noise on the physics capabilities of a neutrino telescope is not as severe as might be feared on the basis of the SuperK experiences with radon (Section 10.1.3). This is because of the completely different energy domains explored by these two instruments. In SuperK, solar neutrino detection is a very important experimental goal. Individual nuclear decay processes, such as the decay of one ^{222}Rn nucleus, can produce a signal that mimics a solar neutrino interaction.

In the case of a neutrino telescope, the experimental signals one is looking for can only be mimicked by a very complicated (and thus unlikely) conspiracy of noise events.

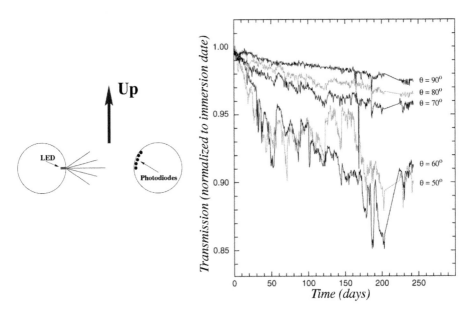

FIG. 10.22. Long-term environmental effects on the performance of the PMTs in the ANTARES underwater neutrino telescope. Shown is the light transmission as a function of time, for five light sensors (photodiodes) oriented at different zenith angles (θ). The measurements were carried out with an LED light source, at 2,400 m depth [Amr 03]. See text for details.

This conspiracy can even be made more unlikely by proper coincidence requirements. For example, the mentioned ANTARES tests revealed that the count rates from ^{40}K decay were reduced by three orders of magnitude when a coincidence was required between signals from two PMTs installed side by side, to ~ 40 Hz. That is about the level that may be expected for purely random coincidences, which is given by

$$N_{cc} = N_1 N_2 \Delta\tau \qquad (10.6)$$

where N_1 and N_2 are the count rates in the two individual PMTs ($\sim 3 \cdot 10^4$) and $\Delta\tau$ is the width of the coincidence gate, 20 ns in these tests.

When the signals produced in two individual PMTs separated by ~ 100 m are compared, the coincidence requirement has to be relaxed considerably. The time difference between the signals produced by one and the same particle in two such PMTs could vary by more than 1 μs, depending on the particle's trajectory. Therefore, in practice $\Delta\tau$ values of ~ 2 μs are used for track searches in neutrino telescopes. The rate at which *ten given PMTs* would produce simultaneous signals due to ^{40}K decays occurring within their fields of view can then be calculated with Equation 10.6 as

$$N_{cc}(10) = 3 \cdot 10^4 \left[3 \cdot 10^4 \times 2 \cdot 10^{-6}\right]^9 \approx 3 \cdot 10^{-7} \text{ Hz}$$

The number of possible PMT configurations for which such a situation could be interpreted as the sign of a passing particle is very large and therefore the radioactive salt is a potentially serious source of fake triggers. However, by grouping the PMTs in pairs and requiring a 20 ns coincident signal from such pairs as a level-0 trigger condition, the rate at which *five given pairs* produce simultaneous signals (again defined by $\Delta\tau = 2~\mu s$) would drop to

$$N_{cc}(2 \times 5) = 40 \left[40 \times 2 \cdot 10^{-6}\right]^4 \approx 2 \cdot 10^{-15}~\text{Hz}$$

This simple arrangement would thus reduce the fake trigger rates caused by the radioactive salt by many orders of magnitude.

In contrast to the ^{40}K signals, the "biospikes" were almost always observed in coincidence between the two PMTs that formed a pair, which indicated that the creatures emitting this light were located nearby. This conclusion was corroborated by the fact that there was almost no correlation observed between the spikes observed in PMT pairs separated by 40 m. To calculate the rate at which these spikes may cause fake track triggers, one should thus use for N_i (Equation 10.6) the rate at which they occur in individual pairs i. In Figure 10.21a, some 30 spikes occur in 600 s, so that $N_i \sim 0.05$. Assuming that the typical duration of one spike is ~ 2 s, the rate at which *five given pairs produce simultaneous biospikes* can be calculated as

$$N_{cc}(2 \times 5) = 0.05 \left[0.05 \times 2\right]^4 = 5 \cdot 10^{-6}~\text{Hz}$$

The biological noise thus seems to be a much more important source of fake track triggers than the radioactive salt.

The consequences of the environmental effects that underwater Čerenkov telescopes have to deal with are not all negative. For example, the natural radioactivity of the salt dissolved in the sea water provides an opportunity to calibrate the detector. Figure 10.23 shows how ANTARES takes advantage of this opportunity [Age 11]. In the decay of ^{40}K, electrons with a kinetic energy up to 1.3 MeV are produced. These electrons may produce up to 150 Čerenkov photons in water, which has a Čerenkov threshold of 0.26 MeV for electrons. If this decay occurs in the vicinity of an optical module, a coincident signal may be recorded by a pair of PMTs in that module.

Figure 10.23a shows the distribution of the measured time differences between hits in neighboring PMTs of a particular optical module. The peak around 0 ns is mainly due to single ^{40}K decays that produce coincident signals. The fit to the data is the sum of a Gaussian distribution and a flat background. The FWHM of the Gaussian function is about 9 ns. This width is mainly due to the spatial distribution of the ^{40}K decays that contribute to the signals. The background is the result of random coincidences with signals from biological activity. The positions of the peaks of the time distributions for different pairs of PMTs in the same optical module are used to cross-check the time offsets computed with the timing calibration. This is illustrated in Figure 10.23b, which shows a comparison of the time offsets calculated from the optical beacon calibration (with LEDs) and those extracted from the analysis of ^{40}K coincidences. The *rms* of this distribution is about 0.6 ns.

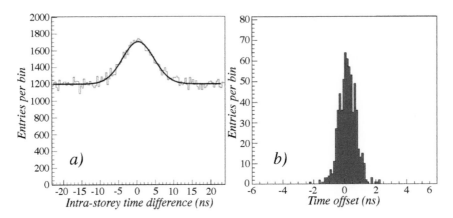

FIG. 10.23. Time difference between a pair of PMTs located in the same optical module of an ANTARES line. The peak is due to single ^{40}K decays (*a*). Differences between the time offsets for all 885 PMTs, determined by the ^{40}K coincidence method (*b*) [Age 11].

Apart from this application for timing purposes, the coincidences induced by ^{40}K decays also provide a powerful tool for monitoring the relative efficiencies of the individual PMTs, with an accuracy of about 5%.

10.2.2.3 *Energy deposit patterns.*   The energy deposit patterns recorded in a deep underwater neutrino telescope are very different from those in SuperK. Even the most minuscule track produces its own characteristic Čerenkov ring in the SuperK detector. This is because every Čerenkov photon typically has a 30% chance of reaching the photocathode of one of the PMTs that records what is happening inside the water tank. However, in a typical 1 km^3 neutrino telescope, a volume the size of the entire SuperK tank would be equipped with, at best, one PMT. Given that the different PMTs are spaced by a distance of the order of the light attenuation length, the observation of a recognizable Čerenkov ring would be an unusual event.

The high-energy neutrino interactions one would like to detect with detectors of this type are characterized by the following energy deposit patterns:

- Charged-current interactions by a ν_e or $\bar{\nu}_e$ produce an energetic electron that develops an em shower with a depth of 30 - 40 X_0, *i.e.*, 10 - 15 m. With PMTs spaced at a multiple of that distance, the signals from such showers would only be recorded in a limited number of PMTs, all located in the same area of the fiducial detector volume. Although most of the light is emitted by shower particles moving in the forward direction (*i.e.*, the direction of the showering electron), there is also an isotropic component. Some of the Čerenkov light may thus even be observed *upstream* of the interaction vertex.

- Similar energy deposit patterns characterize neutral-current events. In these events, the only measurable particles are those produced in the interactions between the Z^0 exchanged in this process and the struck quark. The hadronic showers developing in such events would mainly be visible through their em core, produced by the π^0 shower components.

- Charged-current interactions by a high-energy ν_μ or $\bar{\nu}_\mu$ produce a high-energy muon which, if it carried more than ~ 1 TeV, would not even be fully contained in a 1 km^3 detector. Such muons may produce signals in all the PMTs on a string that spans the entire cross section of the detector. Since the amount of energy emitted in the form of Čerenkov photons is approximately constant per unit track length, the signal pattern in the string of PMTs recording the event is primarily an indication of the trajectory followed by the muon. In general, the following rule holds: the closer the particle passes a certain PMT, the larger the signal it produces in that PMT.

- Charged-current interactions by a ν_τ or $\bar{\nu}_\tau$ produce a τ lepton. This particle has a 17% probability of decaying into a muon, in which case the energy deposit pattern is very similar to that for ν_μ-induced charged-current events. Alternatively, the τ decays either into an electron or hadronically. In those cases, a (em or hadronic) shower develops and the entire energy is deposited in the vicinity of the interaction vertex, as in the case of neutral-current or ν_e-induced charged-current events. Tau neutrinos carrying extremely high energies ($> 10^{18}$ eV) may exhibit a very unique energy deposit pattern, sometimes referred to as the "double-bang topology." At these energies, the τ lepton may travel a substantial distance (~ 100 m) before it decays. If the decay is non-muonic, two energetic showers separated by that distance are expected to be observed in coincidence [Lea 95].

The trajectory of a high-energy muon traversing the neutrino telescope can also be reconstructed with the help of the time information. The arrival time of the light signal in a certain PMT depends on the distance of closest approach. If $d_{\min}$ is the distance of closest approach to a certain PMT, and if the particle passed the PMT at that distance at time $t = 0$, then the Čerenkov light arrives in that PMT after a time Δt, equal to

$$\Delta t = \frac{d_{\min}}{c}\left[\frac{n}{\sin\theta_C} - \cot\theta_C\right] = \frac{d_{\min}}{c}\tan\theta_C \approx \frac{d_{\min}}{c} \tag{10.7}$$

in water. This means that the Čerenkov signal is late by ~ 30 (60) ns if the muon passed the PMT at a distance of 10 (20) m. By comparing the arrival times in all the PMTs contributing to the signal, the muon's trajectory can in principle be reconstructed. This works as follows (see Figure 10.24).

A muon traverses the detector and produces signals in a string of PMTs. If t_0 is some reference time, d_j the distance of closest approach to PMT j and l_j the corresponding track length (starting at t_0), then the arrival time of the Čerenkov light in this PMT is given by

$$t_j - t_0 = \frac{l_j + d_j\tan\theta_C}{\beta c} \tag{10.8}$$

The arrival times t_j for the Čerenkov light in the N PMTs that contribute to the signal in this event constitute the experimental data, β is very close to 1, c and θ_C are constants in this equation. By choosing a trajectory and a point on this trajectory, t_0, l_j

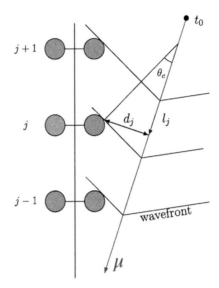

FIG. 10.24. Reconstruction principle of a muon's trajectory in a deep underwater neutrino telescope. The PMTs are indicated by the circles.

and d_j are fixed for all PMTs ($j = 1, 2, ..., N$). The track parameters that best describe the observed t_j values can be found in an iterative procedure.

Not every configuration of hit PMTs leads to a unique solution in this procedure. For example, if all PMTs contributing to the signal are located on a straight line, then rotational symmetry only restricts the solution to the surface of a cone around this line. Also other hit patterns may lead to multiple solutions. It can be demonstrated that signals in at least five PMTs, located on at least three different lines, are needed to produce a unique solution.

The precision with which the direction of the muon (and thus the direction of the ν_μ that produced it) can be determined depends primarily on the position and time resolutions that can be achieved in this telescope. In this type of detector, these resolutions do not depend solely on the electronic precision of the measurements. An angular change of $1°$ for a vertical track traversing a detector over a length of 1 km is achieved when the exit point is shifted by ~ 17 m, keeping the entrance point the same. One would think that a shift of this magnitude should be easy to observe. However, the measurements are always performed *with respect to the position of PMTs*. Underwater currents may change the absolute position of PMTs and the relative position of PMTs with respect to each other. In experiments where the PMTs are connected to cables that are fixed to the ocean floor, the position and orientation of the individual PMTs are usually monitored with a variety of sophisticated instruments (tilt meters, current meters), intended to eliminate any systematic effects due to these environmental factors as much as possible. In Section 10.2.2.5, the importance of this work is illustrated.

10.2.2.4 *Energy measurement in the multi-TeV range.*   The range of a 800 GeV muon
in water is about 1 km. Therefore, charged-current interactions induced by ν_μs with
energies of 1 TeV and above are typically not contained in a 1 km^3 H_2O calorimeter:
the muon produced in such interactions tends to escape from the detector. In those cases
the question arises what energy to assign to this muon and to the (anti-)neutrino that
produced it.

At these extremely high energies, muons are no longer minimum-ionizing particles.
As the energy increases, the energy loss is increasingly dominated by radiative processes. We have already seen that such processes limit the range of high-energy muons
(Figure 10.20b). This phenomenon can also be used to estimate the energy of the muons
traversing the neutrino telescope. There are two possible ways to do this.

The first method is based on measuring the total energy deposited by the muon on
its way through the detector. The combined specific energy loss resulting from ionization and from bremsstrahlung, calculated on the basis of Equation 10.5, is plotted as a
function of the muon energy in Figure 10.25a.

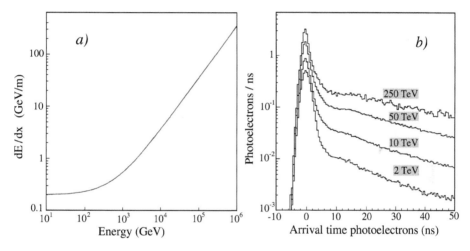

FIG. 10.25. Specific energy loss of muons in sea water, as a function of the muon energy (*a*).
Time structure of the signals generated by muons of different, extremely high energies, in a
PMT of a deep underwater neutrino telescope (*b*) [Hub 99].

This method works particularly well at very high energies, *i.e.*, in the TeV range
and beyond. At these energies, bremsstrahlung completely dominates the energy loss
and thus the PMT signals. The latter are in that case proportional to the muon energy
(Equation 10.5). Over a distance of 1 km, the muon energy decreases significantly in
this energy regime. The associated decrease in the signals observed along the muon's
trajectory may be used to improve the precision of the energy determination.

The second method makes use of a peculiarity of the Čerenkov effect, namely the
feature that the light that constitutes the signal travels at a lower speed than the par-

ticle that produced it. If the muon only lost energy by direct ionization of the water molecules, then a given PMT would only observe light from a very small track segment, defined by the size of the sensitive photocathode surface, the direction of the track and the value of the Čerenkov angle. This is illustrated in Figure 10.26a. Only light from the track segment dl would reach the PMT if direct ionization were the only process that played a role.

However, the muons also lose energy by bremsstrahlung. The bremsstrahlung photons may produce relativistic electrons, for example through Compton scattering. If the Compton electron travels in the proper direction, then Čerenkov photons emitted by this relativistic particle may also reach the photocathode of the PMT. This is also illustrated in Figure 10.26a. In this particular case the Čerenkov light caused by the bremsstrahlung process is emitted about twice as far from the PMT as the light from track element dl. In other cases this difference may be larger or smaller, depending on the production characteristics of the Compton electron. As a matter of fact, the PMT may receive Čerenkov light produced anywhere along the track from this process.

Radiative processes do not only generate Čerenkov light through the simple (Compton scattering) mechanism described above. Sometimes the bremsstrahlung photons carry a lot of energy and develop em showers in the water. Light from these showers can be detected in the PMT if it is emitted in the proper direction. Examples of (simulated) events in which muons with energies of 0.1 TeV and 10 TeV travel through water are shown in Figures 10.26b and 10.26c, respectively [Hub 99].

The track thickness of a muon is an indication of the specific energy loss, with thick segments indicating regions of substantial radiative losses. Only Čerenkov photons reaching the PMT's photocathode surface are shown in these figures. Their number is

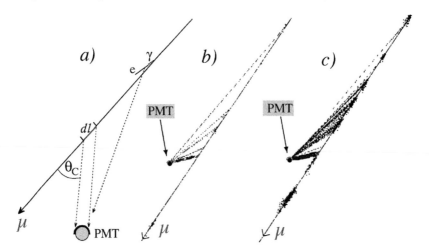

FIG. 10.26. Detection of Čerenkov light produced by direct ionization and by bremsstrahlung in a PMT (*a*). Monte Carlo simulation of Čerenkov photons produced by a 0.1 TeV (*b*) and a 10 TeV (*c*) muon in water. Only photons reaching the sensitive surface of the PMT (*i.e.*, the northern hemisphere) are drawn. From [Hub 99].

considerably larger for the higher-energy muon. Since the Čerenkov light travels at a lower speed than the particles that emit it, the light emitted by bremsstrahlung processes upstream of the track segment dl arrives *later* at the PMT's photocathode surface than the light resulting from direct ionization (emitted from the track segment dl). This time difference may be substantial, given that Čerenkov light takes ~ 11 ns longer than the muon to travel a distance of 10 m in water.

In the events shown in Figures 10.26b and 10.26c, where the photocathode surface of the PMT constitutes the northern hemisphere of the black dot, bremsstrahlung processes *downstream* of dl could not contribute to the signals. The time structure of the events therefore shows a rather sharp peak, representing light emitted from the direct-ionization segment dl, followed by a tail caused by bremsstrahlung processes *upstream* of that segment.

This is illustrated in Figure 10.25b, which shows typical time structures for muons of different energies. The larger the muon energy, the more pronounced the tail becomes. Whereas in the first method discussed above a conclusion about the muon's energy would be based on the *integral* of these pulses, *i.e.*, on the total amount of light collected, in the second method the tail/peak ratio is the determining factor: the larger this ratio, the larger the muon energy.

Figure 10.25b shows that the integral over the peak region alone also depends on the muon energy. This means that the light collected in this peak is not exclusively, or even predominantly, produced in the direct-ionization process of water molecules by the muon itself. As the energy increases, an increasing contribution to the light emitted from the dl track segment comes from electrons produced by Compton scattering or by more complicated em shower development processes initiated by bremsstrahlung in this track segment. Since the bremsstrahlung photons are emitted along the direction of the muon, the relativistic electrons capable of emitting Čerenkov light are also predominantly oriented in the same direction as the muon itself. Therefore, this track segment is also much more likely to produce bremsstrahlung-related Čerenkov photons contributing to the PMT signal than any other track segment.

10.2.2.5 *Other characteristic differences with SuperK.* In spite of the similarities, there are also a number of major differences between the deep underwater neutrino telescopes and SuperK. In the previous subsections we have encountered several of these differences:

- The very different light yield levels, 7 photoelectrons per MeV deposited energy in SuperK, three to four orders of magnitude less in a ν telescope.
- The very different signal patterns. This difference is partially due to the different physics studies, which focus on contained events in SuperK and on non-contained events in the telescope.
- The very different noise levels, caused by the radioactive nuclide ^{40}K that occurs in salt and by the effects of bioluminescence.

Apart from these, there are a number of differences that derive from the environment in which a neutrino telescope has to operate. This environment complicates the experiments in several ways, of which we mention three.

First, the PMTs have to operate at very high *pressures*. Every 10 m of water add 1 atm to the pressure. The PMTs in a ν telescope operating at 3 km depth thus have to withstand a pressure of 300 atm. This requirement affects the design; for example, it limits the possible size of the PMTs.

Second, there are a number of *logistic* problems associated with operating a detector off-shore, at great depth. Power has to be supplied to the detector system and signals have to be extracted and recorded, sometimes under very adverse weather conditions. The nearest shore is typically more than 10 km away from the detector site. Replacing or cleaning PMTs is a major operation that may shut down operations for one year or more.

Finally, the operating conditions in a ν telescope *change* continuously with time, much more so than in SuperK. Underwater currents affect the positions of the PMTs, and bioluminescence levels may change by large factors. This requires extensive monitoring efforts, over and above the monitoring of water quality and PMT performance that is also necessary in SuperK. Let us, as an example, consider the effects of underwater currents.

Even though the strings holding the PMTs are anchored in the bedrock at the bottom of the sea, and held in position with large buoys floating on the water surface, underwater sea currents may cause substantial displacements, as illustrated by Figure 10.27a. Fortunately, these currents affect different nearby strings in similar ways, so that changes in the distances between different optical modules are a second-order effect, much smaller than the absolute changes in position. To monitor these effects, the detec-

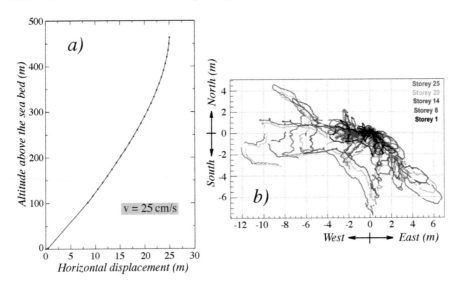

FIG. 10.27. The line shape of a deployed ANTARES string for a sea current velocity of 25 cm s^{-1} (a). The horizontal displacement at the five levels above the sea bed at which the lines were equipped with positioning hydrophones, as determined by the positioning system during a six-month test period [Age 11].

tors are typically equipped with a system of positioning hydrophones. In ANTARES, the signals of a triggered event are accompanied by the (x, y) coordinates of all the PMTs that contributed to the signal. Figure 10.27b shows the (x, y) displacements of hydrophones located at five different altitudes above the seabed during a six-month test period. During that time, displacements of up to 12 m from the nominal position were observed. These tests revealed that the absolute coordinates could be determined at any time with a precision better than 20 cm [Age 11].

10.2.3 The projects

Over the years, a number of different projects for building a neutrino telescope under water have been proposed. Some of these projects were more ambitious than others, ranging from pilot studies to investigate the technological problems to full deployment and exploitation of a multi-string PMT system. Some of these projects are briefly described below.

10.2.3.1 DUMAND.
The first large-scale project of this type was the Deep Underwater Muon And Neutrino Detector that was to be installed off the coast of the main island of the Hawaii archipelago, at a depth of about 4,800 m [Han 98]. The project was started in 1975. Over a period of about 20 years, a variety of technical studies about the deployment of strings of optical modules, and the associated logistic problems of operating a detector of this type in such an environment were carried out. These studies have produced invaluable information for subsequent projects. In 1996, funding for the DUMAND project was terminated.

10.2.3.2 BAIKAL.
The first project in which a three-dimensional detector volume was instrumented was carried out in Lake Baikal, at a depth of $\sim 1,300$ m. Operating since 1993, this project has gradually grown in size and in complexity [Ayt 08]. The original configuration consisted of 36 PMTs. In 1998, the telescope was upgraded and named NT200. An umbrella-like frame carries eight strings, with in total 192 PMTs, arranged in pairs. This configuration surrounds a volume of $\sim 80,000$ m^3, about twice the volume of the SuperK tank. In 2005, three additional strings have been added [Ayt 06]. They are each 200 m long and are placed at 100 m from the center of NT200. Each string contains 12 PMTs, identical to those in NT200 and also grouped in pairs. The new telescope configuration is named NT200+, and the sensitivity for high-energy neutrinos has been increased by a factor of four thanks to this addition.

A schematic overview of the current NT200+ geometry is shown in the right frame of Figure 10.28. For comparison, the earlier version NT200, which now serves as the central part of the telescope, is shown as well in more detail. The 192 PMTs are distributed over eight vertical lines. Seven of these lines enclose a cylindrical heptagon, with sides of 18.6 m, the eighth line forms the axis of this cylinder, which has a total height of ~ 80 m.

The PMTs are grouped in pairs, and the local coincidence requirement between the signals from each pair reduces the noise level (Section 10.2.2.2). Since Lake Baikal is filled with fresh water, ^{40}K is not a problem; the noise is entirely due to bioluminescence. On the other hand, the water quality is not very good, the effective atten-

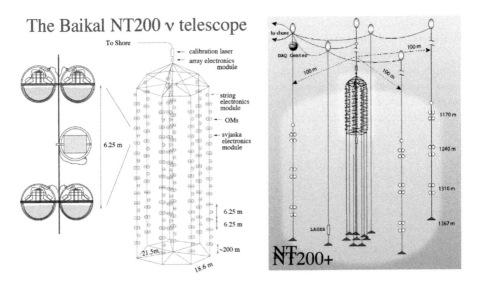

FIG. 10.28. Schematic overview of the NT200 neutrino telescope installed in Lake Baikal. This detector consists of 192 optical modules, distributed over eight strings. The PMTs are grouped in pairs to reduce noise. Details of this arrangement are shown in the blown-up fragment (left). In 2005, three additional strings were added, with 12 PMS each. The new configuration is named NT200+ (right).

uation length is typically of the order of 20 m, and subject to large seasonal variations. Also, sedimentation levels are much higher than in larger bodies of water. In one year, the windows of upward-looking PMTs were found to become 60% less transparent. Mainly for this reason, most of the PMTs in the present setup are oriented downward. One particular advantage of this environment is that the lake is frozen during most of the year. During that period, work on the detector can be performed from a fixed platform installed on the ice overhead.

This detector has taken experimental data in several configurations and results have been presented at various workshops and conferences, and published. For example, upward going muons have been recorded in NT200 at a rate that is in agreement with that expected from atmospheric neutrinos [Ayt 08].

10.2.3.3 *NESTOR.* Not far off the Greek coast, the Mediterranean reaches its deepest point. About 20 km from the Pylos peninsula, an underwater plateau at a depth of $\sim$ 4,000 m is at first sight a very attractive site for an experiment of this type. Some 20 km farther out, the Mediterranean reaches its deepest point, at 5,200 m below sea level (Europe's Abyss).

The NESTOR Collaboration has been investigating the merits of this site since 1991. *In situ* measurements revealed a light attenuation length of 55 ± 10 m at a wavelength of 460 nm. The 15-inch PMTs intended for use in this experiment recorded optical noise at a level of about 75 kHz in these tests [Ana 94].

The telescope envisaged for tests at this site consists of a semi-rigid hexagonal titanium structure with a height of 300 m and a diameter of 32 m. In total, 168 PMTs are distributed over 12 floors (14 PMTs per floor, floors spaced by ~ 25 m). In March 2003, the NESTOR Collaboration successfully deployed one test floor of this detector tower, fully equipped with final electronics and associated environmental sensors to a depth of 3,800 m, situated 80 m above the sea bottom. The 15-inch PMTs were enclosed in a spherical glass housing, designed to be capable of withstanding hydrostatic pressure up to 630 atm. Power and data were transferred through a 30 km long electro-optical cable to the shore laboratory. The deployed detector was continuously operated for more than one month. The monitored experimental parameters, operational and environmental, remained stable within acceptable tolerances. The readout and DAQ chain performed well and with practically zero dead-time. In total, 1.1% of the total observation time was lost due to bioluminescent activity around the detector [Agg 05a]. Some scientific results from this work have been published [Agg 05b].

As far as I know, there have been no further deployments at this site. The Nestor Collaboration participates in KM3NeT, which has made a joint proposal for a km³ water Čerenkov telescope in the Mediterranean (Section 10.2.3.6).

10.2.3.4   *NEMO.*   The Mediterranean Sea is a popular location for neutrino telescopes. Apart from the NESTOR site, at least two other sites are being considered (Figure 10.29). One of these sites is located 60-100 km southeast of Capo Passero (Sicily), a submarine plateau with an average depth of $\sim 3,500$ m. The NEMO Collaboration has been carrying out an evaluation program at this site since 1998, to study its suitability for the construction of a future km³ Čerenkov neutrino telescope. Measurements of the transparency of seawater at this site yielded somewhat better results than at the other mentioned sites, and did not show seasonal variations [Ric 07]. Other important environmental conditions, such as the bioluminescence and underwater currents compared favorably as well. However, of all the Mediterranean sites, this one is located farthest from shore. Yet, in 2008, a 100 km long electro-optical cable was deployed from the

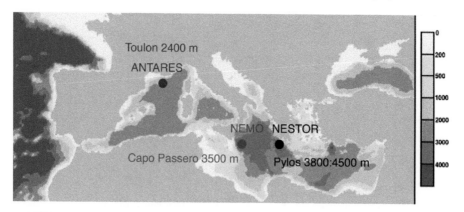

FIG. 10.29.   The three sites in the Mediterranean Sea where large neutrino telescopes are being tested.

envisaged shore station in Portopalo to the deep sea area. It has subsequently been used for tests of prototype KM3NeT modules (Section 10.2.3.6) [Cap 09].

Because of the logistic situation, the NEMO Collaboration initially carried out a research program involving actual detector elements at a location closer to shore, at a depth of about 2,000 m off the Sicilian coast near Catania. A four-floor tower was deployed on the seabed and tested during a period of about five months. These tests involved the power distribution, the data transmission, the timing calibration and the acoustic positioning system, as well as the procedures to use the collected data to reconstruct muon tracks [Aie 10].

10.2.3.5 *ANTARES.*   The third Mediterranean test site is located about 40 km off the French coast, near Toulon, where the sea floor reaches a plateau at a depth of 2,475 m. The ANTARES Collaboration is operating a 885-PMT neutrino telescope at this site since 2008 [Age 11]. This makes it the first ever water Čerenkov telescope installed in the sea. The detector array, which encloses an effective detection volume of ~ 0.1 km^3, consists of vertical strings of PMTs with a length of 500 m. The PMTs are positioned 60–100 m apart. That corresponds to 1.5–2.5 light attenuation lengths.

Before the telescope was installed, the ANTARES Collaboration carried out an extensive R&D program on the feasibility of this project [Age 11]. Several examples of the issues that were studied in this context are addressed in Section 10.2.2 (light yield, the effects of salt, bioluminescence, underwater currents). The expected performance, partially based on the collected data and partially based on simulations, was also the topic of many studies. For example, such studies led to the conclusion that it should be

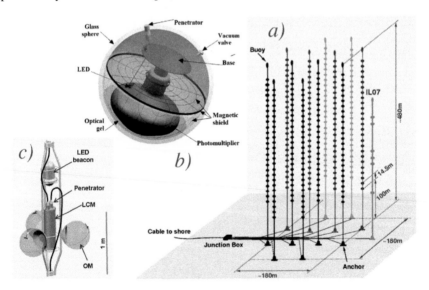

FIG. 10.30.   Schematic overview of ANTARES neutrino telescope (a), the structure of one of the 885 PMTs employed in this instrument (b), and of one of the 295 Optical Module Frames, which holds three PMTs that look downward at 45° (c).

possible to measure the direction of high-energy muons with a precision of ∼ 0.2° with this instrument [Hub 99].

The ANTARES Collaboration has published already a substantial number of scientific papers based on results of the analyses of the collected data, for example on searches for evidence of dark matter annihilation in the Sun [Adr 13] or in the galactic center [Adr 15], a search for magnetic monopoles [Adr 12a] and a study of atmospheric neutrino oscillations [Adr 12b].

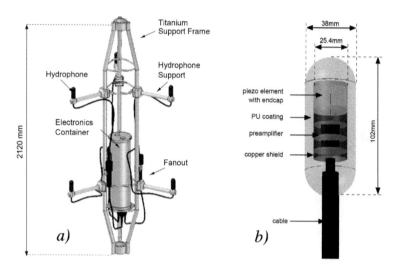

FIG. 10.31. Drawing of a standard acoustic cluster with hydrophones (*a*), and a schematic drawing of one of the hydrophones (*b*) tested in the ANTARES setup [Agu 11]

The telescope infrastructure has also been used to study acoustic particle detection [Agu 11]. The so-called *AMADEUS* system consists of six "acoustic clusters," each comprising six acoustic sensors (hydrophones) that are arranged at distances of ≈ 1 m from each other. Figure 10.31a shows one such cluster. The hydrophones use piezoelectric elements for the broad-band recording of signals with frequencies up to 125 kHz (Figure 10.31b). Measuring acoustic pressure pulses in huge underwater arrays is considered a promising approach for the detection of cosmic neutrinos with ultra-high energies ($> 10^{18}$ eV). The pressure signals are produced by the particle showers that evolve when neutrinos interact with nuclei in water. The resulting energy deposition in a cylindrical volume of a few centimeters in radius and several meters in length leads to a local heating of the medium which is instantaneous on the hydrodynamic time scale. This temperature change induces an expansion or contraction of the medium depending on its volume expansion coefficient. According to the thermo-acoustic model [Ask 79, Lea 79], the accelerated motion of the heated volume forms a pressure pulse of bipolar shape which propagates in the surrounding medium. After propagating several hundreds of meters in sea water, the pulse has a characteristic frequency spectrum that is expected to peak around 10 kHz [Nie 06, Bev 07]. Given the strongly anisotropic propagation pattern of the sound waves, details of the pressure pulse (amplitude, asymmetry, frequency spectrum) depend on the distance and angular position of the observer with respect to the particle shower induced by the neutrino interaction [Bev 07]. The study of acoustic particle detection is motivated by two potentially major advantages over an optical neutrino telescope. First, the attenuation length in sea water is about 5 km (1 km) for 10 kHz (20 kHz) signals. This is one to two orders of magnitude larger than for Čerenkov light detected by the telescope. The second advantage is the more compact sensor design and simpler readout electronics for acoustic measurements. On the other hand, the speed of sound is small compared to the speed of light. Therefore, coincidence windows between two spatially separated sensors have to be correspondingly large. Furthermore, the signal amplitude is relatively small compared to the acoustic background in the sea (whales!). This results in a high trigger rate at the level of individual sensors and makes the implementation of efficient online data reduction techniques essential. Also, it limits the applicability of this technique to the very high end of the investigated energy spectrum of the neutrinos. To particle physicists, $> 10^{18}$ eV is an extremely high energy. However, we are dealing here with a macroscopic phenomenon. AMADEUS is looking for "mini-explosions" in which a total energy of ~ 1 Joule is released (1 J = $6 \cdot 10^{18}$ eV = 0.24 cal).

The AMADEUS project was primarily conceived to perform a feasibility study for a potential future large-scale acoustic neutrino detector. In particular, the knowledge of the rate and correlation length of neutrino-like acoustic background events is a prerequisite for estimating the sensitivity of such a detector.

10.2.3.6 *KM3NeT.* The ultimate goal of each of the three Collaborations discussed in the previous subsections is to build and exploit a water Čerenkov calorimeter with a fiducial volume of the order of 1 km^3. Why? Apart from the fact that it is a nice round number[40], a detector of this size would be the ideal complement to IceCube (Section 10.3), a detector of that size located at the South Pole. Together, these two detectors would cover the entire sky and, importantly, the Northern Hemisphere detector would have a direct view of the center of our Milky Way galaxy, which is suspected to be the origin of several interesting phenomena.

Having realized that a project of this scope is beyond the means of any of the individual groups, NESTOR, NEMO and ANTARES have joined forces and submitted a proposal to their funding agencies for a project they call KM3NeT [KM3 16]. This proposal consists of two parts, called ARCA (Astroparticle Research with Cosmics in the Abyss) and ORCA (Oscillation Research with Cosmics in the Abyss). The neutrino oscillation component of the program needs to be sensitive to neutrinos at the lower end of the spectrum accessible to detectors of this type, and would therefore be better studied with an instrument that has a higher light yield (*i.e.*, a larger photocathode surface per unit detector volume). In the proposal, the ANTARES site is suggested for carrying out ORCA, and the Capo Passero site for ARCA. The equipment used at both sides will be identical, but the optical modules will be much closer spaced in ORCA (9 m *vs.* 36 m). In the context of KM3NeT, a new optical module has been developed. This Digital

---

[40]Ted Turner, commenting on his decision to make a gift of one billion dollars to the United Nations (1997).

FIG. 10.32. Picture of the Digital Optical Module developed for KM3NeT, and its fixation on the detection string [KM3 16]

Optical Module (DOM, Figure 10.32) consists of a transparent 17 inch diameter glass sphere comprising two separate hemispheres, and houses 31 PMTs and their associated readout electronics. This design has several advantages over traditional optical modules that use single large PMTs, as it houses three to four times the photocathode area in a single sphere and has an almost uniform angular coverage. The segmented structure of the photocathode also provides directional information, as well as improved rejection of optical background (Section 10.2.2.2).

In April 2013, the first DOM was deployed at the ANTARES (ORCA) site, and has been operating there for over one year [Adr 14]. From these tests, it was concluded that

- By requesting an eightfold coincidence between signals from the various PMTs in the DOM, an essentially background free sample of muons could be obtained.
- Using this pure muon sample, it was possible to demonstrate the sensitivity of the DOM to the arrival direction of the muons.
- The directionality information of the signals, available because the 31 PMTs all have different orientations in zenith and azimuth, showed that most of the bioluminescent activity emanated from the direction of the support structure and the electronics container. For this reason, in the final KM3NeT string, the DOMs are supported by two thin, 4 mm diameter ropes (Figure 10.32).

In December 2015, a string of 18 DOMs was successfully deployed at the Capo Passero (ARCA) site (Figure 10.33). The unfurling procedure, the deployment on the sea bed, including the submarine connections to the cables already installed there, all proceeded as planned. The tests carried out so far confirm the findings at the ORCA site, and demonstrate the validity of the DOM design for operations at a depth of 3,500 m. The

FIG. 10.33. Photo of the deployment of a launch vehicle for the KM3NeT detector (left). Principle of the unfurling of the launch vehicle (right).

installed equipment also provides a test bench for the software architecture and the data handling for the full-scale KM3NeT detector [Adr 16].

A project of this type is not only valuable for particle physics and astrophysics studies, but also provides great synergetic opportunities for the Earth and Sea Science communities. I mention one example of a spin-off effect of the R&D carried out by the NEMO group on the possibilities of acoustic detection of very-high-energy neutrinos. It turned out that an important contribution to the noise in their hydrophone system was made by sperm whales who passed the area in the Ionian Sea where these tests were conducted. Figure 10.34a shows the pulses in the acoustic signals registered during a period of 5 minutes. It turned out that the details of the waveform of these signals (Figure 10.34b) made it possible to obtain a lot of information about the whales that emitted these signals, such as their size, age, and the structure of the sound producing part of their head [Car 15].

10.3 Arctic Ice

10.3.1 *Ice vs. Water*

Water does not necessarily have to be liquid to make a good Čerenkov calorimeter. Ice is in many ways as good and in some respects better. Many of the factors that make deep seas attractive as large calorimeters apply *a fortiori* to the ice deep underneath the surface of the Antarctic continent. In fact, this environment offers some distinct advantages compared with the "liquid" detectors discussed in the previous section:

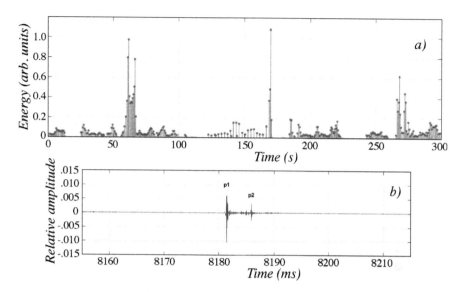

FIG. 10.34. The detection of acoustic clicks emitted by sperm whales. The detection of short pulses in the energy of the signal (Butterworth filter, bandpass 316 kHz) (*a*). Fine structure of the waveform of a single click (*b*) [Car 15].

- The noise level is negligible. The Antarctic ice contains virtually no salt. Light and sound emitting life forms are absent at the depth considered here.

- Once in place, the PMTs stay where they are. Underwater currents that may displace them or worse (DUMAND lost PMTs because of a broken cable) do not play a role.

- There is no need to transport power from and signals to a far-away shore. The control room can be located right above the detector.

Obviously, there are also disadvantages. First and foremost, the Antarctic environment is hostile. There are good reasons why there are no permanent human settlements on this continent. It is also far away from everywhere else, which causes a variety of logistic problems (*e.g.*, transportation of equipment and personnel, maintaining an adequate support structure for the experiments).

The installation of the PMTs at great depth in the ice is far from trivial either. Every string requires a large hole to be drilled, all the way to the depth at which one wants to deploy the instrument. In practice, this "drilling" is done with a technique based on injecting hot water. After the string is installed in the water-filled hole, the PMTs freeze into their final positions. Once installed, the PMTs can no longer be serviced. A failing PMT is lost forever.

There is one essential difference between ice and liquid water that may affect the quality of the measurements and thus the potential capabilities of the instrument. This difference concerns the light attenuation. Light attenuation is caused by two phenomena: absorption and scattering. Both lead to a reduction in the light detected at a certain distance from its source. Both phenomena have their own characteristic length scale (λ_{abs} and λ_{scat}, respectively) and these combine in the following way to the light attenuation length:

$$\frac{1}{\lambda_{att}} = \frac{1}{\lambda_{abs}} + \frac{1}{\lambda_{scat}} \tag{10.9}$$

In liquid water, light attenuation is dominated by absorption. However, air bubbles in the Antarctic ice act as scattering centers and thus may affect the light attenuation considerably. The AMANDA Collaboration, which has operated a neutrino telescope at the Amundsen-Scott research station near the Geographic South Pole since 1994, has studied this phenomenon in considerable detail.

10.3.2   AMANDA/IceCube

A first detector consisting of four strings with a total of 80 PMTs (of which 73 survived the delicate installation procedure) has been operating since 1994 at a depth between 800 and 1,000 m. The performance of this detector was seriously affected by the light diffusion caused by the numerous air bubbles contained in the ice at this depth (λ_{scat} was measured to be < 1 m). As a result, reconstruction of the direction of the observed muons was impossible with this detector. On the other hand, the transparency of the ice was found to be excellent. At a wavelength of 450 nm, λ_{abs} was measured to be 170 m.

The optical properties of the ice encountered with AMANDA-A inspired the Collaboration to explore greater depths, with the desired success. Initially, in 1996, four strings with a total of 86 PMTs (of which 79 survived) were deployed at a depth between 1,500 and 2,000 m. At this depth, diffusion by air bubbles was found to be much less of a problem. The scattering length was measured to be ~ 25 m and the scattering occurred mainly in the forward direction: $\langle \cos \theta \rangle \sim 0.8$. The absorption length was not very different from the value measured before.

Encouraged by these results, six other strings with 36 PMTs each were installed at the same depth (1997). Together with the four earlier ones, these formed the AMANDA-B configuration. A further extension to a 800-PMT array (AMANDA II) was completed in 2001. This array consists of strings with a length of 600 m and covers an effective detection volume of ~ 0.05 km^3.

Based on the scientific successes obtained with these instruments, the researchers convinced the National Science Foundation and other funding agencies to provide the resources for the first (and as yet only) km^3 neutrino detector constructed to date. Construction of this detector, which was appropriately named *IceCube*, started in 2004 and was completed six years later. The international IceCube Collaboration, with more than 40 institutions worldwide, is responsible for the scientific research program.

The in-ice component of IceCube consists of 5,160 digital optical modules (OMs), each containing a 10-inch photomultiplier tube and associated electronics. The OMs are attached to vertical "strings," frozen into 86 boreholes, and arrayed over a cubic kilometer from 1,450 m to 2,450 m depth. The strings are deployed on a hexagonal grid with 125 m spacing and hold 60 optical modules each (Figure 10.35). The vertical separation of these modules is 17 m. Eight of the strings at the center of the array have been deployed more compactly, with a horizontal separation of about 70 m and a

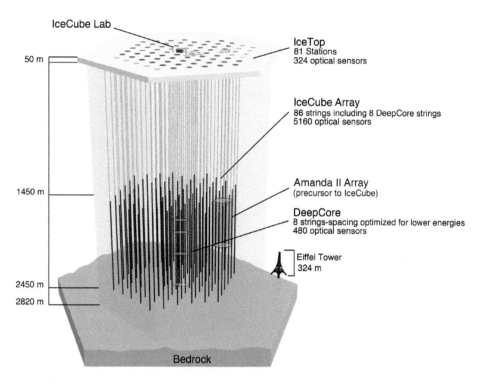

FIG. 10.35. Layout of the IceCube detector complex at the South Pole [Hal 10].

vertical OM spacing of 7 m. This denser configuration forms the *DeepCore* subdetector. The energy threshold for neutrino detection in this part of the array is about 10 GeV, which allows the opportunity to study neutrino oscillations.

To put these numbers in perspective, consider that the total photocathode surface of the IceCube OMs (~ 250 m^2) is about 1/9 of that in SuperK, while the total instrumented volume is 20,000 times larger. In general, the surface area enclosing a volume V scales like $V^{2/3}$. Therefore, while the instrumented volume of HyperK is twenty times larger than that of SuperK, it takes only eight times more PMTs to get the same detection efficiency for Čerenkov photons. Disregarding the effects of light absorption, the total photocathode area of IceCube would thus have to be 737 times *larger* than that of SuperK in order to get the same light yield (7 photoelectrons per MeV deposited energy). The fact that it is, instead, nine times *smaller* than that of SuperK thus means that the expected light yield is, at best, $\sim 7,000$ times smaller, or ~ 1 photoelectron/GeV. A similar calculation shows that a neutrino showering in the DeepCore region is expected to produce, on average, two photoelectrons per GeV deposited energy, or twenty photoelectrons for 10 GeV.

On the surface above IceCube, the *IceTop* array forms an additional component of the detector complex. It consists of 81 stations located on top of the same number of IceCube strings. Each station has two tanks, each equipped with two downward facing

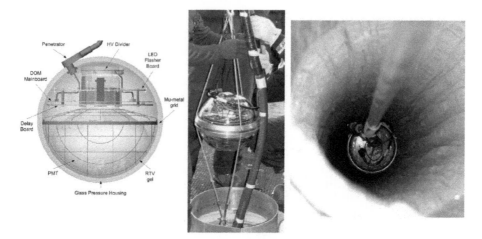

FIG. 10.36. One of the Optical Modules used in the IceCube experiment [Hal 10]. Shown are a schematic drawing of this module (left), the real thing assembled (center) and being lowered to its final position in the ice (right).

OMs. IceTop serves as a veto and calibration detector for IceCube, and also detects air showers from primary cosmic rays in the 300 TeV to 1 EeV energy range. The surface array measures the cosmic-ray arrival directions in the Southern Hemisphere as well as the flux and composition of cosmic rays. Its predecessor, SPASE [Mil 97], had already demonstrated the powerful capabilities added by a surface detector in the AMANDA era. Figure 10.35 shows a schematic overview of the different components of the detector complex described above. To get an impression of the scale of these instruments, the Eiffel Tower is shown as a reference.

10.3.3 Scientific results

IceCube is primarily a neutrino telescope. It is the worlds largest neutrino detector, encompassing a cubic kilometer of ice. Its primary focus is therefore on very-high-energy neutrinos, which are almost certainly of extragalactic origin. They are believed to be produced in violent events: exploding stars, gamma-ray bursts, and cataclysmic phenomena involving black holes and neutron stars. Since these *astrophysical* neutrinos travel from their source to Earth with essentially no attenuation and no deflection by magnetic fields, they might reveal details about the mechanism through which they are produced. This sets them apart from extragalactic charged particles, even those at the very highest energies.

In its first three years of operation, IceCube has detected 106 neutrinos with an energy of more than 10 TeV [Aar 13a, Aar 15a]. Three of these neutrinos had energies of 1 PeV or more. Their event displays are shown in Figure 10.37. Of these high-energy neutrinos, 87^{+14}_{-10} are believed to be of astrophysical origin. The experimental uncertainty derives from the problem that it is not completely clear to what extent high-energy neutrinos produced in the Earth's atmosphere contribute to the event sample. Atmospheric

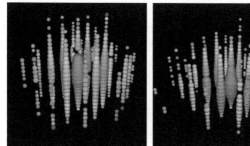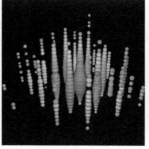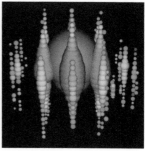

FIG. 10.37. The three highest-energy neutrino events detected in IceCube: Bert (1.0 PeV), Ernie (1.1 PeV) and Big Bird (2.2 PeV) [Aar 13a].

neutrinos originate from the decay of short-lived shower particles produced in the absorption of high-energy cosmic rays in the Earth's atmosphere. Pions and (especially) kaons are the main sources. However, since the decay of these particles competes with them reinteracting in the atmosphere, the neutrinos produced in the decay process carry typically not more energy than a few GeV (Section 11.2). Neutrinos in the TeV regime can only be produced in the atmosphere from the decay of very short-lived particles, *e.g.*, particles that contain a c or b quark. The rate at which such particles are produced in these extensive air showers is uncertain.

Of course, the IceCube researchers have made a major effort to locate the origin of these astrophysical neutrinos. However, there was no significant evidence for clustering [Aar 14b], and the neutrinos were also not associated with suspected sources, such as Active Galactic Nuclei or Gamma Ray Bursts [Aar 15c]. A joint study with scientists from the Auger Collaboration also did not show any correlation between the astrophysical neutrino events and ultra-high-energy cosmic rays detected by Auger [Aar 16].

The rate of atmospheric neutrino interactions in IceCube is very high, but because of the small probability to detect Čerenkov photons, the analysis of neutrino oscillations is limited to particles with energies of at least 10 GeV. Yet, the oscillation phenomena have been clearly observed and the results confirm the findings of SuperK, SNO and other groups [Aar 15b]. In order to become more competitive in this regard, the Collaboration has submitted a proposal to upgrade the detector by installing an additional 40 strings, with 60 OMs each, in the DeepCore region [Aar 14a]. This would increase the number of PMTs (and thus the light yield) in DeepCore by a factor of six, which would make it possible to study atmospheric neutrinos down to energies of a few GeV.

Thanks to IceTop, the Collaboration has also managed to perform very competitive studies of extremely-high-energy cosmic rays. Thanks to its high altitude (2,835 m above sea level) IceTop is sensitive to the em component of cosmic air showers, while IceCube measures the muonic component for energies above ~ 1 TeV. Combination of the signals measured for muons in both IceTop and IceCube has made it possible to determine the track direction of such muons with a resolution of about $0.5°$. Unlike any other experiment, the IceCube Neutrino Observatory is sensitive to cosmic rays of a very large energy range, from ~ 1 PeV, just below the *knee*, to the EeV regime. In one study, they have measured the change in the chemical composition of the cosmic rays in the energy range from 1–30 PeV (Figure 10.38), and confirmed the conclusion of KASCADE-Grande that the knee is a feature that only affects the proton component of the spectrum [Abba 13].

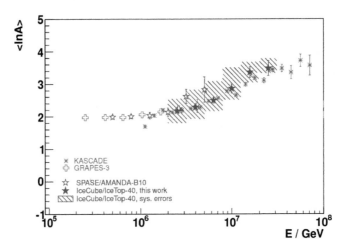

FIG. 10.38. The average atomic number (A) of the particles registered as cosmic rays by the IceCube and KASCADE-Grande Collaborations [Abba 13].

Other studies performed by the IceCube Collaboration in the past six years include a search for dark matter annihilation in the center of the Sun and in the Galactic Center, for magnetic monopoles, both relativistic and non-relativistic, an observation of the shadow of the Moon in the cosmic-ray spectrum (made possible thanks to the very good angular resolution!) and a variety of studies with the collected neutrino data. This and other work completed in the first three years of IceCube operation is nicely summarized in [Hal 14].

Just as in the case of the Mediterranean telescopes, there are synergies with other fields of research, in this case for example with Glaciology. To realize the full potential of IceCube, the properties of light propagation in the Antarctic ice must be well understood. To that end, IceCube has made very detailed measurements of these properties using both light sources deployed on board the IceCube sensors and a dedicated borehole laser probe called the "dust logger." This work has led to some results that are of interest to glaciologists, for example [Aar 13b]:

- It was found that light propagates preferentially in the direction of the movement of the South Pole glacier.
- Evidence was found for the Tova volcano eruption 74,000 years ago, which had never been observed in ice-core studies.
- By comparing the laser data to ice core measurements (which derive temperature information from the $^{18}O/^{16}O$ ratio of the ice), it turned out to be possible to reconstruct a detailed climate record of the last glacial period.

This synergy has played a role in the designation of the South Pole as the site of the next major American ice coring mission.

10.3.4 *Acoustic signals*

At first sight, the South Pole ice cap seems to be a much more favorable environment than the Mediterranean Sea for detecting high-energy neutrinos using acoustic signals. The noise level is much lower, because of the complete absence of sound-emitting life forms, and the penetration of sound by refraction from the surface should play no role at depths of more than 100 m. The only source of noise that may be suspected comes from temperature effects (*cracking*). The speed of sound is much higher in ice than in water, which means that coincidence windows can be much shorter, thus reducing the probability for fake trigger signals. Theoretical estimates put the attenuation length for sound waves with a typical frequency of 20 kHz at 8 km [Pri 06], dominated by absorption rather than scattering.

Given these favorable conditions, the option of installing an array of acoustic detectors, to be operated in conjunction with the light detecting IceCube instruments, seemed very attractive, especially because the fiducial volume could be substantially increased, perhaps by several orders of magnitude. For this reason, a subgroup of IceCube researchers set out to measure the relevant parameters in the IceCube environment. This project became known as *SPATS*, the South Pole Acoustic Test Setup. An array of acoustic transmitters and sensors was set up in some of the IceCube holes, at depths varying from 80 to 500 m below the surface and spaced horizontally up to 690 m [Abba 11]. Some results of the tests include [Lai 12]:

1. The speed of sound quickly increases with depth and reached a maximum value of $\sim 3,900$ m/s at a depth of 200 m.
2. The noise level decreases slightly with increasing depth, it is stable in time and exhibits no correlation with surface conditions (wind, temperature). There is some correlation with human activity, especially if that involves heavy machinery.
3. The relationship between the measured noise floor and the minimum detectable neutrino energy is not clear.
4. The attenuation length was consistently measured, using a variety of methods, sensors, distances and positions, to be ~ 300 m.

The last result was a great surprise. It is not understood where the calculations went so wrong. But in any case, it has important consequences for the design of a hybrid detector system, which would have to be scaled down considerably in size from the originally envisaged 100 km^3 to have any sensitivity at all. In the meantime, more than 35 years after the idea of acoustic neutrino detection was first proposed, not a single neutrino has ever been observed producing both optical and acoustic signals, neither at the South Pole, nor in the Mediterranean Sea, nor in Lake Baikal.

10.4 Calorimetry in the Earth's Atmosphere

The Earth is surrounded by a collar of air, which consists of nitrogen ($\sim 80\%$), oxygen ($\sim 20\%$), and some trace elements (argon, carbon dioxide, *etc.*). The radiation length and the nuclear interaction length of air are 36.66 g cm^{-2} and 90.0 g cm^{-2}, respectively. At sea level, the total thickness of the atmosphere is, on average, about $1{,}030$ g cm^{-2}. Therefore, the atmosphere corresponds to a calorimeter with a thickness of $\sim 11\lambda_{int}$ and $\sim 28X_0$.

As the altitude increases, the atmospheric pressure, and thus the calorimetric thickness of the atmosphere, decreases, to a first approximation, exponentially. This is illustrated in Figure 10.39, which shows the atmospheric pressure as a function of the altitude (above sea level) at which it is measured.

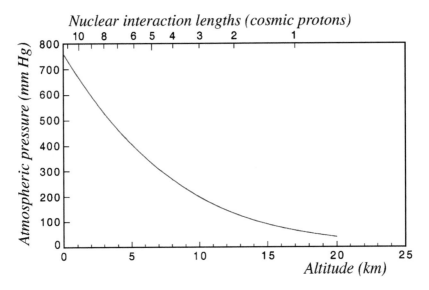

FIG. 10.39. Atmospheric pressure as a function of altitude, measured with respect to sea level. The top horizontal axis indicates the thickness, measured in nuclear interaction lengths, traversed by hadrons entering the Earth's atmosphere from the zenith direction.

The top horizontal axis shows the number of nuclear interaction lengths traversed by cosmic protons as they penetrate the atmosphere. On average, protons entering from the zenith direction interact for the first time with an atomic nucleus in the atmosphere at an altitude of 17 km. The second interaction length is represented by the air between altitudes of 17 and 13 km, the third one between 13 and 10 km, and so on.

10.4.1   Shower development in the atmosphere

Cosmic protons are absorbed in the Earth's atmosphere in ways that are very similar to those discussed in Chapter 2 for much denser absorbing structures. As they traverse the upper layers of the atmosphere, they ionize the air and generate Čerenkov light if they are sufficiently energetic. In their interactions with nuclei, pions and other particles are generated, and the nucleus releases protons, neutrons, αs and/or γs. Neutral pions produced in these processes develop em showers. The charged pions and the other mentioned particles lose their energy by ionizing the air and in nuclear interactions they initiate subsequently themselves.

There are some characteristic differences between hadronic showers developing in the Earth's atmosphere and those in calorimeters of the type discussed in the previous chapters of this book.

- The λ_{int}/X_0 ratio is much smaller than in the dense calorimeters discussed in earlier chapters, 2.45 *vs.* values in excess of 30 for some calorimeters based on high-Z absorber material. This has the unusual consequence that the longitudinal size of em showers of a given energy is larger than that of hadron showers of the same energy. In fact, $28X_0$ is a rather marginal thickness when it comes to containing high-energy em showers.

- Expressed in meters, the nuclear interaction length of the atmosphere is very long, especially in the upper layers. As a result, the fraction of the unstable particles produced in the shower development process that decay before initiating a nuclear interaction is much larger than in dense calorimeters.

  Let us consider, as an example, the case of a 50 GeV pion produced at an altitude of 20 km. The mean free path for this particle, assuming that it equals 1 λ_{int} (which is a lower limit since λ_{int} is defined for protons) is $\sim$ 5 km. The γ value for this pion amounts to $E/m_\pi = 360$, so that the decay length $(\gamma c\tau)$ equals 2.8 km. Therefore, this pion is much more likely to decay than to cause a nuclear interaction, 83% *vs.* 17%.

  Had the same pion been produced in the SPACAL calorimeter (λ_{int} = 20 cm), the decay and interaction probabilities would be 0.007% and 99.993%, respectively.

These properties imply that in practice, most of the charged pions produced in the calorimetric absorption process decay rather than interact with nuclei, primarily through the following processes:

$$\pi^- \to \mu^- + \bar{\nu}_\mu, \qquad \pi^+ \to \mu^+ + \nu_\mu \qquad\qquad (10.10)$$

This is *a fortiori* true for other short-lived mesons, such as kaons. The cosmic muons detected at the Earth's surface (and below) originate from these processes, and so do the so-called *atmospheric neutrinos*.

The muons produced in these decay processes are also unstable ($\tau = 2.2$ μs) and some of them may also decay on their way to the Earth's surface:

$$\mu^- \to e^- + \bar{\nu}_e + \nu_\mu, \qquad \mu^+ \to e^+ + \nu_e + \bar{\nu}_\mu \qquad\qquad (10.11)$$

However, the probability that this actually happens depends strongly on the energy of the muon. The muon's $c\tau$ value is about 660 m. Therefore, the probability that a 10 GeV ($\gamma = 94$) muon produced at an altitude of 10 km decays before reaching the Earth's surface is $\sim$ 15%. For a 100 GeV muon this probability is only 1.5%, while a 1 GeV muon has, on average, a probability of 80% of decaying before reaching the Earth's surface.

Because the baryon number is conserved in nuclear interactions, it is very likely to find an energetic proton or neutron among the reaction products of the nuclear interaction initiated by a cosmic proton. Such a baryon will penetrate the atmosphere farther and initiate, in its turn, a nuclear interaction after having traversed, on average, another nuclear interaction length. The fate of the reaction products of this interaction will be very similar to that of the products of the first interaction: the π^0s initiate em showers, the charged pions decay and the energetic stable baryons will cause new nuclear interactions closer to the Earth's surface.

In a first approximation, showers initiated by energetic protons that penetrate the Earth's atmosphere are thus composed as follows:

- Electromagnetic showers initiated by the π^0s produced in the first nuclear interaction in the upper layers of the atmosphere, and in subsequent interactions of stable baryons produced in this process. Such em showers may easily penetrate the atmosphere all the way down to sea level.

- Muons and muon (anti-)neutrinos from the decay of the charged pions and kaons produced in the various baryon-induced nuclear interactions.

- Electrons, electron (anti-)neutrinos and muon (anti-)neutrinos produced in the decay of the muons mentioned in the previous point. The electrons develop em showers or lose their energy entirely through ionization of the atmosphere.

The "shower leakage" observed at the end of this calorimeter, *i.e.*, the Earth's surface, thus consists of energetic muons, muon (anti-)neutrinos, electron (anti-)neutrinos and possibly some soft photons and electrons from the em shower tails.

The probability that the muons decay in the atmosphere strongly increases at low energies (< 10 GeV). The energy spectrum of the electron (anti-)neutrinos, which almost exclusively originate from μ decay, is softer than that of the muon (anti-)neutrinos, since the latter may also originate from the decay of the (much shorter lived) pions or kaons.

The composition of the (anti-)neutrino sample reaching the Earth's surface thus depends on the energy. At high energies, one expects very few electron (anti-)neutrinos, since only a very small fraction of the high-energy muons decay in the atmosphere. In the low-energy limit, where all muons decay before reaching the Earth's surface, one expects a $(\nu_e + \bar{\nu}_e)/(\nu_\mu + \bar{\nu}_\mu)$ value of 1/2 . Averaged over all energies, the fraction of electron (anti-)neutrinos of atmospheric origin measured on Earth is thus expected to be smaller than one-third of the total number of atmospheric (anti-)neutrinos[41].

In Section 11.2, the measurements of the properties of the atmospheric neutrinos by the SuperKamiokande Collaboration are discussed in some detail.

10.4.2 *Čerenkov light production in atmospheric showers*

The production of Čerenkov light in showers developing in the Earth's atmosphere has very different characteristics from that in the water Čerenkov counters discussed in the previous sections. The reason for this is the very different index of refraction. Using the textbook value for air at standard temperature and pressure ($n - 1 = 2.93 \cdot 10^{-4}$) as the basis for the calculation, it turns out that only particles with $\beta > 0.99971$ are capable

---

[41] In this discussion, the production of charmed particles in the shower development is ignored. Such particles may in principle produce very-high-energy (anti-)neutrinos, both of the electron and muon type (see Section 10.3.3).

of emitting Čerenkov light. This threshold corresponds to a Lorentz factor $\gamma = 41.3$. Therefore, electrons are only capable of emitting Čerenkov light if they carry at least an energy of 21 MeV; the corresponding threshold for muons is 4.4 GeV and for protons 39 GeV.

As a result of the high threshold velocity, only a small fraction of the electrons produced in em showers in the atmosphere are capable of emitting Čerenkov light. It has been estimated [Aha 97] that the useful track length for creating Čerenkov light in high-energy em air showers is only $\sim 25\%$ of the total track length of all the ionizing shower particles. For em showers developing in water (where electrons with kinetic energies as low as 0.26 MeV are capable of emitting Čerenkov light), the corresponding fraction is over 90%.

The shower particles that do emit Čerenkov light in em air showers are predominantly found *before* the shower maximum. Most of the Čerenkov light is actually emitted before the shower maximum is reached. A longitudinal shower profile measured on the basis of the Čerenkov signals would look very different from the energy deposit profile. Even showers initiated by 1 TeV electrons or photons that enter the atmosphere from outside produce almost none of their Čerenkov light at altitudes below ~ 5 km, *i.e.*, beyond $\sim 15X_0$. The maximum of the Čerenkov light production in such showers is reached after $\sim 8X_0$, *i.e.*, at an altitude of ~ 9 km.

Also the Čerenkov angle is very different from that in water: $\theta_C = 1.4°$ for ultra-relativistic particles ($\beta \gg \beta_{\text{threshold}}$), *vs.* 41.4° in water. The Čerenkov light from air showers thus travels in almost the same direction as the particles that emit it. For example, the light cone emitted by an ultra-relativistic shower electron at an altitude of 9 km has a radius of only ~ 200 m by the time it reaches sea level. For softer shower particles, the Čerenkov cone is even more narrow. In addition, the index of refraction decreases with altitude. The value used for calculating the Čerenkov threshold mentioned above is valid for air of 20°C and a pressure of 1 atm, *i.e.*, conditions found at sea level.

Showers that originate at an altitude of ~ 10 km produce Čerenkov cones that have a radius of ~ 100 m by the time they reach the Earth's surface. This is illustrated in Figure 10.40, which shows typical radial light intensity distributions for atmospheric showers initiated by 100 GeV γs.

The total Čerenkov light yield in air showers is somewhat smaller than in water. This can be seen as follows. For ultra-relativistic particles, the specific energy loss through the emission of Čerenkov light is proportional to $\sin^2 \theta_C$ (Equation 4.3). Comparing the θ_C values for air and water, it turns out that this energy loss is about 750 times larger in water. The specific energy loss by ionization is also larger in water, by a factor of ~ 900. And since the effective track length for emitting Čerenkov light is only 25% of the total, *vs.* more than 90% for water, the total Čerenkov light yield is ~ 20–25% of that in water.

Even though the Čerenkov cones are extremely narrow, compared to those in water, a typical detector only observes a small fraction of this light, much less than 1%. This is the result of a geometric factor (the limited acceptance of the light detector), and of scattering and absorption losses in the atmosphere. Assuming that the atmospheric Čerenkov light illuminates an area with a surface of 10^4 m^2, and using the light yield

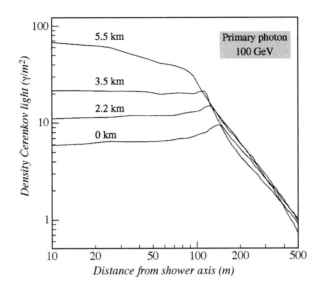

FIG. 10.40. Radial Čerenkov light distributions for 100 GeV γ-induced air showers, measured at different altitudes. Data from [Aha 97].

measured in SuperK ($\sim 7,000$ p.e. GeV^{-1}) as the starting point, we find that a light yield of the order of 0.1 photoelectron per m^2 per GeV of shower energy may be expected, not taking into account scattering and absorption losses. From Figure 10.40, such losses may be estimated at another order of magnitude. A 1 TeV γ-induced atmospheric shower should thus produce a signal of the order of 10 photoelectrons per m^2 in a ground-based Čerenkov telescope.

The Čerenkov signals must be detected against the background illumination of the night sky. This background depends, among other factors, on the charge integration time of the light detectors, which should therefore be kept as short as possible. Unlike in water (see Figure 10.25b), the time structure of the signals themselves is in this respect not a very limiting factor. The difference between the arrival times of Čerenkov photons produced at altitudes of 10 km and 2 km on the surface of the Earth is only ~ 7 ns. A charge integration time of 10 ns would thus include almost all the light produced in the shower.

In practice, most of the Čerenkov light produced in air showers travels to the Earth's surface in approximately the same direction, *i.e.*, the direction of the showering particle. Whereas this feature offers great opportunities to measure the trajectory with great precision, it has two major drawbacks:

1. It limits the geometric acceptance of any instrument intended to detect air showers through the Čerenkov light they generate.

2. It makes it almost impossible to obtain experimental information on the longitudinal shower development.

Yet, atmospheric Čerenkov calorimetry is a widely applied experimental technique, both in observational astronomy and in cosmic-ray physics.

10.4.3 *Atmospheric Čerenkov calorimetry*

The history of observing cosmic-ray showers through the Čerenkov light they produce in the atmosphere started in 1948, when Blackett was the first to suggest this possibility and estimated the light levels involved [Bla 48]. Five years later, Galbraith and Jelley made the first actual observations that confirmed the feasibility of the idea [Gal 53].

The instruments that are nowadays used to measure this Čerenkov light are in many ways very similar to ordinary optical telescopes that are used to detect and analyze the optical signals that reach us from far-away places. Typically, the instrument consists of one or several mirror(s) that collect the light. Detection equipment installed in the focal plane of each mirror is used to record the signals. Figure 10.41 shows an example of a Čerenkov telescope used in the HESS experiment in Namibia [HES 16].

FIG. 10.41. The 800 m^2 Čerenkov telescope used in the HESS experiment in Namibia. One of the smaller (12 m diameter) telescopes is visible to the left [HES 16].

The vast majority of the energetic particles penetrating the Earth's atmosphere are protons and heavier nuclei. High-energy γ-rays form a small, but interesting component, since they are not affected by interstellar magnetic fields. This offers the opportunity to locate their source. Because of the different scientific goals, instruments intended to measure atmospheric showers induced by hadrons are usually quite different from those intended for γ-ray astronomy.

Excellent angular resolution is absolutely essential for locating point sources. More-over, it is very helpful to reduce the background level (see Section 10.3.3.1). For these reasons, the design of Čerenkov telescopes for ground-based γ-ray astronomy usually strongly emphasizes angular resolution, at the expense of the field of view and thus the acceptance of the instrument. These instruments usually have large mirrors, the one pictured in Figure 10.41 has a surface area of almost 800 m^2. A fine-grained imaging camera installed in the focal plane of this mirror provides the needed angular resolution.

On the other hand, Čerenkov telescopes intended for high-energy cosmic-ray studies are designed to maximize the acceptance, and thus the energy reach of the instrument. Some of these telescopes have a field of view close to 2π sr, whereas the aperture of a typical γ-ray telescope is only a few degrees, which corresponds to a field of view that is less than one-thousandth of the visible night sky (0.006 sr). To achieve a large angular acceptance, many small mirrors are needed, each covering a certain fraction of the sky [Ber 12]. Such cosmic-ray telescopes are often used in conjunction with a detector that records the ionizing particles from the shower tails and/or muons produced in the shower development. Such a detector typically consists of an array of plastic-scintillator plates, distributed over a surface area of the order of the size of the showers to be detected, or somewhat larger. A typical diameter of this surface area is several hundred meters. An example of an experiment of this type is located near the town of Yakutsk, in Eastern Siberia [Iva 15].

The first high-quality dedicated Čerenkov telescope for atmospheric shower detec-tion was built on Mt. Hopkins (altitude 1,268 m) in southern Arizona in 1968. This telescope, named after F.L. Whipple, was used until 2013. It contained 248 hexagonal mirrors and provided a total reflecting area of 75 m^2 The imaging camera was located in the focal plane and contained an array of 379 PMTs, which provided a field of view of 2.6° and an angular resolution of 0.117°. It discovered the first TeV γ-ray source, the Crab Nebula [Wee 89], which has since been used as a calibration source by all its successors, as well as several other objects [Vac 91, Ker 95].

The current generation of detectors for ground-based γ-ray astronomy all consist of an array of several closely spaced telescopes with moderate aperture. This is the most economical way to collect more light without sacrificing angular resolution. Examples of such arrays include

- VERITAS, the successor of the Whipple telescope in Arizona, which consists of an array of four 12 m telescopes [Ver 16].

- MAGIC, located at an altitude of 2,200 m on the Canary Island of La Palma, consists of two 17 m telescopes separated by 85 m. These operate in coinci-dence, only events that trigger both telescopes within a time window of 180 ns are recorded [Ale 16].

- HESS, located in Namibia, at a 1,800 m high plateau known for its exceptionally clear skies. It consists of an array of four 12 m telescopes, spread out over an area of 50,000 m^2. In its center, this array is supplemented by the HESS-II tele-scope (Figure 10.41), the largest Čerenkov telescope in the world. Its 32.6×24.3 m^2 mirror consists of 875 hexagonal facets of 90 cm (flat-to-flat) size, made of quartz-coated aluminized glass. Each facet can be positioned independently with a precision of 1 arc second. The detected Čerenkov light is focused on the camera (focal length 36 m). This camera consists of 2,048 "pixels," each subtending an angle of 0.067°, for a total field of view of 3.2°. Each pixel uses a 42-mm PMT to convert the light into electric signals. The total signal integration time is 16 ns [HES 16].

Apart from the fact that the increased light yield reduces the low-energy threshold of the detector, the multiple images obtained from one and the same event constrain the shower trajectory better than a single image. In addition, the discriminative power can be greatly improved [Aha 97]. The pointing precision of these detector systems is illustrated in Figure 10.42, which shows the angular resolution of the MAGIC telescopes as a function of the energy of the γ-rays. These measurements were performed with γs from the Crab Nebula. For energies larger than 1 TeV, the detection rate was about 2 events per minute, with a background of less than 1%.

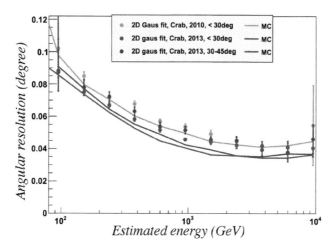

FIG. 10.42. Angular resolution of the MAGIC telescopes as a function of the estimated energy, obtained with a Crab Nebula data sample, and Monte Carlo simulations [Ale 16].

The success of ground-based γ-ray astronomy depends on the telescope's capability to discriminate against the potentially overwhelming flux of hadronic events. One parameter that can be useful in that respect is the apparent shower width. Because of the transverse momentum of the secondary π^0s produced in hadronic interactions, hadronic showers appear to be wider than those induced by γ-rays. Figure 10.43 shows characteristic shower profiles for 100 GeV γs and 300 GeV protons measured with an imaging atmospheric Čerenkov telescope. Taking advantage of these differences, sophisticated algorithms have been developed to discriminate between γ-ray showers and background events. One example is the "ShowerGoodness" parameter used in HESS analyses [Nau 09]. Figure 10.44a shows how this parameter can be used to distinguish γs from cosmic-ray background events in the data sample recorded for a particular

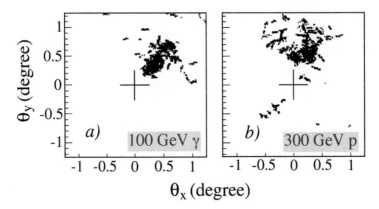

FIG. 10.43. Characteristic shower profiles for 100 GeV γs (*a*) and 300 GeV protons (*b*) measured with an imaging atmospheric Čerenkov telescope. Data from [Aha 97].

blazar. Figure 10.44b shows how the efficiency for γ detection is related to the quality of the background rejection. The results of this particular HESS method are compared to those obtained with a traditionally used method.

It should be noted that background rejection based on image analysis is not the full story. Cherenkov telescope arrays already provide significant background rejection at the trigger level. Typically, a cosmic-ray shower needs to have three times higher energy to trigger a telescope coincidence with the same efficiency as a γ-ray. Since the cosmic-ray spectrum decreases with energy like $E^{-2.7}$, this already implies a background rejection of a factor five to ten at the trigger level.

Even though the telescopes discussed in this subsection were primarily intended for γ-ray astronomy, the rate at which they detect high-energy extensive air showers caused by protons and nuclei absorbed in the atmosphere has allowed them to make

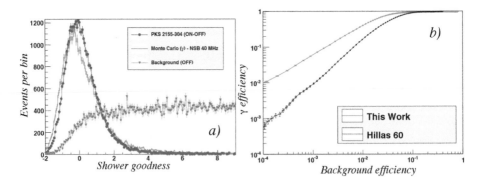

FIG. 10.44. Distribution of the γ/proton discriminating "ShowerGoodness" parameter for HESS data taken on the blazar PKS 2155-304, compared with a simulation with a similar night sky background level (*a*). The efficiency for γ-ray shower identification as a function of background rejection (*b*). See text for details [Nau 09].

important contributions in that field of research as well. For example, HESS recently demonstrated that the center of our galaxy acts as a *PeVatron, i.e.*, as a source of extremely high-energy protons and nuclei [Abr 16]. Until now, it was generally believed that supernova explosions are the accelerators in the PeV regime. In another study, the HESS researchers managed to identify the extremely collimated Čerenkov light emitted by a primary cosmic nucleus *before* it underwent its first nuclear interaction in the Earth's atmosphere. Since the number of Čerenkov photons is proportional to Z^2, this allowed them to identify such nuclei. As a result, they managed to measure the spectrum of iron nuclei in the energy range from 13–200 TeV [Aha 07], something that had not been done before by dedicated experiments. Both studies benefitted greatly from the excellent pointing capability of the HESS telescope array.

The next-generation instrument for ground-based γ-ray astronomy is the CTA (Čerenkov Telescope Array), which will comprise a Northern Hemisphere and a Southern Hemisphere component with in total more than 100 telescopes [CTA 11]. Of these, 99 will be installed at the Cerro Armazones site in Chile (which will also host the future European Extremely Large Telescope), and 19 at La Palma (joining MAGIC). It is expected that CTA will allow the detection of γ-ray induced air showers over a large area on the ground, and increase the number of detected γ-rays dramatically, while at the same time providing a much larger number of views of each cascade. This would result in both improved angular resolution and better suppression of cosmic-ray background events.

10.4.4 *Atmospheric scintillation calorimetry*

Although strongly dominating, the Čerenkov mechanism is not the only source of light production when extended air showers develop in the Earth's atmosphere. Fluorescence of the atmosphere also plays a role, but the light yield from that process is about an order of magnitude smaller. Nearly all fluorescent light in the atmosphere is produced by transitions in the 2P band of molecular nitrogen (N_2) and in the 1N band of the N_2^+ molecular ion. Figure 10.45a shows the fluorescent spectrum. Most of the light is emitted at wavelengths between 300 nm and 450 nm. Figure 10.45b shows the fluorescent

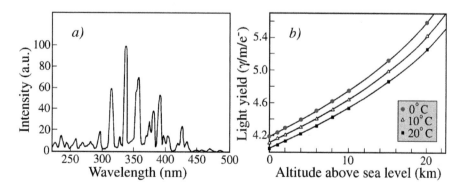

FIG. 10.45. Atmospheric fluorescence spectrum (*a*) and the fluorescent light yield as a function of altitude, for different temperatures (*b*). Data from [Cas 85].

light yield measured with a detector that has made detecting it its specialty: the Fly's Eye observatory [Bal 85, Cas 85], constructed and operated by the University of Utah at the US Army's Dugway Proving Grounds, about 160 km southwest of Salt Lake City, where some of the darkest skies and best visibility conditions in the USA can be found (Figure 10.46).

In its original version, which took data from 1981 to 1993, the Fly's Eye observatory consisted of two measuring stations (FE I and FE II), separated by 3.3 km. Both stations housed a certain number of identical optical units, 67 in the case of FE I, 8 for FE II. Each optical unit contained a 67-in. spherical section mirror. In the focal plane of this mirror, 12–14 hexagonal light-sensing "eyes" were packed together. Each of these eyes consists of a light collector (Winston cone) plus a PMT.

The whole arrangement was made such that each of these eyes observed a designated angular area of the night sky, covering a total solid angle of $\sim 6.5 \cdot 10^{-3}$ sr. The 880 eyes of FE I were arranged to image the entire night sky; the 120 eyes of FE II covered roughly one quadrant, with elevation angles ranging from $2°$ to $38°$ above the horizon.

In 1997, the detectors were replaced by devices that provided a considerably better resolution. Each optical unit was equipped with an array of 256 PMTs, each of which covered an angular area of $\sim 1° \times 1°$ ($5 \cdot 10^{-5}$ sr), and an entire optical unit covered

FIG. 10.46. Aerial view of the Fly's Eye cosmic-ray experiment. Shown is the HiRes II station on Camel's Back Ridge. The experiment was designed to detect extensive air showers developing in the atmosphere by means of nitrogen fluorescence light. Photograph courtesy J. Matthews, Fly's Eye Collaboration.

an angular area of $15° \times 15°$. The detector was renamed HiRes. As before, the optical units were divided between two sites (HiRes I and HiRes II), which were separated by 13 km in the new setup. Figure 10.46 shows an aerial view of the HiRes II site, on Camel's Back Ridge. At this site, 42 optical units were installed, which covered the night sky over the entire $360°$ in azimuth and up to $31°$ in elevation. The HiRes I site (on Little Granite Mountain) housed 22 optical units, which covered elevation angles up to $17°$, over the entire azimuthal range. The experiment has been taking data in this configuration until 2006. In the last few years of this period, HiRes was converted into a true "hybrid" detector system, through the addition of arrays of surface-level detectors (CASA) and muon detectors (MIA). The latter were buried 3 m underground.

In 2007, the entire operation was moved to the southern part of Utah, where it now occupies an area of almost 800 km^2 in the high desert (1,400 m above sea level). The experiment was renamed TAP (Telescope Array Project) and is the largest of its kind on the Northern Hemisphere. Apart from three fluorescence stations located on a triangle with 35 km spacing [Abu 12b], TAP consists of 507 surface detectors (3 m^2 plastic scintillator, [Abu 12a]). The telescopes cover an elevation from $3°$ to $33°$ over the entire azimuthal range. One of the fluorescent stations is equipped with additional telecopes which increase the reachable elevation angle to $59°$. Also, the density of surface detectors has been increased in the vicinity of this station, in order to improve the experiment's sensitivity at the low end of the energy reach (down to 30 PeV). The Sun, the Moon and weather conditions limit the effective duty cycle of the experiment to only 10%. To extend the experimental possibilities, TAP is also equipped with a radar detector system, which should make it possible also to carry out some measurements when optical observations are not possible.

Figure 10.47 shows one of the optical units at the HiRes II observatory. The hexagonal surface structure of the PMTs located in the box in the foreground is reflected in the mirror. The PMTs are equipped with glass that absorbs the UV background light and transmits the nitrogen fluorescence light.

Even though the night sky can be very dark at the site of these observatories, background photons are by no means absent. Even on moonless nights, HiRes recorded, on average, $\sim 5 \cdot 10^5$ photons m^{-2} sr^{-1} μs^{-1}. *Fluctuations* in that rate constitute the noise in the shower measurements. This background (and fluctuations therein) is caused by (scattered) light from stars, galaxies and interplanetary matter, by atmospheric airglow and by man-made light pollution. Obviously, the background levels in all sensing elements need to be continuously monitored, so that proper corrections can be applied to signals from extended air showers.

The fluorescent light yield is more than an order of magnitude smaller than the Čerenkov light yield. However, the angular distribution is completely different. As we saw in Section 10.3.1, Čerenkov light is strongly concentrated in the forward direction; on the other hand, the fluorescent light is isotropically emitted. This means that the relative contributions of these two mechanisms to the light observed at the Earth's surface depend very strongly on the viewing angle.

The geometry of the experimental observations is illustrated in Figure 10.48. The shower–detector plane drawn in this figure is determined by the trajectory of the show-

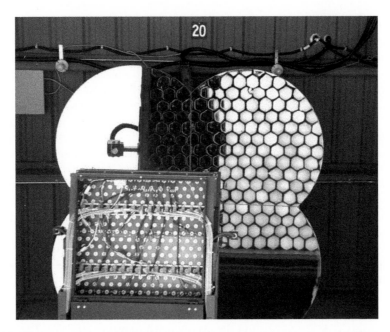

FIG. 10.47. Optical unit of the HiRes II observatory, consisting of a mirror, a PMT cluster and a filter. The detector box consists of an array of 256 hexagonal PMTs. The surface of these PMTs is reflected in the mirror. Photograph courtesy J. Matthews, Fly's Eye Collaboration.

ering particle, and is thus different for each event. The plane contains both this trajectory and the center of the detector. Very relevant in this context are the distance of closest approach, R_p, and the "light emission angle," θ_i. The latter is different for each track segment i along the shower axis.

Shower trajectories of the type shown in Figure 10.48 appear as a sequence of events recorded in different sensing elements of the Fly's Eye detector. Since every such element subtends a specific solid angle in the night sky, the trajectory followed by the developing shower appears to propagate along a great circle projected on the celestial sphere. The precise orientation and distance of the shower trajectory with respect to the detector can be reconstructed from details of the timing sequence of the light signals arriving in the different detector elements. The trajectory geometry may also be reconstructed by stereoscopic viewing of the event from different observatory sites, and/or by using the relative timing information from different stations.

From Figure 10.48 it is clear that for showers that strike within a short distance of the detector ($R_p < 1$ km), the scintillation signals tend to be swamped by the Čerenkov ones. This is because the θ values are small for light produced in the atmosphere high above the detector. In practice, directly beamed Čerenkov light may dominate the isotropically emitted scintillation light for angles θ up to $\sim 25°$, depending on the shower energy. Even at larger viewing angles, the contamination of Čerenkov light may not be completely negligible, since the intense beam of Čerenkov light that

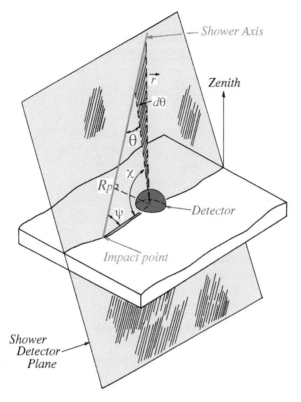

FIG. 10.48. The geometry of an extended air shower as seen by the Fly's Eye detector. The shower–detector plane contains both the shower axis and the center of the detector. The angle ψ and the impact parameter R_{p} are obtained from fits to the observation angles χ_i in individual detector elements i *vs.* the time of observation.

builds up with a propagating shower front may generate enough scattered light at low altitudes to compete with the locally produced scintillation light from the rapidly dying shower.

These effects are illustrated in Figure 10.49, which shows the competition between the different light-generating mechanisms for a shower with an energy of 1 EeV (10^{18} eV) that passes the detector at a distance of 4 km and strikes the Earth's surface at an angle $\psi \sim 40°$. The shower profile, *i.e.*, the number of shower particles, is represented by the curve labeled N_e in this figure.

For this particular event, light emitted at angles θ_i ranging from $\sim 30°$ to 120° was usable for the analysis. That corresponds to light emitted at altitudes between 2 and 7 km above the detector. The light received from the early shower phase, at high altitudes, is more than 95% due to fluorescence, but as the shower propagates further toward the Earth's surface, the relative contribution of Rayleigh-scattered Čerenkov light to the received signals gradually increases, and reaches some 30% near the viewing horizon. In the very last stages of the shower development, in the last kilometer of air traversed,

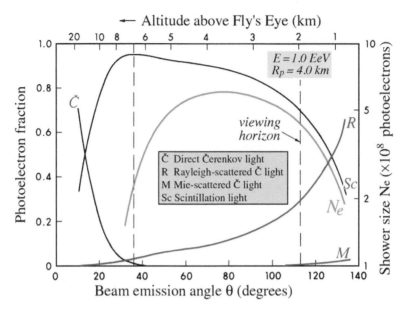

FIG. 10.49. Relative fractions of light received at the Fly's Eye detector, as a function of the light emission angle θ (see Figure 10.48, *lower scale*), or as a function of the altitude above the Fly's Eye (*upper scale*), for an extended air shower with an energy of 10^{18} eV that passes the detector at a distance of 4 km and strikes the Earth's surface at an angle $\psi \sim 40°$. The shower profile is represented by the curve labeled N_e. Data from [Cas 85].

this scattered Čerenkov light rapidly takes over as the dominating light source.

For this particular event, direct Čerenkov light would be the dominant mechanism at viewing angles $\theta_i < 15°$, corresponding to altitudes well above 10 km. However, since the shower development started deeper into the atmosphere, this direct light played no role in this case. In general, scattered light is inconsequential during the early stage of the shower development. To a very good approximation, all the light observed from this early stage is direct. These early signals can thus be used to estimate the Čerenkov light build-up as the shower propagates. This makes it possible to calculate the scattering corrections that have to be applied for light received from later stages of the shower development.

Given the extreme faintness of the fluorescent atmospheric light, and the very low duty cycle, which is imposed by the requirement of clear, moonless nights for the operation of this type of detector with its 2π field of view, one may wonder where the interest in this type of signals comes from. There are two very important advantages of using fluorescent light:

1. The first advantage of atmospheric scintillation calorimetry is that events at large distances can be observed. The geometric acceptance of atmospheric Čerenkov calorimeters is in practice limited to something of the order of 0.1–1 km^2 sr. With existing scintillation calorimeters, this acceptance can be enlarged by up to five

orders of magnitude [Abr 10].

2. A second important advantage is the possibility to measure the longitudinal development of showers in the atmosphere. This longitudinal development may
   provide crucial information about the nature of the showering particle, at least on
   a statistical basis. For example, heavy ions that penetrate the atmosphere from
   outside start showering much earlier than protons. Also, they reach their shower
   maximum earlier than protons of the same energy, counting from the first nuclear
   interaction. Measurements of this type may thus provide insight into the chemical
   composition of the high-energy cosmic nuclei that enter the Earth's atmosphere.

In the next subsection, examples of these advantages are shown. Unfortunately, atmospheric scintillation calorimetry is only applicable for showers of very high energy. This
is a consequence of the low light intensity, in combination with the background level.
For the Fly's Eye detector, the low-energy detection limit was in practice $\sim 10^{17}$ eV,
and at TAP the limit is a factor of three lower. At high energy, the reach of this type
of detector is limited by the rates at which events take place in the Earth's atmosphere.
This rate decreases fast. The primary cosmic-ray spectrum exhibits roughly a power-law
behavior:

$$\frac{dN}{dE} \sim E^{-n} \tag{10.12}$$

with $n \approx 3$ in the energy region of interest. The count rate above a certain energy
threshold thus drops by two orders of magnitude when the threshold is increased by
one order of magnitude. The scientific program of all experiments of this type is mainly
focusing on events of the highest possible energy, in view of the predicted cutoff mechanism [Gre 66] that would prevent cosmic protons with energies in excess of more than
$\sim 10^{19}$ eV from traveling distances of more than ~ 50 Mpc away from their production source. Such protons would be subject to photo-pion production in collisions
with photons constituting the universal microwave background radiation, *i.e.*, the 2.7 K
blackbody spectrum that is believed to be a remnant from the Big Bang.

10.4.5 *The Pierre Auger Observatory*

By far the largest detector based on the principles encountered in the previous subsection
is the Pierre Auger Observatory, a vast complex of diverse instruments installed on an
area of $\sim 3,000$ km^2 in an uninhabited part of western Argentina, not far from the
Andes mountain range. The average altitude amounts to $\sim 1,400$ m, which corresponds
to an atmospheric overburden of 875 g cm^{-2}. The Observatory started collecting data in
2004, and the first physics results were presented one year later. Figure 10.50 shows an
overview of the site, which is covered by 1,660 ground-based tanks of the type shown
in Figure 10.51a [Bau 98], and is overlooked by four fluorescent detector complexes,
located on hills at the edges of the area. Figure 10.51b shows one of these buildings,
each of which houses six telescopes. A single telescope has a field of view of 30° ×
30° in azimuth and elevation, with a minimum elevation of 1.5° above the horizon.
The telescopes face towards the interior of the array so that the combination of the six
telescopes provides 180° coverage in azimuth. These telescopes collect the fluorescent
light produced at altitudes up to 15 km and as far away as 30 km. The concept is the

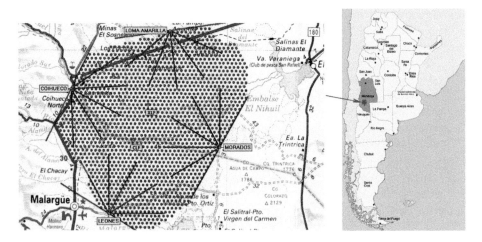

FIG. 10.50. Overview of the site of the Pierre Auger Observatory near the town of Malargüe in Argentina's Mendoza province [Aug 16] .

same as for the HiRes telescopes (Figure 10.47) . The light enters through a circular diaphragm of 1.1 m radius covered with a Schott MUG-6 filter glass window, which reduces the background light flux and thus improves the signal-to-noise ratio of the measured air shower signal. The light is focused by a spherical mirror of $\sim$3,400 mm radius of curvature onto a spherical focal surface with radius of curvature $\sim$1,700 mm, where it is detected by a 80 cm $\times$ 80 cm camera, which consists of 440 PMTs. Each of these PMTs acts as a "pixel" with a field of view of 1.5° $\times$ 1.5°. Because of its large area ($\sim$13 m^2), the primary mirror is segmented to reduce the cost and weight of the optical system.

Each surface detector contains 12,000 liters of ultra-pure water, in which charged shower particles that reach the Earth's surface may generate Čerenkov light that is detected by three PMTs. The water is contained in a tank with a diameter of 3.6 m. Each

FIG. 10.51. One of the 1,660 ground-based tanks (left) and one of the four fluorescence detector buildings (right) of the Pierre Auger Observatory [Aug 16] .

detector is separated from its nearest neighbor by 1,500 m, is powered by a solar panel and communicates with the central data acquisition system using wireless (cell phone) technology.

An essential feature of the hybrid design is the capability of observing air showers simultaneously by two different but complementary techniques. The surface detector operates continuously, it measures the particle densities as the shower strikes the ground. On dark nights, the telescopes record the development of the air shower through the nitrogen fluorescence produced along its path. Since the intensity of fluorescent light is, in first approximation, proportional to the energy dissipated by the shower, integrating the intensity of light produced along the shower axis yields a nearly calorimetric measurement of the energy of the cosmic ray. This energy calibration can then be transferred to the surface array, where the same events have been recorded. A second advantage of the telescope measurements is that the depth at which a shower reaches its maximum size, X_{max}, is observable. This parameter is the most direct of all accessible mass composition indicators. An important advantage of having several fluorescence observation

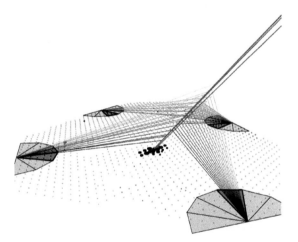

FIG. 10.52. An extensive air shower in view of the detectors of the Pierre Auger Observatory [Aug 16].

sites is that a large number of showers at the higher energies are observed by more than one telescope. At 10^{19} eV, over 60% of the showers are viewed in stereo, increasing to 90% at $3 \cdot 10^{19}$ eV. Stereo observations provide two or more independent hybrid reconstructions of the shower geometry, and of profile parameters such as the energy and X_{max}. This feature allows cross-checks of atmospheric corrections, and of simulations of the detector resolution. It turned out that, in this energy range, the statistical resolutions for these parameters are 10% (E) and 20 g cm^{-2} (X_{max}), for each single observation site [Daw 07]. Figure 10.52 illustrates how the development of an extensive air shower developing in the Auger area may be observed and reconstructed on the basis of observations by several telescopes.

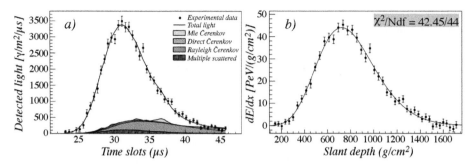

FIG. 10.53. A reconstructed event detected by the Pierre Auger Observatory. Experimental data from [PAC 15].

Figure 10.53 shows an example of an actual air shower observed with the Auger telescopes. The intensity of the detected fluorescent light is shown as a function of time in Figure 10.53a. This information is translated into an energy deposit profile in Figure 10.53b, using a method described in [Ung 08]. The deposited energy is plotted as a function of the *slanted depth* in the atmosphere, *i.e.*, the amount of material encountered by a particle traveling along the shower axis. For this particle, the atmosphere would in total be about 1,700 g cm^{-2} deep, which means that it entered the atmosphere at an angle of about 28° with the Earth's surface. Its energy can be determined by integrating over the curve: $\sim 3 \cdot 10^{19}$ eV.

A key to the success of the hybrid technique is that it allows a precise determination of the position of the shower axis in space, with an accuracy better than could be achieved independently with either the surface array detectors or a single fluorescence telescope. The first step in the geometrical reconstruction makes use of the known orientations of the pixels of the fluorescence detector and of the light intensities registered at the pixels. This enables the shower-detector plane (SDP, see Figure 10.48) to be determined. Timing information obtained from the surface detectors is then used to find the orientation of the shower axis within this plane. Using this hybrid reconstruction method, a directional resolution of 0.5° is routinely achieved.

The surface detector array can also operate in stand-alone mode, *i.e.*, without the assistance of the fluorescence telescopes, as illustrated in Figure 10.54. The wavefront of a shower that hits the surface area at a non-perpendicular angle produces signals at different times in the different surface detectors. These time differences make it possible to estimate the zenith angle of the shower: the smaller the time differences, the steeper the angle of incidence. Additional information about the direction of the shower can be obtained from the *pattern* of the surface detectors that contributed to the signals (Figure 10.54b. The energy of the shower can be estimated from the signals in the individual detectors. In the event displayed in this figure, signals were observed in detectors that were as far apart as 5 km, the time difference between the signals received by the first and last detectors participating in this event may thus be estimated at $\sim 15 \ \mu$s. This is also the time scale of the event depicted in Figure 10.53.

The determination of the energy of these air showers is no trivial matter. In first

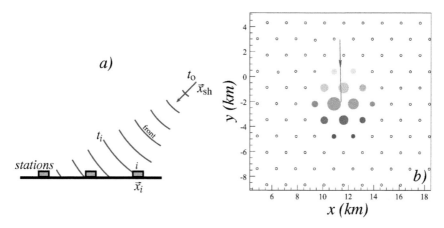

FIG. 10.54. Reconstruction of an air shower event on the basis of surface detector data alone. Schematic representation of the shower front (*a*). The signals induced in the stations of the surface array (*b*), where the arrival time of the shower front is indicated by the color (shading) and the size of the markers is proportional to the logarithm of the energy [PAC 15] .

approximation, the number of detected scintillation photons is a measure of this energy. However, there are many factors that may affect these photons on their way from the spot where they are produced to the telescope. Also, the precise relationship (*i.e.*, the calibration constant) is not a priori clear, in the absence of a calibration source. Atmospheric state variables, including temperature, pressure and humidity, are needed to assess both the longitudinal development of the showers as well as the amount of fluorescence light they induced. Also the surface detector measurements depend on atmospheric conditions such as the local air density, which affects the Molière radius. Furthermore, atmospheric state variables are used to determine effects of Rayleigh scattering of the fluorescent and Čerenkov light. Aerosols and clouds represent the most dynamic monitoring and calibration challenges at the Observatory, since they directly affect the optical transmission properties of the atmosphere. An extensive system of devices has been installed (both on the ground and in weather balloons) to monitor all relevant atmospheric conditions in the vicinity of the Auger site.

Throughout each night of telescope operation, thousands of collimated UV laser pulses are directed into the atmosphere from two facilities located near the center of the surface detector (marked *XLF* and *CLF* in Figure 10.50). Light scattered out of the laser pulses generates tracks in the telescopes. Unlike air showers, the direction, rate, and energy of these laser pulses can be pre-programmed as desired. Laser pulses can be fired at specific directions relative to the ground, for example vertically, or in specific directions relative to the sky, for example aimed at potential sources of cosmic rays. An optical fiber at each laser directs a small amount of light into an adjacent surface detector to provide hybrid laser events. These laser data are used to measure the telescope performance, time offsets between the telescopes and the surface detectors, check telescope pointing, and make hourly measurements of aerosol optical depth vertical profiles for the atmospheric database. Thanks to the described systems, it is believed that the absolute energy scale of the shower measurements is accurate to within 14%, and that the energy resolution varies from 12% at the high-energy end of the accessible spectrum to 16% at the low-energy end [PAC 15].

The Pierre Auger Observatory was primarily constructed to investigate cosmic rays in the energy region of the predicted GZK cutoff [Gre 66]. It was therefore designed to detect air showers with energies of $\sim 10^{18}$ eV (1 EeV) and up. At these energies, the photons from the 2.7 K cosmic microwave background are blue-shifted to the point that they may cause photo-pion production when hit by such a proton ($p+\gamma \to \Delta^+ \to N\pi$). This mechanism[42] would limit the mean free path of such protons to ~ 50 Mpc, and therefore lead to a decrease in the rate at which such protons are observed here on Earth (relative, for example, to the rate at which heavier nuclei would be observed). The Observatory has operated continuously since its inauguration, with a typical efficiency of 85–90%; on average, 15% of the time data are being taken in the hybrid mode (surface detectors + telescopes). The total exposure so far has been $\sim 60,000$ km^2 sr yr. Typically, $\sim 6,000$ quality hybrid events with energies above 1 EeV are recorded each year, 300 of which have energies above 10 EeV.

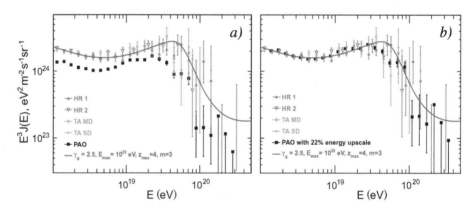

FIG. 10.55. The spectrum of ultra-high-energy cosmic rays measured by the Pierre Auger Collaboration, and by their Northern Hemisphere colleagues from HiRes and TAP. See text for details [Ber 14] .

The most interesting results obtained to date by the Pierre Auger Collaboration are summarized in Figures 10.55 and 10.56. The energy spectrum in the EeV region is shown in Figure 10.55. The Auger results are compared with those published by their Northern hemisphere colleagues from HiRes and TAP. In order to better discern the characteristic features, the differential energy spectrum has been multiplied by E^3. In the left diagram, the published data are shown as such. All sets of data show a very significant steepening of the spectrum beyond 40 EeV, where the GZK cutoff is ex-

---

[42]This process could also be a source of the very-high-energy neutrinos looked for by IceCube and KM3NeT (Sections 10.2/10.3).

pected to kick in (according to the model calculation, represented by the solid curve). However, the Auger data are systematically lower than the other data sets. According to the author who made this plot, this discrepancy could be completely explained from a systematic difference between the calibration constants used by both groups. To check that assumption, he multiplied all Auger data by a factor 1.22 and obtained the results shown in the right diagram after having done so [Ber 14]. Given the relatively large systematic uncertainty in the energy scale (Auger claims 14% [PAC 15]), this is not a far-fetched explanation. We note that a difference in the calibration constants would blow up considerably in a plot such as Figure 10.55, where the energies are cubed.

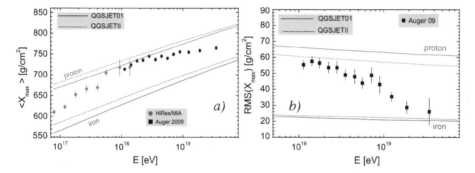

FIG. 10.56. The average depth in the atmosphere at which the shower maximum is reached (a) and the *rms* variation in this depth (b), as a function of the shower energy, together with the predictions from two different Monte Carlo simulation packages [Abr 11, Ber 14] .

A possibly even more interesting result is shown in Figure 10.56, which illustrates the invaluable benefits of the fluorescence measurements and the possibility these offer to measure the depth of the shower maximum in the atmosphere, X_{max}. At a given energy, $\langle X_{max} \rangle$ is larger for protons than for heavier ions, and its value decreases as the nuclear charge Z of the projectiles increases. There are two reasons for these effects:

1. The *nuclear interaction length* (λ_{int}), *i.e.*, the average distance the primary particle penetrates into the atmosphere before undergoing a nuclear interaction, is proportional to $A^{-2/3}$. Therefore, protons penetrate, on average, much deeper into the atmosphere than do heavier nuclei.

2. The *particle multiplicity* is smaller in reactions initiated by protons than in those initiated by heavier nuclei. Therefore, the energy of the incoming proton is transferred to a smaller number of secondaries, which carry thus, on average, more energy than the secondaries produced in reactions initiated by heavier ions of the same primary energy. And since the depth of the shower maximum increases with energy, the showers developed by the secondaries in proton-induced reactions reach their maximum intensity farther away from the primary vertex than in case of showers induced by heavier ions.

In summary, showers induced by protons of a given energy start later and peak at a larger distance from the primary vertex than showers induced by heavier ions of the same energy. These effects can be quantitatively estimated on the basis of the well known characteristics of showers at lower energy [Wig 03]. The much shorter interaction length and the much larger multiplicity of secondaries also imply that event-to-event fluctuations in X_{max} are (much) smaller for heavy nuclei than for protons, by about a factor of three! All these effects are predicted in great detail by the sophisticated Monte Carlo packages that are used to simulate the development of high-energy showers in the atmosphere (Figure 10.56).

The measurements by the Pierre Auger Collaboration show very clearly that there is a transition in the chemical composition of cosmic rays. Between 2 EeV and 20 EeV, the proton component gradually disappears and at the highest energies accessible to Auger the events are almost exclusively caused by iron nuclei [Abr 11]. This observation is consistent with the expected effect, since the GZK mechanism *only* affects the proton component of the cosmic-ray spectrum.

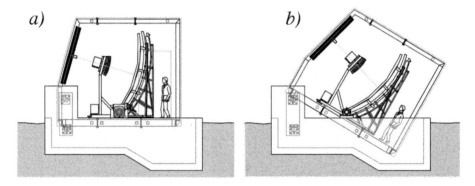

FIG. 10.57. Schematic view of the cross section of one of the HEAT telescopes. Horizontal (downward) mode for service and cross-calibration (*a*). Data taking (upward) mode in tilted orientation (*b*)[PAC 15] .

Subsequent to the completion of the construction of the Pierre Auger Observatory, two significant enhancements have been incorporated into the baseline detectors:

1. The HEAT (High Elevation Auger Telescopes) fluorescence telescopes. The three additional telescopes are identical to the 24 others, but can be tilted upward to cover the elevation range from 30°–58° (Figure 10.57).

2. The AMIGA (Auger Muon and Infilled Ground Array). This is a joint system of additional water Čerenkov detectors and buried scintillators, that spans an area of 23.5 km^2. The detectors are more closely spaced than in the baseline detector (750 m instead of 1,500 m).

Both additions are intended to extend the Observatory's sensitivity down to 10^{17} eV. At lower energies, showers need to enter the atmosphere at a smaller zenith angle to reach the Earth's surface. Therefore, the HEAT telescopes are focused on higher elevation angles, because that is where such showers will reach their maximum intensity (X_{max}). Lower energy also implies a smaller number of shower particles at the level of the surface detectors. Therefore, the density of surface stations had to be increased in order to see adequately large total shower signals. The energy region between 0.1 EeV and 1 EeV is considered interesting because this is where the transition from galactic to extragalactic cosmic rays occurs. It is considered very important to measure the mass composition of the cosmic-ray spectrum in this energy range. AMIGA will allow separate measurements of the electronic and the muonic component of the hadron showers that reach the Earth's surface, which is an important tool in this context, as shown in the next subsection.

10.4.6 *The KASCADE-Grande experiment*

The experiments discussed in the previous subsections are based on detecting light produced when extraterrestrial particles are absorbed in the Earth's atmosphere. However, it is also possible to learn a lot about these particles without detecting such light. In the previous subsection, we saw that the Pierre Auger Observatory also works fine when the fluorescence telescopes are not being used. It only works better when it operates in the hybrid mode.

An example of an experiment that does a great job in cosmic-ray physics *without* detecting atmospheric light is KASCADE-Grande, installed in Karlsruhe, Germany [Ape 10], at 110 m above sea level. This experiment has been taking data continously since 1996, originally as KASCADE, and since 2003 with the addition of the Grande scintillator array. A picture taken during the early stages of this experiment is shown in Figure 1.5. The difference with the experiments discussed in the previous subsections is immediately clear: the grid spacing of the detectors is much smaller, 13 m *vs.* 1,500 m at the Pierre Auger Observatory. The size of the detectors is not very different, the KASCADE-Grande ones measure 2.4 m × 2.4 m each. They are intended for separate measurements of the charged-particle and the muonic shower components of the extended air showers. Each of the 252 detector stations consists, from top to bottom, of 5 cm of liquid scintillator, 10 cm of lead, 4 cm of iron and 3 cm thick plastic scintillator. The latter is intended for the muon measurements, while the liquid scintillator also detects all other charged shower particles, and in particular the tails of the em showers.

Because of the much denser grid spacing, this experiment is intended for much lower energies than Auger and TAP: 0.1–100 PeV, *i.e.*, the region around the *knee* in the cosmic-ray spectrum. The detection principles are the same as for the Auger surface area. The time structure and the pattern of detectors that contribute to the signal provide the direction of the shower, the integrated signal over all counters gives the energy. Because of the lower energy range (and the low altitude of the detector), only events entering the atmosphere at zenith angles $\theta < 40°$ are being considered.

An important goal of the KASCADE-Grande experimental program is the determination of the chemical composition of the cosmic rays in the knee area (1–10 PeV). Having no possibility to measure X_{max}, they have to resort to other means to determine the mass of the particle that initiated the detected showers. The primary tool for that is a comparison between the muonic and charged-particle components of these showers, based on the expectation that the particles that reach the Earth's surface have a larger muon component for showers initiated by heavy nuclei than for proton showers, and

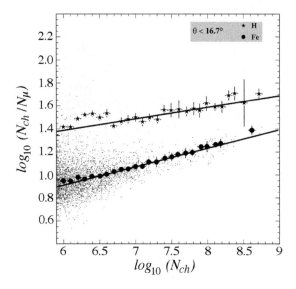

FIG. 10.58. The relationship between the numbers of muons and charged shower particles in high-energy proton and iron air showers detected in KASCADE-Grande [Bert 12] .

that the measured radial shower profile will also, on average, be wider. This expectation is based on the fact that heavy ions interact, on average, at much higher altitude. In addition, the secondaries carry, on average, less momentum because of the much larger multiplicity. Therefore, pions and kaons produced in a heavy-ion shower are more likely to decay, and the muons produced in these decays will hit the Earth surface, on average, at a larger distance from the shower axis than in the case of proton-induced showers.

Figure 10.58 shows a scatter plot, which clearly confirms the relative dominance of the muon component in showers induced by iron nuclei. A very important result of this experimental program of the KASCADE-Grande Collaboration is shown in Figure 10.38, where the average atomic number of the particles that initiate showers in the atmosphere is plotted as a function of the energy. This number increases significantly around the knee. Additional measurements have clearly confirmed that the knee is an exclusive feature of the proton component of the cosmic-ray spectrum [Ant 05].

This observation was one of the reasons for the decision to extend the high-energy reach of the experiment. In 2003, the Grande scintillator array was added, and the detectors now occupy an area of 0.5 km^2. An additional array of digital antennas is intended to study radio emission in air showers above 100 PeV. With these additions, KASCADE-Grande hopes to shed more light on the elemental composition of air showers in the energy range up to the one covered by TAP and Auger.

10.5   Outlook

In the 16 years that have passed since the first edition of this book was published, there has been a tremendous development in the studies of natural phenomena with calorimetric techniques. What was started by inquisitive individuals as a modest attempt to test new ideas has now grown and become a field of research where very large collaborations of scientists exploit huge detectors, and obtain spectacular results in the process. Neutrino detection has been demonstrated from the MeV region (solar and supernova neutrinos), via the few-GeV region (atmospheric neutrinos) all the way to the PeV region and beyond, in search for extragalactic neutrinos produced by unknown mechanisms. Several H_2O-based Čerenkov calorimeters with an instrumented volume of ~ 1 km^3 are continuously on the lookout. Cosmic-ray research has evolved and become a precision science, in which the spectra and the composition are now being studied all the way up to the 10^{20} eV range, where even the 3,000 km^2 Pierre Auger Observatory runs out of steam.

Because of the rate limitations in this interesting energy domain, where elementary particles carry energies in excess of the macroscopic unit of 1 Joule, NASA is considering an experiment (known as OWL) in which showers developing in the atmosphere are viewed from *above*, with satellite-based detectors that record stereoscopic information about these events [Str 98]. In this way, a fiducial detector volume of the order of 10^6 km^3 could be created. This would without a doubt constitute the largest natural calorimeter ever conceived.

11

CONTRIBUTIONS OF CALORIMETRY TO THE ADVANCEMENT OF SCIENCE

In the past decades, calorimeters have acquired a prominent place in particle physics experiments. Some reasons for this were already mentioned in previous chapters. As the energy increases, calorimeters perform better, because of the statistical nature of the processes that generate the signals. Also, the type of physics information that calorimeters can provide (*e.g.*, energy flow data) is better matched to the needs of experiments as the energy increases. And their capability to provide fast triggers is essential as selectivity becomes an increasingly crucial aspect of the experiments.

However, even all these factors combined do not explain the evolution in the design of particle physics experiments that has taken place in the past three decades. The fact that, on several important occasions, the described calorimeter features have turned out to make a decisive difference has contributed in no small part to this evolution as well.

In this final chapter, we describe some important scientific discoveries, all awarded with a Nobel prize, that became possible thanks to the calorimeters used in the experiments in which the discoveries were made. Such discoveries were the driving force behind the development of calorimeters into the sophisticated precision instruments they have become.

And this is how it should be. The continued development of particle detectors into more and more powerful instruments is not a goal in itself, it is intended to increase the capabilities and the precision of the experiments in which they are applied. The possibility to study the properties of matter at increasingly higher temperatures and densities, and at smaller distance scales, thus goes hand in hand with the availability of increasingly sophisticated detectors that make it possible to make the most of these studies.

It is this symbiosis that has made possible the enormous improvement in our understanding of the most elementary structure of matter and of the reasons why it behaves the way it does.

11.1 Discovery of the Intermediate Vector Bosons

The *Standard Model of Particle Physics* is nowadays textbook material. One of the main reasons why this model is universally accepted is the success story of the intermediate vector bosons.

11.1.1 *Historical context*

Ever since the discovery of nuclear β-decay, physicists have been extremely interested in the nature of the weak interaction which mediates this process.

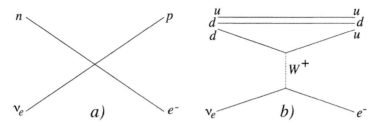

FIG. 11.1. Feynman diagrams for nuclear β-decay as seen by Fermi (a) and in the Standard Model of particle physics (b).

The first complete and coherent theory of nuclear β-decay was formulated by Enrico Fermi, in 1934. The Feynman diagram of the four-fermion interaction, proposed by Fermi, is shown in Figure 11.1a. The four particles involved in the fundamental process

$$n \rightarrow p + e^- + \bar{\nu}_e \qquad (11.1)$$

react with each other at a single point in space–time. Mathematically, this is expressed by saying that, at this point, the wavefunction of the neutron (ψ_n) is transformed into that of a proton and that the wavefunction of the incoming neutrino (which is equivalent to the wavefunction of the outgoing anti-neutrino that is actually observed in this process) is transformed into that of an electron. The probability for the reaction is thus described by

$$M = G_F \left(\bar{\psi}_p \Gamma \psi_n \right) \left(\bar{\psi}_e \Gamma \psi_\nu \right) \qquad (11.2)$$

in which the Fermi coupling constant G_F determines the timescale, and thus the rates, of the weak interaction processes, and the interaction factors Γ contain the fine details of the effects that give rise to the transformation of the particles.

The factors Γ, which take the form of a mixture of vector and axial-vector quantities, could be chosen so as to accommodate the phenomena of parity violation, discovered in 1956 in the β-decay of ^{60}Co.

This description turned out to be remarkably successful. For example, it turned out that the numerical value of G_F, determined from the purely leptonic weak decay process

$$\mu^- \rightarrow e^- + \nu_\mu + \bar{\nu}_e \qquad (11.3)$$

was within experimental errors (few %) equal to that of the nuclear decay process (Equation 11.1), thus proving that these two very different processes are indeed governed by one and the same fundamental interaction.

Fermi's theory was also very useful for calculating the cross sections of interactions induced by neutrinos. Such calculations provided valuable guidance for the experiment in which (anti-)neutrinos were first experimentally observed (by Reines and Cowan, who used the Savannah River nuclear reactor as their source [Rei 59]). Also the rate calculations for Davis' solar neutrino experiment, operating since 1967 [Dav 68], were based on Fermi's model.

Ironically, the neutrino scattering cross sections also made it clear that Fermi's theory was incomplete. This can be illustrated with the purely leptonic neutrino scattering process

$$\nu_e + e^- \rightarrow e^- + \nu_e \qquad (11.4)$$

By inserting the mathematical expression for the wavefunctions and interaction factors into the interaction amplitude M (see Equation 11.2), it is possible to calculate the cross section (σ_{lab}) for this process in the laboratory reference frame. The final answer is particularly simple when the neutrino energy, E_ν, is large compared with the electron mass, m_e:

$$\sigma_{\text{lab}} = \sigma_0 \frac{E_\nu}{m_e} \qquad (11.5)$$

where σ_0 is a constant factor, which sets the scale for the interaction rates that may be expected:

$$\sigma_0 \approx 10^{-44} \text{ cm}^2$$

For incoming neutrinos with energies of a few MeV, such as those in the experiments mentioned above, the cross section is minute, $\sim 10^{-43}$ cm^2. To illustrate how small this is, consider that Ray Davis' solar neutrino experiment, which was based on the transformation

$$\nu_e + {}^{37}\text{Cl} \rightarrow {}^{37}\text{Ar} + e^- \qquad (11.6)$$

registered only of the order of a *few events per week*, in a detector containing 615 tons of C_2Cl_4, while the cross section for this particular reaction is even three orders of magnitude larger ($\sim 10^{-40}$ cm^2).

However, Equation 11.5 shows that the cross section increases linearly with the neutrino energy. For example, at energies of about 50 GeV, easily achievable in modern particle accelerators, the cross section for the process described in Equation 11.4 amounts to $\sim 10^{-39}$ cm^2, and purely leptonic neutrino scattering has indeed been experimentally observed at these energies, albeit with ν_μ as projectiles [Gei 91].

As the neutrino energy increases further, so does the cross section. However, this cannot continue until arbitrarily high energies. We must therefore accept that Equation 11.5 gives only an approximate description of reality, and that this approximation loses its validity at very high energies.

To solve this problem, it is necessary to abandon the idea of a four-fermion point-like interaction and replace it with a particle exchange mechanism. This also puts the description of weak interaction processes on a more common footing with the theories of electromagnetism and of the strong nuclear force.

Electromagnetic interactions between charged particles are, in the field-theoretical formalism known as *Quantum Electro Dynamics* (QED), described as the exchange of a field quantum (a photon) between the interacting particles. Similarly, the strong force is usually described in terms of an exchange of pions (in nuclei) or gluons (inside hadrons) between the interacting particles.

The field quanta of the weak interaction are called the *intermediate vector bosons*, and they have been attributed the symbol W (for weak). The essential difference between this description and Fermi's is that the currents involved in a weak interaction process no longer couple directly to each other in a single space–time point. Instead, each current couples to the W-boson wavefunction at different space–time points and the W boson mediates the interaction between the two currents. This is illustrated in Figure 11.1b, for the nuclear β-decay process.

In order to describe processes such as nuclear β-decay, or the decay of muons, pions, *etc.*, the W boson must come in both negatively and positively charged versions: W^+ and W^-. In these processes, a charged lepton (e, μ) couples to a neutral one $(\nu_e, \bar{\nu}_e, \nu_\mu, \bar{\nu}_\mu)$ in the leptonic interaction vertex, and therefore these are called *charged-current* processes.

In order to describe *neutral-current* processes, a neutral version of the intermediate vector boson is needed as well. This particle, which is exchanged in certain reactions induced by neutrinos $(\nu + p \rightarrow \nu + X)$, became known as the Z^0.

The discovery of the intermediate vector bosons $W^\pm$ and Z^0 was anticipated for many years. Initially, it was generally believed that their masses were so large that the possibility of ever producing and detecting these particles was completely excluded. However, as the theory of weak interactions was further developed, and relevant experimental data, in particular on neutrino scattering, became available, it became clear that these masses could be predicted rather accurately .

In the late 1960s and early 1970s, the efforts of Glashow, Weinberg, Salam, Veltman and 't Hooft led to the formulation of a unified theory for the weak and electromagnetic interactions. This theory describes the interactions of leptons by the exchange of intermediate vector bosons and photons. It incorporates the Higgs mechanism, which describes the phenomenon of spontaneous symmetry breaking, in which all particles acquire their mass (see Section 11.3).

This theory, which became known as the Weinberg–Salam model, contained one free parameter, the so-called Weinberg angle, θ_W. The authors were able to show that the masses of the intermediate vector bosons could be expressed in terms of this parameter, as follows:

$$M_{W^\pm} = \frac{38.5}{\sin \theta_W} \text{GeV}/c^2 , \qquad M_{Z^0} = \frac{M_{W^\pm}}{\cos \theta_W} \qquad (11.7)$$

The parameter θ_W turned out to be also related to measurable experimental quantities, such as the ratios of the cross sections for neutral-current and charged-current interactions between (anti-)neutrinos and nucleons.

As a matter of fact, the Weinberg–Salam model explicitly predicted the occurrence of neutral-current interactions, which had never been observed until then. The experimental discovery of ν_μ-induced neutral-current events (1973) represented a major triumph for the model, which contributed greatly to its general acceptance.

In 1979, Glashow, Weinberg and Salam received the Nobel prize for their work[43], and few people doubted the correctness of their predictions with regard to the masses of the intermediate vector bosons, which had yet to be discovered.

In the meantime, all experimental data combined had led to a fairly precise measurement of the value of the Weinberg angle:

---

[43] In 1999, Veltman and 't Hooft were laureated for their contributions.

$$\sin^2 \theta_W = 0.232 \pm 0.002$$

On the basis of Equation 11.7, the masses of the W and Z bosons were thus expected to be about 80 ± 1 GeV/c^2 and 90 ± 1 GeV/c^2, respectively.

Given these very precise predictions, it was not surprising that in the years between 1976 and 1983 an enormous experimental effort was mounted to detect these eagerly awaited particles. The LEP Collider, by far the largest accelerator ever built, was planned and approved in this period, and this gigantic project was justified exclusively by the prospect of producing intermediate vector bosons, through the reactions

$$e^+ + e^- \rightarrow Z^0 \ (E \approx 90 \text{ GeV})$$
$$e^+ + e^- \rightarrow W^+ W^- \ (E \approx 160 \text{ GeV}) \tag{11.8}$$

However, in the mid-1970s, physicists realized that a less expensive way of searching for the W and Z could be provided by quark–anti-quark annihilation, and that collisions between protons and anti-protons were a promising venue to achieve the high center-of-mass energies needed for the production of these very massive objects.

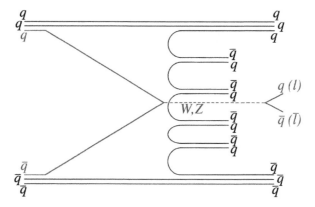

FIG. 11.2. Example of a Feynman diagram for the production of an intermediate vector boson in proton–anti-proton collisions.

In such reactions, one of the quarks inside the proton (which has the quark structure uud) annihilates with one of the anti-quarks contained in an anti-proton ($\bar{u}\bar{u}\bar{d}$). When the quark–anti-quark pair is dissimilar ($u\bar{d}$ or $\bar{u}d$), a W may be produced, while a similar pair ($u\bar{u}$ or $d\bar{d}$) may lead to the production of a Z^0. This is illustrated in Figure 11.2.

Due to the presence of spectator (anti-)quarks which may fragment into a large number of particles, the final states of the $p\bar{p}$ collisions in which these processes take place are usually extremely messy (Figure 11.3). Moreover, the production of intermediate vector bosons takes place in only a very small fraction of the proton–anti-proton collisions, even if the process is energetically possible. This is because the annihilating quark–anti-quark pair must have exactly the right properties for this process to occur. For example, the $q\bar{q}$ center-of-mass energy must equal the boson's rest mass to within a few percent.

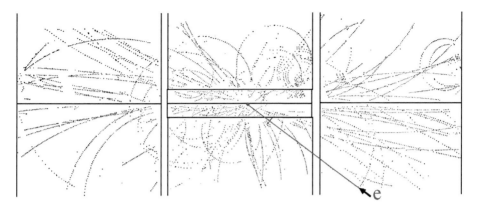

FIG. 11.3. Example of an event in which a W boson is produced in a $p\bar{p}$ collision. This event
was recorded in the UA1 detector at the CERN $p\bar{p}$ collider. The high-$p_\perp$ electron produced
in the W decay is indicated with an arrow [Arn 83a].

Despite these expected difficulties, the idea was too attractive to ignore. In 1976,
C. Rubbia and coworkers suggested that by converting an existing high-energy proton
synchrotron into a proton–anti-proton collider, enough center-of-mass energy could be
amassed at the constituent level to make the production of intermediate vector bosons
energetically possible.

After attempts to sell the idea to the Fermilab management failed, the idea was
picked up by CERN, and the decision was made to use the Super Proton Synchrotron
(SPS) for this purpose. When operating in its "fixed-target" mode, the SPS could accelerate protons to 450 GeV. For operation in the "collider" mode, in which the particles
would be stored in the vacuum ring for extended periods of time (sometimes several
days), and the magnets keeping the particles in orbit would have to operate at a constant
field strength during that period, a lower energy was chosen: 270 GeV.

The elegance of the proton–anti-proton collider idea is that the same machine that
accelerates protons in one direction can also accelerate anti-protons in the opposite direction. This is possible because anti-protons have exactly the same mass as protons,
but the opposite charge. Therefore, the single vacuum ring of the original SPS could accommodate the two counter-rotating beams of protons and anti-protons. The two beams
intersected each other at two points in the ring, and that is where the experiments named
UA1 and UA2 were mounted. The new colliding-beam machine created this way became known as the Sp$\bar{p}$S.

Although the basic idea behind the Sp$\bar{p}$S is very simple, there was one major obstacle on the way to its realization: anti-protons do not occur in nature. In order to
overcome this obstacle, an anti-proton "factory" was built, under the leadership of S.
van der Meer. The anti-protons were produced in collisions induced by protons from a
smaller accelerator at the CERN site, the 26 GeV Proton Synchrotron (PS). The cross
section for $\bar{p}$ production reaches its maximum for particles of about 4 GeV. Anti-protons
of this energy were collected at a rate of about 10^7 per second. These anti-protons were

accumulated in a small storage ring with large aperture, the Anti-proton Accumulator (AA).

Because of the way in which they were produced, the stored anti-protons exhibited a large dispersion in terms of their energies and their momentum vectors. Because of this, only a very small fraction could be expected to survive injection into the main accelerator complex, which only accepted particles with four-vectors within a very narrow range of values.

To overcome this problem, the dispersion in the four-vectors of the anti-protons stored in the AA was considerably reduced, through a process known as *stochastic cooling*. Deviations from the desired ideal orbit were detected with a sophisticated control system. In response, this system generated a signal that flashed across the diameter of the AA ring and applied a correcting magnetic pulse that forced the anti-protons, which arrived only nanoseconds later, closer to the ideal orbit.

Each new bunch of anti-protons was treated this way and could be sufficiently "cooled" during the two seconds between subsequent PS bursts. Following this procedure, the 10^{12} anti-protons needed to achieve the desired luminosity in the Sp$\bar{p}$S were collected in about two days. When enough anti-protons had been collected, they were injected into the PS/Sp$\bar{p}$S complex and accelerated in two steps from 4 GeV via 26 GeV to 270 GeV. At the same time, protons were injected in the opposite direction. This procedure thus resulted in counter-rotating proton and anti-proton bunches which collided head-on at two positions in the ring, with a center-of-mass energy of 540 GeV.

This energy was about six times higher than the threshold energy needed for intermediate vector boson production. However, this excess was very much needed, since the annihilating quarks and anti-quarks typically carried only a small fraction of the $p, \bar{p}$ momentum. In fact, as illustrated in Figure 11.4, the collision energy at the Sp$\bar{p}$S was fairly marginal, especially for the production of the heavier Z^0 particles. At Fermilab's Tevatron, which had a center-of-mass energy of 1.96 TeV (*i.e.*, 3.6 times the Sp$\bar{p}$S value), the cross section for Z^0 production was measured to be larger by a factor of about four. And at present (2016), the LHC produces Ws and Zs at rates that are larger by yet another order of magnitude.

Nevertheless, both the $W^{\pm}$ and the Z^0 were discovered at the first collider of this type, the Sp$\bar{p}$S. This happened in 1983 and may be considered one of the highlights of twentieth-century physics.

11.1.2   *The detectors*

Detecting the W and Z bosons, should these be produced in the $\bar{p}p$ collisions, was the single most important design criterion for both the UA1 and UA2 detectors. Both experiments emphasized electron identification and excellent calorimetry in the central region ($-1 < \eta < 1$). UA1 emphasized, in addition, also muon identification and precise measurements of the four-vectors of these particles.

Figure 11.5 shows a schematic cross section of the UA2 detector, in a plane through the beam line. Unlike almost all current 4π detectors operating at storage rings, UA2 did not have a central magnetic field. This simplified the design considerably and made it possible to construct a very compact detector, with minimal dead space.

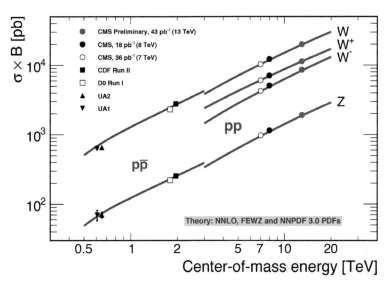

FIG. 11.4. The product of the production cross section and the leptonic branching ratio for Ws and Zs in $p\bar{p}$ and pp collisions, as a function of the center-of-mass energy. Multiplied by the integrated luminosity (in pb^{-1}), the vertical scale gives the production rates of the leptonically decaying vector bosons.

The central region of the detector, *i.e.*, the polar angle interval $40° < \theta < 140°$, was covered over the full azimuth by the central calorimeter [Bee 84]. Since no momentum measurements could be carried out in this region, the energy measurement of the particles relied entirely on the calorimeter. To identify electrons, a preshower detector consisting of a $1.5X_0$ tungsten converter, followed by a multi-wire proportional chamber (C5, see Figure 11.5), was installed in front of this central calorimeter. The tracking system upstream of the PSD/calorimeter combination consisted of four multi-wire proportional chambers and a scintillator hodoscope.

The central calorimeter (see Figure 1.7) was segmented into 240 independent cells, each covering $10°$ in θ and $15°$ in ϕ and built in a tower structure pointing to the center of the interaction region. Each cell was longitudinally segmented into a $17X_0$ deep em section (lead/plastic-scintillator), followed by two hadronic compartments of $2\lambda_{int}$ each (iron/plastic-scintillator). The scintillation light from each compartment was channelled to two PMTs by WLS plates located on opposite sides of each cell. This part of the detector was very hermetic.

The forward regions ($20° < \theta < 37.5°$ and $142.5° < \theta < 160°$) were instrumented as magnetic spectrometers, followed by an em calorimeter. Twelve coils, equally spaced in azimuth, generated a toroidal magnetic field with an average bending power of 0.38 Tm. Each sector was instrumented with 3 drift chambers, a PSD ($1.4X_0$ absorber followed by two layers of proportional tubes) and an em calorimeter. The latter, a lead/plastic-scintillator sandwich structure, was longitudinally subdivided in a $24X_0$ thick part, followed by a $6X_0$ part that was used to provide hadron rejection. Laterally,

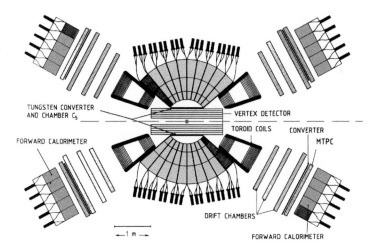

FIG. 11.5. A view of the UA2 detector in a plane containing the beam line [Bag 84]. A picture of the calorimeter is shown in Figure 1.7.

this calorimeter had a tower structure pointing to the vertex region. Each tower covered 3.5° in polar angle and 15° in azimuth.

All calorimeter cells were exposed to 10 GeV electron, pion and muon beams before the final calorimeter was assembled. The calibration constants were derived from these tests. In between Sp$\bar{\text{p}}$S runs, a significant fraction of the central calorimeter cells was recalibrated. Results of this procedure are discussed in Section 6.6.1.

11.1.3   Detecting the bosons

The lifetime of the intermediate vector bosons is very short, far too short to leave measurable tracks in particle detectors. Once produced in the $p\bar{p}$ collisions, the Ws and Zs decay almost instantaneously, either in a quark–anti-quark or a lepton–anti-lepton pair. As is often the case in particle physics, the production of intermediate vector bosons in the collisions can thus only be inferred from the four-vectors of their decay products. These four-vectors make it possible to determine the invariant mass (M) of the decaying object, as follows:

$$M = \sqrt{\left(\sum_{i=1}^{n} E_i\right)^2 - \left(\sum_{i=1}^{n} \mathbf{p}_i\right)^2} \tag{11.9}$$

in which E_i and $\mathbf{p}_i$ denote the energy and the momentum vector for the ith of the n decay products.

It turned out that the easiest way to detect the W and Z in this type of experiment is through their leptonic decay modes:

$$W^+ \rightarrow e^+ + \nu_e, \quad W^+ \rightarrow \mu^+ + \nu_\mu$$
$$W^- \rightarrow e^- + \bar{\nu}_e, \quad W^- \rightarrow \mu^- + \bar{\nu}_\mu \tag{11.10}$$

for the $W^\pm$ bosons, and

$$Z^0 \rightarrow e^+ + e^-, \quad Z^0 \rightarrow \mu^+ + \mu^- \qquad (11.11)$$

for the Z^0.

The rest masses of the decay products are totally insignificant compared with the mass of the decaying objects. The mass difference is thus almost entirely transferred in the form of kinetic energy to the decay products. These therefore acquire kinetic energies of typically 40–50 GeV, for vector bosons produced *at rest* in the laboratory system, more if the vector bosons themselves obtained kinetic energy in the production process.

However, because of the marginal kinematic conditions (see Figure 11.4), most Ws and Zs were produced practically at rest in these initial experiments at the Sp$\bar{p}$S. Therefore, their decay products were almost always emitted back-to-back in the laboratory system. And since the decay is, to a first approximation, isotropic, many of the leptons produced in the decay of these heavy particles carried a large momentum component in the direction perpendicular to the axis of the $p\bar{p}$ collisions.

This *transverse momentum* ($p_\perp$) turned out to be a very powerful tool for selecting events in which vector boson production had occurred, since it represented a unique feature, absent in the more common interactions.

Both UA1 and UA2 had implemented a trigger that was sensitive to electrons of high $p_\perp$ in their data acquisition systems. In UA2, the PMT gains of all calorimeter cells were adjusted so that their signals were proportional to the deposited transverse energy. Since showers initiated by electrons could be shared among (up to four) different calorimeter cells, trigger thresholds were applied to linear sums of signals from matrices of 2×2 cells. Therefore, in the central calorimeter, all possible 2×2 matrices were considered. The trigger threshold was typically set at 10 GeV. Whenever any 2×2 matrix produced a signal in excess of this threshold, the event was recorded, provided that also a signal was observed in the proper segment of the scintillator hodoscope that was part of the tracking system. The latter requirement was made to reject background from sources other than $\bar{p}p$ interactions in the vertex position.

Offline, all recorded events were individually examined to see if the high-$p_\perp$ particle that triggered the data acquisition system was possibly an electron. In UA2, high-$p_\perp$ electrons were required to satisfy the following criteria:

- Almost all energy needed to be deposited in the first, em compartment of the calorimeter, with only little leakage into the deeper compartments.
- There should be a reconstructed track pointing from the vertex region to the calorimeter cell(s) where the energy deposit occurred.
- There should be a signal in the PSD at the position expected on the basis of the extrapolated charged-particle track. This signal should be significantly larger than that from a mip.

If the first and/or third criterion were not fulfilled, then the event was classified as a high-$p_\perp$ *jet*. If there was no charged-particle track pointing to the hit calorimeter region, then the event was classified as a high-$p_\perp$ π^0.

Figure 11.6a shows the $p_\perp$ distribution of 2,444 electron candidates accumulated in this way by UA2 during the years 1982–1984. The integrated luminosity was 452 nb^{-1}

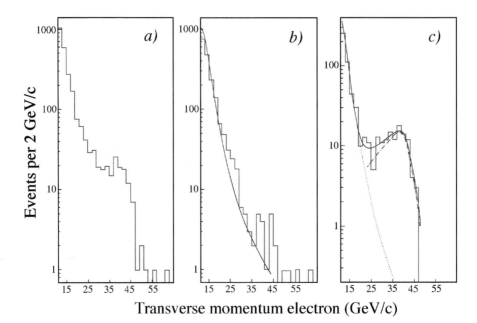

FIG. 11.6. Transverse momentum distributions of all electron candidates with $p_\perp > 11$ GeV/c for the 1982–1984 UA2 data (a), and for electron candidates in events with small (b) or large (c) $p_\perp$ imbalance. The background curve superimposed on histogram b applies to a sample of π^0 events, the dotted curve superimposed on histogram c represents the calculated background from QCD processes. See text for details. Data from [Bag 86].

in this period. This distribution dropped by two orders of magnitude in the region from $p_\perp = 10$ to 30 GeV/c, but then reached a plateau that extended to $p_\perp \sim 45$ GeV/c.

The exponential drop was in excellent agreement with expectations, based on QCD calculations. It was also experimentally verified with a sample of events in which π^0s were selected instead of electrons. In that case, there was *no* charged track pointing to the calorimeter region where the energy deposit took place. The curve in Figure 11.6b describes the results of that measurement. The experiment thus measured a clear excess of high-$p_\perp$ electrons with transverse momenta from 30 to 45 GeV/c. Similar results were obtained by UA1.

Apart from events in which an isolated high-energy electron was produced, UA1 and UA2 also searched for another type of events, namely those with a large missing transverse energy. In the leptonic decay of the Ws, (anti-)neutrinos are produced (Equation 11.10) in conjunction with the charged leptons. Since this is a two-body decay, the transverse momentum distribution of these (anti-)neutrinos is identical to that of the charged leptons. And since the rest mass of the neutrinos is vanishingly small, their transverse energy ($E_\perp$) distribution is almost the same.

However, these (anti-)neutrinos escaped the detector without interacting in it, which led to an imbalance in the transverse energy. Events characterized by a large imbalance in the transverse energy would thus also be indicative of the production and subsequent leptonic decay of W bosons.

The missing transverse energy can be determined from the energy flow, as follows (see also Section 5.5.1). Each energy deposit in a calorimeter cell can be expressed by a vector $(\mathbf{\Delta E})_i$, of which the amplitude is given by the energy deposit $(\Delta E)_i$, and the direction is determined by the spatial position of the cell with respect to the interaction region. By adding all energy deposits vectorially, the total energy flow vector $\mathbf{\Delta E}$ is obtained. Ignoring particle masses, momentum conservation requires $\Delta E = 0$. If there is a momentum imbalance because of the production of an energetic neutrino, then this particle has a momentum vector $-\mathbf{\Delta E}$. In practice, both UA1 and UA2 limited the measurements to the projection of $\mathbf{\Delta E}$ on a plane perpendicular to the beam line, i.e., the missing *transverse* energy, or the transverse energy (momentum) of the escaping ν. That is because in this type of experiment the longitudinal component of the missing energy is prone to fluctuations in undetected energy, caused by particles escaping the detector near the beam line.

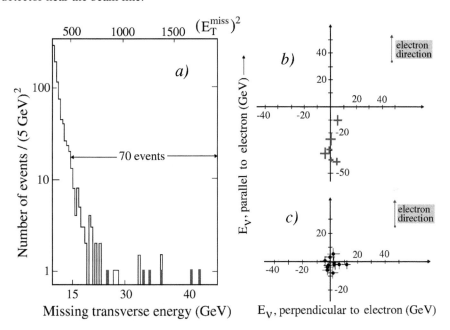

FIG. 11.7. Distribution of the missing transverse energy for events with a high-$p_\perp$ candidate recorded in the UA1 detector (*a*). The missing transverse energy plotted vectorially against the electron direction for events in which a high-$p_\perp$ electron candidate was (*c*) or was not (*b*) accompanied by jets [Arn 83a].

Figure 11.7 shows experimental data from the first UA1 paper on this topic [Arn 83a]. UA1 used criteria very similar to the ones discussed above to select events with a high-$p_\perp$ isolated track that behaved as if it was an electron. They found 39 events that had an electron candidate with $p_\perp > 15$ GeV/c. Each of these events was examined in great detail. It turned out that in six of these events, the electron was *not* accompanied by jets, *i.e.*, there were no other high-$p_\perp$ tracks detected in the same event.

The missing $E_\perp$ distribution for all events with an electron candidate with $p_\perp > 7$ GeV/c is shown in Figure 11.7a. The six "jetless" events mentioned above are highlighted in this plot. They all appeared in the high-end tail of this distribution. These six events were also in other ways strikingly different from the rest.

In Figure 11.7b, the missing transverse energy of these six events is plotted vectorially against the direction of the electron, *i.e.*, $p_\perp^e$. In all six cases, the events had a missing transverse energy that was about equal to the transverse electron energy. Moreover, the two vectors balanced each other almost exactly back-to-back. In other words, these six events were characterized both by the production of an electron with large transverse momentum and an equally large missing transverse energy. The two vectors representing these quantities pointed in opposite azimuthal directions.

The situation was quite different for the events in which the high-$p_\perp$ candidate was accompanied by significant "jet activity." These events exhibited virtually no missing $E_\perp$, as illustrated in Figure 11.7c. Most likely, the electron candidate was in fact a misidentified jet in these events.

The UA2 experiment had similar experiences in analyzing the missing-$E_\perp$ characteristics of their events. They split the event sample shown in Figure 11.6a into two subsamples, according to the measured momentum imbalance. Figure 11.6b shows the $p_\perp$ distribution of the electron candidates in the subsample of events with very little or no momentum imbalance. This subsample represented $\sim 75\%$ of all events from Figure 11.6a. The curve superimposed on the histogram in Figure 11.6b was obtained from a sample of events in which the high-$p_\perp$ particle was classified as a π^0, since no charged track pointed to the calorimeter cell where the energy deposition was recorded.

The $p_\perp$ distribution for the remaining 25% of the events, which did exhibit significant missing transverse energy, is shown in Figure 11.6c. Almost all events that gave rise to the bump between 25 and 45 GeV/c are contained in this plot. The dotted curve superimposed on the histogram represents the background expected from QCD processes, whereas the dash-dotted line describes the expected contribution from $W \to e\nu$ decays.

Also here, the $\bar{p}p$ collisions in which a leptonically decaying W boson was produced were thus characterized by the simultaneous occurrence of a high-$p_\perp$ electron and large missing transverse energy, a very unique and selective signature. Thanks to the calorimeters, these events could be recognized, even though they were outnumbered, by eight orders of magnitude, by other types of $\bar{p}p$ interactions.

Because of the large mass difference between the W and its decay products, the W mass can be determined from the relationship

$$M_W = \sqrt{2p^e p^\nu (1 - \cos \vartheta)} \tag{11.12}$$

where ϑ is the angle between the momentum vectors of the electron and the neutrino. However, since all quantities were measured in a plane perpendicular to the beam line, UA1 and UA2 determined the *transverse mass* of the W:

$$M_\perp(W) = \sqrt{2p_\perp^e p_\perp^\nu (1 - \cos \Delta\phi)} \qquad (11.13)$$

where $\Delta\phi$ represents the azimuthal separation between the electron and neutrino vectors.

A Monte Carlo simulation program was used to generate $M_\perp$ distributions for different values of M_W, and in this way the measured $M_\perp$ distribution could be used to determine the value of M_W. The simulations included the effects of (longitudinal and transverse) W momenta, which depend on the proton structure functions, as well as the effects of experimental cuts that were used to select electrons, missing energy, *etc.* Some of these effects contained intrinsic uncertainties, which led to an irreducible systematic uncertainty in the final result.

In January 1983, UA1 and UA2 simultaneously announced the discovery of the W boson [Arn 83a, Ban 83]. This discovery was based on a combined total of 10 events of the type $W \to e\nu$ (6 for UA1, 4 for UA2). The W mass obtained from reconstructing these decays was 81 ± 5 GeV/c^2 for UA1 and 80^{+10}_{-6} GeV/c^2 for UA2, in excellent agreement with the predictions of the Weinberg–Salam model.

Six months later, both experiments announced the discovery of the Z^0 [Arn 83b, Bag 83]. Because of its larger mass, the cross section for production of the neutral vector boson was smaller, by about a factor of ten, in the Sp$\bar{p}$S (see Figure 11.4). Therefore, it simply took longer to amass enough integrated luminosity to produce a few leptonically decaying Z^0s.

However, the signature of these events was, if possible, even more spectacular than that of the W decays: two electrons, each with very large transverse momentum, emitted back-to-back, so that the transverse momenta canceled each other.

When this discovery was announced, both experiments had observed four such events. In addition, UA1 had observed one event of the type $Z^0 \to \mu^+\mu^-$. The background contribution from "conventional" processes (*i.e.*, other than Z^0 decay) to this signal was estimated to be 0.03 events by UA2. The Z^0 mass obtained from reconstructing these decays was given as 95.2 ± 2.5 GeV/c^2 by UA1 and as 91.9 ± 1.3 (stat.) ± 1.4 (syst.) GeV/c^2 by UA2.

11.1.4 *Epilogue*

In the years following the discovery of the intermediate vector bosons in 1983, the experimental information about these particles has increased to a level that makes them rank among the most detailed studied objects in nature.

The contributions of the Sp$\bar{p}$S to this process were rather limited. As illustrated by Figure 11.4, the available center-of-mass energy was relatively marginal, and therefore the production rates of vector bosons were low.

When Fermilab's Tevatron, a $p\bar{p}$ collider with a center-of-mass energy of 1.8 TeV, became operational in 1988, there were no compelling reasons to extend the life of the Sp$\bar{p}$S machine much longer[44]. It had reaped the fruits of the discovery, and the SPS ring

---

[44] After completion of the Main Ring Injector (1993), the Tevatron's center-of-mass energy was increased to 1.96 TeV.

was needed for other purposes, for example to supply LEP with electrons and positrons.

When the Sp$\bar{p}$S was shut down in 1990, the combined boson harvest of UA1 and UA2 included $\sim 5,000$ Ws and ~ 400 Z^0s. These numbers have since been exceeded by two orders of magnitude by both the CDF and D0 experiments, operating at the Tevatron.

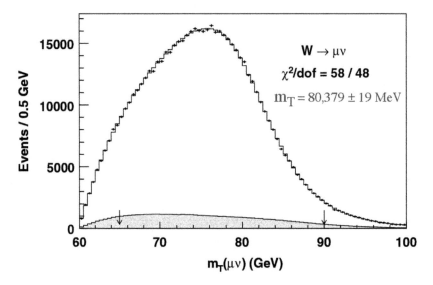

FIG. 11.8. The transverse mass distribution of the W boson, reconstructed from $W \rightarrow \mu\nu$ decays recorded by CDF. The arrows indicate the range over which the fit was made [Aal 12a].

Figure 11.8 shows the transverse mass distribution for $\sim 620,000$ $W \rightarrow \mu\nu$ events obtained with CDF. A W mass value of 80.379 ± 0.019 GeV/c^2 (stat.) was derived from this distribution. This is still the most precise measurement of this parameter to date.

LEP dwarfed all other sources of experimental information with regard to the Z^0. The four experiments operating at the 27 km e^+e^- collider at CERN collected in total some 17 million events of the type $e^+ + e^- \rightarrow Z^0$, and studied all possible aspects of the production and decay of this object [LEP 06]. The mass of this particle is now known with an experimental precision of better than 1 part in 40,000 (91.1876 $\pm$ 0.0021 GeV/c^2). Whereas the $p\bar{p}$ experiments only concentrated on two leptonic decay modes (Equation 11.11), the LEP experiments have studied several hundred other decay modes as well. They even obtained very important information about decay modes that cannot be directly observed:

$$Z^0 \rightarrow \nu\bar{\nu} \tag{11.14}$$

From the total width of the Z^0 resonance, and the associated *invisible width* due to processes of the type 11.14, the number (N) of different neutrino species (with masses smaller than half the Z^0 rest mass) could be measured, with the following result [PDG 14]:

$$N = 2.92 \pm 0.05,$$

while Standard Model fits to all experimental LEP data give

$$N = 2.984 \pm 0.008$$

This is a very strong indication that the three generations of quarks and leptons known today indeed form the complete set.

There is a well known adage in physics: "Yesterday's discovery is today's background and tomorrow's calibration." This most certainly applies to the intermediate vector bosons. As described in Sections 6.5 and 6.6, these particles now play a very important role in the calibration of the ATLAS and CMS calorimeters at the Large Hadron Collider.

11.2 Atmospheric Neutrino Detection

Particle physics started in the 1940s when balloon experiments revealed the existence of particles that did not fit into the scheme of elementary building blocks of nature known at that time. These particles, at that time all termed *mesons*, are now known as pions, muons and kaons. These particles are unstable, they decay on a timescale in the 10^{-6}s to 10^{-10}s range. They are produced in the upper layers of the Earth's atmosphere, in the interactions of highly energetic cosmic protons or heavier nuclei with the nuclei of air molecules. It is quite possible that neutrinos produced in the same processes may lead to another quantum leap in our understanding of the fundamental structure of matter.

In Section 10.3.1, the production of electron and muon (anti-)neutrinos in shower development in the Earth's atmosphere was discussed. Electron neutrinos are almost exclusively produced in the decay of muons, whereas muon neutrinos are, in addition, produced in the decay of the mesons that generate these muons. To a first approximation, one would thus expect that the neutrinos that reach the Earth's surface consist of one-third $(\nu_e + \bar{\nu}_e)$ and of two-thirds $(\nu_\mu + \bar{\nu}_\mu)$. However, in practice, the ratio $(\nu_e + \bar{\nu}_e)/(\nu_\mu + \bar{\nu}_\mu)$ is smaller than 1/2, because of the long decay length of the muons. The value 1/2 thus represents the low-energy limit, and as the energy increases, fewer and fewer electron neutrinos should thus be expected. Since the production processes and the particle properties are well understood, the precise composition and the spectra of the neutrinos and anti-neutrinos reaching the Earth's surface can be accurately modeled.

Figure 11.9a shows the (calculated) energy spectra of the atmospheric neutrinos interacting in the SuperKamiokande detector [Bar 89, Hon 90]. These spectra reach a maximum near 1 GeV. At that energy, the large majority of the muons decay in the atmosphere before reaching the Earth's surface, and the ν_e/ν_μ ratio is thus close to 0.5. As the energy increases, more and more muons reach the Earth's surface before decaying, and the ν_e/ν_μ ratio decreases. This effect is illustrated by Figure 11.9b [Bar 89].

Atmospheric neutrinos are produced in the entire shell of air surrounding our planet. Because of the small interaction cross sections, the probability that these neutrinos traverse the entire Earth without noticing its presence is close to 100%. Therefore, the neutrinos that do interact in a detector installed near the Earth's surface may originate

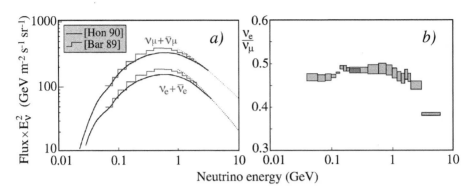

FIG. 11.9. The (calculated) flux of atmospheric neutrinos times E_ν^2 at the SuperKamiokande site (a) and the ν_e/ν_μ ratio (b) as a function of the neutrino energy. Data from [Bar 89] and [Hon 90].

from any region of the atmosphere, ranging from the area located a few kilometers above the detector to the sky above the antipodes, 13,000 km away. This is an ideal situation for studying the eventual occurrence of *neutrino oscillations*.

11.2.1   *Neutrino oscillations*

In the Standard Model, three types of neutrinos are distinguished: ν_e, ν_μ and ν_τ. These neutrinos are eigenstates of the weak interaction. They are produced in conjunction with their charged leptonic counterparts (e, μ, τ), in processes governed by the weak interaction.

However, if lepton number is not strictly conserved (and there is no compelling reason why it should be), mixing between the different neutrino species may occur. If, moreover, the neutrinos have a non-zero rest mass, there is absolutely no reason why the weak eigenstates (ν_e, ν_μ, ν_τ) should be identical to the mass eigenstates. In that case, so-called *neutrino oscillations* may occur.

This possibility was first envisaged in the 1960s [Mak 62, Pon 68], in analogy with phenomena observed in the neutral kaon system. In hadronic processes, where the *strangeness* quantum number is conserved, neutral kaons are produced as eigenstates of the strong interaction: K^0 ($d\bar{s}$ in terms of quark structure) or $\bar{K}^0$ ($s\bar{d}$).

However, these strong interaction eigenstates are not the same as the mass eigenstates, K_S^0 and K_L^0. If the small CP-violating effects are disregarded, K_S^0 and K_L^0 correspond to the CP eigenstates K_1 and K_2, which are distinguished through their lifetimes (90 ps and 52 ns, respectively) and their decay modes ($K_1 \rightarrow \pi\pi$, $K_2 \rightarrow \pi\pi\pi$).

The wave functions of the strong interaction eigenstates, $|K^0\rangle$ and $|\bar{K}^0\rangle$ can be expressed as linear combinations of $|K_1\rangle$ and $|K_2\rangle$, as follows:

$$|K^0\rangle \;=\; \frac{1}{\sqrt{2}}(|K_1\rangle + |K_2\rangle)$$

$$|\bar{K}^0\rangle \;=\; \frac{1}{\sqrt{2}}(|K_1\rangle - |K_2\rangle) \tag{11.15}$$

However, because of the mass difference between K_1 and K_2, the time dependence of the wavefunctions $|K_1\rangle$ and $|K_2\rangle$ is different as well. This means that the wavefunction of a particle produced as a pure K^0 state at time $t = 0$, at any later time t will have developed into a superposition of K^0 and $\bar{K}^0$:

$$|K\rangle(t) = \alpha(t)|K^0\rangle + \beta(t)|\bar{K}^0\rangle \tag{11.16}$$

with $\alpha(0) = 1$ and $\beta(0) = 0$.

This implies that a particle produced in a state with a well-defined strangeness loses this characteristic as it propagates in space and time. If this particle is made to interact in a downstream target, then it may produce either a particle with strangeness −1 (containing an s quark, e.g., a K^-) or with strangeness +1 (containing an $\bar{s}$ quark, e.g., a K^+), with relative probabilities determined by the wavefunction at that point.

A more detailed mathematical analysis of this phenomenon shows that these probabilities oscillate as a function of the time that has elapsed since the particle was produced. The frequency of this oscillation is determined by the mass difference, Δm, between the two mass eigenstates. This mass difference is very small and has been determined experimentally from the oscillating pattern observed in the K^0-induced production of hyperons (baryons containing an s quark) as a function of the distance traveled by the K^0s:

$$\Delta m = 3.52 \cdot 10^{-6} \text{ eV}/c^2$$

The theory of neutrino oscillations was worked out in great detail by Bruno Pontecorvo [Pon 68]. He argued that if neutrinos have a non-zero rest mass, there is no reason why the mass eigenstates (ν_1 and ν_2, the third generation of leptons had yet to be discovered then) would be identical to the weak-interaction eigenstates (ν_e and ν_μ).

The wavefunctions describing these weak-interaction eigenstates can then be expressed as some linear combination of the wavefunctions for the mass eigenstates:

$$\begin{pmatrix} |\nu_\mu\rangle \\ |\nu_e\rangle \end{pmatrix} = \begin{pmatrix} \cos\theta & \sin\theta \\ -\sin\theta & \cos\theta \end{pmatrix} \begin{pmatrix} |\nu_1\rangle \\ |\nu_2\rangle \end{pmatrix} \tag{11.17}$$

i.e., a unitary transformation involving an arbitrary mixing angle θ. The wavefunctions for the neutrinos produced in weak interactions (e.g., pion decay or nuclear β-decay) can thus be written as follows:

$$\begin{aligned} |\nu_\mu\rangle &= |\nu_2\rangle \sin\theta + |\nu_1\rangle \cos\theta \\ |\nu_e\rangle &= |\nu_2\rangle \cos\theta - |\nu_1\rangle \sin\theta \end{aligned} \tag{11.18}$$

This formalism can be extended in a straightforward way to a system involving three different sets of eigenstates.

Because of the different masses, the wavefunctions $|\nu_1\rangle$, $|\nu_2\rangle$ and $|\nu_3\rangle$ develop differently in time. Consider, as an example, a beam of neutrinos produced in nuclear

β-decay at time $t = 0$. The wavefunction of these particles was thus initially $|\nu_e\rangle$. However, as time goes by and the particles travel away from the source, admixtures of $|\nu_\mu\rangle$ and $|\nu_\tau\rangle$ become increasingly important components of the wavefunction:

$$|\nu\rangle(t) = \alpha(t)|\nu_e\rangle + \beta(t)|\nu_\mu\rangle + \gamma(t)|\nu_\tau\rangle \tag{11.19}$$

If we limit the problem, for simplicity's sake, to a situation that involves only two generations, then the probability of finding $|\nu_e\rangle$- or $|\nu_\mu\rangle$-induced (weak) interactions by particles that were produced as $|\nu_e\rangle$ of energy E and that have propagated in space over a distance L can be written as:

$$P(\nu_e \to \nu_e) = 1 - \sin^2 2\theta \sin^2 \left(\frac{1.27 \Delta m^2 L}{E} \right)$$

$$P(\nu_e \to \nu_\mu) = 1 - P(\nu_e \to \nu_e) \tag{11.20}$$

where the mass difference $\Delta m^2 = m_2^2 - m_1^2$ is expressed in units of $(eV/c^2)^2$, the distance L in meters and the neutrino energy E in MeV.

Therefore, if we have a beam of 10 GeV neutrinos and $\Delta m^2 = 0.1$ $(eV/c^2)^2$, then the oscillation length equals $2\pi \cdot 10^5/1.27 = 5 \cdot 10^5$ m, or 500 km. If these particles were produced as muon neutrinos, e.g., through the process

$$\pi^+ \to \mu^+ + \nu_\mu$$

and assuming maximum mixing (i.e., $\theta = 45°$), then the probability of finding ν_e interactions among those induced in a target 10 km away from the production source would be $\sin^2 (10/500) \sim 4 \cdot 10^{-4}$ (0.04%). This probability increases with the distance traveled by the neutrinos.

Next, consider a collection of muon neutrinos with a certain energy spectrum. Each energy corresponds to a different oscillation length. If these particles have traveled a distance corresponding to many oscillation lengths, then the probability of finding $|\nu_e\rangle$-induced interactions among those that occur in a target becomes equal to the average value of the $\sin^2$ function (assuming maximum mixing), i.e., 0.5. In other words, the oscillation is complete and the probabilities of finding $|\nu_e\rangle$- or $|\nu_\mu\rangle$-induced reactions are equal.

In the more general case of three generations, the probability of finding reactions induced by neutrinos of the same type that started the journey is reduced to 1/3, after many oscillation lengths. The fact that the number of nuclear reactions induced by solar neutrinos (e.g., $\nu_e + {}^{37}Cl \to {}^{37}Ar + e^-$) is about a factor of three smaller than predicted by certain (solar) models is sometimes considered experimental evidence for neutrino oscillations.

However, this type of evidence is certainly not conclusive, since it relies on a comparison between experimental data and calculations based on models characterized by certain speculative aspects, such as the temperature in the Sun's center. Also, the rates of reactions induced by the softer, and more abundant, component of the solar neutrino spectrum (e.g., $\nu_e + {}^{71}Ga \to {}^{71}Ge + e^-$) seem to be somewhat less in disagreement with the predicted rates from solar models.

Much more conclusive evidence could be expected if the measurements could, in addition, also be carried out at a much smaller distance from the neutrino source, *i.e.*, at a distance that is short in comparison with the oscillation length. Unfortunately, that is impossible in the case of solar neutrinos. However, for the atmospheric neutrinos this possibility does exist. The (Super)Kamiokande detector has been used to exploit this possibility, and the results are quite exciting.

11.2.2 *The original (Super)Kamiokande results*

The measurements discussed in this section were pioneered with the Kamiokande detector. In 1997, this detector was replaced by SuperKamiokande, which represented an increase in fiducial detector volume by a factor of about thirty.

Both detectors had a crucial performance characteristic in that they were capable of measuring the *direction* of the particles that entered the detector and interacted in its fiducial volume. This made it possible to estimate the distance traveled by the atmospheric neutrinos.

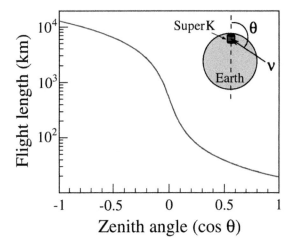

FIG. 11.10. Relationship between the zenith angle (θ) of the incident neutrino and the distance traveled by this particle before interacting in the SuperK detector [Oku 99].

The relationship between this distance and the zenith angle at which the atmospheric neutrino enters the SuperK detector is shown in Figure 11.10. Neutrinos coming from the region right above the detector have $\theta = 0°$, *i.e.*, $\cos\theta = 1$. These are called the *downward going* νs. The neutrinos coming from the opposite side of the Earth have traveled a distance of $\sim 13,000$ km when they interact in SuperK, and their zenith angle is 180° ($\cos\theta = -1$); these are the *upward going* νs.

It is important to realize that the measurements on interactions of atmospheric neutrinos in the SuperK detector do not yield the direction of the incoming neutrino, but rather the direction of the charged lepton produced in these interactions (in the case of charged-current processes). This means that the neutrino direction inferred from these

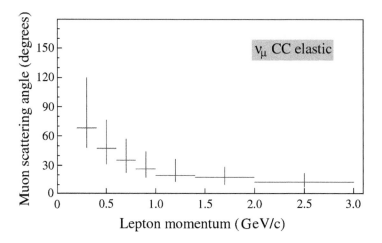

FIG. 11.11. The average angle between the incoming muon neutrino and the muon it produces in elastic charged-current scattering processes [Oku 99].

measurements has an intrinsic, energy-dependent uncertainty.

Figure 11.11 shows the relationship between the directions of the neutrino and the charged lepton for a process that dominates the total cross section at the low energies considered here: elastic charged-current scattering, *i.e.*, the processes $\nu_\mu + n \rightarrow \mu^- + p$ or $\bar{\nu}_\mu + p \rightarrow \mu^+ + n$. The angle between the incoming neutrino and the scattered muon is plotted as a function of the neutrino energy. The cross points correspond to the peak of the distributions: the vertical bars indicate the angular region where 68% of the events are found and the horizontal bars mark the energy bins in this plot. The figure shows that this scattering angle is typically about 25° for events in which 1 GeV/c leptons are produced, and that the angle is somewhere between 15° and 40° in 68% of the events. For lower energies, the intrinsic uncertainty in the neutrino direction increases; for higher energies it decreases.

The SuperKamiokande detector is discussed in considerable detail in Section 10.1. The detector features that are particularly important for the atmospheric-ν analysis are

1. its capability to distinguish muons from electrons, and

2. its capability to determine the direction of these charged leptons.

The distinction between muons and electrons is made on the basis of the characteristics of the Čerenkov ring. The techniques developed by the SuperK Collaboration for this purpose (see Section 10.1.5) are very accurate (Figure 10.14). For energies around 1 GeV, the misidentification probability is only $\sim 1\%$. The possibility to reconstruct the direction of the scattered lepton in SuperK is a direct benefit of the use of Čerenkov light as the source of experimental information. Elaborate Monte Carlo studies have shown that the directionality of the emitted Čerenkov light makes it possible to reconstruct the muon tracks or the axes of the em showers in events induced by atmospheric neutrinos with an average angular resolution of only a few degrees [Oku 99]. This instrumental effect is thus negligible compared with the intrinsic uncertainty in the direction of the neutrino itself.

During the first 1.5 years of operation, SuperK recorded in total about 2,700 one-ring, fully contained charged-current events induced by atmospheric neutrinos in the 22 kiloton fiducial detector volume. Remarkably, this event sample contained about equal numbers of "electron-like" and "muon-like" events, whereas the Monte Carlo simulations predicted that the ν_μ charged-current interactions should outnumber the ν_e ones by almost two to one (63% *vs.* 37%).

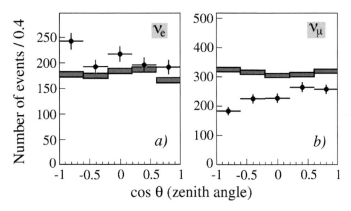

FIG. 11.12. Zenith-angle distribution of charged-current interactions induced by atmospheric ν_e (a) and ν_μ (b). The experimental data are indicated by the black dots; the boxes represent the Monte Carlo predictions, in the absence of ν oscillations [Oku 99].

The apparent "shortage" of ν_μ-induced charged-current events turned out to be much larger for the upward going neutrinos than for the downward ones. This is illustrated in Figure 11.12, which shows the zenith-angle dependence of the charged-current interaction rates of ν_es (Figure 11.12a) and ν_μs (Figure 11.12b). The experimental data are indicated by the black dots in this figure, and the boxes represent Monte Carlo predictions. These Monte Carlo calculations, which did not include any effects of neutrino oscillations, basically predicted the absence of any zenith-angle dependence. However, the measurements appeared to be at variance with this prediction. As the zenith angle, and thus the distance traveled by the neutrinos (Figure 11.10) increased, the rates of ν_μ-induced charged-current interactions gradually dropped. No such effect was observed for the ν_e-induced charged-current interactions. On the contrary, the event rate for these interactions seemed to be significantly larger than that predicted for upward going neutrinos.

Figure 11.12b shows an asymmetry between the rates of charged-current interactions induced by upward-going (here defined as $\cos\theta < -0.2$) and downward-going ($\cos\theta > 0.2$) ν_μs, with the latter outnumbering the former. The degree of this asymmetry turned out to be energy dependent (Figure 11.13): the larger the energy, the larger the asymmetry, expressed as the absolute value of the ratio (up − down)/(up + down),

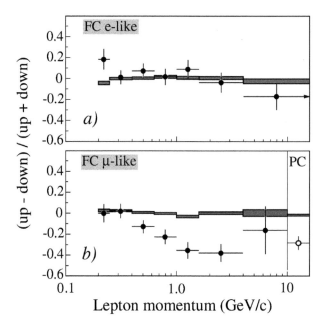

FIG. 11.13. Up/down asymmetry as a function of the charged lepton produced in charged-current interactions induced by atmospheric ν_e (a) and ν_μ (b). The experimental data are indicated by the black dots, and the boxes represent the Monte Carlo predictions, in the absence of ν oscillations. The open circle represents events that were partially contained in the detector; all other data points concern fully contained events [Oku 99].

was found to be. Again, no such effect was found for interactions induced by ν_es.

All these experimental observations could be explained by oscillations of the type $\nu_\mu \rightarrow \nu_\tau$. As the distance traveled by a neutrino produced as a ν_μ increases, so does the $|\nu_\tau\rangle$ component of its wave function (Equation 11.19). Therefore, the probability that it induces a muonic charged-current interaction gradually decreases as the zenith angle increases. At the same time, the probability that it undergoes a weak interaction as a ν_τ increases. However, to induce a charged-current interaction of this type, the energy must be sufficient to create the heavy τ lepton (mass 1.78 GeV/c^2). The τ lepton has an 18% probability of decaying into an electron. If that happened, then a charged-current ν_τ interaction would be classified as an "electron-like" event. A small increase of ν_e-induced charged-current events at large zenith angles, as experimentally observed (Figure 11.12a), would thus be consistent with this scenario.

Even though the SuperK data made it clear that neutrino oscillations do occur, and that neutrinos must thus be massive particles, other aspects of the interpretation of these experimental data were not so clear-cut. The mentioned oscillation scenario is by no means the only one that may explain the experimental data. Other possible schemes involve all three ν species, or even so-called "sterile" neutrinos. The measured energy dependence of the up/down asymmetry is in principle a sensitive tool to distinguish

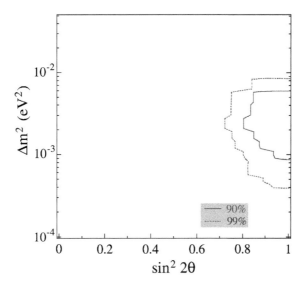

FIG. 11.14. The allowed region for the oscillation parameters Δm^2 and $\sin^2 2\theta$, in the $\nu_\mu \leftrightarrow \nu_\tau$ oscillation mode [Oku 99]. Several slightly different versions of this figure were published in subsequent SuperK analyses [Fuk 98, Kaj 99].

between different oscillation scenarios.

When the experimental SuperK results are interpreted in terms of $\nu_\mu \to \nu_\tau$ oscillations, it is possible to extract the oscillation parameters Δm^2 and $\sin^2 2\theta$ (see Equation 11.20) from the data. The SuperK Collaboration went through this exercise and published the outcome in the form of figures such as Figure 11.14, which shows the allowed region for these parameters.

The first and most important conclusion from this figure is that $\Delta m^2 \neq 0$, which means that at least one neutrino eigenstate must have a non-zero rest mass. The region of $0.002 < \Delta m^2 < 0.006$ (eV)^2 is allowed at a 90% confidence level. Therefore, at least one neutrino must have a mass in excess of ~ 0.04 eV/c^2. The second conclusion is that the mixing between the $|\nu_2\rangle$ and $|\nu_3\rangle$ mass eigenstates is large: $\sin^2 2\theta_{23} > 0.84$, with 90% probability [Kaj 99].

In 2015, the leader of this SuperK program, Dr. Takaaki Kajita, was awarded the Nobel Prize for physics for this important discovery.

11.2.3 Epilogue

After the official announcement of the evidence for neutrino oscillations [Fuk 98], on the basis of 535 days (33 kiloton yr) of data taking, SuperK has continued to collect data from atmospheric neutrinos, even during the enormous setback caused by the accident of November 2001. By now (2016), the total number of events recorded in 220 kiloton yr has increased by more than a factor of six, to $\sim 80,000$. The best reported limits on the parameter values of the $\nu_\mu \leftrightarrow \nu_\tau$ oscillation are, at a 90% confidence level, $\sin^2 2\theta_{23} > 0.92$ and $0.0015 < (\Delta m_{23})^2 < 0.0034$ (eV)^2 [Kaj 16]. The increased statistics is also starting to make it possible to study the oscillation phenomena in more detail, and distinguish between different proposed scenarios. An example of this is shown in Figure 11.15. An expectation for massive neutrino oscillation is that the shape of the oscillation pattern should be truly oscillatory with a frequency proportional to L/E (Equation 11.20). The values found for Δm^2 imply an oscillation length $L/E \sim 500$ km/GeV. When the rate of ν_μ events is plotted as a function of L/E, one would thus expect to see an oscillation in the 100–1,000 km/GeV range, with a minimum near 500 km/GeV, as indicated by the solid histogram in Figure 11.15.

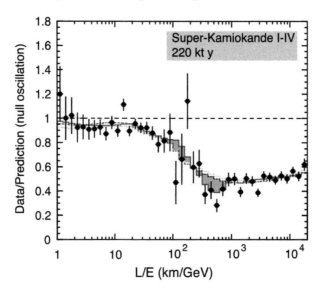

FIG. 11.15. The ratio of experimental ν_μ data to MC events (in the absence of neutrino oscillations) as a function of the reconstructed L/E value, together with the best-fit three-flavor expectation for neutrino oscillation (the solid histogram) and two alternative hypotheses with similar shape. [Kaj 16]

For most atmospheric neutrino events, the baseline distance L and the neutrino energy E are not well determined, due to uncertainty in the direction, the production height in the atmosphere, or the energy transfer to the outgoing lepton. A specialized analysis was developed that minimized these effects by selecting events with good resolution in L/E, largely by excluding low-energy events and events near the horizon. Figure 11.15 shows the measured ν_μ event rate (relative to the expected rate in the absence of oscillations) as a function of L/E for this event sample. The figure shows evidence, claimed to be at the level of four standard deviations, for the expected effect, *i.e.*, a dip in the rate of muon neutrinos near 500 km/GeV that was greater than the 50% deficit found at higher values of L/E. This deficit is due to maximal mixing and averaging of rapid oscillations. To characterize the significance of the dip, this standard neutrino oscillation prediction was compared to two alternate models, neutrino decay and neutrino decoherence. Both predict a more smooth transition to 50% survival at large values of

L/E [Kaj 16].

For an experiment such as SuperK, an oscillation length of a few hundred km/GeV means that the interesting events are concentrated in one of the most difficult kinematic regions, where the neutrinos enter the detector sideways, and a small uncertainty in the arrival direction translates into a much larger uncertainty in the distance traveled by these neutrinos. The large error bars for the data points around 100 km/GeV attest to this problem, which could be greatly alleviated if higher-energy neutrinos were used. For this reason, experiments such as KM3NeT (ORCA) and IceCube aim to improve their sensitivity to atmospheric neutrinos in the 10 GeV range (Sections 10.3/4).

The discovery of neutrino oscillations has also inspired a variety of other experiments. An oscillation length of 500 km/GeV means that long-baseline experiments with accelerator-produced low-energy ν_μ beams are an attractive option to study the oscillation phenomena under controlled conditions. In Europe, CERN has sent neutrino beams to detectors in the Gran Sasso lab (OPERA, ICARUS) since 2006. In the USA, Fermilab is providing such beams for experiments such as MicroBooNE, DUNE, MINOS and NOVA. And in Japan, J-PARC delivers accelerator-produced neutrinos to the detectors in the Kamioka mine. Some of these experiments have obtained results that explicitly support the explanation of the SuperK data as deriving from $\nu_\mu \leftrightarrow \nu_\tau$ oscillations. For example, OPERA has detected five events that were initiated by ν_τs in a neutrino beam that was produced from charged pion and kaon decay [Aga 15].

Other aspects of neutrino mixing are being successfully studied with electron antineutrinos provided by nuclear reactors. In this case, the transition from one ν flavor into another cannot be explicitly observed, since the energies are much smaller than the masses of the charged (μ, τ) leptons. The only way to observe neutrino oscillations in such experiments is through the survival probability of the originally produced $\bar{\nu}_e$. The KAMLAND experiment looked for interactions by $\bar{\nu}_e$s produced at 56 different nuclear reactors, located throughout Japan (and even some in Korea), at an average distance of 180 km from the detector. Their main detector contained one kiloton of liquid scintillator. The observed rate of $\bar{\nu}_e$ interactions, relative to the expected rate based on the on/off status of the different reactors, is plotted as a function of L/E in Figure 11.16a, and exhibits a clear oscillating pattern. The oscillation length is much longer than for the SuperK data and is interpreted as the result of $\nu_e \leftrightarrow \nu_\mu$ oscillations. Therefore, the mass difference $(\Delta m_{12})^2$ is much smaller than the value derived by SuperK for $(\Delta m_{13})^2$: $(\Delta m_{12})^2 = (7.5 \pm 0.2) \cdot 10^{-5}$ eV^2 [Gan 11].

The Daya Bay experiment has performed a series of disappearance measurements at very short distances from the source of the $\bar{\nu}_e$s. Six nuclear reactors (2.9 GW each) generated $\bar{\nu}_e$s that produced events in three different detectors, located at average distances of 470 m, 576 m and 1648 m from the cores. Over a period of 575 days, more than 300,000 events were collected. Figure 11.16b shows the survival probability of the $\bar{\nu}_e$s as a function of L/E. Interestingly, the oscillation that is observed has about the same period as in the case of the SuperK data (0.5 km/MeV $\equiv$ 500 km/GeV). However, the mixing angle is much smaller since the rate drops by less than 10%. These effects are interpreted as evidence for the oscillation $\nu_e \leftrightarrow \nu_\tau$. Thanks to the very high statistics in

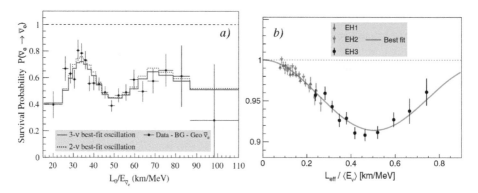

FIG. 11.16. Results of oscillation experiments with reactor anti-neutrinos. The $\bar{\nu}_e$ survival probability is shown as a function of the ratio of the source-detector distance and the $\bar{\nu}_e$ energy (L/E), for experiments carried out by the KAMLAND (a) [Gan 11] and Daya Bay (b) [An 15] Collaborations.

this experiment, the oscillation parameters for this process have the smallest error bars: $(\Delta m_{13})^2 = (2.42 \pm 0.11) \cdot 10^{-3}$ eV^2 and $\sin^2 2\theta_{13} = 0.084 \pm 0.005$ [An 15].

A crucial contribution to further clarification of the neutrino oscillation issues was delivered by the SNO Collaboration, which operated a large water Čerenkov detector in a deep mine (2,092 m below ground level) near Sudbury, Canada [Ahar 06]. This detector [Bog 00] consisted of 10^6 kg of 99.92% isotopically pure heavy water (D$_2$O) contained within a 12 m diameter transparent acrylic vessel (Figure 11.17). Over $7 \cdot 10^6$ kg of H$_2$O between the rock and that vessel shielded the D$_2$O from external radioactive backgrounds. An array of 9,456 inward-facing 20-cm PMTs detected Čerenkov radiation produced in both the D$_2$O and H$_2$O. Extensive purification systems removed radioactive isotopes from the water.

This experiment focused on the effects of neutrino oscillations on solar neutrinos, in particular the ones produced in ^{8}B decay and in the *hep* process (Section 10.1.6). Since the early days of Ray Davis' solar neutrino experiments with ^{37}Cl in the Homestake mine, it was suspected that the discrepancy between the observed and expected rates, which became known as the *solar neutrino deficit*, was due to neutrino oscillations. However, since this deficit was based on a comparison between results from a very difficult experiment and a calculation that depended on several questionable assumptions, this suspicion was not much more than that. The fact that both the experimental rates *and* the theoretical predictions (but not the ratio between the two) decreased substantially over the years did not help either.

We saw earlier that experiments with reactor anti-neutrinos (KAMLAND, Daya Bay) rely on *disappearance* effects, since the surviving particles can only be detected in charged-current reactions. The success of these experiments is due to the availability of experimental L/E information in the relevant domain, *i.e.*, where L/E values are of the order of an oscillation length. That is not the case for solar neutrinos, where L is fixed at about 150 million km. However, compared to the reactor experiments, SNO offered a

FIG. 11.17. The SNO detector in Sudbury, Canada [Ahar 06].

major advantage in that it allowed detection of neutrinos through *neutral-current* events. In a deuteron, the proton and neutron are bound by only 2.2 MeV, which is rather small compared to the typical energy of the ^{8}B neutrinos. These neutrinos may initiate two different types of reactions with a deuteron:

1. They may exchange a W and turn a neutron into a proton: $\nu_e + n \rightarrow e^- + p$. In this case, the reaction products are thus two protons plus an electron, which share the available kinetic energy, *i.e.*, the energy carried by the neutrino minus the 2.2 MeV binding energy of the deuteron.

2. They may exchange a Z^0 and break up the deuteron: $\nu + d \rightarrow \nu + p + n$. In this case, the reaction products are a proton and a neutron, which share the energy of the neutrino minus 2.2 MeV minus the energy carried away by the final-state neutrino.

The first type of reaction can only be caused by ν_es, whereas the second type of reaction can be caused by *all* neutrino flavors. On April 20, 2002, the SNO experiment announced that the total flux of neutrinos, as deduced from the neutral-current processes (# 2) was measured to be larger by a factor of about three than the flux deduced from the charged-current process (#1). This difference could not be explained by any correction, plausible or implausible, to the solar model calculations. Furthermore, it was found that the measured total neutrino flux was almost exactly equal to the predicted flux [Jel 09]. These results are considered the conclusive proof that neutrino oscillations are the cause of the solar neutrino deficit.

The SNO exeriment took data from 1999–2006. Since then, the detector has been refurbished and is now used for studies of $\beta\beta$ decay in ^{130}Te. The leader of the SNO solar neutrino program, Dr. Arthur McDonald, shared the 2015 Nobel Prize for phyics with his SuperK colleague Dr. Kajita.

11.3   The Higgs Boson

11.3.1   *Introduction*

After the discovery of the intermediate vector bosons, which demonstrated the validity of the theory that the electromagnetic and weak nuclear force are two manifestations of the same underlying physical mechanism, the final piece in the puzzle of the electro-weak interaction remained the elusive Higgs boson.

In the 1960s, particle physicists realized that fundamental interactions between particles, such as the electromagnetic and the strong nuclear interaction, could be very successfully described as being mediated by exchange particles. These exchange particles became known as "gauge bosons," the field quanta of the gauge invariant quantum field theories that described the interactions: Quantum Electro Dynamics and Quantum Chromo Dynamics, respectively. An essential ingredient of these theories, known as Goldstone's theorem, implied that these gauge bosons had to be massless. This requirement seemed to exclude the possibility of developing a gauge invariant quantum field theory for the weak interaction, which is mediated by the very massive intermediate vector bosons, W and Z.

The way out of this dilemma turned out to be the mechanism of *spontaneous symmetry breaking*. This mechanism, sometimes referred to as *hidden symmetry breaking*, is a spontaneous process by which a system in symmetrical state ends up in an asymmetrical state. It applies to a system where the Lagrangian that describes the equations of motion is symmetric, but the lowest-energy solutions are not. Often, the "Mexican hat" (Figure 11.18) is used to illustrate it, but nature offers many other examples where this mechanism plays a role.

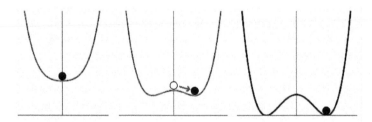

FIG. 11.18. Example of spontaneous symmetry breaking. A ball rolls down an incline and comes to rest at a point where its energy is lowest. Even though the trough may be symmetric in shape, the final solution is not, since the ball ends up in a local minimum, which is not necessarily the center of the trough.

In 1962, Philip Anderson realized that a combination of gauge symmetry and spontaneous symmetry breaking may produce a physical massive vector field [And 63]. This is what happens in superconductivity, a subject about which Anderson was (and is) one of the leading experts. This idea was picked up by Peter Higgs and several others and applied in particle physics [Eng 64, Gur 64, Hig 64]. They developed a quantum field theory in which vector bosons can acquire a non-zero rest mass without explicitly breaking gauge invariance, as a byproduct of spontaneous symmetry breaking. This theory became known as the *Higgs mechanism*. It implies the existence of a scalar field quantum, the Standard-Model Higgs boson.

In contrast to the intermediate vector bosons, whose masses were accurately predicted within the context of the Weinberg–Salam model, theory initially did not provide much guidance with respect to the mass of the Higgs boson. Considerations deriving from unitarity constraints dictated an upper bound of about 1 TeV/c^2. For a long time, theoretical estimates of m_H relied primarily on a precise knowledge of the masses of the W-boson and the top quark, since radiative corrections to the values of these masses depended on it. However, because of the logarithmic nature of these corrections, the predictive effect was rather weak. Nevertheless, as illustrated by Figure 11.19, this type of information (based on results from the Tevatron experiments CDF and D0) suggested that the mass of the Higgs boson was relatively small, probably $\lesssim$ 200 GeV/c^2, at the time when the LHC experiments were designed.

So what should be the strategy for detecting this hypothesized Higgs boson?

11.3.2 *Considerations for the Higgs search*

In the Standard Model, the coupling between the Higgs boson and other particles depends on the mass of these particles. For fermions, the strength of the coupling is proportional to the fermion mass, for bosons it is proportional to the square of the boson mass. The coupling to (massless) gluons is induced by a one-loop Feynman diagram in which the Higgs boson couples to a virtual $t\bar{t}$ pair. Likewise, the Higgs boson's coupling to photons is also generated via loops; in this case, the one-loop diagram with a virtual W^+W^- pair provides the dominant contribution.

Because of these couplings, the relative contributions of different decay modes depend crucially on the mass of the Higgs boson (m_H) itself. The overlap between the wavefunction of the Higgs boson and that of other particles is determined by the mass of these particles: the more massive the particles, the stronger the coupling with the Higgs boson. Therefore, Higgs bosons decay preferentially into the heaviest particles energetically available. A crucial point occurs near 150 GeV/c^2, $\approx 2M_W$. If $m_H > 150$ GeV/c^2, then decay into intermediate vector bosons would completely dominate:

$$H^0 \rightarrow WW \quad \text{or} \quad H^0 \rightarrow ZZ \tag{11.21}$$

and discovery of the Higgs boson would be relatively simple and straightforward, once the required center-of-mass energy and luminosity became available. These processes would also dominate at mass values somewhat below the kinematic limit ($2m_W$). In that case, one of the bosons would be virtual (off the mass shell). In all these cases, by

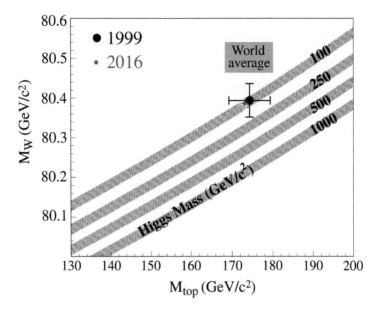

FIG. 11.19. The relationship between the masses of the top quark, the W boson and the Standard-Model Higgs boson. The large dot represents the W, t mass values as they were known around 1999, based on Tevatron data. The small dot, which represents the currently known mass values, illustrates to what extend the knowledge of the W and t mass values has improved in 17 years [Aal 12a].

far the most likely final state consists of four jets, two from fragmenting quarks and the other two from fragmenting anti-quarks:

$$H^0 \to WW^{(*)}, ZZ^{(*)} \to q\bar{q}q\bar{q} \tag{11.22}$$

However, because of the background expected from QCD processes, it was generally believed that such heavy Higgs bosons would be more easily discovered through the leptonic decay modes of the vector bosons.

If, on the other hand, m_H were considerably smaller than $2M_W$, then a variety of other channels would determine the decay. All these channels have one thing in common: discovery of the Higgs boson would be extremely difficult, and present a tremendous challenge for the experimentalists. This turned out to be indeed the case.

Given the lack of strong constraints on the mass, the search for the Higgs boson had to proceed in a way similar to the search for the top quark. And as for the top quark, the decay modes would depend crucially on the new particle's mass. In the case of the top quark, the critical mass value was about 80 GeV/c^2. Top quarks lighter than this could be produced in the decay of W bosons; heavier top quarks would predominantly decay into a W boson plus a b quark.

For the Higgs boson, the critical point lies around 130 GeV/c^2. If m_H turned out to be smaller than $\sim$130 GeV/c^2, decay via intermediate vector bosons would stop playing a significant role, and the dominant decay modes would be those involving the next most massive object, the b quark: $H^0 \rightarrow b\bar{b}$. Such events would contain two very energetic b jets. Since the leading particles of these jets, which carry the fragmenting (anti-)quark, might be highly relativistic ($\gamma \sim 10$), these jets could be recognized by secondary vertices that might easily be displaced from the primary vertex by as much as 5–10 mm. The CDF experience in the top search showed that such events can be successfully recognized at the trigger level.

Nevertheless, discovering the Higgs boson would be much more difficult in this mass regime, especially if m_H was not very different from m_{Z^0}. The production of Z^0s at the LHC is abundant and Z^0s may decay via the same channels as the H^0.

For these reasons, it was widely believed that the best chance for discovering Higgs bosons in the mass window between the lower limit to be set by LEP and 130 GeV/c^2 would be offered by the decay channel

$$H^0 \rightarrow \gamma\gamma \tag{11.23}$$

These were the considerations when the experiments that would explore the new energy regime opened up by the Large Hadron Collider, ATLAS and CMS, designed their detectors. The calorimeter systems were designed such as to be sensitive to Higgs boson decay in the entire mass range between the LEP2 limit and the kinematic limit of the LHC collisions. However, there were some differences in emphasis.

For example, ATLAS put much more emphasis on good hadron calorimetry than CMS. If vector bosons were to play a dominant role in the decay of the H^0, then the leptonic decay modes of these bosons would often involve (anti-)neutrinos. Successful reconstruction of the events would in that case require excellent measurement of the missing transverse energy, carried away by these particles, in order to reduce the large background.

The other LHC experiment, CMS, put most of its eggs in the $H^0 \rightarrow \gamma\gamma$ basket. The primary design goal of the calorimetry for this experiment was to achieve the best possible energy resolution for γ detection, since this energy resolution would directly affect the signal-to-background ratio for this decay mode (Figure 11.20). The CMS calorimeter system is built around an em section consisting of lead tungstate crystals, which were expected to allow γ detection with a resolution better than 1% in the relevant energy range (50-100 GeV). The adverse effects of this choice on the hadronic performance (see Section 9.1.4) were considered an acceptable price to pay.

This example illustrates the gambling aspect in designing detectors for a new experiment. Another example of such gambling was apparent in the LEP experiment L3, which designed its calorimeters in the hope of being able to study the spectroscopy of $t\bar{t}$ bound states. This turned out to be an illusion, since top quarks were too massive to be produced at LEP. At the LHC, CMS gambled on a Higgs mass between 100 and 130 GeV/c^2. If the Higgs boson had a mass outside this range, then CMS would be at a clear disadvantage compared with ATLAS, which has considerably better performance for the detection of hadrons and jets.

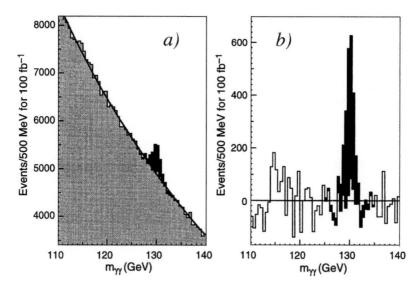

FIG. 11.20. The anticipated $\gamma\gamma$ invariant mass spectrum, measured in an LHC experiment that has a high-resolution calorimeter for γ detection, if $m_H = 130$ GeV/c^2. Results are given before (a) and after (b) the (QCD) background is subtracted. This plot is from the technical design report of the CMS calorimeter system [CMS 97a].

11.3.3   *The discovery*

In the 1990s, an extensive search for the Higgs boson was conducted at the Large Electron Positron Collider at CERN, where this particle could in principle be made by means of the associate production process $e^+ + e^- \rightarrow Z^0 H^0$. At the end of its service in 2000, LEP had found no conclusive evidence for this process. This implied that if the Higgs boson were to exist, it would have to be heavier than 114.4 GeV/c^2 [Hei 03]. The Tevatron experiments CDF and D0 at Fermilab also spent a lot of efforts in the search for the Higgs boson, through its decay into pairs of intermediate vector bosons (Equations 11.21,11.22). Since these experiments had earlier discovered the top quark (mass 173 GeV/c^2), they were in principle also well equipped to discover the Higgs boson if it had a mass in the range 150–200 GeV/c^2. However, since the search was limited to leptonically decaying Ws and Zs, the integrated luminosity was not sufficiently large to obtain very significant results. In a 2010 paper, the absence of a significant signal in the WW invariant mass distribution led to the conclusion that the Higgs mass could not be in the 162–166 GeV/c^2 range, at a 95% confidence level [Aal 10]. The authors also announced not having seen a significant signal in the entire mass range from 130–200 GeV/c^2. In a later publication, they used their very precise measurements of the W and t mass values (Figure 11.19) to exclude mass values for the Higgs boson larger than 145 GeV/c^2 at the 95% confidence level [Aal 12a]. All this left a rather narrow mass window, from 115–145 GeV/c^2 to search for evidence of this elusive particle.

On July 4, 2012, in a joint meeting at CERN, the ATLAS and CMS experiments

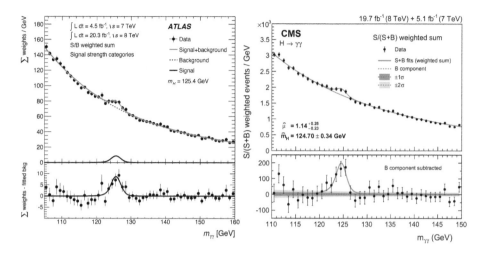

FIG. 11.21. Discovery of the a new resonance around 125 GeV in the $\gamma\gamma$ invariant mass distribution, observed by the ATLAS [Aad 12] and CMS [Chat 12] Collaborations.

announced that they had observed in their 7 and 8 TeV pp collision data a significant resonance in the $\gamma\gamma$ invariant mass distribution, at an energy of $\sim$ 125 GeV (Figure 11.21). This resonance had all the characteristics expected from the Standard-Model Higgs boson. Since the object decayed into two photons, it had to be a boson, and its spin could not be 1 [Lan 48]. The statistical significance of the resonance was better than 5σ, which is in practice the agreed-upon level for a discovery. Both experiments also saw evidence for a resonance at the same energy in invariant mass plots involving vector bosons: $WW^{(*)}$, $ZZ^{(*)}$, albeit at a lower significance level (Figure 11.22a). Searches for a resonance in the decay channels $H^0 \to b\bar{b}$ and $H^0 \to \tau^+\tau^-$ did not yet result in significant signals. The results were clearly limited by event statistics, given that the significance was considerably larger in the sum of all data than in the individual event samples obtained at 7 TeV and 8 TeV (Figure 11.22b). Yet, the collaborations have made enormous efforts to extract the maximum amount of information from the limited event statistics. ATLAS and CMS were in slight disagreement about the mass of the new particle: 126.0 ± 0.6 GeV/c^2 (ATLAS) vs. 124.7 ± 0.4 GeV/c^2. The strength of the observed signal in the $\gamma\gamma$ decay mode is within errors equal to that expected for a Standard-Model Higgs boson. An important point is the spin of the new particle. Measurements of the spin are based on the angular distribution of the photons. In the rest frame of a spin-0 boson, the decay photons are isotropically distributed. The analyses favor the spin-0 hypothesis over the spin-2 one, but would clearly benefit from more statistics [Kha 14]. So far, all experimental data have indicated that the new object is indeed the Standard-Model Higgs boson.

11.3.4 *Epilogue*

At the end of 2012, the LHC shut down operations for an extended period of time, in preparation for running at higher energies and luminosities. In 2015, the experiments

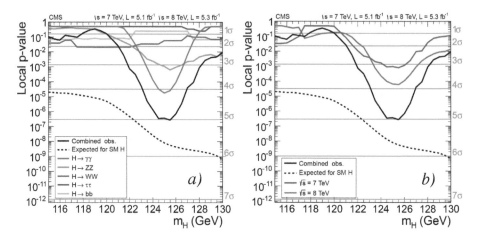

FIG. 11.22. Statistical significance of the Higgs boson discovery. The observed local p-value for the different decay modes and the overall combination as a function of the particle's mass (*a*). The observed local p-value for the $\gamma\gamma$ decay mode for the data collected at 7 TeV and at 8 TeV, as well as for the combination of all data (*b*). In both diagrams, the dashed line shows the expected local p-value for a SM Higgs boson with a mass m_H [Chat 12].

resumed taking data, at a center-of-mass energy of 13 TeV. At the time of this writing (October 2016), the LHC has delivered ~ 40 fb^{-1} worth of collision data at this energy, in addition to the 25 fb^{-1} accumulated at the lower energies. The significance of the Higgs boson peak has further increased, also because the production cross section at 13 TeV is more than twice as large as at 8 TeV [PDG 14].

Based on the 2015 data, there was some excitement about a possible other peak in the $\gamma\gamma$ invariant mass distribution, this time at ~ 750 GeV/c^2. However, this rumor did not survive the additional scrutiny provided by the strongly increased event statistics accumulated in 2016. The LHC is expected to increase the total integrated luminosity accumulated over its lifetime to $\sim 3,000$ fb^{-1}. If the history of accelerators operating at the energy frontier is any guidance, then more such rumors may be expected in the future, and perhaps some real discoveries. In any case, the gradually increasing integrated luminosity will allow increasingly detailed studies of the properties of the Higgs boson. However, this will demand a lot of the detectors, which are venturing deep into uncharted territory in these experiments, both in terms of the radiation levels at which they have to operate and in terms of the complexity of the events, with hundreds of interactions occurring every (25 ns) bunch crossing.

11.4   Outlook

In the previous sections, several experimental milestones in the recent development of the understanding of the fundamental structure of matter are described. Calorimeters played an important role in all these discoveries. The enrichment of the arsenal of experimental techniques brought about by calorimetry is illustrated by the fact that in all these cases, different features of the calorimetric performance made the crucial difference:

- The discovery of the intermediate vector bosons was possible thanks to the capability of the calorimeters to recognize and select events in which large-$p_\perp$ particles were produced on-line, amidst a background that outnumbered these events by some seven orders of magnitude.

- The observation of ν-oscillation effects in SuperKamiokande was possible thanks to some special features that derived from the use of Čerenkov light as the source of experimental information in this calorimeter, namely (a) the possibility to distinguish between low-energy electrons and muons, and (b) the capability to reconstruct the direction of the particles that generated the light cones.

- The experimental observation of the Higgs boson became possible thanks to the capability of the calorimeters to detect neutral particles (γs) and to measure their four-vectors with very high precision.

Other important discoveries in the past half century also benefited in essential ways from calorimetry. For example, the discovery of the top quark by D0 was made possible by the capability of the calorimeter and its auxiliary systems to recognize charged leptons that were part of jets [Abb 99]. And the unraveling of the details of direct CP-violation in the K^0 system was made possible by the excellent mass resolution for π^0s in the calorimeters of the NA31 [Barr 93], KTeV [Ala 99] and KLOE [Des 09] experiments.

In the 1950s and 1960s, a large number of new "elementary particles" were found in bubble chamber experiments, as resonances in invariant-mass plots (see Equation 11.9). Searches for new particles were conducted by carefully measuring the four-vectors of all particles produced in the interactions and combining the four-vector of one particular particle (e.g., a K^0) with that of other particles (e.g., two pions of opposite charge), according to the chosen decay mode. Such searches were often successful (e.g., $K_1(1270) \rightarrow K^0\pi\pi$).

This technique is still being used successfully, for example in the spectroscopy of particles containing a charmed and/or bottom quark, in which the LHCb experiment currently excels [Aai 15]. However, as the energy at which we study the structure of matter is further increased, exclusive hadronic decay modes become very complicated (they involve many final-state particles) and each represent only a tiny fraction of the decays. For example, the decay of some charmed and bottom mesons proceeds through more than one hundred different channels with measurable branching fractions.

Increasingly, the primary process at the quark level is of great interest, rather than the precise composition of the final state. For example, the hadronic decay of a W boson proceeds through the process $W \rightarrow q\bar{q}$. In order to demonstrate that hadronically decaying Ws were produced in certain reactions, it is much more important to be able to reconstruct these two jets and determine their invariant mass than to know the characteristics of all individual particles making up the final state.

In the 1960s, K^0s and Λs could be identified by the fact that the $\pi^+\pi^-$ hypothesis produced the correct mass value for the first one, while the $p\pi^-$ hypothesis reproduced exactly the 1115.68 MeV/c^2 on the books for the Λ. The higher the energy, the more precise the experimental measurements of the four-vectors had to be to distinguish these two particles.

History repeats itself, but now the individual particle tracks have been replaced by jets. Figure 7.70 shows the jet–jet invariant mass distribution measured by the UA2 Collaboration. The peak corresponding to the decay of the intermediate vector bosons W and Z can clearly be distinguished from the QCD background. This is a direct result of the measurement accuracy obtained with the calorimeter used in this experiment. An even better resolution would make it possible to observe W and Z decays as separate peaks in this mass spectrum. Calorimeters are ideal instruments for this type of measurement. The examples discussed in Section 7.7 illustrate that it clearly pays to have a good calorimeter at one's disposal.

Thirty years ago, intermediate vector bosons were the objects of intense experimental searches. Now, they may become the key to discovering new phenomena at the 0.1–1.0 TeV mass scale, and multi-quark spectroscopy may turn out to be an invaluable tool for this.

The quality of the hadronic calorimetry may well turn out to be very important for the study of phenomena beyond the Standard Model, assuming that there is physics beyond this Model that can be studied with the available tools. The discovery of Supersymmetry hinges on detection of the escaping LSP, which manifests itself through missing (transverse) energy in the collisions. Establishing the existence of other hypothesized new objects, such as leptoquarks, requires the explicit detection of fragmenting quarks and precise measurements of their properties.

Calorimeters are instruments for measuring energy. The history of physics in general, and of nuclear and particle physics in particular, is filled with examples that prove that measurement precision pays off. A better, more accurate instrument allows more precise measurements. More precise measurement results make it possible to discover new phenomena, and/or to better understand old ones. Better understanding of the physical world has always been a crucial element in the evolution of mankind.

The history of calorimetry itself illustrates this process in a nutshell. Calorimeters were originally invented as crude, cheap instruments for some specialized applications (for example, detection of neutrino interactions). The original literature is testimony to the fact that their performance was often perceived as somewhat mysterious by their users. Only after the physics on which calorimeters are based was understood in detail did it become possible to develop these detectors into the precision instruments that they are nowadays and which form the centerpiece of many modern experiments in particle physics.

I started this book with a description of the development of nuclear γ-ray detectors. The advent of germanium-based solid-state detectors with their unprecedented energy resolution caused a revolution in nuclear spectroscopy in the 1960s (*cf.* Figure 1.1). We have now entered an era in which calorimetry may allow the measurement of fragmenting quarks and gluons with nuclear-spectroscopic precision. If nature is kind to us, a new world might open up as a result.

APPENDIX A

NOTATION AND ABBREVIATIONS

A.1   Variables and their Units

The following table summarizes the notation for variables used in this book, and the units in which these variables are usually expressed.

Symbol	Definition	Units
A	Number of nucleons in nucleus	dimensionless
c	Speed of light	$2.99792458 \cdot 10^8$ m s^{-1}
e	Elementary charge	$1.6021765 \cdot 10^{-19}$ C
E	Energy	GeV
f_{em}	Electromagnetic fraction of hadron showers	dimensionless
f_{samp}	Sampling fraction for a mip	dimensionless
h	Planck's constant	$6.626075 \cdot 10^{-34}$ J.s
k_B	Birks' constant	g cm^{-2} MeV^{-1}
m_e	Electron rest mass	511.0 keV ($= m_e c^2$)
m_p	Proton rest mass	938.3 MeV ($= m_p c^2$)
n	Index of refraction	dimensionless
N_A	Avogadro's number	$6.02214 \cdot 10^{23}$ mol^{-1}
q^2	Momentum transfer in nuclear reactions	(GeV/c)^2
r_e	Classical electron radius	$2.818 \cdot 10^{-15}$ m
X_0	Radiation length	g cm^{-2} or cm
Z	Number of protons in nucleus	dimensionless
α	Fine-structure constant	1/137.036
β	Velocity as a fraction of the speed of light	dimensionless
γ	Lorentz factor $(1 - \beta^2)^{-1/2}$	dimensionless
ϵ_c	Critical energy	MeV
η	Pseudorapidity ($-\ln \tan(\theta/2)$)	dimensionless
θ	Polar angle in collider experiments	radians
θ_C	Čerenkov angle, $\arccos(1/\beta n)$	radians
λ	Wavelength	μm, nm, Å
λ_{int}	Nuclear interaction length *for protons*	g cm^{-2} or cm
λ_{att}	Attenuation length for *light*	m
ρ_M	Molière radius	g cm^{-2} or cm
ϕ	Azimuthal angle in collider experiments	radians

A.2 Abbreviations

The following table gives an alphabetic list of abbreviations used in this book, and their meaning.

AA	Anti-proton Accumulator
AC	Alternating Current
ADC	Analog-to-Digital Converter
ANL	Argonne National Laboratory
APD	Avalanche Photo Diode
ASIC	Application Specific Integrated Circuit
BEPC	Beijing Electron Positron Collider
BNL	Brookhaven National Laboratory
BGO	Bismuth Germanium Oxide ($Bi_4Ge_3O_{12}$)
BOTEC	Back-Of-The-Envelope Calculation
CC	Charged Current (weak interactions)
CCD	Charge Coupled Device
CEBAF	Continuous Electron Beam Accelerator Facility (Thomas Jefferson Lab)
CERN	Conseil Europeèn pour la Recherche Nucléaire
CESR	Cornell Electron Storage Ring
DC	Direct Current
DESY	Deutsches Electronen Synchrotron
DREAM	Dual-REAdout Method
EFL	Energy Flow Logic
EGS	Electron Gamma Shower
ELSA	Electron Stretcher Accelerator (Bonn, Germany)
ENC	Equivalent Noise Charge
ESS	European Synchrotron Source
EST	Electrostatic Transformer
FAIR	Facility for Anti-proton and Ion Research (Darmstadt, Germany)
FNAL	Fermi National Accelerator Laboratory
FWHM	Full Width at Half Maximum
HPD	Hybrid Photon Detector
HST	Hubble Space Telescope
HERA	Hadron Electron Ring Anlage
IHEP	Institute for High Energy Physics (Serpukhov, Russia)
ILC	International Linear (e^+e^-) Collider
ILL	Institut Laue Langevin (Grenoble, France)
ISR	Intersecting Storage Rings
ISS	International Space Station
J-FET	Junction Field Effect Transistor
J-PARC	Japan Proton Accelerator Research Complex
KEK	Ko Enerugii Kenkyusho (National Lab for High Energy Physics Japan)
LAr	Liquid Argon

LBNL	Lawrence Berkeley National Laboratory
LEAR	Low-Energy Anti-proton Ring
LED	Light Emitting Diode
LEP	Large Electron Positron (collider)
LHC	Large Hadron Collider
LKr	Liquid Krypton
LSP	Lightest Supersymmetric Particle
LXe	Liquid Xenon
MAMI	Mainz Microtron (Mainz, Germany)
MC	Monte Carlo
MWPC	Multi Wire Proportional Chamber
NC	Neutral Current (weak interactions)
PDG	Particle Data Group
PFA	Particle Flow Analysis
PMMA	Polymethylmethacrylate
PMT	Photomultiplier Tube
PS	Proton Synchrotron
PSD	Preshower Detector
PSI	Paul Scherrer Institut (Villigen, Switzerland)
PWO	Lead tungstate ($PbWO_4$)
QCD	Quantum Chromo Dynamics
QED	Quantum Electro Dynamics
QIE	Charge Integrator and Encoder
RF	Radio Frequency (Frequent)
RHIC	Relativistic Heavy Ion Collider
RPC	Resistive Plate Chamber
SiPM	Silicon Photomultiplier
SLAC	Stanford Linear Accelerator Center
SLC	Stanford Linear Collider
SM	Standard Model
SMD	Shower Maximum Detector
SNO	Sudbury Neutrino Observatory (Canada)
SPS	Super Proton Synchrotron
SSC	Superconducting Super Collider
SUSY	Supersymmetry
TDC	Time-to-Digital Converter
TPC	Time Projection Chamber
VPT	Vacuum Photo Triode/Tetrode
WIMP	Weakly Interacting Massive Particle
WLS	Wavelength Shifted (Shifting)

APPENDIX B

PROPERTIES OF MATERIALS USED IN CALORIMETERS

B.1 Calorimetric Scaling Parameters

The following table lists parameters relevant for shower development in materials that are frequently used in sampling calorimeters: the critical energy (ϵ_c), the radiation length (X_0, eq. 2.10), the Molière radius (ρ_M, eq. 2.15), the nuclear interaction length (λ_{int}) and the specific ionization ($\langle dE/dx \rangle$) for mips. The critical energy was determined on the basis of dE/dx data tabulated in [Pag 72], using Rossi's definition (see Section 2.1.1). The nuclear interaction length is given for *protons*.

Table B.1 *Properties of calorimeter materials*

Passive material	Z	Density (g cm^{-3})	ϵ_c (MeV)	X_0 (mm)	ρ_M (mm)	λ_{int} (mm)	$(dE/dx)_{mip}$ (MeV cm^{-1})
C	6	2.27	83	188	48	381	3.95
Al	13	2.70	43	89	44	390	4.36
Fe	26	7.87	22	17.6	16.9	168	11.4
Cu	29	8.96	20	14.3	15.2	151	12.6
Sn	50	7.31	12	12.1	21.6	223	9.24
W	74	19.3	8.0	3.5	9.3	96	22.1
Pb	82	11.3	7.4	5.6	16.0	170	12.7
^{238}U	92	18.95	6.8	3.2	10.0	105	20.5
Concrete	-	2.5	55	107	41	400	4.28
Glass	-	2.23	51	127	53	438	3.78
Marble	-	2.93	56	96	36	362	4.77

Active material	Z	Density (g cm^{-3})	ϵ_c (MeV)	X_0 (mm)	ρ_M (mm)	λ_{int} (mm)	$(dE/dx)_{mip}$ (MeV cm^{-1})
Si	14	2.33	41	93.6	48	455	3.88
Ar (liquid)	18	1.40	37	140	80	837	2.13
Kr (liquid)	36	2.41	18	47	55	607	3.23
Xe (liquid)	54	2.95	12	24.0	42	572	3.71
Polystyrene	-	1.032	94	424	96	795	2.00
Plexiglas	-	1.18	86	344	85	708	2.28
Quartz	-	2.32	51	117	49	428	3.94
Lead-glass	-	4.06	15	25.1	35	330	5.45
Air 20°, 1 atm	-	0.0012	87	304 m	74 m	747 m	0.0022
Water	-	1.00	83	361	92	849	1.99

B.2   Nuclear Properties

Table B.2 *The average binding energy per nucleon (B/A) and the average binding energy of nucleons released in spallation reactions (assuming $\langle \Delta A \rangle = 10$). Data from [Wap 77].*

Target nucleus	Z	A	B/A (MeV)	$\Delta B/\Delta A$ (MeV)
C	6	12.01	7.7	-
Al	13	26.98	8.1	11.3
Ar	18	39.95	8.6	9.6
Fe	26	55.85	8.8	9.6
Cu	29	63.55	8.7	8.5
Sn	50	118.7	8.5	8.0
W	74	183.9	8.0	6.6
Pb	82	207.2	7.9	6.9
^{238}U	92	238.0	7.6	5.9

Table B.3 *Cross sections for elastic neutron scattering*

Neutron energy →	1 keV	10 keV	100 keV	1 MeV
σ_{el} in hydrogen (b)	20	19	12	4.3
σ_{el} in carbon (b)	4.7	4.6	4.4	2.6
σ_{el} in oxygen (b)	3.8	3.8	3.8	8
σ_{el} in argon (b)	0.6	0.5	1.5	2.0
σ_{el} in iron (b)	10	6	4	2.5
σ_{el} in lead (b)	11	11	10	5

Table B.4 *Cross sections for neutron-induced reactions (14 MeV)*

Process	On Carbon	On Oxygen	On Iron
Total inelastic cross section (mb)	500	600	1400
$\sigma (n, \alpha)$ (mb)	70	200	45
Total cross section for α production (mb)	300	200	45
# α's per inelastic reaction	1.5	0.3	0.03

B.3   Scintillating Materials used in Particle Physics Experiments

Table B.5 *Properties of scintillating crystals that are used in particle physics experiments.*

	NaI(Tl)	CsI(Tl)	CsI	BaF$_2$	CeF$_3$	BGO	PbWO$_4$
Density (g cm^{-3})	3.67	4.51	4.51	4.89	6.16	7.13	8.30
Radiation length (cm)	2.59	1.85	1.85	2.06	1.68	1.12	0.89
Molière radius (cm)	4.8	3.5	3.5	3.4	2.6	2.3	2.0
Interaction length (cm)	41.4	37.0	37.0	29.9	26.2	21.8	18.0
$\langle dE/dx \rangle_{\rm mip}$ (MeV cm^{-1})	4.79	5.61	5.61	6.37	8.0	8.92	9.4
Refractive index (at $\lambda_{\rm peak}$)	1.85	1.79	1.95	1.50	1.62	2.15	2.2
Hygroscopicity	Yes	Slight	Slight	No	No	No	No
Emission spectrum, $\lambda_{\rm peak}$							
Slow component (nm)	410	560	420	300	340	480	510
Fast component (nm)			310	220	300		510
Relative light yield							
Slow component	100	45	5.6	21	6.6	9	0.3
Fast component			2.3	2.7	2.0		0.4
Decay time (ns)							
Slow component	230	1300	35	630	30	300	50
Fast component			6	0.9	9		10

Table B.6 *Properties of some plastic scintillators and wavelength shifters applied in particle physics experiments. Listed are the wavelengths at which the absorption (for WLS) and emission spectra peak, the characteristic decay time and, when available, the absorption length (WLS).*

	Absorption (nm)	$\lambda_{\rm abs}$ (μm)	Emission (nm)	Decay time (ns)
Scintillators				
SCSF-38 (Kuraray)			428	2.3
SCSF-81 (Kuraray)			437	2.4
SCSF-3HF (Kuraray)			530	7
BCF-10 (Bicron)			432	2.7
BCF-20 (Bicron)			492	2.7
BCF-60 (Bicron)			530	7
Wavelength shifters				
Y-7 (Kuraray)	440	100	490	10
Y-11 (Kuraray)	435	80	476	10
BCF-91A (Bicron)	420		494	12
BCF-92 (Bicron)	410		492	2.7

B.4 Noble Liquids used in Calorimeters

Table B.7 *Properties of noble liquids applied in particle physics experiments*

Property ↓ Liquid →		LAr	LKr	LXe
Z		18	36	54
Boiling point (K)		87.3	119.8	165.0
Density in liquid phase (g cm^{-3})		1.40	2.41	2.95
Radiation length (cm)		14.0	4.7	2.40
Molière radius (cm)		8.0	5.5	4.2
Nuclear interaction length for protons (cm)		84	61	57
Ionization properties, measured with ^{207}*Bi* [Miy 95]				
Energy needed per electron–ion pair (eV)		24	17	15
Scintillation properties, for 1 MeV e$^-$ *(αs)* [Dok 99]				
Emission spectrum, λ_{peak} *(nm)*		128	147	174
Decay time (ns)				
Fast component		5.0 – 6.3	2.0	2.2
Slow component		860 – 1090	80 – 91	27 – 34
Relative light yield				
Fast component		8% (57%)	1%	5% (31%)
Slow component		92% (43%)	99%	95% (69%)

APPENDIX C

OVERVIEW OF CALORIMETER SYSTEMS

The following pages contain lists of experiments in which calorimeters play(ed) an important role. These experiments have been subdivided into four categories:

1. Fixed-target experiments
2. Colliding-beam experiments
3. Non-accelerator experiments
4. Generic R&D projects aimed at studying new calorimeter techniques, including R&D for experiments that were planned but canceled. Examples of the latter include the experiments at the SSC, the $K^0 \rightarrow \pi^0 \nu \bar{\nu}$ experiment at Brookhaven, the BTeV experiment at Fermilab, the SuperB factory in Frascati and the NOE experiment at the Gran Sasso Laboratory.

The main characteristics of the calorimeters are summarized in a few lines. The granularity is expressed in units of length (mm, cm, X_0), angle (degrees, mr), or in dimensionless numbers, which indicate the segmentation $\Delta\eta \times \Delta\phi$. When the energy resolution is expressed as the sum of two or more terms, these terms usually (but not always) have to be added *in quadrature*. The energy E is expressed in units of GeV, unless stated otherwise.

More information about each of the detectors can be found in the references listed in the last column.

Table C1 Calorimeter Systems in Fixed Target Experiments

Accelerator	Facility	Principal physics goals	Calorimeter structure		Granularity	Energy resolution	Comments	Refs.
			Passive	Active				
SPS (CERN)	WA1 (CDHS)	Neutrino physics Structure functions nucleons	5 cm Fe (105 plates) 15 cm Fe (40 plates) (67 lambda)	1 cm scintillator	145 sampling layers 8x45 cm scint. slabs 15 long. segments	had: 80%/√E (5 cm Fe) 150%/√E (15 cm Fe)	Resolution improvement after off-line weighting (see ch. 4.9)	[Abr 81] [Hol 78a,b]
	CHARM II	(Anti-)neutrino electron scattering. Measurement of Weinberg angle	0.5 Xo glass plates	Plastic streamer tubes (10 mm spacing) Analog strips (30 mm)	150000 digital ch. 9000 analog ch.	em: 28%/√E+3% had: 40%/√E+7%	Optimized for angular resolution em showers: 18 mr/√E	[Dew 86]
	CHORUS	Neutrino oscillations tau neutrino search	Lead (Spaghetti structure)	Scintillating fibers (1 mm)	em: 4 cm x 8 cm had1: 8 cm x 8 cm had2: 10 cm x 20 cm	em: 13.8%/√E had: 32%/√E + 1.4%	Compensating (Pb:plastic = 4:1) em 21Xo, had1 2λ, had2 5λ	[Esk 97] [Dic 96] [Buo 94]
	NOMAD	Same as CHORUS	em: leadglass array (875 blocks)	Strips in had2 (4 mm)	79mm x 112mm (19Xo deep)	3.2%/√E+1.0%+4.3%/E (electrons)	Readout by phototetrodes in magnetic field	[Aut 96,97]
	HELIOS (NA34)	Heavy-ion physics (O-16, S-32 scattering)	had: steel em: 1.7 mm U-238 had: 3.4 mm U-238 3 mm U-238	scintillator 2.5 mm LAr gaps plastic scintillator (2.0/2.5 mm sheets)	2000 channels 700 channels 1200 towers 1200 towers	em: 19%/√E had: 50%/√E em: 17%/√E had: 36%/√E	Optimized for energy flow measurements	[Ake 87]
	WA89	Study of hyperon interactions Production charmed baryons	em: Lead glass (642 blocks) had: lead (80 vol%)	1 mm scintil. fibers	7.5 x 7.5 x 36 cm (642 blocks) 155 hexagonal cells (43 mm apex-to-apex)	em: 6.1%/√E had: 32%/√E + 1%	See Section 8.1	[Bru 92] [Bec 96]
	Omega (WA 69,70,71, 76,81,83)	Various	em calorimeter Lead (50 vol%)	1 mm scintil. fibers	35 mm x 35 mm	em: 10%/√E + 2.2%	wavy fibers to avoid channeling	[Son 87] [Gom 87]
	NA38	Heavy-ion physics Muon pair production	em calorimeter Lead (67 vol%)	1 mm scintil. fibers	30 channels	em: 20%/√E	angle 30-300 mr	[Son 87]
	WA70 (EMC)	Direct photon studies (280 GeV)	em calorimeter 4.2 mm Pb	4.4 mm teflon tubes filled with scint. liquid	3072 channels	em: 12.6%/√E + 3.2%	mass resolution π° 0.3% position resolution 0.3 mm gamma gamma sep. 1.8 cm	[Bon 87]
	WA80	Heavy-ion physics Quark-gluon plasma studies through direct photons	em(had) 2(3) mm U Lead-glass (SAPHIR)	3 mm scintillator	30 cm x 30 cm (10λ) 1278 modules 35 mm x 35 mm		Zero Degree Calorimeter	[You 89] [Bau 90]
	NA31	CP violation in K° system	em calorimeter 2.3 mm Pb/Al	2x2.1 mm LAr	1536 channels 12.5 mm strips 27 Xo	7.5%/√E+10%/E+0.5% (electrons)	position resolution 1 mm	[Bur 88]
	NA48	Same as NA31	Liquid krypton		13500 channels 2 cm x 2 cm 26 Xo	3.5%/√E + 4%/E + 0.4% (electrons)	Mass resolution π° 3.0 MeV Position res. 4.2mm/√E Time res. 260 ps (50 GeV)	[Fan 94] [Barr 96] [Con,Cre 98]
	NA62	Rare K decays	Liquid krypton		Same as NA48	Same as NA48		[Fan 07]
	NA50	Heavy-ion physics Search quark-gluon plasma	30 tantalum slabs 1.5mm, grooved	900 quartz fibers 430 micron diameter	2.5cm x 2.5cm x 65cm (4 cells)	33 TeV Pb: 5% 160 GeV p: 30%	fibers at 0° with beam 1.5 mm fiber-to-fiber	[Arn 98]
	NA58 (COMPASS)	Nucleon structure Spin details, glueballs	em lead glass had1: 20 mm Fe had2: 25 mm Fe	5 mm plastic scintillator 5 mm plastic scintillator	3.8 x 3.8 x 45 cm 4.8 λ 5.5 λ	59%/√E + 7.6% 66%/√E + 5%		[Abb 07]
	COMPASS II		em: 0.8 mm Pb	1.5 mm plastic scint.	4 x 4 cm	7.8%/√E + 2.3%	Shashlyk, MAPD readout	[Anf 13]
	NA61 (SHINE)	Heavy ion + ν experiment	had only: 16 mm Pb	4 mm plastic scintillator	10 x 10 cm, 20 x 20 cm 4.2 λ	56%/√E + 2%	Projectile Spectator Detector	[Abg 14]

Table C1 Continued

Accelerator	Facility	Principal physics goals	Calorimeter structure — Passive	Calorimeter structure — Active	Granularity	Energy resolution	Comments	Refs.
Tevatron accelerator complex (FNAL)	E-494	Production of lepton and hadron pairs (15 - 60 GeV)	Water Cherenkov counter (2 tanks) (15 Xo, 7 lambda)	8-9 WLS bars	30 cm x 3 m	45% (almost independent of energy)		[Bro 78]
	E-594	Neutrino physics	Sand + steel shot 3.7 m x3.7 m x18.3 m (152 Xo, 22 lambda)	4 mm flash chambers 2.5 cm prop. tubes	0.22 Xo sampling 3.5 Xo sampling	em: 8% had: 10%		[Bog 82]
	E-616 (CCFRR)	Neutrino physics	Steel 690 ton target 420 ton toroid magnet	spark chambers liquid scint. (target) plastic scint. (magnet)	10 cm Fe (target) 20 cm Fe (magnet)	had: 78%/√E + 93%/E		[Mac 84]
	E-744/770 (CCFR)	Neutrino physics	Steel	liquid scint. (target)	5 cm Fe (target)	em: 60%/√E had: 85%/√E + 30%/E		[Sak 90]
	E-705	Charm, direct photon studies in 300 GeV π,p beams	Near beam: Scintillating glass (cerium oxide) Farther out: Lead glass	Only em calorimeter	7.5 cm x 7.5 cm 15 cm x 15 cm	em: 1.5%/√E + 1.6%	cerium doped glass 150° more radiation hard than lead glass	[Wag 85] [Jenk 89] [Cox 84] [Ant 93]
	E-706	High transverse momentum photon and jet production in (p,π,K) scattering off Be	em: 2 mm Pb had: 2.5 cm steel	2.5 mm LAr	em: 6600 channels had: 2400 channels	em: 15%/√E had: 70%/√E	Optimized for high rates Sub-cm spatial resolution for photon showers	[Lob 85] [Apa 98]
	E-760	Charmonium production in proton-antiproton collisions (gas jet target)		Lead glass array	20/64 segments in θ/φ 12-16 Xo deep blocks	em: 3.0%/√E + 1.5%		[Bart 90]
	E-731/799/832 (KTeV)	CP violation in the K° system	Forward: lead (1 mm) scintillator (6.4 mm)	Pure CsI crystals	10 cm x 10 cm 50 mm x 50 mm (27 Xo deep)	em: 7.4% at 1 GeV em: <1% for E>5 GeV	WLS plate readout	[Has 90] [Kes 96]
	E-831 (FOCUS)	Photoproduction of charm	4.4 cm Fe	7 mm scintillator	20 cm x 20 cm tiles 7.8 λ deep	had: 85%/√E + 1%		[Are 99]
	E-871 (HyperCP)	CP violation in hyperons	had only: 24.1 mm Fe	5 mm polystyrene	49.5 x 98 cm 16 layers (9.6 λ)	9% @ 70 GeV protons		[Bur 05]
	E-907 (MIPP)	Study of particle production cross sections	em: 5 mm Pb had: 24 mm Fe	MWPCs 5 mm polystyrene	10 Xo deep 9.6λ deep	27%/√E 55%/√E + 2.6%		[Nig 09]
	Muon g-2	Measurement of g-2	PbF₂ Cherenkov crystals		2.5 x 2.5 x 20 cm	3.4 - 4.6%/√E (em)	SiPM readout	[Fie 15]
	Mu2e	Measurement of μ →eγ	Hexagonal BaF₂ Crystals (2 x 930)		20 cm deep 1.65 cm flat-to-flat	5 MeV @ 100 MeV	APD readout	[Pez 15]
	E-875/934/1016 (MINOS)	Neutrino oscillations	2.54 cm Fe 980 t (near), 5400 t (far)	1.0 cm extruded PS	282 planes (near) 249 planes (far)	em: 21.4%/√E + 4% had: 56%/√E + 2%	Magnetized tracking calorimeter	[Mic 08]
	NOvA	Neutrino oscillations	Liquid scintillator, 0.13 t (near), 8.7 kt (far) 20,192 cells (near), 344,064 cells (far)	scintillating strips (triangular)	3.8 x 5.9 x 1550 cm			[Muf 15]
	E-938 (MINERVA)	Low-energy ν interactions	em: 2 mm Pb had: 2.54 cm Fe	scintillating strips (triangular)		Eᵥ: 29%/√E + 13.4%		[Ali 14]
	μBooNE	Neutrino oscillations	LAr TPC imaging calorimeter (170 ton)		5 x 5 x 5 mm		Near detector of Long Baseline experiment	[MB 15]

Table C1 Continued

Accelerator	Facility	Principal physics goals	Calorimeter structure — Passive	Calorimeter structure — Active	Granularity	Energy resolution	Comments	Refs.
AGS (Brookhaven)	E-814	Heavy-ion scattering at 14.6 GeV/nucleon	1.0 cm lead plates (every 6th plate iron)	plastic scintillator (0.3cm)	16 segments in φ, 8 segments in θ, 4 segments in depth	em: 24%/√E, had: 43%/√E	Readout by WLS fibers (1.5 mm diameter)	[Sim 91] [Fox 92] [Sul 93]
	E-852 (in Multiparticle Spectrometer)	Search for mesons with unusual quantum numbers	Lead glass array (3053 crystals)		4 cm x 4 cm x 45 cm (3053 crystals)	em: 6%/√E	Mass resolutions: π° 8%, eta 4%	[Brab 93] [Cri 97]
	E-864	Strangelet search	198 CsI crystals (barrel + ring)		4 Xo deep		Used for soft photon tagging	[Ada 96]
			Lead (Spaghetti structure)	scintillating fibers	754 towers, 10cm x 10cm x 117cm	had: 35%/√E + 3.5%	compensating structure (lead:plastic = 4.55:1)	[Arm 98]
	E-821 (BNL g-2)	Measurement of g-2 (muons)	Lead, 6% Sb	scintillating fibers, 1 mm	22.5 x 14 cm, 24 cells, 13 Xo deep	6.8% @ 2 GeV electrons	high sampling fraction (50% absorber)	[Sed 00]
CEBAF (Jefferson Lab)	CLAS	Study of multi-particle final states of ep, γp reactions	2.2 mm Pb	10 mm plastic scintillator	16 Xo deep	10 - 12% @ 1 GeV	WLS fiber readout	[Amar 01]
	GlueX	Study of the confinement phenomenon	0.5 mm Pb (only 37% absorber)	scintillating fibers, 1 mm high sampling fraction	7.5°, 390 cm wide	5.4%/√E + 2.3%	time resolution 70 ps/√E	[Lev 08]
	Hall A	Measurements with polarized electrons, 1 - 6 GeV	GSO crystal PWO crystals (25)		15 cm deep, 6 cm diam. 2 x 2 x 23 cm		γ detector (GSO replaced PWO matrix)	[Fri 12] [Ney 00]
J-PARC (Tokai, Japan)	T2K	Neutrino oscillations	Near detector (ND280): 3 mm Pb / 10 mm scint. Far detector: Super Kamiokande			3.1%/√E + 1.4% + 0.2%/E	WLS readout, pixel APDs	[Min 07]
	E391a (KOPIO)	Search for K° → π° νν̄	Pure CsI crystals (576)		5 x 5 x 50 cm (24 Xtals), 7 x 7 x 30 cm (496 Xtals)	mass resolution K° → 3π°, 4.3 MeV/c²	time resolution 0.51 ns	[Dor 05]
PSI (Switzerland)	MEG	Search for μ → eγ	Liquid Xenon (900 liters)		14 Xo deep, 846 PMTs (35% wall coverage)	1.56% @ 54.9 MeV γ	time resolution 119 ps	[Ada 13a]
IHEP Serpukhov (Russia)	SIGMA-AIAKS	Study of rare processes in hadron interactions	2 cm steel plates	plastic scintillator (5 mm plates)	20 cm x 20 cm	em: 27%/√E, had: 52%/√E + 2%	Readout with WLS plates, Pos. resolution 2cm @40GeV	[Ant 90]
	GAMS	Search for glueballs and exotic mesons	Lead tungstate crystals (13 x 12 matrix)		13.9 mm x 13.9 mm, 22 - 26 Xo	em: 3% at 9.3 GeV	Mass resolution eta 2.9%	[Bli 97] [Bin 99]
ISIS (Rutherford)	KARMEN	Neutrino oscillations	Liquid scintillator (65000 liters)		18 cm x 18 cm x 3.5 m, 4 PMTs/cell, 512 cells	em: 11.5%/√E (MeV)	Position resolution 5 cm (for mip)	[Drex 90] [Dod 93]
INS-Tokyo (1.3 GeV e)		Coherent photoproduction of π°	Pure CsI crystals (64) (13.5 Xo long)		Hexagonal crystals, 70 mm apex-to-apex	em: 2.5%/√E	Mass resolution π° 7 GeV (at energies 0.5 - 1.0 GeV)	[Oku 95]
ELSA (Bonn)	BGO-OD	Nucleon spectroscopy	BGO crystals (480) arranged as Ball		Δθ = 8.7°, Δφ = 11.3°, 24 Xo deep	3% @ 1 GeV		[Ban 15]
	CB-ELSA / TAPS	Photonuclear reactions	Barrel: 1380 CsI(Tl) crystals (Crystal Barrel) Forward 216 BaF₂ hexagonal crystals		3 x 3 cm, 16 Xo deep, 5.9 cm diameter, 12 Xo	2%/√E, 1.5%/√E + 1.8%	APD readout	[Urb 15] [Die 15]
MAMI (Mainz)	A2	Study of non-perturbative regime strong interactions	Barrel: 720 NaI(Tl) crystals (Crystal Ball) Forward: 384 BaF₂ hexagonal crystals plus 72 PbWO₄ crystals		truncated pyramids, 5.9 cm diameter, 12 Xo, 17 Xo deep	2% @ 1 GeV	Formerly used at SLAC,DESY	[Nei 15]

Table C2 Calorimeter Systems in Collider Experiments

Accelerator	Facility	Principal physics goals	Calorimeter structure Passive	Calorimeter structure Active	Granularity	Energy resolution	Comments	Refs.
SppS (CERN)	UA1	Intermediate vector bosons; Top quark search; QCD tests; Physics beyond Standard Model	em: 1.2 mm Pb had: 5 cm Fe	1.5 mm polystyrene 1 cm plexipop	768 EM channels 416 HAD channels	em: 15%/√E + 3% had: 80%/√E	EM calorimeter was to be replaced by warm-liquid device. Was not realized.	[Cor 85] [Coc 86]
	UA2	Same as UA1	em: 3 mm Pb had: 15 mm Fe	4 mm PVT 5 mm PMMA	Tower structure; 2 or 3 longitudinal segments; 3000 PMT's	em: 16%/√E had: 30%/E**0.25	Good hermeticity; Good electron ID; Good long-term stability	[Bee 84]
Tevatron (FNAL)	CDF	Same as UA1 Also B-physics, CP violation	Central em: 3.2 mm Pb had: 25 mm Fe Forward em: 2.7 mm Pb had: 5.1 cm Fe	5 mm SCSN-38 10 mm PMMA proportional gas tubes proportional gas tubes	956 EM channels 576 HAD channels	em: 13.5%/√E had: 33%/√E + 4% em: 28%/√E had: 140%/√E	Not very hermetic Texas tower effect (ch. 4.8) Replaced in 1999 by a scintillator-based device	[Bal 88] [Bert 88] [Bran 88] [Cih 88] [Fuk 88]
	D0	Same as UA1	em: 3 mm U-238 had1: 6 mm U-238 had2: 46.5 mm Steel	2 x 2.3 mm LAr gaps throughout	.05 x .05, 4 segments 0.1 x 0.1, 4 segments 0.1 x 0.1, 1 segment	em: 157/√E + 0.3% had: 44/√E + 1.3/E + 5%	e/h = 1.08	[Aba 93,94] [Wim 89] [Aln 93]
HERA (DESY) (ep collider)	ZEUS	Proton structure functions, Electroweak interference, Physics beyond Standard Model	3 mm U-238	3.2 mm SCSN-38	em: 5 cm x 20 cm had: 20 cm x 20 cm Total 13000 PMT's	em: 18%/√E had: 35%/√E + 2%	See Section 9.1	[And 91] [Dre 90] [Der 91]
	H1	Same as ZEUS	em: 2.8 mm Pb had: 16 mm Steel Backward Pb/scintillating fibers (em 0.5 mm, had 1 mm)	3 mm LAr gaps	em: 3 or 4 segments, 42000 channels had: 4-6 s (18000 ch) em: 4.05 cm x 4.05 cm had: 11.9 cm x 11.9 cm (in 2 interaction lengths)	em: 10%/√E had: ≤65%/√E 29% for 1-7 GeV hadrons	Use complicated algorithms for "off-line compensation" (see ch. 4.9, 6.4) e/π rejection 300 @ 4 GeV Pb/fiber = 2.27 (em), 3.4 (h)	[Braun88,89] [Fla 90] [Andr 93,94] [Bar 94] [App. 96,97] [Ava 96,98]
	HERMES	Study spin structure nucleon	Lead glass blocks		840 blocks 9 cm x 9 cm x 50 cm	em: 5.1%/√E + 2%		
	HERA-B	CP violation in B0 system	Inner: 8 mm W Middle: 2 mm Pb Outer: 2 mm Pb	4 mm extruded plastic 4 mm extruded plastic 4 mm extruded plastic	2.2 x 2.2 cm, 23 Xo 5.6 x 5.6 cm, 20 Xo 11.2 x 11.2 cm, 20 Xo	20.5%/√E + 1.2% 11.8%/√E + 1.4% 10.8%/√E + 1.0%	Shashlik structure, 2100 cells Shashlik structure, 2128 cells Shashlik structure, 1728 cells	[Gia 06]
LEP (CERN)	ALEPH	Precision tests Standard Model W,Z spectroscopy; Higgs search, B-physics	em: 2 mm Pb (<12 Xo) 4 mm Pb (>12 Xo) had: 5 cm Fe	Proportional gas tubes Plastic streamer tubes	0.94° x 0.93° (3 segments) 3.7° x 3.7° (1 segment)	em: 18%/√E had: 80%/√E	Very good uniformity	[Cat 86] [Dep 89] [Dec 90]
	DELPHI	Same as ALEPH	em: 3 mm Pb had: 5 cm Fe	Proportional gas tubes Plastic streamer tubes	Order 1 cubic cm (9 segments) 3.8° x 3.0° (1 segment)	em: 23%/√E + 1.1% had: 100%/√E	"TPC" readout of em section	[Fis88] [DEL 91] [Abr 96]
	L3	Same as ALEPH	em: BGO crystals had: 5 mm U-238		12000 crystals 24 Xo deep (2 x 2 cm2) 2.5° x 5 cm (22000 ch)	em: 5% at 0.1 GeV, 1% at E>5 GeV had: 55%/√E + 5%		[Bak 85,87] [Sum 88]
	OPAL	Same as ALEPH	em: Lead glass blocks had: 8 cm Fe	3.2 mm MWPC (2 mm anode pitch)	11700 crystals, 20.5 Xo, 9.4 x 9.4 cm2 480+160 cathode pads, 1 longitudinal segment	em: 5%/√E had: 1.05/√E	e/h depends on gas mixture (see ch. 3.3.3) No saturation up to 50 GeV	[Are 89,90] [Adr 91] [Akr 90] [Dad 86] [Cha 88]
SLC (SLAC)	SLD	Same as ALEPH	em: 2 mm Pb had: 6 mm Pb	2.75 mm LAr gaps throughout	Tower structure 4 longitudinal segments 32500 channels	em: 8%/√E had: 45%/√E	Optimized for good hermeticity in barrel region	[Dub 86] [Hal 89] [Axe 93]

Table C2 Continued

Accelerator	Facility	Principal physics goals	Calorimeter structure — Passive	Calorimeter structure — Active	Granularity	Energy resolution	Comments	Refs.
LHC (CERN)	ATLAS	Higgs search; top spectroscopy; Supersymmetry; Physics at 1 TeV mass scale; Physics beyond Standard Model	em: Lead (1.5 mm); had: Copper Iron; forward (η>3.2): Cu/W	Lar (2 x 2.1 mm); Liquid argon (η>1.4); Scintillator (η<1.7); Liquid argon	0.025 x 0.025; 0.1 x 0.1	10%/√E+0.28/E+0.35%; 47%/√E + 2.2%	Ang. resol. 50-60 mr/√E (γ)	[Aub91,92c] [Aub 93a,b] [Aba 09,10] [Adr 09]
	CMS	Same as ATLAS	em: Lead tungstate crystals; had: brass (5 cm); |η|>3.2: steel	plastic scintillator (4 mm); quartz fibers (0.3 mm)	2.2 cm x 2.2 cm; 23 cm (26 Xo) deep; 0.087 x 0.087; 2 depth segments; 5 cm x 5 cm (|η| > 4); 10 cm x 10 cm (|η|<4)	em: 2% at 10 GeV, 1% at 30 GeV, 0.5% at 120 GeV; had: 110%/√E + 5%; had: 20% at 300 GeV	Readout with APDs, VPTs; Ang. resol. 45 mr/√E (PSD); Readout with HPDs (multipixel); Readout with PMTs	[Auf 98] [CMS 08] [Akc 97]
	ALICE	Quark-gluon plasma; Heavy-ion physics	Lead tungstate crystals		2.2 cm x 2.2 cm; 18 cm deep	3.6%/√E + 1.2% + 1.3%/E	Intended as γ spectrometer read out with APDs	[Con 05] [Aam 08]
	LHCb	CP violation, rare b-decays; b quark spectroscopy; exotic hadrons (pentaquarks)	em: 2 mm Pb; had: 10 mm Fe	4 mm extruded PS; 3 mm extruded PS	4x4, 6x6, 12x12 cm; 3 longitudinal segments; 25 Xo + 5.6 λ deep	8.2 - 9.4%/√E + 0.9%; 69%/√E + 9%	Shashlik structure, 25 Xo; longitudinal tile structure, WLS fiber readout	[Mac 09] [Dzh 02] [Alv 08]
	LHCf	pp scattering at η > 8.4	7 mm W (11 layers); 14 mm W (5 layers)	3 mm plastic scintillator; 3 mm plastic scintillator	20 x 20 - 40 x 40 mm; 44 Xo, 1.5 λ deep	5% for electrons > 100 GeV	Position measurements with Si strips or scintillating fibers	[Mas 12]
CESR (Cornell) (e+e-)	CLEO	B-physics; CP violation	CsI(Tl) crystals	CsI(Tl) crystals	7000 crystals, 16 Xo; 5 cm x 5 cm	em: 3.8% for 0.18 GeV; 1.6% for 5 GeV	em calorimetry only; photodiode readout	[Beb 88]
	CUSB	B-physics	Nal(Tl) + Pb glass; (5 segments Nal(Tl) + 1 segment (leadglass)	Nal(Tl) + Pb glass	4° x 8°; 15.8 Xo deep	em: 3.9%/E**0.25	em calorimetry only; PMT readout	[Klo 83] [Scha 84]
SLAC B-factory	BABAR	Same as CLEO		CsI(Tl) crystals (tapered)	6 cm x 5 -> 6 cm; 36 cm deep	2.8% at 100 MeV/c; 1.9% at 405 MeV/c (e-)	Position resolution; 13 (9) mm @ 100 (400) MeV	[Bar 99] [Sta 98]
KEK B-factory	BELLE	Same as CLEO		CsI(Tl) crystals (6624)	6 cm x 6 cm; 30 cm deep (16.2 Xo)	5% at 40 MeV; 1.5% at 800 MeV (γ)	Position resolution; 12 (8) mm @ 100 (450) MeV	[Ohs 96] [Ahn 98]
Super KEK-B	BELLE II	Studies of heavy quarkonia; Lepton violation in τ decay	barrel: CsI(Tl) crystals (6624); endcap: CsI crystals (2212)		6 x 6 cm, 16.2 Xo deep; 6 x 6 cm, 30 cm deep	<3% at E > 50 MeV (γ)	Time resolution < 1 ns; Readout with photopenthodes	[Abe 10] [Auf 15]
SPEAR (SLAC)	MARK II	Charm studies; Later Z decay at SLC	2 mm Pb	3 mm LAr	6 segments in depth; 14 Xo, 3000 channels; (projective strips)	em: 12%/√E		[Abr 78] [Abr 80]
	Crystal Ball	Charm studies; Later B-physics at DORIS		Nal(Tl) crystals	15.7 Xo deep; 672 + 60 crystals	em: 2.7%/E**0.25	Later used at DORIS (DESY) for Upsilon spectroscopy	[Ore 82] [Ner 85]
BEPC (Beijing)	BES	Charm studies; Tau physics	2.8 mm lead plates	aluminium gas tubes; 6 mm x 14-18 mm (Ar/CO2)	azimuth: 560 segments; barrel: 6 depth samples; endcap: 9 depth samples	em: 21%/√E	Position resolution: 15 mm in x, 17 mm in y; 36 mm in z (charge division)	[Bai 94]
	BESS III	Tau/charm physics	CsI(Tl) crystals (6240)	(truncated pyramids)	5.2x5.2 - 6.4x6.4 cm; 28 cm (15.1 Xo) deep	@ 1.5 GeV: 2.3% barrel; 4.1% endcap	2 photodiodes per crystal	[Don 08]
TRISTAN (KEK)	AMY	Top search; QCD tests; Electroweak interference	3.5 mm Pb (<9.4 Xo); 7 mm Pb (last 5 Xo)	Proport. gas tubes; (wire + cathode strip readout)	13 mr x 16 mr (9.4 Xo); 27 x 33 mr (last 5 Xo); 10000 channels	em: 29%/√E + 10%	Resolution improved in 3T magnetic field	[Kaj 86] [Aba 92]
	TOPAZ	Same as AMY	Barrel: Lead glass blocks (20 Xo); Endcap: 2 mm Pb; 3 mm Pb (>3Xo)	Proport. gas tubes	4300 blocks, 130 cm2; 3 longitudinal segments; 2 x 1024 pads	em: 8%/√E; em: 19%/√E		[Kaw 88] [Hay 92]
	VENUS	Same as AMY	Barrel: Lead glass blocks (18 Xo); Endcap: 1.7 mm Pb; Lum. monitor: Lead	3 mm LAr; 1 mm plastic fibers	5160 blocks; 3800 channels	em: 5.2%/√E + 0.7%; em: 11.3%/√E + 1.4%		[Oga 86] [Sumi 88] [Tak 92]

Table C2 Continued

Accelerator	Facility	Principal physics goals	Calorimeter structure		Granularity	Energy resolution	Comments	Refs.
			Passive	Active				
DAΦNE (Frascati)	KLOE	CP violation in Kaon decay	Lead (Spaghetti structure)	1 mm plastic fibers (48% by volume) Orthogonal to particles	3.5 cm x 3.5 cm modules 450 cm in z	em: 4.8%/√E (20 - 100 MeV)	z position determined by timing (50ps/√E)	[Bab 93] [Ant 95]
LEAR (CERN)	Crystal Barrel	Meson spectroscopy in proton-antiproton annihilation	CsI crystals (1400) 16Xo deep		3 cm x 3 cm	em: 2%/√E	Readout with vacuum triodes	[Ake 92]
	JETSET	Study of low-mass resonances in proton-antiproton annihilation	Lead	Scintillating fibers (1 mm diameter) Spaghetti structure	12 Xo deep trapezoidal blocks	em: 6.3%/√E	Pb:plastic = 50:35	[Her 90]
	CPLEAR	CP violation in neutral kaons	1.5 mm lead sheets 18 sampling layers	High-gain gas tubes 17112 in total	4.5 mm wide tubes	13%/√E (photons)	Intended for 50 - 500 MeV photons	[Adi 92,97]
	OBELIX	Exclusive final states in low-E proton-antiproton annihilation	3 mm lead sheets (20)	Limited-streamer tubes	9 mm x 9 mm		Intended for 0.1 - 1.0 GeV photons	[Aff 93]
RHIC (BNL)	PHENIX	Heavy-ion physics Search for quark-gluon plasma	1.5 mm Pb	4.0 mm scintillator	10 cm x 10 cm x 18 Xo; 4 cm x 4 cm x 40 cm	em: 7.8%/√E + 1.4%; em: 7.1%/√E + 0.7%	Shishkebab	[Kis 94]
			Lead-glass array (10000)		3.5cm x 3.5cm x 30 cm; 18 (25) Xo barrel (cap)	em: 1.8%/√E +0.3%/E + 0.8%	For low-Pt γ(< 1 GeV/c)	[Dav 94]
	STAR	Same as PHENIX	5 mm Pb	4 mm scintillator	0.05 x 0.05	em: 16%/√E + 3.5%	Readout with 1 mm WLS fibers	[Bel 96] [Lo 96]
DORIS (DESY)	ARGUS	Upsilon (4S) decay	1 mm Pb in barrel 1.5 mm Pb in endcap	5 mm scintillator	10 cm x 10 cm; 12.4 Xo deep	em: 7%/√E + 7%	Readout with WLS bars	[Hof 79,82] [Dre 83] [Alb 89]
PETRA (DESY)	TASSO	Search for top quark electron-positron annihilation	2 mm Pb	5 mm liquid argon	barrel: 7 cm x 7 cm endcap: 5" x 5-8 cm	em: 10%/√E; barrel: 5% for E > 4 GeV	Barrel 6-8 Xo deep Endcap 13.5 Xo deep	[Ast 81] [Kad 81]
	CELLO	Same as TASSO	1.2 mm Pb	3.6 mm liquid argon	8000 channels	em: 8.5%/√E	20 Xo deep	[Beh 81]
	JADE	Same as TASSO	Lead-glass blocks		8.5 cm x 10 cm x 13 Xo	em: 6%/√E + 3.5%	Angular resolution 7 mr	[Ast, Flu 81]
	MARK J	Same as TASSO	5 mm Pb	5 mm scintillator	em: 2° (φ) x 4° (θ); 6 (4) depth segments in barrel (endcap)	em: 12%/√E (gammas)	Angular resolution 5° - 7°	[Ast, Flu 81]
PEP (SLAC)	MAC	Search for top quark	em: 2.5 mm Pb/Sb/Sn had1: 2.5 cm Fe had2: 10 cm Fe	Proportional Wire Chambers (tubes)	16 Xo deep, 3 segments	em: 23%/√E had: 75%/√E √	Position resolution 0.7° in φ, 1.4° in θ	[Ali 89]
VEPP-4M (Novosibirsk)	KEDR	electron-positron annihilation up to 6 GeV, charm/tau physics		Liquid krypton		6% at 100 MeV 2.2% at 1 GeV (photons)	Position resolution gammas 1.2 (0.8) mm at 0.2 (1) GeV	[Aul 92,97] [Pel 09]
VEPP-2M (Novosibirsk)	SND	CP violation in neutral kaons	3 layers, 2.9Xo + 4.8Xo + 5.7Xo deep	NaI(Tl) crystals	1632 crystals 9° x 9° segments	5.5% for 0.5 GeV photons		[Ach 97]
	CMD-2	Same as SND		CsI(Na) and CsI(Tl) crystals	892 crystals 6 cm x 6 cm x 15 cm	12% for 0.5 GeV electrons		[Aul 93]
	CMD-3	Measure hadronic cross sections in energy range 1.4 - 2 GeV	Barrel: 400 liter LXe (detect ionization), 5 Xo followed by CsI (1152 crystals), 8 Xo Endcaps: BGO (680 crystals), 13.4 Xo		11° x 11° readout cells 6 x 6 x 15 cm; 2.5 x 2.5 x 15 cm			[Fed 06] [Pel 09]
FAIR (Darmstadt)	R³B/CALIFA	Reactions with Relativistic Radioactive Beams, up to 1 GeV/A	CsI(Tl) crystals, 1952 in barrel			5 - 6% @ 1 GeV γ	Readout with APDs	[Alv 14]
	PANDA	Study of charmed meson states and glueballs in pp̄ annihilations	PbWO4 crystals. Barrel: 11360 crystals Forward (backward) endcap: 3895 (600) crystals		21.3 - 27.2 mm diameter 200 mm long	1.95%/√E + 0.48%	Readout: APD in barrel and bEC, VPtetrodes in fEC	[Kav 11]

Table C3 Calorimeter Systems in Non-Accelerator Experiments

Laboratory	Facility	Principal physics goals	Calorimeter structure Passive	Calorimeter structure Active	Granularity	Energy resolution	Comments	Refs.
Morton Salt Mine (USA)	IMB	proton decay / neutrino astrophysics	8000 tonnes of purified water (3300 tonnes fiducial volume)		2048 PMTs (8") separated by 1 m			[Bio 83]
Kamioka (Japan)	Kamiokande	proton decay / neutrino astrophysics	2140 tonnes of purified water (+ 1.5 m thick water veto counter)		948 PMTs (50 cm) separated by 1 m	electrons: 22% at 10 MeV		[Hir 88]
	Superkamiokande	atmospheric neutrinos / solar neutrinos / proton decay	50000 tonnes of purified water (fiducial mass 22000 tonnes)		11460 PMTs (20") ID; 1885 PMTs (8") OD; 40% of ID wall covered	electron: 2.6%/√p + 0.6%; muon: 0.7%/√p + 1.7%	See Section 10.1	[Kos 98] [Oku 99] [Fuk 03]
	KAMLAND	Oscillations reactor neutrinos	1000 tonnes of ultrapure liquid scintillator shielded by 3200 tonnes of ultrapure water		1325 17" + 554 20" PMTs (35% of 4π coverage)		Sensitive to 56 reactors + geoneutrinos	[AbsS 10] [Gan 11]
Daya Bay (China)	Daya Bay Reactor Neutrino Expt.	Measure neutrino oscillations close to nuclear reactor core	v̄ detector: 20 t Gd-doped liquid scintillator inside 20 t of undoped liquid scintillator inside 37 ton of mineral oil		Each detector is viewed by 192 8-in PMTs	7.5%/√E(MeV) + 0.9%	There are 8 such detectors in 3 different halls; There are 6 nuclear reactors	[An 12b]
SNOLAB Sudbury,Canada	SNO	solar neutrinos in particular from Boron-8	1000 tonnes of heavy water shielded by ultrapure light water		9438 inward looking PMT (55% of 4π)	electrons: 15% at 10 MeV	position resol. 10 MeV e: 1m; angular resol. 10 MeV e: 25°	[Aar 87] [Bog 00]
	SNO+	0ν ββ decay (^{130}Te) / ν physics (solar, geo, atm., SN)	780 tonnes of ultrapure liquid scintillator with 800 kg tellurium dissolved in it					[Andr 16]
Grand Sasso (Italy)	MACRO	magnetic monopoles / cosmic rays / neutrino astrophysics	crushed rock; 15 sampling layers; total depth 9.6 m	liquid scintillator; limited streamer tubes; track etch detectors	11.9m x 75 cm x 25 cm		Position resol. 1 cm, 0.2°; Helium/n-pentane; Lexan/CR39	[Ambr 96]
	Borexino	Solar neutrinos especially from Be-7 decay	Liquid scintillator shielded by ultrapure water		20 cm PMTs; 25% coverage		position resol. 1 MeV e:10cm; 300 photoelectrons/MeV	[Ali 98]
	EAS-TOP	cosmic-ray studies	had: 13 cm Fe; 9 layers; 12 m x 12 m x 3 m	streamer tubes; quasiproportional tubes; scintillators (timing)	3 cm x 3 cm	had: 15% at 1 TeV; 25% at 5 TeV (leakage)	Coincident measurements tunnel / mountain top (collaboration with MACRO)	[Agl 93] [Adi 99]
	ICARUS	proton decay, cosmic rays / neutrino oscillations (CERN)	Liquid argon Time Projection Chamber; 600 ton LAr Imaging calorimeter		2 mm x 2 mm x 2 mm	em: 10% at 380 keV	Position resolution 60 μm (results of 3-ton prototype); 600 ton detector	[Ben 93] [Cen 94,99] [Alm 04]
South Pole (~2000 m)	AMANDA	neutrino astrophysics	Arctic ice; Cherenkov light detected with 800 PMTs		PMT spacing Vert: 20 m, Hor: 60 m; 5 mm x 5 mm x 5 mm		Up to 800 PMTs (8 strings) (0.05 km³)	[Mil 97]
	IceCube	neutrino astrophysics / high-energy cosmic rays	Arctic ice; Cherenkov light detected with 5,160 PMTs		86 strings, hexagon grid; Vert: 17 m, Hor: 125 m		5160 Digital Optical Modules; Instrumented volume 1 km³	[Hal 14]
Mediterranean	ANTARES (near Toulon) (~2400 m)	neutrino astrophysics / atmospheric ν oscillations / high-energy cosmic rays	Sea water; Cherenkov light detected with 885 PMTs		13 strings, 295 opt. mod.; Vertical spacing 14.5 m; Horizontal 60 - 100 m		Prototype for KM3NeT-ORCA; instrumented volume 0.1 km³	[Hub 99] [Age 11]
	NESTOR (near Pilos) (~3800 m)	Same as ANTARES	Sea water; Cherenkov light detected with 15" PMTs					[Ana 94] [Agg 05b]
	NEMO (near Sicily) (~3500 m)	Same as ANTARES	Sea water; Cherenkov light detected with PMTs; Acoustic signal possibilities studied				Site for KM3NeT-ARCA (Capo Passero)	[Cap 09] [Aie 10]
Lake Baikal (Siberia)	NT200+ (~1300 m)	neutrino astrophysics	Sweet water; Cherenkov light detected with 229 PMTs		8 strings with 192 PMTs; vert 6.25 m, hor 18.6 m; 3 strings (36 PM), 100m			[Ayt 06,08]

Table C3 Continued

Test site	Built by	Principal physics goals	Calorimeter structure Passive	Active	Granularity	Energy resolution	Comments	Refs.
Dugway (Utah)	Fly's Eye (Hi Res)	high-energy cosmic rays	The Earth's atmosphere	(signals provided by nitrogen scintillation)	1° x 1° (360° x 31° coverage)	had: about 10% (high energy showers)	Only sensitive to showers with E >10 PeV	[Bai 85] [Cas 85]
Southern Utah	Telescope Array Project	high-energy cosmic rays GZK cutoff	The Earth's atmosphere signals from nitrogen scintillation (3 stations) + scintillation surface detectors (muons)		800 km² covered, 3-59° 35 km telescope spacing 507 μ detectors		In total 38 fluorescence telescopes (3 stations)	[Abu 12a,b]
KIZ Karlsruhe (Germany)	KASCADE	high-energy cosmic rays (chemical composition)	The Earth's atmosphere sample electrons, γs, μs, hadrons after 10 λ had: 12-36 cm Fe slabs 1 cm TMS,TMP 16 m x 20 m x 4.5 m scintillators (timing)		252 em stations (20 Xo) 4 0.8 m² cells/station 25 cm x 25 cm	had: 10% at 10 TeV (MC)	Total area covered 200 m x 200 m 11.5 λ deep	[Kla 97] [Eng 99]
	KASCADE-GRANDE	Same as KASCADE (higher energy reach)	Added 370 m² (Grande) + 80 m² (Piccolo) plastic scintillator arrays, Surface area 7 x 8 km				Sensitive from 0.1 - 100 PeV	[Ape 10]
Akeno (Japan)	AGASA	high-energy cosmic rays GZK cutoff	100 km² array of 111 scintillators + 27 μ counters 1 km² array of 156 scintillators + 11 μ counters		S 2.2 m², μ 2.8 - 20 m² S 1 m², μ 25 - 50 m²		Sensitive to E > 100 PeV	[Chi 92]
YangBajing (Tibet)	ARGO-YBJ (4,300 m asl)	γ astronomy high-energy cosmic rays	5,800 m² array of RPCs (93% active area)		61.8 x 6.75 cm		Sensitive from 10 TeV - 1 PeV	[Bar 15]
	LHAASO - KM2A	γ astronomy high-energy cosmic rays	1 km² array of densely packed e, μ, h detectors fluorescence/C telescope, shower core detector		e det:1x1m, spaced 15m μ det:6x6m, spaced 30m		Sensitive to γs with E >30 TeV Hadrons with E = 1 - 10 PeV	[Cui 14]
Yakutsk (Siberia)	Yakutsk array	high-energy cosmic rays	Cherenkov telescope (field-of-view 308 sq.deg.) 10 km² surface detector array (58 plastic scint. of 2 m²), 4 μ counters, 48 PMTs		Multi-anode (16x16) PMT		Sensitive from 1 - 10000 PeV	[Iva 15] [Iva 09]
La Palma Canary Islands	HEGRA	high-energy γ astronomy (gammas with E > 0.1 TeV)		The Earth's atmosphere (signals provided by Cherenkov light)	0.25°/pixel, 271 pixels/telescope 5 telescopes		angular resolution 0.1° field of view telescope 4.3°	[Dau 97]
	MAGIC (2,200 m asl)	high-energy γ astronomy		Two 17 m Cherenkov telescopes, operating in coincidence (stereoscopy)	1039 PMTs / telescope 0.1° pixels, f. of view 3.4°		Sensitive to E > 50 GeV	[Ale 16]
Mt. Hopkins (Arizona)	VERITAS (1,270 m asl)	high-energy γ astronomy		Four 12 m Cherenkov telescopes	399 pixels / telescope field of view 3.5° scope		Sensitivity 50 GeV - 50 TeV	[Ver 16]
Khomas Plateau (Namibia)	HESS (1,800 m asl)	high-energy γ astronomy cosmic-rays in PeV range		Four 12 m Cherenkov telescopes, 120 m apart One 28 m telescope (614 m² mirror) in center	960 pixels, 0.16° fov 5° 2048 pix, 0.067° fov 3.2°		Sensitivity 50 GeV - 1 PeV	[HES 16]
Chili & Namibia	Cherenkov Telescope Array	high energy γ astronomy cosmic rays up to 10 PeV		>100 Cherenkov telescopes			Sensitivity 10 GeV - 10 PeV	[CTA 11]
Malargüe (Argentina)	Pierre Auger Observatory	high-energy cosmic rays GZK cutoff	The Earth's atmosphere signals from nitrogen scintillation (4 stations) 1660 surface detectors (C, muons), 1.5 km apart		4 x 6 telescopes 440 pix/scope, 1.5° x 1.5° 12,000 l H₂O/detector	16% (low E) - 12% (high E)	3,000 km² surface area sensitivity 1 - 1000 EeV	[PAC 15]
Caucasus (3860 m)	Moscow State University	Cosmic rays with E>100 GeV	Iron slabs	Argon ionization chambers	8 slabs of iron 1 lambda each		First experiment to use a sampling calorimeter	[Gri 58]
Sea level and 2900 m	Univ. Maryland	Cosmic rays 80-E-8000 GeV	Iron slabs	Liquid scintillator	2 x 1 m, 7.4 cm deep		Measured shower profiles at high energies	[Sio 79]

Table C3   Continued

Test site	Built by	Principal physics goals	Calorimeter structure Passive	Active	Granularity	Energy resolution	Comments	Refs.
Balloon	MUBEE Collaboration	Cosmic protons/nuclei 10 - 200 TeV	Lead	Photographic emulsion	25 x 0.6 Xo lead plates		Calorimeter serves as target Measure em component	[Zat 93]
	JACEE Collaboration	Cosmic protons/nuclei 1 - 1000 TeV	Lead	Photographic emulsion	22 x 0.3 Xo lead plates	15 - 30% (em component)		[Bur 86]
	Goddard Space Flight Center	Cosmic protons/nuclei 50 - 1000 GeV	Tungsten + iron	Plastic scintillator (6.4 mm thick)	W: 12 x 1 Xo / Fe: 7 x 0.5 lambda	25%	Electronic calorimeter in balloon	[Rya 72]
	MASS Collaboration	Cosmic antiproton flux 3 - 19 GeV	Brass 7.3 Xo (0.75 lambda)	Streamer tubes	40 layers		Calorimeter part of magnetic spectrometer	[Hof 96]
	WiZard/CAPRICE Collaboration	Cosmic (anti)proton and positron fluxes	Tungsten 7 Xo deep	Silicon strips	3.6 mm wide Si strips 7 W layers		Calorimeter part of magnetic spectrometer	[Boc 96]
	HEAT Collaboration	Cosmic (anti)electron fluxes Up to 100 GeV	Lead 10 Xo deep	Plastic scintillator 1 cm thick sheets	10 x 1 Xo lead plates		Imaging calorimeter part of magnetic spectrometer	[Bar 97]
	CREAM	Cosmic protons/nuclei 1 TeV - 1 PeV	3.5 mm W	0.5 mm scint. fibers (ribbons)	20 Xo deep, 50x50 cm readout every Xo 2560 electronic channels	~6% @ 100 GeV electrons	Two 9.5 cm C targets in front 64-pixel HPD readout	[Ahn 07]
	ATIC	Cosmic protons/nuclei 50 GeV - 100 TeV	BGO crystals (320) stacked in 8 layers 50 x 50 x 20 cm		2.5 x 2.5 x 25 cm 18 Xo deep	2% @ 150 GeV electrons (91% containment)	0.75λ C target in front PMT readout	[Gan 05]
	BETS	Cosmic electrons 10 GeV - few 100 GeV	8 Pb plates, 5 mm thick 28 x 28 cm	1 mm scintillating fibers 36 layers of 280 fibers	10,080 fibers read out with CCD camera	14 - 17%	proton rejection power > 2000	[Mur 98] [Tor 01]
Satellite (Proton 1-4)	Moscow State University	Cosmic protons/nuclei 10 GeV - 1 PeV	Iron 2.5 lambda total depth	Plastic scintillator			First calorimeter operated outside Earth's atmosphere	[Aki 69]
Satellite (Sokol 1-2)	Moscow State University	Cosmic protons/nuclei 10 GeV - 1 PeV	Iron 8 x 0.68 lambda	Plastic scintillator				[Gri 90]
Satellite Resurs-Arktika	PAMELA Collaboration	Cosmic antiprotons and positrons, 0.1 - 150 GeV	Tungsten 16 Xo deep	Silicon strips	2.4 mm wide Si strips 23 W layers of 0.7 Xo		Imaging calorimeter part of magnetic spectrometer	[Pam 99]
Satellite (GRO)	EGRET Collaboration	Cosmic gamma rays 20 MeV - 30 GeV	NaI(Tl) crystal 8 Xo deep			15% FWHM (gammas)		[Kan 88]
Satellite	GAMMA-1 Collaboration	Cosmic gamma rays 50 MeV - 5 GeV	Lead Total depth 7.4 Xo	Plastic scintillator 5 mm thick sheets	24 Pb layers of 2 mm	35 - 50% FWHM		[Aki 88]
Satellite	Fermi (formerly GLAST) Large Area Telescope	high-energy cosmic γ rays transients (GRB, SN) dark matter search	CsI(Tl) crystals, 16 modules with 96 crystals each 8 layers, 10 Xo deep W-based gamma converter, 16 layers 0.1 mm (x12), 0.7 mm (x4) 228 μm Si strips		2.7 x 2.0 x 32.6 cm Xtals 1536 channels/layer	2% (196 GeV e-) 4% (5 GeV e-)	Sensitivity 10 MeV - 300 GeV Crystals read out with PIN diodes on both ends (different dynamic ranges)	[Atw 09]
Satellite	DAMPE Collaboration	DArk Matter Particle Explorer high-E e,γ (5 GeV - 10 TeV) cosmic rays (0.1 - 100 TeV)	BGO crystals, 14 layers of 24 crystals each (31 Xo, 1.6 λ deep) W-Si tracker/converter, 6 layers 112 x 112 cm, 3 x 1 mm W, 300k channels	Si strips 9.5 cm long 121 μm pitch	2.5 x 2.5 x 60 cm Xtals	1.5% @ 100 GeV e- 40% @ 800 GeV p	Flies on Chinese satellite Xtals read w PMTs (both ends)	[DAM 16]
International Space Station	AMS-02	Cosmic-ray composition Search for dark and anti-matter	Pb/scintillating fiber calorimeter 65 x 65 cm, 17 Xo deep, 0.7 λ		18 longitudinal samplings 72 lateral samplings	10.4%/√E + 1.4% (e-)	Read out with PMTs (1296 channels)	[Gal 15]
	CALET	e, γ, p, ions at high energy Gamma Ray Bursts Dark matter search	PbWO4 crystals, 12 layersof 16 crystals each 32.6 x 32.6 cm, 27 Xo deep, 1.2λ W/scint. fiber converter + preshower detector 45 x 45 cm, 3 Xo deep		2.0 x 1.9 x 32.6 cm 0.2 Xo (5) + 1 Xo (2) W 14 layers of 1x1 mm fiber	7% @ 10 GeV e- 2% @ 10 TeV e- (MC)	2 layers readout by PMTs rest by Pd/APDs readout by multi-anode PMTs	[Mar 12]

Table C4   Generic Calorimeter Prototypes

Laboratory where tested	Facility	Principal R&D goals	Calorimeter structure Passive	Active	Granularity	Energy resolution	Comments	Refs.
CERN	SPACAL	Develop/test high-resolution compensating lead/scintillator calorimetry	Lead Spaghetti structure	1 mm plastic fibers (20% in volume)	155 hexagonal towers 43 mm apex-to-apex	em: 13%/√E, had: 32%/√E + 1%		[Aco 90-92] [Wig 92] [Liv 95]
	RD1	Develop SPACAL technology for LHC experiments	Lead Spaghetti structure	1 mm, 0.5 mm fibers (20% in volume)	4 cm x 4 cm, 4 towers	em: 9.2%/√E + 0.6%	Modular structure. Also projective structures	[Bad 94a,b]
	Projective Spaghetti	Study projective fiber structures	AF17 alloy (Bi/Pb/Sn)	1 mm plastic fibers (metal:fiber = 3.17:1)	20.5 mm x 20.5 mm (front face)	em:14%/√E	Modular structure with truncated pyramides	[Bert 95]
	SPAKEBAB	Combine advantages SPACAL with excellent em resolution	Perforated lead sheets (0.63 mm, 0.89 mm)	Thin scintillator plates (1 mm)	36 mm x 36 mm	em: 5.5%/√E		[Dub 96]
	RD36	Test performance "shashlik" calorimeter structures	2 m lead	4 mm scintillator	47 mm x 47 mm 27.5 Xo deep	em: 8.4%/√E+.4%/E+.8%	Readout with looping WLS fibers (1.2 mm), 25/cell	[Asp 96] [Bad94c.95]
	NOMAD R&D	Test performance "shashlik" with 2-sided WLS readout	1.5 mm lead plates (20 Xo)	4 mm scintillator	100 mm x 100 mm	em: 7%/√E	Tested with PMT, SPD, APD and phototetrode readout	[Aut 93]
	SICAPO	Study feasibility/benefits of silicon as active material for sampling calorimeters	10 mm uranium plates 60 in total, 5.7 lambda	400 μ m silicon	28 cm2 cells	em: 18%/√E, had: 50%/√E		[Bor 89,93] [Ang 90,92] [Fur 95]
	UA1 Upgrade	Equip UA1 detector with novel em calorimeter	Depleted uranium em: 2 mm, had: 5 mm forward: 10 mm	TMP, 2 x 1.25 mm	14 cm x 14 cm, 15 cm x 12 cm	em: 12%/√E+10%/E+1%, had: 10% at 70 GeV, forward (e): 31%/√E+6%	Studied effect of E-field on e/π ratio	[Bac 90] [Kra 92]
	PPC	Check feasibility of readout with Parallel Plate Capacitors	15 Fe layers with depth from 45-105 mm	Ceramic PPCs, gas: CF4/CO2	5 cm x 5 cm	em: 92%/√E + 3%, had: 192%/√E + 9%		[Ben 97] [Are 95,96] [Biz 93]
	SSC/SDC	Prototype endcap calorimeter	6.35 mm lead sheets	3HF-doped polystyrene (4 mm thick plates)	11 cm x 11 cm	em: 19.5%/√E	Read out with WLS fibers	[Abr 97]
	Crystal Clear	Study of new scintillators	Cerium fluoride		20 mm x 20 mm, 30 mm x 30 mm	em: 0.7% for E > 50 GeV	High-energy tails when readout with SPDs	[Auf 96a,c] [Auf 96b]
	Thin Gap Turbine (TGT)	Study of new LAr structure for high-rate applications	1.65 mm lead 2.0 mm stainless steel	6 x 0.8 mm Lar gaps	Pads 42.5 mm x 30 mm	em: 10 - 13%/√E		[Ber 95] [Braun 96]
	BAYAN (Planar version)	Study of novel sampling calorimeter structure	0.37 mm lead foils	thin scintillator sheets (1.4 mm)	7 cm x 7 cm towers 23 Xo deep	em: 7.7%/√E + 0.2%	WLS fibers embedded in edge	[Dol 95]
	DREAM	Dual-readout calorimetry Test of principles	Cu tubes 4x4 mm (5,130 tubes)	3 scintillating fibers +4 clear fibers per tube	19 hexagonal towers, 80 mm apex-to-apex		Details in Section 8.2	[Akc 04,05]
	RD52	Dual-readout calorimetry Tests of suitable crystals	PbWO4 [Mo] crystals BGO em section, BSO	Individual crystals, or 100-crystal BGO ecal			Details in Section 8.2	[Akc 09-11]
		Fiber modules, high sampling fraction	Cu or Pb absorber	50% fiber filling fraction	4.6 x 4.6 cm towers	em: 13%/√E + 1% @ 1.5°	Details in Section 8.2	[Akc 14c]
	CALICE	Tests of calorimeter modules intended for application in PFA detector systems	em: W sheets 4Xo..8Xo.12Xo (10 each), had: Fe or W sheets 17 mm of 10 mm	Silicon pads, Plastic scintillator strips, Scint. tiles (38 layers), 1x1 cm RPCs	1 x 1 cm pads, 5x45 mm, SiPM readout, 3x3 cm, 7620 tiles, SiPM, 500,000 channels	em: 16.5%/√E + 1.1%	Details in Section 8.3, Details in Section 8.3	[Sef 15]
	NOE	R&D for ν detector in CERN - LNGS beam	Iron bars, 4 x 4 cm 6 m long	2 mm scintillating fibers 14, 23 or 33 fibers/bar			Readout (2 sides) with PMTs. Time resolution ~ 1 ns	[Ale 11]

Table C4 Continued

Laboratory where tested	Facility	Principal R&D goals	Calorimeter structure Passive	Active	Granularity	Energy resolution	Comments	Refs.
FNAL	WALIC	Warm-liquid calorimetry options for SSC	6.35 mm Pb/Fe sheets or multiples of these	TMP, 2 x 1.25 mm electrodes	13 cm x 26 cm electrodes	16%/√E + 0.3/E + 1.3% (electrons)	Studied effect of E-field (Birks' constant) and sampling fraction on e/π	[Aub 92a,b] [Aub 93c]
	Hanging File	Systematic study parameters sampling calorimeters	3.2 mm Pb/Fe/Al or multiples of these	Plastic scintillator (3 mm thick)	1 m x 1 m	Depending on configuration	Readout with WLS fibers embedded in scintillator	[Bere 93]
	CDF miniplug R&D	Study towerless shashlik structure, liquid scintillator	4.8 mm thick Pb plates (15 cm x 15 cm)	Liquid scintillator	2 cm x 2 cm (towerless, Sect. 8.2)	em: 18.1%/√E	Read out by WLS fibers spaced by 1 cm	[Gou 99]
	SSC/em calorimeter	Study cheap high-resolution technology	Brass tubes filled with lead	Brass tubes filled with gas	4.7 mm x 4.7 mm 56 planes	em: 24%/√E	Angular resolution 0.11/√E rad	[Bad 91]
	High-pressure gas	Study radiation hard solution for forward region LHC/SSC	Stainless steel tubes Copper electrodes	Tubes filled with gas at 100 atm	9.5 mm tubes (OD) 7.9 mm ID, 4 mm Cu	em: 32%/√E	Strong dependence response on angle of incidence	[Gio 90] [Bag 95]
	BTeV ECAL	Prototype studies for em calorimeter for BTeV experiment	Array of PWO crystals (5 x 5)		27 x 27 x 220 mm	1.8%/√E + 0.33% + 2.4%/E	σ_x: 2.8/√E + 0.16 mm	[Bat 03]
KEK	Fast-gas EM	Check properties of an em calorimeter with thin-gap wire chamber readout	Brass, 20 plates of 12 mm thickness	Wire chambers (3 mm thick)	12 cm x 12 cm (2 mm wire spacing) 16.5 Xo deep	em: 25%/√E	Rise time 5 ns, width 10 ns	[Che 94]
	Hanging-file Lead/scintillator	Systematic study of e/π ratio and hadronic resolution in lead/scintillator structures	Lead, 4 - 16 mm in steps of 2 mm Total depth 1 m	Plastic scintillator 2, 4 or 6 mm	1m x 1m x1m 22 - 84 PMTs configuration dependent	hadr: 33.6%/√E (best value)	Tested with 1 - 4 GeV pions	[Suz 99]
	Pb/plastic	Compare Pb/fiber & Pb/tile (same sampling fraction)	Lead	1 mm fibers Tiles (0.5, 1, 2 mm)	7 cm x 7 cm (fiber) 10 cm x 10 cm (tile)	em: 13-20%/√E (depending on geometry)	Pb:plastic = 4:1 (all cases) fibers perpendicular to beam	[Ari 92]
	Diamond calorimetry	Test feasibility of diamond as active calorimeter material	Tungsten, 1 Xo thick 20 layers	Diamond, 3 cm x 3 cm	1 cm x 1 cm	em: 19%/√E +8%/E+2.2%		[Tes 94]
(Gadolinium silicate (cerium doped))		Test new crystal for its calorimetric properties	Cerium doped gadolinium silicon oxide (GSO)		Diameter crystal 4.9cm Depth 19.6 cm (14 Xo)	1 GeV e-: 7.7% FWHM		[Kob 91]
(Traditional calorimetry for Linear e⁺e⁻ Collider)		Test Pb/fiber calorimeter at low energies (E <10 GeV)	2 mm lead plates 20x20 cm, 130 cm deep	1 mm scintillating fibers Pb fiber = 4:1	5 x 5 cm cells, readout by PMTs	em: 14%/ E had: 38%/ E + 12%	Excellent e/π separation down to 1 GeV	[Ish 97]
Brookhaven (AGS)	RD3/LKr	Comparison Lar/LKr	1.3 mm lead (accordion structure)	Lar or LKr 2 x 2 mm gaps	2.5 cm x 2.7 cm 45 cm deep	em: 6.7%/√E (LKr) 7.7%/√E (Lar)		[Ben 94a]
		Sampling frequency studies	0.8 mm lead	LAr	same	em: 6.3%/√E		[Ben 94b]
	KOPIO	R&D calorimeter for K⁰→π⁰νν̄ experiment at BNL	0.35 mm Pb plates(240) 15.9 Xo total depth 0.275mm Pb plates (300) 15.9 Xo total depth	1.5 mm PS plates Shashlyk,72 fibers/mod same as above	12 x 12 cm modules (9) readout by PMTs 11 x 11 cm modules (9) readout by APDs	em: 3.8%/√E + 0.8% em: 2.74%/√E + 1.96%	time res: 72/√E + 14/E ps	[Ato 04] [Ato 08]
Berkeley (LBL)	Allene-doped Lar	R&D for heavy-ion detection at RHIC (BNL)	2 mm Fe (40x) 6 mm Fe (110x)	3.5 mm Lar gaps	radius 50 cm, depth 6λ 93 readout channels	had: 81%/√E + 0.9%	Tested with heavy ions 1 - 1.7 GeV/nucleon	[Dok 91]
Serpukhov	High-pressure gas	Study radiation hard solution for forward region UNK/SSC	2 mm thick lead disks (25 Xo)	High-pressure Xe,Ar (20 atm)	radius 40 mm segmented in 4 cells	em: 3.1% at 26.6 GeV		[Kha 90]
	Drift chamber calorimetry	Study dc gas calorimetry for UNK/SSC/LHC	Uranium 6.5 mm (60x) Steel 20 mm (15x)	Various gas mixtures (up to 40 atm)	42 cm x 42 cm x 6.5λ 3 electrodes/drift plane	had: 69%/√E + 300%/E		[Den 93]
Karlsruhe (KfZ) ITEP (Moscow)	Warm-liquid studies	Study feasibility/properties of TMS as active material	Various absorbers (Fe, Pb, U)	Tetramethylsilane	50cm x 50cm chambers 6 and 12 cm electrodes	em: 16%/√E (12 mm Fe) had: 60%/√E (25 mm Fe)	Measured e/mip, e/π Birks' constant = 0.02	[Eng 92]
LNF Frascati	R&D for SuperB	Tests of a LYSO em calorimeter	LYSO crystals (25)	(2 x 3.75 mm)	2.5 x 2.5 x 20 cm	1.1%/√E + 1.2% + 0.4%/E		[Eig 13]
Caltech	New crystals for HEP	Tests of new crystals for applications in particle physics	Various crystals: LSO, GSO, YSO, CeF3, etc.		5 x 5 array of crystals	(for 100 - 500 MeV γs)	Measure decay time, light yield radiation hardness,...	[Zhu 15]

APPENDIX   D

ILLUSTRATION CREDITS

Figure	Reference	Credit
1.1		Measurements courtesy of G. Roubaud, CERN.
1.2	[CMS 13]	Reprinted with permission from IOP Publishing, © 2013.
1.3		Photograph courtesy of CERN.
1.4		Image courtesy of Prof. A. Ereditato, μBooNE Collaboration.
1.5		Photograph courtesy of Kernforschungszentrum Karlsruhe.
1.6		Photograph courtesy of the Tokyo Institute for Cosmic-Ray Research.
1.7		Photograph courtesy of CERN.
1.11		Photograph courtesy of CERN.
1.13	[Pre 00]	Reprinted with permission from Elsevier Science, © 2000.
2.2	[PDG 14]	Reprinted with permission from the Particle Data Group.
2.16	[Bat 70]	Reprinted with permission from Elsevier Science, © 1970.
2.20	[PDG 14]	Reprinted with permission from the Particle Data Group.
2.29		Photograph courtesy of CERN.
2.35	[Adl 13b]	Reprinted with permission from IOP Publishing, © 2013.
2.39	[Bar 90]	Reprinted with permission from Elsevier Science, © 1990.
2.49		Photograph courtesy of CERN.
2.52	[Adl 13b]	Reprinted with permission from IOP Publishing, © 2013.
3.3	[Fla 85]	Reprinted with permission from Elsevier Science, © 1985.
3.21	[Bru 88]	Reprinted with permission from Elsevier Science, © 1988.
3.26	[Wil 74]	Reprinted with permission from Elsevier Science, © 1974.
3.29	[Ata 83]	Reprinted with permission from Elsevier Science, © 1983.
3.32	[Pei 96]	Reprinted with permission from Elsevier Science, © 1996.
3.34	[Adl 09]	Reprinted with permission from Elsevier Science, © 2009.
3.35	[Hug 69]	Reprinted with permission from Elsevier Science, © 1969.
3.36	[Fab 77]	Reprinted with permission from Elsevier Science, © 1977.
3.38	[Ang 90]	Reprinted with permission from Elsevier Science, © 1990.

Figure	Reference	Credit
3.39	[Gal 86]	Reprinted with permission from Elsevier Science, © 1986.
3.40	[Gal 86]	Reprinted with permission from Elsevier Science, © 1986.
3.43	[Suz 99]	Reprinted with permission from Elsevier Science, © 1999.
3.47a	[Ant95]	Reprinted with permission from Elsevier Science, © 1995.
4.2		Measurements courtesy of G. Roubaud, CERN
4.9	[Fis 78]	Reprinted with permission from Elsevier Science, © 1978.
4.13	[Adr 09	Reprinted with permission from Elsevier Science, © 2009.
4.17	[Dre 90]	Reprinted with permission from Elsevier Science, © 1990.
4.18	[Fab 77]	Reprinted with permission from Elsevier Science, © 1977.
4.25	[Beh 90]	Reprinted with permission from Elsevier Science, © 1990.
4.28	[And 91]	Reprinted with permission from Elsevier Science, © 1991.
4.33	[Did 80]	Reprinted with permission from Elsevier Science, © 1980.
4.37	[Adr 10]	Reprinted with permission from Elsevier Science, © 2010.
4.59	[Beh 90]	Reprinted with permission from Elsevier Science, © 1990.
4.60	[Brau 89]	Reprinted with permission from Elsevier Science, © 1989.
4.61	[Brau 89]	Reprinted with permission from Elsevier Science, © 1989.
4.65	[Gal 86]	Reprinted with permission from Elsevier Science, © 1986.
4.67	[Fer 97]	Reprinted with permission from Elsevier Science, © 1997.
4.68	[Abr 81]	Reprinted with permission from Elsevier Science, © 1981.
4.69	[Abr 81]	Reprinted with permission from Elsevier Science, © 1981.
5.19	[Aba 94]	Reprinted with permission from Elsevier Science, © 1994.
5.21	[Ale 97]	Reprinted with permission from Elsevier Science, © 1997.
5.27	[Wil 74]	Reprinted with permission from Elsevier Science, © 1974.
5.28	[Shi 75]	Reprinted with permission from Elsevier Science, © 1975.
5.29	[Ande 88]	Reprinted with permission from Elsevier Science, © 1988.
5.30	[Ere 13]	Reprinted with permission from IOP Publishing, © 2013.
5.32	[Ame 86]	Reprinted with permission from Elsevier Science, © 1986.
5.33	[Ant 14]	Reprinted with permission from IOP Publishing, © 2014.
5.38	[Arn 94]	Reprinted with permission from Elsevier Science, © 1994.
5.39	[Ambr 03]	Reprinted with permission from Elsevier Science, © 2003.
5.40	[Eis 14]	Reprinted with permission from Elsevier Science, © 2014.
5.42	[Yar 93]	Reprinted with permission from Elsevier Science, © 1993.
5.43	[Wil 74]	Reprinted with permission from Elsevier Science, © 1974.
5.51	[Wil 74]	Reprinted with permission from Elsevier Science, © 1974.
5.54	[Aub 93b]	Reprinted with permission from Elsevier Science, © 1993.
5.55	[Arn 94]	Reprinted with permission from Elsevier Science, © 1994.
5.56	[Bert 97]	Reprinted with permission from Elsevier Science, © 1997.

Figure	Reference	Credit
5.59	[Amal 86]	Reprinted with permission from Elsevier Science, © 1986.
5.60	[Joh 92]	Reprinted with permission from Elsevier Science, © 1992.
5.62	[Zhu 98]	Reprinted with permission from Elsevier Science, © 1998.
5.63	[Lec 06]	Reprinted with permission from Elsevier Science, © 2006.
5.64	[Bac 99]	Reprinted with permission from Elsevier Science, © 1999.
6.12	[Aha 06]	Reprinted with permission from Elsevier Science, © 2006.
6.13	[Aha 06]	Reprinted with permission from Elsevier Science, © 2006.
6.14	[Cer 02]	Reprinted with permission from Elsevier Science, © 2002.
6.15	[Cer 02]	Reprinted with permission from Elsevier Science, © 2002.
6.16	[Bere 93]	Reprinted with permission from Elsevier Science, © 1993.
6.17	[Braun 88]	Reprinted with permission from Elsevier Science, © 1988.
6.31	[CMS 08]	Reprinted with permission from IOP Publishing, © 2008.
6.32	[CMS 13]	Reprinted with permission from IOP Publishing, © 2013.
6.33	[CMS 13]	Reprinted with permission from IOP Publishing, © 2013.
6.34	[CMS 13]	Reprinted with permission from IOP Publishing, © 2013.
6.35	[Aad 08]	Reprinted with permission from IOP Publishing, © 2008.
6.36	[Braun 88]	Reprinted with permission from Elsevier Science, © 1988.
6.37	[Jen 88]	Reprinted with permission from Elsevier Science, © 1988.
6.38	[CMS 13]	Reprinted with permission from IOP Publishing, © 2013.
6.39	[CMS 15a]	Reprinted with permission from IOP Publishing, © 2015.
6.40a	[Apa 98]	Reprinted with permission from Elsevier Science, © 1998.
7.1	[Dok 99]	Reprinted with permission from Elsevier Science, © 1999.
7.2	[Mih 11]	Reprinted with permission from IOP Publishing, © 2011.
7.4	[Adz 07]	Reprinted with permission from IOP Publishing, © 2007.
7.5a	[Ale 05]	Reprinted with permission from Elsevier Science, © 2005.
7.5b	[Ike 00]	Reprinted with permission from Elsevier Science, © 2000.
7.6	[Barr 96]	Reprinted with permission from Elsevier Science, © 1996.
7.7	[Ato 08]	Reprinted with permission from Elsevier Science, © 2008.
7.9	[Lev 08]	Reprinted with permission from Elsevier Science, © 2008.
7.24	[Dec 90]	Reprinted with permission from Elsevier Science, © 1990.
7.25	[Aub 93b]	Reprinted with permission from Elsevier Science, © 1993.
7.26	[Ant 95]	Reprinted with permission from Elsevier Science, © 1995.
7.27	[Adi 02]	Reprinted with permission from Elsevier Science, © 2002.
7.31	[Apa 98]	Reprinted with permission from Elsevier Science, © 1998.
7.37	[Adr 10]	Reprinted with permission from Elsevier Science, © 2010.
7.44	[Cob79]	Reprinted with permission from Elsevier Science, © 1979.
7.47	[App 75]	Reprinted with permission from Elsevier Science, © 1975.
7.50	[App 75]	Reprinted with permission from Elsevier Science, © 1975.

Figure	**Reference**	**Credit**
7.51	[App 75]	Reprinted with permission from Elsevier Science, © 1975.
7.52	[App 75]	Reprinted with permission from Elsevier Science, © 1975.
7.53	[Alv 08]	Reprinted with permission from IOP Publishing, © 2008.
7.54		Data courtesy of WA1 Collaboration, CERN
7.58	[Alit 89]	Reprinted with permission from Elsevier Science, © 1989.
7.59	[Alit 89]	Reprinted with permission from Elsevier Science, © 1989.
7.60	[Aad 08]	Reprinted with permission from IOP Publishing, © 2008.
7.61	[CMS 15a]	Reprinted with permission from IOP Publishing, © 2015.
7.62	[CMS 15a]	Reprinted with permission from IOP Publishing, © 2015.
7.63	[Aub 93b]	Reprinted with permission from Elsevier Science, © 1993.
7.66	[Eng 83]	Reprinted with permission from Elsevier Science, © 1983.
7.67	[Berg 92]	Reprinted with permission from Elsevier Science, © 1992.
7.68	[Aad 08]	Reprinted with permission from IOP Publishing, © 2008.
7.69	[Dec 90]	Reprinted with permission from Elsevier Science, © 1990.
7.70	[Jen 88]	Reprinted with permission from Elsevier Science, © 1988.
7.71	[You 89]	Reprinted with permission from Elsevier Science, © 1989.
8.3		Data courtesy of Prof. B. Webber, Cambridge University
8.40	[Boc 01]	Reprinted with permission, World Scientific Publishing, © 2001.
8.43	[Adl 10]	Reprinted with permission from IOP Publishing, © 2010.
8.44	[Sef 15]	Reprinted with permission, American Physical Society, © 2015.
8.45	[Sef 15]	Reprinted with permission, American Physical Society, © 2015.
8.46	[Adl 11a]	Reprinted with permission from IOP Publishing, © 2011.
8.47	[Adl 09]	Reprinted with permission from Elsevier Science, © 2009.
8.48	[Adl 07]	Reprinted with permission from IOP Publishing, © 2007.
8.51	[Adl 07]	Reprinted with permission from IOP Publishing, © 2007.
8.53	[Adl 13a]	Reprinted with permission from IOP Publishing, © 2013.
8.55	[Tho 09]	Reprinted with permission from Elsevier Science, © 2009.
8.56	[Tho 09]	Reprinted with permission from Elsevier Science, © 2009.
8.57	[Adl 11b]	Reprinted with permission from IOP Publishing, © 2011.
8.58	[Adl 11b]	Reprinted with permission from IOP Publishing, © 2011.
9.1	[Fab 77]	Reprinted with permission from Elsevier Science, © 1977.
9.7	[Pet 12]	Reprinted with permission from Elsevier Science, © 2012.
9.10	[Aad 08]	Reprinted with permission from IOP Publishing, © 2008.
9.11	[Cer 02]	Reprinted with permission from Elsevier Science, © 2002.
9.15	[Hee 09]	Reprinted with permission from IOP Publishing, © 2009.
9.16	[Pet 12]	Reprinted with permission from Elsevier Science, © 2012.
9.19	[Ata 83]	Reprinted with permission from Elsevier Science, © 1983.
9.20	[Sef 15]	Reprinted with permission, American Physical Society, © 2015.

Figure	Reference	Credit
10.2		Photograph courtesy of the Tokyo Institute for Cosmic-Ray Research.
10.3a	[Suz 93]	Reprinted with permission from Elsevier Science, © 1993.
10.3b		Photograph courtesy of the Tokyo Institute for Cosmic-Ray Research.
10.4	[Fuk 03]	Reprinted with permission from Elsevier Science, © 2003.
10.5a	[Fuk 03]	Reprinted with permission from Elsevier Science, © 2003.
10.5b,c	[Abe 11]	Reprinted with permission, American Physical Society, © 2011.
10.7	[Fuk 03]	Reprinted with permission from Elsevier Science, © 2003.
10.11	[Abe 14]	Reprinted with permission from Elsevier Science, © 2014.
10.17	[Abe 11]	Reprinted with permission, American Physical Society, © 2011.
10.18	[Abe 11]	Reprinted with permission, American Physical Society, © 2011.
10.19	[Hos 06]	Reprinted with permission, American Physical Society, © 2006.
10.21	[Amr 00]	Reprinted with permission from Elsevier Science, © 2000.
10.22	[Amr 03]	Reprinted with permission from Elsevier Science, © 2003.
10.23	[Age 11]	Reprinted with permission from Elsevier Science, © 2011.
10.27	[Age 11]	Reprinted with permission from Elsevier Science, © 2011.
10.31	[Agu 11]	Reprinted with permission from Elsevier Science, © 2011.
10.32		Photograph courtesy of the KM3NeT Collaboration.
10.33		Photograph courtesy of the KM3NeT Collaboration.
10.35	[Hal 10]	Reprinted with permission from AIP Publishing LLC, © 2010.
10.36	[Hal 10]	Reprinted with permission from AIP Publishing LLC, © 2010.
10.37	[Aar 13a]	Reprinted with permission, American Physical Society, © 2013.
10.38	[Abba 13]	Reprinted with permission from Elsevier Science, © 2013.
10.41		Photograph courtesy of the HESS Collaboration.
10.42	[Ale 16]	Reprinted with permission from Elsevier Science, © 2016.
10.44	[Nau 09]	Reprinted with permission from Elsevier Science, © 2009.
10.46		Phograph courtesy of Dr. J. Matthews, Fly's Eye Collaboration.
10.47		Phograph courtesy of Dr. J. Matthews, Fly's Eye Collaboration.
10.54	[PAC 15]	Reprinted with permission from Elsevier Science, © 2015.
10.55	[Ber 14]	Reprinted with permission from Elsevier Science, © 2014.
10.56	[Ber 14]	Reprinted with permission from Elsevier Science, © 2014.
10.57	[PAC 15]	Reprinted with permission from Elsevier Science, © 2015.
10.58	[Bert 12]	Reprinted with permission from Elsevier Science, © 2012.

Figure	Reference	Credit
11.3	[Arn 83a]	Reprinted with permission from Elsevier Science, © 1983.
11.4		Data courtesy of Dr. S.W. Lee, CDF Collaboration.
11.5	[Bag 84]	Reprinted with permission from Springer-Verlag, © 1984.
11.6	[Bag 86]	Reprinted with permission from Springer-Verlag, © 1986.
11.7	[Arn 83a]	Reprinted with permission from Elsevier Science, © 1983.
11.8	[Aal 12a]	Reprinted with permission, American Physical Society, © 2012.
11.15	[Kaj 16]	Reprinted with permission from Elsevier Science, © 2016.
11.16a	[Gan 11]	Reprinted with permission, American Physical Society, © 2011.
11.16b	[An 15]	Reprinted with permission, American Physical Society, © 2015.
11.17		Photograph courtesy of the SNO Collaboration.
11.19	[Aal 12a]	Reprinted with permission, American Physical Society, © 2012.
11.21a	[Aad 12]	Reprinted with permission from Elsevier Science, © 2012.
11.21b	[Chat 12]	Reprinted with permission from Elsevier Science, © 2012.
11.22	[Chat 12]	Reprinted with permission from Elsevier Science, © 2012.

Frontispiece	Courtesy of P. Loïez, CERN Photography Service.
Picture of author	Courtesy of Dr. M.A. Frautschi.

REFERENCES

[Aab 16]   Aaboud, M. *et al.* ATLAS Collaboration (2016), *Phys. Lett.* **B761**, 350.
[Aad 08]   Aad, G. *et al.* ATLAS Collaboration (2008), *JINST* **3**, S08003.
[Aad 12]   Aad, G. *et al.* ATLAS Collaboration (2012), *Phys. Lett.* **B716**, 1.
[Aai 15]   Aaij, R. *et al.* (2015). *Phys. Rev. Lett.* **114**, 062004.
[Aal 10]   Aaltonen, T. *et al.* (2010). *Phys. Rev. Lett.* **104**, 061802.
[Aal 12a]   Aaltonen, T. *et al.* (2012). *Phys. Rev. Lett.* **108**, 151803.
[Aal 12b]   Aaltonen, T. *et al.* (2012). *Phys. Rev. Lett.* **109**, 152003.
[Aam 08]   Aamodt, K. *et al.* ALICE Collaboration (2008), *JINST* **3** S08002.
[Aar 87]   Aardsma, G. *et al.* (1987). *Phys. Lett.* **194B**, 321.
[Aar 13a]   Aartsen, M.G. *et al.* (2013). *Phys. Rev. Lett.* **111** (2013 021103.
[Aar 13b]   Aartsen, M.G. *et al.* (2013). *J. Glaciology* **59**, 1117.
[Aar 14a]   Aartsen, M.G. *et al.* (2014). arXiv:1410.2046 [physics.ins-det].
[Aar 14b]   Aartsen, M.G. *et al.* (2014). *Astrophys. J.* **796**, 109.
[Aar 15a]   Aartsen, M.G. *et al.* (2015). *Phys. Rev.* **D91**, 022001.
[Aar 15b]   Aartsen, M.G. *et al.* (2015). *Phys. Rev.* **D91**, 072004.
[Aar 15c]   Aartsen, M.G. *et al.* (2015). *Astrophys. J. Lett.* **805**, L5.
[Aar 16]   Aartsen, M.G. *et al.* (2016). *J. Cosm. and Astropart. Phys.* **1**, 37.
[Aba 92]   Abashian, A. *et al.* (1992). *Nucl. Instr. and Meth.* **A317**, 75.
[Aba 93]   Abachi, S. *et al.* (1993). *Nucl. Instr. and Meth.* **A324**, 53.
[Aba 94]   Abachi, S. *et al.* (1994). *Nucl. Instr. and Meth.* **A338**, 185.
[Aba 09]   Abat, E. *et al.* (2010). *Nucl. Instr. and Meth.* **A607**, 372.
[Aba 10]   Abat, E. *et al.* (2010). *Nucl. Instr. and Meth.* **A621**, 134.
[Aba 14]   Abasov, V.M. *et al.* (2014). *Phys. Rev. Lett.* **113**, 032002.
[Abba 11]   Abbasi, R. *et al.* (2011). *Astropart. Phys.* **34**, 382.
[Abba 13]   Abbasi, R. *et al.* (2013). *Astropart. Phys.* **42**, 15.
[Abb 99]   Abbott, B. *et al.* (1999). *Phys. Rev.* **D60**, 012001.
[Abb 07]   Abbon, P. *et al.* (2007). *Nucl. Instr. and Meth.* **A577**, 455.
[Abb 16]   Abbott, B.P. *et al.* (2016). *Phys. Rev. Lett.* **116**, 061102.
[Abd 96]   Abdurashitov, J.N. *et al.* (1996). *Phys. Rev. Lett.* **77**, 4708.
[Abd 08]   Abdullin, S. *et al.* (2008). *Eur. Phys. J.* **C55**, 159.
[Abd 09]   Abdullin, S. *et al.* (2009). *The CMS Barrel Calorimeter Response to Particle Beams from 2 to 350 GeV/c*, CERN-CMS-NOTE-2008-034.
[Abe 95]   Abe, F. *et al.* (1995). *Phys. Rev.* **D52**, 4784.
[Abe 98]   Abe, F. *et al.* (1998). *Phys. Rev.* **D58**, 112004; *Phys. Rev. Lett.* **81**, 2432.
[Abe 10]   Abe, T. *et al.* (2010). arXiv:1011.0352 [physics.ins-det].
[Abe 11]   Abe, K. *et al.* (2011). *Phys. Rev.* **D83**, 052010 and arXiv:1010.0118v3 [hep-ex].
[Abe 14]   Abe, K. *et al.* (2014). *Nucl. Instr. and Meth.* **A737**, 253.

[Abe 16]   Abe, K. *et al.* (2016). *Phys. Rev. Lett.* **116**, 181801.
[AbeS 10]   Abe, S. *et al.* (2010). *Phys. Rev.* **C81**, 025807.
[Abg 14]   Abgrall, N. *et al.* (2014). *JINST* **9**, P06005.
[Abl 10]   Ablikim, M. *et al.* (2010). *Nucl. Instr. and Meth.* **A614**, 345.
[Abo 89]   Abolins, M. *et al.* (1989). *Nucl. Instr. and Meth.* **A280**, 36.
[Abr 78]   Abrams, G.S. *et al.* (1978). *IEEE Trans. Nucl. Sci.* **NS-25**, 309.
[Abr 80]   Abrams, G.S. *et al.* (1978). *IEEE Trans. Nucl. Sci.* **NS-27**, 59.
[Abr 81]   Abramowicz, H. *et al.* (1981). *Nucl. Instr. and Meth.* **180**, 429.
[Abr 92]   Abramowicz, H. *et al.* (1992). *Nucl. Instr. and Meth.* **A313**, 126.
[Abr 96]   Abreu, P. *et al.* (1996). *Nucl. Instr. and Meth.* **A378**, 57.
[Abr 97]   Abrams, G.S. *et al.* (1997). *Nucl. Instr. and Meth.* **A390**, 41.
[Abr 10]   Abraham, J. *et al.* (2010). *Nucl. Instr. and Meth.* **A613**, 29.
[Abr 11]   Abreu, P. *et al.* (2011). arXiv:1107.4804 [astro-ph].
[Abr 16]   Abramowski, A. *et al.* (2016). *Nature* **531**, 476.
[Abt 83]   Abt, I. *et al.* (1983). *Nucl. Instr. and Meth.* **217**, 377.
[Abu 12a]   Abu-Zayyad, T. *et al.* (2012). *Nucl. Instr. and Meth.* **A676**, 54.
[Abu 12b]   Abu-Zayyad, T. *et al.* (2012). *Nucl. Instr. and Meth.* **A689**, 87.
[Ach 97]   Achasov, M.N. *et al.* (1997). *Nucl. Instr. and Meth.* **A401**, 179.
[Ach 08]   Achenbach, R. *et al.* (2008). *JINST* **3**, P03001.
[Aco 90]   Acosta, D. *et al.* (1990). *Nucl. Instr. and Meth.* **A294**, 193.
[Aco 91a]   Acosta, D. *et al.* (1991). *Nucl. Instr. and Meth.* **A302**, 36.
[Aco 91b]   Acosta, D. *et al.* (1991). *Nucl. Instr. and Meth.* **A305**, 55.
[Aco 91c]   Acosta, D. *et al.* (1991). *Nucl. Instr. and Meth.* **A308**, 481.
[Aco 91d]   Acosta, D. *et al.* (1991). *Nucl. Instr. and Meth.* **A309**, 143.
[Aco 91e]   Acosta, D. *et al.* (1991). *Nucl. Instr. and Meth.* **B62**, 116.
[Aco 92a]   Acosta, D. *et al.* (1992). *Nucl. Instr. and Meth.* **A314**, 431.
[Aco 92b]   Acosta, D. *et al.* (1992). *Nucl. Instr. and Meth.* **A316**, 184.
[Aco 92c]   Acosta, D. *et al.* (1992). *Nucl. Instr. and Meth.* **A320**, 128.
[Aco 95]   Acosta, D. *et al.* (1995). *Nucl. Instr. and Meth.* **A354**, 296.
[Ada 96]   Adams, T. *et al.* (1996). *Nucl. Instr. and Meth.* **A368**, 617.
[Ada 05]   Adam, W. *et al.* (2005). *Eur. Phys. J.* **46**, 605.
[Ada 13a]   Adam, J. *et al.* (2013). *Eur. Phys. J.* **73**, 2365.
[Ada 13b]   Adam, J. *et al.* (2013). *Phys. Rev. Lett.* **110**, 201801.
[Adi 99]   Adinolfi Falcone, R.A. *et al.* (1999). *Nucl. Instr. and Meth.* **A420**, 117.
[Adi 02]   Adinolfi, M. *et al.* (2002). *Nucl. Instr. and Meth.* **A482**, 364.
[Adl 92]   Adler, R. *et al.* (1992). *Nucl. Instr. and Meth.* **A321**, 458.
[Adl 97]   Adler, R. *et al.* (1997). *Nucl. Instr. and Meth.* **A390**, 293.
[Adl 07]   Adloff, C. *et al.* (2007). arXiv:0707.1245 [physics.ins-det].
[Adl 09]   Adloff, C. *et al.* (2009). *Nucl. Instr. and Meth.* **A608**, 372.
[Adl 10]   Adloff, C. *et al.* (2010). *JINST* **5**, P05004.
[Adl 11a]   Adloff, C. *et al.* (2011). *JINST* **6**, P04003.
[Adl 11b]   Adloff, C. *et al.* (2011). *JINST* **6**, P07005.
[Adl 12]   Adloff, C. *et al.* (2012). *JINST* **7**, P09017.
[Adl 13a]   Adloff, C. *et al.* (2013). *JINST* **8**, P09001.

[Adl 13b]	Adloff, C. *et al.* (2013). *JINST* **8**, 07005.
[Adl 13c]	Adloff, C. *et al.* (2013). *Nucl. Instr. and Meth.* **A714**, 147.
[Adr 09]	Adragna, P. *et al.* (2009). *Nucl. Instr. and Meth.* **A606**, 362.
[Adr 10]	Adragna, P. *et al.* (2010). *Nucl. Instr. and Meth.* **A615**, 158.
[Adr 12a]	Adrián-Martíinez, S. *et al.* (2012). *Astropart. Phys.* **35**, 634.
[Adr 12b]	Adrián-Martíinez, S. *et al.* (2012). *Phys. Lett.* **B714**, 224.
[Adr 13]	Adrián-Martíinez, S. *et al.* (2013). arXiv:1302.6516 [astro-ph.HE].
[Adr 14]	Adrián-Martíinez, S. *et al.* (2014). *Eur. Phys. J.* **C74**, 3056.
[Adr 15]	Adrián-Martíinez, S. *et al.* (2015). arXiv:1505.04866 [astro-ph.HE].
[Adr 16]	Adrián-Martíinez, S. *et al.* (2014). *Eur. Phys. J.* **C76**, 24.
[Adr 91]	Adriani, O. *et al.* (1991). *Nucl. Instr. and Meth.* **A302**, 53.
[Adz 07]	Adzic, P. *et al.* (2007). *JINST* **2** P04004.
[Aff 93]	Affatato, S. *et al.* (1993). *Nucl. Instr. and Meth.* **A325**, 417.
[Aga 15]	Agafonova, N. *et al.* (2015). *Phys. Rev. Lett.* **115**, 121802.
[Age 11]	Ageron, M. *et al.* (2011). *Nucl. Instr. and Meth.* **A656**, 11.
[Agg 05a]	Aggouras, G. *et al.* (2005). *Nucl. Instr. and Meth.* **A552**, 420.
[Agg 05b]	Aggouras, G. *et al.* (2005). *Astropart. Phys.* **.23**, 377.
[Agl 93]	Aglietta, M. *et al.* (1993). *Nucl. Instr. and Meth.* **A336**, 310.
[Ago 89]	d'Agostini, G. *et al.* (1989). *Nucl. Instr. and Meth.* **A274**, 134.
[Ago 03]	Agostinelli, S. *et al.* (2003). *Nucl. Instr. and Meth.* **A506**, 250.
[Agu 05]	Aguilar, J.A. *et al.* (2005). *Astropart. Phys.* **23**, 131.
[Agu 11]	Aguilar, J.A. *et al.* (2005). *Nucl. Instr. and Meth.* **A626-627**, 128.
[Aha 97]	Aharonian, F.A. and Akerlof, C.W. (1997). *Ann. Rev. Nucl. Part. Sci.* **47**, 273.
[Aha 07]	Aharonian, F. *et al.* (2007). *Phys. Rev.* **D75**, 042004.
[Aha 06]	Aharouche, M. *et al.* (2006). *Nucl. Instr. and Meth.* **A568**, 601.
[Ahar 06]	Aharmim, B. *et al.* (2006). *Phys. Rev.* **C88**, 025501.
[Ahm 02]	Ahmad, Q.R. (2002). *Phys. Rev. Lett.* **89**, 011301.
[Ahn 98]	Ahn, H.S. *et al.* (1998). *Nucl. Instr. and Meth.* **A410**, 179.
[Ahn 06]	Ahn, M.H. *et al.* (2006). *Phys. Rev.* **D74**, 072003.
[Ahn 07]	Ahn, H.S. *et al.* (2007). *Nucl. Instr. and Meth.* **A579**, 1034.
[Aie 10]	Aiello, S. *et al.* (2010). *Astropart. Phys.* **33**, 263.
[Aih 93]	Aihara, H. *et al.* (1993). *Nucl. Instr. and Meth.* **A325**, 393.
[Aja 97b]	Ajaltouni, Z. *et al.* (1997). *Nucl. Instr. and Meth.* **A388**, 64.
[Akc 97]	Akchurin, N. *et al.* (1997). *Nucl. Instr. and Meth.* **A399**, 202.
[Akc 98]	Akchurin, N. *et al.* (1998). *Nucl. Instr. and Meth.* **A408**, 380.
[Akc 03]	Akchurin, N. and Wigmans, R. (2003). *Rev. Sci. Instr.* **74**, 2955.
[Akc 04]	Akchurin, N. *et al.* (2004). *Nucl. Instr. and Meth.* **A533**, 305.
[Akc 05a]	Akchurin, N. *et al.* (2005). *Nucl. Instr. and Meth.* **A536**, 29.
[Akc 05b]	Akchurin, N. *et al.* (2005). *Nucl. Instr. and Meth.* **A537**, 537.
[Akc 07a]	Akchurin, N. *et al.* (2007). *Nucl. Instr. and Meth.* **A581** 643.
[Akc 07b]	Akchurin, N. *et al.* (2007). *Nucl. Instr. and Meth.* **A582**, 474.
[Akc 08a]	Akchurin, N. *et al.* (2008). *Nucl. Instr. and Meth.* **A584**, 304.
[Akc 08b]	Akchurin, N. *et al.* (2008). *Nucl. Instr. and Meth.* **A593**, 530.

[Akc 09a] Akchurin, N. *et al.* (2009). *Nucl. Instr. and Meth.* **A598**, 422.
[Akc 09b] Akchurin, N. *et al.* (2009). *Nucl. Instr. and Meth.* **A604**, 512.
[Akc 09c] Akchurin, N. *et al.* (2009). *Nucl. Instr. and Meth.* **A610**, 488.
[Akc 10] Akchurin, N. *et al.* (2010). *Nucl. Instr. and Meth.* **A621**, 212.
[Akc 11a] Akchurin, N. *et al.* (2011). *Nucl. Instr. and Meth.* **A638**, 47.
[Akc 11b] Akchurin, N. *et al.* (2011). *Nucl. Instr. and Meth.* **A640**, 91.
[Akc 12a] Akchurin, N. and Wigmans, R. (2012). *Nucl. Instr. and Meth.* **A666**, 80.
[Akc 12b] Akchurin, N. *et al.* (2012). *Nucl. Instr. and Meth.* **A686**, 125.
[Akc 14a] Akchurin, N. *et al.* (2014). *Nucl. Instr. and Meth.* **A735**, 120.
[Akc 14b] Akchurin, N. *et al.* (2014). *Nucl. Instr. and Meth.* **A735**, 130.
[Akc 14c] Akchurin, N. *et al.* (2014). *Nucl. Instr. and Meth.* **A762**, 100.
[Akc 16] Akchurin, N. (2016). *Fast-timing capabilities of silicon sensors for the CMS high-granularity calorimeters at the high-luminosity LHC*, Talk at CALOR16, Daegu, Korea, May 15-20, 2016.
[Ake 85] Åkesson, T. *et al.* (1985). *Nucl. Instr. and Meth.* **A241**, 17.
[Ake 87] Åkesson, T. *et al.* (1987). *Nucl. Instr. and Meth.* **A262**, 243.
[Ake 92] Aker, E. *et al.* (1992). *Nucl. Instr. and Meth.* **A321**, 69.
[Akh 00] Akhmadaliev, S. *et al.* (2000). *Nucl. Instr. and Meth.* **A449**, 461.
[Aki 69] Akimov, V.V. *et al.* (1969). *Proc. 11th Int. Conf. on Cosmic Rays*, 06-99.
[Aki 88] Akimov, V.V. *et al.* (1988). *Space Sci. Rev.* **49**, 111.
[Aki 95] Akimenko, S.A. *et al.* (1995). *Nucl. Instr. and Meth.* **A365**, 92.
[Ako 77] Akopdjanov, G.A. *et al.* (1977). *Nucl. Instr. and Meth.* **140**, 441.
[Akr 90] Akrawy, M.A. *et al.* (1990). *Nucl. Instr. and Meth.* **A290**, 76.
[Ala 99] Alavi-Harati, A. *et al.* (1999). *Phys. Rev. Lett.* **83**, 22.
[Alb 88] Albrow, M. *et al.* (1988). *Nucl. Instr. and Meth.* **A265**, 303.
[Alb 89] Albrecht, H. *et al.* (1989). *Nucl. Instr. and Meth.* **A275**, 1.
[Alb 95] Albajar, C. *et al.* (1995). *Nucl. Instr. and Meth.* **A364**, 473.
[Alb 99] Albrow, M. *et al.* (1999). *Nucl. Instr. and Meth.* **A431**, 104.
[Alb 02a] Albrow, M. *et al.* (2002). *Nucl. Instr. and Meth.* **A480**, 524.
[Alb 02b] Albrow, M. *et al.* (2002). *Nucl. Instr. and Meth.* **A487**, 381.
[Ale 63] Alekseeva, K.I. *et al.* (1963). *Proc. of the 8th Int. Conf. on Cosmic Rays*, Jaipur, India, Vol. 5, 356.
[Ale 95] Alexander, G. *et al.* (1995). *Z. Phys.* **C68**, 179.
[Ale 97] Alexeev, G. *et al.* (1997). *Nucl. Instr. and Meth.* **A385**, 425.
[Ale 01] Alexandrov, K.V. *et al.* (2001). *Nucl. Instr. and Meth.* **A459**, 123.
[Ale 05] Aleksandrov, D.V. (2005). *Nucl. Instr. and Meth.* **A550**, 169.
[Ale 16] Aleksic, J. *et al.* (2016). *Astropart. Phys.* **72**, 61 and 76.
[Ali 98] Alimonti, G. *et al.* (1998). *Nucl. Instr. and Meth.* **A406**, 411.
[Ali 14] Aliaga, L. *et al.* (2014). *Nucl. Instr. and Meth.* **A743**, 130.
[Alit 89] Alitti, J. *et al.* (1989). *Nucl. Instr. and Meth.* **A279**, 364.
[Alit 91] Alitti, J. *et al.* (1991). *Z. Phys.* **C49**, 17.
[All 89] Allaby, J.V. *et al.* (1989). *Nucl. Instr. and Meth.* **A281**, 291.
[All 03] Allgower, C.E. *et al.* (2003). *Nucl. Instr. and Meth.* **A499**, 740.
[Alm 91] Almehed, S. *et al.* (1991). *Nucl. Instr. and Meth.* **A305**, 320.

[Alm 04]   Almerio, S. *et al.* (2004). *Nucl. Instr. and Meth.* **A527**, 329.
[Alv 08]   Alves, A.A. *et al.* (2008). *JINST* **3**, S08005.
[Alv 14]   Alvarez-Pol, H. *et al.* (2014). *Nucl. Instr. and Meth.* **A214**, 453.
[Amal 81]   Amaldi, U. (1981). *Phys. Scripta* **23**, 409.
[Amal 86]   Amaldi, U. (1986). *Nucl. Instr. and Meth.* **A243**, 312.
[Amar 01]   Amarian, M. (2001). *Nucl. Instr. and Meth.* **460**, 239.
[Amb 92]   Ambats, I. *et al.* (1992). *Nucl. Instr. and Meth.* **A320**, 161.
[Amb 96]   Ambats, I. *et al.* (1996). *Nucl. Instr. and Meth.* **A368**, 364.
[Ambr 96]   Ambrosio, M. *et al.* (1996). *Astropart. Phys.* **6**, 113.
[Ambr 03]   D'Ambrosio, C. and Leutz, H. (2003). *Nucl. Instr. and Meth.* **A501**, 463.
[Ame 86]   Amendolia, S.R. *et al.* (1986). *Nucl. Instr. and Meth.* **A252**, 399.
[Amr 00]   Amram, P. *et al.* (2000). *Astropart. Phys.* **13**, 127.
[Amr 03]   Amram, P. *et al.* (2003). *Astropart. Phys.* **19**, 253.
[An 12]   An, F.P. *et al.* (2012). *Phys. Rev. Lett.* **108**, 171803.
[An 12b]   An, F.P. *et al.* (2012). *Nucl. Instr. and Meth.* **A685**, 78.
[An 15]   An, F.P. *et al.* (2015). *Phys. Rev. Lett.* **115**, 111802.
[Ana 94]   Anassontzis, E.G. *et al.* (1994). *Nucl. Instr. and Meth.* **A349**, 242.
[And 63]   Anderson, P. *Phys. Rev.* **130**, 439.
[And 78]   Anderson, R.L. *et al.* (1978). *IEEE Trans. Nucl. Sci.* **NS-25**, 340.
[And 90]   Andresen, A. *et al.* (1990). *Nucl. Instr. and Meth.* **A290**, 95.
[And 91]   Andresen, A. *et al.* (1991). *Nucl. Instr. and Meth.* **A309**, 101
[Ande 88]   Anderson, D.F. and Lamb, D.C. (1988). *Nucl. Instr. and Meth.* **A265**, 440.
[Ande 90]   Anderson, D.F. *et al.* (1990). *Nucl. Instr. and Meth.* **A290**, 385.
[Andr 93a] Andrieu, B. *et al.* (1993). *Nucl. Instr. and Meth.* **A336**, 460.
[Andr 93b] Andrieu, B. *et al.* (1993). *Nucl. Instr. and Meth.* **A336**, 499.
[Andr 94a] Andrieu, B. *et al.* (1994). *Nucl. Instr. and Meth.* **A344**, 492.
[Andr 94b] Andrieu, B. *et al.* (1994). *Nucl. Instr. and Meth.* **A350**, 57.
[Andr 02]   Andrieux, M.L. *et al.* (2002). *Nucl. Instr. and Meth.* **A479**, 316.
[Andr 16]   Andringa, S. *et al.* (2016). *Advances in HEP*, 6194250; arXiv:1508.05759
            [physics.ins-det].
[Anf 08]   Anfreville, M. *et al.* (2008). *Nucl. Instr. and Meth.* **A594**, 292.
[Anf 13]   Anfimov, N. *et al.* (2013). *Nucl. Instr. and Meth.* **A718**, 75.
[Ang 90]   Angelis, A.L.S. *et al.* (1990). *Phys. Lett.* **B242**, 293.
[Ang 92]   Angelis, A.L.S. *et al.* (1992). *Nucl. Instr. and Meth.* **A314**, 425.
[Ank 08]   Ankowski, A. *et al.* (2008). arXiv:0812.2373v1 [hep-ex].
[Ant 90]   Antipov, Yu.M. *et al.* (1990). *Nucl. Instr. and Meth.* **A295**, 81.
[Ant 93]   Antoniazzi, L. *et al.* (1993). *Nucl. Instr. and Meth.* **A332**, 57.
[Ant 95]   Antonelli, A. *et al.* (1995). *Nucl. Instr. and Meth.* **A354**, 352.
[Ant 05]   Antoni, T. *et al.* (2005). *Astropart. Phys.* **24**, 1.
[Ant 14]   Antonello, M. *et al.* (2014). *JINST* **9**, P12006.
[Anz 95a]   Anzivino, G. *et al.* (1995). *Nucl. Instr. and Meth.* **A365**, 76.
[Anz 95b]   Anzivino, G. *et al.* (1995). *Nucl. Instr. and Meth.* **A357**, 350.
[Anz 95c]   Anzivino, G. *et al.* (1995). *Nucl. Instr. and Meth.* **A360**, 237.
[Aot 95]   Aota, T. *et al.* (1995). *Nucl. Instr. and Meth.* **A352**, 557.

[Apa 98]   Apanasevich, L. *et al.* (1998). *Nucl. Instr. and Meth.* **A417**, 50.
[Ape 10]   Apel, W.D. *et al.* (2010). *Nucl. Instr. and Meth.* **A620**, 202.
[Aph 03]   Aphecetche, L. *et al.* (2003). *Nucl. Instr. and Meth.* **A499**, 521.
[Apo 92]   Apollinari, G. *et al.* (1992). *Nucl. Instr. and Meth.* **A311**, 520.
[Apo 93]   Apollinari, G. *et al.* (1993). *Nucl. Instr. and Meth.* **A324**, 475.
[Apo 98a]   Apollinari, G. *et al.* (1998). *Nucl. Instr. and Meth.* **A409**, 547.
[Apo 98b]   Apollinari, G. *et al.* (1998). *Nucl. Instr. and Meth.* **A412**, 515.
[App 75]   Appel, J.A. *et al.* (1975). *Nucl. Instr. and Meth.* **127**, 495.
[App 94]   Appuhn, R.-D. *et al.* (1994). *Nucl. Instr. and Meth.* **A350**, 208.
[App 96]   Appuhn, R.-D. *et al.* (1996). *Nucl. Instr. and Meth.* **A382**, 395.
[App 97]   Appuhn, R.-D. *et al.* (1997). *Nucl. Instr. and Meth.* **A386**, 397.
[Apr 85]   Aprile, E., Giboni, K.L. and Rubbia, C. (1985). *Nucl. Instr. and Meth.* **A241**, 62.
[Apr 87]   Aprile, E. *et al.* (1987). *Nucl. Instr. and Meth.* **A261**, 519.
[Aps 91]   Apsimon, R. *et al.* (1991). *Nucl. Instr. and Meth.* **A305**, 331.
[Are 89]   Arefiev, A. *et al.* (1989). *Nucl. Instr. and Meth.* **A285**, 403.
[Are 90]   Arefiev, A. *et al.* (1990). *Nucl. Instr. and Meth.* **A288**, 364.
[Are 95]   Arefiev, A. *et al.* (1995). *Nucl. Instr. and Meth.* **A364**, 133.
[Are 96]   Arefiev, A. *et al.* (1996). *Nucl. Instr. and Meth.* **A376**, 163.
[Are 99]   Arena, V. *et al.* (1999). *Nucl. Instr. and Meth.* **A434**, 271.
[Ari 92]   Arima, T. *et al.* (1992). *Nucl. Instr. and Meth.* **A314**, 417.
[Ari 94]   Ariztizabal, F. *et al.* (1994). *Nucl. Instr. and Meth.* **A349**, 384.
[Arm 98]   Armstrong, T.A. *et al.* (1998). *Nucl. Instr. and Meth.* **A406**, 227.
[Arn 83a]   Arnison, G. *et al.* (1983). *Phys. Lett.* **B122**, 103.
[Arn 83b]   Arnison, G. *et al.* (1983). *Phys. Lett.* **B126**, 398.
[Arn 84]   Arnison, G. *et al.* (1984). *Phys. Lett.* **B139**, 115.
[Arn 94]   Arnaudon, H. *et al.* (1994). *Nucl. Instr. and Meth.* **A342**, 558.
[Arn 98]   Arnaldi, R. *et al.* (1998). *Nucl. Instr. and Meth.* **A411**, 1.
[Aro 88]   Aronson, S *et al.* (1988). *Nucl. Instr. and Meth.* **A269**, 492.
[Asa 80]   Asano, Y. *et al.* (1980). *Nucl. Instr. and Meth.* **174**, 357.
[Ase 92]   Aseev, A.A *et al.* (1992). *Nucl. Instr. and Meth.* **A317**, 143.
[Ask 79]   Askariyan, G.A. *et al.* (1979). *Nucl. Instr. and Meth.* **164**, 267.
[Asp 96]   Aspell, P. *et al.* (1996). *Nucl. Instr. and Meth.* **A376**, 17; *ibid.* **A376**, 361.
[Ast 81]   Astbury, A. (1981). *Phys. Scripta* **23**, 397.
[Ata 81]   Atač, M. *et al.* (1981). *IEEE Trans. Nucl. Sci.* **NS-28**, 500.
[Ata 83]   Atač, M. *et al.* (1983). *Nucl. Instr. and Meth.* **205**, 113.
[Atl 96a]   ATLAS Collaboration (1996). The ATLAS Calorimeter Performance, *report* CERN/LHCC/96-40.
[Atl 96b]   ATLAS Collaboration (1996). The ATLAS Liquid Argon Calorimeter Technical Design Report, *report* CERN/LHCC/96-41.
[Atl 96c]   ATLAS Collaboration (1996). The ATLAS Tile Calorimeter Technical Design Report, *report* CERN/LHCC/96-42.
[Ato 92]   Atoyan, G.S. *et al.* (1992). *Nucl. Instr. and Meth.* **A320**, 144.
[Ato 04]   Atoyan, G.S. *et al.* (2004). *Nucl. Instr. and Meth.* **A531**, 467.

| [Ato 08] | Atoyan, G.S. *et al.* (2008). *Nucl. Instr. and Meth.* **A584**, 291. |

[Ato 08]   Atoyan, G.S. *et al.* (2008). *Nucl. Instr. and Meth.* **A584**, 291.
[Atw 09]   Atwood, W.B. *et al.* (2009). *Astrophys. J.* **697**, 1071.
[Aub 91]   Aubert, B. *et al.* (1991). *Nucl. Instr. and Meth.* **A309**, 438.
[Aub 92a]   Aubert, B. *et al.* (1992). *Nucl. Instr. and Meth.* **A313**, 357.
[Aub 92b]   Aubert, B. *et al.* (1992). *Nucl. Instr. and Meth.* **A316**, 165.
[Aub 92c]   Aubert, B. *et al.* (1992). *Nucl. Instr. and Meth.* **A321**, 467.
[Aub 93a]   Aubert, B. *et al.* (1993). *Nucl. Instr. and Meth.* **A325**, 116.
[Aub 93b]   Aubert, B. *et al.* (1993). *Nucl. Instr. and Meth.* **A330**, 405.
[Aub 93c]   Aubert, B. *et al.* (1993). *Nucl. Instr. and Meth.* **A334**, 383.
[Aub 02]   Aubert, B. *et al.* (2002). *Nucl. Instr. and Meth.* **A479**, 1.
[Auf 96a]   Auffray, E. *et al.* (1996). *Nucl. Instr. and Meth.* **A378**, 171.
[Auf 96b]   Auffray, E. *et al.* (1996). *Nucl. Instr. and Meth.* **A380**, 524.
[Auf 96c]   Auffray, E. *et al.* (1996). *Nucl. Instr. and Meth.* **A383**, 367.
[Auf 98]   Auffray, E. *et al.* (1998). *Nucl. Instr. and Meth.* **A412**, 223.
[Aug 16]   Auger Collaboration. https://www.auger.org
[Aul 90]   Aulchenko, V.M. *et al.* (1990). *Nucl. Instr. and Meth.* **A289**, 468.
[Aul 92]   Aulchenko, V.M. *et al.* (1992). *Nucl. Instr. and Meth.* **A316**, 8.
[Aul 93]   Aulchenko, V.M. *et al.* (1993). *Nucl. Instr. and Meth.* **A336**, 53.
[Aul 97]   Aulchenko, V.M. *et al.* (1997). *Nucl. Instr. and Meth.* **A394**, 35.
[Aul 15]   Aulchenko, V. *et al.* (2015). *J. Phys. Conf. Ser.* **587**, 012045.
[Aut 93]   Autiero, D. *et al.* (1993). *Nucl. Instr. and Meth.* **A336**, 510.
[Aut 96]   Autiero, D. *et al.* (1996). *Nucl. Instr. and Meth.* **A373**, 358.
[Aut 97]   Autiero, D. *et al.* (1997). *Nucl. Instr. and Meth.* **A387**, 352.
[Ava 96]   Avakian, H. *et al.* (1996). *Nucl. Instr. and Meth.* **A378**, 155.
[Ava 98]   Avakian, H. *et al.* (1998). *Nucl. Instr. and Meth.* **A417**, 69.
[Axe 93]   Axen, D. *et al.* (1993). *Nucl. Instr. and Meth.* **A328**, 472.
[Ayt 06]   Aytnutdinov, V. (2006). *Nucl. Instr. and Meth.* **A567**, 433.
[Ayt 08]   Aytnutdinov, V. (2008). *Nucl. Instr. and Meth.* **A588**, 99.
[Bab 93]   Babusci, D. *et al.* (1993). *Nucl. Instr. and Meth.* **A332**, 444.
[Bac 89]   Bacci, C. *et al.* (1989). *Nucl. Instr. and Meth.* **A279**, 169;
[Bac 90]   Bacci, C. *et al.* (1990). *Nucl. Instr. and Meth.* **A292**, 113; *ibid.* **A301**, 445.
[Bac 97]   Baccaro, S. *et al.* (1997). *Nucl. Instr. and Meth.* **A385**, 209.
[Bac 99]   Baccaro, S. *et al.* (1999). *Nucl. Instr. and Meth.* **A426**, 206.
[Bad 91]   Badgett, W. *et al.* (1991). *Nucl. Instr. and Meth.* **A307**, 231.
[Bad 94a]   Badier, J. *et al.* (1994). *Nucl. Instr. and Meth.* **A337**, 314.
[Bad 94b]   Badier, J. *et al.* (1994). *Nucl. Instr. and Meth.* **A337**, 326.
[Bad 94c]   Badier, J. *et al.* (1994). *Nucl. Instr. and Meth.* **A348**, 74.
[Bad 95]   Badier, J. *et al.* (1995). *Nucl. Instr. and Meth.* **A354**, 328.
[Bag 83]   Bagnaia, P. *et al.* (1983). *Phys. Lett.* **B129**, 130.
[Bag 84]   Bagnaia, P. *et al.* (1984). *Z. Phys.* **C24**, 1.
[Bag 86]   Bagnaia, P. *et al.* (1986). *Z. Phys.* **C30**, 1.
[Bag 90]   Bagliesi, G. *et al.* (1990). *Nucl. Instr. and Meth.* **A286**, 61.
[Bag 95]   Bagdasorov, S.L. *et al.* (1995). *Nucl. Instr. and Meth.* **A351**, 336; *ibid.* **A364**, 139.

[Bah 95]   Bahcall, J.N. and Pinsonneault, M.H. (1995). *Rev. Mod. Phys.* **67**, 78.

[Bah 98]   Bahcall, J.N. and Krastev, P. (1998). *Phys. Lett.* **B436**, 243.

[Bah 04]   Bahcall, J.N. and Pinsonneault, M.H. (2004). *Phys. Rev. Lett.* **92**, 121301.

[Bai 94]   Bai, J.Z. *et al.* (1994). *Nucl. Instr. and Meth.* **A344**, 319.

[Baj 00]   Bajanov, N.A. *et al.* (2000). *Nucl. Instr. and Meth.* **A442**, 146.

[Bak 85]   Bakken, J.A. *et al.* (1985). *Nucl. Instr. and Meth.* **228**, 294.

[Bak 87]   Bakken, J.A. *et al.* (1987). *Nucl. Instr. and Meth.* **A254**, 535.

[Bal 85]   Baltrusaitis, R.M. *et al.* (1985). *Nucl. Instr. and Meth.* **A240**, 410.

[Bal 88]   Balka, L. *et al.* (1988). *Nucl. Instr. and Meth.* **A267**, 272.

[Bal 05]   Balatz, M.Y. *et al.* (2005). *Nucl. Instr. and Meth.* **A545**, 114.

[Ban 83]   Banner, M. *et al.* (1983). *Phys. Lett.* **B122**, 476.

[Ban 15]   Bantes, B. *et al.* (2015). *J. Phys. Conf. Ser.* **587**, 012042.

[Bar 69]   Barbier, M. (1969). *Induced Radioactivity* (North-Holland, Amsterdam).

[Bar 70]   Barns, A.V. *et al.* (1970). *Phys. Rev. Lett.* **37**, 76.

[Bar 89]   Barr, G., Gaisser, T.K. and Stanev, T. (1989). *Phys. Rev.* **D39**, 3532.

[Bar 90]   Barreiro, F. *et al.* (1990). *Nucl. Instr. and Meth.* **A292**, 259.

[Bar 94]   Barrelet, E. *et al.* (1994). *Nucl. Instr. and Meth.* **A346**, 137.

[Bar 97]   Barwick, S.W. *et al.* (1997). *Nucl. Instr. and Meth.* **A400**, 34.

[Bar 99]   Barlow, R.J. *et al.* (1999). *Nucl. Instr. and Meth.* **A420**, 162.

[Bar 07]   Bartolini, A. *et al.* (2007). *Nucl. Instr. and Meth.* **A582**, 462.

[Bar 15]   Bartoli, B. *et al.* (2015). *Nucl. Instr. and Meth.* **A783**, 68.

[Bara 90]   Baranov, A. *et al.* (1990). *Nucl. Instr. and Meth.* **A294**, 439.

[Barr 93]   Barr, G.D. *et al.* (1993). *Phys. Lett.* **B317**, 233.

[Barr 96]   Barr, G.D. *et al.* (1996). *Nucl. Instr. and Meth.* **A370**, 413.

[Bart 90]   Bartoszek, L. *et al.* (1990). *Nucl. Instr. and Meth.* **A301**, 47.

[Bat 70]   Bathow, G. *et al.* (1970). *Nucl. Phys.* **B20**, 592.

[Bat 03]   Batarin, V.A. *et al.* (2003). *Nucl. Instr. and Meth.* **A510**, 248.

[Bau 88]   Baumgart, R. *et al.* (1988). *Nucl. Instr. and Meth.* **A272**, 722.

[Bau 90]   Baumeister, H. *et al.* (1990). *Nucl. Instr. and Meth.* **A292**, 81.

[Bau 98]   Bauleo, P. *et al.* (1998). *Nucl. Instr. and Meth.* **A406**, 69.

[Bau 11]   Baunack, S. *et al.* (2011). *Nucl. Instr. and Meth.* **A640**, 58.

[Bau 15]   Baudis, L. (2015). arXiv:1509.00869 [astro-ph.CO].

[Beb 88]   Bebek, C. (1988). *Nucl. Instr. and Meth.* **A265**, 258.

[Bec 96]   Beck, M. *et al.* (1996). *Nucl. Instr. and Meth.* **A381**, 330.

[Bed 95]   Bédérède, D. *et al.* (1995). *Nucl. Instr. and Meth.* **A365**, 117.

[Bed 03]   Beddo, M. *et al.* (2003). *Nucl. Instr. and Meth.* **A499**, 725.

[Bee 84]   Beer, A. *et al.* (1984). *Nucl. Instr. and Meth.* **224**, 360.

[Beh 81]   Behrend, H.-J. *et al.* (1981). *Phys. Scripta* **23**, 610.

[Beh 90]   Behrens, U. *et al.* (1990). *Nucl. Instr. and Meth.* **A289**, 115.

[Beh 13]   Behnke, T.]*et al.* (2013). arXiv:1306.6329 [physics.ins-det].

[Bel 96]   Belousov, V.I. *et al.* (1996). *Nucl. Instr. and Meth.* **A369**, 45.

[Bel 03]   Bell, K.W. *et al.* (2003). *Nucl. Instr. and Meth.* **A504**, 255.

[Ben 86]   Bengtsson, H.U. and Sjöstrand, T. (1986). *Computer Physics Commun.* **46**, 43.

[Ben 92]	Bencheikh, B. *et al.* (1992). *Nucl. Instr. and Meth.* **A315**, 354.
[Ben 93]	Benetti, P. *et al.* (1993). *Nucl. Instr. and Meth.* **A327**, 173; *ibid.* **A332**, 395.
[Ben 94a]	Benary, O. *et al.* (1994). *Nucl. Instr. and Meth.* **A344**, 363.
[Ben 94b]	Benary, O. *et al.* (1994). *Nucl. Instr. and Meth.* **A350**, 131.
[Ben 97]	Bencze, Gy.L. *et al.* (1997). *Nucl. Instr. and Meth.* **A386**, 259.
[Benv 75]	Benvenuti, A. *et al.* (1975). *Nucl. Instr. and Meth.* **125**, 447.
[Ber 95]	Berger, C. *et al.* (1995). *Nucl. Instr. and Meth.* **A357**, 333.
[Ber 12]	Berezhnev, S.F. *et al.* (2012). *Nucl. Instr. and Meth.* **692**, 98.
[Ber 14]	Berezinsky, V. (2014). *Astropart. Phys.* **53**, 120.
[Ber 16]	Berra, A. *et al.* (2016). arXiv:1605.09630v1 [physics.ins-det].
[Bere 93]	Beretvas, A. *et al.* (1993). *Nucl. Instr. and Meth.* **A329**, 50.
[Berg 92]	Berger, F. *et al.* (1992). *Nucl. Instr. and Meth.* **A321**, 152.
[Bern 87]	Bernardi, E. *et al.* (1987). *Nucl. Instr. and Meth.* **A262**, 229.
[Bert 87]	Bertolucci, S. *et al.* (1987). *Nucl. Instr. and Meth.* **A254**, 561.
[Bert 88]	Bertolucci, S. *et al.* (1988). *Nucl. Instr. and Meth.* **A267**, 301.
[Bert 95]	Bertino, M. *et al.* (1995). *Nucl. Instr. and Meth.* **A357**, 363.
[Bert 97]	Bertoldi, M. *et al.* (1997). *Nucl. Instr. and Meth.* **A386**, 301.
[Bert 12]	Bertaina, M. *et al.* (2012). *Nucl. Instr. and Meth.* **A692**. 217.
[Bet 59]	Bethe, H. and Ashkin, J. (1959). Passage of Radiation through Matter, in *Experimental Nuclear Physics*, Vol. 1, Part 2, Segré, E. (ed.), (New York: Wiley).
[Bet 92]	Bethke, S. and Pilcher, J.E. (1992). *Ann. Rev. Nucl. Part. Sci.* **42**, 251.
[Bev 07]	Bevan, S. *et al.* (2007). *Astropart. Phys.* **28**, 366.
[Bez 02]	Bezzubov, V. *et al.* (2002). *Nucl. Instr. and Meth.* **A494**, 369.
[Bin 81]	Binon, F. *et al.* (1981). *Nucl. Instr. and Meth.* **188**, 507.
[Bin 99]	Binon, F.G. *et al.* (1999). *Nucl. Instr. and Meth.* **A428**, 292.
[Bio 83]	Bionta, R.M. *et al.* (1983). *Phys. Rev. Lett.* **51**, 27.
[Bir 64]	Birks, J.B. (1964). *The Theory and Practice of Scintillation Counting* (Oxford: Pergamon).
[Biz 93]	Bizzetti, A. *et al.* (1993). *Nucl. Instr. and Meth.* **A335**, 102.
[Bla 48]	Blackett, P.M.S (1948). *Phys. Soc. of London Gassiot Committee Report* p. 34.
[Bli 97]	Blick, A.M. *et al.* (1997). *Nucl. Instr. and Meth.* **A387**, 365.
[Blo 83]	Bloom, E. and Peck, C. (1983). *Physics with the Crystal Ball detector*, report CALT-88-989 (Caltech), SLAC-PUB-3189 (Stanford).
[Blo 92]	Blömker, D. *et al.* (1992). *Nucl. Instr. and Meth.* **A311**, 505.
[Boc 96]	Bocciolini, M. *et al.* (1996). *Nucl. Instr. and Meth.* **A370**, 403.
[Boc 01]	Bocci, A. *et al.* (2001). *Int. J. of Mod. Phys.* **A16**, suppl. 1A, 255.
[Bog 82]	Bogert, D. *et al.* (1982). *IEEE Trans. Nucl. Sci.* **NS-29**, 363.
[Bog 00]	Boger, J. *et al.* (2000). *Nucl. Instr. and Meth.* **A449**, 172.
[Bon 87]	Bonesini, M. *et al.* (1987). *Nucl. Instr. and Meth.* **A261**, 471.
[Boo 96]	Booth, N.E., Cabrera, B. and Fiorini, E. (1996), *Ann. Rev. Nucl. Part. Sci.* **46**, 471.
[Bor 80]	Böringer, T. *et al.* (1980). *Phys. Rev. Lett.* **44**, 1111.

[Bor 89]	Borchi, E. *et al.* (1989). *Phys. Lett.* **B222**, 525.
[Bor 93]	Borchi, E. *et al.* (1993). *Nucl. Instr. and Meth.* **A332**, 85.
[Bot 81]	Botner, O. (1981). *Phys. Scr.* **23**, 555.
[Brab 93]	Brabson, B.B. *et al.* (1993). *Nucl. Instr. and Meth.* **A332**, 419.
[Bran 88]	Brandenburg, G. *et al.* (1988). *Nucl. Instr. and Meth.* **A267**, 257.
[Brau 85]	Brau, J.E. and Gabriel, T.A. (1985). *Nucl. Instr. and Meth.* **A238**, 489.
[Brau 89]	Brau, J.E. and Gabriel, T.A. (1989). *Nucl. Instr. and Meth.* **A279**, 40.
[Brau 10]	Brau, J.E., Jaros, J.a. and Ma, H. (2010). *Ann. Rev. Nucl. Part. Sci.* **60**, 615.
[Braun 88]	Braunschweig, W. *et al.* (1988). *Nucl. Instr. and Meth.* **A265**, 419.
[Braun 89]	Braunschweig, W. *et al.* (1989). *Nucl. Instr. and Meth.* **A275**, 246.
[Braun 96]	Braunschweig, W. *et al.* (1996). *Nucl. Instr. and Meth.* **A378**, 479.
[Bro 85]	Brown, R.M. *et al.* (1985). *IEEE Trans. Nucl. Sci.* **NS-32**, 736.
[Bro 78]	Brown, B.C. *et al.* (1978). *IEEE Trans. Nucl. Sci.* **NS-25**, 347.
[Bru 88]	Brückmann, H. *et al.* (1988). *Nucl. Instr. and Meth.* **A263**, 136.
[Bru 92]	Brückner, W. *et al.* (1992). *Nucl. Instr. and Meth.* **A313**, 345.
[Buo 94]	Buontempo, S. *et al.* (1994). *Nucl. Instr. and Meth.* **A349**, 70.
[Bur 84]	Burmeister, H. *et al.* (1984). *Nucl. Instr. and Meth.* **225**, 530.
[Bur 86]	Burnett, T.H. *et al.* (1986). *Nucl. Instr. and Meth.* **A251**, 583.
[Bur 88a]	Burkhardt, H. *et al.* (1988). *Nucl. Instr. and Meth.* **A268**, 116.
[Bur 88b]	Burkhardt, H. *et al.* (1988). *Phys. Lett.* **B206**, 169.
[Bur 05]	Burnstein, R.A. *et al.* (2005). *Nucl. Instr. and Meth.* **A541**, 516.
[Bus 95]	Buskulic, D. *et al.* (1995). *Nucl. Instr. and Meth.* **A360**, 481.
[Bus 96]	Buskulic, D. *et al.* (1996). *Phys. Lett.* **B380**, 442.
[Cam 88]	Camporesi, T. *et al.* (1988). *IEEE Trans. Nucl. Sci.* **NS-36**, 90.
[Cap 09]	Capone, A. *et al.* (2009). *Nucl. Instr. and Meth.* **A602**, 47.
[Car 79]	Carrington, R.L. *et al.* (1979). *Nucl. Instr. and Meth.* **163**, 203.
[Car 15]	Caruso, F. *et al.* (2015). http://dx.doi.org/10.1371/journal.pone.0144503
[Car 16]	Cardini, A. *et al.* (2016). *Nucl. Instr. and Meth.* **A808**, 41.
[Cas 85]	Cassiday, G.L. (1985). *Ann. Rev. Nucl. Part. Sci.* **35**, 321.
[Cat 85]	Cattai, A. *et al.* (1985). *Nucl. Instr. and Meth.* **A235**, 310.
[Cat 86]	Catanesi, M.G. *et al.* (1986). *Nucl. Instr. and Meth.* **A247**, 438.
[Cat 87]	Catanesi, M.G. *et al.* (1987). *Nucl. Instr. and Meth.* **A260**, 43.
[Cat 90]	Catanesi, M.G. *et al.* (1990). *Nucl. Instr. and Meth.* **A292**, 97.
[Cen 94]	Cennini, P. *et al.* (1994). *Nucl. Instr. and Meth.* **A345**, 230.
[Cen 99]	Cennini, P. *et al.* (1999). *Nucl. Instr. and Meth.* **A432**, 240.
[Cer 02]	Cervelli, F. *et al.* (2002). *Nucl. Instr. and Meth.* **A490**, 132.
[Chan 78]	Chan, Y. *et al.* (1978). *IEEE Trans. Nucl. Sci.* **NS-25**, 333.
[Cha 88]	Chang, C.Y. *et al.* (1988). *Nucl. Instr. and Meth.* **A264**, 194.
[Char 68]	Charpak, G. *et al.* (1968). *Nucl. Instr. and Meth.* **62**, 235.
[Chat 12]	Chatrchyan, S. *et al.* CMS Collaboration (2012). *Phys. Lett.* **B716**, 30.
[Che 89]	Checchia, P. *et al.* (1989). *Nucl. Instr. and Meth.* **A275**, 49.
[Che 94]	Chen, C. *et al.* (1995). *Nucl. Instr. and Meth.* **A351**, 330.
[Che 13]	Chefdeville, M.A. PhD thesis, NIKHEF Amsterdam (2013). Unpublished. http://www.nikhef.nl/pub/services/biblio/theses_pdf/thesis_MA_Chefdeville.pdf

[Chi 92]   Chiba, N. *et al.* (1992). *Nucl. Instr. and Meth.* **A311**, 338.

[Chi 00]   Chipaux, R. and Géléoc, M. (2000). *Proc. 5th Int. Conf. on scintillators and their applications (SCINT99)*, eds. V.V. Mikhailin and M.V. Lomonosov, 629.

[Cih 88]   Cihangir, S. *et al.* (1988). *Nucl. Instr. and Meth.* **A267**, 249.

[Cih 89]   Cihangir, S. *et al.* (1989). *Neutron-Induced Pulses in the CDF Forward Hadron Calorimeter*, FERMILAB-PUB-89-272-PPD.

[Cle 94]   Cleland, W.E. and Stern, E.G. (1994). *Nucl. Instr. and Meth.* **A338**, 467.

[CMS 97a]   CMS Collaboration (1997). The CMS Electromagnetic Calorimeter Technical Design Report, *report* CERN/LHCC/97-33.

[CMS 97b]   CMS Collaboration (1997). The CMS Hadron Calorimeter Technical Design Report, *report* CERN/LHCC/97-31.

[CMS 05]   CMS Collaboration (2005). *JINST* **5**, P03010.

[CMS 07]   CMS Collaboration (N. Akchurin *et al.*) 2007. *The response of CMS combined calorimeters to single hadrons, electrons and muons*, CERN-CMS-NOTE-2007-012.

[CMS 08]   CMS Collaboration (2008). *JINST* **3**, S08004.

[CMS 09]   CMS Collaboration (2009). *Note* CMS-PAS-PFT-09-001.

[CMS 10]   CMS Collaboration (2010). *Electromagnetic calorimeter commissioning and first results with 7 TeV data*, CMS-NOTE-2010-12.

[CMS 13]   CMS Collaboration (2013). *JINST* **8**, P09009.

[CMS 15a]   CMS Collaboration (2015). *JINST* **10**, P08010.

[CMS 15b]   CMS Collaboration (2015). Technical proposal for the phase-II upgrade of the Compact Muon Solenoid, CERN-LHCC-2015-10.

[Cob 77]   Cobb, J.H. *et al.* (1977). *Nucl. Instr. and Meth.* **140**, 413.

[Cob 79]   Cobb, J.H. *et al.* (1979). *Nucl. Instr. and Meth.* **158**, 93.

[Coc 86]   Cochet, C. *et al.* (1986). *Nucl. Instr. and Meth.* **A243**, 45.

[Col 90]   Colas, J. *et al.* (1990). *Nucl. Instr. and Meth.* **A294**, 583.

[Con 98]   Constantini, F. *et al.* (1998). *Nucl. Instr. and Meth.* **A409**, 570.

[Con 05]   Conesa, G. *et al.* (2005). *Nucl. Instr. and Meth.* **A537**, 363.

[Cor 85]   Corden, M.J. *et al.* (1985). *Nucl. Instr. and Meth.* **A238**, 273.

[Cox 84]   Cox, B. *et al.* (1984). *Nucl. Instr. and Meth.* **219**, 487.

[Cre 98]   Crèpe, S. *et al.* (1998). *Nucl. Instr. and Meth.* **A409**, 575.

[Cri 97]   Crittenden, R.R. *et al.* (1997). *Nucl. Instr. and Meth.* **A387**, 377.

[CTA 11]   The CTA Consortium (2011). *Exp. Astr.* **32**, 193.

[Cui 14]   Cui, S. *et al.* (2014). *Astropart. Phys.* **54**, 86.

[Cum 90]   Cumalat, J.P. *et al.* (1990). *Nucl. Instr. and Meth.* **A293**, 606.

[Dad 86]   Dado, S. *et al.* (1986). *Nucl. Instr. and Meth.* **A252**, 511.

[Daf 07]   Dafni, Th. http://tuprints.ulb.tu-darmstadt.de/epda/000577/Dafni2.pdf, PhD thesis, Darmstadt (2007). Unpublished.

[DAM 16]   DAMPE Collaboration (2016). http://dpnc.unige.ch/dampe/index.html

[Dau 97]   Daum, A. *et al.* (1997). *Astropart. Phys.* **8**, 1.

[Dav 68]   Davis, R. *et al.* (1968). *Phys. Rev. Lett.* **20**, 1205.

[Dav 94]   David, G. *et al.* (1994). *Nucl. Instr. and Meth.* **A348**, 87.

[Daw 07]   Dawson, B.R. (2007). arXiv:0706.1105.

[Dec 90]   Decamp, D. *et al.* (1990). *Nucl. Instr. and Meth.* **A294**, 121.

[Del 89]   Del Peso, J and Ros, E. (1989). *Nucl. Instr. and Meth.* **A276**, 456.

[DEL 91]   DELPHI Collaboration (1991). *Nucl. Instr. and Meth.* **A303**, 233.

[Del 98]   Delbart, A. (1998). *Eur. Phys. J.* **D1**, 109.

[Dem 93]   Demortier, L. *et al.* (1993). *Nucl. Instr. and Meth.* **A324**, 77.

[Den 93]   Denisov, S. *et al.* (1993). *Nucl. Instr. and Meth.* **A335**, 106.

[Dep 89]   DePalma, M. *et al.* (1989). *Nucl. Instr. and Meth.* **A277**, 68.

[Der 91]   Derrick, M. *et al.* (1991). *Nucl. Instr. and Meth.* **A309**, 77.

[Des 89]   DeSalvo, R. *et al.* (1989). *Nucl. Instr. and Meth.* **A279**, 467.

[Des 09]   De Simone, P. (2009). *J. Phys. Conf. Ser.* **171**, 012051.

[Dev 86]   De Vincenzi, M. *et al.* (1986). *Nucl. Instr. and Meth.* **A243**, 348.

[Dew 86]   DeWulf, J.P. *et al.* (1986). *Nucl. Instr. and Meth.* **A252**, 443.

[Dew 89]   De Winter, K. *et al.* (1989). *Nucl. Instr. and Meth.* **A278**, 670.

[Dic 96]   Di Capua, E. *et al.* (1996). *Nucl. Instr. and Meth.* **A378**, 221.

[Did 80]   Diddens, A.N. *et al.* (1980). *Nucl. Instr. and Meth.* **178**, 27.

[Die 88]   Dietrich, S.S. and Berman, B.L. (1988). *Atomic Data and Nuclear Data Tables* **38**, 199.

[Die 15]   Diehl, S. *et al.* (2015). *J. Phys. Conf. Ser.* **587**, 012044.

[Dis 79]   Dishaw, J.P. *et al.* (1979). *Phys. Lett.* **85B**, 142.

[Dod 93]   Dodd, A.C. *et al.* (1993). *Nucl. Instr. and Meth.* **A336**, 136.

[Dok 81]   Doke, T. (1981). *Portugal Phys.*, Vol. 12, fasc. 1-2, p. 9; reprinted in *Experimental Techniques in High Energy Nuclear and Particle Physics*, p. 537, ed. T. Ferbel, 2nd edition, (Singapore: World Scientific, 1987).

[Dok 90]   Doke, T., Masuda, K. and Shibamura, E. (1990). *Nucl. Instr. and Meth.* **A291**, 617.

[Dok 91]   Doke, T. *et al.* (1991). *Nucl. Instr. and Meth.* **A302**, 290.

[Dok 99]   Doke, T. and Masuda, K. (1999). *Nucl. Instr. and Meth.* **A420**, 62.

[Dol 95]   Dolgopolov, A.V. *et al.* (1995). *Nucl. Instr. and Meth.* **A363**, 557.

[Dol 06]   Dolgoshein, B. *et al.* (2006). *Nucl. Instr. and Meth.* **A563**, 590.

[Don 08]   Dong Ming-Yi *et al.* (2008). *Chin. Phys.* **C32**, 11.

[Dor 05]   Doroshenko, M. *et al.* (2005). *Nucl. Instr. and Meth.* **A545**, 278.

[Dra 07]   Drake, G. *et al.* (2007). *Nucl. Instr. and Meth.* **A578**, 88.

[Dre 83]   Drescher, A. *et al.* (1983). *Nucl. Instr. and Meth.* **205**, 125; *ibid.* **216**, 35.

[Dre 90]   Drews, G. *et al.* (1990). *Nucl. Instr. and Meth.* **A290**, 335.

[DRE 13]   DREAM Collaboration, (2013). Internal report CERN-SPSC-2013-012.

[Drex 90] Drexlin, G. *et al.* (1990). *Nucl. Instr. and Meth.* **A289**, 490.

[Dub 86]   Dubois, R. *et al.* (1986). *IEEE Trans. Nucl. Sci.* **NS-33**, 194.

[Dub 96]   Dubois, O. *et al.* (1996). *Nucl. Instr. and Meth.* **A368**, 640.

[Duk 90]   Dukes, E.C. *et al.* (1990). *Proc. 1st Int. Conf. on Calorimetry in High Energy Physics*, Fermilab, eds. D.F. Anderson, M. Derrick and H.E. Fisk (Singapore: World Scientific).

[Dwu 89]   Dwuraźny, A. *et al.* (1989). *Nucl. Instr. and Meth.* **A277**, 76.

[Dzh 02]   Dzhelyadin, R.I. (2002). *Nucl. Instr. and Meth.* **A494**, 332.

[Eig 13]   Eigen, G. *et al.* (2013). *Nucl. Instr. and Meth.* **A718**, 107.
[Eis 14]   Eisenhardt, S. (2014). *Nucl. Instr. and Meth.* **A766**, 217.
[Eng 64]   Englert, F. and Brout, R. (1964). *Phys. Rev. Lett.* **13**, 321.
[Eng 69]   Engler, J. *et al.* (1969). *Phys. Lett.* **29B**, 321.
[Eng 83]   Engelmann, R. *et al.* (1983). *Nucl. Instr. and Meth.* **216**, 45.
[Engl 83] Engler, J. (1983). *Nucl. Instr. and Meth.* **217**, 9.
[Eng 86]   Engler, J., Keim, H. and Wild, B. (1986). *Nucl. Instr. and Meth.* **252**, 29.
[Eng 92]   Engler, J. *et al.* (1992). *Nucl. Instr. and Meth.* **A320**, 460.
[Eng 99]   Engler, J. *et al.* (1999). *Nucl. Instr. and Meth.* **A427**, 528.
[Ens 05]   Enss, C. (ed) (2005). *Cryogenic Particle Detection*, Springer Verlag.
[Ere 13]   Ereditato, A. *et al.* (2013). *JINST* **8**, P07002.
[Esk 97]   Eskut, E. *et al.* (1997). *Nucl. Instr. and Meth.* **401**, 7.
[Fab 77]   Fabjan, C.W. *et al.* (1977). *Nucl. Instr. and Meth.* **141**, 61.
[Fab 82]   Fabjan, C.W. and Ludlam, T. (1982). *Ann. Rev. Nucl. Part. Sci.* **32**, 335.
[Fab 85]   Fabjan, C.W. (1985), in *Experimental Techniques in High Energy Nuclear and Particle Physics*, p. 257, ed. T. Ferbel, 2nd edition, (Singapore: World Scientific).
[Fab 03]   Fabjan, C.W. and Gianotti, F. (2003). *Rev. Mod. Phys.* **75**, 1243.
[Fab 11]   Fabjan, C.W. and Schopper, H. (eds) (2011). *Elementary particles: detectors for particles and radiation. Part 1; principles and methods*, Springer Verlag, Berlin.
[Fan 94]   Fanti, V. *et al.* (1994). *Nucl. Instr. and Meth.* **A344**, 507.
[Fan 07]   Fanti, V. *et al.* (2007). *Nucl. Instr. and Meth.* **A574**, 433.
[Fano 47] Fano, U. (1947). *Phys. Rev.* **72**, 26.
[Fas 93]   Fassò, A. *et al.* (1993). *Nucl. Instr. and Meth.* **A332**, 459.
[Fed 06]   Fedotovich, G.V. (2006). *Nucl. Phys. B (Proc. Suppl.)* **162**, 332.
[Fer 97]   Ferrando, A. *et al.* (1997). *Nucl. Instr. and Meth.* **A390**, 63; *Nucl. Instr. and Meth.* **A400**, 267.
[FF 11]   Fabjan, C.W. and Fournier, D. (2011). Calorimetry, in *Landolt-Börnstein New Series*, Springer Verlag, I 21B1.
[Fie 15]   Fienberg, A. *et al.* (2015). *Nucl. Instr. and Meth.* **A783**, 12.
[Fil 15]   Filha, L.M. de A. *et al.* (2015). *IEEE Trans. Nucl. Sci.* **NS-62**, 3265.
[Fis 78]   Fischer, H.G. (1978). *Nucl. Instr. and Meth.* **156**, 81.
[Fis 88]   Fischer, H.G. *et al.* (1988). *Nucl. Instr. and Meth.* **A265**, 218.
[Fla 85]   Flauger, W. (1985). *Nucl. Instr. and Meth.* **A241**, 72.
[Fla 90]   Flauger, W. (1990). *Nucl. Instr. and Meth.* **A289**, 446.
[Flu 81]   Flügge, G. *et al.* (1981). *Phys. Scripta* **23**, 499.
[Fox 92]   Fox, D. *et al.* (1992). *Nucl. Instr. and Meth.* **A317**, 474.
[Fra 06]   Franzini , P. and Moulson, M. (2006). *Ann. Rev. Nuc. Part. Sci.* **56**, 207.
[Fre 96]   Fretwurst, E. *et al.* (1996). *Nucl. Instr. and Meth.* **A372**, 368.
[Fri 12]   Friend, M. (2012). *Nucl. Instr. and Meth.* **A676**, 96.
[Fuk 88]   Fukui, Y. *et al.* (1988). *Nucl. Instr. and Meth.* **A267**, 280.
[Fuk 98]   Fukuda, Y. *et al.* (1998). *Phys. Rev. Lett.* **81**, 1562.
[Fuk 03]   Fukuda, S. *et al.* (2003). *Nucl. Instr. and Meth.* **A501**, 418.

[Fur 95]   Furetta, C. *et al.* (1995). *Nucl. Instr. and Meth.* **A357**, 64; *ibid.* **A361**, 149; *ibid.* **A368**, 378.

[Gaba 78]   Gabathuler, E. *et al.* (1978). *Nucl. Instr. and Meth.* **157**, 47.

[Gab 85]   Gabriel, T.A. (1985). *Proc. of the Workshop on Compensated Calorimetry*, Pasadena, Internal Report CALTECH–68–1305, p. 238.

[Gab 94]   Gabriel, T.A. *et al.* (1994). *Nucl. Instr. and Meth.* **A338**, 336.

[Gal 86]   Galaktionov, Y. *et al.* (1986). *Nucl. Instr. and Meth.* **A251**, 258.

[Gal 53]   Galbraith, W. and Jelley, J.V (1953). *Nature* **171**, 349.

[Gal 15]   Gallucci, G. (2015). *J. Phys. Conf. Ser.* **587**, 012028.

[Gan 93]   Ganel, O. and Wigmans, R. (1993). A New Approach to Forward Calorimetry. *Internal Report* SDC-93-575, Superconducting Supercollider, Dallas (TX).

[Gan 95]   Ganel, O. and Wigmans, R. (1995). *Nucl. Instr. and Meth.* **A365**, 104.

[Gan 98]   Ganel, O. and Wigmans, R. (1998). *Nucl. Instr. and Meth.* **A409**, 621.

[Gan 05]   Ganel, O. *et al.* (2005). *Nucl. Instr. and Meth.* **A552**, 409.

[Gan 11]   Gando, A. *et al.* (2011). *Phys. Rev.* **D83**, 052002.

[Gei 91]   Geiregat, D. *et al.* (1991). *Phys. Lett.* **B259**, 499.

[Ghe 15]   Ghezzi, A. (2015). *J. Phys. Conf. Ser.* **587**, 012002.

[Gia 06]   Giacobbe, B. (2006). *Nucl. Phys. B (Proc. Suppl.)* **150**, 257.

[Gib 84]   Giboni, K.L. *et al.* (1984). *Nucl. Instr. and Meth.* **225**, 579.

[Gib 93]   Gibbons, L.K. *et al.* (1993). *Phys. Rev. Lett.* **70**, 1203.

[Gin 95]   Gingrich, D.M. *et al.* (1995). *Nucl. Instr. and Meth.* **A364**, 290.

[Gio 90]   Giokaris, N.D. *et al.* (1990). *Nucl. Instr. and Meth.* **A291**, 552.

[Gom 87]   Gomez, J.J. *et al.* (1987). *Nucl. Instr. and Meth.* **A262**, 284.

[Gor 80]   Gorbachev, V.M., Zamyathin, Y.S. and Lbov, A.A. (1980). *Nuclear Reactions in Heavy Elements, a Handbook* (London: Pergamon Press).

[Gou 99]   Goulianos, K. and Lami, S. (1999). *Nucl. Instr. and Meth.* **A430**, 34.

[Gre 66]   Greisen, K. (1966). *Phys. Rev. Lett.* **16**, 748; Zatsepin, G.T. and Kuz'min, V.A.(1966). *JETP Lett.* **4**, 78.

[Gre 90]   Green, D. (1990). *Di-jet spectroscopy at high luminosity*, Fermilab Report Fermilab-Conf-90/151.

[Gre 94]   Green, D. (1994). *Proc. 4th Int. Conf. on Calorimetry in High Energy Physics*, La Biodola, Italy, eds. A. Menzione and A. Scribano, (Singapore: World Scientific), p. 1.

[Gri 58]   Grigorov, N.L., Nurzin, V.S. and Rapoport, I.D. (1958). *J. Exp. Th. Phys. (USSR)* **34**, 506.

[Gri 90]   Grigorov, N.L. (1990). *Sov. J. Nucl. Phys.* **51**, 99.

[Gro 88]   Groom, D.E. *et al.* (1988). *Radiation Levels in the SSC Interaction Regions*, Task Force report SSC Central Design Group, SSC-SR-1033 (Lawrence Berkeley Lab, Berkeley CA).

[Gro 07]   Groom, D.E. (2007). *Nucl. Instr. and Meth.* **A572**, 633.

[Gru 95]   Grunhaus, J., Kananov, S. and Milsténe, C. (1995). *Nucl. Instr. and Meth.* **A354**, 368.

[Gum 08]   Gümüş, K.Z. (2008), *Search for new physics in CMS & Response of CMS calorimeters to particles and jets*, PhD thesis, Texas Tech University (2008), unpublished.

[Gus 04] Gusev, Yu.I. *et al.* (2004). *Nucl. Instr. and Meth.* **A535**, 511.

[Gru 12] Grupen, C. and Buvat, I. (eds) (2012). *Handbook of Particle Detection and Imaging*, Springer Verlag, Berlin. ISBN 978-3-642-14621-3.

[Gur 64] Guralnik, G., Hagen, C.R. and Kibble, T.W.B. (194). *Phys. Rev. Lett.* **13**, 585.

[Hah 88] Hahn, S.R. *et al.* (1988). *Nucl. Instr. and Meth.* **A267**, 351.

[Hal 84] Halzen, F. and Martin, A.D. (1984), *Quarks and Leptons*, Wiley, New York, p. 231.

[Hal 89] Haller, G.M., Fox, J.D. and Smith, S.R. (1989). *IEEE Trans. Nucl. Sci.* **NS-36**, 675.

[Hal 10] Halzen, F. and Klein, S.R. (2010). *Rev. Sci. Instr.* **81**, 081101.

[Hal 14] Halzen, F. and Gaisser, Th. K. (2014). *Ann. Rev. Nucl. Part. Sci.* **64**, 101.

[Ham 96] Hampel, W. *et al.* (1996). *Phys. Lett.* **B388**, 384.

[Ham 07] Hamamatsu (2007). *Photomultiplier Tubes, basics and applications. Version 3a.*

[Han 98] Hanada, H. (1998). *Nucl. Instr. and Meth.* **A408**, 425.

[Har 89] Hartjes, F.G. and Wigmans, R. (1989). *Nucl. Instr. and Meth.* **A277**, 379.

[Has 90] Hasan, M.A. *et al.* (1990). *Nucl. Instr. and Meth.* **A295**, 73.

[Hau 94] Hauger, J.A. *et al.* (1994). *Nucl. Instr. and Meth.* **A337**, 362.

[Hay 92] Hayashi, H. *et al.* (1992). *Nucl. Instr. and Meth.* **A316**, 202.

[Hee 09] Heelan, L. (2009). *J. Phys. Conf. Ser.* **160**, 012058.

[Hei 03] Heister, A. *et al.* (2003). *Phys. Lett.* **B565**, 61.

[Her 77] Herb, S.W. *et al.* (1977). *Phys. Rev. Lett.* **39**, 252.

[Her 90] Hertzog, D. *et al.* (1990). *Nucl. Instr. and Meth.* **A294**, 446.

[HES 16] The High Energy Stereoscopic System (HESS). https://www.mpi-hd.mpg.de/hfm/HESS/

[Hig 64] Higgs, P. (1964). *Phys. Rev. Lett.* **13**, 508; *Phys. Lett.* **12**, 132.

[Hir 87] Hirata, K.S. *et al.* (1987). *Phys. Rev. Lett.* **58**, 1490.

[Hir 88] Hirata, K.S. *et al.* (1988). *Phys. Rev.* **D38**, 448.

[Hir 90] Hirata, K.S. *et al.* (1990). *Phys. Rev. Lett.* **65**, 1297.

[Hir 91] Hirata, K.S. *et al.* (1991). *Phys. Rev.* **D44**, 2241.

[Hira 91] Hirayama, H. *et al.* (1991). *Nucl. Instr. and Meth.* **A302**, 427.

[Hit 76] Hitlin, D. *et al.* (1976). *Nucl. Instr. and Meth.* **137**, 225.

[Hoc 07] Höcker, A. *et al.* (2007). *TMVA - Toolkit for Multivariate Data Analysis*, PoS ACAT 040 [physics/0703039] [inSPIRE].

[Hof 69] Hofstadter, R. *et al.* (1969). *Nature* **221**, 228.

[Hof 79] Hofmann, W. *et al.* (1979). *Nucl. Instr. and Meth.* **163**, 77.

[Hof 82] Hofmann, W. *et al.* (1982). *Nucl. Instr. and Meth.* **195**, 475.

[Hof 96] Hof, M. *et al.* (1996). *Astrophys. J.* **467**, L33.

[Hol 78a] Holder, M. *et al.* (1978). *Nucl. Instr. and Meth.* **148**, 235.

[Hol 78b] Holder, M. *et al.* (1978). *Nucl. Instr. and Meth.* **151**, 69.

[Hol 85]   Holroyd, R.A. and Anderson, D.F. (1985). *Nucl. Instr. and Meth.* **A236**, 294.

[Hol 90]   Holroyd, R.A. (1990). *IEEE Trans. Nucl. Sci.* **NS-37**, 513.

[Hon 90]   Honda, M. *et al.* (1990). *Phys. Lett.* **B248**, 193.

[Hos 06]   Hosaka, J. *et al.* (2006). *Phys. Rev.* **D73**, 112001.

[HSS 06]   *Hadronic Shower Simulations*, 6-8 Sept. 2006, Fermilab; AIP Conference Proceedings **896**, eds. M. Albrow and R. Raja (2007).

[Hub 99]   Hubaut, F. (1999). *Optimisation et caractérisation des performances d'un télescope sous-marin à neutrinos pour le projet ANTARES*, Thesis, Univ. de la Méditerranée, Aix–Marseille II, unpublished.

[Hug 69]   Hughes, E.B. *et al.* (1969). *Nucl. Instr. and Meth.* **75**, 130.

[Hul 93]   Hulbert, M. *et al.* (1993). *Nucl. Instr. and Meth.* **A335**, 427.

[Hut 05]   Huhtinen, M. *et al.* (2005). *Nucl. Instr. and Meth.* **A545**, 63.

[Ike 00]   Ikeda H. *et al.* (2000). *Nucl. Instr. and Meth.* **A441**, 401.

[Ish 97]   Ishii, K. *et al.* (1997). *Nucl. Instr. and Meth.* **A385**, 215.

[Iva 09]   Ivanov, A.A., Knurenko, S.P. and Sleptsov, I.Ye. (2009). *New. J. of Phys.* **11**, 065008.

[Iva 15]   Ivanov, A.A. *et al.* (2015). *Nucl. Instr. and Meth.* **A772**, 34.

[Iwa 80]   Iwata, S. (1980). *Report* DPNU 13-80, Nagoya Univ., Japan.

[Jac 74]   Jackson, J.D. (1974). *Classical Electrodynamics* (New York: Wiley).

[Jan 82]   Janni, J.F. (1982). *Atomic Data and Nuclear Data Tables* **27**, 147.

[Jef 85]   Jeffreys, P.W. *et al.* (1985). *Rutherford Lab report* RAL-85-058.

[Jel 09]   Jelley, N., McDonald, A.B. and Robertson, R.G.H. (2009). *Ann. Rev. Nucl. Part. Sci.* **59**, 431.

[Jen 88]   Jenni, P. (1988). *Nucl. Phys. B (Proc. Suppl.)* **3**, 341.

[Jenk 89]   Jenkins, C.M. *et al.* (1989). *IEEE Trans. Nucl. Sci.* **NS-36**, 117.

[Joh 92]   Johnson, K.F. *et al.* (1992). *Nucl. Instr. and Meth.* **A317**, 506.

[Kad 81]   Kadansky, V. *et al.* (1981). *Phys. Scripta* **23**, 680.

[Kaj 86]   Kajino, F. *et al.* (1986). *Nucl. Instr. and Meth.* **A245**, 507.

[Kaj 99]   Kajita, T. (1999). *Proc. of the 6th Topical Seminar on Neutrino and Astroparticle Physics*, San Miniato (I), eds. F. Navarria and G. Pelfer, *Nucl. Phys. B (Proc. Suppl.)* 85 (2000) 44.

[Kaj 16]   Kajita, Y., Kearns, E. and Shiozawa, M. (2016). *Nucl. Phys.* **B908**, 14.

[Kam 94]   Kampert, K.-H. *et al.* (1994). *Nucl. Instr. and Meth.* **A349**, 81.

[Kan 88]   Kanbach, G. *et al.* (1988). *Space Sci. Rev.* **49**, 69.

[KAS 16]   https://kcdc.ikp.kit.edu; http://www-ik.fzk.de/KASCADE_home.html

[Kav 11]   Kavatsyuk, M. *et al.* (2011). *Nucl. Instr. and Meth.* **A648**, 77.

[Kaw 88]   Kawabata, S. *et al.* (1988). *Nucl. Instr. and Meth.* **A270**, 11.

[Ker 95]   Kerrick, A.D. *et al.* (1995). *Astrophys. J.* **A452**, 588.

[Kes 96]   Kessler, R.S. *et al.* (1996). *Nucl. Instr. and Meth.* **A368**, 653.

[Kha 90]   Khazins, D.M. *et al.* (1990). *Nucl. Instr. and Meth.* **A300**, 281.

[Kha 14]   Khachatryan, V. *et al.* CMS Collaboration (2014). *Eur. Phys. J.* **C74**, 3076.

[Kha 16]   Khachatryan, V. *et al.* CMS Collaboration (2014). *Phys. Rev.* **D93**, 072004.

[Kir 88]   Kirby, J. (1988). *Proc. Workshop for the INFN Eloisatron Project – Vertex Detectors*, ed. F. Villa (New York: Plenum Press), p. 225.

[Kis 94]   Kistenev, E.P. (1994). *Proc. 5th Int. Conf. on Calorimetry in High Energy Physics*, Brookhaven National Laborotory, eds. H. Gordon and D. Rueger (World Scientific, Singapore, 1995), p. 211.

[Klo 83]   Klopfenstein, C. (1983). *Phys. Lett.* **130B**, 444.

[KM3 16]   The KM3 Collaboration (2016). arXiv:1601.07459 [astro-ph.IM].

[Kob 87]   Kobayashi, M. and Kobayashi, S. (1987). *Nucl. Instr. and Meth.* **A262**, 264.

[Kob 91]   Kobayashi, M. *et al.* (1991). *Nucl. Instr. and Meth.* **A306**, 139.

[Kob 95]   Kobayashi, S. *et al.* (1995). *Nucl. Instr. and Meth.* **A364**, 95.

[Kop 85]   Kopp, R. *et al.* (1985). *Z. Phys.* **C28**, 171.

[Kos 98]   Koshio, Y. (1998). *Study of Solar Neutrinos at Super Kamiokande*, thesis Inst. for Cosmic-Ray Research, Univ. of Tokyo, ICRR-Report-426-98-22.

[Kra 92]   Krammer, M. (1992). *Nucl. Instr. and Meth.* **A315**, 294.

[Kri 99]   Křivková, P. and Leitner, R. http://cds.cern.ch/record/683812/files/tilecal-99-007.pdf

[Kru 92]   Krüger, J. (1992). The Uranium/Scintillator Calorimeter for the ZEUS Detector at the Electron-Proton Collider HERA, *Internal Report* DESY F35-92-02, DESY Hamburg.

[Kub 92]   Kubota, Y. *et al.* (1992). *Nucl. Instr. and Meth.* **A320**, 66.

[Kud 90]   Kudla, I. *et al.* (1990). *Nucl. Instr. and Meth.* **A300**, 480.

[Kul 06]   Kulchitskii, Yu.A., Tsiareshka, P.V. and Vinogradov, V.B. (2006). *Electron Energy Resolution of the ATLAS TileCal Modules with Fit Filter Method*, report ATLAS-TILECAL-PUB-2006-004.

[Kun 97]   Kunori, S. (1997). *Proc. 7th Int. Conf. on Calorimetry in High Energy Physics*, Tucson, Arizona (Singapore: World Scientific, Singapore), p. 224.

[Lac 74]   Lachkar, J. *et al.* (1974). *Nucl. Phys.* **A222**, 333.

[Lai 12]   Laihem, K. (2012). *Nucl. Instr. and Meth.* **A692**, 192.

[Lan 48]   Landau, L.D. (1948). *On the angular momentum of a two-photo system*, Dokl. Akad. Nauk Ser. Fiz. **60**, 207.

[Lea 79]   Learned, J.G. (1979). *Phys. Rev.* **D19**, 3293.

[Lea 95]   Learned, J.G. and Pakvasa, S. (1995). *Astropart. Phys.* **3**, 267.

[Lec 06]   Lecomte, P. *et al.* (2006). *Nucl. Instr. and Meth.* **A564**, 164.

[Lee 16a] Lee, S. *et al.* (2016). *Characteristics of the light produced in a dual-readout fiber calorimeter*. Internal report, RD52 Collaboration.

[Lee 17]   Lee, S. *et al.* (2017). *Nucl. Instr. and Meth.* **A866**, 76.

[Leh 78]   Lehraus, I. *et al.* (1978). *Nucl. Instr. and Meth.* **153**, 347.

[Lem 89]   Lemeilleur, F. *et al.* (1989). *Phys. Lett.* **B222**, 518.

[Leo 87]   Leo, W.R. (1987). *Techniques for Nuclear and Particle Physics Experiments* (Berlin/Heidelberg: Springer-Verlag).

[LEP 06]   The LEP & SLC Collaborations (2006). *Phys. Rep.* **427**, 257.

[Ler 86]   Leroy, C., Sirois, Y. and Wigmans, R. (1986). *Nucl. Instr. and Meth.* **A252**, 4.

[Ler 00]   Leroy, C. and Rancoita, P.G. (2000). *Rep. Progr. Phys.* **63**, 5005.

[Ler 11]   Leroy, C. and Rancoita, P.G. (2011). *Principles of Radiation Interaction in Matter and Detection* (3rd ed.), World Scientific, Singapore.

[Lev 08]   Leverington, B.D. *et al.* (2008). *Nucl. Instr. and Meth.* **A596**, 327.

[Lig 08]   Lightfoot, P.K. *et al.* (2008). arXiv:0807.3220 [physics.ins-det].

[Lin 12]   Linssen, L. (2012). arXiv:1202.5940 [physics.ins-det].

[Liu 96]   Liu, L.-I. *et al.* (1996). *Appl. Spectr.* **50**, nr. 12, 1545.

[Liv 95]   Livan, M., Vercesi, V. and Wigmans, R. (1995). Scintillating-fibre Calorimetry, *CERN Yellow Report*, CERN 95-02, Genève, Switzerland.

[Llo 96]   Llope, W.J. *et al.* (1996). *Proc. 6th Int. Conf. on Calorimetry in High Energy Physics*, Frascati, Italy, eds. A. Antonelli, S. Bianco, A. Calceterra and F. Fabbri (Frascati Physics Series, 1997), p. 187.

[Lob 85]   Lobkowicz, F. *et al.* (1985). *Nucl. Instr. and Meth.* **A235**, 332.

[Lob 02]   Lobban, O., Sriharan, A. and Wigmans R. (2002). *Nucl. Instr. and Meth.* **A495**, 107.

[Loh 85]   Lohmann, W., Kopp, R. and Voss, R. (1985). Energy loss of muons in the energy range 1 GeV to 10 TeV, *CERN yellow report*, CERN 85-03, Genève, Switzerland.

[Lor 94]   Lorenz, E. *et al.* (1994). *Nucl. Instr. and Meth.* **A344**, 64.

[Lor 02]   Loredo, T.J. and Lamb, D.Q. (2002). *Phys. Rev.* **D65**, 063002.

[Lub 92]   Lu, B., Mo, L.W. and Nunamaker, T.A. (1992). *Nucl. Instr. and Meth.* **A313**, 135.

[Lud 81]   Ludlam, T. *et al.* (1981). *IEEE Trans. Nucl. Sci.* **NS-28**, 517.

[Mac 09]   Machikhiliyan, I. (2009). *J. Phys. Conf. Ser.* **160**, 012047.

[Mag 81]   Mageras, G. *et al.* (1981). *Phys. Rev. Lett.* **46**, 1115.

[Mak 62]   Maki, Z. *et al.* (1962), *Prog. Theor. Phys.* **28**, 870.

[Man 92]   Manuisch, J. *et al.* (1992). *Nucl. Instr. and Meth.* **A312**, 451.

[Mar 12]   Marrocchesi, P.S. (2012). *Nucl. Instr. and Meth.* **A692**, 293.

[Mas 12]   Mase, T. *et al.* (2012). *Nucl. Instr. and Meth.* **A671**, 129.

[MB 15]   MicroBooNE, http://www-microboone.fnal.gov

[McN 09]   McNabb, R. *et al.* (2009). *Nucl. Instr. and Meth.* **A602**, 396.

[Mel 92]   Melèse, P. *et al.* (1992). *Nucl. Instr. and Meth.* **A322**, 189.

[Mic 84]   Micke, U. *et al.* (1984). *Nucl. Instr. and Meth.* **221**, 495.

[Mic 08]   Michael, D.G. *et al.* (2008). *Nucl. Instr. and Meth.* **A596**, 190.

[Mie 95]   Mielke, H.H. *et al.* (1995). *Nucl. Instr. and Meth.* **A360**, 367.

[Mih 11]   Mihara, S. (2011). *J. Phys. Conf. Ser.* **308**, 012009.

[Mik 88]   Mikenberg, G. (1988). *Nucl. Instr. and Meth.* **A265**, 223.

[Mil 97]   Miller, T.C. *et al.* (1997). *Proc. 26th Int. Cosmic Ray Conf.*, Durban, South Africa (1997).

[Min 07]   Mineev, O. *et al.* (2007). *Nucl. Instr. and Meth.* **A577**, 540.

[Miy 95]   Miyajima, M., Sasaki, S. and Shibamura, E. (1995). *Nucl. Instr. and Meth.* **A352**, 548.

[Mkr 13]   Mkrtchyan, H. *et al.* (2013). *Nucl. Instr. and Meth.* **A719**, 85.

[Mue 81]   Mueller, J.J. *et al.* (1981). *IEEE Trans. Nucl. Sci.* **NS-28**, 496.

[Muf 15]   Mufson, F. *et al.* (2015). *Nucl. Instr. and Meth.* **A799**, 1.

[Mur 98]   Murakami, H. *et al.* (1998). *Adv. Space Res.* **21**, 1029.

[Nag 01]   Nagaslaev, V.P., Sill, A.F. and Wigmans, R. (2001). *Nucl. Instr. and Meth.* **A462**, 411.

[Nau 09]   De Naurois, M. and Rolland, L. (2009). *Astropart. Phys.* **32**, 231.

[Nei 15]   Neiser, A. (2015). *J. Phys. Conf. Ser.* **587**, 012041.

[Nel 78]   Nelson, W.R., Hirayama, H. and Rogers, D.W.O. (1985). EGS 4, *SLAC Report* 165, Stanford.

[Ner 85]   Nernst, R. *et al.* (1985). *Phys. Rev. Lett.* **54**, 2195.

[Ney 00]   Neyret, D. *et al.* (2000). *Nucl. Instr. and Meth.* **A443**, 231.

[Nie 06]   Niess, V. and Bertin, V. (2006). *Astropart. Phys.* **26**, 243.

[Nig 09]   Nigmanov, T.S. *et al.* (2009). *Nucl. Instr. and Meth.* **A598**, 394.

[Nik 13]   Nikiforou, N. (2013). arXiv:1306.6756 [physics.ins-det].

[Nyg 81]   Nygren, D.R. (1981). *Phys. Scripta* **23**, 584.

[Oga 86]   Ogawa, K. *et al.* (1986). *Nucl. Instr. and Meth.* **A243**, 58.

[Ohs 96]   Ohshima, Y. *et al.* (1996). *Nucl. Instr. and Meth.* **A380**, 517.

[Oko 82]   Okoshi, T. (1982). *Optical Fibers* (San Diego, CA: Academic Press).

[Oku 95]   Okuno, H. *et al.* (1995). *Nucl. Instr. and Meth.* **A365**, 352.

[Oku 99]   Okumura, K. (1999). *Observation of Atmospheric Neutrinos in Super-Kamiokande and a Neutrino oscillation Analysis*, thesis Inst. for Cosmic-Ray Research, Univ. of Tokyo, ICRR-Report-450-99-8.

[Ons 38]   Onsager, L. (1938). *Phys. Rev.* **54**, 554.

[Ore 82]   Oreglia, M. *et al.* (1982). *Phys. Rev.* **D25**, 2259.

[Ori 83]   Orito, S. *et al.* (1983). *Nucl. Instr. and Meth.* **216**, 439.

[Ott 06]   Otte, N. (2006), *The silicon photomultiplier - a new device for high energy physics, astrophysics, industrial and medical applications*, SNIC Symposium, Stanford (CA), 3-6 April 2006.

[PAC 15]   The Pierre Auger Collaboration (2015). *Nucl. Instr. and Meth.* **798**, 172.

[Pag 72]   Pages, L. *et al.* (1972). *At. Data* **4**, 1.

[Pam 99]   PAMELA Collaboration (1999). *Proc. 26th Int. Conf. on Cosmic Rays* **4**, 187.

[Pat 13]   Patterson, R.B. *et al.* (2013). *Nucl. Phys. B (Proc. Suppl.)* **235-236**, 151.

[PDG 14]   Particle Data Group, Olive, K.A. *et al.* (2014 and 2015 update). *Chin. Phys.* **C38**, 090001.

[PDH 90]   Philips Components (1990). *Photomultiplier Data Handbook*, PC04 (Eindhoven, the Netherlands: Philips).

[Pei 96]   Peigneux, J.P. *et al.* (1996). *Nucl. Instr. and Meth.* **A378**, 410.

[Pel 09]   Peleganchuk, S. (2009). *Nucl. Instr. and Meth.* **A598**, 248.

[Per 00]   Perkins, D.H. (2000). *Introduction to High Energy Physics*, 4th ed. (Cambridge: Cambridge University Press).

[Pet 12]   Petyt, D.A. (2012). *Nucl. Instr. and Meth.* **A695**, 293.

[Pez 15]   Pezzullo, G. *et al.* (2015). *J. Phys. Conf. Ser.* **587**, 012047.

[Pin 65]   Pinkau, K. (1965). *Phys. Rev.* **139B**, 1548.

[Pit 92]   Pitzl, D. *et al.* (1992). *Nucl. Instr. and Meth.* **A311**, 98.

[Ple 06]   Plewnia, S. *et al.* (2006). *Nucl. Instr. and Meth.* **A566**, 422.

[Pon 68]   Pontecorvo, B. (1968). *Sov. Phys. JETP* **26**, 984.

[Pre 87]   Pretzl, K., Schmitz, N. and Stodolsky, L. (eds.) (1987). *Low-Temperature Detectors for Neutrinos and Dark Matter* (Berlin: Springer Verlag).

[Pre 00]   Pretzl, K. (2000). *Nucl. Instr. and Meth.* **A454**, 114.

[Pri 81]   Price, J. and Ambats, I. (1981). *IEEE Trans. Nucl. Sci.* **NS-28**, 506.

[Pri 06]   Price, B.J. (2006). *J. Geophys. Res.* **111**, B02201.

[Pro 79]   Prokoshkin, Yu.D. (1979). *Proc. of the Second ICFA Workshop on the Possibilities and Limitations of Accelerators and Detectors*, Les Diablerets, October 1979, U. Amaldi (ed.), p. 405.

[Rad 88a]   Radeka, V. (1988). *Ann. Rev. Nucl. Part. Sci.* **38**, 217.

[Rad 88b]   Radeka, V. and Rescia, S. (1988). *Nucl. Instr. and Meth.* **A265**, 228.

[Rad 92]   Radermacher, E. *et al.* (1992). Liquid detectors for Precision Calorimetry, in *Instrumentation in High Energy Physics*, p. 387, ed. F. Sauli (Singapore: World Scientific).

[Rei 59]   Reines, F. and Cowan, C. (1959). *Phys. Rev.* **113**, 273.

[Rep 08]   Repond, J. *et al.* (2008). *JINST* **3**, P08001.

[Rep 12]   Repond, J. *et al.* (2012). https://twiki.cern.ch/twiki/pub/CALICE/CaliceAnalysisNotes/CAN-039.pdf

[Rep 13]   Repond, J. (2013). *Nucl. Instr. and Meth.* **A732**, 466.

[Ric 07]   Riccobene, G. *et al.* (2007). *Astropart. Phys.* **27**, 1.

[Rit 10]   Ritt, S. *et al.* (2010). *Nucl. Instr. and Meth.* **A623**, 486.

[Ros 52]   Rossi, B. (1952). *High-Energy Particles* (Englewood Cliffs, N.J: Prentice Hall).

[Rua 14]   Ruan, M. and Videau, H. (2014). arXiv:1403.4784 [physics.ins-det].

[Rud 66]   Rudstam, G. (1966). *Z. Naturforsch.* **21a**, 1027.

[Rya 72]   Ryan, M.J., Ormes, J.F. and Balasubrahmanyan, V.K. (1972). *Phys. Rev. Lett.* **28**, 985.

[Sad 01]   Sadrozinski, H.F.-W. (2001), *Nucl. Instr. and Meth.* **A466**, 292.

[Sak 90]   Sakumoto, W.K. *et al.* (1990). *Nucl. Instr. and Meth.* **A294**, 179.

[Sau 77]   Sauli, F. (1977). Principles of Operation of Multiwire Proportional and Drift Chambers, *Report* CERN 77-09, Genève, Switzerland; reprinted in *Experimental Techniques in High Energy Nuclear and Particle Physics*, p. 79, ed. T. Ferbel, 2nd edition, (Singapore: World Scientific, 1987).

[Saw07]   Sawada, R. (2007). *Nucl. Instr. and Meth.* **A581**, 522.

[Scha 84]   Schamberger, R.D. *et al.* (1984). *Phys. Lett.* **138B**, 225.

[Schm 69]   Schmidt, W.F. and Allen, A.O. (1969). *J. Chem. Phys.* **50**, 5037.

[Schm 77]   Schmidt, W.F. (1977). *Can. J. Chem* **55**, 2197.

[Scho 70]   Schonkeren, J.M. (1970). *Photomultipliers*, Philips Application Book Series, ed. by H. Kater and L.J. Thompson (Eindhoven, the Netherlands: Philips).

[Sed 00]   Sedykh, S.A. *et al.* (2000). *Nucl. Instr. and Meth.* **A455**, 346.

[Sef 15]   Sefkow, F. *et al.* (2016). *Rev. Mod. Phys.* **88**, 1.

[Seg 92]   Séguinot, J. *et al.* (1992). *Nucl. Instr. and Meth.* **A323**, 583.

[Seg 94]	Séguinot, J. and Ypsilantis, T. (1994). *Nucl. Instr. and Meth.* **A343**, 1.
[Ses 79]	Sessoms, A.L. *et al.* (1979). *Nucl. Instr. and Meth.* **161**, 371.
[Shi 75]	Shibamura, E. *et al.* (1975). *Nucl. Instr. and Meth.* **131**, 249.
[Shi 05]	Shimizu, H. *et al.* (2005). *Nucl. Instr. and Meth.* **A550**, 258.
[Sic 13]	Sicking, E. *et al.* (2013). https://twiki.cern.ch/twiki/pub/CALICE/CaliceAnalysisNotes/CAN-044.pdf
[Sim 91]	Simon-Gillo, J. *et al.* (1991). *Nucl. Instr. and Meth.* **A309**, 427.
[Sim 93]	Simon, D.A. *et al.* (1993). *Nucl. Instr. and Meth.* **A335**, 86.
[Sim 08]	Simonyan, M. (2008). *Performance of the ATLAS Tile Calorimeter to pions and protons.* CERN-THESIS-2008-032.
[Sio 79]	Siohan, F. *et al.* (1979). *Nucl. Instr. and Meth.* **167**, 371.
[Sir 85]	Sirois, Y. and Wigmans, R. (1985). *Nucl. Instr. and Meth.* **A240**, 262.
[Sni 88]	Snider, F. (1988). *Nucl. Instr. and Meth.* **268**, 75.
[Sod 81]	Söding, P. and Wolf, G. (1981). *Ann. Rev. Nucl. Part. Sci.* **31**, 231.
[Son 87]	Sonderegger, P. (1987). *Nucl. Instr. and Meth.* **A257**, 523.
[Sta 98]	Stahl, A. *et al.* (1998). *Nucl. Instr. and Meth.* **A409**, 615.
[Sta 02]	Starchenko, E. *et al.* (2002). *Nucl. Instr. and Meth.* **A494**, 381.
[Ste 14]	Steen, A. *et al.* (2014), arXiv:1403.8097 [physics.ins-det].
[Sto 70]	Storm, E. and Israel, H.I. (1970). *Nucl. Data Tables* **7**, 565.
[Sto 78]	Stone, S.L. *et al.* (1978). *Nucl. Instr. and Meth.* **151**, 387.
[Str 98]	Streitmatter, R.E. (1998). *Proc. Workshop on Observing Giant Cosmic-Ray Air Showers from $> 10^{20}$ eV Particles from Space*, AIP Conf. Proc. **433**, eds. J.F. Krizmanic, J.F. Ormes and R.E. Streitmatter (Woodbury, NY: AIP Press, 1998), 95.
[Sul 93]	Sullivan, J.P. *et al.* (1993). *Nucl. Instr. and Meth.* **A324**, 441.
[Sumi 88]	Sumiyoshi, T. *et al.* (1988). *Nucl. Instr. and Meth.* **A271**, 432.
[Sum 88]	Sumner, R. (1988). *Nucl. Instr. and Meth.* **A265**, 252.
[Suz 93]	Suzuki, A. *et al.* (1993). *Nucl. Instr. and Meth.* **A329**, 299.
[Suz 99]	Suzuki, T. *et al.* (1999). *Nucl. Instr. and Meth.* **A432**, 48.
[Tak 92]	Takasaki, F. *et al.* (1992). *Nucl. Instr. and Meth.* **A322**, 211.
[Tay 79]	Taylor, J.H., Fowler, L.A. and Weisberg, J.M. (1979). *Nature* **277**, 437.
[Tes 94]	Tesaret, R.J. *et al.* (1994). *Nucl. Instr. and Meth.* **A349**, 96.
[Tho 97]	Thomson, J.J. (1897). *Nature* **55**, 606.
[Tho 09]	Thomson, M.A. (2009). *Nucl. Instr. and Meth.* **A611**, 25.
[Tor 01]	Torii, S. *et al.* (2001). *Astrophys. J.* **559**, 973.
[Tsa 74]	Tsai, Y.S. (1974). *Rev. Mod. Phys.* **46**, 815.
[Twe 96]	Twerenbold, D. (1996). *Rep. Progr. Phys.* **59**, 349.
[Ung 08]	Unger, M. *et al.* (2008). *Nucl. Instr. and Meth.* **A588**, 443.
[Uoz 02]	Uozumi, S. *et al.* (2002). *Nucl. Instr. and Meth.* **A487**, 291.
[Urb 15]	Urban, M. *et al.* (2015). *J. Phys. Conf. Ser.* **587**, 012043.
[Vac 91]	Vacanti, G. *et al.* (1991). *Astrophys. J.* **A377**, 467.
[Ver 16]	VERITAS (2016). http://veritas.sao.arizona.edu
[Wag 85]	Wagoner, D.E. *et al.* (1985). *Nucl. Instr. and Meth.* **A238**, 315.

[Wap 77]   Wapstra, A.H. and Bos, K. (1977). *Atomic Data and Nuclear Data Tables*
            19, 177.
[Web 15]   Webber, B. (Cambridge University). *Private communication.*
[Wee 89]   Weekes, T.C. *et al.* (1989). *Astrophys. J.* **342**, 379.
[Whi 88]   White, T.O. (1988). *Nucl. Instr. and Meth.* **A273**, 820.
[Wig 87]   Wigmans, R. (1987). *Nucl. Instr. and Meth.* **A259**, 389.
[Wig 88]   Wigmans, R. (1988). *Nucl. Instr. and Meth.* **A265**, 273. Reprinted in *Experi-
            mental Techniques in High Energy Nuclear and Particle Physics*, p. 679,
            ed. T. Ferbel, 2nd ed. (Singapore: World Scientific).
[Wig 91]   Wigmans, R. (1991). *Ann. Rev. Nucl. Part. Sci.* **41**, 133.
[Wig 97]   Wigmans, R. (1997). *Proc. 7th Int. Conf. on Calorimetry in High Energy
            Physics*, Tucson, Arizona (Singapore: World Scientific, 1998), p. 182.
[Wig 98]   Wigmans, R. (1998). *Rev. Sci. Instr.* **69**, 3723.
[Wig 02]   Wigmans, R. and Zeyrek, M. (2002). *Nucl. Instr. and Meth.* **A485**, 385.
[Wig 03]   Wigmans, R. (2003). *Astropart. Phys.* **19**, 379.
[Wig 06a]   Wigmans, R. (2006). CALOR06, AIP Conf. Proc. **867**, eds. S. Magill and
            R. Yoshida, 90.
[Wig 06b]   Wigmans, R. (2006). *Proc. of the workshop on hadronic shower simula-
            tion*, Fermilab (2006). AIP Conf. Proc. **896**, eds. M. Albrow and R. Raja
            (2007), 123.
[Wig 07]   Wigmans, R. (2007). *Nucl. Instr. and Meth.* **A572**, 215.
[Wig 08a]   Wigmans, R. (2008). *New Journal of Physics* **10**, 025003.
[Wig 08b]   Wigmans, R. (2008). *Scientifica Acta* **2**, nr. 1, 20.
[Wig 11]   Wigmans, R. (2011). Calorimeters, in *Handbook of Particle Detection and
            Imaging*, eds. C. Grupen and I. Buvat, Springer Verlag, vol. 1, 497.
[Wig 13]   Wigmans, R. (2013). *Nucl. Instr. and Meth.* **A718**, 43.
[Wig 16]   Wigmans, R. (2016). *Nucl. Instr. and Meth.* **A824**, 721.
[Wil 74]   Willis, W.J. and Radeka, V. (1974). *Nucl. Instr. and Meth.* **120**, 221.
[Wim 89]   Wimpenny, S.J. *et al.* (1989). *Nucl. Instr. and Meth.* **A279**, 107.
[Win 99]   Winn, D.R. (Fairfield University). *Private communication.*
[Yar 93]   Yarema, R.J. *et al.* (1993) , *IEEE Trans. Nucl. Sci.* **NS-40**, 750; *ibid. Nucl.
            Instr. and Meth.* **A360**, 150.
[You 89]   Young, G.R. *et al.* (1989). *Nucl. Instr. and Meth.* **A279**, 503.
[Yps 95]   Ypsilantis, T. and Séguinot, J. (1995). *Nucl. Instr. and Meth.* **A368**, 229.
[Zat 93]   Zatsepin, V.I. *et al.* (1993). *Proc. 23rd Int. Conf. on Cosmic Rays* **2**, 13.
[Zeu 86]   The ZEUS Collaboration (1986). *Technical Proposal for the ZEUS Detec-
            tor*, DESY, Hamburg, p. 5.
[Zhu 91]   Zhu, R.Y. (1991). *Nucl. Instr. and Meth.* **A302**, 69.
[Zhu 94]   Zhu, R.Y. (1994). *Nucl. Instr. and Meth.* **A340**, 442.
[Zhu 96]   Zhu, R.Y. (1996). *Nucl. Instr. and Meth.* **A376**, 319.
[Zhu 98]   Zhu, R.Y. (1998). *Nucl. Instr. and Meth.* **A413**, 297.
[Zhu 11]   Zhu, R.Y. (2011). *J. Phys. Conf. Ser.* **293**, 012004.
[Zhu 15]   Zhu, R.Y. (2015). *J. Phys. Conf. Ser.* **587**, 012055.
[Zio 94]   Ziock, H. *et al.* (1994). *Nucl. Instr. and Meth.* **A342**, 96.

INDEX

About the author

Dr. Richard Wigmans (1948), a native of the Netherlands, obtained academic degrees in Physics, Mathematics, Astronomy and Economics at the Vrije Universiteit in Amsterdam. In 1975 he was awarded a Ph.D. in Physics for an experimental study of the production and decay of short-lived nuclides in the $Z \sim 50$ region. As a graduate student, he developed a facility for the very fast (~ 1 minute) production of chemically and isotopically pure sources. Thanks to this facility, he was able to determine the decay properties of some ten β^+-unstable nuclei in great detail. Moreover, he discovered several previously unknown tellurium isotopes (^{112}Te, ^{113}Te and ^{115m}Te).

In 1975 Dr. Wigmans switched from nuclear to particle physics. In the next seventeen years, he worked alternatingly at the Dutch National Institute for High Energy Physics (NIKHEF) in Amsterdam and at the European Center for Nuclear Research (CERN) in Genève, in different capacities. Early in his tenure at CERN, Dr. Wigmans was primarily involved in research on heavy-ion scattering in the framework of the HELIOS experiment. Inspired by problems encountered in the HELIOS calorimeters, he engaged in theoretical and experimental studies of the fundamental aspects of calorimetry, in particular the *compensation* mechanism. In 1987, he initiated SPACAL, a detector R&D project in which the compensating lead/scintillating-fiber technology was developed. Under his leadership, an international collaboration of ~ 50 physicists, engineers and technicians built a 20-ton generic prototype calorimeter and studied its (record setting) performance in great detail. In 1990 he initiated RD1, a detector R&D project intended to study the merits of this type of detector in high-luminosity experiments.

In 1992 Dr. Wigmans relocated to the United States. Attracted by the prospect of the Superconducting Supercollider (SSC), under construction in Texas at that time, and driven by the desire to work in a teaching environment after having spent twenty years in full-time research, he accepted a faculty position at Texas Tech University, where he holds the J. Fred and Odetta Greer Bucy Chair in Physics. The research group he started at TTU is currently involved in the CMS experiment at CERN, and in RD52, a detector R&D project in which the merits of calorimeters based on the simultaneous detection of scintillation and Čerenkov light produced in the absorption process, are being studied and optimized. This technique is known as the *dual-readout method* (DREAM).

Based on his personal contributions to the development of the field, Dr. Wigmans is considered one of the world's leading experts on calorimetry.